# modern
# WELDING

## complete coverage of the welding field in one easy-to-use volume!

by
**Andrew D. Althouse**
Technical-Vocational Education Consultant
Member of American Welding Society

and
**Carl H. Turnquist**
Career Education Consultant
Member of American Welding Society

and
**William A. Bowditch**
Career Education Consultant
Life Member of American Welding Society

and
**Kevin E. Bowditch**
Hughes Aircraft Company
Industrial Products Division
Assembly and Test Products
Carlsbad, California
Member of American Welding Society

South Holland, Illinois
**THE GOODHEART-WILLCOX COMPANY, INC.**
Publishers

Library of Congress Catalog Card Number 91-40908
International Standard Book Number 0-87006-966-7

2 3 4 5 6 7 8 9 10    92    96 95 94 93

**Library of Congress Cataloging in Publication Data**

Modern welding: complete coverage of the welding
field in one easy-to-use volume! / by Andrew D.
Althouse . . . [et al.].

p.   cm.
Includes index.
ISBN 0-87006-966-7
1. Welding. I. Althouse, Andrew Daniel.
TS227.M557  1992
671.5'2--dc20                                91-40908
                                              CIP

# INTRODUCTION

MODERN WELDING is an authoritative text written for secondary and post secondary students, apprentices and journeyman welders, technical students, instructors, and those who wish to teach themselves about welding.

MODERN WELDING provides you with an understanding of the entire welding field. The text covers the theory, fundamentals, and basic processes along with the practical applications that build skills and techniques for the welder.

The text is organized into nine logical parts which allow you to build from the solid welding basics to advanced processes and specialized study. Part 1, Welding Fundamentals, explains and colorfully illustrates the various welding and cutting processes. Reading, interpreting, and using welding symbols complete this part.

Oxyfuel Gas Processes in Part 2 describes welding, cutting, soldering, and brazing. Part 3 details Shielded Metal Arc Welding principles and practices.

Gas Tungsten Arc Welding and Gas Metal Arc Welding are covered in Part 4. The techniques and procedures used in GTAW and GMAW are described and shown in numerous easy-to-understand illustrations.

Arc Cutting information is provided in Part 5, while Resistance Welding is the subject of Part 6.

Parts 7 through 9 will provide you with coverage of Special Welding and Cutting Processes, Metal Technology, and Professional Welding. These interesting and informative parts are designed to help you have a thorough understanding of the entire field of welding and to have a head start on the job.

MODERN WELDING has many outstanding features which make it easy to use and to understand. It is highly illustrated with photographs and special line drawings to help you visualize the welding and cutting operations. Color is used throughout to aid your learning. Most measurements are provided in both conventional and SI metric units. The terms used are those recommended by the American Welding Society, with the terms used in the trade also provided when necessary. The logical sequence of the topics is designed to make it easier for learners to learn and for teachers to teach.

**William A. Bowditch**
**Kevin E. Bowditch**

# CONTENTS

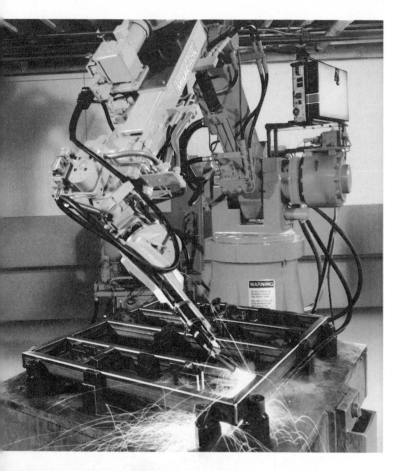

Industry photos of various welding applications. (Miller Electric Mfg. Co., Cincinnati Milacron)

part
1

# WELDING FUNDAMENTALS

Industry photo of freight car positioner. Large assembly is rotated by positioner to achieve flat or horizontal weld position as a means of producing quality welds economically under the production line conditions. Welders use semi-automatic, gas shielded, flux cored arc welding equipment to make finish welds. (Thrall Car Manufacturing Co.)

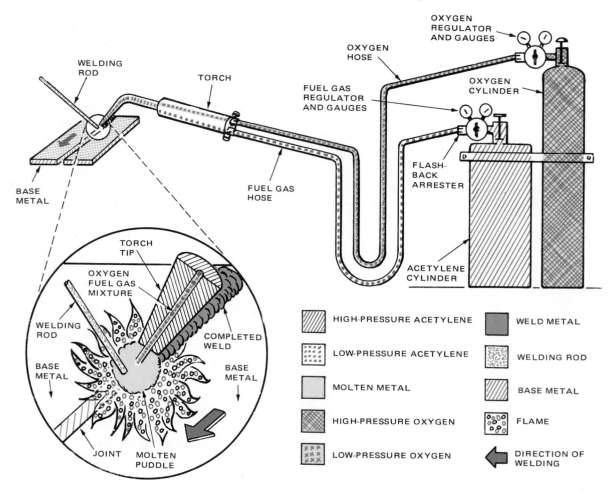

Fig. 1-1. Oxyacetylene welding (OAW). The oxygen and acetylene gas are mixed in a torch. The mixture burns at the torch tip. The heat from this flame is used to melt the base metal and welding rod. This melted material forms a welded joint.

# 1 WELDING AND CUTTING PROCESSES

This introductory section illustrates the many welding and cutting processes used in modern industry. The parts and materials used in each process are color keyed for clarity. To more easily understand each process, the student should refer to the drawings frequently while reading the text.

American Welding Society (AWS) standard terminology is used in these descriptions.

Each welding process is explained as follows:
1. Its application and purpose.
2. Colored illustration(s) of the process.
3. The energy source used to produce heat.
4. Controls.
5. Operation of the process.
6. Safety.
7. References to text chapters and paragraphs where more detailed explanations are given.

The color illustrations in this introductory section are intended to show only the general process. In each illustration the welding station is diagrammed. Some details have been enlarged to help explain the process.

## 1-1. OXYACETYLENE WELDING (OAW)

The oxyacetylene welding process combines oxygen and acetylene gases to provide a high temperature flame for welding. This flame provides enough heat to melt most metals. The oxyacetylene flame may also be used for all types of brazing.

An oxyacetylene outfit, as shown in Fig. 1-1, is used in this process. Oxyacetylene welding is a manual process. The welder must personally control the torch movement and welding rod. Acetylene is supplied from one cylinder; compressed oxygen is supplied from another cylinder. Both cylinders must be equipped with a pressure reducing regulator. Each regulator is fitted with two gauges.

One pressure gauge (HIGH) indicates the pressure in the cylinder. The other pressure gauge (LOW) indicates the pressure of the gas being fed to the torch.

Separate flexible hoses carry the gases to the torch. The torch has two needle valves. One torch valve controls the rate of flow of the oxygen; the other controls the rate of flow of the acetylene to the torch tip. The mixed gases burn at the torch tip orifice (opening).

As shown in Fig. 1-2, acetylene burns in the atmosphere with a yellow-red flame. A carburizing flame (excess acetylene with oxygen) is blue with an orange and red end. It may release black smoke. A neutral flame (perfect mixture of oxygen and acetylene) has a quiet, blue-white inner cone. This is the flame used in most welding processes. An oxidizing flame (excess of oxygen) results in a short, noisy, hissing inner cone. It tends to burn the metal being welded.

Other fuel gases can be used in place of acetylene. These include LP (liquefied petroleum), MAPP (methylacetylene propadiene), natural gas and hydrogen.

Welding goggles should be worn for eye protection. Gloves, nonflammable clothing, and all other required safety clothing should be worn to protect against burns. Good fire safety and prevention techniques should be employed. Good ventilation must be provided.

See Chapters 3 and 4 for additional oxyacetylene welding information.

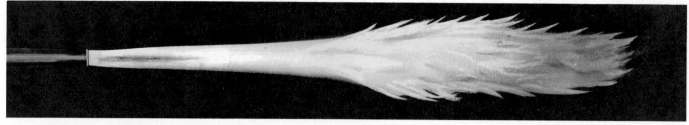

**Acetylene Burning in Atmosphere**
Open fuel gas valve until smoke clears from flame.

**Carburizing Flame**
(Excess acetylene with oxygen.) Used for hard-facing and welding white metal.

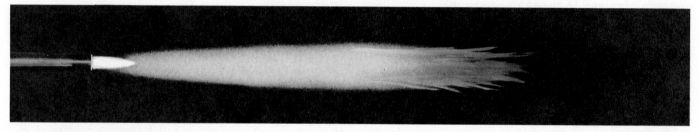

**Neutral Flame**
(Acetylene and oxygen.) Temp. 5720°F (3160°C). For fusion welding of steel and cast iron.

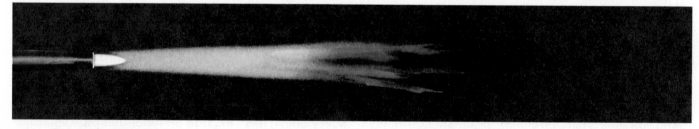

**Oxidizing Flame**
(Acetylene and excess oxygen.) For braze welding with bronze rod.

Fig. 1-2. Color appearance of oxyacetylene flames.
(Smith Equip., Div. of Tescom Corp.)

## 1-2. OXYFUEL GAS CUTTING (OFC)

It is possible to burn (rapidly oxidize) iron or steel. The oxyfuel gas flame raises the temperature of the metal to a cherry-red color of 1472°-1832°F (800°-1000°C). Then, a high-pressure jet of oxygen from the cutting torch is directed at the metal. This causes the metal to burn (oxidize) and blow away very rapidly. This is why the term ''burning'' is sometimes used in connection with the oxyfuel gas cutting process.

The process requires cylinders of oxygen and fuel gas, as shown in Fig. 1-3. Each cylinder has a regulator and two pressure gauges. One pressure gauge indicates the pressure in the cylinder. The other gauge indicates the pressure of the gas being fed to the torch. Flexible hoses carry the gases to the torch. Construction details of a typical oxyfuel gas cutting torch are shown in Fig. 1-4.

Several different fuel gases may be used in this process. The flame adjustments used when acetylene is the fuel gas are shown in Fig. 1-5. The flame adjustments for liquefied petroleum (LP) are shown in Fig. 1-6.

Welding goggles should be worn for eye protection. Approved gloves and proper clothing must be worn to prevent burns. It is important that the area of work be cleared of combustible material. Good fire prevention practices should be followed. Good ventilation should be provided. See Chapters 5 and 6 for more detailed instructions concerning oxyfuel gas cutting.

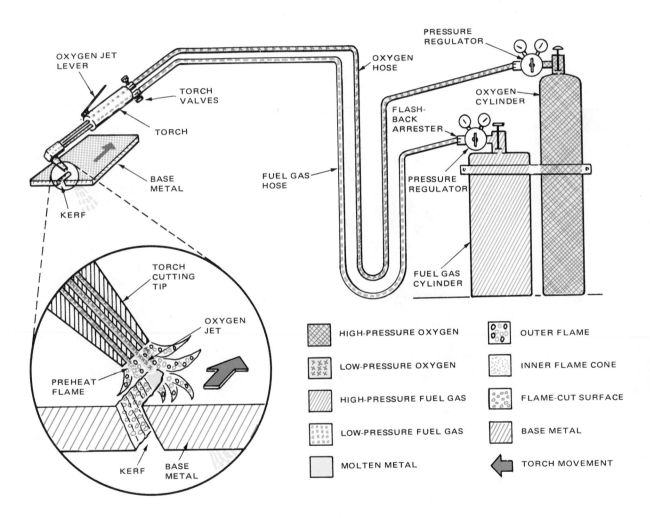

Fig. 1-3. Oxyfuel gas cutting (OFC). Oxygen and a fuel gas are mixed in the torch. The mixture burns at several orifices (openings) in the torch tip. When the flame has heated the base metal to a dull cherry red, a lever on the torch is pressed. This allows a jet of oxygen to rush out of an orifice. The oxygen jet quickly oxidizes the heated base metal and blows it away. This removal of material leaves a cut (kerf) in the base metal.

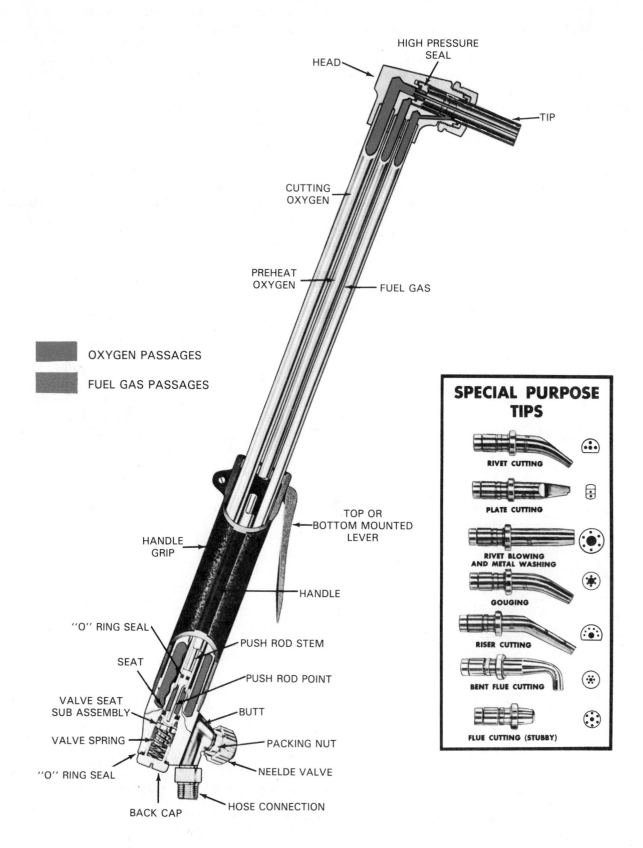

HIGH PRESSURE
SEAL

HEAD

TIP

CUTTING
OXYGEN

PREHEAT
OXYGEN

FUEL GAS

OXYGEN PASSAGES

FUEL GAS PASSAGES

TOP OR
BOTTOM MOUNTED
LEVER

HANDLE
GRIP

HANDLE

"O" RING SEAL

PUSH ROD STEM

SEAT

PUSH ROD POINT

VALVE SEAT
SUB ASSEMBLY

BUTT

VALVE SPRING

PACKING NUT

"O" RING SEAL

NEELDE VALVE

BACK CAP

HOSE CONNECTION

SPECIAL PURPOSE
TIPS

RIVET CUTTING

PLATE CUTTING

RIVET BLOWING
AND METAL WASHING

GOUGING

RISER CUTTING

BENT FLUE CUTTING

FLUE CUTTING (STUBBY)

Fig. 1-4. Cutting torch. Oxygen and a fuel gas are mixed and then carried to the tip orifice(s) to form preheat flames. Oxygen, carried directly to the tip, oxidizes the metal and blows it away to form the cut. (Smith Equip., Div. of Tescom Corp.)

**Acetylene Burning in Atmosphere**
Open fuel gas valve until smoke clears from flame.

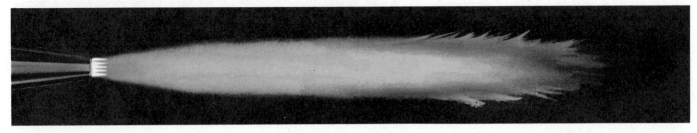

**Carburizing Flame**
(Excess acetylene with oxygen.) Preheat flames require more oxygen.

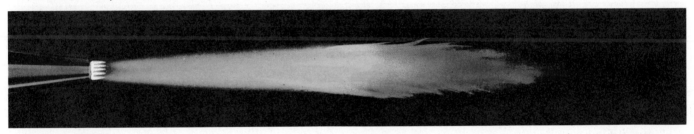

**Neutral Flame**
(Acetylene with oxygen.) Temp. 5720°F (3160°C). Proper preheat adjustment for cutting.

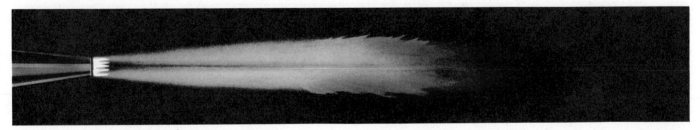

**Neutral Flame with Cutting Jet Open**
Cutting jet must be straight and clear.

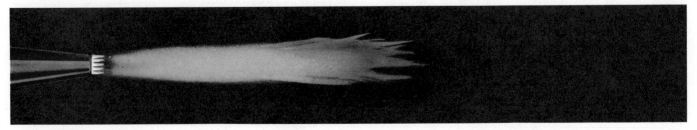

**Oxidizing Flame**
(Acetylene with excess oxygen.) Note recommended for average cutting.

Fig. 1-5. Conditions of the oxyacetylene cutting flame when adjusting the torch.
(Smith Equip., Div. of Tescom Corp.)

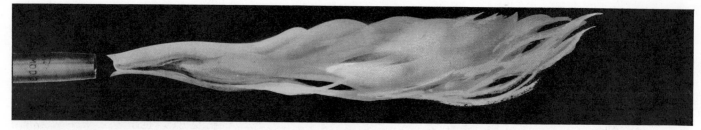

**LP Gas Burning in Atmosphere**
Open fuel gas valve until flame begins to leave tip end.

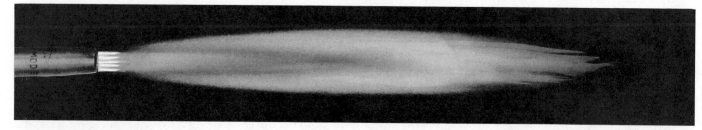

**Reducing Flame**
(Excess LP-Gas with oxygen.) Not hot enough for cutting.

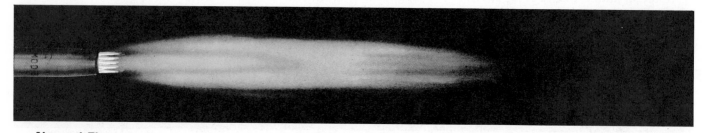

**Neutral Flame**
(LP-Gas with oxygen.) For preheating 1/8 in. and under prior to cutting.

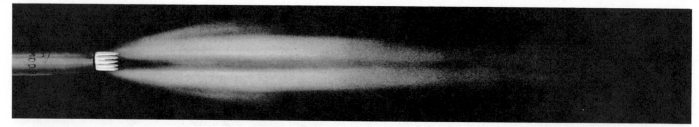

**Oxidizing Flame with Cutting Jet Open**
Cutting jet stream must be straight and clean.

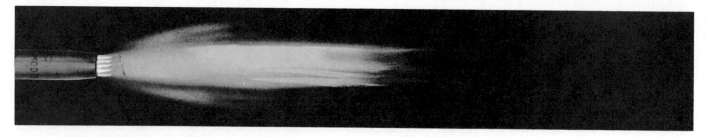

**Oxidizing Flame without Cutting Jet Open**
(LP-Gas with excess oxygen.) The highest temperature flame for fast starts and high cutting speeds.

Fig. 1-6. Conditions of the oxygen-LP Gas cutting flame when adjusting the torch.
(Smith Equip., Div. of Tescom Corp.)

## 1-3. TORCH SOLDERING (TS)

Torch soldering, as shown in Fig. 1-7, uses an air-fuel gas flame. The oxyfuel gas flame referred to in Fig. 1-1 may also be used for soldering. The base metal is heated enough to allow the solder to melt and then bond to the base metal. Soldering occurs at temperatures below 840 °F (449 °C). This is a popular method of joining metals in manufacturing and service operations. Torch soldering is used to fill a seam or to make an air-tight joint.

The air-fuel or oxyfuel gas torch can be used for soft soldering. It can also be used to braze small metal parts.

In the soldering process, the amount of heat is controlled by the amount of the gases flowing through the torch. Larger torch tip orifices are used when more heat is needed. The rate of gas flow is usually controlled by a needle valve on the torch. When air-fuel soldering, atmospheric air is drawn into the torch through holes at the base of the torch tip, as shown in Fig. 1-7.

A regulator is mounted on the cylinder to control the gas pressure to the torch. The final flame adjustment is made with the torch valve(s).

MAPP gas (methylacetylene propadiene) or LP (liquefied petroleum) may be used as the fuel gas. Some torches may burn any of these gases. Some torches, however, can only be used with one type of fuel. Be sure to check the recommended gas for the type torch being used.

Safety goggles or flash goggles are recommended to protect the eyes. Use pliers to handle the hot metal. Keep moisture away from molten solder. Moisture in contact with molten solder instantly changes to steam. This may cause molten solder to fly in all directions.

See Chapter 7 for more detailed instructions about soldering.

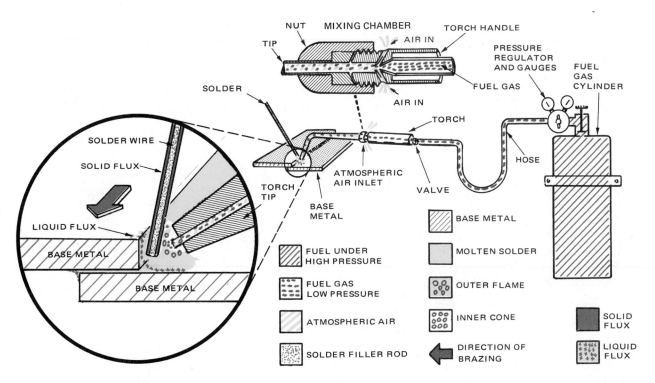

Fig. 1-7. Torch soldering (TS) with the air-fuel gas flame. A mixture of air and fuel gas is burned at the end of tip. The flame provides the heat to warm the base metal and to melt the solder. A flux is needed to keep the base metal and solder clean enough to allow the solder to adhere (stick). The oxyfuel gas station may also be used for torch soldering. See Fig. 1-1.

## 1-4. TORCH BRAZING (TB)

In torch brazing, as shown in Fig. 1-8, an oxy-fuel flame heats the base metal, and the heated base metal melts the brazing rod. Brazing is done at temperatures above 840°F (449°C). Brass and bronze are used as brazing rods. Their melting temperatures are much lower than that of steel. When torch brazing, a thin layer of brass or bronze is used. The process of brazing is similar to soldering in this respect. See Fig. 1-8. There is usually less warping of the base metal when brazing than when welding. This is due to the lower temperature involved in brazing or braze welding. Fig. 1-8 illustrates a typical torch brazing or braze welding station.

The oxyfuel gas flame is adjusted to provide a neutral, oxidizing, or reducing flame, depending on the behavior of the metal in the joint. This means that flame adjustment is very important. See Fig. 1-9.

Metal parts to be brazed or braze welded are first carefully cleaned. A flux material is used to keep the metal clean. Flux is added in powder or paste form or as a coating on the brazing filler metal.

The oxyfuel gas flame is used to heat the base metal. The brazing filler metal is touched to the base metal and it is heated by the base metal. It then melts and flows into the joint. If the mating surfaces were properly cleaned, a good brazed joint should result.

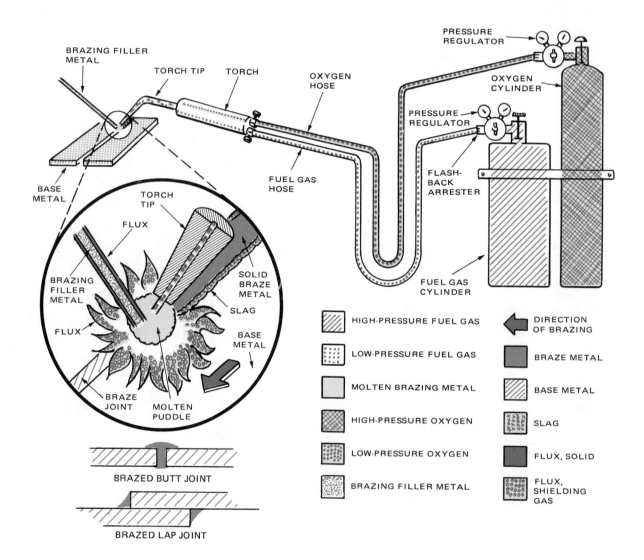

Fig. 1-8. Torch brazing (TB). Oxygen and a fuel gas are mixed in a torch. The mixture is burned at the torch tip. Heat raises the temperature of a spot on the base metal until the brass or bronze filler metal melts and adheres to the base metal. A flux is used to keep the base metal and filler metal clean during the operation.

Torch brazing can also be used to produce thick, beveled joints similar to oxyfuel gas welding. The process used on these thicker joints is called braze welding. The base metal is not melted.

The welder should wear goggles, gloves, and fire-resistant clothing. There may be a severe health hazard if brazing filler metal contains zinc, cadmium, phosphorous, or beryllium alloys. Overheating the brazing alloy should, therefore, be avoided. The brazing operation must be well ventilated to remove the fumes from the brazing alloys and flux.

See Chapter 8 for more detailed instructions about brazing.

**LP Gas Burning in Atmosphere**
Open fuel gas valve until flame begins to leave tip end.

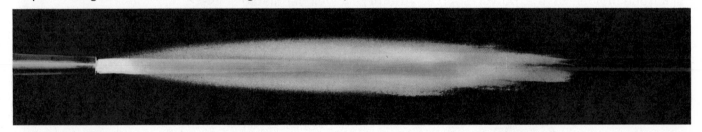

**Reducing Flame**
(Excess LP-Gas with oxygen.) For heating and soft soldering or silver brazing.

**Neutral Flame**
(LP-Gas with oxygen.) For brazing light material.

**Oxidizing Flame**
(LP-Gas with excess oxygen.) Hottest flame about 5300 °F (2927 °C). For fusion welding and heavy braze welding.

Fig. 1-9. Color appearance of oxygen-LP gas flames.
(Smith Equip., Div. of Tescom Corp.)

## 1-5. SHIELDED METAL ARC WELDING (SMAW)

The shielded metal arc welding process, Fig. 1-10, uses an electric arc between a flux covered metal electrode and the metal being welded (base metal). Heat from the electric arc melts both the end of the electrode and the base metal to be joined. This process is often used for maintenance and small production welding. Heavy pipe welding is done almost exclusively with shielded metal arc welding.

Equipment used in this welding process provides an electric current for welding. The electric current may be either alternating (AC) or direct (DC). Current adjustment controls on the welding machine allow the welder to set the desired current. Movement of the hand-held electrode holder is controlled by the operator. The electrode is a flux covered metal wire. A cable connects the electrode holder to the power source. Another cable connects the work to the power source to complete the circuit.

Fig. 1-10 illustrates a typical station for shielded metal arc (stick) welding.

The heat of the electric arc may be controlled by the current setting and by the arc length. Electrode diameter and flux material will determine the type (AC or DC) and amount of welding current required.

The arc between the welding electrode and the base metal is struck (initiated) by the operator. Correct arc length must also be controlled by the operator.

Some of the covering on the electrode turns into a protective gas shield which surrounds the arc as the electrode melts. Some of the covering melts to form a slag which covers the completed weld. The slag layer protects the hot metal from oxidizing while it cools.

Welders must wear an approved helmet, gloves, and protective clothing. The work station must be well ventilated.

Chapters 9 and 10 give more detailed information on shielded metal arc welding.

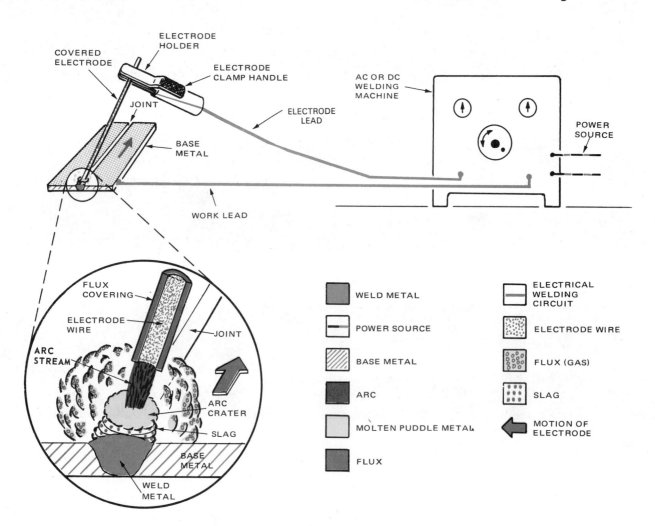

Fig. 1-10. Shielded metal arc welding (SMAW). An electric arc is drawn between the covered electrode and the base metal. The heat of the arc melts the end of the electrode and the base metal where the arc contacts it. The metal from the electrode provides the filler metal for the weld.

## 1-6. GAS TUNGSTEN ARC WELDING (GTAW)

Gas tungsten arc welding uses the heat of an electric arc between a tungsten electrode and the base metal, Fig. 1-11. A separate welding filler rod is fed into the molten base metal, if needed. A shielding gas flows around the arc to keep away air and other harmful materials. This process is also called TIG (Tungsten Inert Gas) welding.

Gas tungsten arc welding is particularly desirable when welding stainless steel, aluminum, titanium, and many other nonferrous metals.

Fig. 1-11 illustrates a typical station for gas tungsten arc welding. An AC-DC welding machine may be used with a regulated flow of a shielding gas such as argon or helium. The shielding gas flows from a cylinder through a regulator, flow meter, and a hose to the GTAW torch.

The welder normally manually operates the torch (electrode holder) and the filler metal. This torch has a gripping device to hold the tungsten electrode. A heat-resistant gas flow cup or nozzle surrounds the electrode. Some small capacity torches are air-cooled. Large torches are water cooled.

Heating properties of the arc may be controlled by changing current and arc length. The diameter of the tungsten electrode, thickness, and kind of base metal, will determine welding amperage.

GTAW welding may be done in any position with excellent results.

Gas tungsten arc welding generates intense heat and light, with no metal spatter. The welder must wear a welding helmet, gloves, and welder's clothing. See Chapters 11 and 12 for additional information.

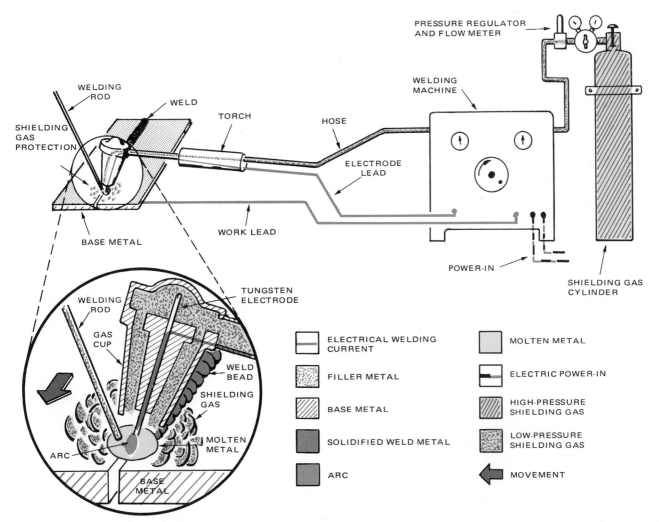

Fig. 1-11. Gas tungsten arc welding (GTAW). An electric arc is drawn between the end of the tungsten electrode and a spot on the base metal. Only the base metal melts. If filler metal is needed in the joint, a separate welding rod is used. The welding rod is added to the molten puddle as needed. Shielding gas flows out a nozzle around the tungsten electrode.

## 1-7. GAS METAL ARC WELDING (GMAW)

In gas metal arc welding, Fig. 1-12, an electric arc between a continuously fed metal electrode and the base metal produces heat. The arc is shielded by a gas. This process is popular in production and repair shops. It is often called MIG (Metal Inert Gas) welding.

As shown in Fig. 1-12, a shielding gas cylinder, regulator, flow meter, and hose provide a flow of shielding gas to the arc. Shielding gases such as carbon dioxide, argon, or helium may be used. An electrode feeding device supplies metal electrode continuously. A torch and cable carry the electrode wire, current and shielding gas to the arc. The torch usually has a trigger switch for starting and stopping the electrode feed and gas flow.

A constant voltage DC welder is used with this process. The desired voltage is set on the welding machine. Current is changed by adjusting the wire feed speed. Speed controls for the wire are usually mounted in the wire feed mechanism.

Shielded gas volume adjustments are made at a gas flow meter on the regulator. The kind of shielding gas used usually depends on the metals being welded.

The welder:

1. Selects the electrode size.
2. Sets the desired voltage.
3. Adjusts the shielding gas flow.
4. Adjusts the rate of electrode feed.
5. Controls the torch movement and electrode extension. (The electrode extension is the distance from the torch tip to the arc.)

The welder must wear an approved helmet, gloves, and welder's clothing. The welding area must have good ventilation. See Chapters 11 and 13 for additional information.

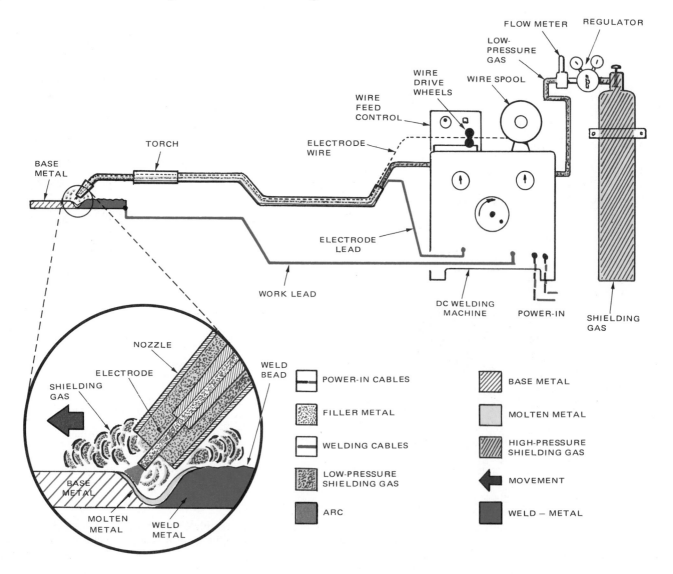

Fig. 1-12. Gas metal arc welding (GMAW). An electric arc is drawn between a metal electrode and the base metal. Heat from the arc melts the end of the electrode wire and a spot on the base metal. A shielding gas flows out of the torch nozzle. This gas keeps the oxygen and impurities in the air from contacting the weld.

## 1-8. FLUX CORED ARC WELDING (FCAW)

In flux cored arc welding, Fig. 1-13, heat comes from an arc between a flux cored electrode and the base metal. This process is particularly desirable for welding structural steel and in other low-carbon applications.

A constant voltage (constant potential) DC arc welding machine is used.

The electrode is a hollow metal tube with the center (core) filled with a flux material. The electrode wire is fed to the torch from a large spool on the electrode drive mechanism. Some flux cored wires are used without $CO_2$ shielding gas. Others use $CO_2$.

The heat of the arc depends upon the arc length, voltage setting, and the wire feed speed. The speed of the electrode feed mechanism may be adjusted. Adjusting the wire speed varies the current at the arc. The higher the wire feed speed, the higher the current and, thus, the greater the heat of the arc.

The welder is exposed to heat and light from the arc. An arc welder's helmet, leather gloves, and protective clothing should be worn. Excellent ventilation should also be provided.

Chapters 11 and 13 give more detailed information concerning the operation of this process.

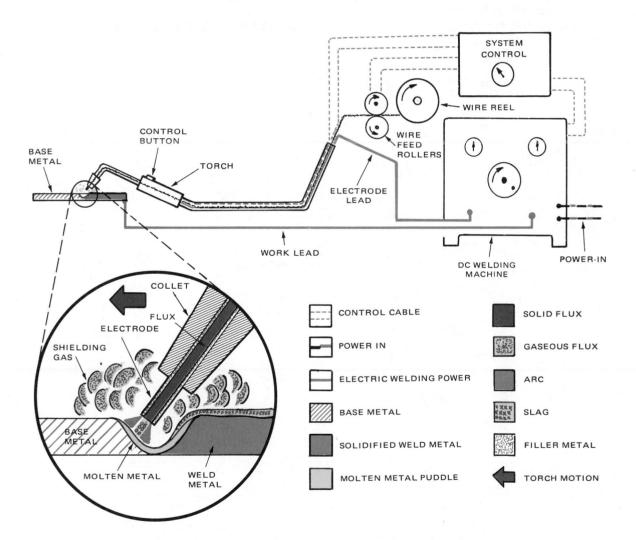

Fig. 1-13. Flux cored arc welding station. (FCAW). The heat energy comes from an electric arc. The end of the electrode and a spot on the base metal are melted to form the weld. The flux core provides a gaseous shield around the arc and also provides a slag covering to keep the air away from the weld until it cools.

## 1-9. AIR CARBON ARC CUTTING (AAC)

The air carbon arc cutting process uses an electric arc to melt the base metal. A jet of air then blows the melted metal away. Air carbon arc cutting may be used on many metals.

Fig. 1-14 shows a typical station for air carbon arc cutting and gouging. The recommended electrical supply is direct current (DC) or alternating current (AC). An electrode lead (flexible cable) connects the electrode holder to the welding machine. A work lead (ground cable) connects the base metal to the welding machine.

The air jet may be supplied from either a compressed air cylinder or an air compressor. The air line is attached to the electrode holder. A lever-operated valve in the electrode holder controls the air flow. The welder operates the electrode holder manually. This process can be used for either cutting or gouging metal.

The length of the carbon electrode between the air jet nozzle and the arc must be maintained at such a distance that the air jet will be effective in blowing away the molten metal.

Current is regulated by adjustments on the welding machine. The arc length is controlled by the welder.

This cutting process produces considerable sparking. The welder must be protected by gloves, helmet, and clothing. Excellent ventilation is needed. Good fire prevention practices must be followed.

See Chapters 14 and 15 for more details.

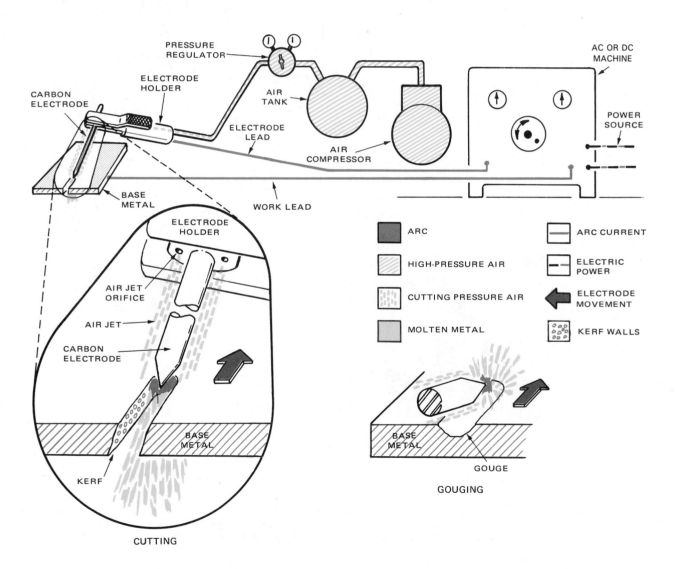

Fig. 1-14. Air carbon arc cutting (AAC). The electric arc between the carbon electrode and the base metal melts the base metal. Manually operated air jets attached to the electrode holder blow the molten metal away. This process is used for gouging base metal as well as for cutting.

## 1-10. OXYGEN ARC CUTTING (AOC)

The oxygen arc cutting process uses an electric arc to heat the base metal. Then a jet of oxygen cuts the heated metal. It is used for cutting cast iron, steel, and many other metals. It is a rapid metal-cutting process. Fig. 1-15 illustrates a typical station for oxygen arc cutting.

The equipment includes a special electrode holder and a hollow metal electrode. An oxygen cylinder and regulator provide a controlled flow of pressurized oxygen through the hollow electrode. A flexible hose carries the oxygen to the electrode holder. The operator controls the oxygen through a hand valve on the electrode holder. An AC or DC welding machine supplies an arc current to the electrode. Current is conducted to the electrode holder through the electrode lead. A work lead connects the metal workpiece (base metal) to the welding machine.

To make the cut, the operator strikes an arc and, as soon as the metal surface to be cut has a molten spot, the operator opens the oxygen valve. The flow of oxygen into the arc and against the metal very rapidly oxidizes and blows away the metal. This process may be used either in air or under water.

The heat of the arc is adjusted by a current adjustment control on the welding machine.

This cutting process may produce a great shower of sparks. There is also a danger from sparks falling around the legs and feet. The operator must wear a helmet, gloves, and protective clothing. Good ventilation and fire protection are important.

See Chapters 14 and 15 for more details on this process.

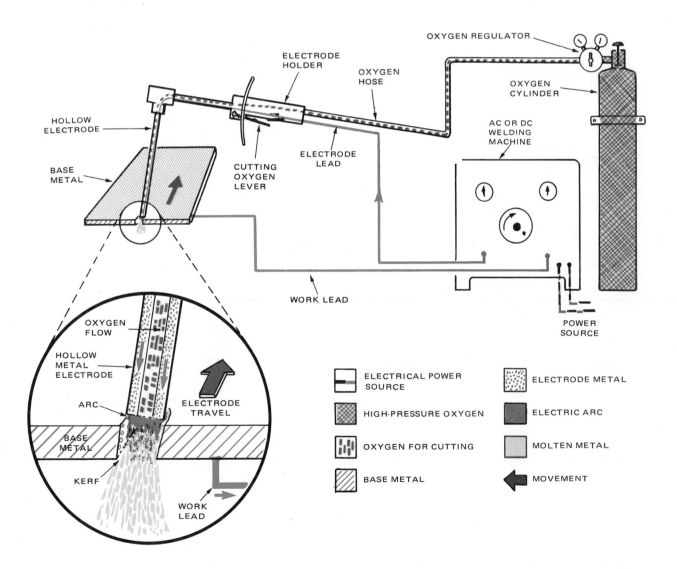

Fig. 1-15. Oxygen arc cutting (AOC). An electric arc is drawn between a hollow electrode and the base metal. A lever on the holder allows a jet of oxygen to flow through the electrode. This jet cuts the base metal by rapid oxidization (combining the metal with oxygen).

## 1-11. GAS TUNGSTEN ARC CUTTING (GTAC)

In gas tungsten arc cutting, Fig. 1-16, an arc between a tungsten electrode and the base metal heats the metal. The shielding gas then blows the melted metal away. This process is used on aluminum, stainless steel, nickel and many other metals. Since the electrode being used to melt the metal is tungsten, the electrode does not burn away rapidly.

Fig. 1-16 illustrates a typical station for gas tungsten arc cutting. Equipment includes a water-cooled electrode holder and a current supply. A regulated supply of shielding gas is required.

Gases used are argon, helium, or hydrogen.

The current supply is usually direct current straight polarity. The DCSP is furnished by an arc welder generator or rectifier. Current may be adjusted to meet job requirements.

The operator adjusts the flow of current, water, and shielding gas. The cutting torch can be either manually or automatically operated.

Shielding gas serves two purposes. It blows the molten metal away from the cutting area and keeps the surfaces of the cut from oxidizing.

Gloves, helmet, and protective clothing must be worn. There is considerable sparking with this type of cutting.

See Chapters 14 and 15 for more details.

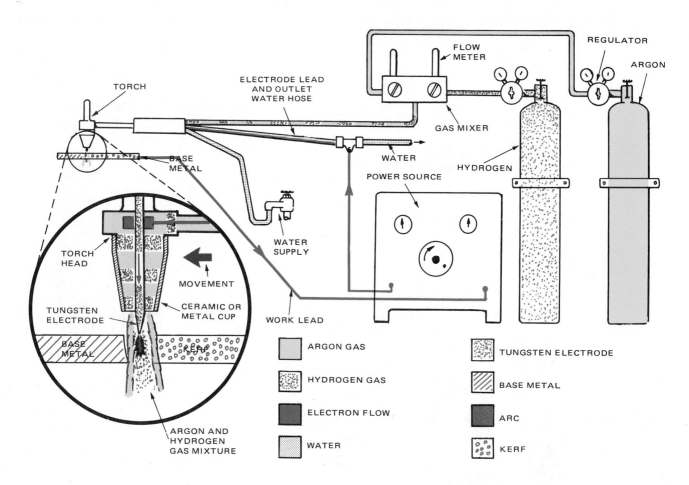

Fig. 1-16. Gas tungsten arc cutting (GTAC). An arc between a tungsten electrode and the base metal heats and melts the base metal. A mixture of argon and hydrogen gases is used to blow the melted metal away to form a kerf (cut) in the base metal.

## 1-12. SHIELDED METAL ARC CUTTING (SMAC)

The shielded metal arc cutting process, Fig. 1-17, uses an arc between a metal electrode and the base metal. This arc melts the base metal. The electrode is heavily covered with flux. The molten metal drops away from the base metal to form a kerf (cut). This process is used mainly for small maintenance jobs.

The arc cutting process requires a current source, either AC or DC. A manually operated electrode holder provides a grip for controlling the electrode. Electric current from the arc welding machine flows through the electrode holder lead and arcs between the electrode and the workpiece. A work lead (ground cable) between the workpiece and the power source completes the circuit in this process.

Fig. 1-17 illustrates an arc cutting station. Heat from the arc is controlled by the arc length, current, and electrode material.

In operation, both electrical and mechanical equipment is adjusted to provide the desired arc. The operator strikes the arc between the electrode and the base metal and the cutting is started. The operator moves the electrode as the cut progresses through the base metal.

Protection is needed from the intense heat, the light of the arc and the sparks. This requires wearing an approved helmet, gloves, and welder's clothing. The equipment used must include the necessary shielding and safety devices. Excellent ventilation is needed. Good fire prevention practices must be followed.

Refer to Chapters 14 and 15 for more detailed instructions on SMAC.

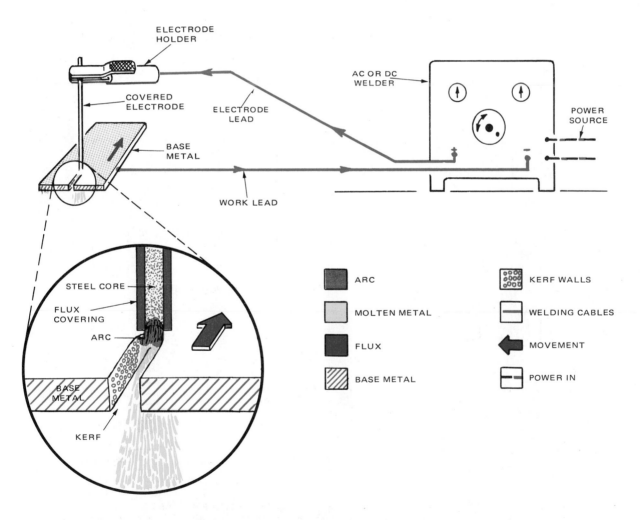

Fig. 1-17. Shielded metal arc cutting (SMAC). An arc between a very heavily covered metal electrode and the base metal melts the end of electrode and the base metal. The metal electrode melts far back into the covering, and this produces a jet action of gases which blows the molten base metal away.

## 1-13. PLASMA ARC CUTTING (PAC)

Plasma arc cutting uses an electric arc and fast-flowing ionized gases to melt and cut metals. This process cuts aluminum, stainless steel, and most other metals rapidly. Plasma arc cutting can be used to cut non-metals such as concrete. Fig. 1-18 illustrates a station for plasma arc cutting.

The metal is melted by the heat of the plasma arc. Then the molten metal is blown away by the high velocity of the shielding gas.

Plasma arc cutting requires a special water-cooled cutting nozzle. It makes use of a tungsten electrode connected to a source of DC power, compressed gas, and suitable controls.

Plasma arc cutting is usually used along with automatic cutting devices. The current is controlled by devices on the power source. Water flow to cool the torch is usually manually adjusted by the operator.

Plasma arc cutting is a very noisy process. The operators must be protected from the noise by the use of ear plugs or industrial "ear muffs." It is sometimes necessary to use a "walkie-talkie" type of communication system where plasma arc cutting is being done. The operator must also be protected with an approved helmet, gloves, protective clothing, and other required safety equipment.

See Chapters 14 and 15 for more details.

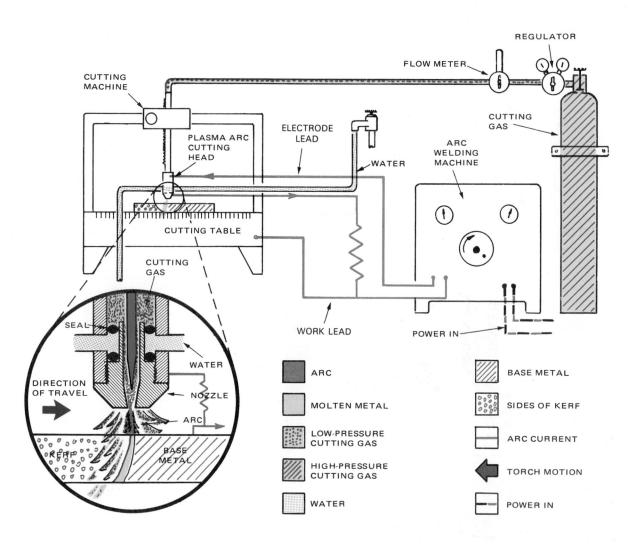

Fig. 1-18. Plasma arc cutting (PAC). An electric arc between a tungsten electrode and the base metal ionizes some of the cutting gas. This ionized gas (plasma) leaves the torch and hits the base metal. The very high plasma temperatures superheat the base metal and very rapidly forms a kerf (cut) in the base metal.

## 1-14. RESISTANCE SPOT WELDING (RSW)

Resistance spot welding, Fig. 1-19, passes an electric current through the metal. Resistance to the electrical flow heats the metal to welding temperature. The process is used to weld together two or more overlapping pieces. It is well suited to automatic welding. Spot welding is commonly used to join auto body sections, cabinets, and other sheet metal assemblies.

A step-down transformer converts fairly high voltage-low amperage current to a low voltage-high amperage current. The weld is made between two electrodes which press the metals together. A large electrical current flows from one electrode through the metals to be welded together to the second electrode.

These electrodes are special copper alloys which can carry the high current and still have physical strength to operate under high pressures. The electrodes on small spot welders, used to weld thin materials, may be air-cooled. Electrodes for welding thicker metals are water-cooled.

Resistance spot welding is controlled by the amperage, the electrode pressure, and the length of time the current flows.

In an automatic spot welder, the operator sets the electrode current, electrode pressure, and the timing. The electronic controller repeats the desired weld cycle each time the start switch is pushed.

The operator must wear flash goggles. If the metals must be manually handled, gloves are needed to prevent cuts and burns to the hands.

See Chapters 16 and 17 for detailed information concerning spot welding.

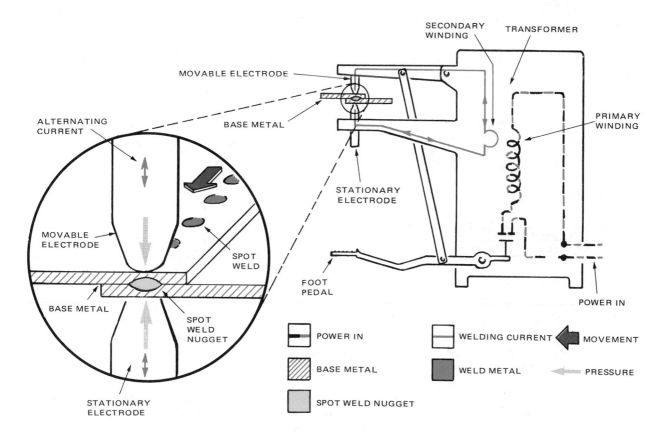

Fig. 1-19. Resistance spot welding (RSW). A step-down transformer provides a low voltage, high current electrical flow. Current through the electrodes heats a small area on two sheets of metal as these sheets are pressed together between the spot welding electrodes. The metals become hot enough to fuse together.

## 1-15. PROJECTION WELDING (RPW)

The projection welding process uses resistance to the flow of electricity to create heat for welding. See Fig. 1-20. It is similar to spot welding. It is commonly used in production welding.

One of the two pieces of metal is run through a machine which makes bumps or projections of a designed shape and size in the metal.

The welding machine electrodes are flat plates called platens. The two pieces of base metal are placed together between the platens. They touch only at the projections or pumps on the one piece.

Welding current is supplied by a resistance welder transformer. The welding current flows through the pieces to be welded while they are clamped between the platen plates. Due to the projections, the current is concentrated at the points of contact. These points heat up and fuse.

The welding current flows for a short time and, at the same time, pressure is applied between the two platens. This completes the weld. Timing of the current flow and application of welding pressure are important parts of this welding process.

Some flash and sparking may take place during the welding. The operator should wear flash goggles, approved clothing, safety shoes, and leather gloves.

A more complete explanation of this process is given in Chapters 16 and 17.

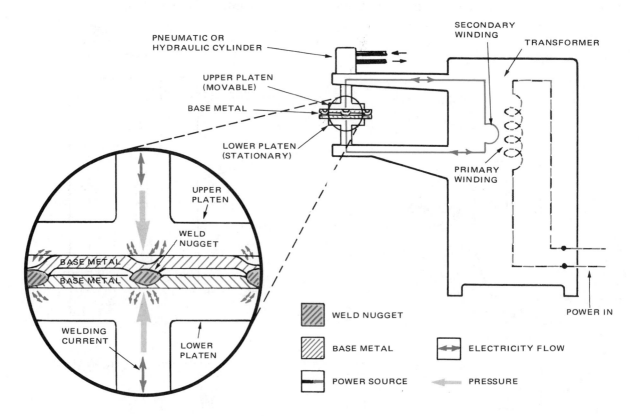

Fig. 1-20. Projection welding (RPW). Electrical energy heats the projections in one sheet of base metal as these projections touch another sheet of base metal. Both pieces become hot enough at the contact spots to fuse together as pressure is applied.

## 1-16. FLASH WELDING (FW)

The flash welding process uses an electric arc to heat the base metals. It provides a strong, clean weld joint. Its chief use is in production welding. It combines resistance welding, arc welding, and pressure welding.

The process, as shown in Fig. 1-21, uses a step-down transformer to provide the welding current. The two pieces to be welded are held in current-conducting movable clamps.

To make a flash weld, the workpieces are brought together under light pressure. A low voltage, high current travels between the two base metals.

As soon as the current is established, the two pieces of metal are drawn apart very slightly. At this point an electric arc passes between them. The arc heats the surfaces of the two metals. When the surfaces are sufficiently heated, they are forced together under very high pressure. This pressure causes a slight outward flow of heated and somewhat dirty metal from the joining surfaces. The clean, heated subsurface metal brought into contact produces a good weld. The finished weld will have a flash or enlargement at the joint.

Welding heat is controlled by the current adjustment setting. The quality of the weld is controlled by the current, the length of time of the arc, and finally, the pressure at the time the two surfaces are brought together.

When flash welding manually, the operator must be able to judge the metal temperatures, time that the welding surfaces must be brought together, and the proper welding pressure.

It is necessary that protective clothing, a face shield, and gloves be worn. Flying sparks (metal expelled at joint) are produced during this process.

See Chapters 16 and 17 for more detailed information.

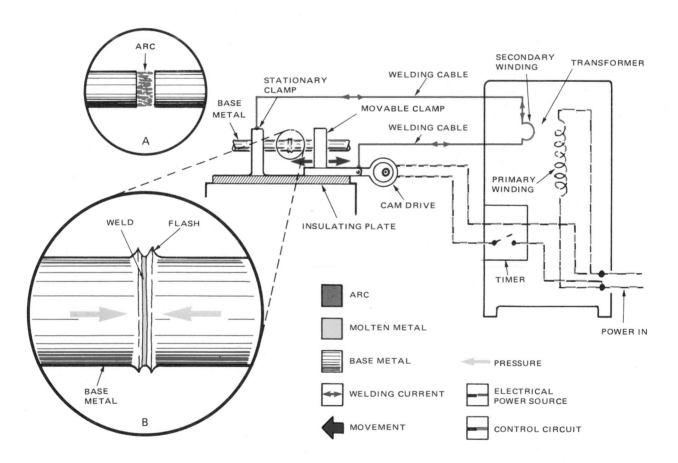

Fig. 1-21. Flash welding (FW). A—Electrical energy creates an arc which melts the ends of two pieces of base metal. B—When the ends are molten, they are pushed together to fuse into one piece.

## 1-17. RESISTANCE SEAM WELDING (RSEW)

Resistance seam welding, Fig. 1-22, is a special application of spot welding. It is often used to weld joints in containers and other products which require an airtight or vapor-tight seam.

The electrodes are wheels. The work to be welded is passed between the revolving wheel electrodes. A timing device turns on the welding current at controlled but rapidly repeating intervals. The rapidly repeating current flow makes a series of overlapping spot welds which appear to be a continuous line of welding. These machines are usually automatic. The current, pressure on the electrodes, and sequence times are set by a welding operator. The timing of the process is regulated by an electronic controller.

The operator must wear flash goggles and all other required safety clothing. If the metal must be handled, the operator should wear approved gloves.

Refer to Chapters 16 and 17 for additional information.

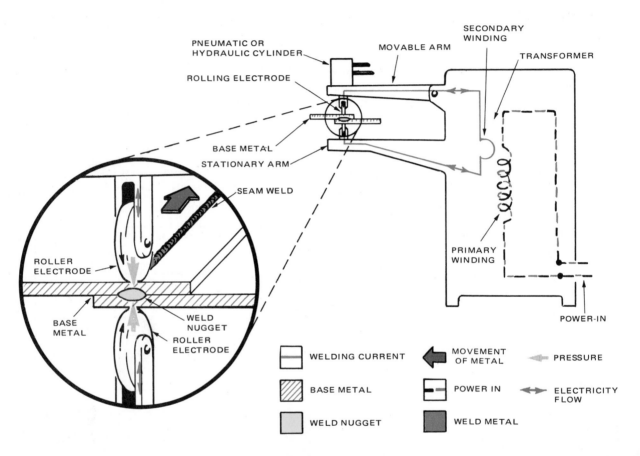

Fig. 1-22. Resistance seam welding (RSEW). Electrical energy travels between two electrode wheels. As the wheels travel, they clamp two sheets of base metal together. As the electricity travels through the base metal, the metal becomes hot enough to fuse the two sheets together and form a seam weld.

## 1-18. SUBMERGED ARC WELDING (SAW)

In submerged arc welding, an electric arc between an electrode and the base metal produces welding heat. The arc is submerged in a granular flux. Some of the flux melts and forms a shielding slag over the weld. This process is often used when welding thick plate joints. The equipment is usually automatically or semiautomatically operated. The electrode also feeds automatically into the arc. See Fig. 1-23.

A hopper and feeding mechanism are used to provide a flow of flux over the joint being welded. The arc, generated by AC or DC current, is submerged in the flux. See Chapter 18.

The chemical composition of the flux will affect the composition of the completed weld. Alloy elements can be added to the weld by controlling the chemical composition of the flux.

Welding heat is regulated by changing the voltage and wire feed speed on the welding machine.

The electrode, which is usually power fed and made of mild steel, extends through the flux to a point just above the base metal. The flux material shields the base metal from oxidation while it cools. The correct arc length is automatically maintained underneath the shielding material.

As the weld progresses, some of the flux melts and forms a slag over the weld. A vacuum machine may be used to pick up the flux for reuse.

Since the arc is submerged in the shielding material, it cannot be seen during welding. This reduces, to some extent, the hazard of burns and flying sparks. Still, the operator should wear approved goggles, gloves, clothing, and provide good ventilation. See Chapter 18 for more details on submerged arc welding.

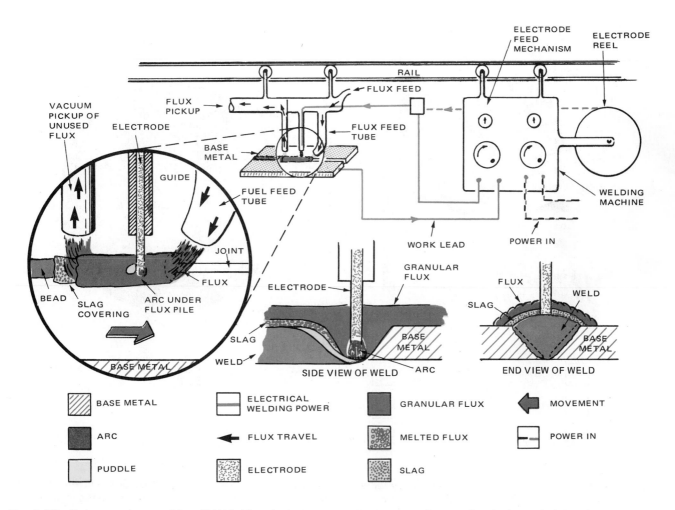

Fig. 1-23. Submerged arc welding (SAW). Electrical current creates an arc to heat and melt the end of the electrode wire and the base metal. The arc is covered with grains of flux. Some flux melts and forms a slag over the weld.

## 1-19. ELECTROSLAG WELDING (ESW)

Electroslag welding, Fig. 1-24, is used to weld butt joints on thick metal. The weld is done vertically and it moves upward.

Prior to starting the weld, a flux material several inches deep is placed between the two base metals. The flux is able to conduct electricity.

To start the weld an electric arc is struck between one or more electrodes and the base metal. The electrodes are continuously fed by a drive mechanism. The flux is melted by the heat of the arc. After the flux becomes molten, the arc is stopped, but electricity continues to flow. The flux (slag) is kept molten by its resistance to the flow of the electricity through the electrode and flux to the base metal.

The molten flux melts the base metal and the continuously fed filler metal to form a weld.

Movable, water-cooled molds (shoes) are used on each side of the joint. These shoes (molds) keep the molten flux, filler metal, and base metal in place as the weld moves up.

The shoes are water cooled to prevent them from melting into the weld.

Molten flux (slag) acts as a shield from the atmosphere. The electrode is often automatically oscillated (moved back and forth) when extremely thick plates are welded.

As the weld moves up, the shoes (molds) are moved up also. Fig. 1-24 illustrates a typical set-up for electroslag welding.

The heat source is controlled by the amount of current and the physical characteristics of the flux. Operators must wear protective clothing. The heat developed in this process is very intense. See Chapter 18 for more details.

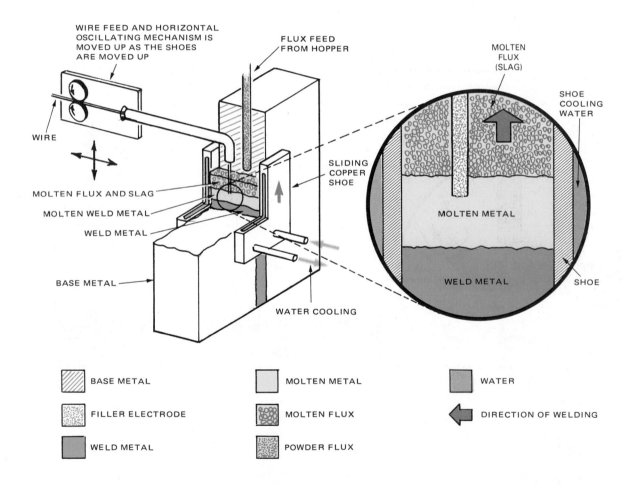

Fig. 1-24. Electroslag welding (ESW). The energy for welding is an electric current. This current keeps the flux molten. The molten flux melts the base metal and filler metal. The weld is formed vertically. Water-cooled shoes contain the molten metal until it solidifies.

## 1-20. STUD ARC WELDING (SW)

Stud arc welding is a semiautomatic welding process. It is used to attach metal fastening devices to metal plates or beams without drilling and tapping. Bolts, screws, rivets, and spikes may be attached in this way.

Fig. 1-25 shows a stud arc welding station. The heat source is an electric arc. The enlargement shows the stud, stud chuck, and ceramic ferrule. Another sequence of line drawings shows the four steps in making the welded joint.

The energy source (welding machine) is an elec-tric welding transformer. The control on the welding machine determines the current in the electric arc. Current settings vary with the size of the stud and the kind of metal. The control unit has a timer which controls the duration of the arc.

The operator installs the stud in the gun. The gun is then positioned on the base metal. A switch on the gun starts the stud welding operation cycle.

The operator must wear gloves, a face shield with flash goggles, and fire-resistant clothing.

A more complete explanation of stud welding is given in Chapter 18.

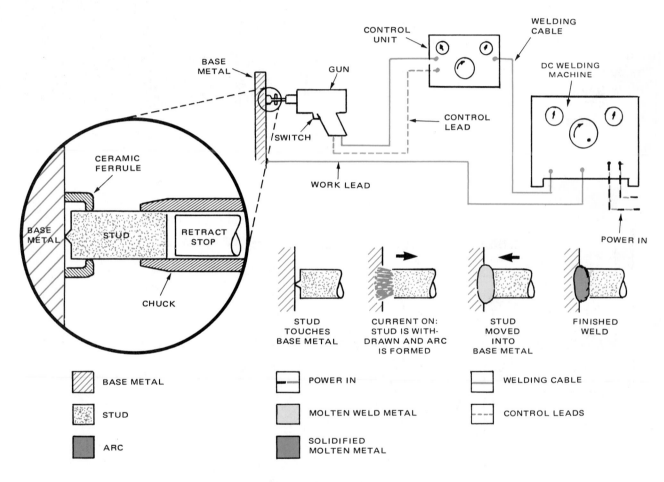

Fig. 1-25. Stud arc welding (SW). Electrical energy creates an arc which melts the end of a special electrode (a bolt, a pin, a clip, or a stud) and the base metal. The special electrode is then moved into the base metal puddle. The assembly is held until the weld metal cools.

## 1-21. COLD WELDING (CW)

No outside heat source is used. Cold welding uses very high pressure to force metals together. Only the surface molecules are heated and fused to form a weld. The method is used mainly on softer metals such as aluminum-to-aluminum, copper-to-copper, and aluminum-to-copper. Good fusion occurs, resulting in a strong weld. Butt welds and lap welds may be made with this process. Metal surfaces being joined must be very, very clean.

The source of energy to produce the weld is a tremendous pressure usually produced by using hydraulic cylinders. See Fig. 1-26. The weld is controlled by the size of the die surfaces in contact with the metal and the amount of hydraulic pressure.

The operator should wear gloves, a face shield or safety goggles, and approved clothing.

See Chapter 18 for more details on the cold welding process.

Semiautomatic flux-core arc welding of box columns on a major construction project. (Lincoln Electric Co.)

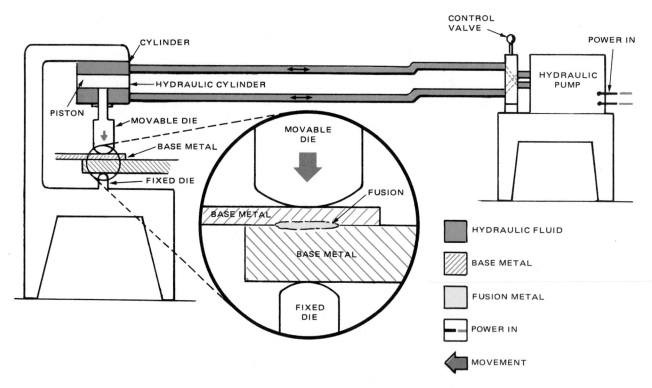

Fig. 1-26. Cold welding (CW). Pressure creates the welding energy. In this illustration a thin sheet of base metal is lap welded to a thicker sheet of base metal. Note that the upper die is larger than the bottom die when the top base metal is softer than the bottom base metal. A hydraulic piston is used to force the upper die against the metal. The hydraulic pressure used is only enough to fuse the surfaces of the base metals being welded.

## 1-22. EXPLOSION WELDING (EXW)

Explosion welding, Fig. 1-27, joins metals together by using a powerful shock wave. This creates enough pressure between two metals to cause surface flow and cohension. It is often used to weld large sheets together. In one common application, it is used to weld thin stainless steel sheet to mild steel sheet. It is also used for welding aluminum and molybdenum together.

The energy souce is the tremendous shock wave caused by igniting an explosive material. The operation requires very careful setup. Bonding takes place in an instant. Such welding is done either in a safety chamber, or less frequently, under water.

Safety is very important! Both the explosives and fixtures must conform to approved written specifications. The welder must be protected from the sound of the explosion by wearing industrial "ear muffs" and/or ear plugs. Face shields, safety helmets, and approved clothing should be used. Special permits are required from government authorities because of the explosives used.

Chapter 18 explains this special type of welding in more detail.

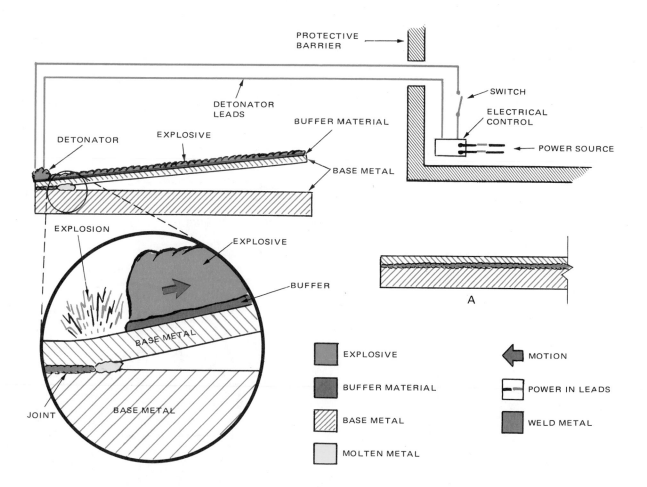

Fig. 1-27. Explosion welding (EXW). A buffer material is applied to the top surface of one of the metal sheets. Welding energy comes from explosive material placed on top of buffer material. An igniter (detonator) is operated from behind a barrier. The explosion proceeds from left to right and welds the top plate to the bottom plate almost at once, without deforming either piece of metal. A—Completed weld.

## 1-23. FORGE WELDING (FOW)

In forge welding a furnace heats two pieces of metal to a plastic temperature. Pounding (hammer blows) fuses the two pieces together. This process is used to join wrought iron, low carbon steel, and medium carbon steel pieces. It can be used in places where there is no electricity or fuel gas available. Fig. 1-28 illustrates a typical blacksmith's forge welding station.

The forge has a cast iron pan or tub. A blower supplies air at the bottom of the forge through an opening called a tuyere. The fuel is usually a good grade of soft coal. Heat changes the burning coal to coke near the center of the fire. The fire is carefully tended by feeding coal from the outside toward the center. Coking means driving out the combustible gases from the coal. As this coal is being coked, the gas that is released will burn in an open flame over the fire. A thick bed of coked coal should be made before any welding is attempted by this process.

Heat is controlled by the airflow through the tuyere (air inlet). Increasing airflow increases the rate of combustion.

Metal parts to be welded are placed down in the coke at the point of combustion. When the parts reach the forging temperature indicated by the bright red color of the metal, they are withdrawn. When heated, the parts are placed together on an anvil and pounded together to make a weld. Heating may also be done in a furnace burning a fuel gas. See Chapter 18.

The blacksmith welder should wear eye protection, safety shoes, and flame-resistant clothing and gloves. This process produces a great deal of heat. Sparks are often created when the hot metal is pounded.

See Chapters 18 and 30 for more details.

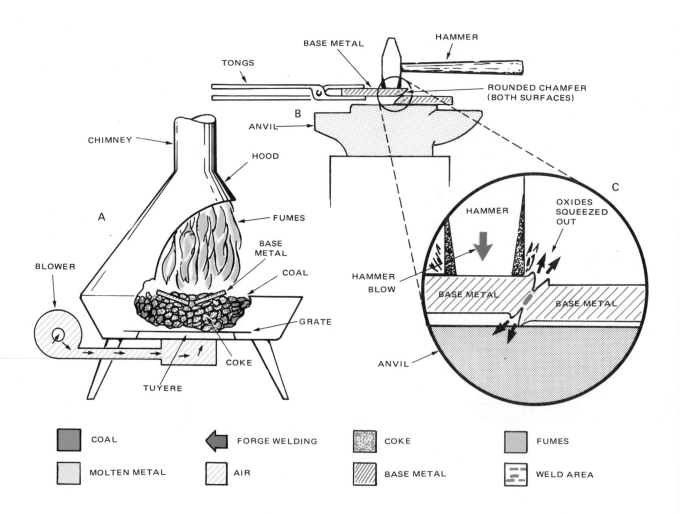

Fig. 1-28. A forge welding (FOW) station. A—Cross section through blacksmith's coal fired forge. B—Hammer and anvil being used to make forged weld. C—Enlarged view of welding action.

## 1-24. FRICTION WELDING (FRW)

Friction welding uses friction to create enough heat to fuse two pieces of metal together. This process is used chiefly in butt welding rather large, round rods or cylinders. Fig. 1-29 illustrates a friction welding station.

No outside heat is supplied. One of the pieces is made to revolve. The ends of the parts to be joined are then brought together under a light pressure. The resulting friction between the stationary and revolving part develops the heat needed to form the weld. As the metal surfaces reach the plastic state, they are forced together under a much greater pressure. The process creates a clean metal-to-metal welded surface.

Equipment includes the necessary clamping devices, a mechanism for revolving one part and a method for subjecting the friction surfaces to a high pressure.

The control of this type of weld is based on the pressure and the speed at which the surfaces rotate against one another.

Friction welding produces considerable sparking. The operator needs to wear goggles or a face shield to avoid injury, and approved clothing.

See Chapter 18.

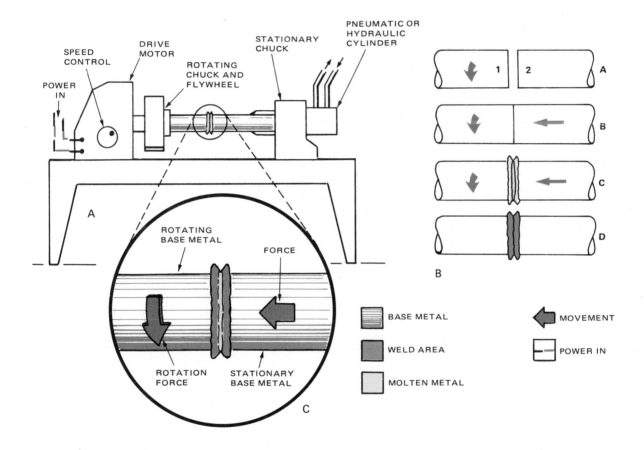

Fig. 1-29. Friction welding (FRW). A—Schematic of friction welding station. B—Four steps in producing friction weld. C—Close-up of a friction weld in process.

## 1-25. ULTRASONIC WELDING (USW)

In ultrasonic welding, very high sound frequencies excite metal surface molecules. This movement among the molecules produces fusion. This process is most often used to join very light materials. For example, ultrasonic welding is used for attaching fine wires to foil or wires to wires.

Fig. 1-30 illustrates the ultrasonic welding station. Since no outside heat is applied, this process is particularly desirable where the control of the heat affected zone is important.

Small rods contact the two metals to be welded.

One of these rods vibrates at a very high frequency (ultrasonic), much above the normal sound range. This causes the material being joined to vibrate at a corresponding rate. During the vibrations, some molecules of the two surfaces become intermixed and form a strong joint.

This type of welding is controlled by the rate of vibration and the pressure exerted by the vibrating elements on the parts being welded. Surfaces to be joined by ultrasonic welding must be very clean and free of oxidation.

Operators should wear gloves and goggles. Fine particles could be thrown off. See Chapter 18 for more details.

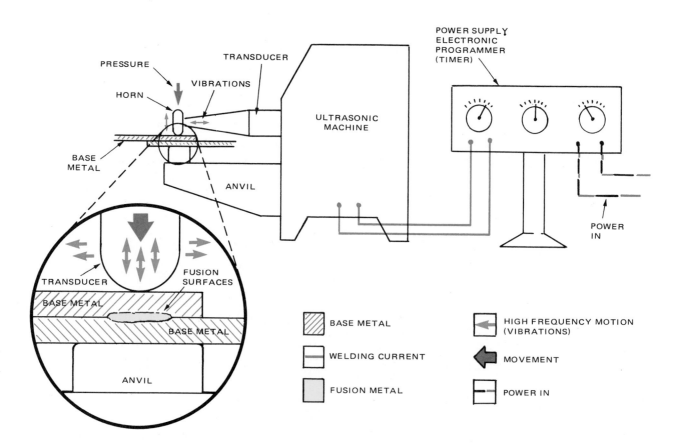

Fig. 1-30. Ultrasonic welding (USW). Welding energy is created by very, very high frequency vibration of the machine horn. The molecules at the surfaces of the two pieces of base metal are made to move so rapidly that the surfaces fuse.

## 1-26. LASER BEAM WELDING (LBW)

Laser beam welding uses a single frequency light beam. This beam puts energy into a metal causing it to heat up to its melting temperature. The laser beam is useful in welding small, light materials, particularly in locations where it is very difficult to weld with any other process. Fig. 1-31 shows a typical laser beam welding setup.

The beam is created by putting light or heat energy into a single molecule of a substance (ruby or carbon dioxide). The single frequency energy of the single molecule substance increases in intensity by traveling between two mirrors until it passes through the weaker or poorer of the mirrors. The release of the laser beam is controlled by the operator.

The heat of the laser beam is very intense and is easily directed to the spot where needed. Since the laser beam can be reflected, it can be directed to the weld joint by any combination of mirrors. It can be either a continuous heat source or a pulsed beam. There is an instantaneous heat release when the beam contacts the base metals. Controlling the input to the laser beam source controls the resulting heat.

Laser beams must be very carefully guarded from being directed toward any part of the body or anything of a heat-sensitive nature. Operators must wear special goggles designed for laser operators. See Chapter 18 for more details.

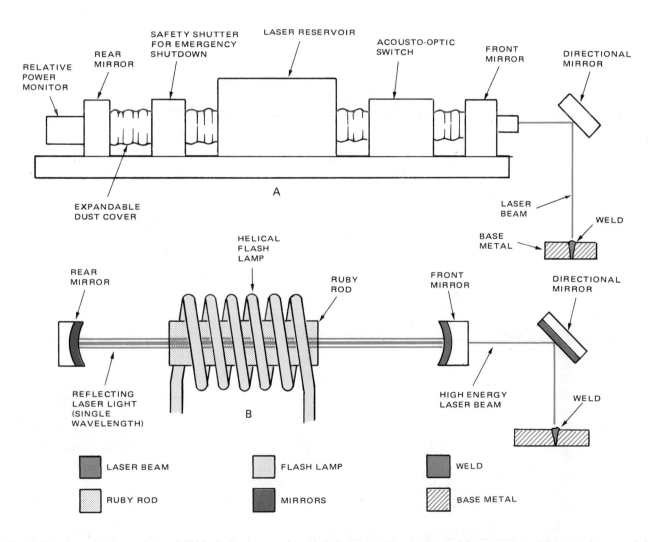

Fig. 1-31. Laser beam welding (LBW). A single wavelength light beam creates the welding energy as it hits the base metal. The laser beam can also be used for cutting and piercing base metals. A—Laser beam station. B—Laser beam light source.

## 1-27. ELECTRON BEAM WELDING (EBW)

Electron beam welding uses energy from a focused stream of electrons to heat and fuse metals. It is a good process for welding thick parts when the distance between them is small. Welds can be made in deep, narrow spaces with an extremely narrow weld zone. Fig. 1-32 shows an electron beam welding setup. This process is used to join metals difficult to weld by other processes. It may also be used to weld metals at very high speeds.

Because the equipment is large and expensive, it is used only where other processes cannot do the job.

The machine uses a filament which emits (gives off) electrons. These streams of electrons are controlled (focused and concentrated) by electromagnets called a magnetic lens. The electron beam is generated in much the same way as the

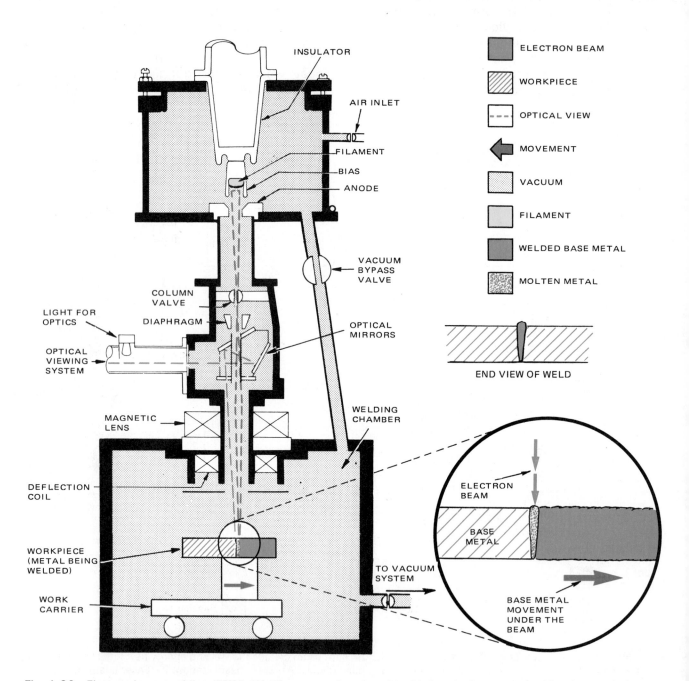

Fig. 1-32. Electron beam welding (EBW). Welding energy is created by hitting the base metal with a beam of electrons. The heat energy is very intense. Welds formed are very narrow.

light beam in a television receiver. It can be bent or directed by a magnetic field as in the television set. Weld energy is regulated by the current in the electron gun filament.

The electron beam weld is usually made in a vacuum, since air molecules tend to interfere with the beam. The vacuum chambers are also a shield against radiation.

The operator watches the weld through a safe optical system and directs the beam with remote controls. Some modern electron beam welding equipment may be used to produce welds under nonvacuum conditions.

The surfaces to be joined should be cleaned prior to welding. The parts being joined must fit together very tightly.

The operator and other persons near the machine must be protected from the radiation given off by the beam. Most machines use lead as a shielding material.

See Chapter 18 for more details on electron beam welding processes.

Industry photo of fully automatic production welding for freight car sides. This photo shows two submerged arc, solid wire, fully automatic welding machines depositing fillet welds. The welding machines traveling on a common beam join side posts to side sheets.   (Thrall Car Manufacturing Co.)

## 1-28. TORCH PLASTIC WELDING

In torch plastic welding, heated air or heated shielding gas melts and fuses plastic base materials and plastic filler materials together. The use of plastics in modern industry makes considerable amounts of plastic welding necessary. Fig. 1-33 shows a typical torch plastic welding setup.

The gas or air is heated by an electrical heating coil. The heated gas flows through the heating nozzles and heats the parts to be welded. The welding tip is designed to press the heated plastic filler material into the weld area. Plastics are joined at between 400° and 800°F (204°-427°C).

The heat source is controlled by adjusting the resistance unit and/or gas flow rate. The operator must manipulate both the torch and the filler material.

The temperature of the filler material during welding, is hot enough to cause severe burns. It is advisable to wear gloves and safety goggles. Good ventilation is recommended.

See Chapter 20 for more details.

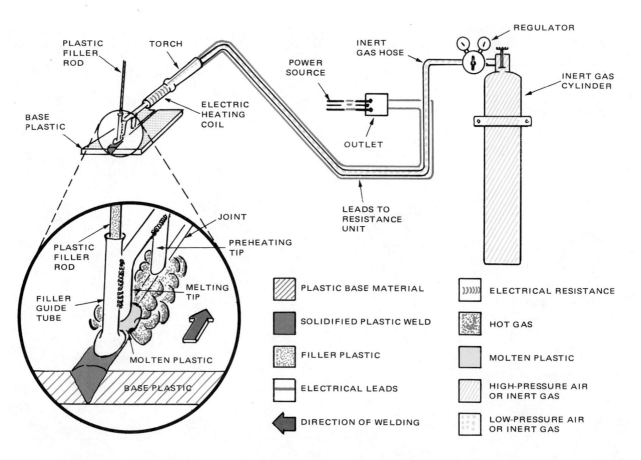

Fig. 1-33. Torch plastic welding. The heat source is heated air or a heated inert gas. The heated gas first preheats the joint, then a second orifice melts the plastic as a plastic filler is fed through a tube to the molten plastic.

## 1-29. OXYGEN LANCE CUTTING (LOC)

In the oxygen lance cutting process, an oxyfuel gas flame heats the base metal while a jet of oxygen is directed at the heated metal to cut (burn) it. This method has been used for many years to cut heavy steel sections. Fig. 1-34 illustrates a typical oxygen lance cutting station.

The oxygen lance is a straight piece of iron pipe with a hand valve. The oxygen lance is attached by a hose to one or more oxygen cylinders equipped with regulators and gauges.

The oxygen lance is used along with an oxyfuel gas cutting torch (see Heading 1-2). After the cut is started with the cutting torch, the lance oxygen valve is opened. The stream of oxygen from the iron pipe flows into the kerf started by the cutting torch. It can cut sections of great thicknesses. The lance is slowly melted away during the cutting and must be replaced frequently.

As with all cutting operations, wear proper eye protection, gloves, and clothing to prevent personal injury. It is also important to cover or clear the work area of combustible materials. Follow all fire prevention practices. Good ventilation is required.

See Chapter 22 for more information.

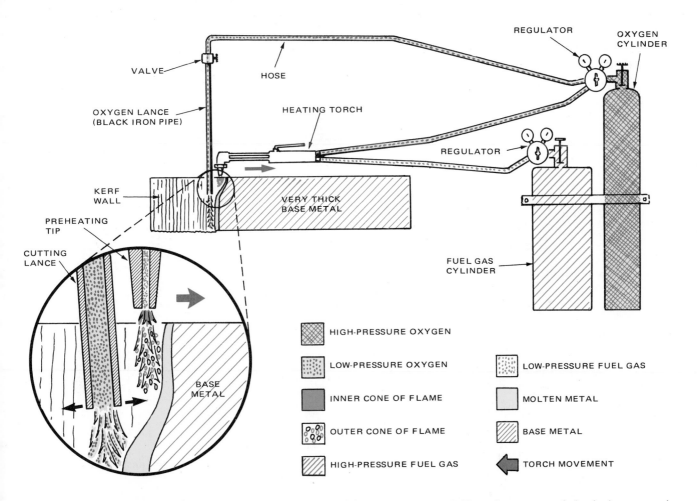

Fig. 1-34. Oxygen lance cutting (LOC). An oxyfuel gas torch heats the base metal. The valve on a metal pipe is then opened and a jet of oxygen rapidly oxidizes the heated base metal and blows it away. The pipe is slowly consumed.

## 1-30. OXYFUEL GAS UNDERWATER CUTTING

To heat the base metal, the underwater oxyfuel gas cutting process uses an oxyfuel gas flame, surrounded by compressed air. An oxygen jet oxidizes the heated metal and blows it away. The oxyhydrogen flame plus an oxygen jet has long been used to cut steel under water. The process is used for salvage operations and underwater construction.

Fig. 1-35 shows an underwater cutting outfit. Acetylene gas cannot be used as a fuel gas because acetylene is unstable at pressures above 15 psi (pounds per square inch) or 103.4 kilopascals (kPa). This means that acetylene could not be used safely in water deeper than 10 or 12 ft. (3.05-3.66 m). Oxyhydrogen can be used at any depth up to 200 ft. (61 m).

The proprietary gas called MAPP, manufactured by Airco Industrial Gases, may also be used for this purpose. It is stable under high pressure.

In addition to oxygen and a fuel gas, compressed air is required for underwater cutting. This air forms an air pocket in which the flame burns under water.

The equipment for using either oxyhydrogen or oxygen and MAPP is much the same. However, different cutting tips are used.

The oxygen and the fuel gas cylinders are connected to the torch with regulators and suitable lengths of hose. The valves on the underwater cutting torch control the torch flame.

The fuel gas is usually lighted under water by an electric ignitor built into the air jacket. Gas cutting under water requires suitable diving gear and extensive training as a diver.

See Chapter 22 for more details.

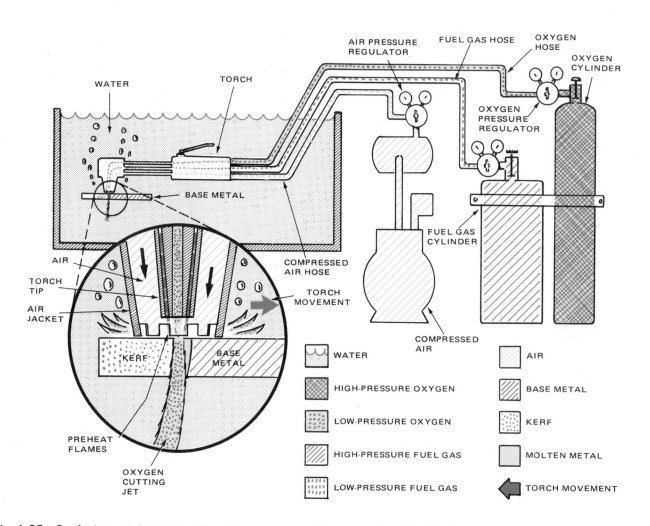

Fig. 1-35. Oxyfuel gas underwater cutting. The oxyfuel gas flames burn in a "bubble" of compressed air which keeps the water away from the flame. The heated metal is rapidly oxidized and is blown away by the oxygen jet.

## 1-31. CHEMICAL FLUX CUTTING (FOC)

Chemical flux cutting uses an oxyfuel gas torch to heat the base metal. This process is used for cutting alloy steels, cast iron, and nonferrous metals which form oxides with a very high melting temperature. Fig. 1-36 shows a basic chemical flux cutting station.

Chemical flux cutting requires an oxygen cylinder and a fuel gas cylinder, both fitted with regulators and gauges. Compressed air carries the chemical flux powder to the torch tip. An oxygen jet flows against the heated metal to cut it.

In chemical flux cutting, flux powder is introduced into an oxyfuel gas cutting torch flame. The flux powder decreases the formation of refractory oxides (solids). The molten metal is then easily removed to form a kerf.

A very similar process uses an iron powder. The iron powder increases the total heat of the flame. The iron in the chemical flux powder also absorbs the oxygen in the area of the cut, reducing the alloy oxides.

Because of the hazard of sparks, the operator must wear goggles, gloves, high shoes, and protective clothing. Be sure to follow all fire safety practices. The area must be cleared of combustible material.

See Chapter 22 for more details.

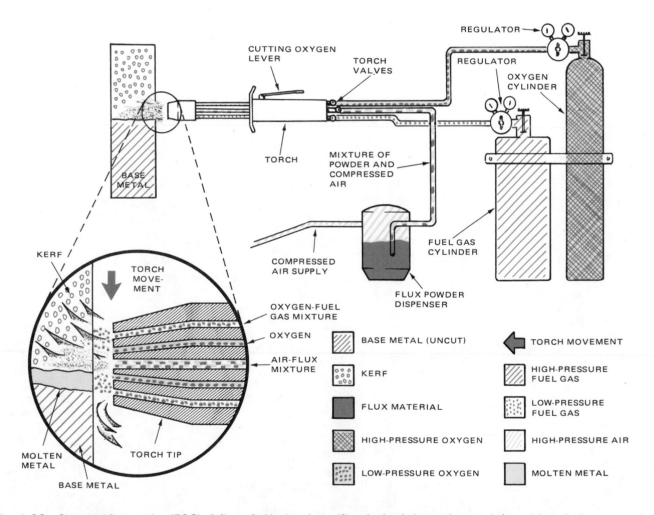

Fig. 1-36. Chemical flux cutting (FOC). A flame fed by heating orifices in the tip heats the metal. A torch lever is then pressed and an air jet mixed with a powdered chemical rapidly melts the base metal. The flux makes this melted metal very fluid and the air jet, plus the oxygen jet, blows the metal and flux away to form a kerf (cut) in the base metal. This process is used mainly for alloy steels.

## 1-32. STEEL TEMPERATURE AND COLOR RELATIONSHIPS

It is important to understand what happens when metals are heated, melted, and then cooled. The physical properties of the steel are determined by these temperatures and actions.

There are thermometers which can accurately measure surface temperatures of steels (melting crayons, pyrometers, and thermocouples). However, the color of the steel is a very popular and quite accurate way to judge the temperature. Fig. 1-37 illustrates the color range for steel.

At low temperatures, steel looks gray to the naked eye. When heated the steel surface starts to oxidize. First the oxide coating is gray from 0° to 100°F (-18° to 38°C). From 400° to 475°F (205° to 245°C) the color changes from a faint straw color to a deep straw color. At 520°F (270°C) the surface becomes bronze in color.

Between 540° and 700°F (280° to 370°C) the surface oxide turns purple, blue, then black. The surface is dark red at 1000°F (538°C). The red becomes brighter as the temperature increases to 1500°F (815°C) where the color is an orange red.

The color finally changes to a bright yellow at 2400°F (1316°C). At 2400°F (1316°C), the welding temperature, steel begins to melt.

Note that the temperature affects steel properties differently as the steel's carbon content varies.

Because metals are good conductors of heat, always use extreme caution when handling heated pieces of steel. Temperatures above 120°F (49°C) will cause severe burns. Protect the body by wearing approved goggles, leather gloves, and protective clothing. Handle hot metals with pliers or tongs.

See Chapter 27 for more details.

Industry photo shows student practicing gas metal arc welding.　　(Miller Electric Mfg. Co.)

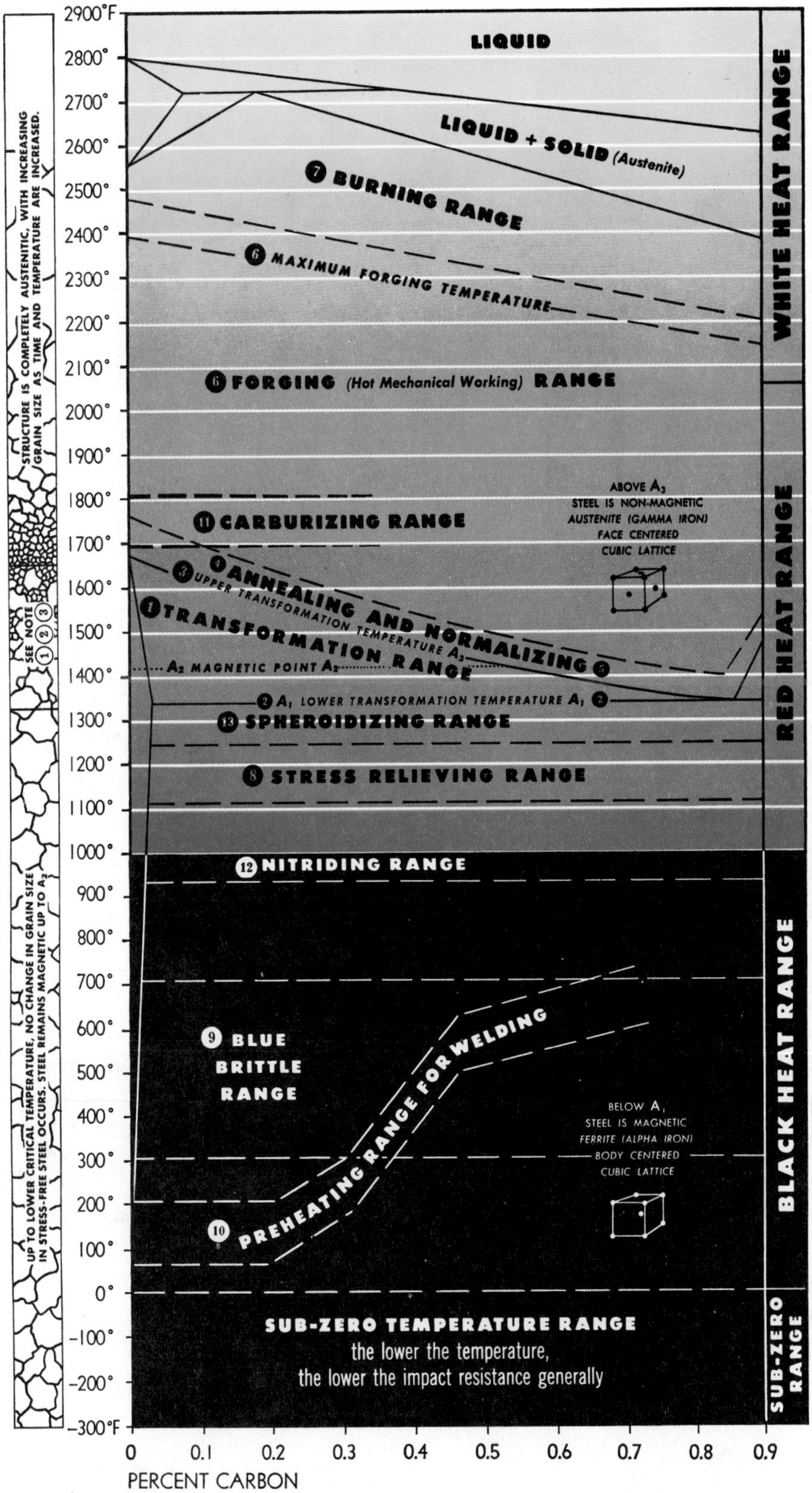

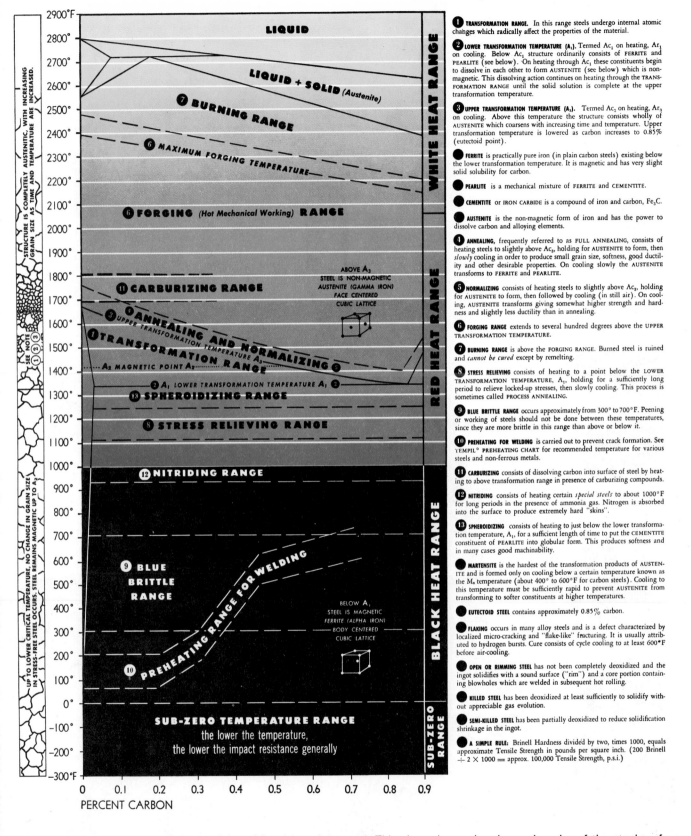

LIQUID
LIQUID + SOLID (Austenite)
7 BURNING RANGE
6 MAXIMUM FORGING TEMPERATURE
6 FORGING (Hot Mechanical Working) RANGE
11 CARBURIZING RANGE
ABOVE $A_3$
STEEL IS NON-MAGNETIC
AUSTENITE (GAMMA IRON)
FACE CENTERED
CUBIC LATTICE
4 ANNEALING AND NORMALIZING 5
3 UPPER TRANSFORMATION TEMPERATURE $A_3$
1 TRANSFORMATION RANGE
$A_2$ MAGNETIC POINT $A_2$
2 $A_1$ LOWER TRANSFORMATION TEMPERATURE $A_1$ 2
13 SPHEROIDIZING RANGE
8 STRESS RELIEVING RANGE
12 NITRIDING RANGE
9 BLUE BRITTLE RANGE
10 PREHEATING RANGE FOR WELDING
BELOW $A_1$
STEEL IS MAGNETIC
FERRITE (ALPHA IRON)
BODY CENTERED
CUBIC LATTICE
SUB-ZERO TEMPERATURE RANGE
the lower the temperature,
the lower the impact resistance generally
WHITE HEAT RANGE
RED HEAT RANGE
BLACK HEAT RANGE
SUB-ZERO RANGE
STRUCTURE IS COMPLETELY AUSTENITIC, WITH INCREASING GRAIN SIZE AS TIME AND TEMPERATURE ARE INCREASED.
SEE NOTE 1 2 3
UP TO LOWER CRITICAL TEMPERATURE, NO CHANGE IN GRAIN SIZE IN STRESS-FREE STEEL OCCURS. STEEL REMAINS MAGNETIC UP TO $A_2$
2900°F
2800°
2700°
2600°
2500°
2400°
2300°
2200°
2100°
2000°
1900°
1800°
1700°
1600°
1500°
1400°
1300°
1200°
1100°
1000°
900°
800°
700°
600°
500°
400°
300°
200°
100°
0°
-100°
-200°
-300°F
0 0.1 0.2 0.3 0.4 0.5 0.6 0.7 0.8 0.9
PERCENT CARBON


**1 TRANSFORMATION RANGE.** In this range steels undergo internal atomic changes which radically affect the properties of the material.

**2 LOWER TRANSFORMATION TEMPERATURE ($A_1$).** Termed $Ac_1$ on heating, $Ar_1$ on cooling. Below $Ac_1$ structure ordinarily consists of FERRITE and PEARLITE (see below). On heating through $Ac_1$ these constituents begin to dissolve in each other to form AUSTENITE (see below) which is non-magnetic. This dissolving action continues on heating through the TRANSFORMATION RANGE until the solid solution is complete at the upper transformation temperature.

**3 UPPER TRANSFORMATION TEMPERATURE ($A_3$).** Termed $Ac_3$ on heating, $Ar_3$ on cooling. Above this temperature the structure consists wholly of AUSTENITE which coarsens with increasing time and temperature. Upper transformation temperature is lowered as carbon increases to 0.85% (eutectoid point).

● **FERRITE** is practically pure iron (in plain carbon steels) existing below the lower transformation temperature. It is magnetic and has very slight solid solubility for carbon.

● **PEARLITE** is a mechanical mixture of FERRITE and CEMENTITE.

● **CEMENTITE** or IRON CARBIDE is a compound of iron and carbon, $Fe_3C$.

● **AUSTENITE** is the non-magnetic form of iron and has the power to dissolve carbon and alloying elements.

**4 ANNEALING,** frequently referred to as FULL ANNEALING, consists of heating steels to slightly above $Ac_3$, holding for AUSTENITE to form, then *slowly* cooling in order to produce small grain size, softness, good ductility and other desirable properties. On cooling slowly the AUSTENITE transforms to FERRITE and PEARLITE.

**5 NORMALIZING** consists of heating steels to slightly above $Ac_3$, holding for AUSTENITE to form, then followed by cooling (in still air). On cooling, AUSTENITE transforms giving somewhat higher strength and hardness and slightly less ductility than in annealing.

**6 FORGING RANGE** extends to several hundred degrees above the UPPER TRANSFORMATION TEMPERATURE.

**7 BURNING RANGE** is above the FORGING RANGE. Burned steel is ruined and *cannot be cured* except by remelting.

**8 STRESS RELIEVING** consists of heating to a point below the LOWER TRANSFORMATION TEMPERATURE, $A_1$, holding for a sufficiently long period to relieve locked-up stresses, then slowly cooling. This process is sometimes called PROCESS ANNEALING.

**9 BLUE BRITTLE RANGE** occurs approximately from 300° to 700°F. Peening or working of steels should not be done between these temperatures, since they are more brittle in this range than above or below it.

**10 PREHEATING FOR WELDING** is carried out to prevent crack formation. See TEMPIL° PREHEATING CHART for recommended temperature for various steels and non-ferrous metals.

**11 CARBURIZING** consists of dissolving carbon into surface of steel by heating to above transformation range in presence of carburizing compounds.

**12 NITRIDING** consists of heating certain *special steels* to about 1000°F for long periods in the presence of ammonia gas. Nitrogen is absorbed into the surface to produce extremely hard "skins".

**13 SPHEROIDIZING** consists of heating to just below the lower transformation temperature, $A_1$, for a sufficient length of time to put the CEMENTITE constituent of PEARLITE into globular form. This produces softness and in many cases good machinability.

● **MARTENSITE** is the hardest of the transformation products of AUSTENITE and is formed only on cooling below a certain temperature known as the $M_s$ temperature (about 400° to 600°F for carbon steels). Cooling to this temperature must be sufficiently rapid to prevent AUSTENITE from transforming to softer constituents at higher temperatures.

● **EUTECTOID STEEL** contains approximately 0.85% carbon.

● **FLAKING** occurs in many alloy steels and is a defect characterized by localized micro-cracking and "flake-like" fracturing. It is usually attributed to hydrogen bursts. Cure consists of cycle cooling to at least 600°F before air-cooling.

● **OPEN OR RIMMING STEEL** has not been completely deoxidized and the ingot solidifies with a sound surface ("rim") and a core portion containing blowholes which are welded in subsequent hot rolling.

● **KILLED STEEL** has been deoxidized at least sufficiently to solidify without appreciable gas evolution.

● **SEMI-KILLED STEEL** has been partially deoxidized to reduce solidification shrinkage in the ingot.

● **A SIMPLE RULE:** Brinell Hardness divided by two, times 1000, equals approximate Tensile Strength in pounds per square inch. (200 Brinell ÷ 2 × 1000 = approx. 100,000 Tensile Strength, p.s.i.)

Fig. 1-37. Color-temperature chart of 0 to 90 point carbon steel. This chart shows the change in color of the steel surface as the temperature changes from 1000° to 2900°F (538° to 1539°C). (Tempil°, Big Three Industries, Inc.)

# 1-33. FLOWLINE OF STEELMAKING

From iron ore, limestone, and coal in the earth's crust to space-age steels—this fundamental flowline shows only major steps in an intricate progression of processes with their many options.

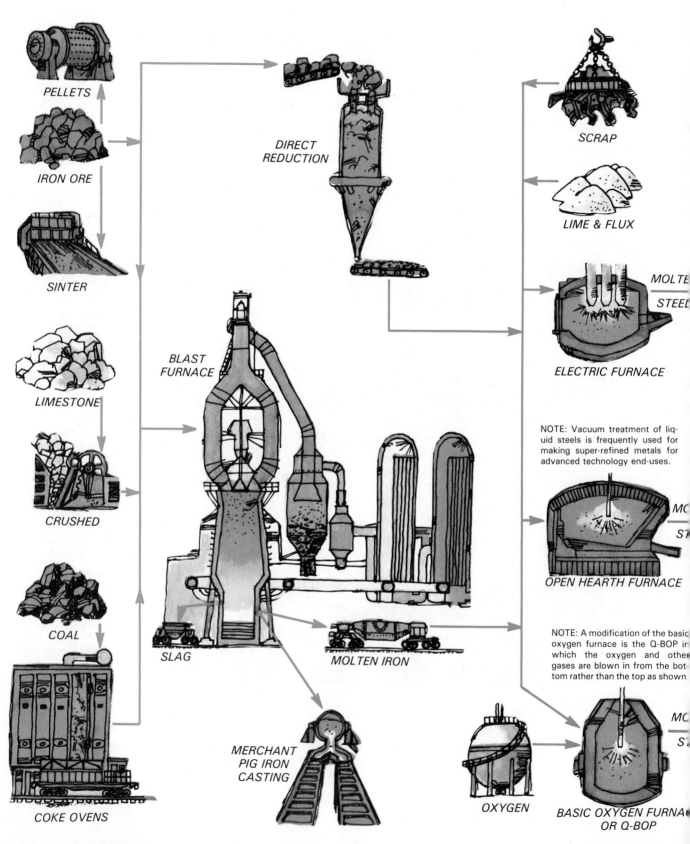

PELLETS

IRON ORE

SINTER

LIMESTONE

CRUSHED

COAL

COKE OVENS

DIRECT REDUCTION

BLAST FURNACE

SLAG

MOLTEN IRON

MERCHANT PIG IRON CASTING

SCRAP

LIME & FLUX

MOLTEN STEEL

ELECTRIC FURNACE

NOTE: Vacuum treatment of liquid steels is frequently used for making super-refined metals for advanced technology end-uses.

OPEN HEARTH FURNACE

NOTE: A modification of the basic oxygen furnace is the Q-BOP in which the oxygen and other gases are blown in from the bottom rather than the top as shown

OXYGEN

BASIC OXYGEN FURNACE OR Q-BOP

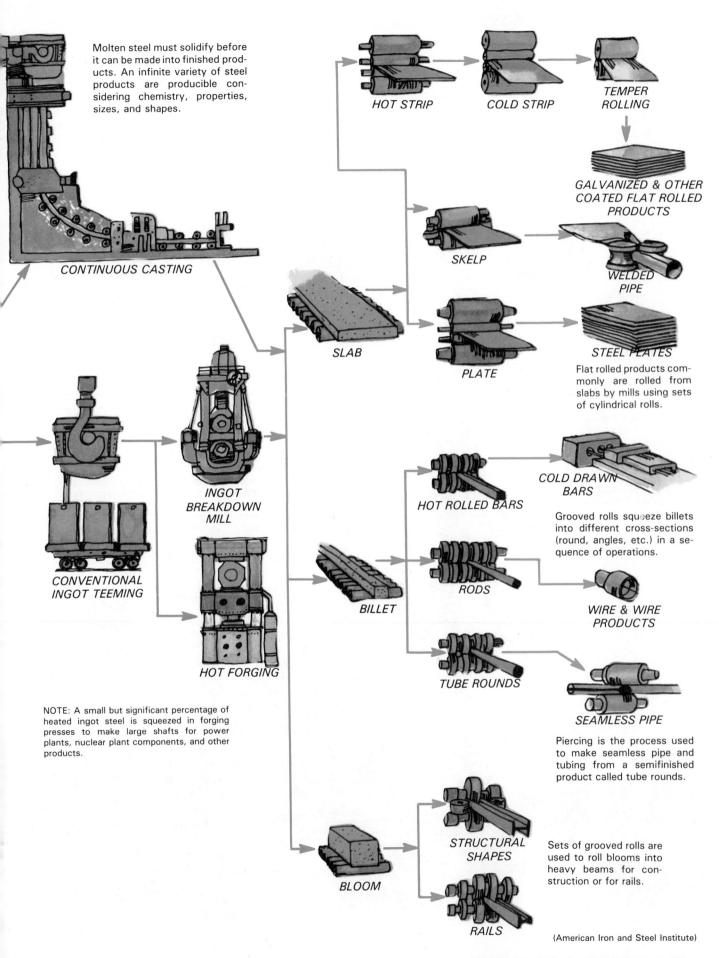

Molten steel must solidify before it can be made into finished products. An infinite variety of steel products are producible considering chemistry, properties, sizes, and shapes.

HOT STRIP

COLD STRIP

TEMPER ROLLING

GALVANIZED & OTHER COATED FLAT ROLLED PRODUCTS

CONTINUOUS CASTING

SKELP

WELDED PIPE

SLAB

PLATE

STEEL PLATES

Flat rolled products commonly are rolled from slabs by mills using sets of cylindrical rolls.

INGOT BREAKDOWN MILL

HOT ROLLED BARS

COLD DRAWN BARS

Grooved rolls squeeze billets into different cross-sections (round, angles, etc.) in a sequence of operations.

CONVENTIONAL INGOT TEEMING

BILLET

RODS

WIRE & WIRE PRODUCTS

HOT FORGING

TUBE ROUNDS

SEAMLESS PIPE

NOTE: A small but significant percentage of heated ingot steel is squeezed in forging presses to make large shafts for power plants, nuclear plant components, and other products.

Piercing is the process used to make seamless pipe and tubing from a semifinished product called tube rounds.

STRUCTURAL SHAPES

Sets of grooved rolls are used to roll blooms into heavy beams for construction or for rails.

BLOOM

RAILS

(American Iron and Steel Institute)

# 1-34. BLAST FURNACE IRONMAKING

A blast furnace is a cylindrical steel vessel, often as tall as a ten-story building, lined inside with a heat-resistant brick. Once it is fired up, the furnace runs continuously until this lining is worn out and must be replaced. Coke, iron ore, and limestone are charged into the top of the furnace. They proceed slowly down through the furnace being heated as they are exposed to a blast of super-hot air that blows upward from the bottom of the furnace. The blast of air burns the coke, releasing heat and gas which remove oxygen from the ore. The limestone acts as a cleansing agent.

Freed from its impurities, the molten iron collects in the bottom where it is drawn off every three-to-five hours as a white hot stream of liquid iron. From there, the molten iron will be further refined in a steelmaking furnace.

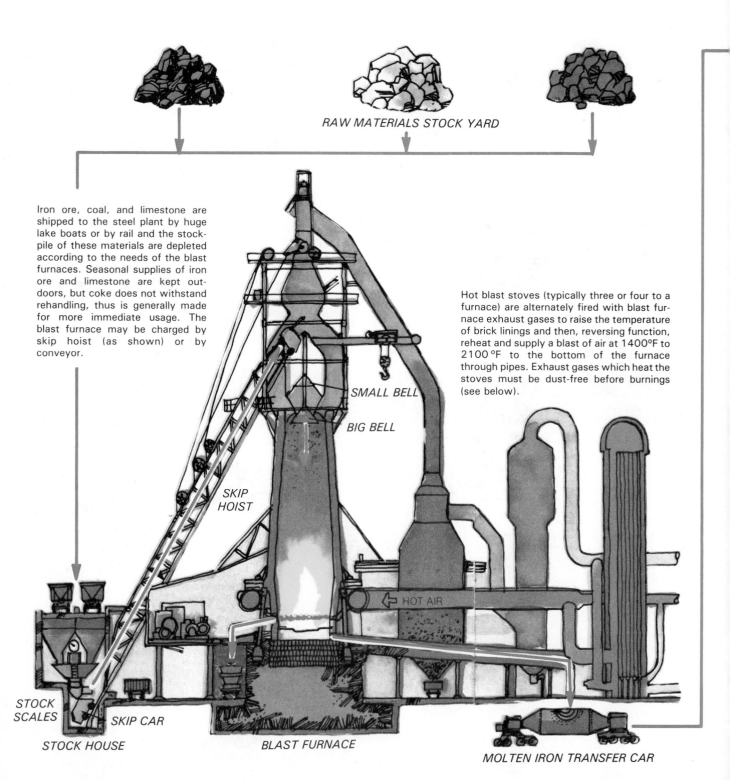

RAW MATERIALS STOCK YARD

Iron ore, coal, and limestone are shipped to the steel plant by huge lake boats or by rail and the stockpile of these materials are depleted according to the needs of the blast furnaces. Seasonal supplies of iron ore and limestone are kept outdoors, but coke does not withstand rehandling, thus is generally made for more immediate usage. The blast furnace may be charged by skip hoist (as shown) or by conveyor.

Hot blast stoves (typically three or four to a furnace) are alternately fired with blast furnace exhaust gases to raise the temperature of brick linings and then, reversing function, reheat and supply a blast of air at 1400°F to 2100 °F to the bottom of the furnace through pipes. Exhaust gases which heat the stoves must be dust-free before burnings (see below).

SMALL BELL

BIG BELL

SKIP HOIST

HOT AIR

STOCK SCALES

SKIP CAR

STOCK HOUSE

BLAST FURNACE

MOLTEN IRON TRANSFER CAR

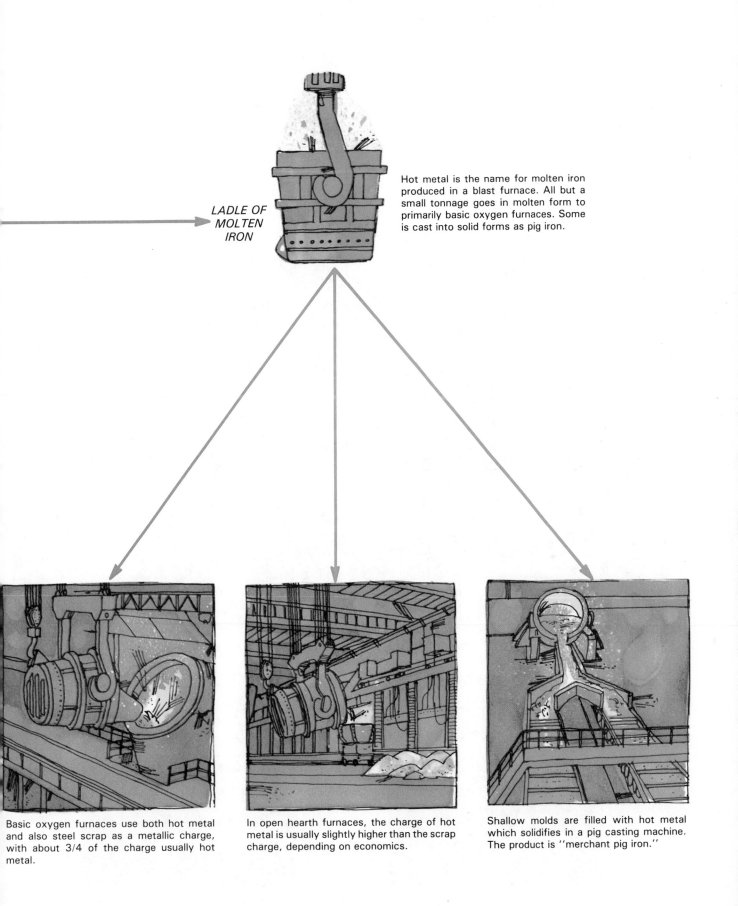

LADLE OF
MOLTEN
IRON

Hot metal is the name for molten iron produced in a blast furnace. All but a small tonnage goes in molten form to primarily basic oxygen furnaces. Some is cast into solid forms as pig iron.

Basic oxygen furnaces use both hot metal and also steel scrap as a metallic charge, with about 3/4 of the charge usually hot metal.

In open hearth furnaces, the charge of hot metal is usually slightly higher than the scrap charge, depending on economics.

Shallow molds are filled with hot metal which solidifies in a pig casting machine. The product is "merchant pig iron."

(American Iron and Steel Institute)

# 1-35. BASIC OXYGEN STEELMAKING

In the United States more steel is produced by basic oxygen furnaces (BOF's) than by any other means. They consume large amounts of oxygen to support the combustion of unwanted elements and so eliminate them. No other gases or fuels are used. BOF's make steel very quickly compared with the other major methods now in use — for example, 300 tons in 45 minutes as against several hours in open hearth and electric furnaces. Most grades of steel can be made in BOF's

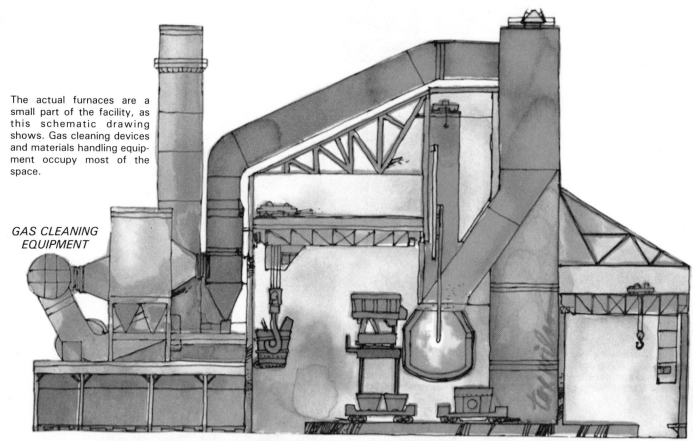

The actual furnaces are a small part of the facility, as this schematic drawing shows. Gas cleaning devices and materials handling equipment occupy most of the space.

*GAS CLEANING EQUIPMENT*

*DIAGRAM OF BASIC OXYGEN FURNACE FACILITY*

The first step for making a heat of steel in a BOF is to tilt the furnace and charge it with scrap. The furnaces are mounted on trunnions and can be rotated through a full circle.

Hot metal from the blast furnace accounts for up to 80 percent of the metallic charge and is poured from a ladle into the top of the tilted furnace.

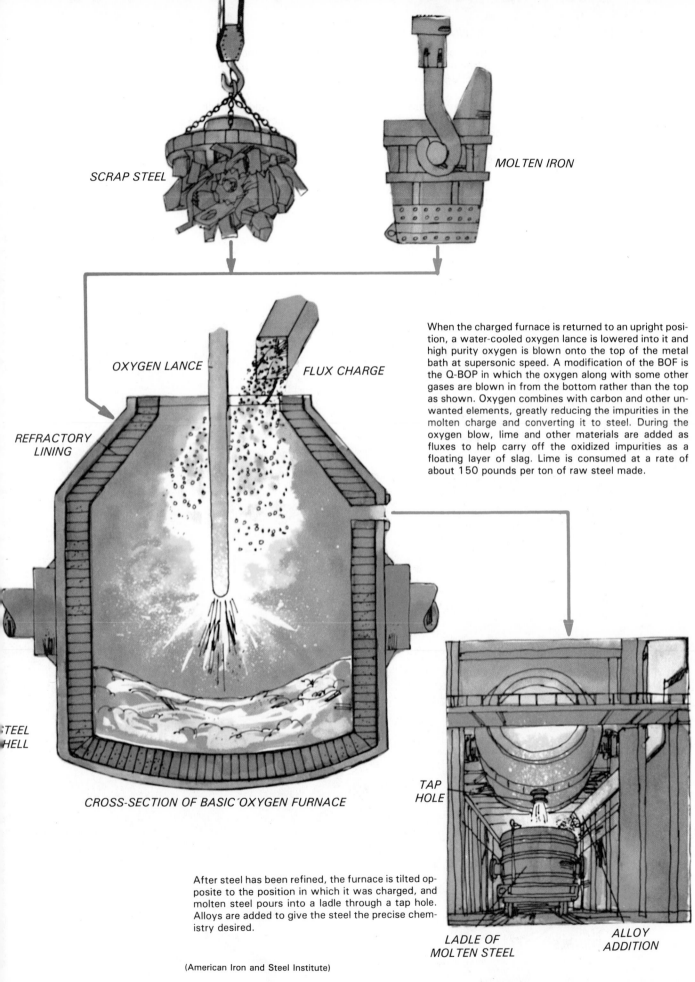

SCRAP STEEL

MOLTEN IRON

OXYGEN LANCE

FLUX CHARGE

REFRACTORY
LINING

STEEL
SHELL

CROSS-SECTION OF BASIC OXYGEN FURNACE

When the charged furnace is returned to an upright position, a water-cooled oxygen lance is lowered into it and high purity oxygen is blown onto the top of the metal bath at supersonic speed. A modification of the BOF is the Q-BOP in which the oxygen along with some other gases are blown in from the bottom rather than the top as shown. Oxygen combines with carbon and other unwanted elements, greatly reducing the impurities in the molten charge and converting it to steel. During the oxygen blow, lime and other materials are added as fluxes to help carry off the oxidized impurities as a floating layer of slag. Lime is consumed at a rate of about 150 pounds per ton of raw steel made.

TAP
HOLE

After steel has been refined, the furnace is tilted opposite to the position in which it was charged, and molten steel pours into a ladle through a tap hole. Alloys are added to give the steel the precise chemistry desired.

LADLE OF
MOLTEN STEEL

ALLOY
ADDITION

(American Iron and Steel Institute)

## 1-36. OPEN HEARTH STEELMAKING

Open hearth furnaces are so named because the limestone, scrap steel and molten iron charged into the shallow steelmaking area (the hearth) are exposed (open) to the sweep of flames. A furnace that will produce a fairly typical 350 tons of steel in five to eight hours may be about 90 feet long and 30 feet wide.

The cutaway drawing below shows several steps simultaneously that would normally occur in sequence. First the long-armed charging machine picks up boxes of limestone and steel scrap, thrusts them through the furnace doors and dumps the contents. The flame of burning fuel oil, tar, or gases partially melts the solid charge, after which molten iron (lower right) is poured into the furnace. High-temperature reactions cause several unwanted elements to combine with the limestone to form a slag.

When tests of samples show the steel to be of specified chemistry, the tap hole is opened by an explosive charge and the steel runs into a ladle. The slag, which is lighter than steel, floats on the metal and overflows into a slag thimble during pouring. Alloy additions are made to the steel in the ladle.

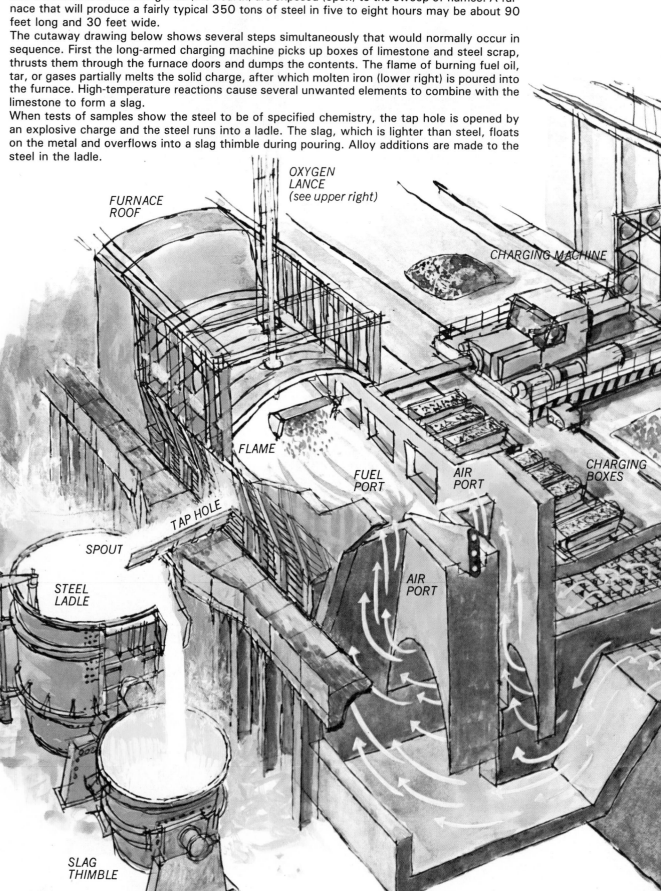

In recent years practically all open hearth furnaces have been converted to the use of oxygen. The gas is fed into the open hearth through the roof by means of retractable lances. The use of gaseous oxygen in the open hearth increases flame temperature, and thereby speeds the melting process. Less than one percent of the steel produced in the U.S.A. is produced in open-hearth furnaces.

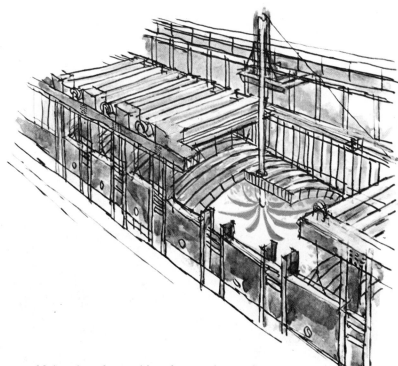

CONTROL PANELS

Molten iron from a blast furnace is a major raw material for the open hearth furnaces. A massive "funnel" is wheeled to an open hearth door and the contents of a ladle or iron are poured through it into the furnace hearth. The principal addition of molten metal is made after the original scrap charge has begun to melt.

Brick checker chambers are located on both ends of the furnace. The bricks are arranged to leave a great number of passages through which the hot waste gases from the furnace pass and heat the brickwork prior to going through the cleaner and stack. Later on, the flow is reversed and the air for combustion passes through the heated bricks and is itself heated on its way to the hearth.

*BRICK CHECKER CHAMBERS*

(American Iron and Steel Institute)

# 1-37. ELECTRIC FURNACE STEELMAKING

Traditionally used to produce alloy, stainless, tool, and specialty steels, electric furnaces have been developed in size and capability to become high-tonnage makers of carbon steel, too.

The heat within the furnace is precisely controlled as the electric current arcs from one electrode (of the three inserted through the furnace roof) to the metallic charge and back to another electrode.

ELECTRODES

At left, this cutaway drawing shows an electric furnace with its carbon electrodes attached to support arms and electrical cables and extending into the furnace. Molten steel and the rocker mounting on which the furnace may be tilted is also shown.

*ELECTRIC FURNACE FACILITY*

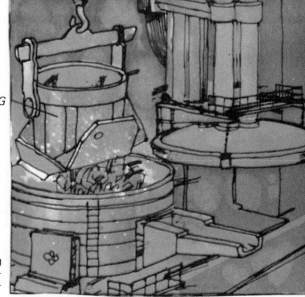

CHARGING
BUCKET

At right, the roof of a furnace is pivoted aside so that a charging bucket of scrap may be lowered into position for bottom-dumping. Alternatively, direct reduced pellets may be fed continuously during the meltdown.

Steel scrap is the principal metallic charge to electric furnaces. It is classified as "home scrap" (croppings originating in steel mills), "prompt industrial scrap" (trimmings returned by steel users), and "dormant scrap" (the materials collected and processed by dealers).

Direct reduction of iron ore produces pellets rich enough in iron to be used as a metallic charge, also.

*SCRAP STEEL*

*FLUXES*

*CHARGING BOX*

Alloying elements from many parts of the world are added to the molten metal, usually in the form of ferroalloys.

*TAPPING SPOUT*

*CROSS SECTION OF ELECTRIC FURNACE*

When the roof of the furnace is in place, the three carbon electrodes are lowered until they approach the cold scrap. Electric arcs produce heat to melt the scrap.

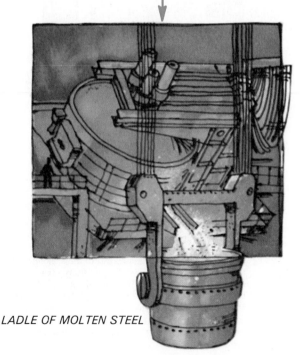

*LADLE OF MOLTEN STEEL*

*SLAG*

*SLAG THIMBLE*

Limestone and flux are charged after the scrap becomes molten. Impurities in the steel rise into a floating layer of slag, some or most of which can be poured off.

(American Iron and Steel Institute)

When the chemical composition of the steel meets specifications, the furnace is tilted backward. Molten steel pours out the spout into a ladle. The slag follows the steel and serves as an insulating layer.

FURNACE LADLE

## 1-38. VACUUM PROCESS OF STEELMAKING

Steels for special applications are often processed in a vacuum to give them pr◖ ties not otherwise obtainable. The primary purpose of vacuum processing is to re⬤ such gases as oxygen, nitrogen, and hydrogen from molten metal to make higher-p⬤ steel.

Many grades of steel are degassed by processes similar to those shown on this ⬤ Even greater purity and uniformity of steel chemistry than available by degassi⬤ obtained by subjecting the metal to vacuum melting processes.

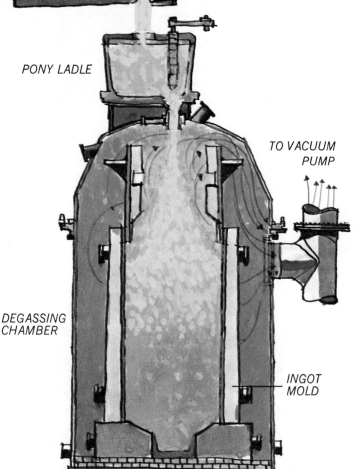

PONY LADLE

TO VACUUM PUMP

DEGASSING CHAMBER

INGOT MOLD

### The Vacuum Degasses

In vacuum stream degassing (left), a ladle of molten steel from ⬤ ventional furnace is taken to a vacuum chamber. An ingot m⬤ shown within the chamber. Larger chambers designed to contain ⬤ are also used. The conventionally melted steel goes into a pony ⬤ and from there into the chamber. The stream of steel is broken ⬤ to droplets when it is exposed to vacuum within the chamber. D⬤ the droplet phase, undesirable gases escape from the steel an⬤ drawn off before the metal solidifies in the mold.

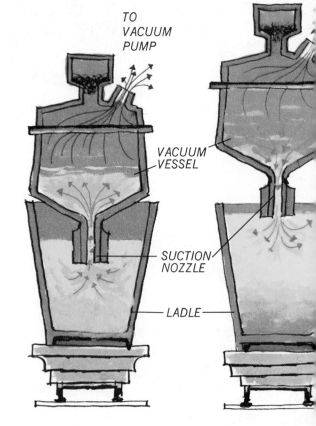

TO VACUUM PUMP

VACUUM VESSEL

SUCTION NOZZLE

LADLE

Ladle degassing facilities (right) of several kinds are in current use. In the left-hand facility, molten steel is forced by atmospheric pressure into the heated vacuum chamber. Gases are removed in this pressure chamber, which is then raised so that the molten steel returns by gravity into the ladle. Since not all of the steel enters the vacuum chamber at one time, this process is repeated until essentially all the steel in the ladle has been processed.

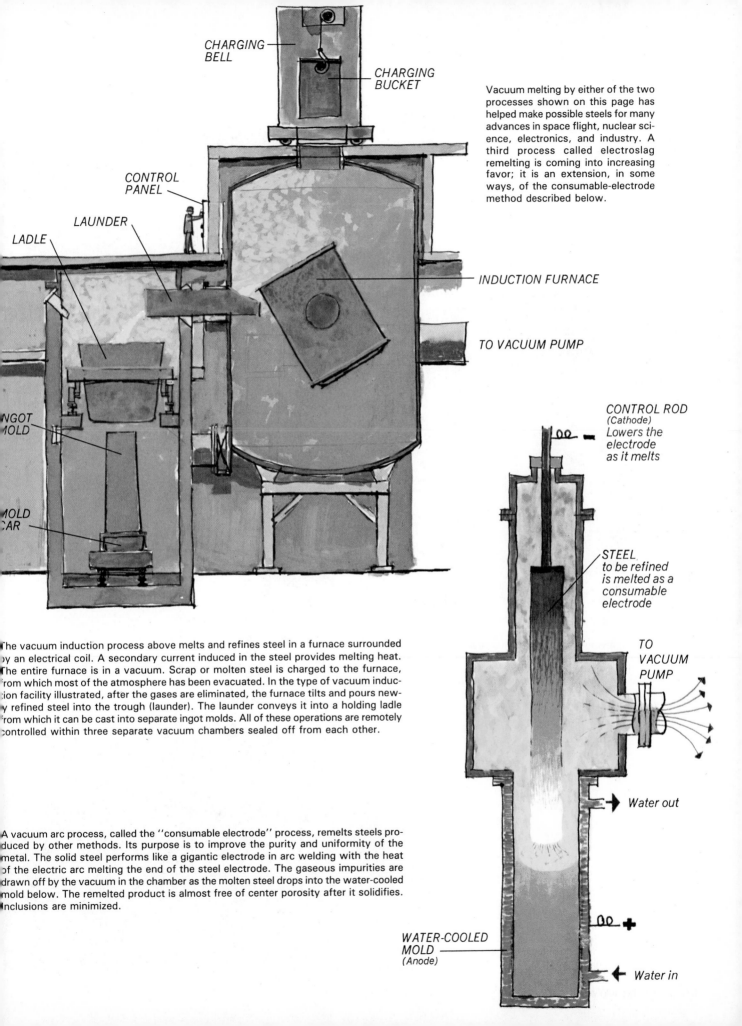

CHARGING BELL

CHARGING BUCKET

CONTROL PANEL

LAUNDER

LADLE

INGOT MOLD

MOLD CAR

INDUCTION FURNACE

TO VACUUM PUMP

Vacuum melting by either of the two processes shown on this page has helped make possible steels for many advances in space flight, nuclear science, electronics, and industry. A third process called electroslag remelting is coming into increasing favor; it is an extension, in some ways, of the consumable-electrode method described below.

The vacuum induction process above melts and refines steel in a furnace surrounded by an electrical coil. A secondary current induced in the steel provides melting heat. The entire furnace is in a vacuum. Scrap or molten steel is charged to the furnace, from which most of the atmosphere has been evacuated. In the type of vacuum induction facility illustrated, after the gases are eliminated, the furnace tilts and pours newly refined steel into the trough (launder). The launder conveys it into a holding ladle from which it can be cast into separate ingot molds. All of these operations are remotely controlled within three separate vacuum chambers sealed off from each other.

A vacuum arc process, called the "consumable electrode" process, remelts steels produced by other methods. Its purpose is to improve the purity and uniformity of the metal. The solid steel performs like a gigantic electrode in arc welding with the heat of the electric arc melting the end of the steel electrode. The gaseous impurities are drawn off by the vacuum in the chamber as the molten steel drops into the water-cooled mold below. The remelted product is almost free of center porosity after it solidifies. Inclusions are minimized.

CONTROL ROD
(Cathode)
Lowers the electrode as it melts

STEEL
to be refined is melted as a consumable electrode

TO VACUUM PUMP

Water out

WATER-COOLED MOLD
(Anode)

Water in

## 1-39. THE FIRST SOLID FORMS OF STEEL

Molten steel from basic oxygen, electric, and open hearth furnaces flows into ladles and then follows one or the other of two major routes to the rolling mills that form most of the industry's finished products. Both processes shown on these pages produce solid semifinished products—one, ingots (below); the other, cast slabs (far right).

*SOAKING PIT*

*BUGGY*

*LADLE*

*STRIPPER CRANE*

Stripped ingots are taken to f▢ naces called soaking pits. The▢ they are "soaked" in heat u▢ they reach uniform temperatu▢ throughout. Then the reheated ▢ gots are lifted out of the pit a▢ carried to the roughing mill on ▢ buggy.

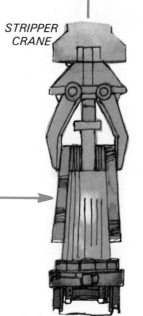

*INGOT MOLDS*

The traditional method of handling molten steel is to position the ladle via an overhead crane above a line of ingot molds (left). Then the operator opens a stopper rod within the ladle, and a stream of steal flows through a hole in the bottom of the ladle to "teem" or fill the cast iron molds which rest on special ingot railroad cars.

Molten steel in an ingot mold cools and solidifies from the outside towards the center. When the steel is solid enough, a stripper crane lifts away the mold while a plunger holds the steel ingot down on the ingot car.

*ROUGHING MILL*

Roughing mills are the first stage in shaping the hot steel ingot into semifinished steel—usually blooms, billets, or slabs. Some roughing mills are the first in a series of continuous mills, feeding sequences of finishing rolls.

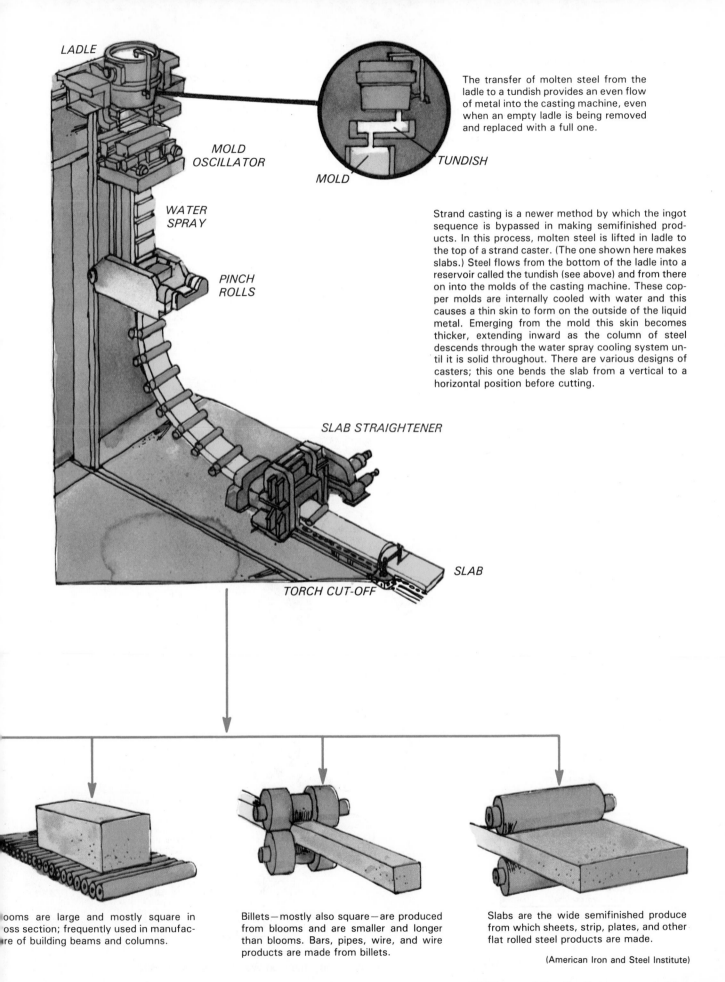

LADLE

MOLD
OSCILLATOR

WATER
SPRAY

PINCH
ROLLS

The transfer of molten steel from the ladle to a tundish provides an even flow of metal into the casting machine, even when an empty ladle is being removed and replaced with a full one.

MOLD

TUNDISH

Strand casting is a newer method by which the ingot sequence is bypassed in making semifinished products. In this process, molten steel is lifted in ladle to the top of a strand caster. (The one shown here makes slabs.) Steel flows from the bottom of the ladle into a reservoir called the tundish (see above) and from there on into the molds of the casting machine. These copper molds are internally cooled with water and this causes a thin skin to form on the outside of the liquid metal. Emerging from the mold this skin becomes thicker, extending inward as the column of steel descends through the water spray cooling system until it is solid throughout. There are various designs of casters; this one bends the slab from a vertical to a horizontal position before cutting.

SLAB STRAIGHTENER

SLAB

TORCH CUT-OFF

looms are large and mostly square in oss section; frequently used in manufac- re of building beams and columns.

Billets—mostly also square—are produced from blooms and are smaller and longer than blooms. Bars, pipes, wire, and wire products are made from billets.

Slabs are the wide semifinished produce from which sheets, strip, plates, and other flat rolled steel products are made.

(American Iron and Steel Institute)

## 1-40. WELDING AND ALLIED PROCESSES

The American Welding Society recognizes 12 major headings for welding and allied processes. See Fig. 1-38. Each process is explained in detail elsewhere. Refer to Chapter 2 for information on welding symbols.

## 1-41. TEST YOUR KNOWLEDGE

Write your answer to these questions on a separate sheet of paper. Do not write in this book.

1. What color is the neutral oxyacetylene flame?
2. In the OFC (oxyfuel gas cutting) process, after the metal is heated to a melting temperature, what causes the rapid oxida- tion or burning of the steel?
3. What is the temperature of the oxidizing, oxygen-LP gas flame?
4. Why is submerged arc welding referred to as submerged?
5. What prevents the shoes in ESW (electroslag welding) from becoming welded to the base metal?
6. In ultrasonic welding, what creates the heat required for welding?
7. Why is most EBW (electron beam welding) done in a vacuum?
8. While cutting under water with oxyfuel gas, what keeps the water away from the preheating flames?
9. In the oxygen cutting process, what fuel gas is used with OFC-H?
10. What process uses the initials AOC?

## MASTER CHART OF WELDING AND ALLIED PROCESSES

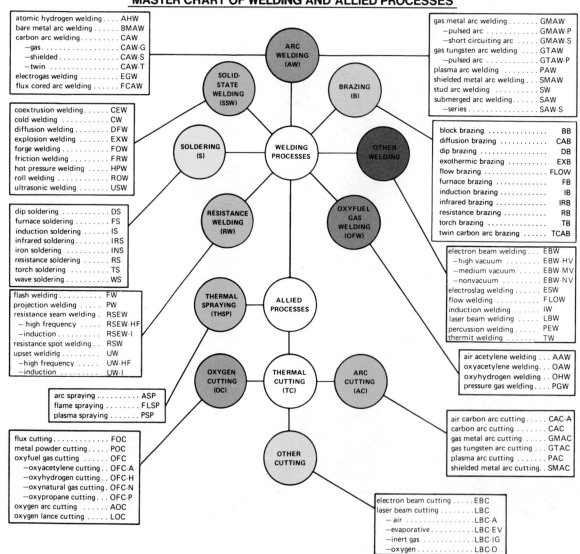

Fig. 1-38. Welding processes defined by the American Welding Society (AWS).

# READING WELDING SYMBOLS

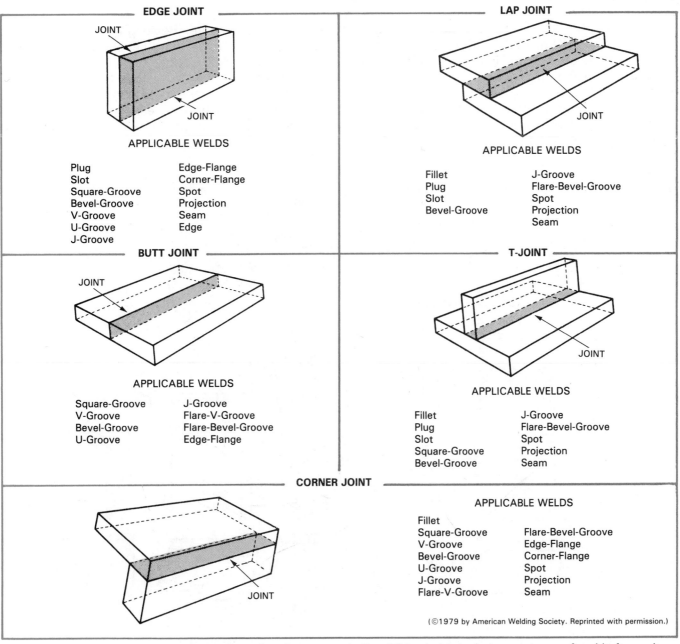

**EDGE JOINT**

JOINT

JOINT

APPLICABLE WELDS

| | |
|---|---|
| Plug | Edge-Flange |
| Slot | Corner-Flange |
| Square-Groove | Spot |
| Bevel-Groove | Projection |
| V-Groove | Seam |
| U-Groove | Edge |
| J-Groove | |

**LAP JOINT**

JOINT

APPLICABLE WELDS

| | |
|---|---|
| Fillet | J-Groove |
| Plug | Flare-Bevel-Groove |
| Slot | Spot |
| Bevel-Groove | Projection |
| | Seam |

**BUTT JOINT**

JOINT

APPLICABLE WELDS

| | |
|---|---|
| Square-Groove | J-Groove |
| V-Groove | Flare-V-Groove |
| Bevel-Groove | Flare-Bevel-Groove |
| U-Groove | Edge-Flange |

**T-JOINT**

JOINT

APPLICABLE WELDS

| | |
|---|---|
| Fillet | J-Groove |
| Plug | Flare-Bevel-Groove |
| Slot | Spot |
| Square-Groove | Projection |
| Bevel-Groove | Seam |

**CORNER JOINT**

JOINT

APPLICABLE WELDS

| | |
|---|---|
| Fillet | |
| Square-Groove | Flare-Bevel-Groove |
| V-Groove | Edge-Flange |
| Bevel-Groove | Corner-Flange |
| U-Groove | Spot |
| J-Groove | Projection |
| Flare-V-Groove | Seam |

(©1979 by American Welding Society. Reprinted with permission.)

Fig. 2-1. Basic welding joint designs. Note that there are only five (5) basic joints, but many types of welds for each joint. (American Welding Society)

## 2-1. BASIC WELDING JOINTS

There are five basic types of welded joints. They are the BUTT, CORNER, TEE, LAP, and EDGE joint. See Fig. 2-1. The five basic types of welded joints can be made in the four different welding positions shown in Fig. 2-2. For illustration purposes, most of the illustrations in this chapter will be in the flat position. Remember that you may be called upon to weld in any position.

The names or labels given to the various parts of a completed weld are called out in Fig. 2-3. Study these names and make them a part of your vocabulary.

The edges of the metal to be welded (base metal) are often prepared for welding by cutting, machining, or grinding. This preparation is done to insure that the base metal is welded through its entire thickness. Edge preparation is also done on thick metal to open up the joint area. This provides a space large enough to permit welding at the bottom of the joint.

The butt joint may be welded from the top, bottom, or both. The butt joint may be welded using one of the following types of welds shown in Fig.

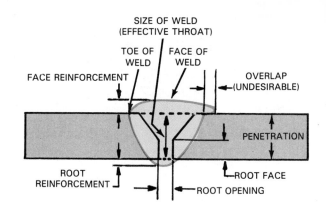

Fig. 2-3. Labels used to describe various parts of completed welds.

2-4: Square groove, Bevel groove, V groove, J groove, U groove, Flare-bevel-groove, Flare V groove, and Edge Flange.

Corner joints may be welded from the inside or outside of the corner. Occasionally the corner may be welded from both sides. Fig. 2-5 shows the methods used to prepare an inside or outside corner joint for welding.

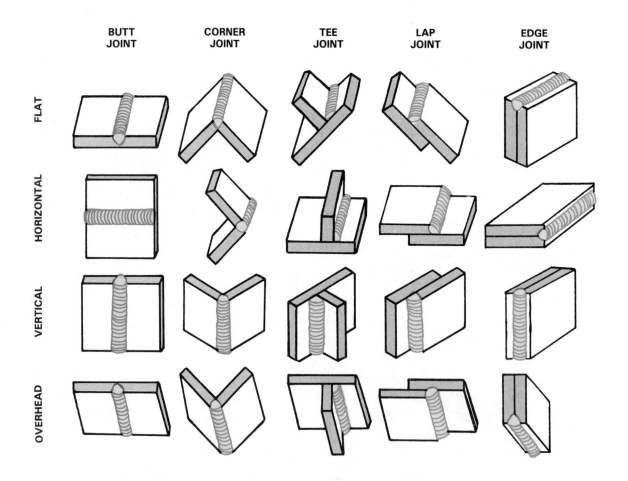

Fig. 2-2. The five basic weld joints may be made in four different welding positions.

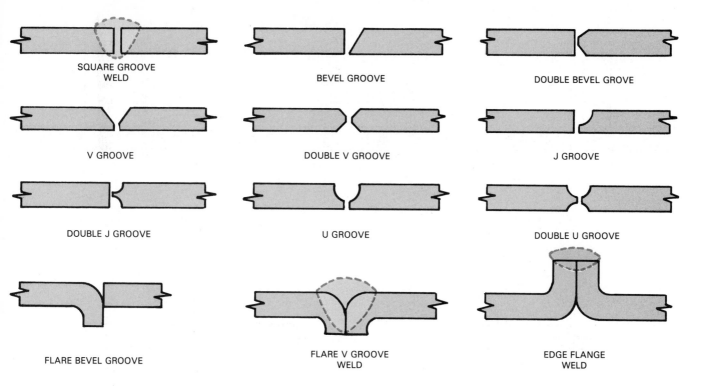

Fig. 2-4. Butt joint edge preparation methods. Double grooves are often used on thick metal and welded from both sides. Note that the metal is bent to form the flare bevel, flare V groove, and edge flange.

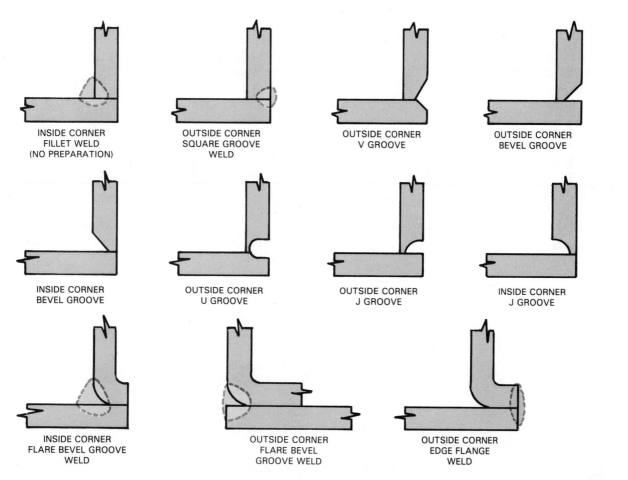

Fig. 2-5. Corner joint edge preparation methods. The placement of the weld is shown in five places where metal preparation is not used.

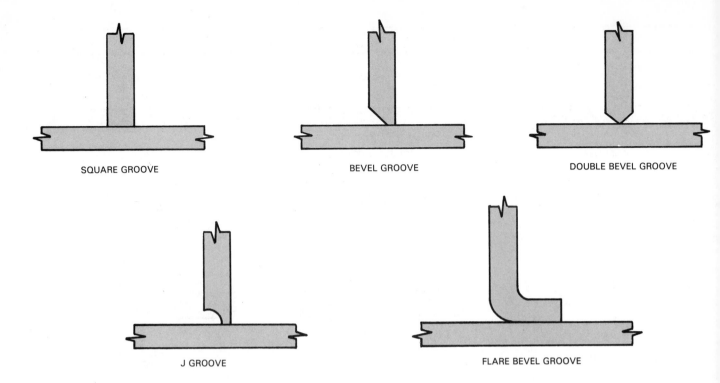

Fig. 2-6. T joint edge preparation methods. The double bevel and double J groove are often used on thick metal.

The T joint obtains its name from the placement of the base metals to form a T shape. See Fig. 2-6 for methods of preparing the metal edges for several types of T joint welds.

The metal which forms the lap joint is seldom altered in preparation for welding.

Edge joints may be prepared in a number of ways also as seen in Fig. 2-7.

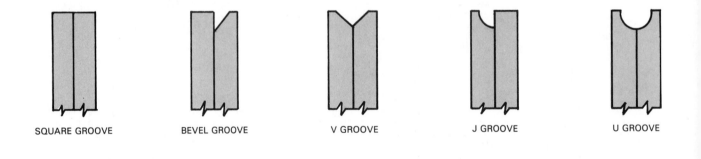

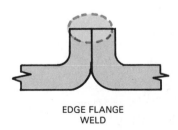

Fig. 2-7. Methods of preparing the edge joint for welding.

## 2-2. THE WELDING SYMBOL

The welding symbol adopted by the American Welding Society (AWS) is used internationally. Currently, all structural, bridge, government, and nuclear fabricators use the symbols.

The symbol shown in Fig. 2-8 is used on drawings of parts to be welded. Whenever two or more pieces of a welded part (weldment) are placed together, their surfaces and edges form a joint. The drawing of the part to be welded indicates how the parts will be assembled and what type of

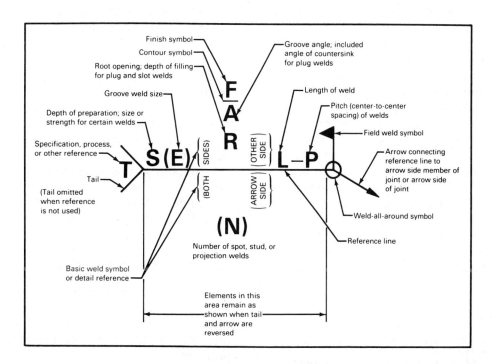

Fig. 2-8. The American Welding Society (AWS) welding symbol. Standard locations of information on a welding symbol are marked. (From ANSI/AWS A2.4-86)

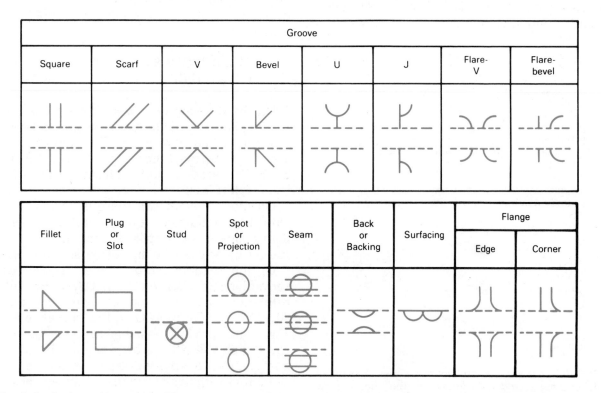

Fig. 2-9. Basic weld symbols. These are a part of the complete welding symbol. (From ANSI/AWS A2.4-86)

welded joint will be made. Refer to Fig. 2-1 for the types of welded joints and the types of welds used on the various joints.

The complete welding symbol will tell the welder how to prepare the base metal, the welding process to use, the method of finishing, and much more information regarding each weld. Dimensions on a welding symbol may be in SI metric units or conventional U.S. units.

A complete welding symbol contains all the information about a welded joint. Part of the welding symbol is the "basic weld symbol" which shows what type of weld to use. Fig. 2-9 shows the basic weld symbols used by AWS.

Information given in each part or area of the welding symbol will be explained in later paragraphs. A number of weld drawings, with their corresponding welding symbols, will be shown to illustrate the information given in the various areas of the complete welding symbol. The edges of the weld joint will be shown in red as they would be prepared and fitted up prior to welding. A completed weld for the welding symbol will also be shown to illustrate the use of the welding symbol.

## 2-3. THE REFERENCE LINE, ARROWHEAD, AND TAIL

The REFERENCE LINE, shown in Fig. 2-10, is always drawn as a horizontal line. It is placed on the drawing near the joint to be welded. All other information to be given on the welding symbol is shown above or below this horizontal reference line. All information shown on a complete welding

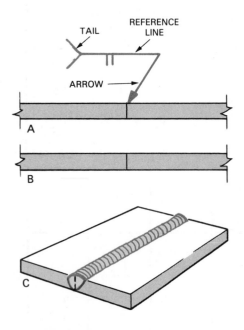

Fig. 2-10. Reference line, arrow, and tail of welding symbol. A—Welding drawing and welding symbol. B—Preparation of metal. C—Completed weld.

symbol is always shown in the same position as indicated in Fig. 2-8 and reads from left to right.

The ARROW may be drawn from either end of the reference line. The welding symbol may appear in any view of the welding drawing. The arrow always touches the line which represents the welded joint.

The TAIL is used only when necessary. If used, it may give information on specifications, the welding process used, or other details required but not shown on the welding symbol. A number, such as 1, 2, or 3, etc., may be used in the tail to refer the user to a note elsewhere on the drawing.

Companies may use their own number or letter codes in the tail to indicate the welding process, procedure, finishing method, or company specification.

If no tail is used, somewhere on the drawing there is a note such as, "Unless otherwise specified, all welds will be made in accordance with Specification No. XXXX."

## 2-4. BASIC WELD SYMBOLS

The basic weld symbol shown on the complete welding symbol indicates the type of weld made on a weld joint. It is also a miniature drawing of any metal edge preparation required prior to welding.

Fig. 2-11 illustrates how some of the various types of weld symbols shown in Fig. 2-9 are used on a welding symbol. The vertical line used with a fillet, bevel, or J-groove weld is ALWAYS drawn to the left.

## 2-5. THE ARROW SIDE AND OTHER SIDE

On the drawing of a welded part, the arrow of the welding symbol touches the line to be welded. The metal has two sides. The side of the metal which the arrow touches is always the ARROW SIDE. The opposite surface from the arrow is called the OTHER SIDE.

On many weldments there is no inside or outside, top or bottom, left or right because of the joint position. To simplify the location of the weld, the terms arrow side and other side are used.

On the welding symbol the arrow side weld information is ALWAYS SHOWN BELOW the reference line. The other side weld information is ALWAYS SHOWN ABOVE the reference line.

It is not always possible to place the welding symbol on the side to be welded. The drawing is sometimes too crowded and complicated. See Fig. 2-12 for examples of the use of the arrow side and other side on the welding symbol.

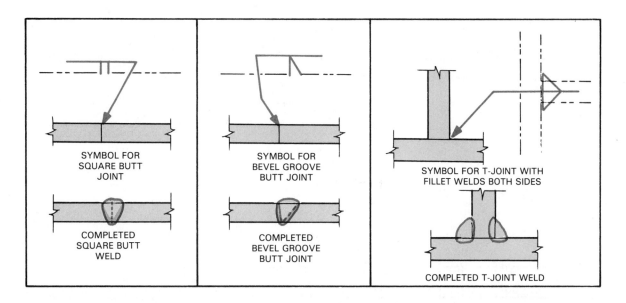

2-11. Comparing some welding symbols and actual welds. Phantom lines are not shown on a basic weld symbol. They are used here, however, to illustrate that the basic weld symbol is a miniature drawing of the edge preparation and/or type of weld used. The vertical line in the bevel groove and fillet weld symbols are drawn to the left.

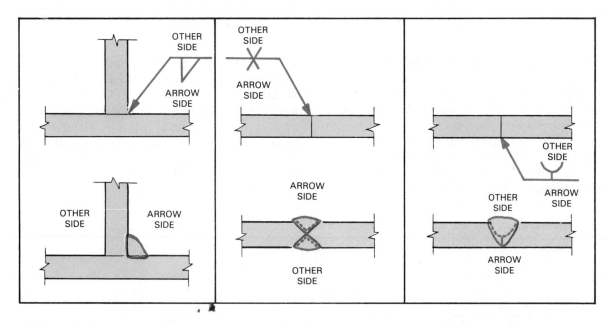

Fig. 2-12. Placement of the weld symbol for welding on the arrow side and on the other side. The side of the metal that the arrow touches is always the arrow side.

## 2-6. ROOT OPENING AND GROOVE ANGLE

The ROOT OPENING is the space between the metals at the bottom of the joint (root). This root opening may be specified on the drawing in metric units, in fractions of an inch, or as a single-place decimal of an inch. The root opening size appears inside the basic weld symbol on the complete welding symbol.

The included angle or total angle of a groove weld is shown beyond the basic weld symbol. See Fig. 2-13. When preparing the edges for welding, half the groove angle is cut on each piece so that when placed together, the combined angles will total the angle shown.

When a bevel or J-groove weld is used, only one piece of metal is cut or ground. The arrow of the welding symbol is bent to point to the piece to be cut or ground. See Fig. 2-13, view D, and Fig. 2-14, view A.

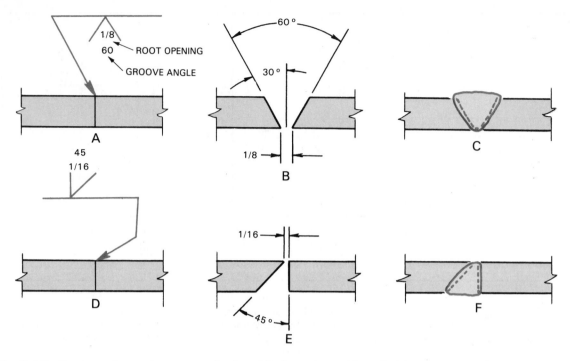

Fig. 2-13. Root opening and groove angle. A and D show the weld symbol for a groove weld. B and E show the pieces cut and set up for welding. C and F show the completed weld. Note that the bend in arrow at D points to left piece, which is the part to be cut or machined.

## 2-7. CONTOUR AND FINISH SYMBOLS

The shape or contour of the completed weld bead is shown on the welding symbol as a straight or curved line between the basic weld symbol and the finish symbol. The straight contour line indicates that the weld bead is to be made as flat as possible. The curved contour line indicates a normal convex or concave weld bead. See Fig. 2-14.

If the weld is not to remain in an "as welded" condition, a FINISH SYMBOL is used on the welding symbol. See Fig. 2-14. The finish symbol indicates the method of finishing. A surface texture or degree of finishing may also be added if required. If all welds are to be finished in the same manner, a note on the drawing may indicate the finish used. Users of the finish symbol may create their own finish symbols.

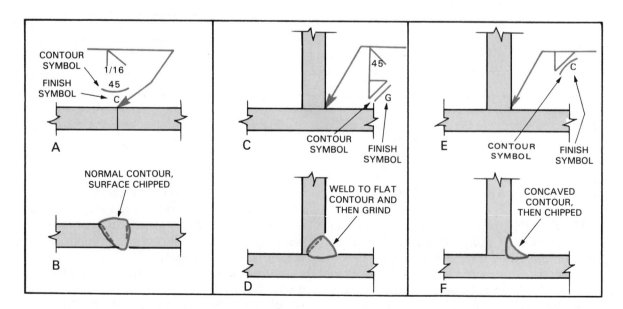

Fig. 2-14. Weld contour and finish symbols. In A, C, and E contour and finish symbols are shown on welding symbol. B, D, and F illustrate shape and finish of completed weld.

The American Welding Society lists the following finish symbols: C—Chipping; G—Grinding; M—Machining; R—Rolling; H—Hammering.

## 2-8. DEPTH OF PREPARATION, GROOVE WELD SIZE, AND EFFECTIVE THROAT

The "S" position on the welding symbol indicates the DEPTH OF PREPARATION for groove type welds. Groove weld size is given in position "(E)." GROOVE WELD SIZE is the depth to which the weld penetrates into the base metal. See Fig. 2-15. The depth of preparation and depth of weld penetration are generally determined by welding codes or specification, or by a welding engineer.

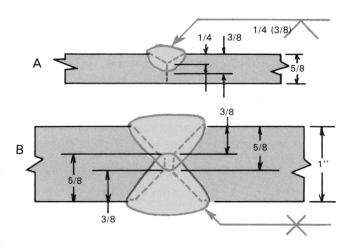

Fig. 2-15. Depth of preparation and groove weld size. In A, the depth of preparation is 1/4" and the groove weld size or depth of penetration is 3/8". The weld at B is a double groove weld. The depth of preparation is 3/8" and the weld groove size or depth of penetration is 5/8".

The EFFECTIVE THROAT of a fillet weld is the distance, minus any convexity, between the weld face and the root of the weld. The effective throat can never be greater than the metal thickness. See Fig. 2-16.

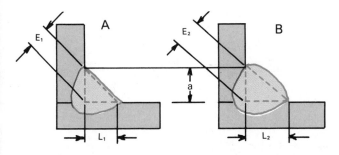

Fig. 2-16. Effective throat and fillet weld size. The apparent size (a) of the fillet welds at "A" and "B" appear to be the same. However, the weld size (L) or triangle drawn within the weld contour and the effective throat (E) size at "A" are smaller than those at "B.".

Weld strength or weld size for the welds other than groove type welds may also be given in the "S" position. See Fig. 2-17.

The size of a fillet weld is the length of each leg (side) of the triangle that can be drawn inside the cross section of the finished weld, as shown in Fig. 2-16. The length of the legs on a typical fillet weld are equal, so only one dimension is given in the "S" position on the welding symbol. If the legs are unequal in size, two dimensions are given in the "S" position within parentheses (). The weld shape is shown on the welding drawing to indicate which is the short leg. See Fig. 2-17. The fillet weld size is shown to the left of the basic weld symbol. The fillet weld size is placed in parentheses only when two dimensions are necessary. The effective throat size of a fillet weld is shown to the left of the basic weld symbol and it is always given in parentheses, as shown in Fig. 2-17.

When no groove weld size is shown for the single groove and double groove welds, complete penetration is required. See Figs. 2-15 and 2-18 for examples of groove weld size and effective throat size.

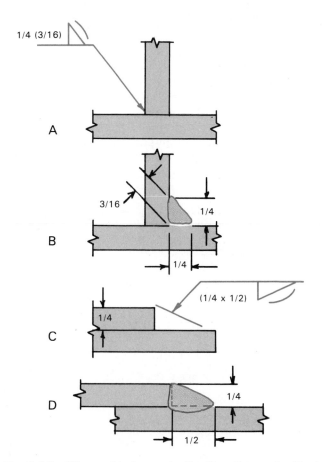

Fig. 2-17. Fillet weld size and effective throat. A—Single dimension indicates both legs are equal and 1/4" in size. The effective throat is 3/16". B—How finished weld will look. C—Two dimensions indicate unequal leg sizes. Relative size of legs is shown on welding drawing. D—How the finished weld will look.

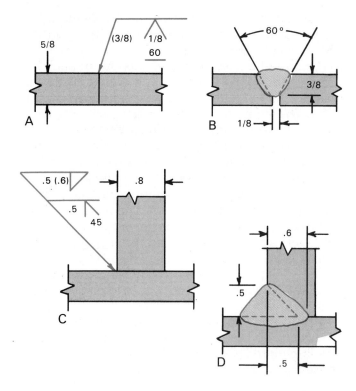

Fig. 2-18. Groove weld size. Groove weld size or depth of penetration is shown in parentheses at ''A'' and ''C.''Note at ''B'' that the groove weld size is less than the metal thickness. At ''D,'' it is greater than the depth of preparation.

## 2-9. LENGTH AND PITCH OF THE WELD

In many welded parts it is not necessary to weld continuously from one end of the joint to the other. To save time and expense, where strength is not affected, short sections of weld may be spaced across the joint. This is called INTERMITTENT WELDING.

On intermittent welds, the LENGTH DIMENSION is used to indicate the length of each weld. The PITCH DIMENSION indicates the distance from the center of one weld segment to the center of the next. See Fig. 2-19 for examples of such welds. The length and pitch dimension is always shown to the right of the basic weld symbol on the welding symbol.

When intermittent fillet welds are required on both sides of a welding joint, they may be one of two types. One type is chain intermittent welding; the other is staggered intermittent welding. The welds on either side of a CHAIN INTERMITTENT WELD begin and end at the same spot. So does the welding symbol. STAGGERED INTERMITTENT WELDS are offset so the welded segments do not line up on each side of the joint. This is shown on the welding symbol by offsetting the fillet weld symbols. See Fig. 2-19, views C and D.

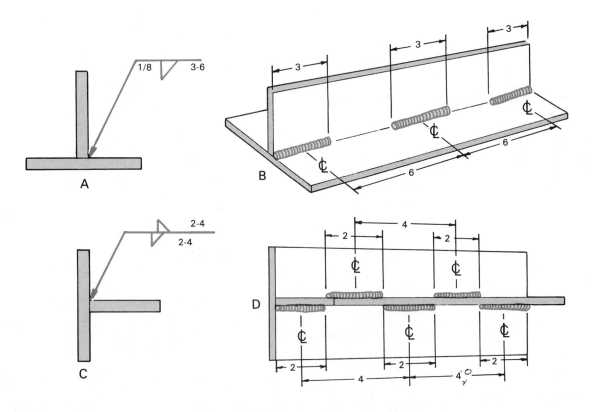

Fig. 2-19. Length and pitch dimensions of weld. A—Note placement of length (3) and pitch (6) on the welding symbol. B—Weld shows a series of 3 in. long welds which are 6 in. apart from center to center of the welds. C and D—Staggered weld. Notice staggered fillet symbols in C.

Continuous and intermittent welds may be made on the same joint. In such a case, the drawing will use dimensions to show where each weld symbol's effectiveness begins and ends. See Fig. 2-20, view A.

A spacing different from the regular pitch is used between the end of the continuous weld and the beginning of the intermittent weld. See 4'' dimension in Fig. 2-20, view B. This spacing is equal to the intermittent pitch minus the length of one intermittent weld. The spacing between the continuous and intermittent welds in Fig. 2-20, view B, equals the pitch minus the length, or 6'' − 2'' = 4'', as shown.

## 2.10 BACKING WELDS AND BURN-THROUGH SYMBOLS

Weld joints that require complete penetration may be welded from both sides. A STRINGER BEAD (single pass weld without a weaving motion) may be all that is required on the side opposite a groove weld to insure complete penetration. In such cases, a backing weld symbol may be used, Fig. 2-21.

The burn-through symbol is used when 100 percent penetration is required on one-side welds, Fig. 2-21.

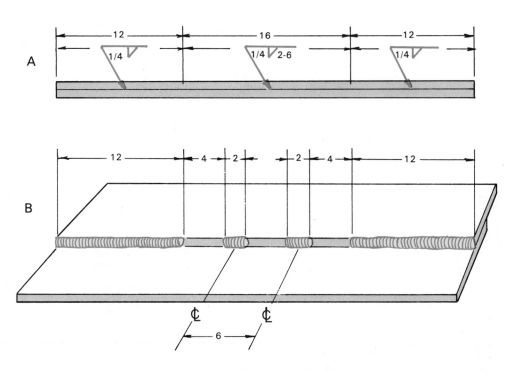

Fig. 2-20. Continuous and intermittent welds. Note that the dimensions on the top welding drawing limits use of welding symbol to distance shown. Note also that the spacing between continuous and intermittent weld is equal to pitch minus length of one intermittent weld (4 in. in this application).

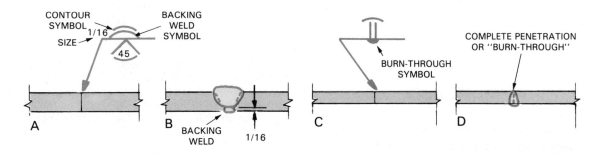

Fig. 2-21. Backing welds and burn-through symbols. The burn-through symbol is used on welds which are welded from one side only and which require 100 percent penetration, as shown in C and D. A backing weld may be used to obtain 100 percent penetration when welding is possible on both sides, as shown in A and B. Note that contour symbols and dimensions may be used with these symbols, as in A and C.

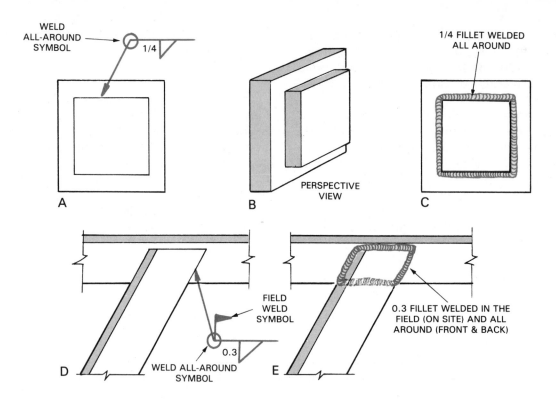

Fig. 2-22. Weld all-around and field weld symbols. Note the 0.3 fillet in D and E is welded in the field. It is welded all around the angle iron both front and back.

## 2-11. WELD ALL-AROUND AND FIELD WELD SYMBOLS

Directions given on a welding symbol are no longer of any value when the weld joint makes a sharp change in direction such as going around a corner. When the joint changes direction sharply, a new welding symbol must be used or a weld-all-around symbol may be used.

The WELD-ALL-AROUND SYMBOL is used when the same type weld joint is used on all edges of a box or cylindrical part. See Fig. 2-22.

Some parts are assembled and welded in the shop. It is often necessary to take parts into the field to make final assembly and welds.

When welds are to be made in the field away from the shop, a FIELD WELD SYMBOL is used, as in Fig. 2-22, view D. If a weld is to be made in the shop, the field weld symbol is not used. Refer to Fig. 2-8.

## 2-12. MULTIPLE REFERENCE LINES

When a sequence of operations are to be performed two or more reference lines may be used.

The reference line nearest the arrow indicates the first operation. The last operation is on the reference line furthest from the arrow. See Fig. 2-23.

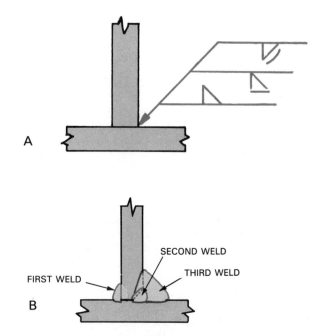

Fig. 2-23. Use of multiple reference lines. In A, three reference lines are used. The first weld is a backing weld. The second operation is to perform the bevel weld. The final operation is a flat contour fillet weld. Note that no sizes were used in this example. The completed joint is shown at B.

## 2-13. PLUG AND SLOT WELDS

Occasionally it is necessary to weld two pieces together at places away from the edges. This is done by creating a hole in one piece and welding the two pieces together through this hole as shown in Fig. 2-24. The holes may be round or of any other shape (generally an elongated slot). The hole may be drilled, flame cut, or machined.

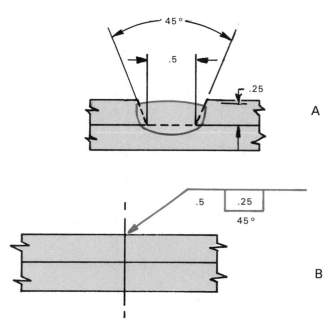

Fig. 2-25. Plug weld symbol. A—Cross-section of a completed plug weld. The desired dimensions are also shown. B—The weld symbol used to complete the desired weld.

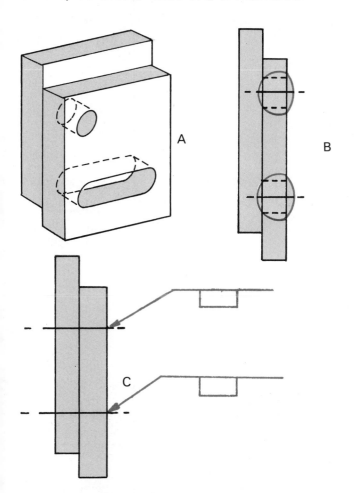

Fig. 2-24. Plug and slot welds. In A, the hole and slot are shown as cut in one piece. Cross-section of the completed welds are shown at B. The basic welding symbols for the plug and slot weld are shown at C.

A PLUG WELD is used if the hole is round. A SLOT WELD is used if the hole is elongated. The sides of the hole may be slanted. The welding symbol used for the plug weld is shown in Fig. 2-25. The size of the plug weld is shown to the left of the weld symbol. The angle of the countersink is shown below the weld symbol (arrow side weld). Inside the weld symbol is shown the depth of weld. The location of a plug weld is shown on the assembly drawing.

For a slot weld, the length, width, angle of the countersink, and the location and spacing of the slots are not shown on the basic weld symbol. These dimensions are shown on the assembly drawing. The depth of filling is shown on the welding symbol. If there are a series of plug or slot welds,

the center to center distance is shown to the right of the basic weld symbol. See Fig. 2-26.

## 2-14. SPOT WELDS

A spot weld may be accomplished using resistance welding, gas tungsten arc welding, electron beam welding, and ultrasonic welding as described in Chapter 1.

The SPOT WELD SYMBOL is a small circle. The circle may be on either side of the reference line, or it may straddle the reference line. If the weld is accomplished from the arrow side, the weld symbol should be below the reference line as in all other welding symbols. If the welding is done on both sides, as in resistance spot welding, the circle straddles the reference line. See Fig. 2-27.

Projection welding is another process to produce spot welds. To indicate which piece has the projections on it, the circle is placed above or below the reference line.

The following information is given for a spot weld: size, strength, spacing, number of spot welds.

The weld size is given to the left of the weld symbol. Weld strength is also shown to the left in pounds or newtons per spot. To the right of welding symbol is found the weld spacing. Centered above or below the spot welding symbol, in parenthesis, is the number of welds desired.

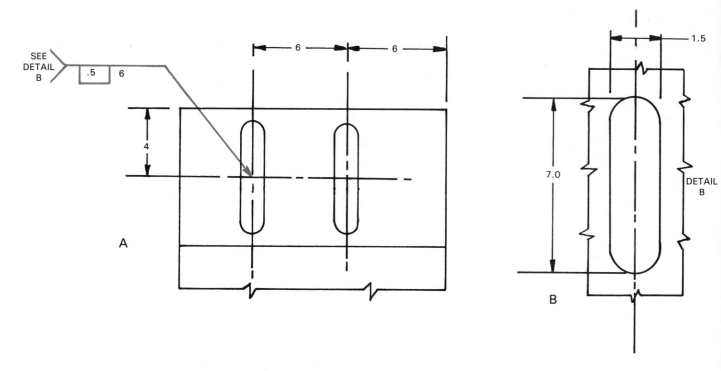

Fig. 2-26. Slot weld drawing and welding symbol. The location of the slot welds is shown at A. Also shown in the assembly drawing at A is the welding symbol. Details for the size of the slot weld are shown in the detail at B.

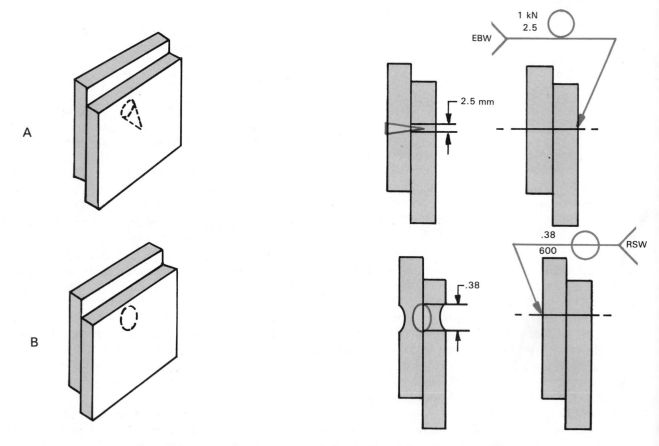

Fig. 2-27. The weld at A is an electron beam spot weld. Its size at the point of fusion is 2.5 mm. Its required strength is 1 kilonewton. The weld is made from the other side. The weld at B is a resistance spot weld. Its size is 3/8 in. and its strength is 600 pounds. The weld is made from both sides and the symbol straddles the reference line.

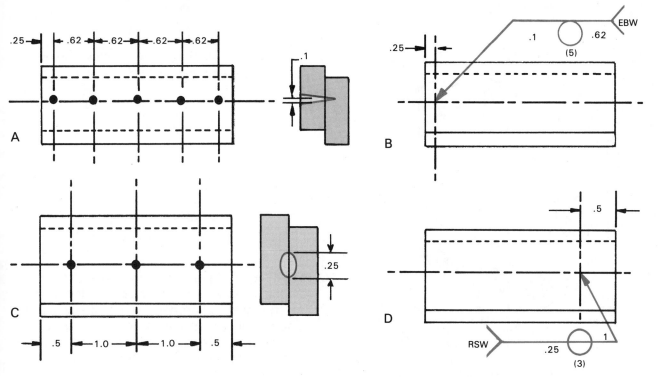

Fig. 2-28.  A series of electron beam spot welds are shown in A and B. In A, the desired weld and spacing is shown. B shows the correct welding drawing and symbol. Drawing D indicates a resistance spot welding symbol and welding drawing for the weld desired at C.

The welding process used is shown in the tail of the welding symbol. See Fig. 2-28 for examples of the welding symbols used for spot welding. The welding symbol may be placed in any view of a drawing as shown in Fig. 2-28.

## 2-15.  SEAM WELDS

Seam welds may be made with a number of processes, such as electron beam, resistance, or gas tungsten arc.

The process used is shown in the tail. The size (width) of the weld and strength of the weld are shown to the left of the weld symbol. The strength is given in pounds per linear inch or in newtons per millimeter.

The length of the seam may be shown to the right of the weld symbol.

The weld symbol may straddle the reference line if welded from both sides as in resistance seam welding. For electron beam and gas tungsten arc welding, the symbol is placed above or below the reference line. This indicates from which side of the part the weld is made. See Fig. 2-29.

For more details see AWS publication A2.4 - 79.

Fig. 2-30 shows examples of various welds and

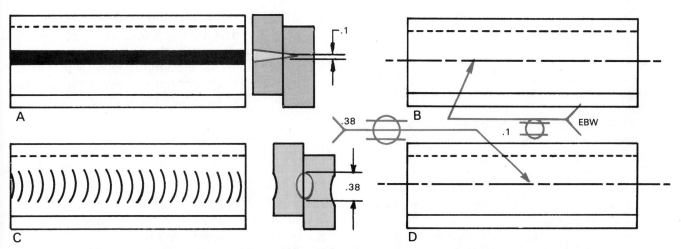

Fig. 2-29.  ''A'' illustrates a seam weld made with the electron beam. Its size at the fusion point is .1 in. The weld symbol and welding drawing for this electron beam seam weld is shown at B. The welding drawing and weld symbol shown in D will produce the weld shown in C.

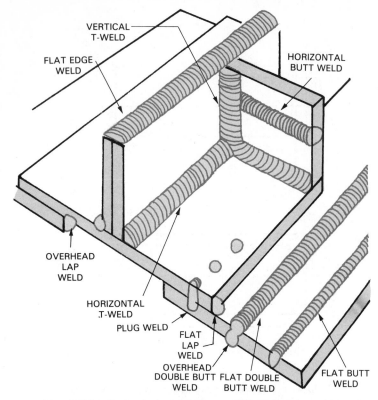

VERTICAL T-WELD

FLAT EDGE WELD

HORIZONTAL BUTT WELD

OVERHEAD LAP WELD

HORIZONTAL T-WELD

PLUG WELD

FLAT LAP WELD

OVERHEAD DOUBLE BUTT WELD

FLAT DOUBLE BUTT WELD

FLAT BUTT WELD

Fig. 2-30. Examples of various welds and the positions in which they may be made.

positions of welding. The ability to visualize the resulting weldment after reading the welding symbol is a very important skill for the welder to master. If you are not sure of the symbol used in prints to specify the welds shown in Fig. 2-30, go back over this chapter in review. A periodic review of this chapter as you move through the material covered in this book will help build your print reading confidence.

## 2-16. TEST YOUR KNOWLEDGE

Answer these questions on a separate sheet of paper. Do not write in this book.

1. List the five basic welding joints.
2. List five things the welding symbol will tell the welder about the welds that is to be made.
3. Why is the tail used on the welding symbol?
4. What does the basic weld symbol tell the welder when used on the welding symbol?
5. The information below the reference line refers to the _____ side of the weld.
6. Where does the root opening appear on a groove weld symbol?
7. Is the weld size or the effective throat placed in parenthesis?
8. A chain intermittent welding symbol has the fillet weld symbols aligned or offset?

9. An intermittent fillet weld with the length and pitch dimensions of 4-10 has how much space between the weld segments?

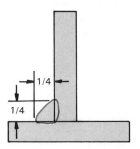

10. On a sheet of paper (do not write in the textbook), sketch the complete welding symbol for the weld sketched in the space above. The metal part shown is not to be ground or cut prior to welding, and the weld is to be continuous rather than intermittent.

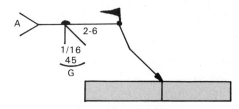

Refer to the welding symbol above when answering Questions 11 through 16:

11. Which piece of metal is to be ground, machined, or cut, prior to welding?
12. What does the small, black, half-round symbol means?
13. What shape is the weld face to be and what method is used to finish it?
14. Is the weld made in the shop or on site (in the field)?
15. At what angle is the one piece ground, and how far apart are the pieces at the root of the weld?
16. Is the weld made continuously, or is it intermittent? If intermittent, how long is each weld? How far apart center to center?

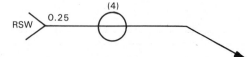

Refer to the welding symbol above when answering Questions 17-19.

17. What type of weld is to be made?
18. How many welds will be made?
19. What size are the welds?
20. Can a weld symbol ever straddle (cross) the reference line?

# part 2

# OXYFUEL GAS PROCESSES

*Chapters 7 and 8 continued on page 80.*

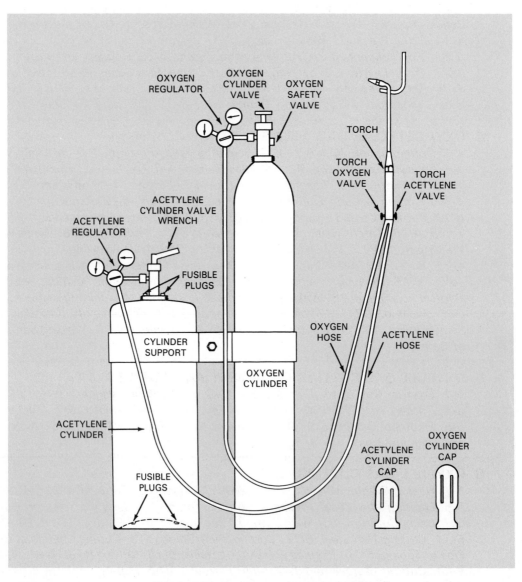

Fig. 3-1. An oxyacetylene welding outfit.

# 3 OXYFUEL GAS WELDING EQUIPMENT AND SUPPLIES

The operator should have a thorough knowledge of the purpose, design, construction, and operation of welding equipment and supplies. This knowledge is necessary to insure safe, conscientious use of the equipment, and to obtain the best results from the materials on hand.

## 3-1. COMPLETE OXYACETYLENE WELDING OUTFIT

The term OXYACETYLENE OUTFIT refers to the basic equipment needed to weld. An oxyacetylene station includes an oxyacetylene outfit, a welding table, ventilation, lighting, and other necessary room equipment.

The equipment necessary for an oxyacetylene outfit may vary depending upon the welding operations performed.

In general, a welding outfit will consist of the following:
1. GAS SUPPLY
   a. Oxygen cylinder (compressed gas) or tank (bulk liquid).
   b. Acetylene cylinder or generator.
2. REGULATORS with high and low pressure indicators or gauges. Cylinder and hose fittings are also required. One regulator with gauges and fittings is required for the oxygen cylinder and one regulator with gauges and fittings for the acetylene cylinder.
3. HOSES with appropriate fittings.
4. Complete WELDING TORCH with welding tip, mixing chamber, needle valves, and hose fittings.
5. WELDING GOGGLES with either a No. 4, 5, or 6 tinted filter shade.
6. SPARK LIGHTER.

The equipment for a complete outfit is shown connected and ready for use in Fig. 3-1.

## 3-2. OXYGEN SUPPLY

Oxygen used for welding is stored in cylinders of various sizes, which are usually painted. Since there is no standard national code, the cylinders are painted a color selected by the manufacturer.

Oxygen is stored in cylinders at a pressure of from 2,000 to 2,640 psig (pounds per square inch gauge), or 13 790 to 18 202 kPa (kilopascal). The pressure varies depending on the cylinder material and size. The pressure in a cylinder will vary also according to room temperature.

Oxygen is obtained by three different processes. One process consists of the liquefying of atmospheric air by compression and cooling. The atmospheric air consists of approximately 21 percent oxygen, 78 percent nitrogen, 1 percent other gases (by volume).

Oxygen and nitrogen have different boiling temperatures so it is easy to separate the oxygen from the nitrogen. The nitrogen will boil off first because oxygen boils at a higher temperature. The boiling point of liquid oxygen is $-297\,°F$ ($-183\,°C$). Nitrogen boils at $-320\,°F$ ($-196\,°C$) at atmospheric pressure.

The one percent by volume of other gases consists mainly of water vapor, carbon dioxide, argon, hydrogen, neon, and helium. Water vapor and carbon dioxide are removed during the compression and liquefying process. The other gases are removed at the time the oxygen and nitrogen are separated by evaporation. Nitrogen and the other gases have a lower boiling temperature than oxygen. As these gases boil away, the oxygen remains and is stored either as a gas or as a liquid, depending on how it is to be used.

A second method of producing oxygen is by electrolysis of water. In this process, an electric current is passed through water causing the water ($H_2O$) to separate into its elements, which are oxygen (O) and hydrogen (H). In the electrolytic process, oxygen will collect at the positive electrode, and hydrogen will collect at the negative electrode. Fig. 3-2 illustrates an experimental schematic of the electrolysis of water.

A third method used to produce oxygen for

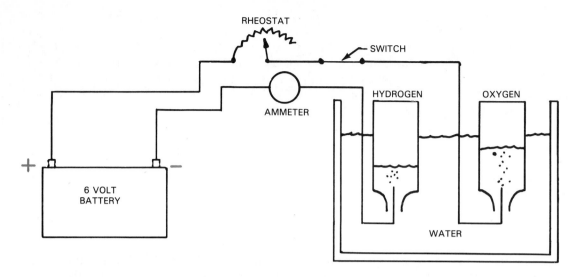

Fig. 3-2. A schematic illustrating the electrolysis of water. With DC current flowing, bubbles of oxygen will appear in one bottle and hydrogen in the other bottle. Oxygen is collected at the positive ( + ) terminal and hydrogen at the negative ( − ) terminal.

welding requires the heating of an oxygen-bearing pellet.

Commercial oxygen sold on market is close to being 100 percent pure. A popular method of distribution consists of shipping oxygen as a liquid. Liquid oxygen installations are chiefly used in steel mills and by steel fabricators that use large quantities of oxygen.

Oxygen sold in liquid form is in large thermos bottle-like tanks or vessels. These vessels are also known as Dewar flasks. Liquid oxygen is not held under very high pressure. The pressure in a liquid oxygen vessel is seldom greater than 240 psig (1655 kPa). The evaporation of some of the liquid keeps the temperature of the liquid very low, approximately −297°F (−183°C). At this low temperature, the oxygen remains a liquid under normal atmospheric pressure. Liquid air in a container will rapidly evaporate if the pressure is reduced.

Some portable liquid oxygen vessels have been developed and experience indicates that their use will increase. The advantage of liquid oxygen storage is chiefly with the saving in size and weight of the container. Fig. 3-3 illustrates a liquid oxygen storage vessel or Dewar flask.

## 3-3. OXYGEN CYLINDERS

Most oxyacetylene welding operators will be using oxygen from gaseous cylinders. Therefore, they must know how to use these cylinders and the various SAFETY PRECAUTIONS which must be taken. The oxygen stored in cylinders is under very high pressure. The gauge pressure is about 2,200 pounds per square inch of tank area (15 170 kPa). For safety sake, cylinders must be of very sturdy construction. The Interstate Commerce Commission (I.C.C.) has prepared specifications

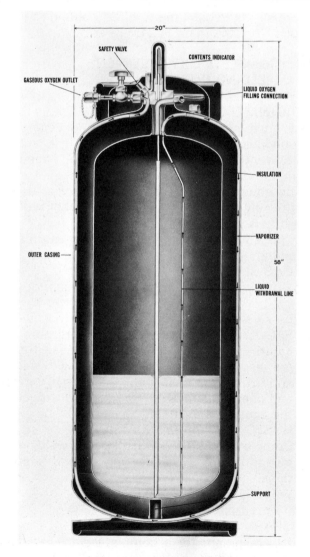

Fig. 3-3. Liquid oxygen vessel or Dewar flask (sectional view). Note that liquid oxygen drawn from the inner cylinder vaporizes as it passes through the withdrawal line in the insulated space surrounding the inner cylinder.
(Linde Div., Union Carbide Corp.)

for the construction of these cylinders. They are forged in one piece. No part of the cylinder is less than 1/4 inch thick. The steel used is armor plate type, high carbon steel. Fig. 3-4 illustrates an oxygen cylinder which has a section cut away to show construction of the cylinder and the valve. Oxygen cylinders are tested regularly. They must withstand hydrostatic (water) pressure of over 3,300 psig (22 753 kPa). These cylinders are periodically heated and annealed. This annealing relieves stresses created during on-the-job handling. They are also periodically cleaned using a caustic (eating away by chemical action) solution.

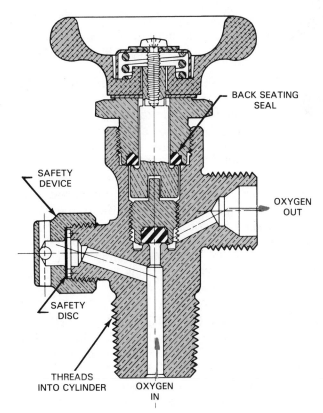

Fig. 3-5. Internal construction of an oxygen cylinder valve. Note that the valve must be opened completely to engage the back seating seal.

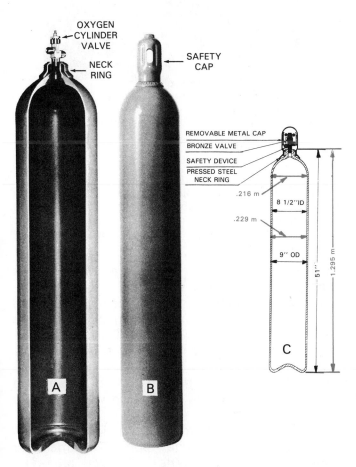

Fig. 3-4. A typical oxygen cylinder with a 244 cu. ft. (6.91 m³) capacity. A—Internal construction of cylinder. Note the one piece forged construction. B—Exterior of the cylinder. Note the cap over the valve. C—Dimensions of a 244 cu. ft. (6.91 m³) cylinder.    (Pressed Steel Tank Co.)

shows the internal construction of an oxygen cylinder valve. The valve is fastened to the oxygen cylinder by 1/2 or 3/4 inch pipe threads. The 3/4 inch size is the most popular. The valve, located in the upper end of the cylinder, incorporates a pressure safety device. The safety device consists of a pressure disc, Fig. 3-5. This thin metal disc will burst, as shown in Fig. 3-6, before the pressure becomes great enough to rupture the cylinder. The valve outlet fitting is a standard male

The cylinder valve is of special design to withstand high pressure and is made of forged brass. The oxygen cylinder valve is a "back seating valve." When the valve is turned all the way open, the stem is sealed to prevent the leakage of oxygen around the stem. When in use, this valve should be turned all the way open and NEVER USED IN A PARTLY OPEN POSITION. Fig. 3-5

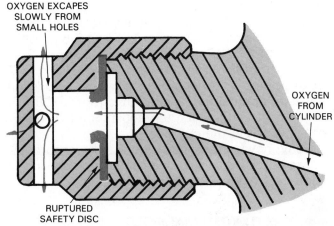

Fig. 3-6. A schematic of the oxygen cylinder safety plug. Note that when the disc ruptures, oxygen escapes from several drilled holes. See Fig. 3-5 also.

**Oxyfuel Gas Welding Equipment and Supplies / 83**

thread to which all standard U.S. made pressure regulators may be attached. A handwheel for operating the valve is securely attached to the valve stem.

Threads on the upper body of the cylinder are provided for the safety cap. The heavy steel safety cap is screwed over the valve to protect it from injury during movement or shipment. The thread size is 3 1/8 inch diameter (79.4 mm), 7 or 11 threads per inch. The thread pitch in SI units is 3.62 mm or 2.31 mm. If the cylinder valve should ever be broken off, the very high pressure oxygen will escape rapidly. The escaping oxygen may help any material it contacts to burn. It could also tend to give the cylinder rocket propulsion. Because of this danger, it is RECOMMENDED THAT TWO PERSONS MOVE THE CYLINDER. WHEN TRANSPORTING CYLINDERS, THEY SHOULD BE SECURED IN A VERTICAL POSITION TO ELIMINATE THE DANGER OF ITS BEING TIPPED OVER OR DROPPED. THE CYLINDERS, WHEN FULL OR PARTIALLY FULL, SHOULD NEVER BE ALLOWED TO STAND BY THEMSELVES WITHOUT ADEQUATE SUPPORT. The pressure in the oxygen cylinder will indicate the amount of oxygen remaining in the cylinder.

CYLINDERS SHOULD ALWAYS BE KEPT VALVE END UP, AND THE VALVE SHOULD BE CLOSED WHEN THE CYLINDER IS NOT IN USE, WHETHER THE CYLINDER IS FULL OR EMPTY.

Oxygen is purchased by the cubic foot measured at atmospheric pressure. The cylinders usually remain the property of the oxygen manufacturer and are loaned to the consumer. The price of an empty oxygen cylinder is quite high. It is generally to the advantage of the consumer to use the rental system. Most companies will not charge any rental for a cylinder if it is returned within thirty days from date of delivery. After thirty days, a fee is charged on a per day basis, often referred to as a demurrage charge.

Oxygen cylinders come in many sizes. The most common sizes are: 20 (.57 m³), 55 (1.56 m³), 80 (2.27 m³), 122 (3.45 m³), 220 (6.23 m³), 244 (6.91 m³), 300 (8.50 m³) cubic feet. See Fig. 3-7 for examples of various cylinder sizes. Under full pressure, approximately one cubic foot (.028 m³) of gas is stored in each ten cubic inch (.000164 m³) of space in the cylinder.

Many newer cylinders permit a pressure increase of ten percent with an equal increase in the amount of oxygen in the cylinder.

## 3-4. THE OXYGEN MANIFOLD

In many industries and in many school welding shops, an oxygen cylinder is not made a part of

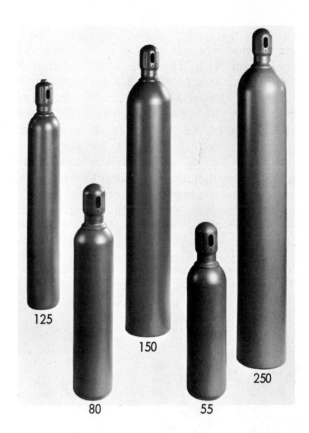

Fig. 3-7. Five different capacity oxygen cylinders. They range from 55 cu. ft. to 250 cu. ft. (1.56 m³ - 7.08 m³) capacity, and under certain conditions they may be charged with ten percent more oxygen. (Pressed Steel Tank Co., Inc.)

each welding station. In such cases, oxygen is piped to the welding stations. In such installations, one or more oxygen cylinders are attached to a manifold. The oxygen is piped from the manifold to the welding stations. The oxygen leaving the manifold is reduced in pressure. Pressures between 30 and 100 psig (206.8 - 689.5 kPa) are common. The length and size of the pipe and the amount of oxygen being used will determine the pressure needed.

The oxygen manifold is usually located in the oxygen cylinder storage room. The manifold cylinders are generally outside the work area. The shop is safer and space is saved by not having the cylinders in the shop.

With a manifold system, an oxygen line regulator is used at each welding station in order to control the oxygen pressure to each torch. A line regulator has only one pressure gauge and this gauge indicates the pressure on the delivery (torch) side of the system. Fig. 3-8 shows a typical oxygen manifold installation. Note that the manifold has pressure up to the manual shut-off valve. This valve is kept closed when the oxygen system is not in use. If the manifold is to be shut down completely, it is recommended that the cylinder valves be closed also. A master regulator

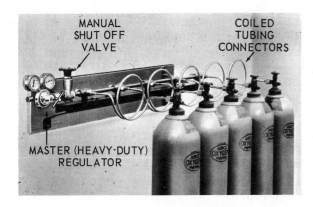

MANUAL
SHUT OFF
VALVE

COILED
TUBING
CONNECTORS

MASTER (HEAVY-DUTY)
REGULATOR

Fig. 3-8. An oxygen manifold for five cylinders. A master shutoff valve is located between the regulator and the cylinders. Be sure to check local and national fire safety codes prior to installation. (Airco Welding Products)

controls the pressure of the oxygen in the piping system after it leaves the manifold. This regulator has two gauges. One gauge shows the pressure in the manifold, and the other gauge shows the pressure in the piping. See Heading 3-15 for information concerning the construction and operation of these regulators.

In manifold installations, the copper tubing (pigtail) connecting the cylinders to the manifold should be frequently annealed (heated and cooled to soften and make less brittle). This is because the tubing is subjected to high cylinder pressure, and it therefore may become brittle and more subject to breakage. LOCAL AND NATIONAL SAFETY AND FIRE CODES MUST BE CHECKED TO PROPERLY INSTALL AN OXYGEN MANIFOLD.

## 3-5. OXYGEN SAFETY PRECAUTIONS

1. If a label on a cylinder is not legible or missing, do not assume that a cylinder contains a particular gas. Return it to the supplier.
2. Do not permit smoking or open flames in any area where oxygen is stored, handled, or used.
3. Liquid oxygen at −297°F (−183°C) may cause freeze burns to the eyes or skin if it comes in contact with them.
4. Keep all organic materials such as oil, grease, kerosene, cloth, wood, tar, and coal dust away from contact with oxygen.
5. Do not place liquid oxygen equipment on asphalt or surfaces with oil or grease deposits.
6. Remove all clothing which has been splashed or otherwise saturated with oxygen gas. Such clothing is highly flammable. It is not safe to wear for at least 30 minutes or until no oxygen remains.

The following are some safety codes which may be referred to:

ANSI (American National Standard Institute)

Z49.1 "Safety in Welding and Cutting."

NFPA (National Fire Protection Association) Standard No. 50, "Bulk Oxygen Systems at Consumer Sites."

NFPA Standard No. 51, "Oxygen-Fuel Gas Systems for Cutting and Welding."

Linde Form 9888, "Precautions and Safe Practices - Liquid Atmospheric Gases."

Linde Form 2035, "Precautions and Safety Practices in Welding and Cutting with Oxygen-Fuel Gas Equipment."

Local and National Safety Codes.

## 3-6. ACETYLENE SUPPLY

Acetylene is produced by the chemical combination of calcium carbide with water. The chemical formula for acetylene is $C_2H_2$. (See Heading 31-4 for further technical information concerning the chemical structure and nature of acetylene.)

Acetylene is made available for oxyacetylene welding using two different methods:
1. Acetylene storage cylinder.
2. Acetylene generator.

## 3-7. ACETYLENE CYLINDERS

Acetylene gas may be stored in cylinders specially designed for this purpose. The gas is first passed through filters and purifiers. THE STORAGE OF ACETYLENE IN ITS GASEOUS FORM UNDER PRESSURE IS NOT SAFE AT PRESSURES ABOVE 15 psig (103.4 kPa). The method used to safely store acetylene in cylinders is as follows:
1. The cylinders are filled with a monolithic (massive uniformity) filler which cures to a porosity of 85 percent as required by federal safety regulations. This prevents large acetylene gas accumulations.
2. The cylinders are then charged with acetone. Acetone absorbs acetylene.

The theory is that the acetylene molecules fit in between the acetone molecules. Using both of these techniques prevents the accumulation of a pocket of high pressure acetylene.

These cylinders, like oxygen cylinders, are fabricated according to Interstate Commerce Commission (I.C.C.) specifications.

The base of this cylinder is concave and it usually has two plugs threaded into it (pipe threads). These threaded plugs (fuse plugs) have a center made of a special metal alloy which will melt at a temperature of approximately 212°F (100°C). Fuse plugs may also be threaded into the top of the cylinder. Fig. 3-9 illustrates the fuse plug. If

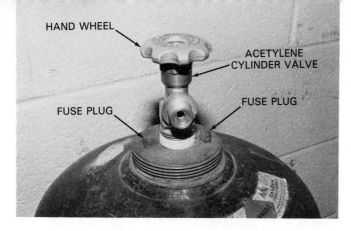

Fig. 3-9. Acetylene cylinder fuse plugs. The body of the plug is usually made of brass. The center is a fusible material which melts at about 212 °F (100 °C).

the cylinder should be subjected to a high temperature, the plugs will melt and allow the gas to escape before the pressure builds up enough to burst the cylinder. In a fire, the acetylene from the fuse plug will burn, but at a relatively slow rate.

These precautions are necessary as the pressure in an acetylene cylinder builds up rapidly with an increase of temperature. Fig. 3-10 shows the construction of an acetylene cylinder.

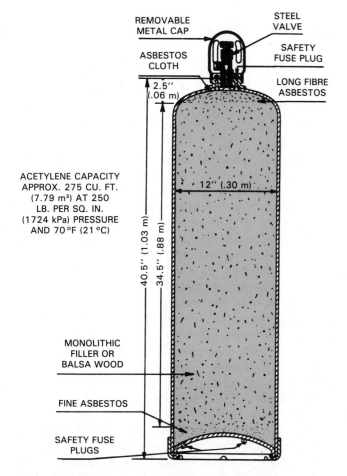

Fig. 3-10. Cutaway view of acetylene cylinder showing the porous filler. Note the fuse plugs at the top and bottom of cylinder.

Acetylene cylinder valves come in two types. One has a hand wheel which should remain attached at all times. The other is provided with a 3/8 inch (9.5 mm) square shank. The square shank is turned by means of a 3/8 inch (9.5 mm) square box end wrench.

It is recommended that the cylinder valve be opened only 1/4 to 1/2 turn. The wrench should be left on the valve stem at all times that the valve is open. This is done so that the valve may be closed quickly in case a hose or some other part catches fire.

The regulator fitting is a female fitting. Fig. 3-11 shows the construction of the hand wheel type of acetylene cylinder valve.

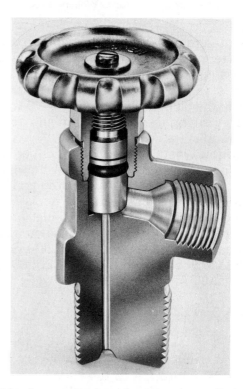

Fig. 3-11. A cutaway view of an acetylene cylinder valve.

The amount of acetylene in a cylinder cannot be estimated by the pressure in the cylinder because the pressure of the acetylene gas coming out of the acetone solution will remain fairly constant (depending on the temperature) until most of the gas is consumed. The amount of acetylene in a cylinder can be determined accurately by weighing the cylinder. The tare weight (weight without the acetylene gas) of the cylinder is always stamped on the cylinder. Subtract the tare weight from the weighed cylinder to give the weight of the gas remaining. The gas weighs one pound per 14 1/2 cubic feet (1.10 kg per m³).

Acetylene cylinders should always be used in an upright position, otherwise, some of the acetone is likely to escape with the acetylene and contaminate the equipment and the flame. Each time that an acetylene cylinder is refilled, the tare weight is checked and acetone is added if necessary.

Acetylene cylinders are available in a variety of sizes from 10 ft.³ to 380 ft.³ (.28 m³ - 10.8 m³). The most common sizes used in welding are: 130 cubic feet (3.68 m³), 290 cubic feet (8.21 m³), and 330 cubic feet (9.34 m³) capacity.

Two small size acetylene cylinders are available for portable welding and cutting equipment. These are: B size - 40 cubic feet (1.13 m³) and MC size - 10 cubic feet (2.83 m³). The fittings on these two small size cylinders (sometimes called tanks) are Presto-O-Lite (P.O.L.) fittings.

All acetylene cylinders are designed and constructed according to Interstate Commerce Commission, (I.C.C.) Specifications No. 8. All designs must be tested by the Bureau of Explosives and must pass these tests before they can be used commercially.

Acetylene is dissolved in acetone. Therefore, acetylene cannot be drawn from the cylinder any faster than it can be released from the acetone. This release of acetylene is a kind of boiling action. The boiling action or release of acetylene from acetone occurs efficiently at room temperature. If acetylene is drawn out of a cylinder too rapidly, a considerable amount of acetone may be drawn from the cylinder along with the acetylene.

The maximum safe rate for drawing acetylene from a cylinder is one-seventh of the cylinder's capacity per hour. Thus, a single 250 cubic foot (7.08 m³) cylinder can supply acetylene at about 36 cubic feet (1.0 m³) per hour.

AN OXYACETYLENE FLAME CONSUMING SOME ACETONE WILL BURN WITH A PURPLE COLOR.

Acetone in the flame is not desirable. It lowers the flame temperature and increases gas consumption. Moreover, the quality of the weld is affected.

Low or freezing temperatures can cause the flow of acetylene to decrease as heat is needed to boil off the acetylene. Another factor causing a slowdown in acetylene flow is a nearly exhausted cylinder. As the cylinder approaches the empty or discharged condition, the acetylene is released more slowly. If a greater flow rate is needed, a number of cylinders may be connected to a manifold to supply the required flow. This kind of installation makes it possible to use up more of the acetylene from each cylinder. This may also cut down the cost of the acetylene used.

The rate at which acetylene may be drawn off depends on the three following conditions:

1. The temperature of the cylinder.

2. The amount of charge remaining in the cylinder.
3. The number of cylinders providing the flow.

Too rapid removal of acetylene can be dangerous. A pressure drop could occur which would cause a FLASHBACK. A flashback is when the torch flame burns inside the torch or hoses. If this occurs, shut off the gas supply immediately.

Fig. 3-12 illustrates several sizes of acetylene cylinders. Some have a recessed top. This recess protects the cylinder valve which has a female regulator connection or fitting. The cylinder on the extreme right is the MC size, while the one next to it is the B size. The valves in the larger cylinders shown in Fig. 3-12 are known as P.O.L. Commercial.

Fig. 3-12. Acetylene cylinders come in a variety of sizes to meet the needs of most users.

## 3-8. ACETYLENE MANIFOLD

As explained in Heading 3-4, there are several advantages in a manifold system in which the fuel gas and oxygen cylinders are not located at the welding stations in the room. In the case of acetylene or other fuel gases, fire safety is one additional reason for having a manifold installation.

Fig. 3-13 illustrates a modern acetylene cylinder manifold installation. It should be noted that the acetylene manifold system incorporates some features not included in an oxygen manifold system. In addition to regulator and control valves, the acetylene manifold also has a waterseal type flash arrester to prevent flashback. Local and national safety fire codes must be checked to properly install an acetylene or fuel gas manifold.

BRASS PIPING MAY BE USED WITH ACETYLENE. COPPER PIPING MUST NOT BE USED IN THE PRESENCE OF ACETYLENE. COPPER AND ACETYLENE WILL FORM COPPER ACETYLIDE,

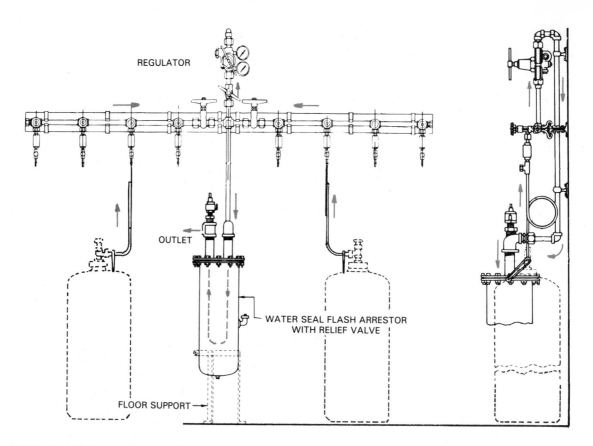

Fig. 3-13. A typical acetylene manifold installation. Be sure to check local or national codes for proper installation.

AN UNSTABLE COMPOUND THAT DISASSOCI—ATES VIOLENTLY AT THE SLIGHTEST SHOCK.

## 3-9. ACETYLENE GENERATOR

Acetylene is produced by the chemical action of water with calcium carbide. A generator brings the proper amounts of water and calcium carbide together at the rate required to safely generate the acetylene needed.

Acetylene generators are available in two types: a low pressure type, and a medium pressure type. The low pressure type generates acetylene at approximately 1/4 psig (4 to 6 inches of water column) or 1.72 kPa. With this type of generator, it is necessary to use an injector type torch. See Heading 3-21.

The medium pressure type acetylene generator generates acetylene up to 15 psig (103.42 kPa) which is the maximum safe pressure at which free acetylene gas may be stored. With this type of generator, equal pressure type torches may be used. See Heading 3-20 for a description of the equal pressure type torch. Fig. 3-14 illustrates an acetylene generator. The use and installation of an acetylene generator is controlled by safety codes which should be carefully studied and rigidly followed. See Heading 3-10. A generator installation will require regular care which includes check-

ing and replenishing the water and calcium carbide supply, and removal of the calcium hydroxide sludge. The removal and disposing of the sludge is perhaps the greatest problem in the use of the acetylene generator.

Referring to Fig. 3-14, note that the generator is partially filled with water. The calcium carbide is stored in the hopper at the top of the generator. A feed mechanism feeds the calcium carbide into the water. Acetylene is generated by the chemical action of the water on the calcium carbide. As soon as the predetermined acetylene pressure has been generated, the feed of calcium carbide is stopped. No more can be fed into the water until the acetylene has been drawn off and the acetylene pressure in the generator decreases. A water level supply device stops the feeding of calcium carbide if insufficient water is in the base of the generator.

## 3-10. ACETYLENE SAFETY PRECAUTIONS

1. Concentrations of acetylene between 2.5 percent and 81 percent by volume in air are easy to ignite and may cause an explosion.
2. Smoking, open flames, unapproved electrical equipment, or other sources of ignition must not be permitted in acetylene storage areas.
3. Under certain conditions, acetylene forms

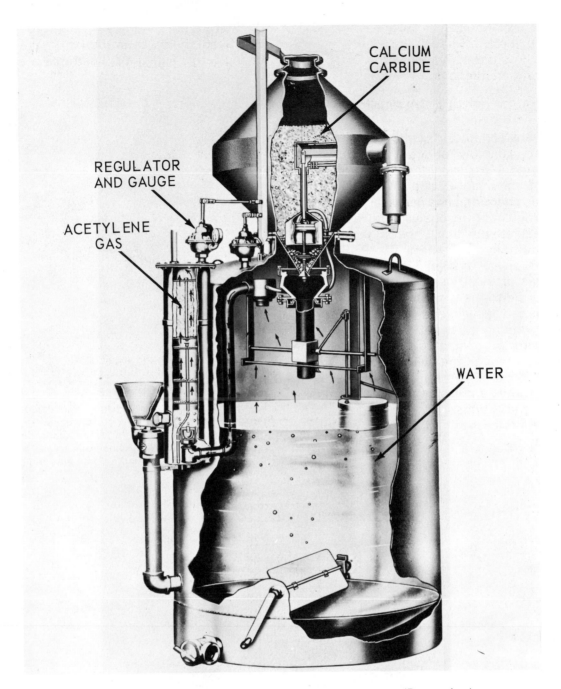

REGULATOR AND GAUGE

ACETYLENE GAS

CALCIUM CARBIDE

WATER

Fig. 3.14. A cutaway view of an acetylene generator. (Rexarc, Inc.)

readily explosive compounds with copper, silver, and mercury. Acetylene must be kept away from these metals, their salts, compounds, and high concentration alloys.

4. The fusible plugs used on acetylene cylinders will melt at 212 °F (100 °C).

5. Keep cylinders away from overhead welding or cutting. Hot slag may fall on a cylinder and melt a fusible plug.

6. Adequate ventilation in welding areas is essential. Acetylene has a garlic-like odor. Acetylene may displace air in a poorly ventilated space. Any atmosphere which does not contain at least 18 percent oxygen may cause dizziness, unconsciousness, or even death.

7. Leave the handwheel, wrench, or key on the cylinder valve for emergency shut off.

8. Keep the torch flame away from the cylinder and fuse plugs.

The following are some safety codes which may be referred to:

ANSI (American National Standards Institute) Standard Z 49.1, "Safety in Welding and Cutting."

NFPA (National Fire Protection Association) Standard No. 51, "Oxygen-Fuel Gas Systems for Welding and Cutting."

Linde Form 2035, "Precautions and Safety

Practices in Welding and Cutting with Oxy-Acetylene Equipment.''

Also, it is important to follow the equipment manufacturer's instructions.

## 3-11. PRESSURE REGULATOR PRINCIPLES

All gases are commonly stored in cylinders at pressures considerably above the working or flame pressures. Most welding torches operate at pressures of between 0 and 30 psig (0 and 206.84 kPa). Therefore, it is necessary to provide a pressure regulating mechanism to reduce and otherwise regulate the pressure from the cylinder. This mechanism is called a PRESSURE REGULATOR. Every system which requires the control of gas pressure uses this type regulator. The regulator performs two functions:

1. It reduces the high storage cylinder pressure to suitable working pressure.
2. It maintains a constant gas pressure at the torch (even though the cylinder pressure may vary).

Fig. 3-15 shows a pressure regulator complete with gauges and fittings.

There are two basic types of regulator mechanisms:
1. Nozzle type.

2. Stem type.

Fig. 3-16 shows construction of the nozzle type of regulator. Fig. 3-17 illustrates a cross section

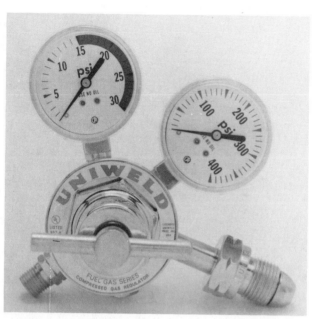

Fig. 3-15. Single stage acetylene cylinder regulator. The high pressure gauge shows cylinder pressure. The low pressure gauge shows the working pressure. Note that low pressure gauge indicates that it is dangerous to use acetylene above 15 psig (103.4 kPa). (Uniweld Products, Inc.)

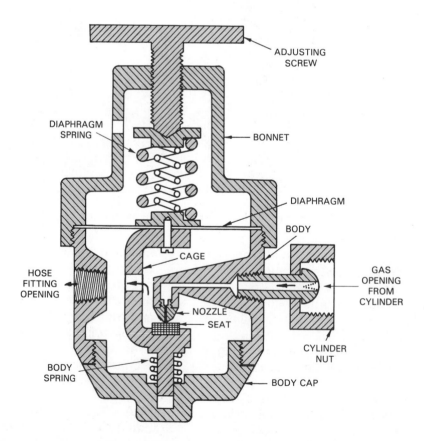

Fig. 3-16. Schematic cross section drawing of a nozzle type pressure regulator.

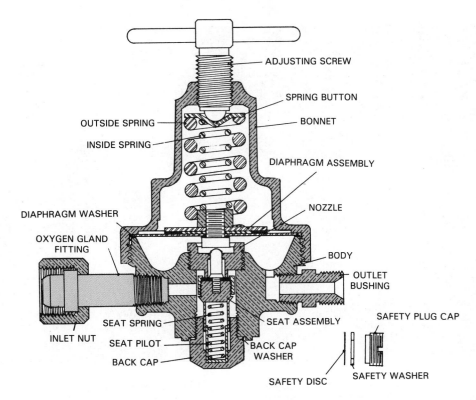

Fig. 3-17. Cross section drawing of nozzle type pressure regulator. Note shape of oxygen gland fitting which, with the inlet nut, connects regulator to cylinder.

of a stem type regulator.

Most regulators have two gauges. The HIGH PRESSURE GAUGE shows the pressure in the cylinder. The LOW PRESSURE GAUGE shows the pressure of gas being delivered to the torch.

Regulators are also obtainable as:

1. Single stage regulators which reduce supply pressure to working pressure in one step. Example — 2000 psi to 5 psi (13 790 kPa - 34.5 kPa).
2. Two stage regulators which reduce supply pressure to working pressure in two steps. Example — 2000 psi to 200 psi (13 790 kPa - 1379 kPa), then 200 psi to 5 psi (1379 kPa - 34.5 kPa).

## 3-12. NOZZLE TYPE PRESSURE REGULATOR

See Fig. 3-16 for an illustration of a nozzle type pressure regulator.

This regulator consists of a forged, die cast, brass or aluminum body. A fitting to attach the regulator to the cylinder is threaded into the body. The body also has openings for both a high pressure and a low pressure gauge. A third opening connects the regulated, low pressure gas to the torch hose. A diaphragm or flexible disc separates and seals the body from the bonnet. The diaphragm spring is mounted between the diaphragm and the adjusting screw. Force on the diaphragm is adjusted by means of an adjusting screw. A cage or carrier is attached to the body side of the diaphragm. This cage curves down into the regulator body chamber. The seat is attached to the cage near the bottom of the body chamber. This seat is constantly pushed up against the nozzle by the body spring to stop the gas flow into the regulator.

The opening of the nozzle and seat is controlled by the position of the diaphragm. The line that leads to the nozzle comes from the cylinder and to this line is attached the high pressure gauge. A fine mesh screen or ceramic filter is commonly located in this line. This filter keeps dirt from entering and injuring the regulator. The screen also serves as a flame arrester and should be always left in place.

If the adjusting screw in the body is turned "in" (clockwise), the diaphragm spring on the bonnet side of the diaphragm will overcome the body spring. The diaphragm spring will then move the seat away from the nozzle, and allow some gas to pass from the cylinder into the regulator body. As this gas enters the regulator body, it tends to build up a pressure in the body. The force created by this pressure tends to push the diaphragm up against the diaphragm spring. The upward movement of the diaphragm moves the cage and the seat up. This closes off the nozzle opening. The pressure tends to fall as the gas is released from the regulator and flows through the hose to the torch. This action will allow the diaphragm spring

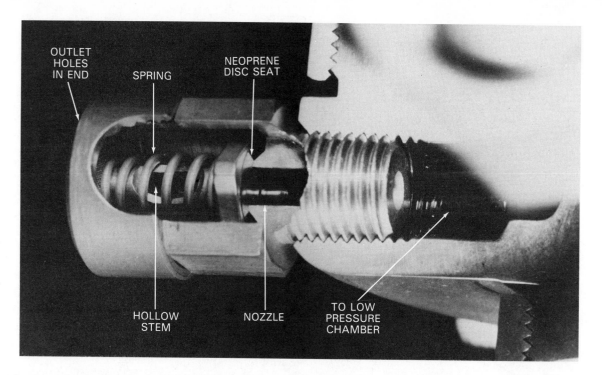

Fig. 3-18. A cutaway view of a regulator pressure relief valve. Over pressure may open and reset this valve. A sudden high pressure or fire will rupture the neoprene seat and vent the regulator through the hollow stem. (Air Products and Chemicals, Inc.)

to move the diaphragm down slightly. The nozzle and seat valve open and allow more gas to come into the regulator body. The balance of the diaphragm spring pushing the diaphragm and seat open, and the pressure under the diaphragm pushing the seat closed, tends to keep the pressure flowing through the regulator at a constant pressure.

The action of the diaphragm will maintain a constant preset pressure regardless of the pressure in the cylinder.

Note: When the adjusting screw on the regulator is turned all the way out, the flow of gas from the cylinder is stopped completely.

These regulators come in various gas-flow capacities and nozzle orifice sizes. The diaphragm size and the spring size are designed to provide the volume of gas desired. Master regulators are designed to allow a large amount of gas to flow. Line regulators allow the flow of a relatively small amount of gas.

The springs are made of a good grade of spring steel, while the diaphragm may be made of brass, phosphor bronze, sheet spring steel, stainless steel, or rubber.

The diaphragm is sealed at the joint between the diaphragm and the regulator body by means of suitable gaskets and the clamping action between the body and the bonnet.

The nozzle is usually made of bronze, while the seat may be made of various materials such as ca-

sein, hard rubber, plastic, or fiber. A cross section of a nozzle type regulator is shown in Fig. 3-17. A regulator pressure relief valve is shown in Fig. 3-18.

## 3-13. STEM TYPE PRESSURE REGULATOR

The stem type regulator works on the same principle as the nozzle type, but instead of using a nozzle and seat, it uses a poppet valve and seat. Fig. 3-19 illustrates a typical stem type regulator. The operation of this type regulator is described as follows.

High pressure gas enters the chamber below the seat when the cylinder valve is opened. The construction is such that the high pressure and the body spring tend to force the valve against its seat. When the pressure adjusting screw is turned in, it forces the diaphragm spring and diaphragm down. The diaphragm pushes the stem down. This downward movement forces the valve away from the seat.

As pressure builds below the diaphragm, the diaphragm moves up and the valve closes. Gas from the low pressure chamber is fed to the torch.

As the pressure drops in the low pressure chamber, the diaphragm spring forces the diaphragm, stem, and valve down again. This allows more high pressure gas to enter the low pressure chamber and the valve shut off again. This constant opening and closing of the valve as

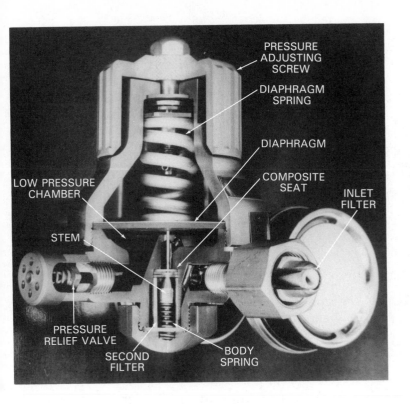

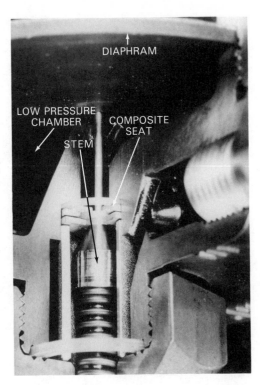

Fig. 3-19. A cutaway view of a stem type, single stage regulator. Note the filters at the high pressure inlet and around the valve and seat. The details of the stem valve, and seat may be seen in the close-up view. (Air Products and Chemicals, Inc.)

the diaphragm moves up and down controls the outlet pressure within close preset limits.

The stem type regulator lends itself to installations requiring a rather high rate of flow, and is commonly used on manifolds and flame cutting machines.

The materials in a stem type pressure regulator are similar to the nozzle type. The seat may be constructed of casein, hard rubber, plastic, or fiber. The stem (pin) is usually made of stainless steel. The stem and seat are designed to enable the complete assembly to be removed as a cage, permitting easy servicing. Fig. 3-20 shows an exploded view of a stem type regulator.

## 3-14. TWO STAGE PRESSURE REGULATOR

The two stage pressure regulator may be considered to be two regulators in one. In the first stage, the high pressure is reduced and regulated to an intermediate pressure. This is done by a valve and diaphragm mechanism, which has a fixed pressure adjustment. Figs. 3-21 and 3-22 illustrate the external appearance of typical two stage oxygen regulators. Fig. 3-23 illustrates a gaugeless two stage regulator. The second stage, which is adjustable, regulates the pressure and flow of the gas to the torch. A two stage regulator is operated in the same manner as a single stage regulator.

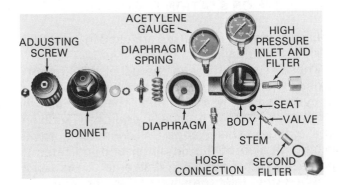

Fig. 3-20. An exploded view showing the parts of a stem type pressure regulator similar to that in Fig. 3-19. (Air Products and Chemicals, Inc.)

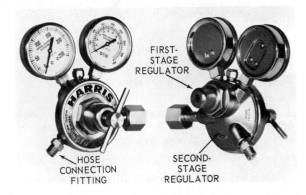

Fig. 3-21. Two views of a two stage oxygen pressure regulator. (Harris Calorific Div. of The Lincoln Electric Co.)

**Oxyfuel Gas Welding Equipment and Supplies / 93**

Fig. 3-22. Two stage regulator. Adjustment is on second stage. (L-TEC Welding & Cutting Systems)

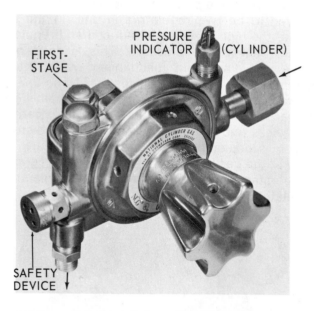

Fig. 3-23. An illustration of a gaugeless two stage regulator. The pressure is indicated by the height of a pin pushed up by oxygen pressure.

In a two stage oxygen regulator, the pressure is reduced through the first stage to about 200 psig (1379 kPa) in the intermediate stage. This means that in the second stage, the pressure need only be regulated from 200 psig (1379 kPa) to the torch pressure.

It is claimed that the two stage regulator will provide a more constant torch pressure, especially when large volumes of gas are being consumed. Fig. 3-24 illustrates, using color, the mechanical operation and the pressure conditions in this type regulator. A high performance single stage regulator is shown in Fig. 3-25. This illustration also shows the details of construction of this regulator.

In a two stage acetylene regulator, the first stage reduces the pressure to approximately 50 psig (344.74 kPa).

The two stage regulator may be constructed to use either the nozzle type or stem type mechanism. Fig. 3-26 illustrates a two stage stem type regulator. Some two stage regulators have been made which are a combination of the two styles, as shown in Fig. 3-27.

### 3-15. MASTER SERVICE REGULATOR

As stated in Headings 3-4 and 3-8, manifold systems for both oxygen and acetylene require the use of large master regulators which control the flow of the gases from the manifold to the welding station line. These regulators are basically of the same construction as described in Headings 3-12, 3-13, and 3-14; however, some modifications are made to adapt the regulator to its particular job. The regulator must be capable of controlling large volumes of gas through the regulator even when the difference between the cylinder pressure and the line pressure is small. These regulators always have two gauges; a high pressure gauge which indicates the manifold (cylinder) pressure, and a low pressure gauge which indicates the line pressure. The typical cylinder regulator usually has insufficient gas flow capacity to be used as a master regulator on a manifold installation.

### 3-16. LINE OR STATION REGULATOR

Line or station regulators are used in connection with manifold systems.

These regulators have the usual adjustment in order to control the gas pressure to the torch, and in every way they are handled the same as other regulators. They are usually equipped with only one gauge. This gauge is connected to the discharge side of the regulator and indicates the pressure of the gas being delivered to the torch.

Since the inlet pressure is lower than with regulators attached to cylinders, the regulator will have larger orifice nozzles and a more flexible, sensitive diaphragm. These regulators usually

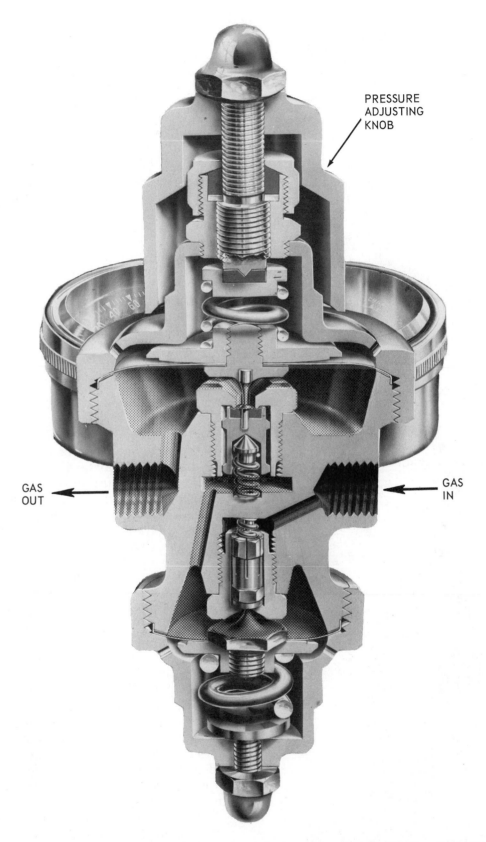

Fig. 3-24. Two stage regulator provides two diaphragms, two needles, and two seats. The first stage reduces the high gas pressure (solid red) as it comes from the cylinder to some intermediate pressure (dark pink). The second stage is the low pressure stage which reduces the intermediate pressure to some constant pressure needed by the torch (light pink).

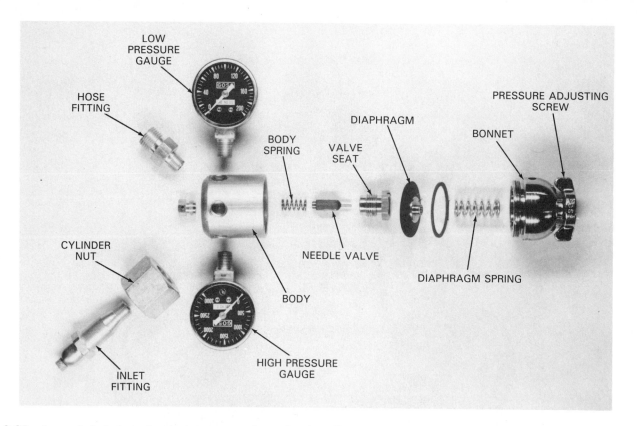

Fig. 3-25. An exploded view of a single stage regulator showing all the parts in their correct relative positions. (Goss, Inc.)

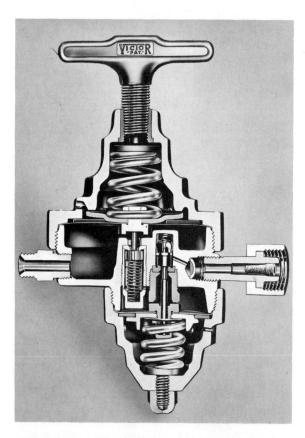

Fig. 3-26. Cross section view of a two stage regulator which uses stem type valves in both stages. (Victor Equip. Co.)

connect to the gas distribution pipe, therefore the inlet connection may be a standard pipe fitting and not the usual tank fitting. The discharge fitting will be the same as the usual regulator. A RIGHT-HAND THREAD FITTING IS USED FOR OXYGEN and a LEFT-HAND THREAD FITTING FOR ACETYLENE OR FUEL GASES.

## 3-17. PRESSURE GAUGES

As mentioned in a previous paragraph, the gauges are mounted on the regulators. The high pressure gauge is connected into the regulator between the regulator nozzle and the cylinder valve. It registers the cylinder pressure when the cylinder valve is opened. The low-pressure gauge is connected into the diaphragm chamber of the regulator, and registers the pressure of the gas flowing to the torch. The gauges are built with gears and springs similar to a pocket watch. Being of delicate construction, they must be handled accordingly. Refer to Fig. 3-28 for details of the construction of pressure gauges.

The basic principle of operation of the pressure gauge depends on the Bourdon tube. This tube which is made of phosphor bronze, is flat in cross section and is bent to fit inside the circular case. It is closed at one end and is connected with the pressure to be measured at the other. As the

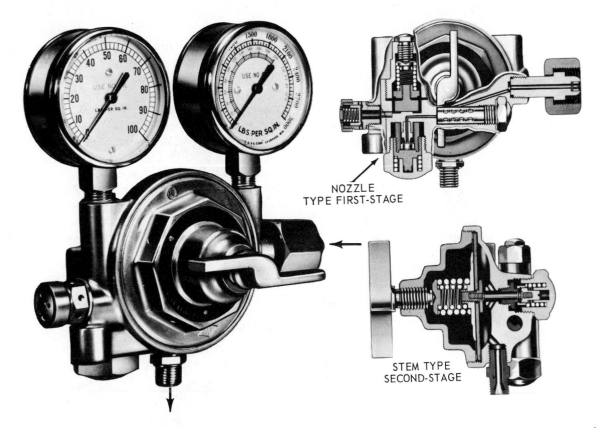

Fig. 3-27. Cross section view of a two stage regulator which has a nozzle type first stage and a stem type second stage.

pressure in the tube increases, it tends to straighten. As it straightens, it operates the gear and pointer mechanisms. The dial is calibrated to indicate corresponding pressures. The gauge is usually fastened to the regulator body using 1/4 inch diameter national pipe thread.

A heavy glass is used to cover the dial face and

needle, and is held in the body by means of a large threaded clamp ring, called a bezel. The gauges come in various sizes. The 2 1/2 in. (63.5 mm), 3 in. (76.2 mm), and 3 1/2 in. (88.9 mm) diameter dials are the most popular. The calibration of the gauges depends entirely upon the pressure to be used. It is generally recommended that the gauge

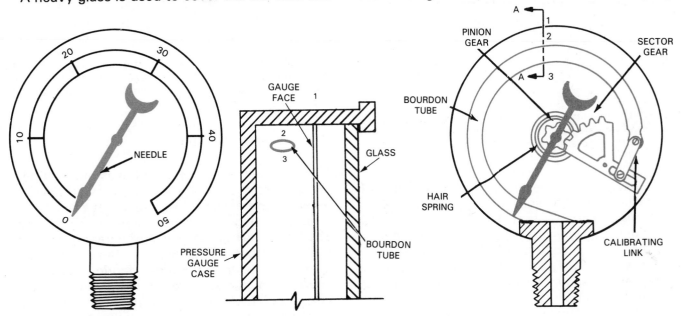

Fig. 3-28. Pressure gauge. Left—Exterior view. Right—Internal view. The thread on the gauge is a 1/4 in. national pipe thread. Section A-A shows the shape of the Bourdon tube.

reading be 50 percent higher than the highest pressure to be used. If 100 psig (689.5 kPa) is the highest pressure to be used, a gauge with a 150 psig (1034.2 kPa) or greater reading should be purchased.

The oxygen high pressure gauge usually has a 3 1/2 in. (88.9 mm) diameter dial, and is calibrated from 0 psig to 3,000 psig (0 - 20 684 kPa). There are sometimes two scales on the high pressure oxygen gauge dial. One is the pressure scale. The other scale is calibrated in cubic feet ($m^3$). This scale indicates the amount of gas left in the cylinder under various pressures as indicated on the dial. This type of scale is suitable for the oxygen. Oxygen is stored under direct pressure, and the amount of gas remaining in the cylinder is proportional to the pressure (Boyle's Law). Fig. 3-29 illustrates some typical pressure gauge calibrations.

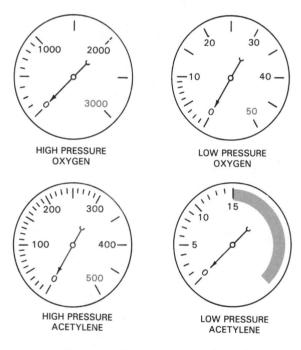

Fig. 3-29. Some typical regulator pressure gauge calibrations. Note red area above 15 psig (103.4 kPa) on the low pressure acetylene gauge.

The oxygen low pressure gauge has a variety of dial calibrations. For light welding, the 2 1/2 in. (63.5 mm) dial is calibrated up to 50 psig (345 kPa). For heavy welding and for cutting, the gauge may read as high as 200 (1379 kPa), 400 (2758 kPa), or even 1000 psig (6895 kPa). The diameter of the gauge used in cutting is usually 3 in. (76.2 mm) in diameter.

The acetylene high pressure gauge is usually of 3 in. (76.2 mm) diameter size, and is calibrated up to 400 or 500 psig (2758 - 3447 kPa).

The acetylene low pressure gauge is usually of 2 1/2 in. (63.5 mm) dial size and is calibrated from 0 to 30 psig (0 - 207 kPa) or from 0 to 50 psig (0 - 345 kPa). Many gauges are only calibrated up to 15 psig (103.4 kPa), leaving the space from 15 to 50 psig (103.4 - 345 kPa) red in color and unmarked.

REMEMBER, ALWAYS KEEP THE ACETYLENE TORCH PRESSURE BELOW 15 psig.

Some regulators use a calibrated spring-loaded diaphragm in place of a gauge; Fig. 3-30. These devices use less space than gauges, are more rugged, need less service, but are usually not as sensitive or accurate as a Bourdon tube type gauge.

Fig. 3-30. A single stage regulator with a calibrated diaphragm spring. Note the large readings on the bonnet and the fewer calibrations on the dial that turns with the adjusting screw.   (Dockson Corp.)

## 3-18. WELDING HOSE

The oxygen and acetylene hoses must be flexible and strong, and the gases must have no deteriorating effect on the materials of construction. Each hose is built of three principal layers. The inner layer is composed of a very good grade of gum rubber. The inner layer is surrounded by layers of rubber-impregnated fabric. The outside cover, or wearing cover, is made of a colored vulcanized rubber, plain or ribbed, to furnish a long-wearing surface. See Fig. 3-31. Hoses are manufactured in three common colors: black, green, and red. The use of these colors is not stan-

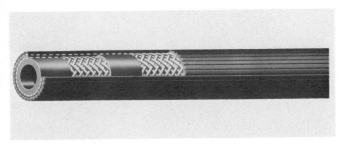

Fig. 3-31. A cutaway of a single welding hose. Note the inner rubber, the two layers of fabric and the ribbed outer rubber cover. (Anchor Swan Corp.)

dardized. However, the red is usually used for carrying acetylene or other fuel gases. Either the green or the black is used to carry the oxygen. The hose is specified according to its inside diameter and it comes in several sizes. The most common are 3/16, 1/4, and 5/16 in. (4.8, 6.4, 7.9 mm) ID (inside diameter). The size to be used depends on the size of the torch, and the length of the hose to be used. The 3/16 in. (4.8 mm) ID hose is very flexible and light, and is used extensively for light duty welding. The ID of the hoses should be increased if the hose is excessively long. Hoses can also be obtained as double hose as shown in Fig. 3-32 to minimize entanglement. Suppliers usually furnish hoses in 25 foot (7.62 m) lengths.

Fig. 3-32. A dual or siamese welding hose. One hose (green) carries the oxygen and the other hose (red) carries the fuel gas. (Anchor Swan Corp.)

HOSES SHOULD NEVER BE INTERCHANGED, CARRYING FIRST ONE GAS AND THEN ANOTHER GAS. If oxygen were to pass down a used acetylene hose, a combustible mixture might form. To prevent this, special precautions are used when attaching the hose to the regulators and torch. The hose is clamped to a nipple by means of a hose clamp. This nipple is fastened to a regulator or torch by means of a nut.

The nut and fitting have RIGHT-HAND THREADS when they are to be used with oxygen.

The oxygen nut may also be marked "OXY." The six sides of the oxygen nut are flat and smooth.

The fitting and nut used for acetylene, are LEFT-HAND THREADS, the nut may be marked "ACE."

The acetylene nut also has a groove machined around its six sides. Both the oxygen nipple and the acetylene nipple use a rounded face as a sealing surface. Fig. 3-33 illustrates some typical hose fittings.

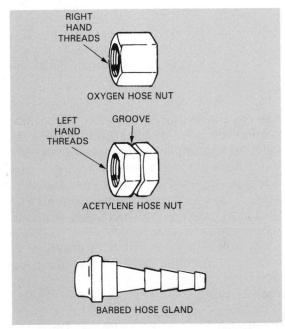

Fig. 3-33. Hose fittings. Fuel gas (acetylene) hose nut has groove cut into it. It also has a left hand thread (L.H.). (Airco Welding Products, Div. of Airco, Inc.)

The hose must be carefully handled to prevent accidents. It should not be allowed to come in contact with any flame or hot metal. Care should be taken that the hose is not kinked sharply. Kinking the hose might crack the fabric and permit the pressure to burst the hose. A kink will also hinder the gas flow. When the equipment is not being used, the hose should be hung away from the floor and away from other things that might injure it. When welding, the hose should be protected from falling articles, from vehicles running over it, and from being stepped on, as these actions might injure the hose.

Hose reels are available. They are usually spring loaded and roll up the hose into the container when the station is not being used.

## 3-19. OXYACETYLENE TORCH TYPES

The WELDING TORCH (sometimes called a blowpipe) and the CUTTING TORCH (sometimes incorrectly called a burning torch) are each designed for a specific purpose. Both of these

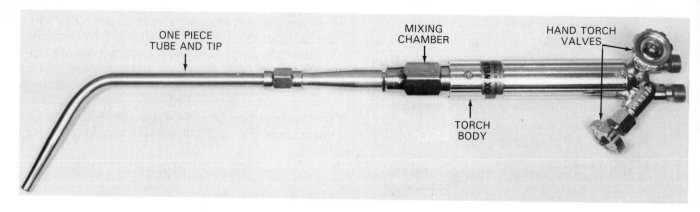

Fig. 3-34. An oxyacetylene equal pressure welding torch.
(L-TEC Welding & Cutting Systems)

torches are somewhat similar in construction, but the cutting torch is provided with a separate control valve to regulate an oxygen jet which does the cutting. These torches come in a variety of sizes and designs, depending on their intended use. See Chapters 5 and 6 for additional information.

There are two types of welding and cutting torches in common use:
1. Equal pressure type (medium pressure type). This is sometimes called balanced pressure type.
2. Injector type.

## 3-20. EQUAL PRESSURE TYPE WELDING TORCH

The gases are mixed in the equal pressure type welding torch, and the gases are burned at the end of the torch tip. The welding torch consists of four main parts, as shown in Fig. 3-34.
1. Body.
2. Hand torch valves.
3. Mixing chamber.
4. Tip.
The equal pressure torch is used with cylinder gases. Its construction necessitates that each gas

be supplied under enough pressure to force it into the mixing chamber, as illustrated in Fig. 3-35. Torches are made of several materials; including brass, aluminum, and/or stainless steel. The various parts are threaded and silver brazed together.

The hand valves are located either at the end of the torch where the hoses are attached to the handle, as shown in Fig. 3-36, or at the tip end of the handle as shown in Fig. 3-37. They are of the needle-type design. They are usually made of yellow brass with a packing of asbestos twine, or impregnated leather, as shown in Fig. 3-38. These hand shut-off valves are used chiefly for shutting the gas off and turning it on; however, many operators use these valves to throttle (make the final adjustment) the gases being fed to the torch.

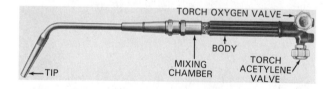

Fig. 3-36. A medium-duty equal pressure type welding torch.

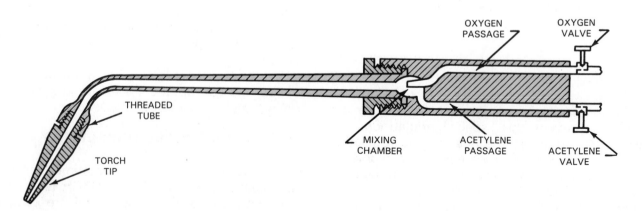

Fig. 3-35. A schematic drawing of an oxyacetylene welding torch.

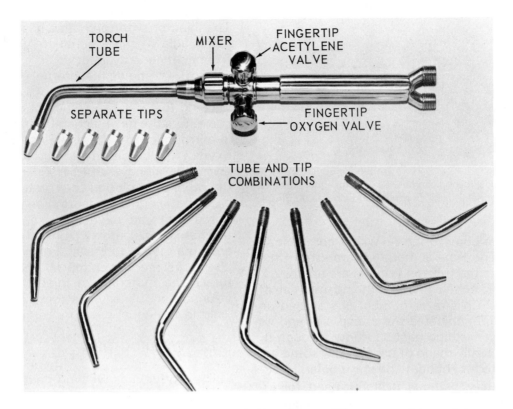

Fig. 3-37.  A light-duty equal pressure torch. Note the two different sets of tips and also note the location of the torch valves. One mixer is used for all tips.
(Air Products and Chemicals, Inc.)

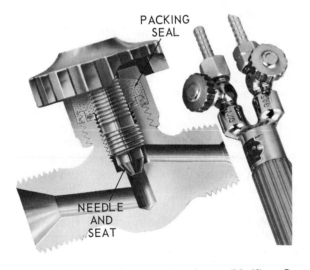

Fig. 3-38.  A torch valve cross section.  (Veriflow Corp.)

Figs. 3-36 and 3-39 illustrate the construction of a medium-duty equal pressure type welding torch, while Fig. 3-37 shows a light-duty torch with two different sets of tips.

The mixing chamber is usually located inside the torch body, although some torches incorporate the mixing chamber in the torch tube. Gases are fed to this chamber through two brass or stainless steel tubes leading from the torch valves. The size and design of the mixing chamber depends on the size of the torch. Some torch designs change the size and shape of the mixing chamber at the same time the tip is changed. The size and shape of the torch tip holes and the chambers should never be altered, nor should the parts be abused. Fig. 3-40 illustrates a well designed mixing chamber. The gases, after being mixed, are fed through the tube

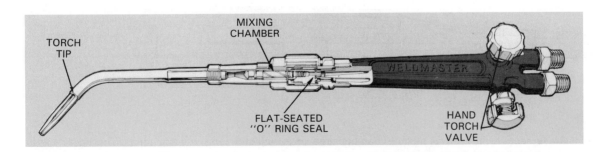

Fig. 3-39.  A cutaway of a medium-duty equal pressure oxyacetylene welding torch.  (Modern Engineering Co., Inc.)

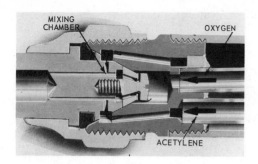

Fig. 3-40. A typical mixing chamber as used on equal pressure type welding torches. (Air Products and Chemicals, Inc.)

or barrel of the body to the tip where combustion takes place. The tube is made separate from the tip. Sometimes the tube and the tip are one piece.

The ORIFICE, or hole drilled in the tip, must be of accurate size. Tip size or number is stamped on the tip. TIP SIZE indicates the orifice size and its ability to allow welding gases to flow through it. The tips are usually made of copper, but some are nickel-plated to reflect heat and stay cooler. Heading 4-10 explains various systems used for indicating torch tip size. In this text, torch tip size is indicated in number drill sizes. Refer to Fig. 4-6 on page 122 for the drill size and tip number used by various tip manufacturers.

## 3-21. INJECTOR TYPE WELDING TORCH

The injector (low pressure) type welding torch looks much like the equal pressure type torch. However, the internal construction of the injector type torch is somewhat different. The chief characteristic of the injector type torch is its ability to operate using very low acetylene pressure. In general, the acetylene pressure remains practically constant regardless of the size tip or thickness of the metal being welded.

The ability of this torch to operate on low acetylene pressure has certain advantages. It is particularly desirable for use in connection with the low pressure acetylene generator which supplies acetylene at a pressure of 1/4 psig (1.7 kPa).

The injector-type torch also has the advantage of being able to more completely draw the charge from acetylene cylinders.

Fig. 3-41 shows the internal construction of the mixing chamber portion of an injector type welding torch. It should be noted that the oxygen line enters the mixing chamber through a jet which is surrounded by the acetylene passage. As the oxygen flows from the jet, it draws (or injects) the acetylene along with it. Handling the valves and the other operation of the torch is much the same as with the equal pressure type torch. It should be noted that the oxygen pressure used in these torches is considerably higher than with the equal-pressure type torches. Follow the torch adjustments recommended by the torch manufacturer. The materials of injector type torch construction are usually the same as those materials used in the equal pressure torch.

## 3-22. WELDING TIPS

The oxyfuel gas welding flame is maintained at the end of a solid copper welding tip. Since copper conducts heat rapidly, there is little danger of overheating and backfiring. Tip size and condition are most important for proper welding. There are two types of tips:
1. One piece (tip and tip tube are one piece).
2. Two piece (tip and tip tube are separate).
Fig. 3-42 shows the basic differences in the two types of tips.

The tip is subjected to both mechanical wear and flame erosion. As tips are removed and installed in a tip tube, the attaching threads may be subjected to considerable wear and abuse. Wrenches used on these tips should be of the box-end type. Pliers should never be used. Do not try to remove a hot tip from a tip tube. Allow the tip and tip tube to cool first. Also, do not install a cold tip in a hot tip tube.

While welding, the molten metal may ''pop'' and throw molten droplets into the tip orifice where they may remain until removed with a tip cleaner (described in the next Heading 3-23).

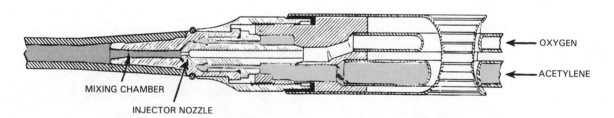

Fig. 3-41. Cross section drawing of mixing chamber area of an injector type welding torch. The acetylene is injected (drawn) into the mixing chambers by the pulling action (suction) of the oxygen jet. Injector torches are particularly adaptable for use with acetylene generators which operate under low pressure.

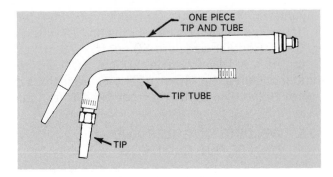

Fig. 3-42. Two commonly used welding tip designs.

Also, the heat from the flame will cause some tip erosion. Always use tips which are made for a particular torch as shown in Fig. 3-43. Some makes of torches use a pliable, heat resistant synthetic gasket to seal the joint between the tip tube and the torch. With this type of torch, a wrench is

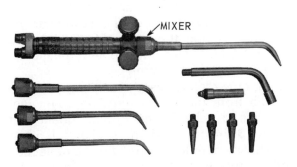

Fig. 3-43. Light-duty equal pressure torch. A separate mixer is provided with each one piece tip. (CONCOA)

not needed to install or service a tip. Hand tightening is sufficient. Fig. 3-44 illustrates a popular type welding tip with the tip and tube brazed together on a one piece unit.

Avoid dropping a tip as the seat which seals the joint may be damaged. The flame end of the tip may receive mechanical damage by being allowed to come in contact with the welding work, the bench, or firebricks. This damage may roughen the end of the tip and cause the flame to burn with a "fishtail" appearance.

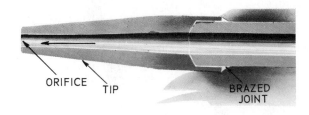

Fig. 3-44. Welding torch tube and tip brazed to form a one piece unit.

## 3-23. WELDING TIP CLEANERS

As mentioned before, welding tips are sometimes subject to considerable abuse. The orifice must be kept smooth and clean if the tip is to perform satisfactorily. When cleaning a welding tip, the orifice must not be enlarged nor scarred. Carbon deposits and slag must be removed regularly, if good performance is expected.

Special welding tip cleaners have been developed which perform this service operation satisfactorily. The cleaner consists of a series of broach-like (tool to shape or bore) wires which correspond in diameter to the diameter of the tip orifices. See Fig. 3-45. These wires are packaged in a holder which makes their use safe and convenient. Fig. 3-46 illustrates a tip cleaner in use.

Some welders prefer to use the correct size number drill to clean welding tip orifices. If a

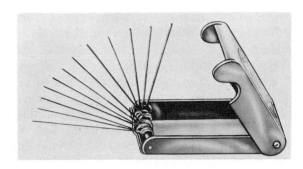

Fig. 3-45. Welding tip orifice cleaner. Notice the various sizes of broaching wires. (Thermacote-Welco Co.)

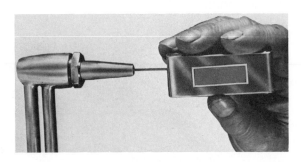

Fig. 3-46. A tip cleaner used to clean cutting tip orifices. (Maitlen & Benson)

number drill is used, it must be used very carefully so that the orifice diameter is not enlarged, bell-mouthed, reamed out-of-round, or otherwise damaged.

The flame end of the tip must be clean and smooth. Its surface must be at right angles to the center line of the tip orifice, if a correctly shaped flame is desired. A four inch mill file is commonly used to recondition the orifice end of the tip. Fig. 3-47 shows a mill file in use on a torch tip.

Fig. 3-47. Reconditioning the orifice end of a cutting torch tip. (Maitlen & Benson)

## 3-24. AIR-ACETYLENE TORCH

The air-acetylene torch, Fig. 3-48, is often used where a light portable flame of a medium temperature of 2500 °F (1370 °C) is required. This torch is used extensively on copper plumbing (soft soldering), refrigeration lines (silver brazing), and to solder or braze small parts. If large parts are to be silver brazed, the oxyacetylene torch is the recommended torch.

The air-acetylene torch receives its acetylene from a cylinder through a regulator and hose. As the acetylene flows through the torch, it draws air into it from the atmosphere, in order to supply the oxygen necessary for combustion. The torch operates on the same principle as a Bunsen burner used in chemistry laboratories.

The same precautions should be observed when using the acetylene cylinders, regulators, and torches as required when handling the oxyacetylene torch.

Cylinders for small portable air-acetylene torches come in various capacities as illustrated in Fig. 3-12. Regular sizes are 10 cu. ft. (.28 m³) and 40 cu. ft. (1.14 m³). The 10 cu. ft. (.28 m³) cylinder is called the MC size. The valve fittings on both of these cylinders are Prest-O-Lite fittings.

## 3-25. WELDING GOGGLES AND PROTECTIVE CLOTHING

The operator must wear suitable goggles when doing oxyacetylene welding. The flame and puddle of molten metal emit both ultraviolet and infrared rays. Both of these rays may cause eye injury if viewed at a close distance. Goggles also protect eyes from flying sparks. The glare is reduced too, and the operator will be able to see the weld puddle more clearly.

The common welding goggle has a sparkproof frame for the lenses. An elastic band holds the goggles securely on the operator's head. The welding lens may be round or rectangular. If the lenses are round, they are 50 millimeters in diameter. If the lens is rectangular, it measures 2 x 4 1/4 in. (50.8 x 108 mm).

A clear cover lens is used to protect the inner or filter lens from metal spatter. The cover lens is clear glass or plastic, of optical quality. It is generally 3/64 to 1/16 in. (1.19 - 1.59 mm) thick. These cover lenses need to be replaced frequently; otherwise, the operator's view of the weld will be dimmed.

The filter lenses are tinted either green or brown and are made in a variety of shade intensities. A

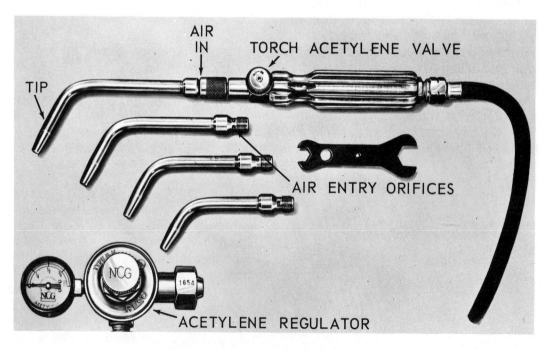

Fig. 3-48. An air-acetylene torch.

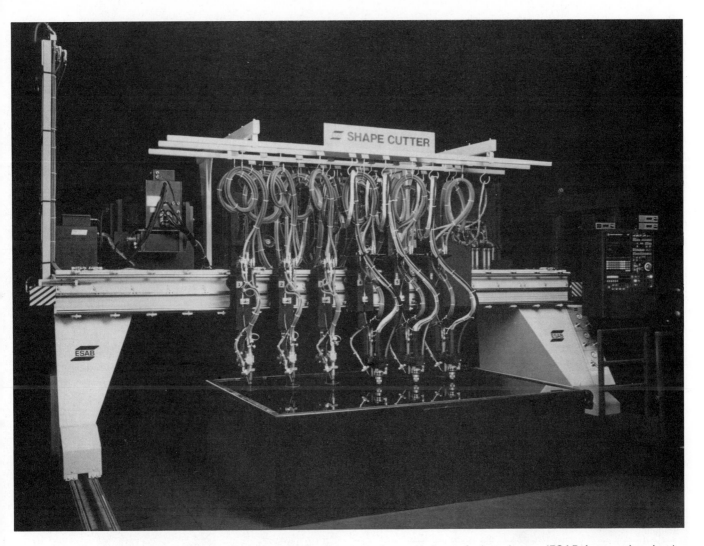

A computer-controlled gantry shape cutter with three plasma torches and six oxyfuel torches.     (ESAB Automation, Inc.)

more recent filter lens, which has a gold reflective layer on the outside, has been found to filter harmful rays exceptionally well. The shade intensities are indicated by shade number. These range from No. 1 to No. 14; the higher the number, the darker the shade.

Filter lenses must conform to the American National Standards Institute (ANSI) requirements for

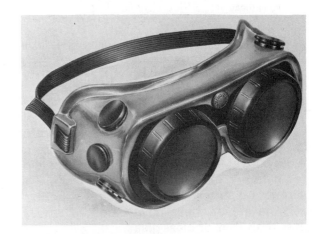

Fig. 3-49.  Two types of 50 mm round welding goggles. Both will fit over prescription glasses. The goggles at right have a soft flexible one piece frame.    (Jackson Products)

eye protection for welding (ANSI Z87.1-1968). Approved lenses carry a shade number and manufacturer's mark.

Welding goggles, Fig. 3-49, are often designed to fit over glasses. Fig. 3-50 lists recommended shades for various welding applications. If operations are of short duration, the lighter shade indicated may be used. For longer or continuous operation, the darker shade should be used. In general, shade number should increase with tip size.

| APPLICATION | BASE METAL THICKNESS | SUGGESTED SHADE NO. |
|---|---|---|
| Shielded metal-arc welding 1/16, 3/32, 1/8, 5/32 in. dia. electrodes | | 10 |
| Gas-shielded arc welding (nonferrous) 1/16, 3/32, 1/8, 5/32 in. dia. electrodes | | 11 |
| Gas-shielded arc welding (ferrous) 1/16, 3/32, 1/8, 5/32 in. dia. electrodes | | 12 |
| Shielded metal-arc welding 3/16, 7/32, 1/4 in. diameter electrodes 5/16, 3/8 in. dia. electrodes | | 12<br>14 |
| Atomic hydrogen welding | | 10-14 |
| Carbon-arc welding | | 14 |
| Soldering | | 2 |
| Torch brazing | | 3 or 4 |
| Light cutting | Up to 1 in. | 3 or 4 |
| Medium cutting | 1 in. to 6 in. | 4 or 5 |
| Heavy cutting | Over 6 in. | 5 or 6 |
| Gas welding (light) | Up to 1/8 in. | 4 or 5 |
| Gas welding (medium) | 1/8 in. to 1/2 in. | 5 or 6 |
| Gas welding (heavy) | Over 1/2 in. | 6 or 8 |

Fig. 3-50. Recommended lens shade numbers for various arc welding and oxyfuel gas cutting and welding applications. (ANSI Z87.1-1968)

Cover lenses are rather inexpensive. Filter lenses are quite expensive. It is, therefore, necessary that the filter lenses ALWAYS be protected by cover lenses. Many cover lenses are protected by a thin layer of transparent plastic which keeps metal spatter from pitting and adhering to them. The life of these plastic coated lenses is greater than that of clear glass.

Some oxyfuel gas welding operators prefer to use the rectangular eye shield type of eye protec-

tion. The lenses are the same size as those used in arc welding helmets. These shields not only fit over spectacles well, but also give a good range of vision. Fig. 3-51 illustrates the eye shield type of eye protection. Fig. 3-52 shows an eye shield which permits the filter lens and cover lens to be flipped up for a view of the weld through a clear lens.

Fig. 3-51. This woman is wearing goggles with a rectangular lens as she makes an overhead weld.

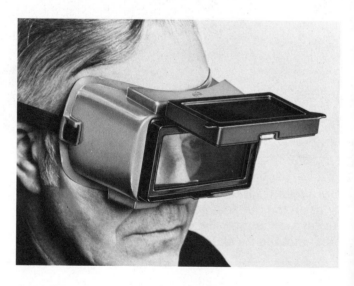

Fig. 3-52. This eye shield permits a view of the weld through a clear lens when the filter lens and cover lens are flipped up. (Jackson Products)

Face shields with large, flexible, filter quality lenses are available, Fig. 3-53. ALWAYS WEAR EYE PROTECTION. EYES CAN NEVER BE REPLACED.

The welding operator must wear protective clothing. The hands should be protected with leather or fabric gloves. The cuffs on the gloves should be either the gauntlet type, or an elastic band which makes a tight seal between the glove and the coat sleeve. Jackets should be either leather or of a fabric treated to be nonflammable or slow burning. Trousers should be without

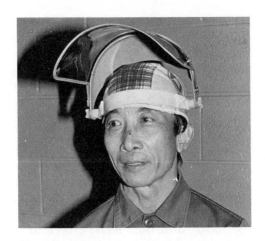

Fig. 3-53. A filter quality face shield is worn by this welder. It provides excellent visibility.

cuffs and the fabric should be treated to resist burning.

## 3-26. TORCH LIGHTERS AND ECONOMIZERS

Matches or burning paper should not be used for lighting a welding torch. CARRYING MATCHES OR OTHER COMBUSTIBLE MATERIAL, SUCH AS COMBS OR PENS IN POCKETS, WHILE WELDING IS DANGEROUS. IF A SPARK SHOULD ENTER A POCKET, A SERIOUS BURN MIGHT RESULT BEFORE THE FIRE CAN BE EXTINGUISHED.

A flint and steel spark lighter is perhaps the most popular type torch lighter. See Fig. 3-54. Pistol grip spark lighters are also available.

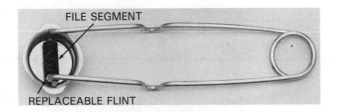

Fig. 3-54. Flint and steel spark lighter. The steel cup tends to trap the gas. When the flint is rubbed on the file segment, the spark quickly and safely ignites the fuel gas.

Many establishments which use a number of gas welding stations, provide pilot lights which use either city gas or acetylene. The city gas is piped to an outlet near the welding station, and a small flame is kept burning continuously. The flame outlet should be located overhead where it will not have any chance of igniting anything on the living level of the room.

There are two types of acetylene pilot lights. One leaves a very small acetylene flame burning at the torch tip when the torch valves are turned off (not the cylinder valves). When the operator turns the acetylene on, the flame grows to the desired size immediately.

The other type of acetylene pilot light provides a torch holder which incorporates the pilot light. Fig. 3-55 shows a combination economizer and lighter. It consists of a mechanism through which the oxygen and acetylene are first fed before going to the torch. This mechanism is also used as the torch holder, when the torch is not being used.

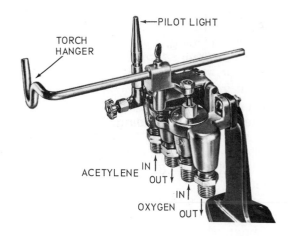

Fig. 3-55. Combination economizer and torch lighter.

Before the torch is placed in this holder, it is lighted and adjusted. When put in the holder, it presses a lever which turns off both the oxygen and acetylene, leaving a very small acetylene flame burning at a special outlet, or pilot light, as shown in Fig. 3-56. When the torch is lifted from this holder, gas flow starts, the gas is ignited by the pilot light, and the torch is ready to be used for welding. The device saves considerable gas and time. Safety is also improved, since the chance of having the torch laid aside while still lighted is minimized.

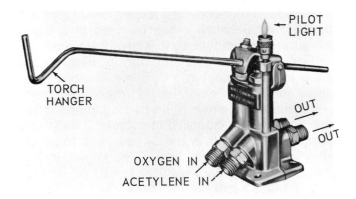

Fig. 3-56. Combination economizer and torch lighter.

## 3-27. OXYFUEL GAS WELDING SUPPLIES

Many supply items are needed in order to perform the usual oxyfuel gas welding operations. The more common supplies needed are:

1. Welding gases.
    a. Oxygen.
    b. Fuel gas.
2. Welding rod (filler metal) for:
    a. Steel.
    b. Stainless steel.
    c. Cast iron.
    d. Aluminum.
    e. Hard surfacing.
       (See Chapters 18 and 26)
3. Fluxes for:
    a. Cast iron welding.
    b. Aluminum welding.
    c. Stainless steel welding.
    d. Refer to Chapters 7 and 8 for lists of supplies needed for soldering and brazing. (See Chapter 18).
4. Firebrick.
5. Carbon paste and forms.
6. Noncombustible materials.
    a. Sheet.
    b. Powder.
7. Glycerine.
8. Litharge.

## 3-28. OTHER FUEL GASES

Handling oxygen and acetylene is explained in Headings 3-3 and 3-6. Other fuel gases in common use are:

1. Hydrogen.
2. LP (liquefied petroleum), propane, and butane.
3. Natural gas.
4. Methylacetylene-propadiene (MAPP).
5. Polypropalene based fuel gas (FG-2).

No fuel gas mixed with oxygen has a flame temperature as high as the oxyacetylene flame. The OXYHYDROGEN flame, however, is very clean and is recommended for welding aluminum and magnesium. Because it can be used at a higher pressure than acetylene, it is also recommended for underwater welding and cutting. Since hydrogen is a reducing agent, this flame, if properly adjusted, minimizes oxidation. A regular oxyacetylene torch may be used with hydrogen as the fuel gas. Hydrogen is supplied in cylinders, as is oxygen. The pressures in the oxygen and hydrogen cylinders are about the same. The standard sizes of hydrogen cylinders are 200 ft.³ (5.66 m³) and 100 ft.³ (2.83 m³).

Hydrogen cylinders are fitted with special fittings and the regulators used on these cylinders must be provided with proper mating attachments. Hydrogen has no odor, and when combined with either air or oxygen, it forms a possible powerful explosive mixture. Hydrogen connections should be regularly checked for leaks using a soap-and-water solution.

LIQUEFIED PETROLEUM (LP) is sold under a variety of names. It may have some variations in chemical analysis. For welding use, the general title of liquefied petroleum (LP) gas is used. This fuel is supplied in liquid form and is under a positive pressure which varies with the temperature. LP gas is used mostly for cutting, soldering, and brazing. Most oxygen cutting torches can use LP fuel, but special LP cutting tips are required. An air-fuel gas type torch is commonly used for general heating purposes using this fuel. Refer to Fig. 3-48.

Portable oxyfuel welding kit for general repair work.
(L-TEC Welding & Cutting Systems)

LP gas is sold by the pound (kilogram). Common sizes of tanks are: 4 1/4 lbs. (1.9 kg), 11 lbs. (5 kg), 20 lbs. (9.1 kg), 30 lbs. (13.6 kg), 40 lbs. (18.1 kg), 60 lbs. (27.2 kg), and 100 lbs. (45.4 kg).

LP gas is also provided by the gallon. Sizes are: 120 gal. (454 L), 250 gal. (946 L), 500 gal. (1893 L), and 1000 gal. (3785 L).

The customer usually purchases the required tank size and returns them to the dealer for refilling. The larger sizes are usually leased. Industries which use large quantities of LP gas provide their own bulk storage. The fuel is delivered to them from bulk tank cars or trucks.

When using LP gas, the pressure must be regulated using an LP gas regulator. These regulators are most often supplied with two gauges. One gauge shows the pressure in the storage tank, and the other indicates the pressure in the torch or burner line. These regulators are usually attached to the tank using a standard P.O.L. Commercial fitting for this purpose.

NATURAL GAS, as now piped to most communities, is an excellent fuel for certain uses. It is particularly adaptable for cutting, soldering, brazing, and preheating.

Because natural gas is delivered at a rather low pressure, injector type torches are used both for cutting and for general heating. Some small torches have been developed which use compressed air and natural gas particularly for soldering and brazing. Natural gas serves very nicely for many preheating operations. Natural gas should be protected by a water seal or a blow back valve to keep air and oxygen from backfiring into the gas supply line, refer to Fig. 3-13 for the water seal flash arrestor. Always consult all required safety authorities and regulations on this matter before proceeding.

Stabilized METHYLACETYLENE-PROPADIENE is a fuel gas sold under the trade name of MAPP. The fuel has the safety and ease of handling of liquefied petroleum gas (LP) with a heating value approaching that of acetylene.

MAPP may be stored and shipped in the liquefied state. The cylinders are available in various sizes. The usual acetylene regulator may be used with this gas. MAPP gas cylinders have the same thread as acetylene cylinders.

This fuel has some advantages over acetylene when used for cutting. It has a narrower explosive range. The explosive range is 3.4 to 10.8 percent in air compared to a range of 2.5 to 80 percent for acetylene.

MAPP gas is used for underwater cutting since it may be used at pressures of over 15 psig (103.4 kPa) and acetylene cannot.

Oxyacetylene cutting torches may be used with this fuel gas. However, special tips made for MAPP gas must be used. These tips are available for all common cutting torches.

OXY-MAPP cutting is rapid and a very liquefied slag is formed which flows away leaving a clean cut.

The customer usually owns the storage cylinders so there are no demurrage (charge per day) costs. Since the fuel is in the liquid form, a cylinder of MAPP fuel contains many more cubic feet or cubic meters of gas than an acetylene cylinder of equal size.

## 3-29. CHECK VALVES AND FLASHBACK ARRESTORS

Check valves are used to prevent the reverse flow of gases through the torch, hoses, or regulators. See Fig. 3-57. Fig. 3-58 shows the check valve in the open and closed positions. It also explains how the check valve works.

A check valve allows a gas to flow in only one direction. Gas pressure opens the valve to allow gas flow. The valve closes when the flow stops or tries to flow in a reverse direction. Check valves are generally installed at the torch inlet, or placed at the regulator outlet. Check valves may also be placed at both the torch and regulator.

Flashback arrestors are designed to eliminate the possibility of an explosion in the regulator or cylinder. Built into the flashback arrestors are the

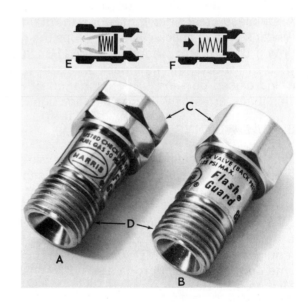

Fig. 3-57. Check valves prevent mixtures of fuel gas and oxygen from backing up and possibly burning in the welding hose and regulator. A—Fuel gas check valve (note groove on hex). B—Oxygen check valve. C—Torch or regulator connection. D—Welding hose connection. E—Normal flow; valve open. F—Reverse flow; valve closes.
(Harris Calorific Div. of The Lincoln Electric Co.)

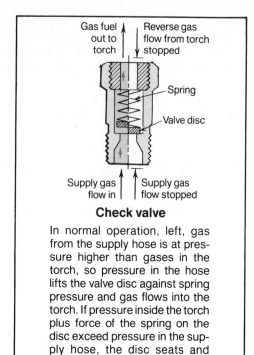

Fig. 3-58. Check valve.
(Welding Design & Fabrication)

**Check valve**

In normal operation, left, gas from the supply hose is at pressure higher than gases in the torch, so pressure in the hose lifts the valve disc against spring pressure and gas flows into the torch. If pressure inside the torch plus force of the spring on the disc exceed pressure in the supply hose, the disc seats and shuts off flow of gas from the torch into the hose. Not reusable after a flashback.

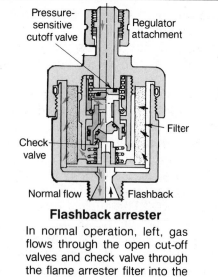

Fig. 3-59. Flashback arrestor
(Welding Design & Fabrication)

**Flashback arrester**

In normal operation, left, gas flows through the open cut-off valves and check valve through the flame arrester filter into the hose.
In the event of a flashback, right, the stainless steel filter stops the flame and the pressure wave activates the cut-off valve, stopping the flow of gas to extinguish the flame. The check valve operates when gas flows towards the cylinder. If the arrester is exposed to fire, the thermal cut-off valve shuts the gas supply. Reusable after a flashback.

following safety valves as shown in Fig. 3-59.
1. A reverse flow check valve. This valve stops the flow of gas in the wrong direction.
2. A pressure sensitive cut-off valve that cuts off gas flow if an explosion occurs.
3. A stainless steel filter that stops the flame.
4. A heat sensitive check valve that stops the gas flow if the arrestor reaches 220°F (104°C).
   Fig. 3-59 shows the flashback arrestor in the open and closed position. An explanation of how the flashback arrestor works is also included.

## 3-30. GAS WELDING ROD (Filler Metal)

The American Welding Society defines welding rod as follows: ''A form of filler metal used for welding or brazing which does not conduct the electric current.'' Welding rods used in GTAW (Gas Tungsten Arc Welding) will be discussed in more detail in Chapter 11 covering the supplies and equipment for use in GTAW.
Some common welding rods are:
1. Mild steel.
2. Cast iron.
3. Stainless steel.
4. Braze welding alloys.
5. Aluminum.
   a. Drawn.
   b. Extruded.
   c. Cast.

Mild steel, braze welding alloys, stainless steel, and some aluminum rods are made in 36 in. (.91 m) lengths and are available in the following diameters: 1/16 in. (1.6 mm), 3/32 in. (2.38 mm), 1/8 in. (3.18 mm), 5/32 in. (3.97 mm), 3/16 in. (4.76 mm), 1/4 in. (6.35 mm), 5/16 in. (7.94 mm), and 3/8 in. (9.53 mm). Metric sized welding rods may not be available as shown above.
They are packaged in 50 lb. (22.7 kg) bundles. Mild steel welding rods are copper coated to keep them from rusting. Aluminum welding rods or wires are packed in 36 in. (.91 m) lengths or in coils. Some aluminum rods are flux coated. The coated rods are sold in 28 in. (.71 m) lengths.
Iron and steel welding rod specifications are sometimes confusing. The problem has been somewhat simplified by the fact that the old Army Air Corps, Navy, and Federal specifications are now combined into one name and one series of numbers under military (MIL) specifications. However, both the MIL and the American Welding Society (AWS) numbers are still used. In this text, AWS specification numbers are used.
The AWS numbers for oxyacetylene steel welding rods are shown in Fig. 3-60. In these

| NUMBER | USE | TENSILE STRENGTH |
|--------|-----|------------------|
| RG 45 | mild steel rod | 45,000 psi (310.3 MPa) |
| RG 60 | low alloy steel rod | 65,000 psi (448.2 MPa) |
| RG 65 | low alloy high strength steel rod | 105,000 psi (723.9 MPa) |

Fig. 3-60. AWS numbers for oxyacetylene steel welding rods.

| TENSIL STRENGTH LBS./SQ. IN. | CARBON MAX. PERCENT | MANGANESE MAX. PERCENT | SULPHUR MAX. PERCENT | PHOSPHORUS MAX. PERCENT | SILICON MAX. PERCENT |
|---|---|---|---|---|---|
| 45,000 | 0.06 | 0.25 | 0.035 | 0.025 | 0.03 |

Fig. 3-61. Characteristics of the AWS RG 45 Gas Welding Rod.

numbers, the letter "R" stands for a welding rod. The letter "G" stands for gas welding. The numbers 45, 60, and 65 indicate the approximate tensile strength of the weld in thousands of pounds per square inch (45 = 45,000 psi).

Fig. 3-61 lists the characteristics of the RG-45 welding rod.

## 3-31. WELDING FLUXES

The American Welding Society defines FLUX as follows: "Material used to prevent, dissolve, or facilitate removal of oxides and other undesirable surface substances."

Fluxes are required when heating certain metals with the intent to join them with solder, silver alloys, or copper alloys. Fluxes are also used in many welding operations. The composition of a flux is determined by the specific application for which it is made. The general classification for fluxes is shown in Fig. 3-62.

Note from the flux classifications that no flux is required for mild steel. Protection of the weld from oxidation is not so critical with mild steel. The iron

oxides melt at a much lower temperature than the mild steel and float to the surface during welding.

## 3-32. FIREBRICKS

FIREBRICKS are used to form welding table tops, to build forms around articles, to aid with preheating, and to relieve stress. They are useful for building up supports for articles to be welded or brazed. They can be used to protect combustible materials which may become heated when exposed to the welding or cutting flame.

Firebricks are made of refractory (difficult to burn) materials. The size of firebricks in common use is 8 3/4" x 4 1/2" x 2 1/2" (222.2 mm x 111.1 mm x 63.5 mm).

## 3-33. CARBON PASTE AND FORMS

There are many places where the welder will find carbon paste materials useful. A few typical uses for carbon paste or carbon forms are:

1. For building dams to contain the molten metal when building up a broken section.
2. For protecting drilled or threaded holes which are in or adjacent to a weld or braze.
3. For building up a support for an uneven surface which is to be welded.
4. As a protecting cover over metal which is adjacent to a weld but which might be injured by spatter or heat.

The paste form is available in various size cans. The formed carbon is available in round or square rods and in plate form. See Chapter 18 for more information.

## 3-34. ASBESTOS

CAUTION: THERE MAY BE DANGER IN HANDLING ASBESTOS IN ANY FORM. INHALED

| SOLDERING | |
|---|---|
| Aluminum | Sheet steel |
| Copper alloys | Stainless steel |
| Galvanized sheet | |
| **BRAZING** | |
| Aluminum | Cast iron |
| Steel | Copper alloy |
| Steel alloy (such as stainless steel) | |
| **WELDING** | |
| Aluminum welding | Cast iron welding |
| Braze welding | Stainless steel welding |

Fig. 3-62. General classification for fluxes.

POWDERED ASBESTOS MAY CAUSE CANCER OF THE LUNG. This danger may be avoided by wearing a respirator. A temporary furnace can be constructed using sheet asbestos and firebricks. Sheet asbestos is supplied in 1/16 in. (1.59 mm) and 1/8 in. (3.18 mm) thickness and in 36 in. (.91 m) rolls.

Powdered asbestos is sometimes used in a metal box where small malleable iron castings are placed after welding or braze welding, in order that they may cool slowly. Powdered asbestos may be mixed with water and molded over surfaces to be protected during welding or brazing. This asbestos paste serves the same purpose as carbon paste.

Asbestos sheet or molded Transite board is often used to support welding exercises as they are being welded. The asbestos acts as an insulator and permits better heat control of the metal.

## 3-35. SEALING COMPOUNDS

To make an airtight seal on threaded pipe joints, certain proprietary (private brand) compounds are recommended. Apply these sealing compounds on the male threads only. In this way, there is little danger of compound entering the pipe.

Sealing compounds are available in either paste form or tape form. One should avoid using sealing compounds that have a lead content. SEALING COMPOUNDS WITH AN OIL CONTENT SHOULD NEVER BE USED ON OXYGEN LINES.

## 3-36. CLAMPS AND CLAMPING FIXTURES

The success or failure of many welding operations depends on how the metals are held in place during the welding operation. Many varieties of clamps and clamping devices have been developed for this purpose.

Pliers with clamping jaws of special design for holding and aligning parts are shown in Fig. 3-63. The pliers have deep jaws to enable clamping around obstructions. Fig. 3-64 shows various applications for this type of clamping pliers. Very complex shapes may be clamped for welding with the chain-clamp pliers, shown in Fig. 3-65. C-clamp pliers have also been developed to hold parts firmly and in alignment as shown in the illustration in Fig. 3-64.

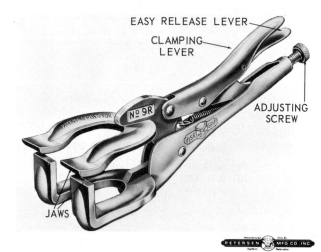

Fig. 3-63. Clamping pliers for holding stock being welded or brazed.

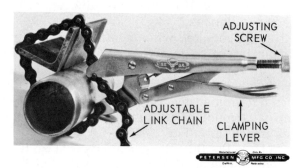

Fig. 3-65. Quick release chain-clamp pliers. This clamp is capable of clamping material of almost any shape.

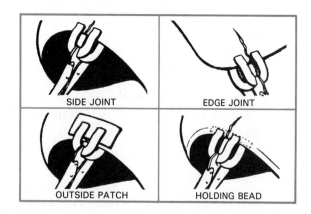

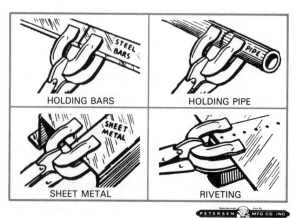

Fig. 3-64. Application of the clamping pliers shown in Fig. 3-63.

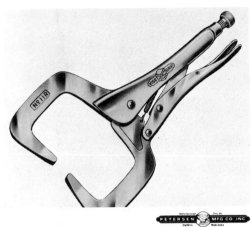

Fig. 3-66. C-clamp pliers.

Special fixtures are very convenient for aligning, holding, and positioning various shaped metals, as shown in Fig. 3-67. No. 1 in the illustration shows a special double fixture with a protractor scale which makes quick and accurate aligning of the parts possible.

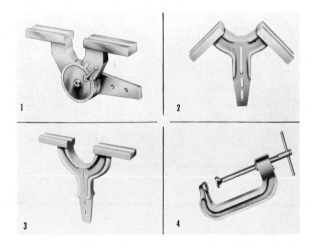

Fig. 3-67. Special alignment fixtures for holding stock which . is being welded. (Strippit, Div. of Houdaille Industries, Inc.)

## 3-37. TEST YOUR KNOWLEDGE

Write your answers on a separate sheet of paper. Do not write in this book.

1. What parts and equipment make up a complete oxyacetylene welding station?
2. What type safety device is used in an oxygen cylinder?
3. What type safety device is used in an acetylene cylinder?
4. Completely describe the acetylene hose and regulator nut and its threads.
5. What materials are used in regulator diaphragms?
6. What factors determine the rate at which acetylene gas can be drawn from the cylinders?
7. Why must oil not be used on welding station fittings?
8. What colors are used on oxygen and acetylene hoses?
9. What is the maximum pressure which should be used with a gauge whose highest number is 400 psig (2758 kPa)?
10. What is a two stage regulator?
11. Name two types of welding torches and describe the difference between them.
12. What is the purpose of a flux?
13. What is the maximum working gauge pressure for acetylene gas?
14. How does a flashback arrestor work?
15. What is MAPP?
16. List five (5) standard sizes for oxygen cylinders.
17. Why must clothing saturated with oxygen be removed?
18. What is the tensile strength of the RG 60 welding rod?
19. At what temperature in degrees Fahrenheit and degrees Celsius do the acetylene cylinder fuse plugs melt?
20. Is the regulator open or closed when the regulator adjusting screw is turned out?

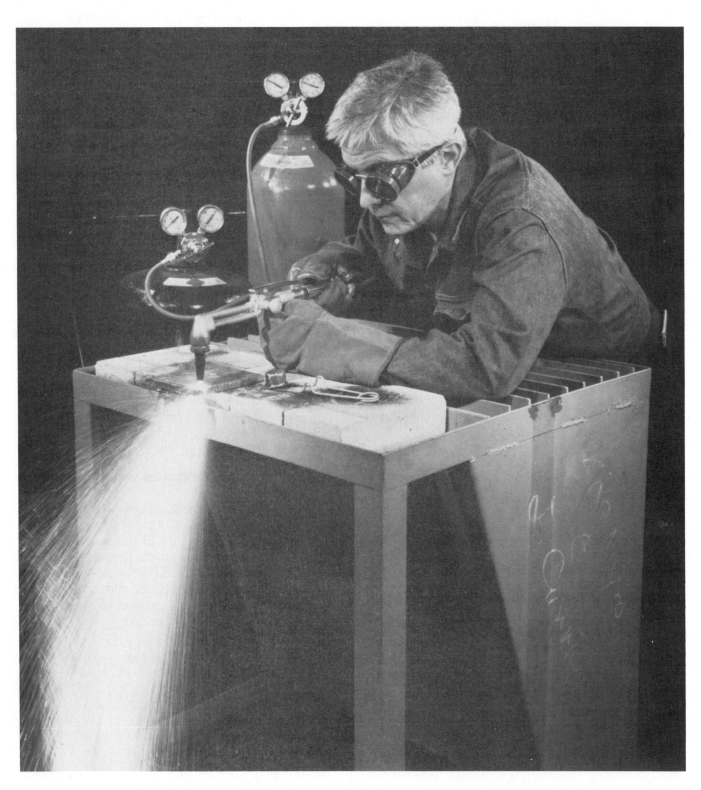

Industry photo showing oxyfuel gas cutting. The fundamentals of oxyfuel gas cutting are similar to the fundamentals of oxyfuel gas welding.

# 4  OXYFUEL GAS WELDING

## 4-1. DEFINITION OF WELDING

OXYFUEL GAS WELDING may be described as a metal-working process in which metals are joined by heating them to the melting point, and allowing the molten portions to fuse or flow together. Review Heading 1-1. Before attempting to make any welds, study the safety Headings 4-33, 4-34, 4-35, and 4-36.

## 4-2. SOLDERING AND BRAZING

Two other metal joining processes which are often confused with oxyfuel gas welding are soldering and brazing.

SOLDERING occurs when two metals, which are not melted, are joined by a third metal which has a melting point below 840 °F (450 °C). Review Heading 1-3. An example of soldering is the joining of copper to steel using a tin-lead alloy. For more information on soldering see Chapter 7.

BRAZING is done when two metals, which are not melted, are joined with a third metal which melts at temperatures above 840 °F (450 °C). Review Heading 1-4. An example of brazing is the joining of two pieces of steel with a silver alloy. For more information on brazing see Chapter 8.

Manual welding or soldering may be defined as an art. The skill to weld and solder metals together can only be obtained after a diligent study of the methods and after careful and correct practice. In classifying welding as an art, it is meant that some persons can do welding better than others. This is due to a seemingly natural gift. However, it has been found that any person can become a successful welder with good instruction and by following correct procedures. Continuous practice is necessary in order to maintain a high standard of skill as a welder. It is, therefore, recommended that only proper equipment and metals be used while learning welding.

A thorough, fundamental procedure should be followed. To correct any possible mistakes early, trainees should work under close supervision.

## 4-3. DIFFERENT TYPES OF WELDING AND CUTTING

The most common types of welding are: oxyfuel gas welding, shielded metal arc welding, gas tungsten arc welding, gas metal arc welding, and resistance welding. Other types include: thermit welding, cold welding, ultrasonic welding, electron beam welding, friction welding, laser beam welding, and electroslag welding. Refer to the various welding procedures in Chapter 1.

Two popular types of thermal cutting are oxygen cutting and arc cutting. All of these processes will be explained in detail in the chapters of this book.

The oxyfuel gas process will be studied first because:
1. The fundamentals of gas welding include fundamentals important to most other forms of welding.
2. The oxyfuel gas process is a popular manual welding process. It is slower and easier to control than some other processes.
3. Since it is slower, it is easier for a beginner to observe the fusion process. The shape and flow of the oxyfuel gas welding puddle and bead is similar to the various arc welding craters and beads.

## 4-4. OXYFUEL GAS WELDING

One of the most popular welding methods is to use the oxyfuel gas flame as the source of heat. This flame is produced by burning a fuel gas in the presence of the oxygen from the air, from a pure oxygen source, or from both. Oxyfuel gas flames may receive their oxygen in three ways:

1. From the surrounding atmosphere which:
   a. Gives lowest flame temperature.
   b. Is the least clean.
   c. Produces the least heat.
2. From air, containing oxygen, drawn in from the atmosphere through holes in the torch. This process:
   a. Gives a higher flame temperature.
   b. Is cleaner.
   c. Gives more heat.
3. From a supply of pure oxygen under pressure mixed with the fuel gases before they burn which:
   a. Gives the highest flame temperature.
   b. Is cleanest.
   c. Gives the most heat.

Fig. 4-1 illustrates three methods of adding oxygen to a fuel gas to support combustion.

ful materials to the metal.
6. The products of combustion should not be toxic (poisonous).

The quantity of heat is determined by the type and number of cubic feet per hour (liters per minute) of gases burned. To obtain more heat, a torch tip with a larger orifice (hole) is used. More pressure must also be supplied to feed sufficient gas to the larger tip. Whether a large torch tip or a small torch tip is used, the temperature of the flame is the same with a given fuel gas.

It should be remembered that the amount of heat generated, and therefore the thickness of the metal which may be welded, will depend on the amount of fuel gas burned per unit of time. Therefore, the amount of heat depends on the size of the torch orifice (hole).

There are several commercially used oxyfuel

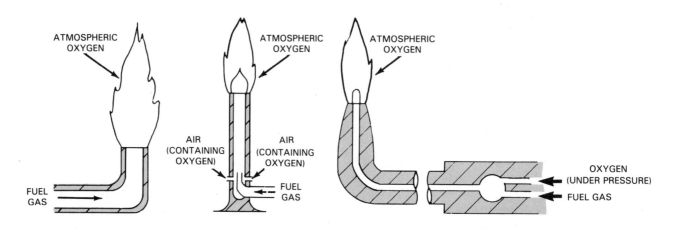

Fig. 4-1. Three methods of combining oxygen and fuel gases to produce fuel gas flames.

## 4-5. WELDING FLAMES

In oxyfuel gas welding, metals are joined by melting and fusing. A very intense, concentrated flame is applied to the metal until a spot under the flame becomes molten and forms a liquid puddle. When two metals melt or puddle, and the molten pools run together and solidify (fuse), the edges of the two pieces become one. This process must be performed carefully to minimize damage to the metals.

Certain conditions are necessary for a good weld. They are:
1. The temperature of the flame must be high enough to melt the metals.
2. Enough heat must be supplied to overcome the heat losses.
3. The flame must not burn (oxidize) the metal.
4. The flame must not add dirt or foreign material to the metal.
5. The flame must not add carbon or other harm-

gas welding and cutting flames. They are:
1. OXYACETYLENE, (oxygen and acetylene) used for welding and cutting.
2. OXYHYDROGEN (oxygen and hydrogen) used for welding and cutting.
3. OXYNATURAL GAS (oxygen and natural gas) or artificial gas, used for cutting.
4. OXYPROPANE (oxygen and propane), a liquefied petroleum gas, used for cutting.
5. MAPP (oxygen and methylacetylene-propadiene) used for welding and cutting.
6. Oxygen and Linde FG-2, used for cutting.

For flame temperatures of various oxygen and fuel gas mixtures, see Heading 31-4.

## 4-6. THE OXYFUEL GAS FLAME

Harmful oxidizing or carburizing flames result when the wrong proportions of two gases, oxygen and a fuel gas, are used. If too much oxygen is used, an oxidizing flame results which will burn the metal. If too much fuel gas is used, a carburizing flame results. This flame can add harmful carbon to the metal being welded. The oxyhydrogen flame will not become carburizing. Fig. 4-2 shows the various flame adjustments. See also Figs. 1-2 and 1-9. The two harmful flames are easily recognized. See Headings 4-11 and 4-13 for flame adjustments. The correct flame heats the metal and does not carburize (add carbon) or oxidize (burn) it. The correct flame is called a ''NEUTRAL FLAME.'' A NEUTRAL FLAME is the result of a perfect proportion and mixture of the fuel gas and oxygen. In a neutral flame, these two gases unite so that the oxygen burns up the carbon and the hydrogen in the fuel gas, then releases only heat and harmless gases. The colors of the flames are shown in Fig. 1-2. (Refer to Chapter 31 for more on the chemistry of the welding flame.)

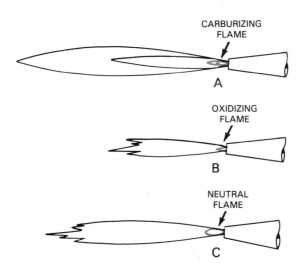

Fig. 4-2. Three oxyfuel gas flame adjustments. Top—A carburizing flame has three distinct flame sections. Center—The inner cone on an oxidizing flame is sharp and pointed. Bottom—The inner cone on a neutral flame is smooth and rounded. The neutral flame is the correct flame for most welding operations.

Chemically, when oxygen and acetylene burn they create carbon dioxide, water vapor, and heat. Oxygen and all fuel gases with the exception of hydrogen will form the following gases when burned: carbon dioxide ($CO_2$) and water vapor ($H_2O$). These are considered harmless gases.

Oxygen in the air surrounding the flame is also used to complete the burning process. In corners, where the air has difficulty getting to the flame, additional cylinder oxygen must be fed to the flame.

The effect of an improper mixture of gases in the welding flame is easily recognized, and the final test for a neutral flame is determined by the manner in which the melting metal reacts to the flame.

Dirt in a welding flame may come from two sources:
1. Dirty gases.
2. Dirty equipment.

Good quality gases should always be used. The purity of the gases made by the manufacturers should be noted and taken into consideration. A neutral oxyacetylene welding flame will produce a temperature of approximately 5720 °F (3160 °C). A neutral oxypropane flame reaches 5300 °F (2927 °C). An oxidizing flame will produce a slightly higher temperature.

Temperatures required to melt various metals are listed in the table, Fig. 4-3. An oxyacetylene welding flame is the hottest oxyfuel gas flame. It is hot enough to melt the common metals.

| METAL | MELTING TEMP. | |
|---|---|---|
| | °F | (°C) |
| Aluminum | 1215 | (657) |
| Brass (yellow) | 1640 | (893) |
| Bronze (cast) | 1650 | (899) |
| Copper | 1920 | (1049) |
| Iron, Gray cast | 2200 | (1204) |
| Lead | 620 | (327) |
| Steel (.20%)SAE 1020 | 2800 | (1538) |
| Solder (50-50) | 420 | (216) |
| Tin | 450 | (232) |
| Zinc | 785 | (418) |

Fig. 4-3. Melting temperatures of some of the commonly used metals.

Acetylene is generally the only fuel gas used for welding.

Propylene gas and MAPP gas (methylacetylene-propadiene) are used for preheating, cutting, torch brazing, flame spraying, and flame hardening.

Propane is used for preheating, soldering, and brazing.

Natural gas (methane) is generally used for heating, cutting, soldering, and brazing.

Hydrogen gas is used in welding nonferrous metals and in some brazing operations.

## 4-7. THE OXYFUEL GAS WELDING OUTFIT

Before discussing how to make a good weld, it is advisable to know the welding equipment used, its limitations, and its proper use.

Basically, the oxyfuel gas welding outfit consists of a source of supply of the two gases; oxygen and fuel gas regulators; hose fittings; and a torch. A welding station includes the welding outfit plus a welding table, ventilation, and possibly a booth. Fig. 4-4 shows a complete oxyacetylene gas welding outfit. In the order of

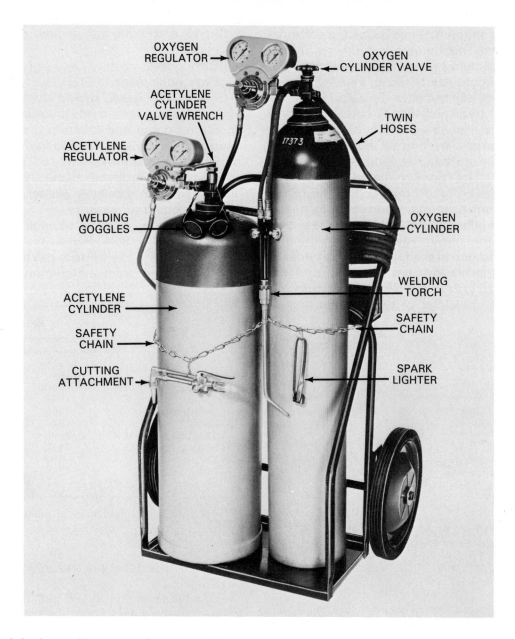

Fig. 4-4. A portable oxyacetylene gas welding and cutting station. (Modern Engineering Co., Inc.)

flow of the gases through the outfit, the following items will be found in common use:

1. Gas cylinders; oxygen cylinder, fuel gas cylinder.
2. Pressure regulators and gauges; oxygen regulator, fuel gas regulator.
3. Hoses; oxygen hose, fuel gas hose.
4. Welding torch.

There are two types of oxyfuel gas welding torches in common use, the equal pressure type torch, and the injector type torch. As the name implies, the equal pressure type torch operates on approximately equal, or the same, pressure for both oxygen and the fuel gas. This type equipment is the most common.

The injector type torch is usually used in connection with an acetylene generator. The torch operates on a relatively low acetylene pressure and a much higher oxygen pressure.

See Chapter 3 for further details concerning the construction and operation of each of these torches.

Study Heading 4-36 before attempting any welding practice.

## 4-8. ASSEMBLING THE OXYFUEL GAS WELDING OUTFIT

Proper handling of welding equipment is essential to safety, and to obtaining good welds and a reasonable amount of service from the equipment.

Oxygen and most fuel gas cylinders are usually owned by the companies which furnish the gases contained in the cylinders. A small rental, or demurrage, is charged for the use of the cylinders after a reasonable rent-free period. The remainder of the equipment, however, is usually the property of the operator or of the company employing the operator.

Because of the high pressure in the oxygen cylinder and the ease of combustion of the fuel gases, especially acetylene, it is necessary that great care be used when handling cylinders. See Heading 4-9 for further information on handling cylinders.

Always wear approved welding goggles when welding. See Heading 3-25 for specifications of welding goggles.

Before welding equipment is used, it is very important to make sure the equipment is properly assembled. Check to see if the gas cylinders are in good condition. The cylinders should be fastened securely to a wall, post, or column by means of steel straps or chains. This will eliminate the possibility of them falling or being knocked over.

If the outfit is portable, the cylinders should be fastened to an approved cylinder truck by means of steel straps or chains. The cylinder truck should be so designed that it is almost impossible to knock it over accidentally.

Whatever method is used to secure the cylinders, it should permit the cylinders to be changed easily.

Before attaching a regulator to a cylinder, the cylinder opening should be cleaned out. The cylinder opening is cleaned by briefly opening the cylinder valve slightly (cracking) and then closing it. This allows some of the high pressure gas to blow through the valve opening to remove dirt and other particles. Inspect the sealing surfaces and the fittings. Avoid using damaged or worn parts. The regulators may then be attached to the cylinder.

Only fixed end wrenches, having wide jaws, provided for the purpose, should be used on any fittings. Be sure that the regulator nut fits the cylinder valve fitting properly. FUEL CYLINDER VALVES ARE USUALLY FITTED WITH LEFT HAND THREADS. OXYGEN CYLINDER VALVES ARE FITTED WITH RIGHT HAND THREADS. THE THREAD DIAMETERS ON THE TWO CYLINDER VALVES ARE ALSO DIFFERENT. THESE ARE SAFETY PRECAUTIONS. By using different threads, it is impossible to change the regulators from one type gas cylinder to another. This prevents the mixing of gases accidentally. Many different varieties of cylinder and regulator fittings are in common use.

The hose connecting the regulators to the torch should be fastened firmly to the fittings at each end of the hose. The hose should be installed so that it is not twisted when the torch is held in the welding position. If the hose is twisted, it will strain the welder's hand and cause fatigue.

Before attaching the hose to the torch, the hose should be blown out or purged. With the regulators attached, open the cylinder valves, then gently open and close the regulator valves, first the fuel gas regulator, then the oxygen regulator. This brief purging will clear the hose lines. Where pipe threads are used, they should be sealed with pipe thread sealing compound (such as glycerine and litharge paste) or Teflon tape when assembling.

Following the hose purging, the torch should be attached to the hose. Note that on oxyfuel gas welding equipment, the fuel gas hose nipple nuts have left hand threads, and the oxygen hose fittings have right hand threads. After the welding equipment is assembled, test for leaks using soapsuds. Testing for leaks must be done when installing any new cylinder or any new part of apparatus.

To test for leaks, the recommended procedure is to put SOAPSUDS on the outside of the joints suspected of leaking. OIL OR FLAMES OF ANY KIND SHOULD NEVER BE USED. Before testing for leaks, also read Heading 4-11 on turning on the oxyacetylene welding outfit. To test for leaks, turn the regulator screws out all the way, open the cylinder valves, and build up from 5 to 15 psi of pressure in the regulators and hoses by turning the regulator screws in (clockwise) slowly. Then, apply the soapsuds solution to all joints. Leaks, if any, will be indicated by bubbles.

When first using welding equipment, remember that the procedure is:
1. Learn the proper ways to prepare the station for use.
2. Learn the proper methods of igniting the torch.
3. Learn how to adjust gas flows to produce a proper flame.
4. Learn the proper method of shutting down the equipment.

The proper procedure for handling the torch and selecting the proper size tip to use are dependent on several factors. These factors are the type of weld desired, kind of metal used, thickness of the metal, and shape and position of the metal. Instructions on tip selection are found in Heading 4-10. Heading 4-11 covers the turning on and lighting of an oxyacetylene torch. Heading 4-36 reviews safety in oxyfuel gas welding.

## 4-9. SAFETY IN HANDLING FUEL AND OXYGEN CYLINDERS

Welding cylinders, when properly handled, are quite safe. Improperly handled, they may be very dangerous. Cylinders should never be dropped or allowed to tip over. The cylinder cap, enclosing and protecting the cylinder valve, should always be screwed on the cylinder when the cylinder is not in use, or when it is being moved.

To move a cylinder, it is advisable to use a cylinder truck. The cylinder must be secured on the truck so that it cannot tip or fall from the truck.

A cylinder in use should be firmly anchored in an upright position, and in such a way that it cannot be tipped. Avoid the storing or use of cylinders in extremely hot locations. High temperatures may cause the cylinder pressure to reach dangerously high levels. Check on local community building and fire codes and be certain that the cylinders are used and stored according to those codes.

## 4-10. SELECTING THE CORRECT SIZE WELDING TORCH TIP

Welding torch tip size is designated by a number stamped on the tip. The tip size is determined by the size of the orifice (opening). There is no standard system of numbering welding torch tip sizes. Manufacturers have their own numbering systems. For this reason, in this text, tip size instructions are given in orifice "number drill" size. Number drills consist of a series of eighty drills numbered one through 80. The diameter of a number one drill is .2280 in. (5.79 mm). The diameter of a number 80 drill is .0135 in. (.34 mm). Note that the larger the number, the smaller the drill diameter. See Chapter 31 for a table of number drill sizes. See Fig. 4-5 for a list of manufacturers' welding tip numbers corresponding to number drill sizes. When a welder becomes familiar with the operation of a certain manufacturer's torches and numbering system, it is seldom necessary to refer to orifice number drill sizes.

The orifice size determines the amount of fuel gas and oxygen fed to the flame. The orifice therefore determines the amount of heat produced by the torch. The larger the orifice, the greater the amount of heat generated.

With a balanced pressure type torch, the tip sizes shown in Fig. 4-6 should give satisfactory results for oxyacetylene welding.

If the torch tip orifice is too small, not enough heat will be available to bring the metal to its melting and flowing temperature. If the torch tip is too large, poor welds will result for the following reasons:

1. The weld will have to be made too fast.
2. The welding rod will melt too quickly and the weld puddle will be hard to control.
3. The appearance and quality of the weld bead will be generally poor.

## 4-11. TURNING ON THE OXYACETYLENE OUTFIT AND LIGHTING THE EQUAL PRESSURE TYPE TORCH

To light the torch, the gases must be turned on

and adjusted to the proper pressures. To do this, proceed as follows:

1. Visually check the equipment for condition.
2. Inspect regulators.
3. Turn the regulator adjusting screws all the way out (counterclockwise) before opening the cylinder valves. This prevents damage to the regulator diaphragm.
4. Stand to one side of the regulator when opening the cylinder valves. A burst regulator or gauge could cause severe injury.
5. Slowly open the acetylene cylinder valve 1/4 to 1/2 turn counterclockwise. This will usually provide adequate flow and will permit rapid closure in an emergency. Use the proper size wrench and leave it on the valve in case an emergency shutoff is needed.
6. Open the oxygen cylinder valve very slowly (counterclockwise) to prevent damage to the regulator diaphragm from the pressure and heat of 2000 psi. When the regulator high pressure gauge reaches its highest reading, turn the cylinder valve all the way open. This is necessary because the oxygen clyinder valve has a double seat or a back seated valve. In the all out position, this seat closes any possible opening along the valve stem through which the high pressure oxygen could escape.
7. Open the acetylene torch valve one turn. Turn the acetylene regulator adjusting screw in slowly (clockwise) until the low pressure acetylene gauge indicates a pressure which is correct for the tip size. (Never use acetylene gas at a gauge pressure above 15 psig (103.4 kPa). An approximate setting may be arrived at as shown in the table, Fig. 4-6. Turn off the acetylene torch valve using finger tip force only. Important - check the low pressure gauge for an indication of possible regulator defects. See Heading 4-12.
8. Adjust the oxygen torch pressure. Open the torch oxygen valve one turn. Turn the oxygen regulator adjusting screw in (clockwise) until the low pressure oxygen gauge indicates the pressure which is correct for the tip orifice. Then turn off the oxygen torch valve. The regulator pressures have now been adjusted to approximately the proper levels. Important - check the low pressure gauge for an indication of possible regulator defects.
   To prevent damage to the needle valves and seats, use only fingertip force to open and close the torch valves.
9. PURGE THE SYSTEM BEFORE LIGHTING THE TORCH. To ensure that the proper gases

Table of manufacturers' gas welding tips (drill sizes in header; upper-row and lower-row drill numbers combined into single columns in decreasing order).

| Trade name | Series | 71 | 70 | 69 | 68 | 67 | 66 | 56 | 55 | 54 | 53 | 52 | 51 | 50 | 49 | 48 | 45 | 44 | 43 | 42 | 41 | 40 | 36 | 35 | 34 | 1/8 | 30 | 29 | 28 | 27 | 26 |
|---|---|---|---|---|---|---|---|---|---|---|---|---|---|---|---|---|---|---|---|---|---|---|---|---|---|---|---|---|---|---|---|
| Airco | All |  |  |  | 1 |  |  | 3 |  | 4 |  |  |  | 5 |  | 6 |  | 7 |  |  |  | 8 | 9 |  |  |  | 10[5] |  |  |  |  |
| Canadian Liquid Air | All |  |  |  | 1 |  |  | 3 |  | 4 |  |  |  | 5 |  |  |  |  |  | 7 |  |  |  |  | 9 |  | 10 |  |  |  |  |
| Craftsman | AA |  | 1 |  |  |  |  |  |  | 3 |  |  |  |  |  |  |  |  |  | 5 |  |  |  |  |  |  |  |  | 8[6] |  |  |
| Dockson | 4EC, 4SC, 7SC / 3EC, 5EC, 6EC, 7EC |  |  |  |  |  | 3 | 5 |  |  |  | 6 |  |  |  | 7 |  |  |  |  |  |  |  |  |  |  |  |  |  |  |  |
| Dockson | All |  |  |  |  |  |  |  |  |  |  |  |  |  |  |  |  |  |  | 4 |  |  | 5 |  |  |  |  |  |  | 7 | 8 |
| Gasweld | G25, G35 |  |  |  | 1 |  |  | 4 |  | 5 |  | 6 |  |  |  | 7 |  | 8 |  | 9 |  | 10 | 12 |  | 13 |  | 16 |  |  |  |  |
| Gasweld | G55 |  |  |  | 2 |  |  | 4 |  | 5 |  | 6 |  |  |  |  |  | 8 |  |  |  |  |  |  |  |  |  |  |  |  |  |
| Gasweld | AVG |  |  |  | 2 |  |  |  |  | 6 |  | 7 |  | 8 |  |  |  |  |  |  |  |  |  |  |  |  |  |  |  |  |  |
| Harris | 2890-F | 00 |  |  |  | 1 |  |  |  |  |  |  | 2 |  |  |  |  |  |  |  |  |  |  |  |  |  |  |  |  |  |  |
| Harris | 6290 |  |  | 00 |  | 1 |  |  |  |  |  |  |  |  | 2 |  |  |  |  |  |  |  |  |  |  |  |  |  |  |  |  |
| Harris | 7490-A |  |  |  |  | 1 |  |  |  |  | 2 |  |  |  |  | 3 |  |  |  |  |  |  |  |  |  |  |  |  |  |  |  |
| Harris | 23, 13-F, 23A swedged / 17F swedged |  |  |  |  |  |  |  |  |  |  |  |  |  |  |  |  |  |  | 9 |  |  | 10 |  |  |  | 15[7] |  |  | 19 |  |
| K-G | AP, APM, APL |  |  |  | 1 |  |  | 3 |  | 4 |  |  |  | 5[1] |  | 6 |  | 7 |  |  |  | 8[8] |  |  | 9 |  | 10[9] |  |  |  |  |
| Liquidweld | 90, 70, 72 |  | 00 |  |  |  |  | 2 |  |  | 3 |  |  |  | 4 |  |  |  |  | 5 |  |  | 6 |  |  |  | 7 | 8 | 9 | 10 | 11 |
| Liquidweld | 80, 82 |  | 00 |  |  |  |  | 2 |  |  | 3 |  |  |  | 4 |  |  |  |  | 5 |  |  | 6 |  |  |  | 7 | 8 |  |  |  |
| Marquette | A |  |  |  | 0 | 1 |  |  |  | 4 |  |  |  |  | 5 |  |  |  |  | 7 |  | 8 | 9 |  |  |  | 11 |  |  | 12 |  |
| Marquette | B |  |  |  | 2 |  |  | 4 |  | 5 |  |  |  | 6 |  |  |  | 7 |  |  |  | 8 | 9 |  |  |  | 11 |  |  |  |  |
| Marquette | F |  |  | 0 | 2 |  | 1 | 4 |  | 5 |  |  |  | 6 |  |  |  | 7 |  |  |  | 8 | 9 |  |  |  |  |  |  |  |  |
| Marquette | G |  |  | 0 |  |  | 1 | 3 |  | 4 |  |  |  | 5 | 6 |  |  |  |  |  |  | 8 | 9 |  |  |  |  |  |  | 12 |  |
| Marquette | H & J |  |  |  | 2 |  |  | 4 |  | 5 |  |  |  | 6 |  |  |  |  |  |  |  |  |  |  |  |  |  |  |  |  |  |
| Meco | All |  |  |  | 1 |  |  |  |  | 4 |  |  |  | 5 |  |  |  |  |  | 7 |  |  | 8 |  |  |  |  |  | 9 | 10 |  |
| National | B | 2 |  |  |  | 3 |  |  | 7 | 8 |  | 9 |  | 10 |  | 11 |  | 12 |  |  |  | 13 | 14 |  |  |  |  |  |  |  |  |
| National | G, P |  |  |  |  |  | 2 |  |  | 3 |  |  |  |  | 4 |  |  |  |  |  |  |  | 6 | 8 |  |  | 10 |  |  |  |  |
| National | R |  |  | 00 |  |  |  | 2 |  |  | 3 |  |  |  |  |  |  |  |  | 5 |  |  |  |  |  |  | 7 | 8 | 9 | 10 | 11 |
| Oxweld | W-15 |  |  |  |  | 2 |  | 4 |  | 5 |  | 6 |  | 7 |  |  |  |  |  |  |  |  |  |  |  |  |  |  |  |  |  |
| Oxweld | W-29 |  |  | 2 |  |  |  | 6 |  | 9 | 12 |  |  | 15 |  |  |  |  |  |  | 30 |  |  |  |  |  |  |  |  |  |  |
| Oxweld | W-17, W-22, W-26 |  |  |  |  |  |  | 6 |  | 9 | 12 | 15 |  | 20 |  |  | 30 |  |  |  |  | 40 |  |  |  |  | 55 | 70 |  |  | 100 |
| Oxweld | W-45, W-47 |  |  |  |  |  |  |  |  | 9 | 12 | 15 |  | 20 |  |  |  |  |  |  |  | 40 |  |  |  |  | 70 |  | 85 |  |  |
| Powr-Craft (Montgomery Ward) | 84-5881 |  |  |  |  | 1 |  | 5 |  |  |  |  |  |  | 7 |  |  |  |  | 9 |  |  |  |  |  |  |  |  |  |  |  |
| Prest-O-Lite | 420 |  |  |  |  |  |  |  |  |  | 15 |  |  | 20 |  |  |  | 30 |  |  |  |  |  |  |  |  |  |  |  |  |  |
| Prest-O-Weld | W-109 |  |  |  | 2 |  |  | 4 |  | 5 |  |  |  | 6 |  |  |  | 8 |  |  |  |  |  |  |  |  |  |  |  |  |  |
| Prest-O-Weld | W-110, W-111 | 2 |  |  | 3 |  |  |  | 6[2] | 7 |  |  |  | 8 |  |  |  |  |  |  | 9 | 10 | 11 |  |  |  |  | 29 |  |  |  |
| Prest-O-Weld | W-120 |  |  |  |  |  |  |  |  | 9 | 12 | 15 |  | 20 |  |  |  | 30 |  |  |  |  | 40 |  |  |  | 70 |  |  |  |  |
| Prest-O-Weld | W-121, W-122 |  |  |  |  |  |  |  |  | 9 | 12 | 15 |  | 20 |  |  |  | 30 |  |  |  |  | 40 |  |  |  | 70 |  | 85 |  |  |
| Purox | 33 |  |  |  | 2 |  |  | 4 |  |  |  |  |  | 5 | 6 |  |  |  | 8 |  |  |  | 10 |  |  |  |  |  |  |  |  |
| Purox | 34 |  |  |  |  |  | 2 | 4 |  | 4 |  |  |  |  | 6 |  |  |  |  |  |  |  |  |  |  |  |  |  |  |  |  |
| Purox | 35 |  |  |  |  |  | 2 |  |  |  |  |  |  | 5 |  |  |  |  |  |  |  |  |  |  |  |  |  | 13 |  |  |  |
| Purox | W-200 |  |  |  |  |  |  |  |  | 9 | 12 | 15 |  | 20 |  |  | 30 |  |  |  |  |  |  |  |  |  |  |  |  |  |  |
| Purox | W-201, W-202 |  |  |  |  |  |  |  |  | 9 | 12 | 15 |  | 20 |  |  | 30 |  |  |  |  | 40 |  |  |  |  | 70 |  | 85 |  |  |
| Purox | 00-D |  |  |  | 2 |  |  | 4 |  | 5 |  |  |  |  |  |  |  |  |  |  |  |  |  |  |  |  |  |  |  |  |  |
| Rego | GX, GXU, SX |  |  |  | 68 |  |  |  | 55 |  | 53 |  |  | 50[10] |  |  |  |  |  | 42 |  |  | 36[11] |  |  |  |  |  |  |  |  |
| Smith | Pipeliner MW[3] | 101 |  | 102 | 103 |  | 106 |  |  | 107 |  | 108 |  |  | 109 |  |  | 110 |  |  |  | 111 |  |  | 112 |  |  |  |  |  |  |
| Smith | Airline AW[3] | 101 |  | 102 | 103 |  | 106 |  |  | 107 |  | 108 |  |  | 109 |  |  | 110 |  |  |  | 111 |  |  | 112 |  |  |  |  |  |  |
| Smith | Silver Star LW[3] | 101 |  | 102 | 103 |  | 106 |  |  | 107 |  | 108 |  |  | 109 |  |  | 110 |  |  |  | 111 |  |  |  | 113 |  |  |  | 114 |  |
| Smith | LW[4] | 700 |  |  |  |  |  |  |  |  | 703 |  | 704 |  |  | 705 |  | 706 |  |  |  | 707 | 708 |  |  |  |  |  | 710 |  | 711 |
| Torchweld | GP 570, 870 |  |  |  | 68 |  |  |  | 55 |  | 53 |  |  | 50 |  |  |  |  |  | 42 |  |  | 36[12] |  |  |  |  |  |  |  |  |
| Torchweld | 71, 370, 170 |  |  |  | 68 |  |  |  | 55 |  | 53 |  |  | 50 |  |  |  |  |  | 42 |  |  | 36 |  |  |  |  |  |  |  |  |
| Victor | All |  |  | 00 |  |  | 00½ |  | 2 | 2½ |  | 3 | 3½ |  | 4 |  |  |  |  | 5 |  |  | 6 |  |  |  | 7 | 8 | 9 | 10 | 11 |
| Weldit | All | 1 |  |  |  |  |  |  |  | 4 | 5 | 6 |  |  |  | 7 |  | 8 | 9 |  |  | 10 | 12 |  |  |  | 16 |  |  |  |  |

[1]APL to No. 5 only. [2]W-110 to No. 6 Tip only. [3]Soft-flame tip. [4]Heavy-duty Tip. [5]Airco Tip. No. 11-Drill Size No. 25; No. 12-Drill Size No. 20; No. 13-Drill Size No. 10; No. 14-Drill No. 2; No. 15-1/4 in. Drill. [6]Tips 7 and 8 require special gooseneck. [7]13-F and 17-F Swedged to No. 15 Tip only. [8]APM to No. 8 Tip only. [9]AP to No. 10 only. [10]SX to No. 46 Tip only. [11]GXU to No. 31 only. [12]GP 570 to No. 31 only. (Welding Engineer)

Fig. 4-5. A table of manufacturers' gas welding tips with tip numbers and orifice drill sizes.

are in the respective hoses (no air or oxygen in the acetylene hose and no fuel gas in the oxygen hose), the system must be purged. This is done by allowing the acetylene to flow through the acetylene hose and oxygen to flow through the oxygen hose for a short time before lighting. Note: If Steps 1 through 8 are followed, the system will have been purged and the torch is ready for lighting.

10. After purging, crack the acetylene torch

| METAL THICKNESS | SIZE* WELDING TIP ORIFICE | WELDING ROD DIAMETER | OXYGEN | | ACETYLENE | | WELDING SPEED FT/HR. |
| | | | PSIG PRESSURE | CU. FT./HR. | PSIG PRESSURE | CU. FT./HR. | |
| --- | --- | --- | --- | --- | --- | --- | --- |
| 1/32 | 74 | 1/16 in. | 1 | 1.1 | 1 | 1 | |
| 1/16 | 69 | 1/16 in. | 1 | 2.2 | 1 | 2 | |
| 3/32 | 64 | 1/16 in. or 3/32 in. | 2 | 5.5 | 2 | 5 | 20 |
| 1/8 | 57 | 3/32 in. or 1/8 in. | 3 | 9.9 | 3 | 9 | 16 |
| 3/16 | 55 | 1/8 in. | 4 | 17.6 | 4 | 16 | 14 |
| 1/4 | 52 | 1/8 in. or 3/16 in. | 5 | 27.5 | 5 | 25 | 12 |
| 5/16 | 49 | 1/8 in. or 3/16 in. | 6 | 33. | 6 | 30 | 10 |
| 3/8 | 45 | 3/16 in. | 7 | 44. | 7 | 40 | 9 |
| 1/2 | 42 | 3/16 in. | 7 | 66. | 7 | 60 | 8 |

*Note the tip orifice size as shown is the number drill size. These recommendations are approximate. The torch manufacturers recommendations should be carefully followed.

Fig. 4-6. A table showing the relationships between welding tip size, gas pressures, welding rod diameters, and metal thicknesses. This table is for oxyacetylene welding with an equal pressure type torch.

valve no more than 1/16 of a turn. Using a flint lighter, ignite the acetylene gas coming out of the tip.

11. Continue to turn on the acetylene torch valve slowly until the acetylene flame jumps away from the end of the tip slightly. This indicates that the proper amount of acetylene is being fed to the tip. A quick flip of the torch should make the flame leap away from the tip 1/16 in. and come back again. If the flame will not move back to the tip, too much acetylene has been turned on. (If the tip is worn, it may be difficult to make the flame jump away from the tip.)

Another method for determining the correct amount of acetylene is to increase the flow until the flame becomes turbulant (rough) a distance of 3/4 to 1 in. from the torch tip. With the right amount of acetylene, the flame will no longer smoke, or release soot. Look at Fig. 4-7 and compare the flames. Refer to Fig. 1-2 on page 10 for a color illustration of the flames.

12. After the acetylene is regulated, slowly open the oxygen valve on the torch. As the oxygen is fed into the flame, the brilliant acetylene flame turns purple and a small inner cone starts to form. This inner cone is light green in color. When first formed, the extremity of this inner cone will have a blurred and irregular contour. As one continues to turn on oxygen, the inner cone loses its blurred edge and becomes a round, smooth cone. Stop the adjustment at this point. Any increase in

oxygen will result in an "OXIDIZING FLAME." (Too much oxygen will burn or oxidize the metal being heated.) The tip of this inner cone is the hottest part of the flame. See Fig. 1-2 for a color illustration.

The correct quantities of gases for the smaller tip sizes may also be detected by listening to the torch flame. It should emit a soft purr, not a sharp irritating hiss, when correctly adjusted.

13. If the torch burns with an irregular contour (feather) to the cone, the flame is called a "CARBURIZING FLAME." There is an excess of acetylene. See Fig. 1-2.

If the inner cone has a very sharp point and if it hisses excessively, it usually means that too much oxygen is being used. If the flame has a smooth inner cone, the flame is called "neutral." See Fig. 1-2.

There is another method of adjusting the welding torch:

1. Turn on the cylinder gases as described previously.

2. Open the acetylene torch valve one turn. Slowly turn in the adjusting screw on the acetylene regulator. When the acetylene starts flowing, light the acetylene. Resume turning the acetylene regulator screw in until the acetylene is made to jump away from the torch, or the turbulence is correct as in the first method. Refer back to Fig. 4-7.

3. Open the oxygen torch valve one turn. Turn the oxygen regulator adjusting screw in slowly until enough oxygen is being fed to the torch

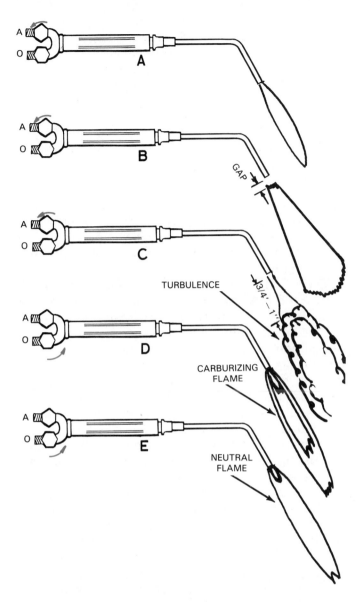

Fig. 4-7. Recommended steps for lighting the oxyacetylene welding torch. A—Open the acetylene torch valve slightly and light the acetylene with a spark lighter. B—The correct amount of acetylene is flowing if the flame jumps away from the tip when the torch is shaken, or, C—As shown here, a turbulence is created in the acetylene flame and the sooty smoke is eliminated. D—Begin turning on the oxygen by opening the torch oxygen valve. E—Continue to turn on the oxygen torch valve until the middle flame is eliminated and a rounded inner cone is seen.

to completely consume all the acetylene and a neutral flame is obtained. (The neutral flame is described earlier.) Also see the color illustration in Fig. 1-2.

4. This method may be used in place of the first method. In cases where the operator uses a long hose, the second method compensates for the pressure drop in the hoses. The operator may choose either of the two methods. Results of either method are generally satisfactory.

## 4-12. REGULATOR GAUGE READINGS

After adjusting the desired pressure on the regulator low pressure gauge, the torch valve is closed. You will see a slight rise in the pressure on the low pressure gauge. The pressure reading will rise and stop. This is normal. The flowing pressure (dynamic pressure) is always lower than the pressure when the torch valve is closed (static pressure).

If the low pressure gauge reading continues to rise (creeps) after the torch valve is turned off, this indicates a leaking regulator nozzle and seat. Turn off the cylinder valve immediately or the low pressure gauge may receive the full cylinder pressure. This will cause the gauge to rupture.

Replace the regulator or have it repaired before using the station again if a continuous rise (creep) occurs.

Review Heading 4-34 on regulator safety before attempting to weld.

## 4-13. TURNING ON THE OXYACETYLENE OUTFIT AND LIGHTING THE INJECTOR TYPE WELDING TORCH

The steps for lighting an injector type torch are:
1. Inspect the equipment to make sure all parts are in good operating condition.
2. Inspect the regulators. The adjusting screws of the regulators should be turned all the way out (counterclockwise), before the cylinder valves are opened. This prevents damage to the regulator diaphragm when the cylinder valve is opened. Never stand in front of the regulator when opening the cylinder valves. Injury may result if the regulator or gauges rupture.
3. Open the oxygen cylinder valve very slowly until the regulator high pressure gauge reaches its maximum reading, then turn the valve all the way open. The oxygen cylinder valve is turned all the way out because this valve has a double seat or back seat. In the all-out position, this seat closes any possible opening through which the high pressure oxygen might escape along the valve stem.
4. With an acetylene cylinder wrench, slowly open the acetylene cylinder valve a quarter to a half turn. Leave the wrench on the acetylene cylinder valve stem so the cylinder may be shut off quickly in an emergency.
5. To adjust the oxygen torch pressure, open the torch oxygen valve wide open, about 1 1/2 turns. Turn the pressure adjusting screw in on the oxygen regulator until the low pressure (delivery) gauge registers the approximate pressure shown in Fig. 4-8. Then close the oxygen valve. Important - check the low pressure

| METAL THICKNESS | TORCH TIP NUMBER DRILL SIZE | OXYGEN REGULATOR PRESSURE PSI | ACETYLENE REGULATOR PRESSURE PSI |
|---|---|---|---|
| 1/16 (1.59 mm) | 56 | 8-20 (55.2-137.9 kPa) | 5 (34.5 kPa) |
| 1/8 (3.18 mm) | 53 | 11-25 (75.8-172.4 kPa) | 5 (34.5 kPa) |
| 1/4 (6.35 mm) | 48 | 12-23 (82.7-158.6 kPa) | 5 (34.5 kPa) |

Fig. 4-8. Table shows correct oxygen and acetylene pressures for welding different thicknesses of metal using an injector type torch.

gauge for an indication of regulator failure. See Heading 4-12. Use only fingertip force to close the torch valves. Too much force will damage the needle valves.

6. Open the torch acetylene valve 1/2 turn. Turn in the pressure adjusting screw on the acetylene regulator until the low pressure (delivery) gauge shows a pressure of 5 psi (34.5 kPa). Close the torch acetylene valve.

By following this procedure, the system will have been purged.

7. To light the torch after purging, open the torch oxygen valve 1/4 turn. Open the torch acetylene valve 1/2 turn. Light the gases at the tip with a lighter. Open the torch oxygen valve wide (1 1/2 turns) and adjust the torch acetylene valve to the desired flame.

8. Should the acetylene delivery pressure become so low that a delivery pressure of 5 psi (34.5 kPa) can no longer be obtained, proceed as follows:
   a. Open the torch acetylene valve all the way.
   b. Turn the oxygen regulator pressure adjusting screw to the left until the flame shows an excess acetylene feather about four times as long as the inner cone.
   c. Adjust the torch acetylene valve to give the desired flame.

## 4-14. TORCH ADJUSTMENTS

The torch may be adjusted to produce the following flame characteristics:
1. Neutral flame.
2. Carburizing flame.
3. Oxidizing flame.

In general, the neutral flame is the one desired. However, in welding aluminum, in brazing, and in some other operations where oxidizing of the metals would interfere with welding, a slightly carburizing flame is often used. Fig. 1-2 and Fig. 4-2 illustrate the appearance of each of these flames. While a slightly carburizing flame may be recommended for certain work, usually a neutral flame will do just as well. However, because of the slight fluctuation in gas pressures, it is difficult to maintain a perfectly neutral flame and it may vary from neutral to slightly oxidizing or carburizing. Therefore, in order to avoid the possibility of running into an oxidizing flame, a slightly carburizing flame is usually safer.

The torch may occasionally backfire. (pop). A BACKFIRE is a small explosion of the flame at the torch tip which may be the result of several avoidable conditions. The most frequent cause is preignition of the gases. Some causes of backfiring are:

1. The gas is flowing out too slowly and the pressures are too low for the size tip (orifice) used. The gases are therefore burning faster (flame propagation) than they can flow out of the tip. This trouble may be corrected by adjusting to a slightly higher pressure for both the oxygen and the acetylene.

2. The tip may become overheated from overuse, from operating in a hot corner, or from being too close to the weld. Cool the tip.

3. The inside of the tip may have carbon deposits or a hot metal particle may be lodged inside the orifice. These particles become overheated and act as ignitors. Correct this by cleaning the tip. See Headings 3-23 and 4-36. Backfiring happens rarely but could occur when the inner cone of the flame is submerged in the puddle.

A FLASHBACK is the burning of the gases in the mixing chamber or beyond toward the regulator. Flashback is an extremely dangerous occurrence! Action must be taken immediately to extinguish the flame.

Generally when flashback occurs, a squealing or sharp hissing noise is heard. If flashback occurs, the torch oxygen valve should be turned off and then the fuel gas torch valve. This is done in an attempt to keep the flashback from going beyond the torch.

Close the fuel gas and oxygen cylinder valves to cut off the supply of the gases.

Inspect for damages. If flashback occurs, the hose, torch, and regulators are usually damaged and must be replaced or overhauled.

Flashbacks occur for the following reasons:
1. Failure to purge the system before lighting the

torch. Purging insures that the proper gas is flowing in each hose and torch passage. Any oxygen in the fuel gas passages is cleared out and the fuel gas is cleared from the oxygen passages prior to lighting the flame.

2. An overheated tip.

Review Heading 4-35 on torch safety before attempting to weld.

## 4-15. TURNING OFF THE TORCH AND SHUTTING DOWN THE OXYACETYLENE OUTFIT

If the operator wishes to leave the welding station for just a few minutes, it is only necessary to close the torch valves and lay the torch aside. To extinguish the flame, turn off the acetylene (fuel gas) torch valve first. Then turn off the oxygen torch valve.

If the equipment is not to be used for some time, the outfit should be completely shut down as follows:

1. Close the hand valves on the torch; the acetylene valve first. This extinguishes the flame and eliminates the soot.
2. Close the cylinder valves (tightly).
3. Open the hand valves on the torch.
4. Wait until BOTH the high and low pressure gauges on BOTH the acetylene and oxygen regulators read zero.
5. Close both hand valves on the torch (lightly) and hang up the torch.
6. Turn the adjusting screws on both the acetylene and oxygen regulators all the way out

These instructions may be followed for both the balanced pressure and the injector type torch.

Caution: If turning out the regulator screws is not the last step, pressure will remain in the high pressure gauge and regulator.

## 4-16. TORCH POSITIONS AND MOVEMENTS

Moving the torch in the direction that the tip is pointing is called FOREHAND WELDING. The torch is held at an angle of 15 - 75 deg. to the work. This will depend on the tip size used, metal thickness, and other welding conditions. A 30 to 45 deg. angle is typical. See Fig. 4-9. In forehand welding the flame spreads over the work ahead of the weld. This preheats the metal before it comes under the high temperature flame.

Use either the OSCILLATING or the CIRCULAR TORCH MOTION. Fig. 4-10 shows different torch motions used by welders. In either case, the cone of the flame should never go outside of the weld pool. The tip of the inner flame cone should be about 1/16 in. (1.6 mm) to 1/8 in. (3.2 mm)

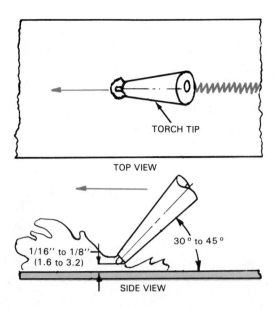

Fig. 4-9. Recommended torch angle, direction, and flame distance from the metal when running a continuous weld pool in the flat position. Angles between 15° - 75° are used to compensate for metal thickness and tip size.

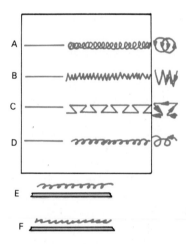

Fig. 4-10. Common types of flame movement patterns used when gas welding. All motion must remain within the size of the weld pool.

above the metal. Fig. 4-11 shows a butt weld in progress.

## 4-17. RUNNING A CONTINUOUS WELD POOL

Before attempting a weld of any kind, it is recommended that the beginning welder practice running a continuous weld pool. Once described by the term "puddling," this is the creation and control of the molten pool of metal which is carried along the seam of the parts to be welded together.

An experienced welder can, by watching the

Fig. 4-11. Oxyacetylene welding a butt joint. Steel plates are 5/16 in. (7.94 mm) thick. Note position and angle of torch flame and filler rod.

appearance of the weld pool carefully, tell the following about the weld:
1. The amount of weld penetration.
2. The torch adjustment and heat provided.
3. How and when to move the torch.
4. When and how often to add filler metal.

The size (diameter) of the weld pool will be in proportion to its depth. The operator may judge the depth, or penetration, of a weld by watching and controlling the size of the pool of molten metal. On very thin metal, the penetration or depth of the weld pool will be greater, in proportion to the width, than with thicker metal.

The appearance of the surface of the pool will indicate the condition of adjustment of the torch. The neutral flame, when melting a good grade of metal will give a smooth, glossy appearance to the weld pool. The edge of the pool away from the torch will have a small bright incandescent spot. This spot will move actively around the edge of the pool. If this spot is oversize, the flame is NOT NEUTRAL. If the weld pool bubbles and sparks excessively, an oxidizing flame is in use. A poor quality and/or dirty metal will also spark when it is being welded. The weld pool will have a dull and dirty (sooty) appearance, if the flame is carburizing to any great extent.

The tip of the inner cone of the torch flame must be held within the boundary of the weld pool at all times. A correctly adjusted flame prevents the oxygen in the atmosphere from coming in contact with the surface of the pool and causing an oxidizing condition. The hottest area of the flame is

1/16 in. (1.6 mm) to 1/8 in. (3.2 mm) from the end of the inner flame cone. See Fig. 4-12. If the weld pool sinks or sags too far, indicating too much penetration, lower the angle of the torch or increase the speed of movement rather than draw the torch away from the surface. Fig. 4-13 shows

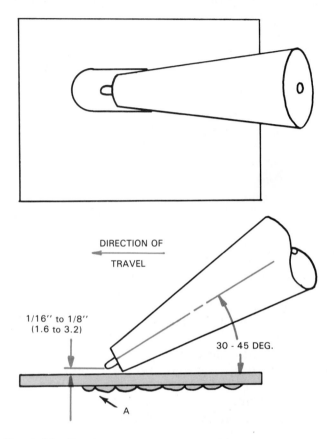

Fig. 4-12. Continuous weld pool procedure. This illustration shows the correct position of the torch in relation to the base metal. Detail A shows the penetration (sag) below the under surface of the base metal.

a continuous weld pool in progress. Note the width of the pool and the torch movement which controls the width of the pool. Note also the eye of the pool, a bright flake of oxide that indicates the movement of the molten metal. A partially

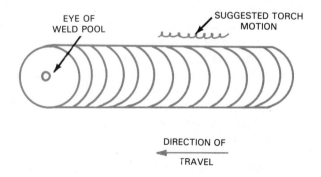

Fig. 4-13. This illustration shows the continuous weld pool in progress.

completed weld pool practice piece is shown in Fig. 4-14. Note the side view showing the penetration.

Before starting to practice with welding rod, the beginning welder should be able to produce five consecutive passes by running a continuous weld pool. Each should be at least 5 in. long (130 mm), without any holes melted through the metal and with good penetration. The passes must also be straight (in line) and even in width. Beginners who can do this have become familiar with torch operation. They now know about the theory and practice of the weld pool sufficiently to proceed with learning the manipulation of the welding rod.

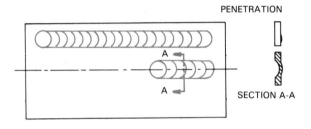

Fig. 4-14. Weld pool procedure. This illustrates a partly completed exercise piece. Note the penetration. This operator has obtained uniform width and has carried the weld pool in a straight line. The penetration is uniform.

## 4-18. TYPE OF WELDS MADE WITHOUT THE USE OF A WELDING ROD

The outside corner joint can be welded without the use of a welding rod. It therefore makes an excellent welding exercise for beginners. This exercise teaches how to weld by using some of the parent metal as the filler metal. The pieces are placed one against the other, at right angles so the vertical piece extends beyond the surface of the horizontal sheet approximately 1/32 to 1/16 in. (.8 - 1.6 mm) as shown in Fig. 4-15. The two pieces are then tacked together at their ends. A TACK WELD is a small weld used to hold parts together prior to making the completed weld. The extended metal serves as the filler metal.

The weld must have good penetration, but the penetration should not show on the inside corner. The operator will find that very little torch motion is needed for this exercise. The torch tip should be slightly tilted, making the flame point inward toward the flat or horizontal surface. The weld should be all on the horizontal surface. None of it should run over on the vertical edge. This is necessary, because in many cases of metal finishing, the weld is made into a right angle corner by grinding the excess metal from the one side. After checking the weld for appearance, the penetration may be tested by bending the two

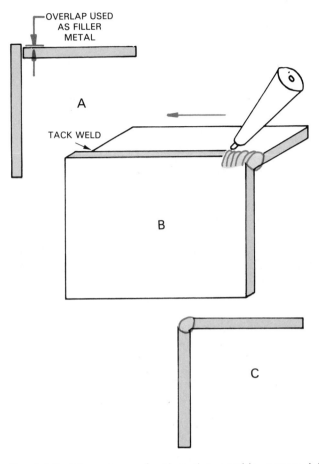

Fig. 4-15. Steps in performing the outside corner joint without a welding rod. A—Metal in position to weld. B—Weld in progress. C—Appearance of the finished weld.

pieces of metal open like the pages of a book. Any cracking or breaking of the metals at the joint will indicate a lack of penetration or fusion.

Another good exercise to help in learning the use of a welding torch when welding without welding rod is the flange weld.

To prepare the metal for this weld, bend a 90 deg. flange about 1/4 - 1/2 in. (6.4 - 12.7 mm) long on two pieces of sheet steel about 1/32 to 1/16 in. (.8 - 1.6 mm) thick. Be sure the lengths of the two flanges are equal. Place the flanges together along their lengths and tack weld several times to hold the metals in position. Fuse the two flanges with the welding torch using the flanges as the filler metal in the same manner as was done when welding the outside corner joint. Fig. 4-16 shows a flange joint weld in progress.

## 4-19. TYPES OF WELDS MADE WITH A WELDING ROD

To become proficient in the art of gas welding, certain fundamental exercises must be planned and practiced until satisfactory welds can be performed consistently. The different fundamental

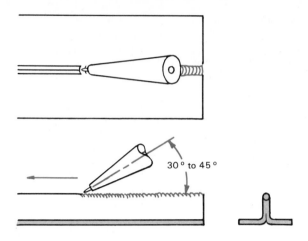

Fig. 4-16. A flange joint weld in progress. Note that the base metal is used as the filler metal, therefore a welding rod is not needed. The joint should be tack welded several times before starting the main weld.

gas welding operations may be classified according to the type of joint and position of the weld.

The basic joints are:
1. Butt.
2. Lap (fillet weld).
3. Outside corner.
4. Inside corner (fillet weld).

The basic welding positions are:
1. Flat or downhand - A horizonal weld on a horizontal surface.
2. Horizontal - A horizontal weld on a vertical surface.
3. Vertical - A vertical, weld on a vertical surface.
4. Overhead - A horizontal weld on a horizontal surface over one's head.

The welding of each of the above joints should be practiced in each of the above positions. Fig. 4-17 illustrates some typical types of joints. Welding of these joints should be performed on both thin sheet steel and finally on steel plate of at least 3/8 in. (9.53 mm) thickness. After obtaining the necessary skill on steel plate with these exercises, the welder may then proceed to study special welding applications such as pipe welding, aluminum welding, or cast iron welding.

## 4-20. USE OF WELDING ROD IN RUNNING A BEAD

Running a continuous weld pool without the use of welding rod is used mainly on outside corner or flange joints. Welding rod (filler metal) is added when extra metal is needed to create the correct weld shape and strength. Welding thicker metal joints with the weld pool method only causes a thinning of the metal in the weld area. To obtain a strong weld, metal from a welding rod is fed into the pool to decrease its depth and to increase the thickness

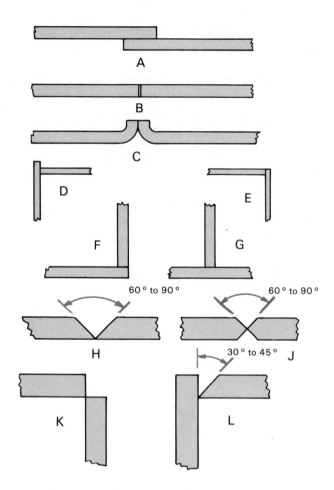

Fig. 4-17. Some typical welding joint designs: A—Sheet steel lap joint in the flat position. B—Sheet steel butt joint in the flat position. C—Flange joint in the flat position. D and E—Outside corner joints. F and G—Inside corner joints (G is sometimes called a T-joint.) H, J, K, and L—Joint designs for metal plate. Note that when welding joints A, B, and E through L, welding rod is used as the filler metal. When welding the joints at C and D on sheet metal, no welding rod is required as a filler metal because the metal pieces themselves are melted to form the bead and to join the pieces together.

of the metal in the weld area. See Heading 3-30 for specifications of welding rods. The bead in a correctly made weld should be slightly convex (crowned) for extra thickness and strength of the weld. Fillet welds are often made with a concave bead. The fillet must be made large enough to provide sufficient strength to the joint. Fig. 4-18 illustrates how welding rod material is added to make a slightly convex bead.

To weld using welding rod, bring the torch to that part of the joint where the weld is to start. Melt a small weld pool on the surfaces of the two pieces and allow the pools to flow together. At the same time, with the other hand, bring the welding rod to within 3/8 in. (9.5 mm) of the torch flame and 1/16 to 1/8 in. (1.6 - 3.2 mm) from the surface of the weld pool. In this position, the rod will become preheated, and melt sooner when dipped

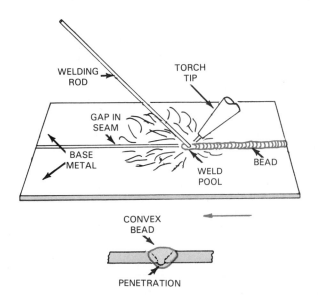

Fig. 4-18. Recommended torch and welding rod positions for welding a butt joint in the flat position.

into the weld pool. When the operator judges that the weld pool needs additional metal, the end of the welding rod is inserted into the pool, and some of the welding rod is melted and mixed with the molten parent metal. Continue to dip the rod into the pool until enough welding rod metal is added to raise the pool to a slight crown. At the same time, continue the torch forward motion without interruption. Torch control at this time is very important. Control over weld pool condition and welding rod melting can be done by small changes of torch position. As soon as enough welding rod metal has been added, withdraw the welding rod. Keep the welding rod close to the pool and flame to maintain the end of the welding rod in a preheated condition.

If the welding rod is withdrawn too far from the torch, it will become too cool. This cool rod will cool and chill the pool when it is again inserted into it. On the other hand, if the welding rod is held too close to the flame of the torch, it will become too hot. If it should become molten, drops of molten welding rod will be blown by the flame onto the cooler parts of the metal being welded. Such a condition will result in a very uneven bead, poor fusion, and probably poor penetration. As the welding rod melts, it becomes necessary to change the position of one's hand on the rod. To change the hand position, lay the welding rod down with the hot end away from the body. Then pick it up at a new position. DO NOT place the rod against the body to slide the hand to a new position. Many welders have been injured when the tip of the welding rod burned through clothing and touched the body.

The beginner is sometimes tempted to use dif-

ferent size welding rods for the same weld. This change should be avoided because:

1. If a 3/32 in. (2.38 mm) welding rod is correct, and if a change is made to 1/16 in. (1.59 mm) welding rod, it will be extremely difficult to add enough welding rod to obtain a good weld.
2. The smaller rods will make it more difficult to control the weld pool.
3. There will be a tendency to burn (oxidize) the smaller size welding rod.

If the operator changes to 1/8 in. (3.18 mm) welding rod when 3/32 in. (2.38 mm) welding rod is correct, the following troubles might result:

1. The larger welding rod will cool the weld pool too much while it is being added and prevent consistent welding.
2. There will be a tendency to add too much welding rod, destroying penetration and building the weld higher than it should be on the top surface.

Whether to use a 1/16 (1.59 mm), 3/32 (2.38 mm), or 1/8 in. (3.18 mm) welding rod is not so important as it is to adhere to one size after becoming used to it for a certain thickness of metal and a certain size tip. See Fig. 4-6 for a table of recommended welding rod sizes and Chapter 3 for more detailed information on welding rod choices.

A good weld with good fusion, good bead, and good penetration, is obtainable only by attaining skill in the handling of the welding torch and the welding rod in harmony with each other. The torch motion should be constant in forward speed and in the width of the motion. The proper distance between the flame cone and the metal must also be maintained. The slant of the torch in respect to the surface should always be the same. The filler metal additions should be made at regular intervals to keep the bead a uniform size and shape.

### 4-21. BUTT JOINT WELDING

The butt joint is one of the most common welds made with the oxyfuel gas torch. The instructions which follow will aid the beginner in making this weld on thin steel.

Procure two pieces of mild steel approximately 1 in. wide and 5 in. (130 mm) long. The pieces should be clean, flat, and the edges should be straight. Place the two pieces of metal across two firebricks, permitting the bricks to support the ends of the metal. Place the edges of the two pieces of metal together at the end where the weld is to start. As the weld proceeds along the joint, the metal shrinks. As it cools from a molten state, it tends to pull the two pieces of metal together. This shrinkage may cause the edges to lap one over the other, or warp the metal. The operator may prepare the metal for this expansion and contraction by:

1. Tack welding (fusing) the two pieces of the metal together before proceeding with the welding as in Fig. 4-19, Part A. This method will produce some internal strain, but will keep the ends sufficiently in line to enable the operator to make a good weld. Tack weld every two inches or so along the joint.
2. Tapering the gap between the two pieces of metal to allow for contraction as in Fig. 4-19, Part B. The approximate contraction is from 1/8 in. to 1/4 in. per foot, or from 10 mm to 20 mm per meter of length. As the weld pool gets wider, the contraction increases.
3. Using especially prepared wedges which may be placed between the two pieces of the joint to prevent the contraction of the metals as the weld cools, as shown in Fig. 4-19, Part C. This method is more generally used with long joints.
4. Clamping the metal in a heavy fixture to minimize movement.

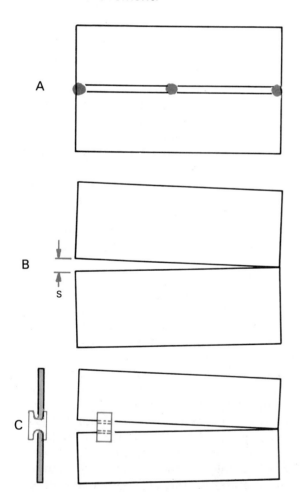

Fig. 4-19. Some methods used to maintain correct position of welded pieces, since the weld metal shrinks as it solidifies and cools: A—"Tacking" pieces together before welding. B—Allowing for shrinkage (S). C—Use of special wedges.

After the metal is prepared, light the torch, adjust it to a neutral flame, and proceed as follows: Bring the torch to the point where the weld is to start, holding the torch at a 30-45 deg. angle with the tip pointing along the direction of the joint. Hold the inner cone approximately 1/16 in. to 1/8 in. (1.6 - 3.2 mm) away from the metal. With the other hand, bring the end of the welding rod approximately 3/8 in. (10 mm) away from the welding torch and about 1/8 in. (3 mm) above the metal. The torch flame will melt a pool on the edges of each of the pieces of metal. The pool should spread equally over the two pieces of metal. Apply the welding rod (filler metal) as directed in Heading 4-19.

Advance the torch a very short distance until the weld pool again reaches the size of the previous pool. The tip motion should continue as the torch moves forward. The welding rod is again dipped into the pool and the weld bead is built up to a crown. Continue this procedure throughout the length of the weld joint, using a continuous torch motion. See Fig. 4-10 for suggested torch motions.

The tip should be kept at a uniform distance from the weld, and the torch angle with the metal should remain unchanged. The welding rod should be added in uniform amounts, and at regular intervals. After the weld has been finished, allow it to cool, and then inspect it. Fig. 4-20 is a magnified photograph of a butt weld in mild steel.

Fig. 4-20. Macrographs (4X) of a butt joint weld in mild steel. The metal has been etched to show the grain and fusion. Note that the penetration is excessive in relation to the thickness of the base metal.

There is a procedure in butt joint welding in which the heat source penetrates completely through the workpiece before the welding rod is added. This process is called KEYHOLE WELDING. A hole is formed at the leading edge of the weld metal. As the heat source progresses, the molten filler metal and base metal fills in behind the hole to form the weld bead and excellent penetration. The keyhole procedure is often used on thicker sections of metal that have been beveled (ground). See Fig. 4-18. As long as the keyhole continues to appear, penetration is certain. See Fig. 4-21 for an example of the keyhole procedure.

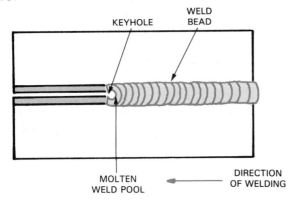

Fig. 4-21. The keyhole welding procedure. Note the hole (keyhole) at the leading edge of the molten weld pool. The continual appearance of the keyhole will ensure good, uniform penetration.

## 4-22. LAP JOINT WELDING

The lap joint is one of the four basic joint designs. The joint consists of lapping one piece of sheet metal over the one to which it is to be welded, as shown in Fig. 4-22. This weld should be first performed in the flat position. Although the welding technique is typical, several things must be kept in mind in order to obtain a satisfactory lap weld.

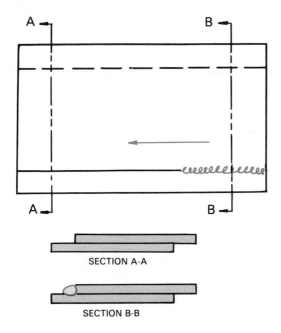

Fig. 4-22. Lap joint welding. Section A—A shows the position of the two pieces in relation to each other. The two pieces should be placed as close together as possible (no gaps). Section B—B shows the finished weld.

1. It will be found difficult to heat the bottom piece of metal to a molten state before the top metal edge melts too much, making the weld very ragged. The way to prevent this is to concentrate the torch flame on the lower surface. The bottom piece of metal requires two-thirds of the total heat, as shown in Fig. 4-23.

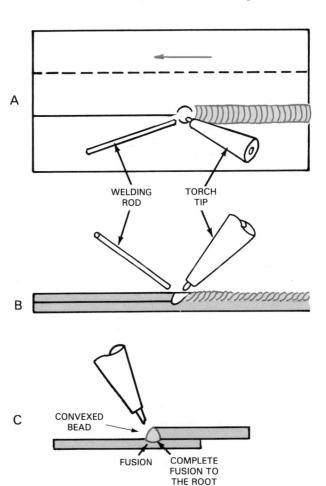

Fig. 4-23. Recommended position of the torch when fillet welding a lap joint. Note the position of the torch and welding rod in relation to the base metal. In C, the torch points more directly at the bottom surface than at the edge.

2. The welded portion (the weld nugget) must be at least as thick as the original metal. To provide this thickness, add enough welding rod metal to make a convexed (curved outward) bead.

The beginner usually has a tendency to perform this weld without heating the bottom metal sufficiently to obtain fusion. The destructive separation test (Chapter 22) will quickly show the lack of fusion at this point. Before testing the specimen, the appearance of the weld should be closely inspected for such things as an even bead, consistency of width, cleanliness, and good general appearance. The metal should not sag on the reverse side of the bottom piece (too much penetration), and the bead should be straight.

A special form of lap welding is known as plug welding. The American Welding Society defines a PLUG WELD as follows: ''A circular weld made through a hole in one member of a lap or T-joint, fusing that member to the other. The walls of the hole may or may not be parallel and they may be partially or completely filled with weld metal---.'' See Fig. 4-24 for an example. The torch flame should be concentrated on the base plate in order to bring it to its melting temperature at the same time that the edges of the hole melt.

In aircraft tubing repairs, the plug weld is known as a ROSETTE WELD.

A SLOT WELD is the same in all respects as a plug weld, except that instead of a round hole being drilled, an elongated hole or slot is cut.

method of performing an outside corner weld is shown in Fig. 4-25. In this method, the two pieces of metal do not overlap. While welding the outside corner joint, watch the outside edges of the metal. When the weld pool touches the outside edges of the metal, the welding rod should be placed in the pool. Enough filler metal should be melted to form a convexed bead. The torch is moved forward and the welding rod is again placed in the weld pool when it touches the outside edges of the joint. This operation is repeated until the joint is completely welded. The keyhole welding technique may be used on this joint. Fig. 4-2 shows a macrograph of a completed outside corner joint made with welding rod.

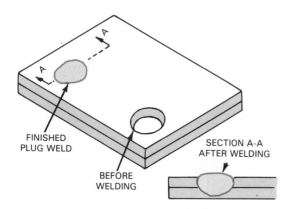

Fig. 4-24. A plug weld. This is a special application of lap welding.

## 4-23. OUTSIDE CORNER JOINT WELDING

Outside corner joints may be made without using welding rod; see Heading 4-17. An alternate

Fig. 4-26. Cross sections of the outside corner welds (a 4X macrograph). The weld at the top was tightly fitted prior to welding. The weld at the bottom has a space between the metals which creates a weaker joint. The metal has been etched to emphasize grain lines and fusion.

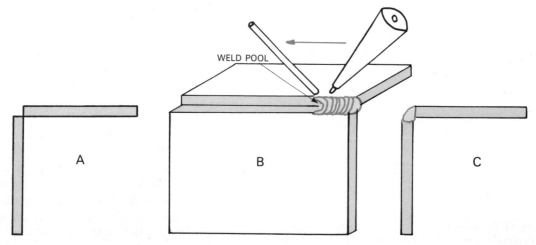

Fig. 4-25. Recommended steps when welding an outside corner joint using a welding rod: A—Metal is in position. B—Weld is in progress. C—Weld is completed.

## 4-24. INSIDE CORNER AND T-JOINT WELDING

This is a fairly easy exercise to perform, but one in which the operator may find it difficult to obtain sufficient penetration. Two pieces of stock are placed with their surfaces at right angles to each other, either in an L-formation (corner joint) or in an inverted T-formation (T-joint). The inside corner joint or T-joint is welded as shown in Fig. 4-27. When welding in this position, the torch flame is placed close into the corner. It is difficult to obtain additional oxygen from the air to complete the combustion of acetylene when welding in a corner. Therefore, it is sometimes necessary to open the torch oxygen valve slightly to provide enough oxygen for complete combustion. This adjustment would result in an oxidizing flame under normal circumstances, but in this special case, it produces a neutral flame. The exercise may be set up in two different ways.

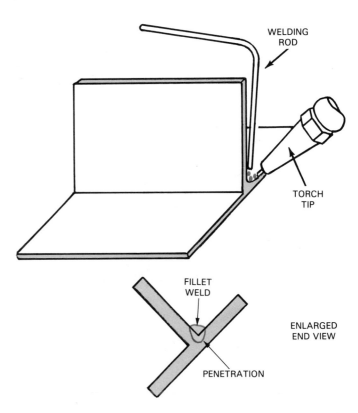

Fig. 4-28. Fillet welding of a T-joint in flat (downhand) position.

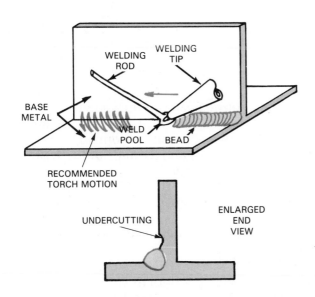

Fig. 4-27. Fillet welding on a T-joint in the horizontal position. Note the lack of filler metal on the vertical piece. This is a welding defect called undercutting. The recommended inclined half moon torch motion will help eliminate undercutting.

The preferred method is to have the two pieces set at an angle of 45 degrees from the horizontal, forming a trough as shown in Fig. 4-28. This is the flat (downhand) position. The operator will find that the exercise is easier with the specimen set up in this manner. This position means that the joint is in a 45 degree position, and that the face (top) of the weld is in the flat position. The two pieces should be tacked at points along the joint before making the weld.

A second method is to have one piece horizontal and the other vertical, with the edge of the vertical piece touching the middle of the surface of the horizontal piece. This is called the horizontal posi-

tion. Before welding the pieces in this position, the operator should tack weld the joint to keep it in line while welding. It will not be necessary to provide for expansion and contraction since the metals are pulled one against the other as the weld cools and solidifies.

The technique of handling the torch and welding rod will be almost the same in this exercise as in the lap weld. The operator will find that very little torch motion is necessary. It is very important to produce a weld pool before attempting to add the welding rod; otherwise insufficient penetration will result. To secure fusion at the root (bottom) of the weld, the torch should be held as close as possible without allowing the inner cone of the flame to touch the metal. The torch is then drawn back slightly when adding the welding rod. Before adding the filler metal in an inside corner joint, both surfaces must be melted. The surfaces are molten when the edges of the weld pool appear to run ahead, forming a "C" shape. The welding rod should be placed into the pool at the time that the "C" shaped pool forms. See Fig. 4-29. After the weld is completed, inspection should show a good bead, good fusion, consistent width of the weld, a clean appearance. The weld must also have equal distribution, meaning that half of the weld is on one piece, and half on the other. Also, the vertical piece of metal especially should not show any indications of having some

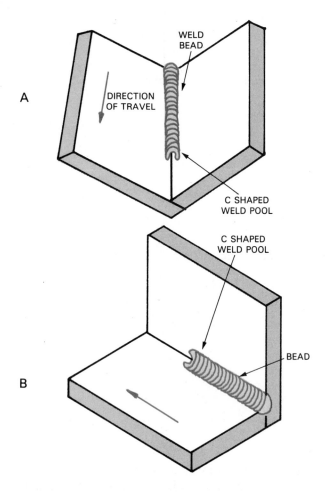

Fig. 4-29. Welding in inside corner joint. At A, the joint is set up in the flat (downhand) position. At B, the joint is set up in the horizontal position. The C shaped weld pool must be seen to ensure that both surfaces are melted before the filler metal is added.

of its metal melted away, leaving it thinner (undercutting). To eliminate undercutting on the vertical surface of a horizontal inside corner joint, use an inclined half moon motion. Stop momentarily at the edge of the weld pool. The inclined half moon motion will help to push molten metal into the undercutting area. See Fig. 4-27. In addition, adding the welding rod into the pool on the vertical surface will help prevent undercutting.

To test the weld for penetration, the two pieces of metal are closed together like the pages of a book. Lack of penetration, or fusion, is indicated by the added metal peeling away from the parent metal. Fig. 4-30 shows some alternate setups for inside corner joint designs. Fig. 4-31 illustrates some typical T-joint designs.

## 4-25. WELDING POSITIONS

Welds made in the flat position (downhand) are the easiest to make. However, it is often necessary to make welds in various other positions.

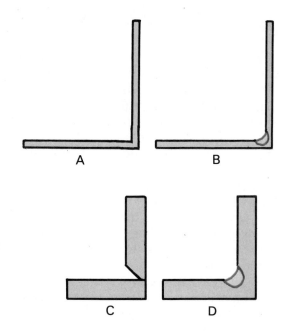

Fig. 4-30. End views of some popular corner joint preparations for fillet welds. A—Thin metal corner joint in position for horizontal fillet welding. B—Completed fillet weld on a thin metal corner joint. C—Thick metal single bevel groove corner joint. D—Completed fillet weld on a thick metal single bevel groove corner joint.

The recognized positions are shown in Fig. 4-32.

FLAT POSITION WELDING—The position of welding where welding is performed from the upper side of the joint, and the face of the weld is approximately horizontal.

HORIZONTAL POSITION FILLET WELD— The position of welding where welding is performed on the upper side of an approximately horizontal surface, and against a surface which is approximately vertical.

HORIZONTAL POSITION GROOVE WELD—The position of welding where the axis of the weld lies in a plane that is approximately horizontal and the face of the weld lies in a plane that is approximately vertical.

VERTICAL POSITION WELDING—The position of welding where the axis of the weld is approximately vertical.

OVERHEAD POSITION WELDING—The position of welding where welding is performed from the underside of the joint and where the axis of the joint and base metal are both approximately horizontal.

## 4-26. OXYFUEL GAS WELDING IN THE HORIZONTAL POSITION

This position consists of welding a horizontal joint on a vertical surface as shown in Fig. 4-33. Welding in this position should be practiced on butt joints, inside corner joints, and lap joints. A

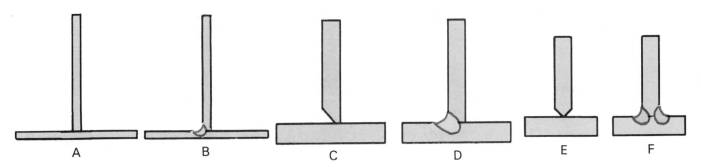

Fig. 4-31. End views of some popular T-joint preparations for fillet welds. A—Thin metal T-joint. B—Completed thin metal T-joint fillet weld. C—Single bevel groove T-joint on thick metal. D—Completed fillet weld on a single bevel groove T-joint. E—Double bevel groove T-joint in thick metal. F-Completed fillet welds in double bevel groove T-joint.

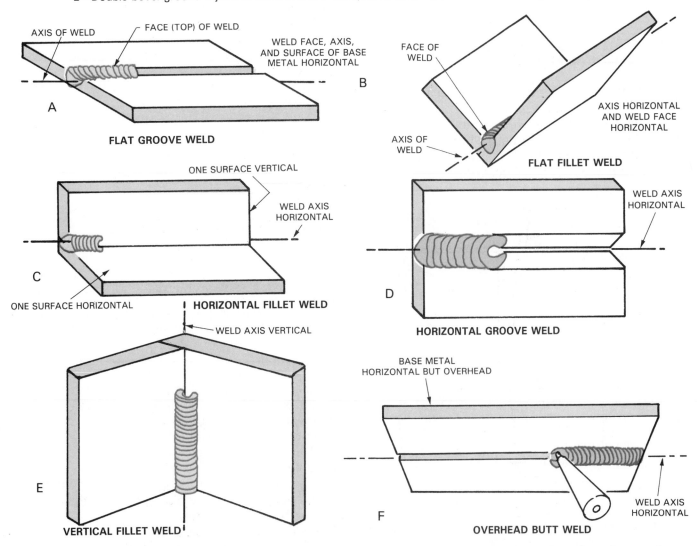

Fig. 4-32. Welding positions. Notice the position of the weld axis, weld face, and base metal surfaces in each example.

precaution to be noted in order to produce an excellent weld is that the torch tip, instead of pointing directly along the weld joint, should point at a slightly upward angle. This tip position will enable the force from the velocity of the gases to keep the molten metal from sagging. The welding rod should be placed into the upper edge of the weld pool. This will result in a more evenly distributed bead. It will be found that it is easier to obtain more consistent penetration, in both the horizontal position and the vertical position welding, than in flat position welding. The operator will have little difficulty in obtaining a good looking bead, but it will be more difficult to make a vertical or a horizontal weld with as good an appearance as a weld in the flat (downhand) position.

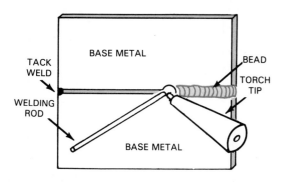

Fig. 4-33. Horizontal butt joint weld. The base metal surfaces are vertical. Note that the torch flame is directed upward to help hold the molten metal in place.

## 4-27. OXYFUEL GAS WELDING IN THE VERTICAL POSITION

Vertical welding consists of welding in a vertical direction on a vertical surface. Exercises should be performed on various weld joints including the butt joint, Fig. 4-34. Vertical welding will not be found difficult after the first few attempts if the following precautions are observed.

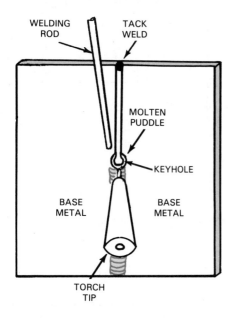

Fig. 4-34. A butt joint being welded while in a vertical position. Note the position and angle of the welding rod and torch tip with respect to the weld.

The weld should proceed upward. The torch is inclined from the surface of the metal at an angle of approximately 15 to 30 degrees. The torch tip is pointed up. This tip position enables the force from the gas velocities to keep the molten metal from falling, or sagging (due to gravitational pull).

A very small torch motion should be used so that the gas velocities keep the molten metal continually in position, and not allow it to sag.

Bead, fusion, and penetration should be checked according to the methods previously recommended for flat position welding.

## 4-28. OXYFUEL GAS WELDING IN THE OVERHEAD POSITION

Overhead welding is considered by many welders as the most difficult welding position. The operator should be skilled in welding in all other positions before starting overhead welding as shown in Fig. 4-35. This requires consistent, diligent practice to reach a satisfactory degree of skill. The operator's body, hands, and head should be protected with fire resistant work clothes, a cap, and gauntlet gloves. He or she should wear high-top shoes or welding spats when performing this exercise. Trousers should have covered pockets and no cuffs. The top botton of the shirt or coveralls must be buttoned to prevent burns.

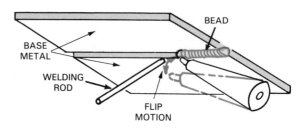

Fig. 4-35. A butt joint being welded while in the overhead position. Note the flip motion used to control the puddle temperature. CAUTION: WEAR GLOVES, CAP, AND LONG SLEEVES. BE SURE ALL CLOTHING IS FIRE RESISTANT.

Overhead welding should be practiced on all the standard joints. There should be at least two excellent samples of each type of joint before practice welding is concluded. The exercise should be mounted approximately 6 to 12 in. (150 - 300 mm) above the operator's head to be most comfortable, and the welder should stand to one side of the seam, welding parallel to the shoulders.

The position of the welding rod and the torch is almost the same in reference to the weld pool as in welding in other positions. The force of the torch flame gas velocities, plus the surface tension of the metal in the pool (the attraction of the molecules for each other), helps to overcome the pull of gravity on the molten metal. However, be careful to keep the metal as close to its lowest flow temperature as possible. Any superheating of the metal produces a more fluid condition and may cause the molten metal to fall.

One method of keeping the molten puddle from becoming too hot is to use a flip motion of the torch tip. The FLIP MOTION is a quick movement of the tip from the weld pool and back again. The momentary removal of the flame allows the pool and bead to cool slightly. Control of the weld pool and good timing when inserting the filler rod will result in welds of comparable quality to flat position welds.

## 4-29. JOINT PREPARATION FOR PLATE WELDING

Welders should be able to successfully weld both thin and thick metal (plate). The American Welding Society has standardized the names of the joints and the names of those parts of the joint which are in the weld zone. Fig. 4-36 shows standard terms for various parts of a groove weld.

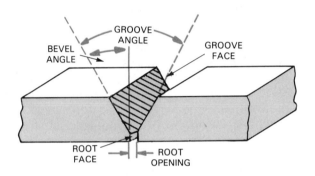

Fig. 4-36. Standard terms for the various parts of a plate joint.

Several edge preparation designs have been developed to prepare the edges of thick metal sections for welding. The edges are prepared in these ways to insure complete penetration. Fig. 4-37 shows edges prepared for various types of butt joints. The straight bevel and the V-groove joint preparation are the most common. The bevel or V-groove may be ground, machined, or flame cut. J and/or U-groove preparation may be machined or flame cut (gouged) with special cutting tips. The bevel and V-grooves are more economical in respect to quantity of welding rod metal used, gas or electricity cost, and in time saved.

Fig. 4-38 shows some suggested angles and measurements used when preparing metal edges for butt joint groove welds.

When oxyfuel gas welding thick plate joints, welding rod motion and torch motion must be used in a coordinated fashion. Fig. 4-39 shows how the welding rod is moved to one side of the pool as the torch is brought to the other side. The oscillating motion is then reversed. This permits excellent control of the large weld pool and provides adequate fusion and build up.

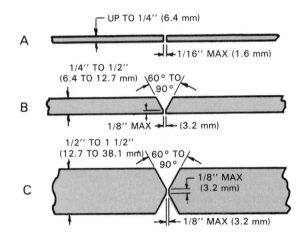

Fig. 4-38. Suggested angles and dimensions for butt joints.

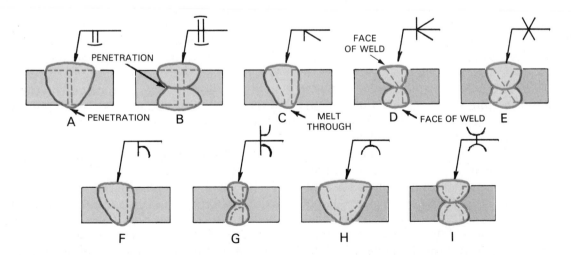

Fig. 4-37. Methods used to prepare metal edges before butt welding. The method used depends on metal thickness and whether it can be welded from both sides. A—Square groove welded from one side. B—Double square groove welded from both sides. C—Single bevel groove. D—Double bevel groove welded from both sides. E—Double V-groove welded from sides. F—Single J-groove welded from one side. G—Double J-groove welded from both sides. H—U-groove welded from one side. I—Double U-groove welded from both sides. See Chapter 2 for explanation of weld symbols shown.

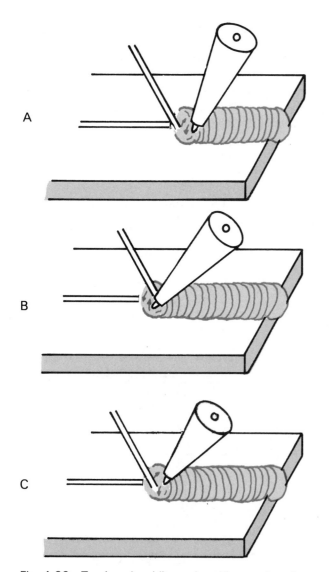

A

B

C

Fig. 4-39. Torch and welding rod positions and motions used when welding thicker metals where wide weld pools are required. The flame tip and welding rod are on opposite sides of the pool. Their positions are continually changed as the weld proceeds.

## 4-30. BACKHAND WELDING

BACKHAND WELDING, as shown in Figs. 4-40 and 4-41 requires that the torch be held at an angle of 30 to 45 degrees with the work, and the flame directed back over the portion of the work that has been welded. Directing the flame in this manner tends to anneal the completed weld and relieves the welding stresses to a great extent. In addition, the backhand direction of the flame tends to help the welder form a higher bead, and attain good penetration of the weld. Backhand welding is commonly used in welding cast iron or drain pipe, and in welding thick, heavy sections. Relieving the welding stresses created by welding is necessary in these heavier sections. In backhand welding, the continued spread of the

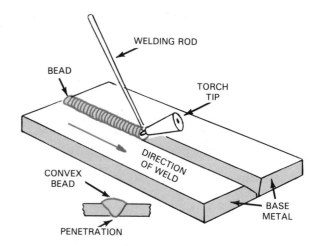

Fig. 4-40. Welding a single V-groove joint using the backhand method. Note the torch tip points away from the direction of the weld progress.

torch flame on the hot portion just welded tends to maintain a rather large pool of molten metal. In order that the edge of the pool may solidify into a bead, it is necessary to move the flame upward (flip motion), at frequent regular intervals. This action allows the edge of the pool to cool slightly and solidify.

## 4-31. MULTIPLE PASS WELDING

Welding a number of thin passes (layers of weld) one on top of the next is known as MULTIPLE PASS WELDING.

Multiple pass welding is well suited to welds on thick metal. It may be used on thin sections also where necessary.

The determining factor of whether to use a single or multiple pass is the thickness and width of the puddle that a welder can handle. Also, if the pass is too thick, impurities may not be able to rise to the surface before the puddle cools.

The layers of the weld may be made continuously for the full length of the joint or the welding may be done by the "step" method illustrated in Fig. 4-42. The continuous pass, multiple layer, type weld is generally preferred.

## 4-32. APPEARANCE OF A GOOD WELD

The weld is usually inspected by careful visual examination. Some inspectors use a magnifying glass of two to ten diameters magnification.

The weld should be of consistent width throughout its length. It should be straight so the two edges form two straight parallel lines.

The weld should be slightly crowned or convex (built up above the surface of the parent metal). This crown should be consistent (even). A gauge

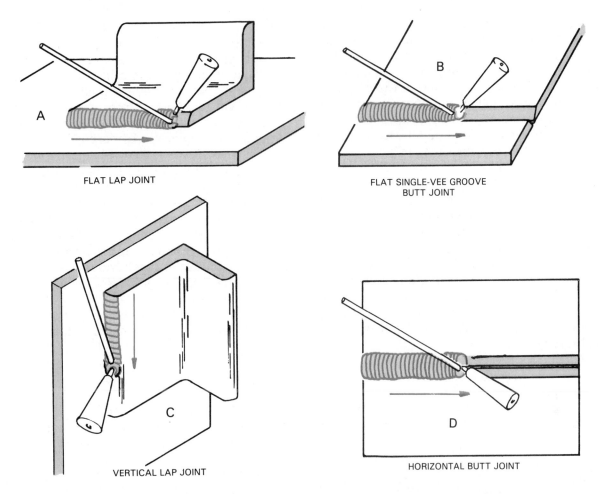

Fig. 4-41. Positions of welding rod and flame when welding, using backhand method.

FLAT LAP JOINT

FLAT SINGLE-VEE GROOVE BUTT JOINT

VERTICAL LAP JOINT

HORIZONTAL BUTT JOINT

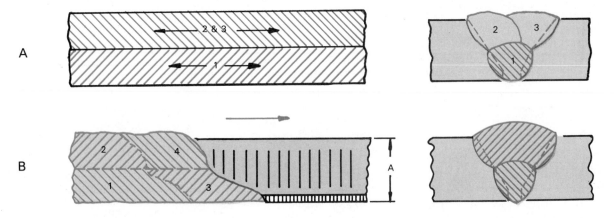

Fig. 4-42. The multiple pass and "step" pass. A—Shows several continuous passes used to weld a thick piece. B—Shows a step weld. The weld progress is in small steps from bottom to top; bottom to top, etc.

is sometimes used to check the weld contour.

The weld should have the appearance of being fused into the base metal, have a blended appearance, and not have a distinct edge between it and the parent metal.

The surface of the weld should have a ripple throughout its length. The ripples should be evenly spaced.

The weld should have a clean appearance. There should be no color spots, no scale on the weld, and no rough pitty appearance to the weld. Lap welds and corner welds should normally show no visible penetration on the side opposite the bead. On a butt weld, turn the specimen over and check the penetration. The degree of penetration will be indicated by the sag of the lower surface of the weld. The sag should be slight, and yet penetration should be obtained along the complete length of the weld. The amount of sag should be approximately 1/64 to 1/32 in. (.4 - .8 mm). Complete penetration and fusion is hard to determine; the easiest way to test for it is to place the specimen in a vise, or jig, with the weld held at the edges of the jaws. The upper half of the specimen is then bent, closing the two edges upon the welded part like a book. If the weld has not penetrated or fused with the edges satisfactorily, it will crack open at the joint as it is being bent.

The welder should obtain consistent and efficient penetration and fusion. It is possible to secure a very good weld with excessive penetration, and still build the upper surfaces of the weld up to the correct height. The weld will be of sufficient strength, but a weld of this nature will cost more than the weld described previously. It is only necessary to produce a weld as strong, or a little stronger, than the original metal. It is not possible to produce sufficiently strong butt joints if full penetration is not obtained.

Fig. 4-43 shows the cross section of both butt joint and lap joint welds. Note that B and D show poor penetration.

Fig. 4-44 shows the cross section of both good and poor lap and butt joints. In a lap joint, the main

fault is lack of fusion along the toe of the bead, as shown at B. To correct, direct more heat toward the bottom metal. The overlap at C is caused when too much welding rod is added to the puddle.

Fig. 4-45 shows severe undercutting in cross section. On an inside corner weld, this condition should be improved by stopping momentarily at the edges of the weld. Undercutting may also be decreased by placing the welding rod in the upper edge of the puddle. Another method used to eliminate undercutting is to use a torch motion which pushes the molten puddle up toward the undercutting area. See Fig. 4-27. In the lap weld, undercutting is caused by overheating the weld metal by poor positioning of the torch. It is also caused by improper torch movement, too small a tip with too high a gas velocity or by using an undersized welding rod.

A specially designed welding bench may be used to practice welding under the best conditions. See Figs. 4-46 and 4-47.

## 4-33. METAL FUME HAZARD

When heated, many metals release irritating and toxic fumes. The metals which release fumes that

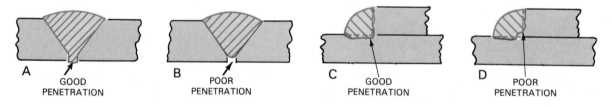

Fig. 4-43. A and B. Examples of butt welds in cross section. C and D. Examples of lap joint welds in cross section. Note in B and D that the fusion has not been completed to the root (bottom) of the joint.

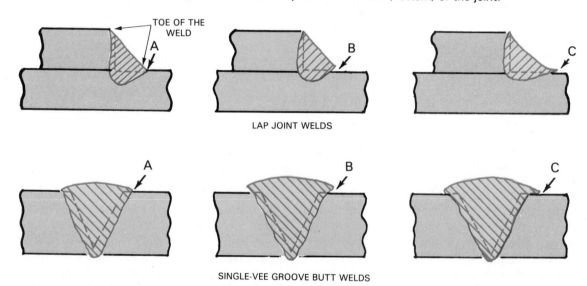

Fig. 4-44. Cross sections of good and poor butt and lap joints. A—Properly made joint with proper fusion at the toe (edges) of weld. B—Poor fusion at the toe (edges) of the weld. C—Poor fusion and overlapping of filler metal at the toe (edges of the weld.

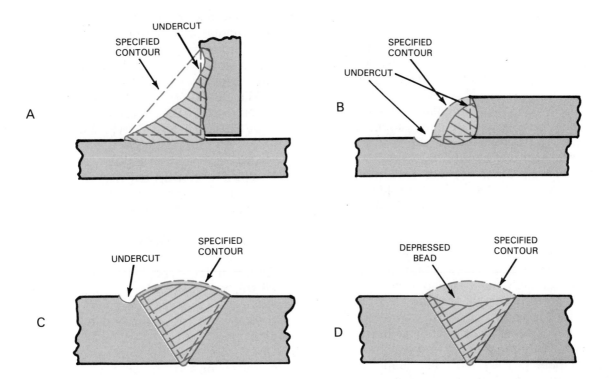

Fig. 4-45. Cross sections of some typical faults in finished welds. The shape of the bead contour is usually specified. When undercutting occurs, or if the face of the weld is not built up to specifications, the welded joint is weakened.

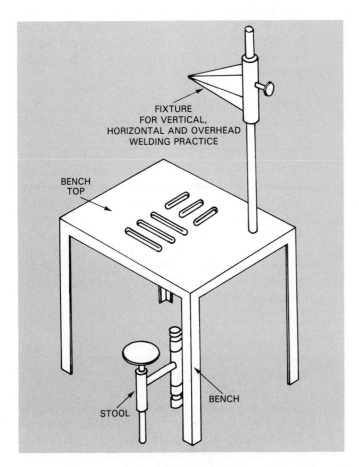

Fig. 4-46. A well planned welding bench or table. Note the facilities for holding work when welding in the overhead position and the built-in stool.

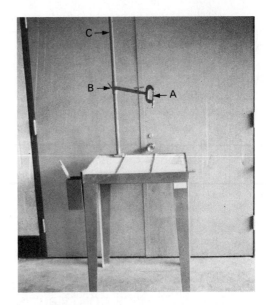

Fig. 4-47. Well designed bench enables a person to practice welding in any position and at various heights. A—Clamp is used for holding welding exercises flat, horizontal, or vertical. B—Height adjustment enables one to position welding exercise at various heights above the bench. C—This fixture may be removed from the bench. (Greene Mfg., Inc.)

are most dangerous are: cadmium, zinc, lead, and beryllium.

Cadmium is widely used as a rust protective plating over steel. It is white in color. The fumes generated by welding, brazing, or cutting cadmium coated metals are very dangerous. If it is necessary to perform welding operations on cadmium coated metals, very thorough ventilation must be provided. Cadmium coatings may usually be identified by the fact that the metal, when gently heated with a torch, will turn to a yellow-gold color.

Zinc is widely used in die-cast metal parts. Galvanized sheet metal is an example of zinc coated metal. It is white in color and melts at a rather low temperature, about 787 °F (419 °C). When alloyed with other metals, this melting temperature may be either raised or lowered. When heated with either a welding torch or an electric arc, zinc gives off a white vapor which is very irritating to the respiratory system. Very thorough ventilation must be used when welding or heating zinc parts or zinc coated materials.

Lead is not as widely used as cadmium or zinc. However, some plating tanks are lead lined. Lead is also used for tanks and pipes for handling certain liquids and gases. Most storage batteries (electric) have lead plates, posts and cell connectors. Some paints may contain lead oxide pigments. Many solders contain lead in varying amounts.

When heating any substances containing lead, thorough ventilation is required. Lead poisoning may occur without the operator being warned in any way of the danger since its fumes are not irritating. The human body does not appear to be able to throw off lead taken into the body either through the lungs or the digestive system. Therefore a person exposed to lead fumes or lead in any form may slowly build up an accumulation of lead until acute lead poisoning develops.

Beryllium has a high modulus of elasticity. It is used in aerospace when metals are needed which are light but stiff. It is also used in the atomic energy field. Its vapors are toxic and excellent ventilation must be provided when beryllium is welded.

## 4-34. WELDING REGULATOR SAFETY

THE REGULATOR IS PERHAPS THE SINGLE MOST IMPORTANT DEVICE CONTRIBUTING TO SAFETY IN OXYFUEL GAS WELDING. If handled properly, its safety qualities will be preserved. See Headings 3-12 and 3-13 for the operation and use of regulators. If the regulator is abused, it may fail in its safe operation. To get both life and safety from a regulator, observe the following rules:

1. Clean the cylinder passages prior to fitting a regulator to a replacement cylinder. This is done by slightly and quickly opening and closing the cylinder valve (cracking the cylinder valve). Aim the fitting away from the operator during this operation.

2. Examine the condition of the threads on both the regulator and the cylinder fittings. The regulator fitting should screw on the cylinder valve easily. Have the fittings repaired rather than use great force to assemble.
   NOTE: FUEL GAS CYLINDERS HAVE LEFT-HAND THREADS. ALL OXYGEN CYLINDERS HAVE RIGHT-HAND THREADS. REGULATORS, HOSES, TORCHES, ETC., HAVE LEFT-HAND THREADS FOR FUEL GAS AND RIGHT-HAND THREADS FOR OXYGEN.

3. Use the proper wrench to tighten the regulator fitting to the cylinder. Be sure that the wrench fits the regulator nut properly. NEVER USE COMMON PLIERS OR PIPE WRENCHES.

4. Be sure that the regulator adjustment is turned all the way out before opening the cylinder valve. In this position no gas will flow through the regulator into the low-pressure side.

5. Open the cylinder valve slowly. A full cylinder, opened quickly, may cause a high enough pressure and a resulting heat of compression to damage the regulator. The temperature at the regulator seat may approach 1000 °F (538 °C) and the seat could fail. If the full cylinder pressure is allowed to enter the regulator too rapidly, the gauge may be damaged or burst. This may cause the gauge glass to blow outward.

   All approved type pressure regulators incorporate a safety pressure disc on the low pressure side. This disc is designed to burst at a pressure between 100 and 200 psig. This is generally below the pressure at which the diaphragm would burst. If the nozzle or seat should leak severely, such a safety device will keep the regulator diaphragm from bursting. However, even with all these safety devices, regulators have exploded!

   If the seat is in poor condition or leaking, the heat of friction along with the heat of compression may cause the regulator seat to melt or decompose. Do not stand in front of a regulator as the cylinder valve is opened, even when opening it very slowly.

6. The pressure to be used in a gauge should always be ONE-HALF to TWO-THIRDS of the maximum calibration of the dial. This means that if the gauge is calibrated up to 300 psig, a 200 psig reading should be the maximum used.

7. Teflon® tape or a paste made of glycerine and litharge should be used for sealing the threads that connect the gauge to the regulator. Do not use pipe compounds containing oil.

8. Never use acetylene at a gauge pressure above 15 psig (103.4 kPa).
9. Never use oil or petroleum based products on oxyfuel gas equipment. Do not allow oil or grease to come in contact with any part of oxyfuel gas welding equipment. Never work with welding equipment when wearing greasy or oily gloves or clothing.
10. Test for leaks ONLY with a soap and water solution.
11. Never interchange oxygen and fuel gas regulators or gauges.
12. Around electrical equipment, insulate the tanks with wool or rubber to prevent ground. A grounded cylinder may allow an electrical arc to form and the resulting heating or burning of the cylinder may cause an explosion.

## 4-35. SAFETY WHEN HANDLING TORCHES

The following are some pointers concerning safety when handling torches. These recommendations concern safety both to the equipment and to the operator:
1. Do not put a cold tip in a hot tip tube since the hot tip tube will produce shrinking action on the tip as it cools.
2. Use only a clean wood surface or a leather surface to clean the end of a tip. Keep the oxygen flowing during this operation to prevent plugging the orifice.
3. Be careful when cleaning a tip with a tip cleaner not to increase the size of the orifice, or to cause it to become out-of-round or tapered.
4. Always extinguish a torch whenever it is not in your hands.
5. The torch hand valves should only be turned with the fingertips.
6. If a torch backfires, find the trouble and remedy it before continuing to use the torch.
7. Each welding station should be provided with a hook upon which to hang the unlighted torch.
8. Be careful that a torch is not directed toward another person while it is being lighted.
9. Be sure no flammable material is near the welding station.

## 4-36. SAFETY IN OXYFUEL GAS WELDING

Oxyfuel gas welding equipment is safe to use if it is properly used. It does, however, possess great potential destructive power if carelessly used. Therefore, it is important that the operator be familiar with all of the potential dangers in the welding processes. Most of the safety hazards of oxyfuel gas welding were pointed out as the operation of the equipment was explained in this Chapter. Welders should know all these precautions, and follow them for their own safety, for the safety of fellow workers, and to protect the equipment. The following precautions are reviewed and elaborated on, in the interest of safety:
1. PROTECTIVE CLOTHING AND SHIELDS
    a. Wear goggles. Various shades are required for various welding applications. When welding or cutting heavier metals, a darker (high numbered) shade of lens is required.
    b. Wear a pair of heavy leather or asbestos gloves.
    c. Wear a leather or asbestos apron or jacket for overhead welding, and for other applications where clothing is in danger of being exposed to sparks.
    d. Remove combustibles from pockets, especially matches.
    e. Avoid wearing trousers with cuffs and open pockets into which sparks may fall. Avoid oily articles.
    f. Use only fireproof materials for the support of articles being welded or cut.
2. WELDING OR CUTTING OF TANKS AND CONTAINERS
    a. Do not weld or cut a tank or container, unless the device has been processed to be safe for such operations. See the American Welding Society recommendations describing procedures to be followed in preparing for welding and/or cutting certain types of containers that have held combustibles.
    b. Remember, many substances which are not usually considered flammable or explosive become vaporized and therefore explosive when heated to a high temperature.
3. HANDLING OXYGEN AND FUEL GAS CYLINDERS AND EQUIPMENT
    a. All equipment used for oxygen such as cylinders, regulators, valves, and torches, must be kept free of oil.
    b. Check for leaks, using soap and water. Never use gas welding equipment with leaks.
    c. Open oxygen and fuel gas cylinder valves slowly. Open oxygen cylinders fully when in use, to eliminate possible leakage around the cylinder valve stem.
    d. Purge oxygen valves, regulators, lines, and torches before use.
    e. Most fabrics, in an atmosphere of oxygen, will burn with explosive force.
    f. Support pressurized cylinders so they cannot tip over. A valve broken from an oxygen cylinder will cause it to become a rocket with tremendous force. A fuel gas cylinder

will behave like a flame thrower, if the cylinder valve is broken off and if the gas is ignited.

g. Stand to one side of oxygen fuel gas regulators when opening the cylinder valve.

h. Always call oxygen "oxygen." It should never be called air. Call acetylene "acetylene," not "gas." Identify the content of cylinders by the name marked on the cylinder. If a cylinder is unnamed, do not use it.

i. Keep cylinders away from exposure to high temperatures. Remember the pressure in any cylinder increases with the temperature.

j. Store full and empty cylinders separately in a well ventilated space.

k. Mark empty cylinders "M T" with chalk as soon as they are taken out of service.

l. If a cylinder leaks around a valve or a fuse plug, tag it to indicate the fault, move it to a safety area, and immediately notify the supplier to pick it up.

m. Keep cylinder caps screwed on all cylinders that are not in use and particularly while they are being moved.

n. Keep the regulators in good repair. Do not use a leaking (creeping) regulator.

4. THE FLAME

a. If the torch is to be laid down, the flame must be turned off.

b. Be sure that no combustible material is in the area in which an oxyacetylene torch is to be lighted.

5. HANDLING THE OXYFUEL GAS WELDING STATION

a. Blow out the cylinder valves before attaching the regulators. Caution: When blowing out fuel gas valves, regulator and hoses, be sure no open flames are near because fuel gases are flammable.

b. Use a well fitting wrench to attach regulators and hoses.

c. Blow out the hose before attaching a torch.

d. Fuel gas hose fittings generally use left hand threads. A groove on the periphery (around the sides) of the acetylene hose nut indicates that it has a left hand thread. Oxygen hose fittings have right hand threads. Do not interchange fittings on hoses.

e. Use a spark lighter for lighting a torch.

f. Keep all hoses away from oil and grease. Examine the hoses regularly for leaks. Leaking fuel gas may cause a severe explosion or fire. If a hose is burned or injured by a flashback, replace it. A flashback usually burns the inner wall of the hose and makes it unsafe to use. Never repair a hose by binding it with tape.

g. The adjusting screw on the regulator must be turned out (counterclockwise, regulator closed) before opening the cylinder valve. Open the cylinder valve slowly. Generally the fuel gas cylinder needs to be opened only 1/4 to 1/2 turn. The wrench should be left on the valve stem to permit quick closure of the valve in an emergency. Adjust oxygen and fuel gas regulator pressures as recommended by the manufacturer of the torch.

h. Never use acetylene pressure in excess of 15 psig (pounds per square inch gauge) or 103.4 kPa (kilopascals).

i. A backfire is caused by an instantaneous extinguishing and reignition of the flame at the torch tip. Usually the trouble will clear itself immediately. If it does not, carefully inspect the equipment, purge the lines and relight. A backfire is usually caused by overheating the tip.

j. A flashback is a burning back of the flame into the tip, torch, or even into the hose. It is characterized by a squealing or sharp hissing sound and by a smoky or sharp pointed flame. In case of a flashback, immediately extinguish the flame by first closing the torch oxygen valve and then the torch fuel gas valve. Also close the cylinders and regulators. Flashbacks indicate something radically wrong with the equipment. Before relighting, purge each line separately, and readjust the regulator pressures to the recommended pressures.

The Occupational Safety and Health Administration (OSHA) has established many compulsory safety requirements for the welding industry. The safety precautions in this text will conform to OSHA requirements.

## 4-37. TEST YOUR KNOWLEDGE OF SAFETY

Write your answers on a separate sheet of paper. Do not write in this book.

1. Why must the regulator adjusting screws be turned out until loose before opening the cylinder valves?

2. Name at least two things which must be done with cylinders to protect them from accidents or harm.

3. The operator should stand to one side of the regulator and gauges when the cylinder valve is turned on. Why?

4. What is the maximum gauge pressure which may ever be set for acetylene?

5. Why is it necessary that a spark lighter or gas economizer flame be used to light the oxyacetylene flame and NOT a match?

6. Why must the oxygen and acetylene regulator, hose, and torch passages be purged before lighting the torch?
7. What is flashback? What must be done immediately if a hissing or squealing sound is heard?
8. Good ventilation is always required when welding. Name three metals which when welded require especially good ventilation.
9. What personal precautions must a welder take when welding in the overhead position?
10. When the welding rod melts down, what procedure should be followed in order to change the position of one's hand on the rod?

## 4-38. TEST YOUR KNOWLEDGE

Write answers to questions on a separate sheet of paper. Do not write in this book.
1. Give a definition of the term "welding."
2. What size torch tip and welding rod is suggested for welding 1/8 in. (3.18 mm) steel?
3. What is meant by undercutting a welded joint?
4. What are the gases produced by the combustion of the oxygen-acetylene flame?
5. What is the approximate temperature of the oxyacetylene flame?
6. Does oxygen burn?
7. Name the four main parts of an oxygen-fuel gas welding station.
8. What should be used when checking an oxy-fuel gas station for leaks?
9. What is a number drill?
10. How are the sizes of welding tip orifices designated?
11. What are two popular types of welding torches?
12. List the steps required for starting an oxy-acetylene station and lighting the torch.
13. What is meant by purging the system?
14. List, in the proper order, the six steps to be followed when shutting down an oxy-acetylene welding station.
15. How does the "puddle" indicate the penetration being obtained?
16. What is the approximate angle between the welding tip and the weld when welding flat stock in the flat position?
17. How is welding rod metal added to a weld?
18. Name three things which may cause a torch to "pop" or backfire.
19. What is the position of the torch in forehand welding? Backhand welding?
20. Name four basic types of welded joints.
21. Name or illustrate four methods used to compensate for shrinkage when welding a butt joint.
22. When welding an inside corner joint, a flame adjustment of a slightly oxidizing nature is used. Why?
23. What are the four recognized welding positions?
24. How does the welding position affect torch position?
25. How may undercutting be prevented on an inside corner joint which is welded in the horizontal position.

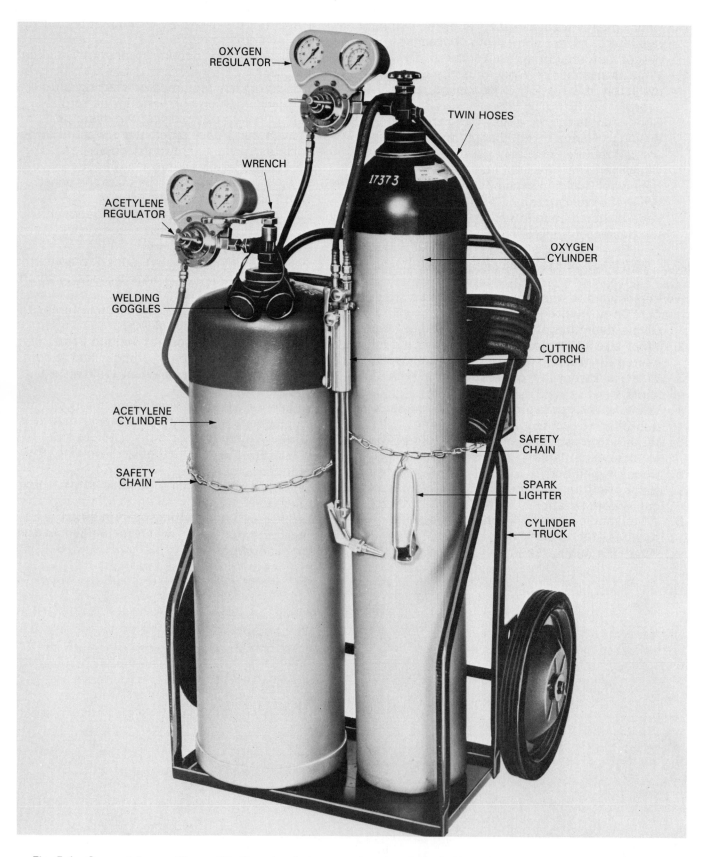

Fig. 5-1. Oxyacetylene cutting outfit. Note the chains around each cylinder holding them securely to the cylinder truck. (Modern Engineering Co., Inc.)

# 5 OXYFUEL GAS CUTTING EQUIPMENT AND SUPPLIES

Oxyfuel gas cutting deals with high pressures, flammable gases, flying sparks, possible rough handling, and other severe services. Therefore, the equipment must be reliable and rugged, and yet quick to respond to changes in demand.

The design and construction of an oxyfuel gas cutting outfit is similar to an oxyfuel gas welding outfit. There is a major difference, however, in the design of the torch and in the oxygen regulator for heavy-duty cutting.

It is very important, for the sake of safety, for quality of results and for economical operation, that the operator become thoroughly familiar with the design and construction of this equipment, and its correct use.

## 5-1. COMPLETE OXYFUEL GAS CUTTING OUTFIT

The complete oxyfuel gas cutting outfit consists of:
Cylinder truck
Oxygen cylinder
Oxygen regulator
Oxygen hose and fittings
Fuel gas cylinder
Fuel gas regulator
Fuel gas hose and fittings
Cutting torch.
This equipment is shown in Fig. 5-1. Each part of the cutting outfit, with the exception of the cylinders, hoses and fittings, will be described in this chapter. The parts will be described in the order in which they are assembled on the complete outfit.

## 5-2. CYLINDER RACK OR CYLINDER TRUCK

The cylinders must be securely fastened in an upright position with a chain or clamping device. If the cutting outfit is portable, the cylinder truck must be designed in such a way that it is resistant to tipping. The cylinders must be securely fastened to the cylinder truck as shown in Fig. 5-1. The possibility of injury to the cylinders must be reduced to a minimum.

## 5-3. REGULATORS FOR OXYFUEL GAS CUTTING

A standard fuel gas regulator is usable for cutting stations. However, the oxygen regulator is quite often a heavy-duty type designed for cutting applications. Since the oxygen cutting pressures sometimes reach as much as 100 to 150 psig and because the gas flow is quite high, the oxygen regulators usually have heavy-duty springs and a high capacity regulator orifice. The two-stage regulator is preferred for cutting applications. The regulator fittings used to attach the regulator to the cylinders are standard and so are the hose fittings. If heavy cutting is to be done it is advisable to check the oxygen regulator by model number and manufacturer's specifications to be sure the correct regulator is being used. See Headings 3-11 to 3-16 for more information regarding regulators.

## 5-4. THE CUTTING TORCH

The cutting torch carries the fuel gas and oxygen in separate tubes to the mixing chamber within the torch head or torch body. In the mixing chamber the two gases combine. The combined gases then travel to the torch preheat orifices (holes) in the tip to produce the preheat flames. The torch must also carry the cutting oxygen to a separate orifice in the tip. As it emerges from the tip it oxidizes the metal and blows it away to form a clean KERF (cut).

The torch is usually constructed from a yellow brass body, stainless steel tubes which carry the gases, and a brass head or tip holder. All these

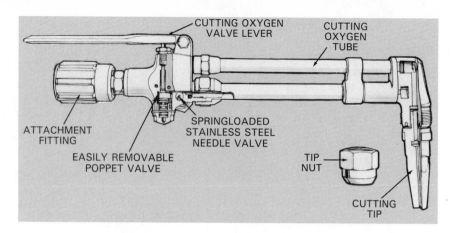

Fig. 5-2.  A cutaway of a cutting torch attachment. This unit attaches to the welding torch body after the welding tip is removed.     (Modern Engineering Co., Inc.)

| NUMBER OF PREHEAT ORIFICES | DEGREE OF PREHEAT | APPLICATION |
|---|---|---|
| 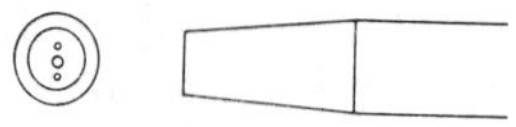 2 | Medium | For straight line or circular cutting of clean plate. |
| 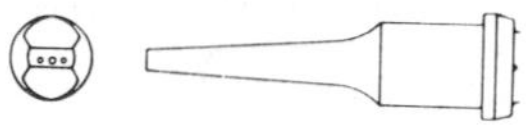 2 | Light | For splitting angle iron, trimming plate and sheet metal cutting. |
| 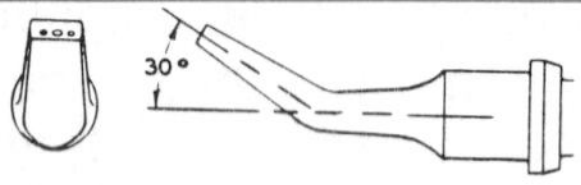 2 | Light | For hand cutting rivet heads and machine cutting 30 deg. bevels. |
| 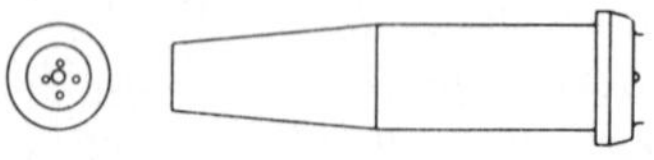 4 | Light | For straight line and shape cutting clean plate. |
| 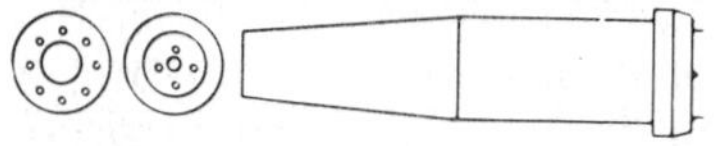 4, 6, 8 | Medium | For rusty or painted surfaces. |
| 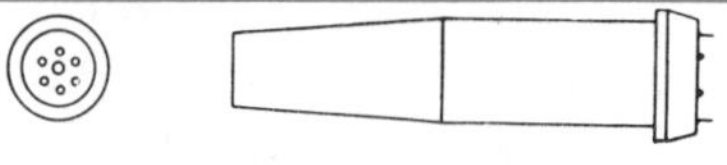 6 | Heavy | For cast iron cutting and preparing welding V's. |
| 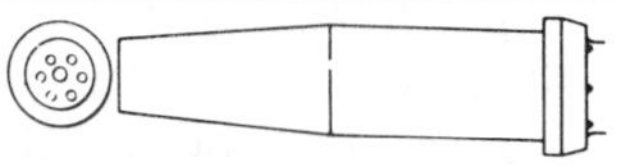 6 | Very Heavy | For general cutting also for cutting cast iron and stainless steel. |
| 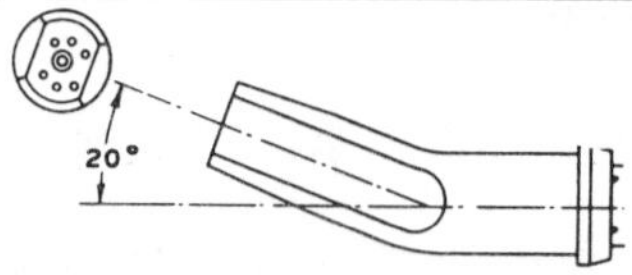 6 | Medium | For grooving, flame machining, gouging and removing imperfect welds. |
| 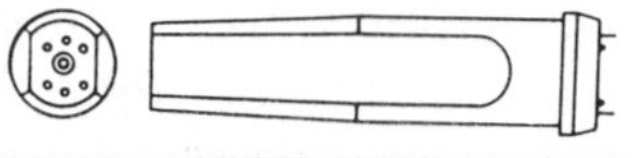 6 | Medium | For grooving, gouging or removing imperfect welds. |
| 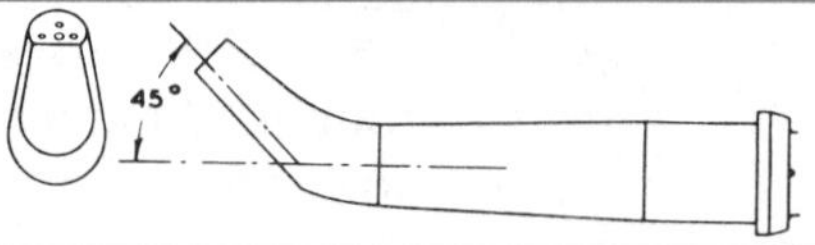 3 | Medium | For machine cutting 45 deg. bevel or hand cutting rivet heads. |
| 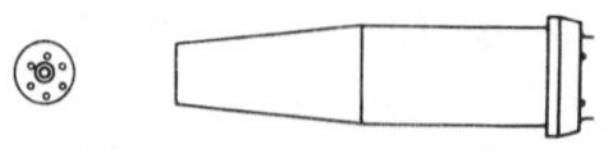 6 | Heavy | Flared cutting orifices provides large oxygen stream of low velocity for rivet head removal (washing). |

Fig. 5-3.  Table of some common cutting torch tips and their uses.

parts are silver soldered together. The copper tip is held into the torch head by a threaded nut. The torch is equipped with three valves:

1. A fuel gas valve similar to the welding torch.
2. An oxygen valve similar to the welding torch.
3. A cutting oxygen valve, button or lever operated, with an automatic spring closing device.

A cutting torch may consist of an attachment which may be fastened to a welding torch body as shown in Fig. 5-2, or it may be a torch designed for cutting operations only, as shown in Fig. 1-4 on page 12.

The combination welding and cutting torch is most popular in small shops where welding and cutting are auxiliary operations to the work in the shop.

## 5-5. CUTTING TORCH TIPS

Just as in welding, the proper size cutting tip is very important if quality work is to be done. The preheat flames must furnish just the right amount of heat, and the oxygen jet orifice must deliver the correct amount of oxygen at just the right pressure and velocity to produce a clean kerf (cut). All of this must be done with a minimum consumption of oxygen and fuel gases. Careless workers or ones not acquainted with the correct procedure may waste both oxygen and fuel gas.

Each manufacturer has cutting tips of different designs. The orifice arrangements and the copper

Fig. 5-4. Cutting tip designs. Notice the various preheat and cutting orifice arrangements and the sizes of the cutting orifices used.

alloy tip material are much the same among various manufacturers. The part of the tip which fits into the torch head, however, often differs in design among manufacturers. Fig. 5-3 is a table showing several different orifice arrangements and their uses. Fig. 5-4 also illustrates several cutting tip orifice arrangements.

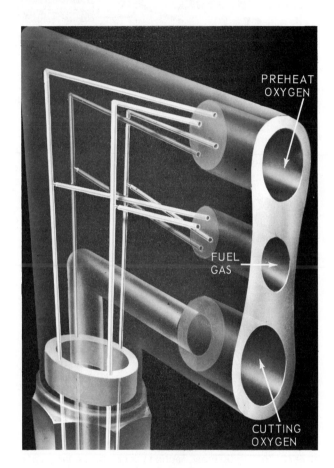

Fig. 5-5. Phanton view of cutting torch head showing gas passageways.

The internal construction of one type cutting tip and head which is designed to mix the preheat gases at the tip is shown in Fig. 5-5. A typical tip design is shown in Fig. 5-6.

Some cutting torches are designed to mix the preheat flame gases in the handle (body). The tips and seats are designed and constructed to produce a good flow of gases, to keep the tips as cool as possible, and to produce leakproof joints. The sealing areas are generally metal to metal between the tip and the torch head. These surfaces must be kept clean and free from damage. If these joints leak, the preheat gases may mix with the cutting oxygen or they may escape to the atmosphere. Fig. 5-7 shows the details of a cutting tip seating design.

If is very important that the orifices and passages be kept clean and free of burrs, to permit

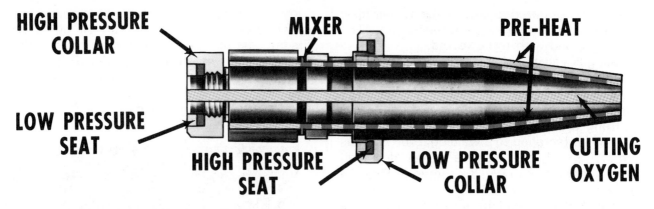

Fig. 5-6. A cutting tip cross section. Also study Fig. 1-4 to see how the gases enter this type of tip.

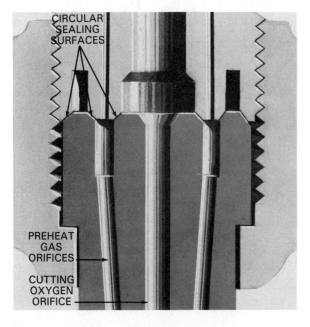

Fig. 5-7. Cutting tip cross section which shows details of the cutting tip seating design. These are metal to metal sealing surfaces.

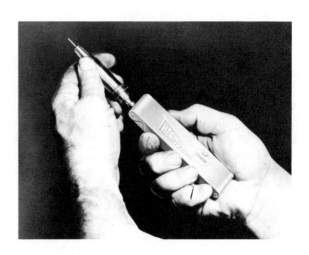

Fig. 5-8. Cutting oxygen orifice being cleaned. A different sized wire broach may be required to clean the preheat orifices. (Thermacote-Welco Co.)

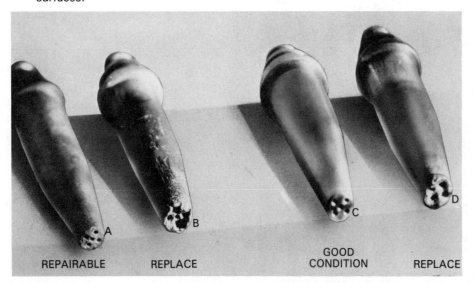

Fig. 5-9 Four cutting tips. Good condition in C. Repairable shown in A. Tips at B and D must be discarded.

free gas flow and to form a well shaped flame. Fig. 5-8 shows a cutting tip orifice being cleaned. Cleaning a cutting tip is done in the same manner as a welding tip. With a cutting tip there may be two or more orifice sizes. Move the tip cleaner straight in and out of the orifice or a bell-mouthed hole may result. Study Fig. 5-9 which illustrates several tips. Fig. 5-9A is in need of repair. Fig. 5-9B and D are beyond repair and must be replaced. Fig. 5-9C is in good condition. Fig. 5-10 shows a tool used to recondition the flame end of a cutting tip. Since it is extremely important that the sealing surfaces be kept clean and free of scratches or burrs, the tips should be stored in a container that cannot scratch the seats, preferably an aluminum or wood rack.

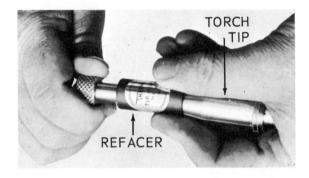

Fig. 5-10. Special tool used to reface the flame end of the cutting tip. This could be used on the tip in Fig. 5-9A to make it flat again. (Thermacote-Welco Co.)

## 5-6. TORCH GUIDES

The operator should always try to cut a smooth kerf and to cut accurately to a dimension. Freehand cutting (holding the torch in your hands) makes both of these objectives very difficult. Many mechanical, electrical and electronic devices have been developed to help produce clean cuts, accurate size cuts, and exact duplicate pieces. Fig. 5-11 shows a simple torch guide which you can easily use.

## 5-7. MECHANICAL GUIDES

Mechanical guides are used to help control the position of the torch. They do not control the speed of the cutting operation. Therefore, the operator must be very skilled; otherwise rough, ragged cuts may result. To cut straight edges, a piece of angle iron may be clamped to the work and the tip moved along the edge of the angle iron to insure a straight cut as shown in Fig. 5-12.

Another popular type guide makes use of a small steel wheel mounted on the torch tip to reduce the friction. The wheel can be used to control the height of the tip above the metal, or it can

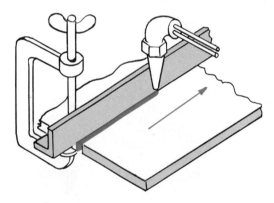

Fig. 5-12. Straightedge used as a cutting torch guide. Use two or more clamps to hold the angle iron.

Fig. 5-11. Aid to cutting straight lines with a torch. Put a 3/4 in. band type clamp on the cutting torch tip, as shown. Fasten the clamp so the end of the band will ride on top of a piece of angle iron clamped onto the stock to be cut. Locate the clamp so that the tip of the torch is the right distance above the metal. to cut in a straight line, hold the torch tip against the angle iron as you cut.

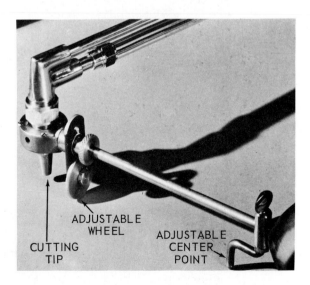

Fig. 5-13. Circle cutting guide. The center of the circle to be cut should be marked with a center punch. (Victor Equip. Co.)

be used to allow the tip to move more smoothly along the guide.

A metal, plaster, or wood template is often used when the cut is to be irregular.

To cut arcs or circles, a circle guide may be used. This device consists of a rod, adjustable in length, with a center pivoting point and a means of holding the tip. Fig. 5-13 illustrates such a device. Fig. 5-14 shows the circle guide in use.

Fig. 5-14. Circle cutting guide in use. Note safety clothing worn by this operator. (Smith Equip. Div., Tescom Corp.)

Guides are also available which may be used to hold a cutting torch and allow the operator to duplicate a pattern of the same size, enlarged, or reduced. With these guides, torches are mounted on a PANTOGRAPH FRAME. See Fig. 5-15. The

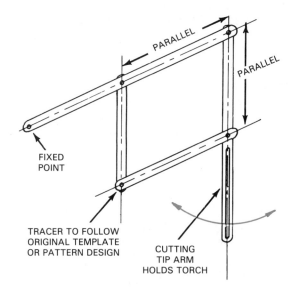

Fig. 5-15. Pantograph cutting torch holder used with a cutting torch.

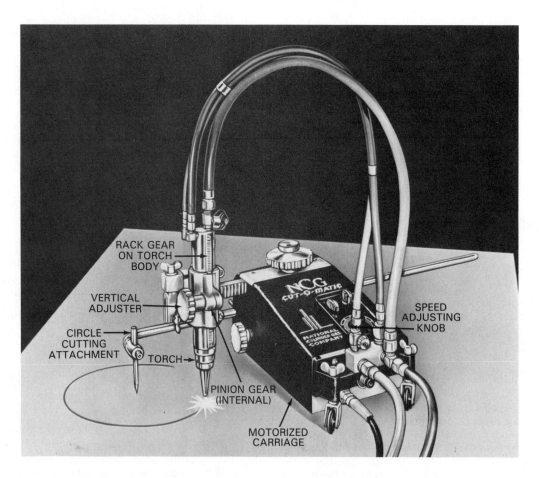

Fig. 5-16. Electric motor driven carriage being used to cut circle in steel plate.

operator moves a tracer around a template, pattern, or sample. The torch moves to produce a duplicate shape in some ratio to the size of the template. For example, the cut piece may be one half the size of the template. Or the pantograph machine can be adjusted to produce a cut piece twice the size of the template.

## 5-8. ELECTRICAL GUIDES

An ELECTRIC MOTOR DRIVEN GUIDE is an improvement over the mechanical guides. This unit has a variable speed motor that enables the operator to cut at any desired set speed. It may be mounted on a track or used with a circle cutting attachment to enable the operator to cut to a dimension. This device usually has four wheels, one driven by a reduction gear, two on swivels (caster style) and one freewheeling. The torch is mounted on the side of the motorized carriage and is adjusted up and down by a rack and pinion gear.

The rack gear is a part of the special torch. The pinion gear is built into the torch holder on the motorized carriage. The torch can also be tilted to cut bevels. The apparatus is obtainable with a straight two-groove track. It also has a radius arm for use when cutting circles and arcs. A motor driven cutting torch guide being used to cut arcs is shown in Fig. 5-16. There is an off-and-on switch, a reversing switch, a clutch, and a speed adjusting dial calibrated in feet per minute or meters per minute.

Fig. 5-17 shows an electrically driven carriage being used on a straight track. The operator must be sure the electric cord and the hose will not become caught on any obstruction during the cutting operation. The best way to check hose and electric cord clearance, and the clearance of the torch, is to freewheel the carriage the full length of the track by hand with the clutch released. This will permit a check of the smoothness of operation.

## 5-9. TRACER DEVICES

In an ELECTRICALLY GUIDED CARRIAGE, a light source is used to follow a trace line on a pattern or drawing. The trace line may be as thin as .040 of an inch.

A pinpoint of light is directed onto the trace line on a template drawing. Some of this light is reflected into a PHOTOELECTRIC CELL which produces a small current. This small current is made larger through an amplifier and signals a SERVOMOTOR (steering motor). The servomotor will cause the steerable wheel on the carriage or a number of specially mounted cutting torches to turn and follow the pattern. See Fig. 5-18. The pattern used may be a drawing or it may be cut from black colored material and mounted on a white background.

Fig. 5-18. An electronic pattern tracer uses a light beam to sense changes in the line direction on a black and white line pattern. The unit electronically controls the movements of the cutting torch to cut the base metal to the shape of the pattern.

The electronic circuit is so designed that the direction the servomotor (steering motor) turns is determined by the intensity of the light reflected from the template surface as shown in Fig. 5-19. When the light source reflects off a dark surface to

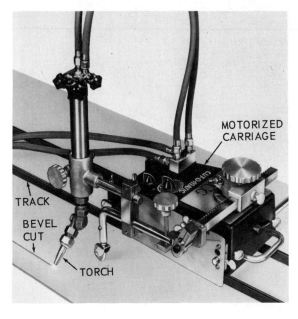

Fig. 5-17. Electrically driven cutting torch carriage being used on a straight track to cut a beveled edge on steel plate.

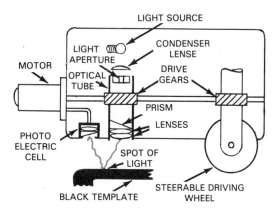

Fig. 5-19. Diagrammatic sketch of an electronic pattern tracer which uses an optical system in the scanning head for controlling torch movements.

the photoelectric cell, the servomotor turns in one direction. When the light is reflected from a white surface, the servomotor turns in the opposite direction. The servomotor does not change direction when the light source reflects equally from a black surface and a white surface, as when the light source is right on the edge of the black template or line.

The light spot which follows the template is aimed to lead the position of the carriage slightly. This gives the servomotor time to correct the position of the steering wheel as changes in the pattern occur.

If the light was following a straight line on the template, the reflected light would be equal from the black surface of the template and the white background surface. The servomotor would not be called to change direction, it would be neutral. When the light source reaches a curve on the template, it leaves the black pattern. The photoelectric cell now picks up a more intense reflected light from the white background. This causes the servomotor to energize and change the position of the steerable wheel to bring the light source back to the edge of the black pattern. As the curve continues or changes, the steerable wheel is constantly bringing the carriage and light source back to the neutral position at the edge of the pattern. A complex or irregular pattern can be followed to a very precise degree with this type of electronic carriage.

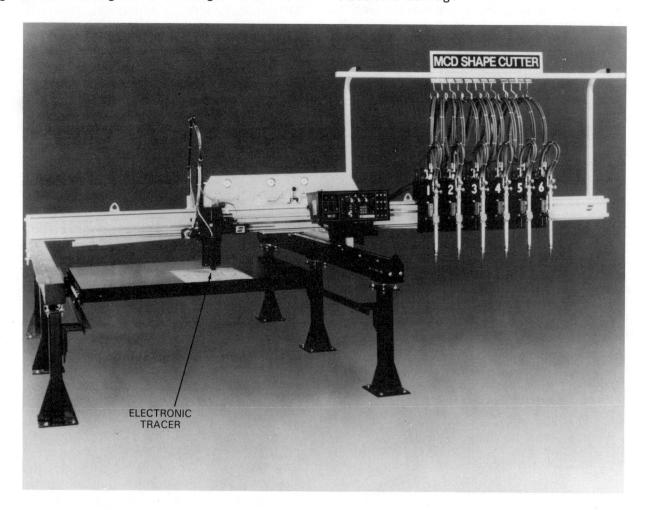

Fig. 5-20. An electronic optical digital tracer used to control the operation of six cutting torches.
(ESAB Automation, Inc.)

Two devices used to guide a cutting torch are the MAGNETIC FOLLOWER and the ELECTRONIC PATTERN TRACER described above. These devices depend on a very sensitive response to the movement of a MAGNETIC STYLUS (tracer roller) or photoelectric cell.

Using the ELECTRONIC PATTERN TRACER, Fig. 5-20, a small current is developed and varies with each change of direction of the pattern tracer. This small current is amplified (made larger) and is used to operate servomotors. The servomotors move a rail which holds the cutting heads so that they duplicate the movement of the pattern tracer. The cutting heads can be at some distance from the electronic pattern tracer. See Fig. 5-20. Two servomotors are used, one for lateral movement and one for longitudinal movement. The pattern tracer unit is either mounted on its own stand or can be set up on the cutting table.

With the MAGNETIC TRACER, the tracer roller is rotated by a small motor. The tracer and cutting torch or torches are mechanically connected. The roller is magnetized to keep the roller in contact with the steel pattern. As the tracer is rotated, it follows the exact shape of the pattern and causes the cutting torch or torches to follow this pattern on the metal being cut. See Figs. 5-21 and 5-22.

A control unit is used to set the speed of rotation of the tracer. As the speed of rotation increases, the speed of movement of the tracer over the pattern increases. The speed of the cutting torch increases also since it moves at the same speed as the tracer.

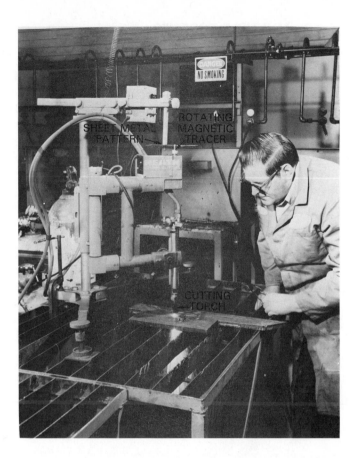

Fig. 5-22. A magnetic tracing machine used to control the movement of a cutting torch. Note the sheet metal pattern used.    (ESAB Automation, Inc.)

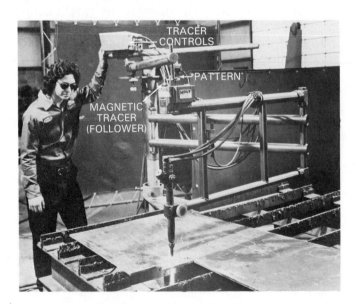

Fig. 5-21. Magnetic tracer which mounts the magnetic pattern follower above the cutting tip. As the tracer rotates and follows the pattern, the cutting torch travels a duplicate path. (ESAB Automation, Inc.)

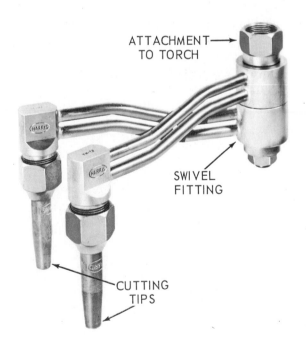

Fig. 5-23. Attachment for manual flame cutting torch which enables two cuts to be made at one time. Swivel fitting permits placing two tips in various positions.
(Harris Calorific Div., The Lincoln Electric Co.)

## 5-10. MULTIPLE TORCHES

There are many cases when it is necessary to make more than one piece of a specially shaped metal part. To do this, two or more torches may be mounted on the cutting machine. Fig. 5-20 shows such an arrangement. The torches have to be carefully mounted to insure that the cuts are accurate. Fig. 5-23 illustrates a device which allows two cuts to be made at the same time using a hand or mechanically operated cutting torch.

## 5-11. SPECIAL CUTTING AND GOUGING TIPS

Many special cutting tips have been designed for special applications. Special tips are made for use with various fuel gases such as acetylene, liquefied petroleum (LP), MAPP, and propane. Tips for cutting rivets or for gouging metal are other examples of special tips. See Fig. 5-3 for a table of special cutting and gouging tips and their suggested uses. Local welding supply companies can provide information regarding special cutting tip sources.

## 5-12. TEST YOUR KNOWLEDGE

Write your answers on a separate sheet of paper. Do not write in this book.

1. How are the fuel gas and oxygen cylinders fastened to the cutting outfit or portable truck?
2. What type of fuel gas regulator is used for a cutting station?
3. The oxygen pressure required in oxyfuel gas cutting can reach _____ to _____ psig.
4. Does the oxygen regulator for a cutting outfit differ from the one used on a welding outfit? How?
5. The fuel gas and oxygen are combined in the _____ _____.
6. What is another name for the slot made in the base metal while cutting?
7. The cutting torch is usually constructed from the following types of metals: a _____ _____ body, _____ _____ tubes which carry the gases, and a _____ head or tip holder.
8. Of what material is a cutting tip made?
9. How is the cutting tip connected to the torch?
10. What are the three valves on a cutting torch?
11. What is a cutting attachment?
12. Are all cutting tips manufactured in the United States constructed the same?
13. If a cutting torch tip has three orifices, what are the names of the orifices?
14. Why is it so important that the orifices and passages of the cutting tip be free of dirt, scratches, or dents?
15. What is the best way to store cutting tips when they are not in use?
16. How may a welder cut a straighter kerf without using a cutting machine?
17. Two types of guides are available to make cutting easier and better. Two types are _____ guides and _____ guides.
18. How does an electronic tracer follow a pattern?
19. When using a multiple torch cutting machine with an electronic tracer, how many patterns must be used to operate all the cutting torches?
20. List three fuel gases that can be used for cutting

# 6 OXYFUEL GAS CUTTING

In oxyfuel gas cutting of metal, an oxyfuel gas flame is used to heat the metal, and an oxygen jet is used to perform the cutting. Oxyfuel gas cutting and oxygen cutting are approved AWS terms.

The art of oxyfuel gas cutting has progressed rapidly. It is now possible to accurately flame cut both very thin, and very thick steel sections. For production work, many layers of metal may be cut at the same time (stack cutting). This process greatly reduces both time and costs.

Oxyfuel gas cutting is particularly useful when shape cutting metal parts. Oxyfuel gas cutting can be done with great accuracy. It leaves the edges smooth enough to satisfy most finished job requirements. Oxygen cutting can be done much faster than machine saw cutting. In many fabrications in shipbuilding, machine frames, and building structures, standard rolled plates or sections may be used. They are cut to size, and then welded together to form a solid steel structure. Such structures known as WELDMENTS are strong, economical to build, and present an attractive appearance.

## 6-1. THE HEAT OF COMBUSTION OF STEEL

Steel may be thought of as a combustible material if sufficiently heated. During the process of burning, it releases a considerable amount of heat measured in British thermal units or joules. This is called the heat of combustion. The burning metal helps maintain the high temperature in the metal which is required in the oxyfuel gas cutting process.

## 6-2. OXYFUEL GAS CUTTING PROCESS

The oxyfuel gas cutting process consists of using one or more oxyfuel gas flames to heat a spot on a piece of steel to a cherry red temperature which is approximately 1300-1400 °F (704-760 °C). The oxyfuel gas flames are adjusted and used in the same manner as when these flames are used for welding. When the spot in the metal reaches the cherry red temperature, the oxygen jet is turned on. As rapidly as the jet action cuts the metal, the torch is moved in the direction the operator wishes the cut to travel. The preheat flames are kept operating during the cutting action. The cutting action may cease unless the heat from the preheat flames provides extra heat. The oxidation action alone will not usually supply enough heat to permit cutting to continue. As the cutting torch is moved along the line to be cut, a kerf (slot) is created behind the tip.

## 6-3. CUTTING OUTFIT

An outfit used for manual oxyfuel gas cutting is similar to the oxyacetylene welding outfit shown in Fig. 6-1. They differ only in the torch and possibly the oxygen regulator used. The term cutting outfit is used to include all equipment required to perform a cut. A cutting station would include the outfit, lighting, ventilation, a cutting table, and possibly a booth.

Since the cutting torch must provide an oxygen cutting jet, it is quite different than a welding torch.

Because the oxygen pressures for cutting are usually higher than the pressures used when welding, an oxygen regulator with a higher working pressure should be used. Also, use a heavier duty oxygen hose.

Carefully review Chapters 3 and 5 before connecting and operating the oxyfuel gas cutting outfit. Information concerning cylinders, manifolds, regulators, hoses, torches, and tips applies to the oxyfuel gas cutting outfit.

## 6-4. CUTTING TORCH

A cutting torch is similar to a welding torch but, in addition, has a separate passageway for the

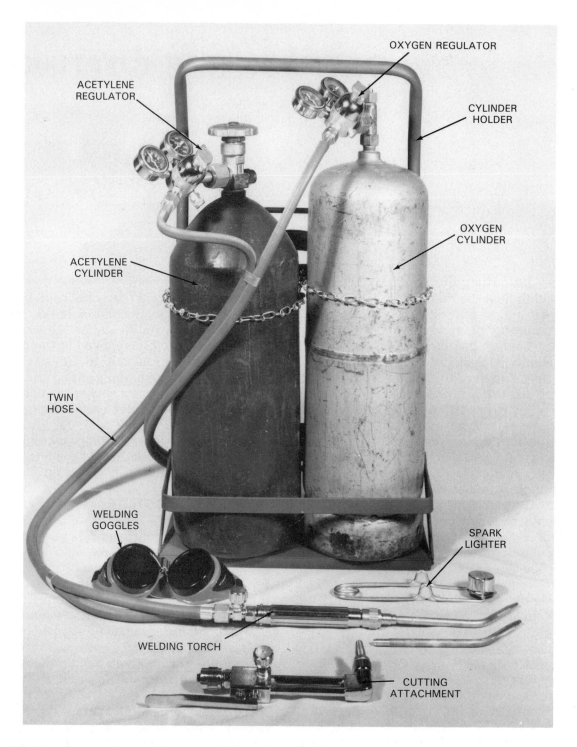

Fig. 6-1. Oxyacetylene welding and cutting outfit. Note the small cylinders used on this portable unit. (Goss, Inc.)

oxygen jet. Review Chapter 5 for a better description of the oxyfuel gas cutting torch.

This chapter will deal with only the procedure for oxyfuel gas cutting. Also see Chapter 22 for other methods of cutting.

In an oxyfuel gas cutting torch the heating flame comes from one or more orifices arranged around a center oxygen orifice. Figs. 6-2 and 6-3 show two different designs of cutting torches. The operator controls the cutting operation through the use of a CUTTING OXYGEN LEVER. In operation, a preheating flame is maintained at the tip through small orifices arranged around the cutting oxygen orifice. Cuts are made by depressing the cutting oxygen lever on the torch which controls the flow of oxygen from the center orifice as shown in Fig. 6-4. As in welding, the cutting torch is connected to oxygen and fuel gas cylinders. See

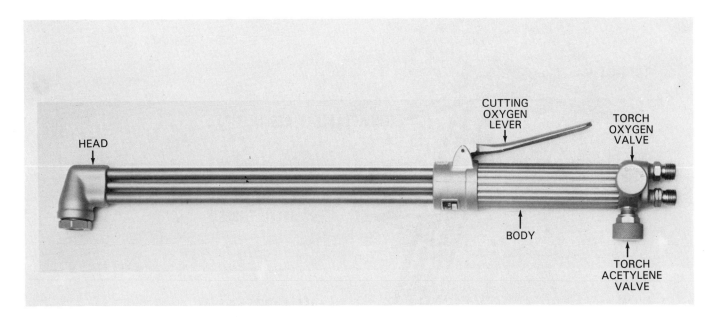

Fig. 6-2. Cutting torch with three tubes. The preheat flame mixing chamber is in the torch head. Note the cutting tip is not installed. (Air Products & Chemicals, Inc.)

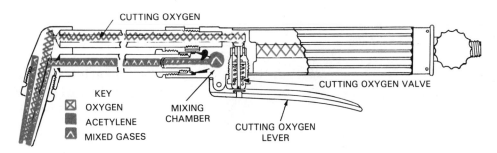

Fig. 6-3. Schematic cross sectional drawing of cutting torch with preheat mixing chamber located in the torch body. (L-TEC Welding & Cutting Systems)

Headings 4-7 and 4-8.

Two different types of flame cutting torches are in use:
1. Equal Pressure Torch.
2. Injector Type Torch.

## 6-5. LIGHTING THE EQUAL PRESSURE TYPE OXYACETYLENE CUTTING TORCH

To turn on the outfit, purge the system, and light the oxyacetylene equal pressure type hand cutting torch, proceed as follows:
1. Check the equipment for condition.
2. Inspect the regulators. Turn adjusting screws all the way out (regulator closed).
3. Open the oxygen cylinder valve very slowly until the regulator high pressure gauge reaches its maximum reading. Then, turn the cylinder valve all the way open to close the double seating valve. While doing this, the operator should stand to one side of the gauges.
4. Open the acetylene cylinder valve slowly 1/4 to 1/2 turn. Leave the acetylene cylinder valve wrench in place so the cylinder valve may be shut off quickly if necessary.
5. Open the torch oxygen valve one turn. With this valve open, next open the oxygen cutting valve and adjust the oxygen regulator to give the desired operating cutting pressure. If the oxygen cutting valve is not opened while adjusting the oxygen pressure, a drop in oxygen pressure will occur when this valve is opened during the cutting operation. This will result in a reduced preheat flame, and possibly, a poor cut. Close the torch oxygen valve and the oxygen cutting valve. See the table in Fig. 6-5 for oxygen and acetylene cutting pressure.
6. Open acetylene torch valve one turn. Slowly turn in acetylene regulator adjusting screw until the low pressure acetylene gauge indicates pressure corresponding to tip size. Refer to Fig. 6-5. Close torch acetylene valve. Regulator pressures are now adjusted. Fig. 6-6 shows a single stage oxygen regulator.

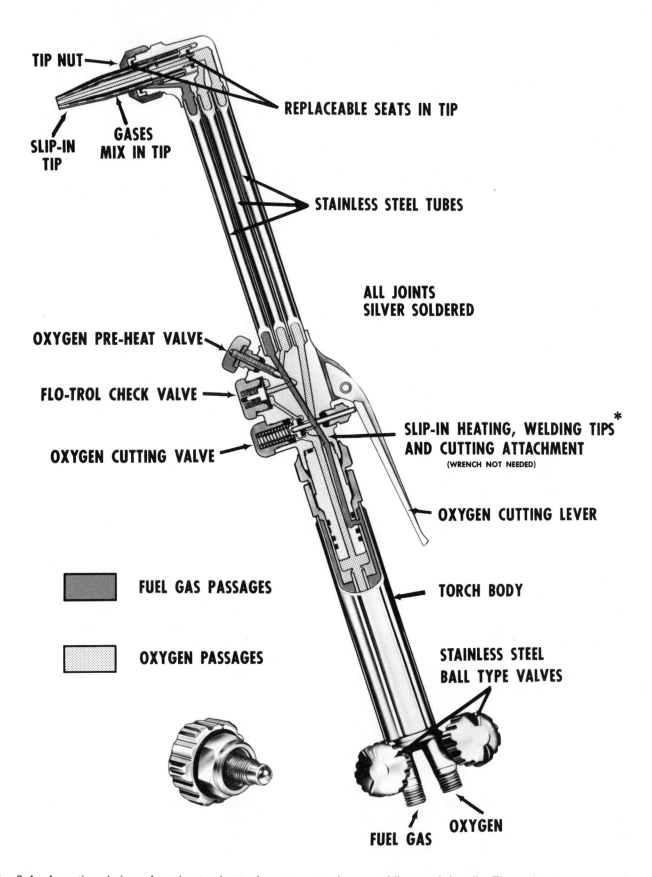

TIP NUT

SLIP-IN TIP

GASES MIX IN TIP

REPLACEABLE SEATS IN TIP

STAINLESS STEEL TUBES

ALL JOINTS SILVER SOLDERED

OXYGEN PRE-HEAT VALVE

FLO-TROL CHECK VALVE

OXYGEN CUTTING VALVE

SLIP-IN HEATING, WELDING TIPS *
AND CUTTING ATTACHMENT
(WRENCH NOT NEEDED)

OXYGEN CUTTING LEVER

FUEL GAS PASSAGES

OXYGEN PASSAGES

TORCH BODY

STAINLESS STEEL BALL TYPE VALVES

FUEL GAS

OXYGEN

Fig. 6-4. A sectioned view of cutting torch attachment mounted on a welding torch handle. The preheat gases are mixed in the tip of the torch where it connects into the torch head. (Smith Equipment, Div. of Tescom Corp.)

| Metal Thickness | | Preheat Orifice Drill Size | Cutting Orifice Drill Size | Oxygen Pressure | | Acetylene Pressure | | Speed | |
|---|---|---|---|---|---|---|---|---|---|
| in. | mm. | | | psig | kPa | psig | kPa | in/min. | mm/min. |
| 1/8-3/8 | 3.2- 9.5 | 70 | 67 | 20-30 | 138-207 | 3 | 21 | 14-18 | 356-457 |
| 3/8-3/4 | 9.5-19.1 | 58 | 62 | 30-40 | 207-276 | 5 | 34 | 12-15 | 305-381 |
| 3/4-1 1/2 | 19.1-38.1 | 57 | 54 | 40-45 | 276-310 | 5 | 34 | 10-12 | 254-305 |
| 1 1/2-2 | 38.1-50.8 | 68 | 51 | 45-50 | 310-345 | 5 | 34 | 9-10 | 229-254 |

Fig. 6-5. Table showing approximate oxygen and acetylene pressures used when cutting steel sheet and plate with the equal pressure type cutting torch. Most cutting tip manufacturers will recommend a range of at least three sizes for any cutting condition. One end of range gives most economy, other end maximum speed. For the learner, the middle size is usually the best. Table is for torch using tip which provides four preheat orifices. For heavier or faster cutting a tip which provides 6-8-10-12 preheat orifices may be used.

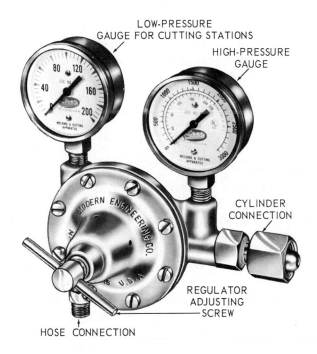

Fig. 6-6. Single stage nozzle type oxygen regulator. The low pressure gauge is calibrated up to 200 psig indicating this regulator is used in cutting operations.
(Modern Engineering Co., Inc.)

7. To light the torch, open the torch acetylene valve approximately 1/16 turn. Then, use a flint lighter to ignite the acetylene.

8. Open the torch acetylene valve until the acetylene flames jump away from the end of the tip slightly and back again when the torch is given a shake or whipping action. An alternate method of adjusting the acetylene, after the torch is lighted, is to turn on the acetylene until most of the smoke clears from the flame. See Figs. 1-5 and 1-6 for illustrations of various conditions of the flame adjustment.

9. Now open the torch oxygen valve, and adjust it to obtain a NEUTRAL FLAME. Open the cutting oxygen valve, and readjust the preheat flame if necessary. The neutral flame may be altered when the cutting oxygen valve is opened. The torch is now adjusted and is ready to be used as a cutting torch.

## 6-6. LIGHTING INJECTOR TYPE OXYACETYLENE CUTTING TORCH

Fig. 6-7 shows typical gas flow through injector type oxyacetylene cutting torch. The following procedure is used to turn on the oxyacetylene cutting outfit, purge the system, and light the torch:

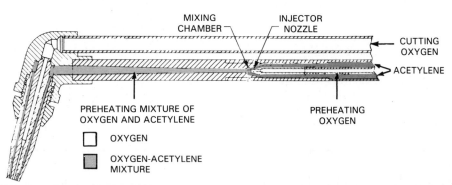

Fig. 6-7. Cross section of an injection type cutting torch. The acetylene is induced into the mixing chambers by the pulling action (suction) of the oxygen jet. This injector type cutting torch is particularly adaptable for use with acetylene generators which operate under low pressure.

| Metal Thickness | | Preheat Orifice Drill Size | Cutting Orifice Drill Size | Oxygen Regulator Pressure | | Acetylene Regulator Pressure | |
|---|---|---|---|---|---|---|---|
| inches | millimeters | | | psig | kPa | psig | kPa |
| 1/8-1/4 | 3.2- 6.4 | 75 | 67 | 15-20 | 103-138 | 1 | 7 |
| 1/4-3/8 | 6.4- 9.5 | 74 | 62 | 20-25 | 138-172 | 1 | 7 |
| 3/8-1/2 | 9.5-12.7 | 72 | 59 | 25-30 | 172-207 | 1 | 7 |
| 1/2-3/4 | 12.7-19.1 | 71 | 55 | 30-35 | 207-241 | 1 | 7 |
| 3/4-1 | 19.1-25.4 | 70 | 54 | 35-40 | 241-276 | 1 | 7 |
| 1 1/2-2 | 38.1-50.8 | 68 | 51 | 45-50 | 310-345 | 1 | 7 |

Fig. 6-8. Table of the approximate oxygen and acetylene pressures used when cutting steel with injector type cutting torch.

1. Check the equipment to make sure all parts are in good operating condition.
2. Inspect the regulators. The adjusting screws of the regulators should be turned all the way out (regulator closed).
3. Open the oxygen cylinder valve very slowly until the regulator high-pressure gauge reaches its maximum reading. Then turn the valve all the way open.
4. Using an acetylene cylinder wrench, slowly open the acetylene cylinder valve 1/4 to 1/2 turn. Leave the wrench in place on the acetylene cylinder valve.
5. Open the torch oxygen valve 1/4 turn. Open the torch oxygen cutting orifice lever wide open. Adjust the oxygen regulator screw to give the correct delivery pressure, as shown in the table in Fig. 6-8. Close the torch oxygen valves.
6. Open the torch acetylene valve fully. Adjust the acetylene regulator delivery pressure to give the pressure required in the table in Fig. 6-8. Close the torch acetylene valve.
7. The pressures are now adjusted, and the cutting torch is ready to be lighted.

8. To light the torch, open the torch oxygen valve 1/4 turn. Open the torch acetylene valve fully. Use a spark lighter to ignite the fuel gas.
9. Press down the oxygen cutting lever and adjust the torch acetylene valve until the preheating flames are neutral.
10. The torch is now ready for oxyacetylene cutting operations.

## 6-7. USING A CUTTING TORCH

The cutting torch must be carefully used if cuts are to be clean and accurate. The tip must be in excellent condition, the preheat flames must be correctly adjusted, and the cutting oxygen pressure must be correct.

To cut, bring the tip of inner cone of the preheating flames to the edge of the metal to be cut. The cutting torch should be held so that the inner cone on the preheat flames are about 1/16 to 1/8 in. (1.6-3.2 mm) from the surface of the metal being cut, as shown in Fig. 6-9.

As soon as the surface of the metal has been heated to a cherry red or white heat color, open

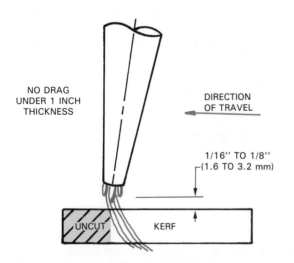
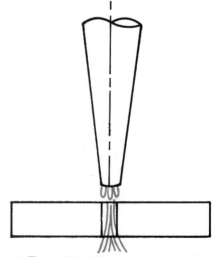

Fig. 6-9. The cutting torch position for cutting with the oxyfuel gas cutting torch. The tip is held perpendicular or inclined slightly in the direction of travel.

the oxygen cutting valve all the way. The jet of oxygen coming through the center of the tip (oxygen jet) will cause the heated metal to burn (oxidize) away, forming the kerf (cut).

One of the best indications of a good cutting operation is the appearance of the slag stream at the bottom of the cut. The following may be noted:

1. The ideal slag stream passes directly through a plate which is less than one inch thick. For economic use of oxygen, when cutting thicker steel sections, some drag is desirable. (DRAG is a measurement, made in the direction of travel, between the entry and exit points of the cutting jet.)
2. If the slag stream lags excessively behind the torch tip travel:
   a. The flame adjustment may be incorrect.
   b. The cutting oxygen pressure adjustment may be too low.
   c. The tip travel is too fast and the metal is not being preheated enough.

## 6-8. CUTTING ATTACHMENTS

Most manufacturers of oxyfuel gas welding and cutting torches market a cutting attachment. The attachment is connected to the welding torch to change it into a cutting torch. The cost of such an attachment and the welding torch is usually less than the cost of a separate welding torch and a cutting torch. For portable kits, such an attachment saves space. To connect a cutting attachment, it is only necessary to remove the welding tip tube, and screw on the cutting attachment.

The operation of the torch with the cutting attachment is the same as the operation of a regular cutting torch. Figs. 5-2 and 6-10 illustrate a cutting attachment. There may be two oxygen and fuel gas torch valves when using a cutting attachment: one set on the torch body and one set on the cutting attachment.

To make flame adjustments the torch valves on the cutting attachment should be opened one or more turns. The torch valves on the torch body are then used to adjust the flames.

Caution: The torch valves on the torch body must be turned off before the cutting attachment is disconnected.

## 6-9. CUTTING TIPS

The cutting tip will normally have at least two orifices. One orifice, which is usually in the center

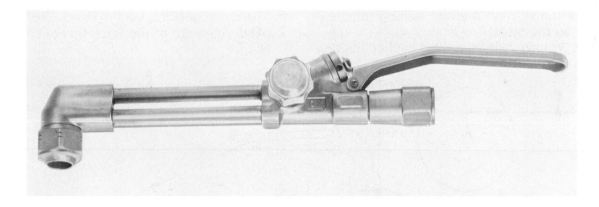

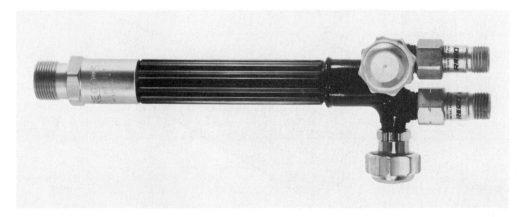

Fig. 6-10. Cutting attachment (top) which is used to convert an oxyacetylene welding torch (bottom) into a cutting torch.

of the tip, is for the cutting oxygen and one or more smaller orifices are for preheating the metal to be cut, as shown in Fig. 6-11 (upper left corner), also Figs. 5-3 and 5-4.

Cutting tips may be of one or two piece construction. One piece tips are used for oxyacetylene cutting only. Two piece tips are used for all other gas cutting: natural gas, propane, MAPP, etc.

For satisfactory service, tips must be kept in good condition. The orifice end must be clean. The surface must be at right angles to the orifices so the preheat flame is shaped and aimed properly. The sealing faces of the tip, (which fastens into the torch body) must be clean and free from scratches, burrs, nicks, etc., or these joints may leak.

Clean the tips as described in Heading 5-5.

Always store the extra tips in soft holders. A wooden block with holes drilled in it works well.

See Chapter 5 for more detailed information.

## 6-10. CUTTING STEEL WITH OXYFUEL GAS CUTTING TORCH

Metals which may be cut with the oxyfuel gas cutting torch may be divided into two classes:
1. Metals whose oxides (compound of oxygen and the metal) have a lower melting temperature than the metal.
2. Metals whose oxides have a higher melting

temperature than the metal.

Practically all steels fall under the first classification and, therefore, cutting presents little difficulty. When the cutting jet is turned on, the iron oxides which form melt at a lower temperature than the base metal. These are easily blown away by the cutting oxygen jet, leaving a clean and straight cut. Fig. 6-11 shows a cut in progress. With an expert handling a cutting torch, or in automatic machine cutting, the kerf formed during the cut has a smoothness of machine-like quality.

The second group includes cast iron, some alloy steels, stainless steel, and nonferrous metals. These metals present a complication in cutting because the oxide has a higher melting temperature than the metal. It is almost impossible to cut an even kerf. It is very important that these oxides, called REFRACTORY OXIDES, be reduced by chemical action or be prevented from forming. See Chapter 22 for special cutting processes used with refractory oxides.

Items of importance to be watched in cutting are:
1. Pressure of the oxygen fed to the cut.
2. Size of the oxygen jet orifice.
3. Speed of the cutting torch across the metal.
4. Distance of the preheat flame from the metal.
5. Size of the preheat flames or the amount of heat delivered to the base metal.
6. Torch tip position (angle) relative to the metal.
7. The alignment of the torch tip orifices with the kerf.

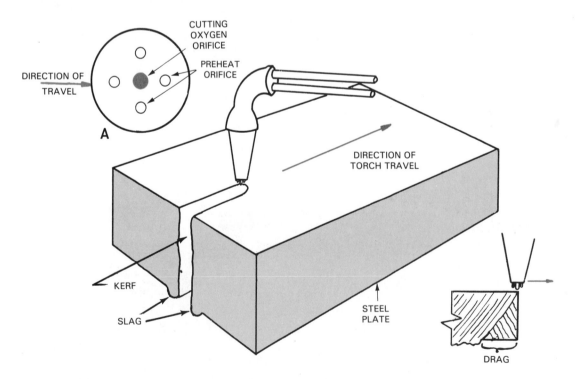

Fig. 6-11. An oxyfuel gas cutting torch being used to cut a steel plate. Detail A shows an end view of a typical cutting tip with a center oxygen orifice and four preheat orifices.

It should be noted that the oxygen pressure will determine the velocity of the oxygen jet. The orifice size will determine the amount of oxygen delivered in CFH (cubic feet per hour) or L/min (liters per minute) at any particular pressure.

The cut should proceed just fast enough to provide a slight amount of drag at the line of cutting. If the drag is too small, the oxygen consumption is too great. If the drag is large, the cutting tip orifices may be too small for the job.

Fig. 6-12 shows the result if too much oxygen is fed to the steel being cut. The cut widens out as the jet penetrates the thickness of the metal. This leaves a bell-mouthed kerf on the side of the metal away from the torch.

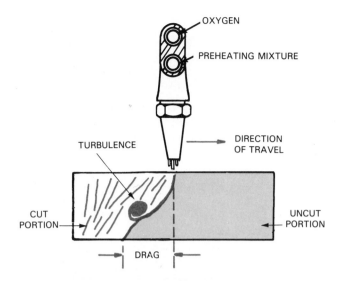

Fig. 6-13. The effect of moving a cutting torch too rapidly across the work. The drag becomes too large.

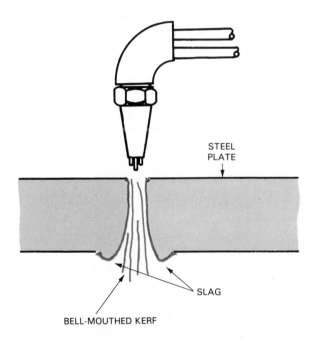

Fig. 6-12. The effect of using too much oxygen when cutting steel. Note how the kerf widens at the bottom of the plate to create a bell-mouthed kerf.

Fig. 6-13 shows the result if the torch is moved too rapidly across the work. When the torch is moved too rapidly, the metal at the bottom, or far side of the cut, will not be burned away. This is because it does not receive enough heating and oxygen to complete the cut. The large drag results in a turbulent action of the torch gases which will leave a very rough and irregular shaped kerf.

If the torch is moved too slowly across the work, the preheated metal will be completely burned away, and the preheating flame wasted. If the metal preheat temperature is lost, the oxygen to the cutting orifice should be closed off. The metal should again be preheated to the proper temperature by the preheating flames. When a cherry red color is obtained, the cutting oxygen should be turned on and the cut continued. These

starting and stopping actions may also cause an irregular cut.

Metal which is very dirty and rusty should be cleaned before starting the cutting operation. The impurities on the metal will slow the cutting speed, and may cause a rough and irregular kerf. Fig. 6-14 shows a number of completed cuts. The cause of each poor cut is explained in the captions.

The torch motion to be used in cutting is a matter of the operator's own experience. Usually no motion is used. In some cases the thickness of the metal requires an oscillating (side to side) motion in order to obtain the necessary width of cut.

The operator, when cutting, should stand in a comfortable position, where he or she can look into the cut as it is being formed. The torch movement should be away from the operator, rather than toward the operator, in order to see into the kerf. The cut may be made from right to left with a good view of the kerf. The torch is usually held with both hands for best control.

Normally the tip is perpendicular to the surface being cut. The end of the preheating flame inner cone should be held just above the metal. If the cutting tip has 4, 6, or more preheating orifices, one orifice should precede (lead) the cutting orifice. One orifice should follow the cut. The other two, four, or more orifices should be aligned to heat each side of the kerf equally. See Fig. 6-15.

The operator should wear leggings, safety boots with high tops, and trousers without cuffs. (Cuffs of trousers should be covered to keep them from catching the white hot metal slag as it drops from the cut.) A container should be placed under the cut, to catch the very hot liquid slag. It is preferred that this container be lined with a heat resistant (refractory) material.

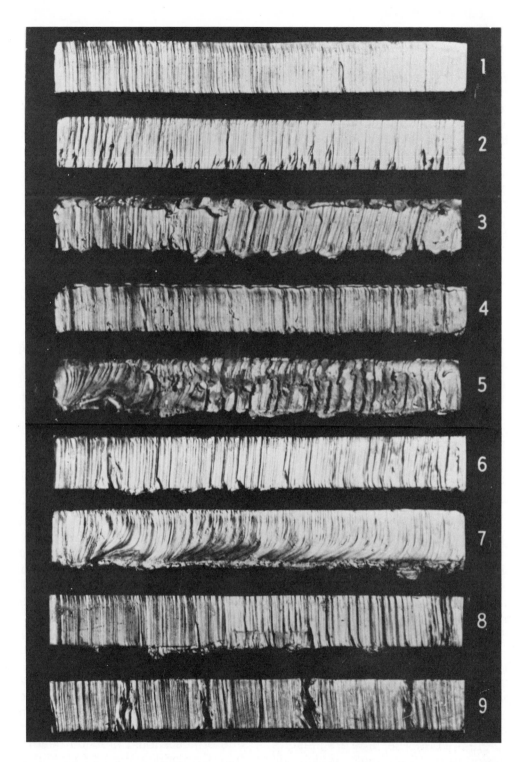

Fig. 6-14. Typical edge conditions resulting from oxyfuel gas cutting operations: (1) good cut in 1 in. (25 mm) plate—the edge is square, and the drag lines are essentially vertical and not too pronounced; (2) preheat flames were too small for this cut, and the cutting speed was too slow, causing bad gouging at the bottom; (3) preheating flames were too long, with the result that the top surface melted over, the cut edge is irregular, and there is an excessive amount of adhering slag; (4) oxygen pressure was too low, with the result that the top edge melted over because of the slow cutting speed; (5) oxygen pressure was too high and the nozzle size too small, with the result that control of the cut was lost; (6) cutting speed was too slow, with the result that the ir-regularities of the drag lines are emphasized; (7) cutting speed was too fast, with the result that there is a pronounced break in the dragline, and the cut edge is irregular; (8) torch travel was unsteady, with the result that the cut edge is wavy and irregular; (9) cut was lost and not carefully restarted, causing bad gouges at the restarting point.

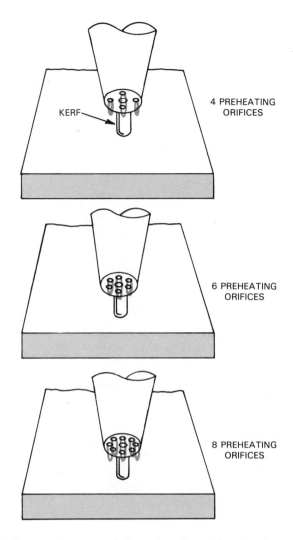

4 PREHEATING ORIFICES

KERF

6 PREHEATING ORIFICES

8 PREHEATING ORIFICES

Fig. 6-15. Proper alignment of the preheating orifices on the cutting tip. Notice one orifice in each case leads the oxygen jet and one follows.

## 6-11. CUTTING THIN STEEL

Cutting steel of 1/8 in. (3.18 mm) thickness or less requires the use of the smallest cutting tip available. A tip with few preheat holes is often used. In addition, the tip is usually pointed in the direction the torch is traveling. If even a small tip size seems too large, change the tip angle to 15°-20° as shown in Fig. 6-16. On very thin metal, holding the tip near vertical will produce too much preheating. The resulting cut will be very

poor. Many welders actually rest the edge of the tip on the metal during this process. Be careful to keep the end of the preheating inner cone just above the metal.

## 6-12. CUTTING THICK STEEL

Steel over 1/2 in. (13 mm) thickness should be cut by holding the torch so the tip is perpendicular to the surface of the base metal being cut. Fig. 6-17 shows the position of the cutting torch tip orifice when cutting thick steel.

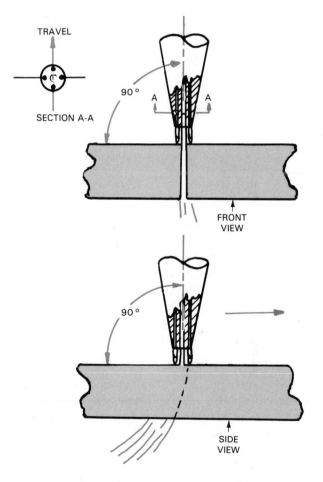

TRAVEL

SECTION A-A

90°

A    A

FRONT VIEW

90°

SIDE VIEW

Fig. 6-17. A recommended technique for cutting thick steel. Note the position of the torch tip preheat orifices in relation to the line of the cut (kerf). Two preheat flames are in the line of the torch progress. This position enables one preheat flame to be ahead of the cut, two flames to heat the sides of the cut, and one flame to heat down in the kerf.

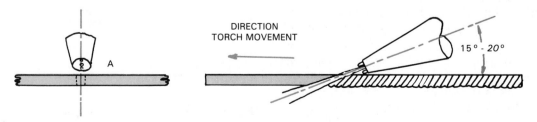

A

DIRECTION TORCH MOVEMENT

15° - 20°

Fig. 6-16. A recommended procedure for cutting thin steel. Notice that the two preheat flames are in line with the cut (kerf).

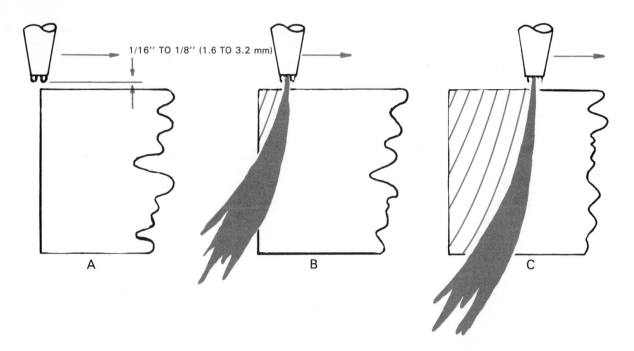

1/16″ TO 1/8″ (1.6 TO 3.2 mm)

A    B    C

Fig. 6-18. Progress of a cut in thick steel. A—Preheat flames are 1/16 to 1/8 in. (1.6 to 3.2 mm) from the metal surface. The torch is held in this spot until the metal becomes cherry red. B—Note that the torch is moved slowly to maintain the rapid oxidation, even though the cut is only part way through the metal. C—Note that as the cut is made through the entire thickness, the bottom of the kerf lags behind the top edge slightly.

The cut is normally started at the edge of the stock. The torch may be moved from left to right, or right to left. Either direction is good if it permits the operator to look into the kerf and check cutting progress. If a line of travel is chalked on the metal, torch movement from right to left enables a right-handed operator to follow the guide lines on the metal best. Fig. 6-18 shows the progress of a cut in thick steel.

After the edge has been heated to a dull cherry red, the oxygen jet should be opened all the way, by pressing on the cutting lever. As soon as the cutting action starts, move the torch tip at a steady rate. Avoid an unsteady movement of the torch, or the cut will be irregular and the cutting action may stop.

To start a cut faster in thick plate, the operator may start at the corner of the metal by slanting the torch in a direction opposite the direction of travel, as shown in Fig. 6-19. As the corner is cut, the operator moves the torch to a vertical position, until finally the total thickness is cut, also shown in Fig. 6-19, and the cut may proceed.

Two other methods used to start cuts are described. One is to nick the edge of the metal, where the cut is to start, with a cold chisel. The sharp edges of the metal upset by the chisel will preheat and oxidize rapidly under the cutting torch. This makes it possible to start without preheating the entire edge of a thick plate.

The second method is to place an iron filler rod under the preheating flames at the edge of a thick plate. The filler rod will reach the cherry red temperature quickly and when the cutting oxygen is turned on the rod will oxidize and cause the thicker plate to start oxidizing. Fig. 6-20 shows a welder cutting a thick section.

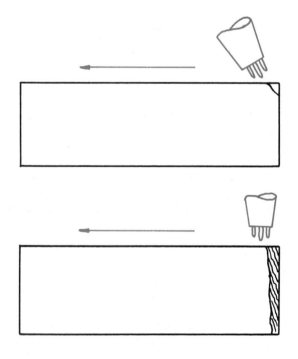

Fig. 6-19. The correct oxyfuel gas cutting torch position for starting cut; also the required change in the torch position as the cut progresses.

Fig. 6-20. A cut in progress on thick steel.

## 6-13. CUTTING CHAMFERS (BEVELS)

Another important torch cutting operation is the cutting of chamfers for bevel and V-joints on the edges of steel plate prior to welding, as shown in Fig. 6-21. Thicker pieces of steel must have the chamfered edge preparation so the weld will penetrate through the thickness of the metal.

Bevel angles may be cut at the same time the metal is being cut to size and shape. The chamfer may be cut as a separate operation.

Instructions relative to cutting thick metal, given previously, are usable when cutting bevel angles.

It is important to obtain a high quality cut when making the bevel. This will minimize any further plate preparation and insure a good fit up of the joint to be welded.

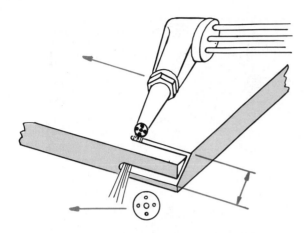

Fig. 6-21. A method of obtaining a beveled edge on thick metal plate using a cutting torch. Note the alignment of the cutting tip holes. Two are in line with the kerf and one on each side.

## 6-14. CUTTING PIPE OR TUBING

One of the most popular uses of the oxyfuel gas cutting torch is the preparation of pipe for joining by welding. See Chapter 21 for more pipe welding instructions. The cutting torch is especially useful for preparing odd-shaped joints on the job. Also, it may be used extensively to chamfer the edges of thick pipe to provide a V-joint for the welder.

The exact procedure to follow when cutting pipe depends on the diameter. For small diameter pipe, it is best to keep the tip almost tangent to the circumference of the pipe. (Note a similar approach to cutting very thin metal in Fig. 6-16.) This will prevent cutting through both sides of the pipe at once. A poor cut usually results when attempting to cut through the two thicknesses of the pipe simultaneously.

With a pipe diameter of approximately four inches and larger, it is possible to keep the torch tip perpendicular to the pipe surfaces while cutting without burning through the other side. Of the two methods, the perpendicular position permits a cleaner and straighter cut. If the welder's helper rotates the pipe as it is being cut, a very clean cut can be obtained. Most welders start the cut at the extreme edge of the pipe, and then cut back to the marked chamfer ring.

When chamfering pipe by hand, it is best to point the torch toward the end of the pipe. This procedure provides a clean chamfer, permits more accurate cutting, and should not leave any excessive oxide clinging to the pipe when the cut is completed. Cutting machines which use a mechanism to revolve the torch around the pipe produce excellent chamfered edges.

Fig. 6-22 shows how to use a cutting torch to bevel or chamfer the end of the pipe. It must be remembered that when cutting pipe, it is not the

diameter of the pipe that determines the size of the cutting torch tip. The thickness of the pipe wall is the controlling factor. The proper welding codes and procedures must be followed whenever pipe is cut for use in structural or pressure vessel applications.

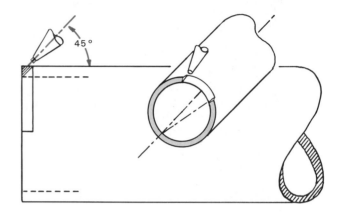

Fig. 6-22. A cutting torch being used to cut a bevel on a steel pipe.

## 6-15. PIERCING AND CUTTING HOLES

Holes may be pierced in steel plates rapidly and with accurate results. To PIERCE is to produce a relatively small hole (in comparison to the size of the metal surface) through a steel plate. The process consists of holding the cutting torch with the nozzle perpendicular to the surface of the metal, and preheating the spot to be cut until it is a bright, cherry red, 1400 °F (760 °C). After the metal is brought up to the proper temperature, the oxygen jet may be turned on very slowly. At the same time, the nozzle should be raised enough to eliminate the slag being blown back into the nozzle orifices. A greater amount of heat is required to preheat the surface for piercing than when starting on an edge. It is, therefore, recommended that the operator use at least the next larger tip in relation to the thickness of the metal than recommended in Figs. 6-5 and 6-8. See Heading 22-3 for use of the oxygen lance for cutting holes in thick plate.

To cut larger holes in steel plate, the typical steel cutting method described in this Chapter is recommended. It is good practice to outline the hole first, using special chalk. This outline is used as a guide to permit the operator to cut an accurate hole. If the size of the hole warrants it, it is best to do the cutting with an automatic machine, or with a radius bar attachment clamped to the torch head.

## 6-16. CUTTING AND REMOVING RIVET HEADS

The cutting torch is frequently used in salvage operations for removing rivet heads when dismantling large fabricated structures which have been assembled with rivets.

Two typical rivet shapes are:
1. Round head.
2. Countersunk head.

The procedure for removing these heads by cutting is fundamentally the same as for any cutting operation, but one additional precaution should be observed. If possible, the operator should do the cutting without damaging the steel plate. To do this, it is very important that the size of the tip is carefully selected. If too large a tip is used, the steel plate will be injured at the same time the rivet head is being removed. If too small a tip is used, the method becomes too slow. Practically all welding equipment companies recommend special shaped cutting tips for rivet cutting.

The procedure for cutting the round head rivet is to preheat the head of the rivet to a bright cherry red. The steel plate is usually adequately protected from the preheating flame by a coating of oxide (scale). The special cutting tip is placed on the base metal and the rivet head cut in the usual manner. See Fig. 6-23. The appearance of an accurately performed job shows a clean removal of the head without any score marks from the cutting jet on the steel plate.

Removal of a countersunk rivet head is more difficult than the round head rivet. The rivet is usually tightly embedded in the plate. Countersunk rivet heads may be removed with very little, if any, damage to the steel plate. This is done by carefully selecting the tip size and by carefully cutting around the countersunk angle.

## 6-17. GOUGING WITH CUTTING TORCH

The cutting torch may be used to cut a GOUGE or a curved groove on the edge or surface of a plate. It may also be used to remove bad areas in welds to prepare for rewelding. The principle of gouging or groove cutting is that instead of using a high-velocity jet as in a cutting tip, a large orifice low-velocity jet is used. Fig. 6-24 illustrates a typical oxygen gouging operation. The slower moving oxygen stream oxidizes the surface metal only and penetrates very slowly enabling the operator to gouge or groove with considerable accuracy.

Fig. 6-24. A typical oxygen gouging operation. A low velocity cutting jet is used to enable better control of the gouge width and depth.

In a GOUGING TIP, there are usually five or six preheat orifices to provide for even preheat distribution. In an automatic machine, a gouging tip will do very accurate gouging to exact depths. It can be used to remove bad spots and to quickly and conveniently prepare the edges of metal for welding.

If the gouging cut is not started properly, it is possible to cut too deeply and actually cut through the entire metal thickness. It is also possible to cut too shallow and cause the gouging operation to stop. The speed at which the torch is moved along the gouge line is important. Too rapid a torch

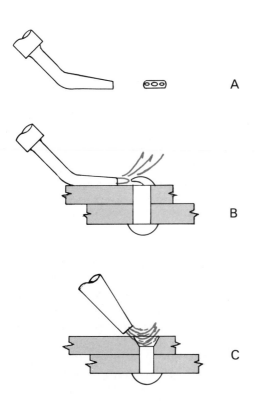

Fig. 6-23. Recommended methods of removing rivet heads using a cutting torch. A—Special rivet cutting tip. B—Special tip used to cut a round head rivet. C—Regular cutting tip used in cutting out a countersunk rivet.

movement will create a narrow, shallow gouge. Too slow a torch movement will create a gouge which is too deep and wide.

## 6-18. CUTTING FERROUS ALLOY METALS

The introduction of many alloy steels into industry has made it necessary for new cutting techniques to be developed for these metals to be successfully and economically cut. Of the alloy steels, stainless steel is perhaps the most widely used. Stainless steel consists of chromium, nickel, and mild steel. See Chapters 19, 25, and 26 for more information about alloy metals.

Many of these alloy metals have melting temperatures below that of steel. The oxides formed when cutting have a melting temperature higher than that of the original metal. These high melting point oxides must be reduced and/or removed from the cut as it proceeds, or the cutting action will stop. It has been found that for the same relative thickness of metal, stainless steels need approximately 20 percent more preheating flame and 20 percent more oxygen for the cutting. It has also been found that it is a good practice to use a slightly carburizing flame when preheating stainless steel.

The metal to be cut should be placed so that the cutting tip and flame are in a horizontal position. The cut should start at the top of the metal and proceed downward in a vertical line. A slight, but quick, up and down motion of the torch facilitates the removal of the slag. Fig. 6-25 shows how the torch should be moved up and down to facilitate the slag removal. Using a regular cutting torch, it is

difficult to obtain as clean and narrow a kerf when cutting alloy metals as when cutting straight carbon steels.

As in the case of steel, the alloy metals must be preheated before the cutting operation is started. Stainless steels, especially, must be preheated to a white heat before the cutting oxygen is turned on. The cutting action is much more violent with stainless steels than with straight carbon steels. Cutting takes place with considerable sparking and blowing of the slag.

In situations where the progress of the cutting is frequently interrupted by the presence of unmeltable slag, the operator may find it advisable to hold a mild steel welding rod in the kerf of the metal. A mild steel rod, mixed with the alloy steel, dilutes or reduces the percentage of the alloys in the area of the cut. The cutting properties of the alloy metal in the area of the cut thus become more like those of mild steel and the cut proceeds more smoothly.

Adding welding rod to the cut is also useful when cutting poor grade steels, cast irons, and old, oxidized steel castings. Powder cutting, plasma arc, and inert gas arc cutting have proven much more practical for cutting steel alloys and nonferrous metals. For full details on these processes refer to Chapter 22.

## 6-19. CUTTING CAST IRON

As mentioned previously, it is more difficult to cut cast iron than steel. This is because iron oxides of cast iron melt at a higher temperature than the cast iron itself. However, successful cutting has been performed on cast iron in many salvage shops and foundries. When cutting cast iron it is important to preheat the whole casting before the cutting is started. The metal should not be heated to a temperature that is too high, as this will oxidize the surface and make cutting difficult. A preheat temperature of about 500 °F (260 °C) is usually satisfactory.

When cutting cast iron, a carburizing flame is recommended to prevent oxides from forming on the surface before the cutting starts. The cast iron kerf is always wider than a steel kerf because of the oxidation difficulties. After completing a cut on cast iron, the casting should be cooled very slowly if a gray cast iron is desired. Rapid cooling will create a white cast iron grain structure.

It is difficult to cut cast iron with the usual oxy-fuel gas cutting torch. Other more satisfactory methods of cutting have been developed, such as the oxygen arc, plasma arc, chemical flux, metal powder, and gas tungsten arc cutting. See Chapter 22 for a description of these cutting processes.

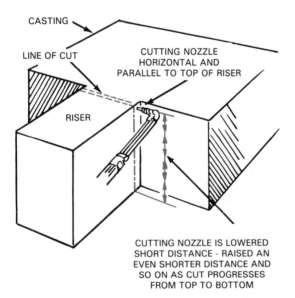

CASTING

LINE OF CUT

RISER

CUTTING NOZZLE HORIZONTAL AND PARALLEL TO TOP OF RISER

CUTTING NOZZLE IS LOWERED SHORT DISTANCE - RAISED AN EVEN SHORTER DISTANCE AND SO ON AS CUT PROGRESSES FROM TOP TO BOTTOM

Fig. 6-25. A technique used for the oxyacetylene cutting of chromium steel. Note how the torch is raised and lowered to assist in removing the slag from the cut.

## 6-20. AUTOMATIC CUTTING

Automatic cutting machines are being constantly improved. Metal powder, chemical flux, plasma arc, and gas tungsten arc methods of cutting stainless steels, low alloy steels, and cast iron have been automated by the use of cutting machines. The accuracy and quality of the cut have been improved by the use of electric solenoid valves to control gas flow. Servomotors (steering motors) are used to control the torch movement more precisely. Electronic and magnetic traces for controlling the torch movement have also been developed. The electronic tracers enable extremely accurate following of a pattern or template, and consequently the torch will produce almost perfect duplicate shapes. Electronic tracers are described in Heading 5-9. Electronic devices have also been developed to maintain a constant tip height over the metal being cut to insure good quality cuts.

There are many automatic mechanisms available, which perform automatic cutting operations, using the oxyfuel gas cutting torch. Automatic, multiple torch cutting machines are used extensively in industry. They will cut several exact copies of a desired part at one time. Inexpensive metal templates or line drawings can be used to guide the mechanical or electronic tracers. See Fig. 6-26. Chapter 22 describes semi-automatic and automatic cutting processes.

Practically all automatic cutting torches are driven by servomotors and variable speed electric motors.

A magnetic tracer with steel patterns can be used as the guide device. A variable speed electric motor on the magnetic tracer permits the tracer to follow the pattern at varying speeds. The cutting torches controlled by this tracer move at varying speeds also. This variable speed device enables the machine to adapt to different thicknesses of metal.

The cutting torches are mounted on a light rail. Both rail and torches are moved in the X-Y (lateral and longitudinal) axis by electric servomotors. As the tracer moves around the pattern, the servomotors move the torches to duplicate the pattern. A servomotor gets feedback information from the

Fig. 6-26. A gantry shape cutter using a computer to control a variety of cutting torches. (ESAB Automation, Inc.)

pattern tracer. This feedback causes the servo-motor to move the torches in the right direction to duplicate the pattern.

The operation of these automatic machines necessitates four important adjustments:

1. Adjustment for the rapidity of the cut.
2. Gas pressure must be carefully adjusted to insure a clean cut through the thickness of the metal without wasting gas.
3. Flame adjustments on each torch must be carefully made.
4. Distance of the torch tip from the metal being cut must be carefully adjusted to obtain the best results. This adjusting is done by means of a graduated scale on the torch body, and a vernier gear mechanism (measuring device which permits fine adjustment), to raise and/or lower the torch.

Once set up, the apparatus is self-operating. However, the initial adjustments must be very carefully made.

## 6-21. SAFETY IN OXYFUEL GAS CUTTING

Oxyfuel gas cutting may be safely performed. However, as in all oxyfuel gas welding and cutting, proper procedures must be followed in order to eliminate dangers which might be caused by incorrect handling or carelessness. Certain potential hazards exist and the operator should always observe approved procedures to eliminate these dangers. The following points should always be carefully observed. Observe the correct handling procedures for oxygen and fuel gas equipment. See also Headings 4-33, 4-34, 4-35, and 4-36.

Considerable sparking and flying of sparks, which are globules (round-shaped particles) of molten metal, accompany cutting operations. In order to avoid accidents from this hazard:

A. Floors on which cutting is done should be concrete or other fireproof material.
B. Workbenches and other necessary furniture should be of metal or other fire resistant material.
C. Oil, paper, wood shavings, gasoline, lint, flammable materials should not be in the room in which flame cutting is performed.
D. Leather or other slow burning fabrics should be worn. Trousers should be without cuffs. Pockets and clothing should be inspected for possible flammable materials such as buttons, combs, celluloid rules, matches, pencils, and other items.
E. Objects being flame cut may present inherent hazards. Tanks and containers may be welded or cut only by an experienced welder. It is generally required to steam the tank or pass an inert gas through the tank as it is being cut.

Steam or inert gas flow is intended to displace any combustible gases in the tank. This process is often required for hours prior to any attempted cutting operation. The tank may also be filled with water except in the area of the work. In all cases the tank must be vented to prevent the entrapment (holding) of potentially explosive gases. This work should never be done except under the supervision of a qualified safety engineer. A small amount of flammable material in such a container or tank may cause a powerful explosion.

F. Certain metals such as magnesium may burn with an explosive force if flame cut. Be certain of what is being cut.
G. Face and hands must be protected from metal splatter.
H. A fire extinguisher should always be at hand while flame cutting for use in possible emergencies.
I. In Chapter 4, the correct procedures for setting up and handling cylinders, regulators, hoses, and torches were explained. Be sure to know these procedures. It is a good idea to review Chapter 4.
J. Review Heading 4-33 on metal fume hazards, and be sure to know what metal is being worked on before performing any oxyfuel gas cutting operation.
K. The normal atmosphere contains about 21 percent oxygen by volume. As the oxygen content in an enclosed space is increased above this percentage, there is an increasing danger of a spark or flame causing a fire or explosion.

## 6-22. TEST YOUR KNOWLEDGE

Write your answers on a separate sheet of paper. Do not write in this book. Carefully review these questions. Answering them will check your knowledge of flame cutting as covered in this Chapter.

1. Which is the correct word to describe this process ''flame cutting,'' ''burning,'' or ''oxygen cutting''?
2. Is an oxyfuel gas cut smooth or rough when complete?
3. Are the preheat flames used after the cutting has begun? Why?
4. What are the two types of flame cutting torches called?
5. In adjusting the pressures before lighting a cutting torch, why it is necessary to check the pressures with the torch cutting lever in the fully open position?
6. What oxygen and acetylene pressures are used when cutting 1/2 in. (12.7 mm) sheet steel using an equal pressure torch?

7. Why is it necessary to provide a higher oxygen pressure with the cutting torch than with a welding torch?

8. What is the purpose of the preheating jets or orifices in a cutting torch tip? When is the oxygen cutting valve opened?

9. List two conditions which may cause the slag stream to lag behind the torch tip (drag).

10. When are two piece cutting tips used?

11. If too much oxygen is used when cutting, the cut widens out and leaves a _____ kerf.

12. What is meant by kerf and drag?

13. Should the cutting torch be moved toward, from side to side, or away from the operator?

14. How should the cutting torch be held, in relation to the work, when cutting thin sheet steel?

15. Is it possible to cut a piece to shape and to place a bevel on the metal at the same time?

16. What controls the size tip to be used when cutting pipe?

17. a. When piercing a hole in a plate, how is the torch held in relation to the base metal.

    b. What are the requirements for preheat compared to starting a cut on the edge of a plate?

18. How is it possible to cut off rivet heads without injuring metal plates riveted together?

19. What procedure is recommended when it is necessary to cut stainless steel or cast iron with the oxyfuel gas cutting torch?

20. What safety precautions must be considered before starting a flame cutting operation?

A plasma arc cutting torch being used to cut a "nest" of shapes. The nest was created by a computer program that positions the pieces to reduce scrap. (ESAB Automation, Inc.)

Fig. 7-1. This television repair technician is making a soldered repair with a small electric soldering iron.

# 7 SOLDERING

SOLDERING is a term applied to fastening two like or unlike metals together with another metal entirely different from either or both of the base metals. Soldering fastens two metals together without melting either one of them. The theory of soldering is that the molten binding or joining metal adheres to the clean surfaces of the parent metal by means of molecular attraction. The molecules of solder entwine (wrap around) with the parent metal molecules and form a very strong bond. This process is called ADHESION. In some cases the metals in the solder may form a surface alloy with one or both of the parent metals. Refer to Heading 1-3 and to Fig. 1-7.

In soldering, the joining metal melts and flows at temperatures less than 840 °F (450 °C). The biggest advantage of soldering is minimum warpage and minimum disturbance of the heat treatment of the parent metals being joined. If the joining metal melts and flows above 840 °F (450 °C) the process is called BRAZING. Brazing is described in Chapter 8. This method of identifying soldering and brazing was established by the American Welding Society.

## 7-1. SOLDERING PRINCIPLES

Soldering is used where a leakproof joint, neatness, a low resistance electrical joint, and sanitation are desired. The joint produced by means of soldering is not as strong as a brazed or welded joint, and in many cases a mechanical joint is used together with the solder seam. The solder most commonly used is a tin and lead alloy. Other solders contain such metals as antimony, silver, zinc, cadmium, indium, and aluminum. The proportions of the metals in a soldering alloy are varied, producing a variety of properties. Different solders are produced to meet various needs. There are different solders for soldering tin, copper, brass, aluminum, bronze, sheet iron, sheet steel, and even glass to glass. Soldered joints have ex-

cellent heat conductivity and electrical conductivity. See Fig. 7-1. The soldered joint usually has less strength than the metals being joined. Soldered assemblies must be kept at a low operating temperature to keep the soldered joint from failing.

## 7-2. CLEANING METHODS PRIOR TO SOLDERING

Metal surfaces should be cleaned prior to soldering to insure a good, strong soldered joint.

Cleaning may be done mechanically by machining, sanding, or wire brushing. The wire brush should be clean. On most metals it is advisable to use a stainless steel wire brush.

Chemical cleaning is also done. A different chemical solution is generally required for each metal or alloy which is to be soldered.

Some of the recommended cleaning solutions and the metals they are used on are listed:

For copper and copper alloys, solvents such as toluene, mixed acetates and trichloroethylene may be used. Alkaline cleaning solutions may also be used.

PICKLING is a process for removing scale and oxides from metals using acids or other chemical solutions. To pickle copper or copper alloys, a 5 - 10 percent solution of sulfuric acid is used at a temperature of 125 ° - 150 °F (51 ° - 65 °C).

Aluminum may be degreased with a degreasing solvent. Mechanical cleaning is also advised.

Stainless steel should be mechanically cleaned. Pickling is recommended using 15 - 20 percent nitric acid and 2 - 5 percent hydrofluoric acid at 140 ° - 180 °F (60 ° - 82 °C).

For most surfaces, degreasing solvents or alkaline solutions can be used to remove grease and oil. Hydrochloric, nitric, phosphoric, sulfuric, and hydrofluoric acids can be used for pickling.

Parts that have been pickled should be dried quickly and soldered immediately while the surfaces are clean.

Always wear chemical type eye goggles, rubber gloves, and long sleeves when using cleaning solutions, pickling solutions, or acids. Chemical type eye goggles do not have holes in the shield area above the eyes.

If any of these chemical solutions, acids, or fluxes come in contact with the skin, wash the area thoroughly under running water. If such chemicals contact the eyes, wash them thoroughly under running water and have a doctor check the eyes immediately.

## 7-3. SOLDER ALLOYS

An ALLOY is a mixture of one metal with one or more metals and/or nonmetals. A metal alloy has different characteristics such as hardness, melting temperature, strength, and others.

For example: Alloying tin and lead creates a lower melting point metal. Lead melts at 621 °F (327 °C) and tin melts at 450 °F (232 °C). If they are combined in a 50-50 percent alloy, the melting temperature is 361° - 421 °F (183° - 217 °C). The alloy melts at a temperature lower than pure tin or pure lead.

Soldering alloys must melt below 840 °F (450 °C) and below the melting point of the base metals on which they are used.

Tin-lead solders are the most commonly used solders. They may be alloyed in percentages from 5 percent tin and 95 percent lead to 70 percent tin and 30 percent lead. The tin content is always given first.

Fig. 7-2 is a graph which shows the solidus and liquidus lines for tin-lead solders. The SOLIDUS LINE (A, B, C, D, E) indicates the temperature at

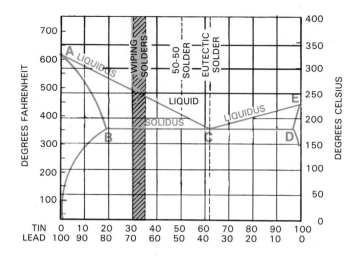

Fig. 7-2. A graph of the solidus and liquidus temperatures for alloys of tin and lead. The solidus temperature is the point at which the metal begins to melt. At the liquidus temperatures all the metal is a liquid.

which the metal or alloy begins to melt. The LIQUIDUS LINE (A, C, E) indicates the temperature at which the metal or alloy is completely liquid. The graph covers the range from 0 percent tin and 100 percent lead to 100 percent tin and 0 percent lead. This graph is called the TIN LEAD PHASE DIAGRAM.

For a 40/60 tin-lead solder, the solidus point is at 361 °F (183 °C). The liquidus point is at 455 °F (235 °C). This alloy therefore melts between these two temperatures.

An alloy of 62/38 tin-lead falls at the eutectic point for tin-lead alloys. The EUTECTIC POINT is the one point on the graph where the alloy melts at one temperature. All other alloys melt over a range of temperatures from solidus to liquidus.

| ASTM SOLDER CLASSI-FICATION | COMPOSITION (% by weight) | | SOLIDUS TEMPERATURE | | LIQUIDUS TEMPERATURE | |
|---|---|---|---|---|---|---|
| | TIN | LEAD | °F | °C | °F | °C |
| 5 | 5 | 95 | 572 | 300 | 596 | 314 |
| 10 | 10 | 90 | 514 | 268 | 573 | 301 |
| 15 | 15 | 85 | 437 | 225 | 553 | 290 |
| 20 | 20 | 80 | 361 | 183 | 535 | 280 |
| 25 | 25 | 75 | 361 | 183 | 511 | 267 |
| 30 | 30 | 70 | 361 | 183 | 491 | 255 |
| 35 | 35 | 65 | 361 | 183 | 477 | 247 |
| 40 | 40 | 60 | 361 | 183 | 455 | 235 |
| 45 | 45 | 55 | 361 | 183 | 441 | 228 |
| 50 | 50 | 50 | 361 | 183 | 421 | 217 |
| 60 | 60 | 40 | 361 | 183 | 374 | 190 |
| * | 62 | 38 | 361 | 183 | 361 | 183 |
| 70 | 70 | 30 | 361 | 183 | 378 | 192 |

Fig. 7-3. The solidus and liquidus temperatures for various tin-lead solders. *This is the eutectic alloy. Its liquidus and solidus temperatures are the same at 361 °F (183 °C).

See Fig. 7-3 for a table which shows the solidus and liquidus temperatures for tin-lead alloys.

Tin-lead solders may be used to join most metals. The alloys 35/65, 40/60, and 50/50 have the best wetting (flowing) properties, strength, and economy. They are therefore widely used.

High lead content solders like 10/90, 15/85, and 20/80 have the best mechanical (strength) properties of all the alloys.

Other soldering alloys include tin-antimony, tin-antimony-lead, tin-silver and tin-silver-lead, tin-zinc, cadmium-silver, cadmium-zinc, zinc-aluminum, and indium.

A 95 percent tin and 5 percent antimony solder has a melting range between 450° - 464°F (232° - 240°C). It is often used for soldering in refrigeration, air conditioning, and plumbing. See Fig. 7-4.

Antimony may be added to tin-lead solder. The mechanical properties of a tin-lead solder are improved by adding up to 6 percent antimony. The antimony percentage replaces the tin in the alloy. Antimony alloy solders are not advised for use on galvanized iron or steel. The antimony may mix with the zinc coating and produce a gritty, brittle solder mixture.

Tin-silver alloys, which contain no lead, are used to join stainless steel on food handling equipment. This solder alloy is 96 percent tin and 4 percent silver. See Fig. 7-4.

When soldering silver surfaces in electronic equipment, a tin-lead-silver alloy is used. This alloy is 62 percent tin, 36 percent lead, and 2 percent silver.

For high strength and cryogenic (extremely cold) applications, a high lead content alloy is used. Such an alloy contains 5 percent tin, 94.5 percent lead, and 0.5 percent silver. See Fig. 7-4 for other tin-lead-silver alloy combinations.

Only inorganic fluxes are recommended for use with these solders. This is due to the high melting temperatures involved. Fluxes will be studied in Heading 7-4.

Tin-zinc solders are used to solder aluminum. Zinc contents as high as 40 percent or the addition of 1 - 2 percent aluminum improves corrosion resistance.

A 95 percent cadmium and 5 percent silver

## SOLDER ALLOY COMPOSITIONS AND MELTING TEMPERATURES

| ALLOY | COMPOSITION (% BY WEIGHT) | | | | | | | | SOLIDUS TEMPERATURE | | LIQUIDUS TEMPERATURE | |
|---|---|---|---|---|---|---|---|---|---|---|---|---|
| | Tin | Lead | Silver | Antimony | Cadmium | Zinc | Aluminum | Indium | °F | °C | °F | °C |
| Tin-Antimony | 95 | | | 5 | | | | | 450 | 232 | 464 | 240 |
| Lead-Tin-Silver | 96 | | 4 | | | | | | 430 | 221 | 430 | 221 |
| | 62 | 36 | 2 | | | | | | 354 | 180 | 372 | 190 |
| | 5 | 94.5 | 0.5 | | | | | | 561 | 294 | 574 | 301 |
| | 2.5 | 97 | 0.5 | | | | | | 577 | 303 | 590 | 310 |
| | 1.0 | 97.5 | 1.5 | | | | | | 588 | 309 | 588 | 309 |
| Tin-Zinc | 91 | | | | | 9 | | | 390 | 199 | 390 | 199 |
| | 80 | | | | | 20 | | | 390 | 199 | 518 | 269 |
| | 70 | | | | | 30 | | | 390 | 199 | 592 | 311 |
| | 60 | | | | | 40 | | | 390 | 199 | 645 | 340 |
| | 30 | | | | | 70 | | | 390 | 199 | 708 | 375 |
| Silver-Cadmium | | | 5 | | 95 | | | | 640 | 338 | 740 | 393 |
| Cadmium-Zinc | | | | | 82.5 | 17.5 | | | 509 | 265 | 509 | 265 |
| | | | | | 40 | 60 | | | 509 | 265 | 635 | 335 |
| | | | | | 10 | 90 | | | 509 | 265 | 750 | 399 |
| Zinc-Aluminum | | | | | | 95 | 5 | | 720 | 382 | 720 | 382 |
| Tin-Lead-Indium | 50 | | | | | | | 50 | 243 | 117 | 257 | 125 |
| | 37.5 | 37.5 | | | | | | 25 | 230 | 138 | 230 | 138 |
| | | 50 | | | | | | 50 | 356 | 180 | 408 | 209 |

Fig. 7-4. Solder alloy compositions and melting temperatures. Alloys containing cadmium are indicated in red since they are a health hazard. Note that the melting temperature of an alloy occurs in a range between the solidus and liquidus temperatures.

solder is used to solder parts which operate at high temperatures. This alloy melts between 640° - 740°F (338° - 393°C). See Fig. 7-4. Aluminum to aluminum and aluminum to other metal joints can be accomplished with this solder. High strength joints are also possible with this alloy.

Alloys containing cadmium pose a health hazard. Extremely good ventilation is necessary. Care must be taken with respect to breathing cadmium fumes.

Cadmium and zinc alloy solders are used to solder aluminum. They provide good corrosion resistance when used with the proper flux. See Fig. 7-4.

A 95 percent zinc and 5 percent aluminum alloy was developed for use in soldering aluminum. Its melting point is very high at 720°F (382°C).

Indium solder, containing 50 percent indium and 50 percent tin, adheres to glass. This solder may be used to solder glass to glass and glass to metal. Indium solder does not require any special techniques in its use. See Fig. 7-4.

Fusible alloys are used in soldering operations where the temperature should remain below 361°F (183°C). These alloys contain bismuth as about 50 percent of the alloying metal. See Fig. 7-5.

flux must be used. If the surfaces are plated with tin or tin-lead a rosin flux can be used. More information about fluxes is found in Heading 7-4.

Solder comes in a variety of solid shapes including ingots, ribbons, foils, solid and flux cored wires, and in paste forms. Any size, alloy, weight, or form is available on special order.

For further information regarding solder specifications refer to the following:

American Society for Testing and Materials publications:
    Solder Metal; ASTM B32
    Rosin Flux Core Solder; ASTM B284
    Paste Solder; ASTM B486
United States Government publication:
    Solders; QQ-S-571.

## 7-4. SOLDERING FLUXES

Surfaces to be soldered must be cleaned of oxides and other surface compounds prior to soldering. This cleaning action is done by a SOLDERING FLUX. A good flux will also keep the solder and base metal from oxidizing during the soldering operation.

WETTING is an action in which a liquid solder or flux flows or spreads evenly and adheres in a thin, continuous layer on the base metal. Soldering flux

| FUSIBLE ALLOY SOLDERS | | | | | | | | | |
|---|---|---|---|---|---|---|---|---|---|
| COMPOSITION (% BY WEIGHT) | | | | | | SOLIDUS TEMPERATURE | | LIQUIDUS TEMPERATURE | |
| Alloy | Bismuth | Lead | Tin | Cadmium | Antimony | °F | °C | °F | °C |
| Eutectic | 52 | 40 | | 8 | | 197 | 91 | 197 | 91 |
| Eutectic | 52.5 | 32 | 15.5 | | | 203 | 95 | 203 | 95 |
| Bending | 50 | 25 | 12.5 | 12.5 | | 158 | 70 | 165 | 74 |
| Rose's | 50 | 28 | 22 | | | 204 | 96 | 225 | 107 |
| Lipowitz | 50 | 26.7 | 13.3 | 10 | | 158 | 70 | 158 | 70 |
| Matrix | 48 | 28.5 | 14.5 | | 9 | 217 | 102 | 440 | 227 |
| Mold or Pattern | 55.5 | 44.5 | | | | 255 | 124 | 255 | 124 |

Fig. 7-5. Fusible alloy solders. Cadmium alloy numbers are in red. Cadmium soldering must be done with excellent ventilation.

The low melting temperatures of these alloys make them useful when soldering on metals or in areas where higher temperatures cannot be tolerated. Such applications are:
1. Soldering heat treated areas.
2. Soldering on fire sprinkler links where the solder must melt at a low temperature to permit sprinklers to actuate in a fire.
3. Soldering near heat sensitive devices.

When using bismuth alloy solders a corrosive

should have a good wetting action.

There are three classifications of fluxes: INORGANIC, ORGANIC, and ROSIN BASED.

Organic compounds in chemistry are compounds (mixtures) which contain carbon.

INORGANIC FLUXES do not contain carbon compounds. They are considered the most active because they clean surfaces better than other types of flux. Torch, oven, resistance, or induction soldering can be done with these fluxes because they do not char or burn easily. Inorganic fluxes are highly corrosive. All areas which have had contact with an inorganic flux must be cleaned after soldering to stop any possible corrosive action. This flux is not recommended for soldering electrical joints.

ORGANIC FLUXES are moderately active and have a medium level cleaning ability. Organic fluxes are corrosive during the soldering operation. After soldering they generally become non-corrosive or inert. Water soluble organic fluxes are ideal for use on electronic assemblies. The flux residue can be removed by washing with water. These fluxes do char or burn easily. The temperatures at which an organic flux is most efficient range between 200° - 600°F (93° - 316°C). These fluxes consist of organic acids and bases.

The fumes produced during soldering can be corrosive. Care must be taken to prevent these corrosive fumes from damaging parts in the area. Organic flux residues are easily cleaned and removed after soldering is completed.

ROSIN FLUXES are the least active of the flux types. They are the least effective in cleaning off metal oxides or tarnishes. All types of electrical and electronic soldering is best done with a rosin based flux. Rosin fluxes are classified by their activity as:
1. Non-active.
2. Mildly active.
3. Fully active.

Federal specifications designate rosin fluxes as R, RMA, and RA.

NON-ACTIVE (R) FLUXES are rosins dissolved in alcohol or turpentine. They are only used on highly solderable surfaces.

MILDLY ACTIVE (RMA) FLUXES are used on highly solderable surfaces. They clean better and faster than non-active fluxes.

FULLY ACTIVE (RA) FLUXES are most commonly used. They are most active and clean best. They may be corrosive but only during the soldering operation. After the soldering is completed this flux is generally non-corrosive. Flux residues which are known to be corrosive must be removed from the joint after soldering.

Rosin fluxes are easily cleaned from parts after soldering, and they generally leave no corrosive

| FLUXES FOR ELECTRONIC SOLDERING | | | | | |
|---|---|---|---|---|---|
| | | ROSEN FLUXES | | | ORGANIC FLUXES Water Soluble |
| METALS | SOLDERABILITY | Non-Activated | Mildly Activated | Activated | |
| Platinum Gold Copper Silver Cadmium Plate Tin (Hot Dipped) Tin Plate Solder Plate | Easy to Solder | X | X | X | X |
| Lead Nickel Plate Brass Bronze Rhodium Beryllium Copper | Less easy to Solder | Not Suitable | | X | X |
| Galvanized Iron Tin-Nickel Nickel-Iron Mild Steel | Difficult to Solder | Not Suitable | | | X |
| Chromium Nickel-Chromium Nickel-Copper Stainless Steel | Very Difficult to Solder | Requires Precoating for Electronic Applications | | | Not Suitable |
| Aluminum Aluminum-Bronze | Most Difficult to Solder | | | | Not Suitable |
| Beryllium Titanium | Not Solderable | | | | |

Fig. 7-6. Suggested fluxes for use in electronic soldering on a variety of metals. Some metals are easier to solder than others. Beryllium and titanium are not solderable.

residue. See Fig. 7-6 for suggested fluxes for use in soldering electric and electronic connections.

Fluxes may be purchased as dry powder, cores within soldering wires, pastes, and liquids. See Fig. 7-7 for a table which shows the flux requirements for soldering various metals.

### RECOMMEND FLUXES FOR VARIOUS METALS

| Base Metal, Alloy or Applied Finish | Flux Recommendations | | |
|---|---|---|---|
| | Corrosive | Non-Corrosive | Special Flux and/ or Solder |
| Aluminum | | | X |
| Aluminum-Bronze | | | X |
| *Beryllium | | | |
| Beryllium Copper | X | | |
| Brass | X | X | |
| Cadmium | X | X | |
| Cast Iron | | | X |
| *Chromium | | | |
| Copper | X | X | |
| Copper-Chromium | X | | |
| Copper-Nickel | X | | |
| Copper-Silicon | X | | |
| Gold | | X | |
| Inconel | | | X |
| Lead | X | X | |
| Magnesium | | | X |
| *Manganese-Bronze | | | |
| Monel | X | | |
| Nickel | X | | |
| Nichrome | | | X |
| Palladium | | X | |
| Platinum | | X | |
| Rhodium | X | | |
| Silver | X | X | |
| Stainless Steel | | | X |
| Steel | X | | |
| Tin | X | X | |
| Tin-Bronze | X | X | |
| Tin-Lead | X | X | |
| Tin-Nickel | X | X | |
| Tin-Zinc | X | X | |
| *Titanium | | | |
| Zinc | X | | |
| Zinc Die Castings | | | X |

Fig. 7-7. Flux recommendations for soldering various metals.
*Soldering not recommended.

## 7-5. FLUX RESIDUE REMOVAL

Inorganic fluxes contain inorganic acids and salts. These are highly corrosive and must be removed completely.

Organic fluxes which contain mild organic acids or urea compounds should also be removed to prevent continuing corrosion.

Rosin fluxes may generally be left on the joints. However, active rosin fluxes which contain organic compounds must be removed. Rosin fluxes should be removed if the joint is to be painted or otherwise finished.

Inorganic fluxes are chloride salts, such as sodium chloride and zinc-aluminum chloride. Sodium and zinc chloride based fluxes are corrosive. Their residue will absorb water and cause rust to form. Sodium and zinc chloride based fluxes are best removed with hot water containing two percent hydrochloric acid. Follow all safety procedures listed in Heading 7-2. This application should be followed with a hot water and mild detergent rinse. If necessary, this cleaning should be followed by washing with hot water and sodium carbonate (washing soda) and a hot water rinse.

Aluminum joints soldered with active type fluxes may generally be cleaned with hot water. If this is not effective, then the joint must be scrubbed with hot water. The joint must then be dipped in a two percent sulfuric acid solution. This is followed by immersion into a one percent solution of nitric acid. Follow all safety recommendations.

Organic fluxes are generally easily removed with hot water. An organic solvent may be required to remove greasy or oily flux residues.

If rosin flux residues must be removed, alcohol or chlorinated hydrocarbons may be used. Certain rosin residues must be removed with an organic solvent followed with a hot water rinse.

## 7-6. SOLDERING PROCEDURES

Many methods may be used to solder single or muliple parts, using manual or automatic soldering equipment. Some of these methods are:
1. Soldering iron (copper).
2. Torch soldering.
3. Dip soldering.
4. Wave soldering.
5. Oven soldering.
6. Resistance soldering.
7. Induction soldering.
8. Infrared soldering.

Before describing soldering procedures in detail, it should be pointed out that several things must be done in order to produce successful soldering. These are:
1. Metals to be soldered together must be chemically and/or mechanically cleaned. All the oxides, grease, and dirt must be removed.
2. Metals to be soldered together must be heated.
3. Metals to be soldered must be firmly supported during the soldering operation and until they cool.
4. The proper flux must be used. This flux must be fresh and it must be as chemically pure (CP) as possible.
5. The solder should be melted only by the heat in the metals to be soldered together.

6. The soldering operation should be done as quickly as possible.

7. An excess of solder is useless and unsightly.

8. Most solder fluxes should be thoroughly removed from the joint as soon as possible after the soldering operation is completed to prevent continuing corrosion.

The metals may be cleaned chemically if they are then thoroughly rinsed and dried. The metals may also be cleaned by filing, or wire brushing. Use only clean tools, clean steel wool, and/or clean stainless steel wool. The metals should be heated just above the flow temperature of the solder. Clean heat should be used.

If either piece of parent metal moves while the metal is cooling from its liquid temperature to its solidification temperature, the solder will probably contain cracks and will fail. It is therefore necessary to firmly support the metals with a fixture, clamps, etc., to make sure they do not move while the soldering operation is in process.

Be sure the main flux container is sealed when not in use in order to keep the flux clean. Remove only that quantity of flux needed for the particular job. Apply the flux with a clean brush, paddle, injector, or automatic applicator. Brushes and paddles should be thoroughly washed each day. Waterbase fluxes should be used immediately. With most paste type fluxes, you may wait as long as an hour before using the flux. The soldered joint should be completed as soon as possible once the flux is heated, as any delay will cause flux salts to form.

One method used to properly proportion the flux to the joint to be soldered is to use a paste made of the flux and of the filler metal in a powder form. Such a paste is shown in Fig. 7-8. This is obtainable in a variety of solders and brazing alloys mixed with the proper fluxes. These fluxes are applied by using dispensers as shown in Figs. 7-9 and 7-10.

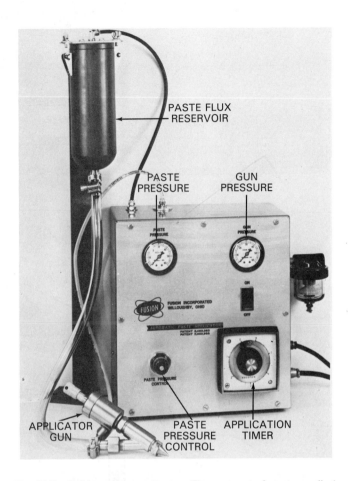

Fig. 7-9. Solder paste applicator. The amount of paste applied is controlled by the paste pressure and by the timer to 1/20 sec. The quantity applied by weight can vary from .001 oz. (0.128 g) to 1 oz. (137.8 g).

Fig. 7-8. Kit of various special paste solders. These compounds contain both a paste flux and a filler metal. A syringe is provided for applying the mixture to the soldered joint. (Fusion, Inc.)

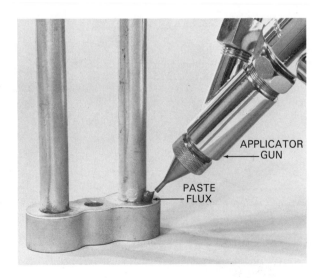

Fig. 7-10. A close-up of soldering flux paste being applied to a part. The amount applied is very precise.

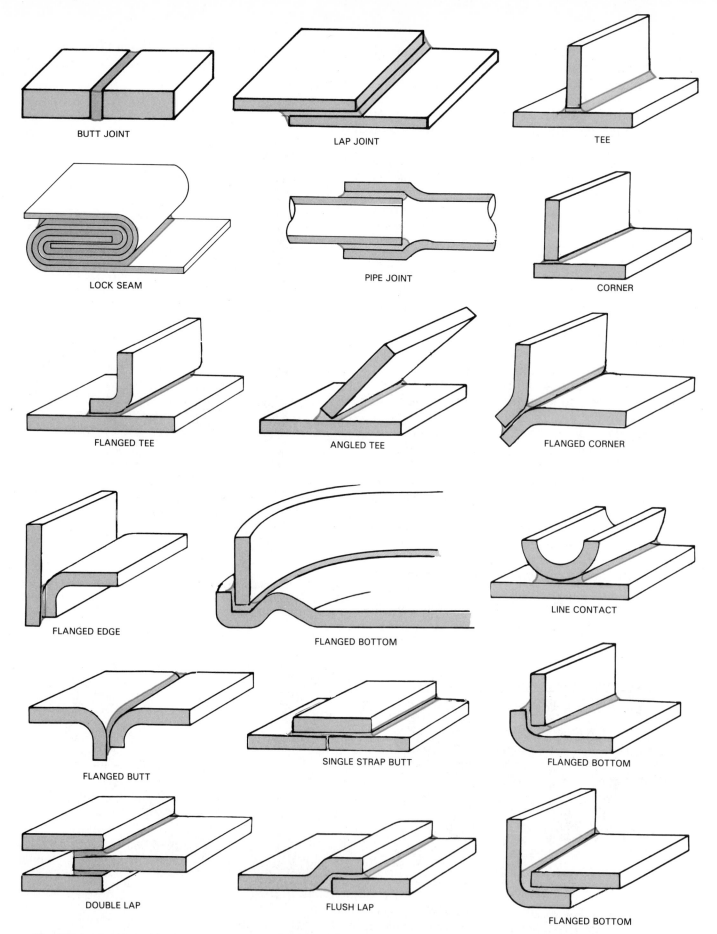

BUTT JOINT

LAP JOINT

TEE

LOCK SEAM

PIPE JOINT

CORNER

FLANGED TEE

ANGLED TEE

FLANGED CORNER

FLANGED EDGE

FLANGED BOTTOM

LINE CONTACT

FLANGED BUTT

SINGLE STRAP BUTT

FLANGED BOTTOM

DOUBLE LAP

FLUSH LAP

FLANGED BOTTOM

Fig. 7-11. Suggested joint designs for soldered and brazed joints. The spacing is exaggerated for purposes of illustration. Normal spacing and filler metal thickness is approximately .003 in. (.076 mm). (Continued)

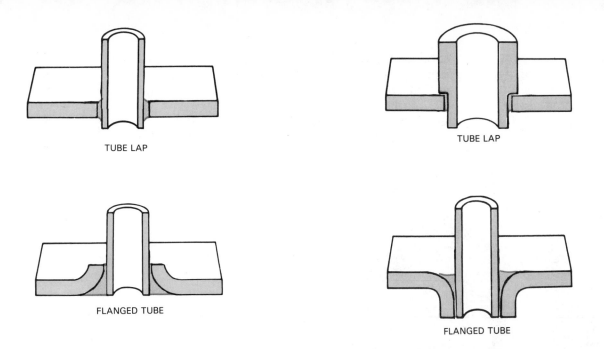

TUBE LAP

TUBE LAP

FLANGED TUBE

FLANGED TUBE

Fig. 7-11. (Continued) Suggested joint designs for soldered and brazed joints.

## 7-7. JOINT DESIGN FOR SOLDERING OR BRAZING

The design for joints to be soldered or brazed is considerably different than for welding.

Solder and brazing alloys are weaker than the base metals joined. Therefore, joints must be designed so that strong joints occur without a dependence on the strength of the filler metal used.

There are many joint designs used for soldering and brazing. Examples of these joints are shown in Fig. 7-11. Clearance between the parts of a soldered or brazed joint must be very small. The clearance between the parts must be small enough to allow capillary action to take place. CAPILLARY ACTION is an action which causes a liquid to be drawn into the space between tightly fitted parts. This desired clearance is about .003 in. (.076 mm). Too large a clearance may stop the capillary drawing of solder into the joint. Large clearances, if filled with solder or brazing alloy, will be weak. Forces on the joint may cause the joint to fail in the weaker filler metal.

## 7-8. TINNING

Adhering a very thin layer or film of solder to a metal surface is called TINNING. The word tinning is from an old sheet metal term used when thin layers of tin were applied to steel. This tinning was done with heat to prevent corrosion. Tin cans are actually not tin cans but are made of steel with an extremely thin coating of tin or a tin alloy on the surface of the steel. Copper wire is also frequently tinned as it is manufactured. In all soldering opera-

tions, the solder tins the surfaces as the process travels along the joint. Some assemblies are easier to solder if each part is first tinned. The parts are assembled after tinning. They are then reheated to cause the tinning solder to flow and complete the joint. This process for soldering is called SWEATING a joint. It can be done with a torch or soldering iron. Tinning normally requires an active corrosive flux. Noncorrosive flux can be used where the final assembly is difficult to clean.

## 7-9. SOLDERING IRON (COPPER) METHOD

The soldering iron method of soldering is a popular method for doing certain types of tin-lead alloy soldering. This tool is called a soldering iron because it "irons" the solder along a seam. The term soldering iron is preferred over the often used term soldering copper. A soldering iron usually consists of a square or octagonal solid copper bar with a four-sided tapered point. This copper bar is attached by a steel rod to a heat resistant handle. Soldering irons come in several sizes and shapes. One pound to sixteen pound soldering irons are available. Shapes available include pointed, flat bottom, blunt pointed, tapered flat bottom, and hatchet. The efficiency of heat transmission from the copper bar to the work makes copper the ideal metal for the purpose. Also, copper is easily tinned (coated with solder) so that the molten solder will adhere to the soldering iron making the handling of the solder less difficult. The soldering iron may be heated by an external or internal gas flame, by a carbon arc, or by an electrical resistance heating element. A 200 watt electric soldering iron is suitable for most sheet metal soldering. Small electric soldering irons are used for electronics applications. Refer to Fig. 7-1. Fig. 7-12 shows a closed flame air-fuel gas torch heating a soldering copper. Fig. 7-13 shows soldering coppers being heated by the twin-carbon arc method.

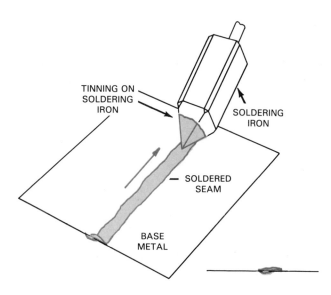

Fig. 7-14. Soldering with a soldering iron. Note the direction in which the copper is moved as the soldered seam progresses.

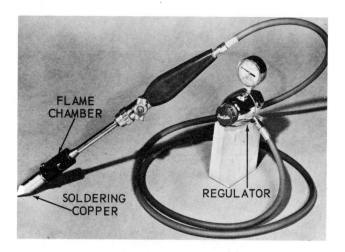

Fig. 7-12. A closed flame air-fuel gas torch equipped with a soldering copper attachment. The regulator is attached to the fuel gas cylinder. (L-TEC Welding & Cutting Systems)

Advantages of the soldering iron are that it produces a concentrated heat, and the copper is not likely to be heated to such a high temperature that it will injure the base metals or solder. If the soldering iron is heated too much the tinning will burn off the copper and the copper will blacken. The soldering iron also acts as a means of spreading, or smoothing (ironing) the solder at the same time it is melting and adhering the solder to the metals, as shown in Fig. 7-14. A soldering iron heated by an external source has the disadvantage of requiring reheating quite frequently. However, electrically heated soldering irons and internal flame heated coppers do not have this problem. The tip of the soldering iron must be kept clean and tinned at all times.

The process of cleaning and solder coating the soldering iron is called "tinning." To "tin" a soldering copper, it is necessary to file the point with a clean file to a smooth coppery finish without leaving any dirt or pits. The tip is then cleaned chemically by dipping it in a cleaning compound. A thin coat of solder is then applied to the tip in the presence of sal ammoniac.

Fig. 7-15 shows a soldering iron used to "sweat" a joint. The resultant joint is strong and neat. It is very important that the two metals fit together snugly.

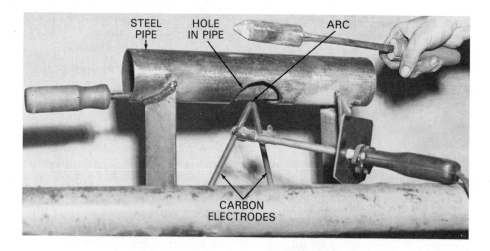

Fig. 7-13. A twin-carbon electric arc being used to heat soldering coppers in a specially built fixture. (The Lincoln Electric Co.)

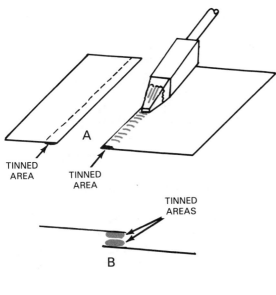

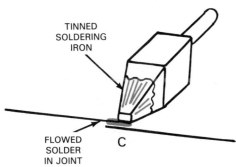

Fig. 7-15. Steps required to make a "sweated" soldered joint. A—Solder is applied in a thin film (tinning). B—Surfaces are lapped to form the joint. C—Soldering iron is applied and moved along the joint to flow the solder on the previously tinned surfaces.

## 7-10. TORCH SOLDERING METHOD

Soldering torches provide a fast and flexible method for providing heat for soldering. Several types of soldering torches are the:
1. Oxynatural gas torch.
2. Compressed air torch and natural gas.
3. Air-fuel gas torch, as shown in Figs. 7-16 and 7-17.
4. Compressed air-fuel gas torch.
5. Oxyfuel gas torch.
6. Propane torch.

In order to solder satisfactorily with a torch, the following conditions must be met:
1. The flame must heat the metals to be soldered.
2. The flame must be clean so that the surfaces heated will not be corroded by the flame gases.
3. Heat from the flame must be concentrated.
4. The amount of heat must be easily adjusted. Fig. 7-18 shows a variety of tips used for soldering with the air-fuel gas flame.

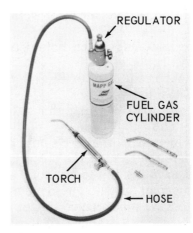

Fig. 7-16. An air-fuel gas soldering and brazing outfit. The regulator is attached to the fuel gas cylinder.

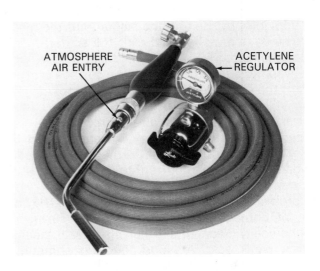

Fig. 7-17. Air-fuel gas torch and regulator. This torch is equipped with a "swirljet" tip. The air and fuel gas is swirled through a set of propeller-like vanes at the rear of the flame-tube. This produces a high heat output.
(L-TEC Welding & Cutting Systems)

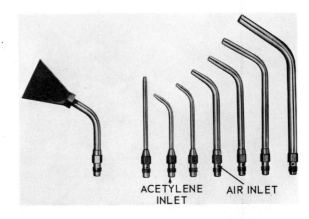

Fig. 7-18. Variety of tips used with an air-fuel gas torch. More heat may be obtained by using a larger tip size; however, each torch tip flame operates at the same temperature.
(L-TEC Welding & Cutting Systems)

The general procedure for torch soldering is as follows:

1. Clean the surfaces to be soldered. Refer to Heading 7-2.
2. Apply the correct flux to the joint surfaces. Refer to Heading 7-4.
3. Support the joint through the soldering and cooling period.
4. Heat the joint with the torch until the joint is just hot enough to melt the solder alloy used.
5. Continue to move the torch close to the joint to keep the joint hot enough to melt the solder. Then withdraw the torch so the joint does not overheat. Touch the solder to the joint. Continue to add solder to the joint until it is completely filled.
6. If the solder does not adhere, stop the operation and reclean and reflux the joint.
7. A common soldering error is to add solder to a cool joint by melting the solder with the torch. The base metal should melt the solder.

Another error is to use too much solder. Only the properly adhered solder at the surface is effective. If too much solder is used, a neat looking joint may still be obtained by wiping off the excess molten solder, using a clean, thick cloth. The solder should be wiped away while it is in the liquid condition rather than when it is nearly solidified. It is very important to avoid overheating the metal.

Torch soldering is often used to restore the contour of damaged parts. The irregular surface is mechanically cleaned. It is then chemically cleaned with a weak acid. A wood paddle is sometimes used to help apply the solder to the torch heated surfaces. Fig. 7-19 shows a metal joint being filled in by the torch soldering method. The solder is then "dressed" by filing and sanding to match the sheet metal surface.

The refrigeration, air conditioning, and plumbing industries use a great number of soldered fittings. See Fig. 7-20. The joint must be mechanically cleaned, fluxed, the tubing firmly supported, and the soldering done as quickly as possible. A thin layer of flux is usually put on the outside of the tubing and the inside of the fitting. A weaving action of the torch flame will heat the metal evenly and prevent overheated spots. This type joint may be soldered successfully with the tubing and the fittings in any position. A change of color of the secondary flame usually indicates that a copper fitting is being overheated.

Soldered pipe and tube fittings use the principle of capillary action to draw the molten solder into the joint. A small clearance is provided between a pipe or tube and the fitting. This allows solder to flow into the joint. Tube or pipe joints may be either soldered or brazed. See Chapter 8 for brazing procedures.

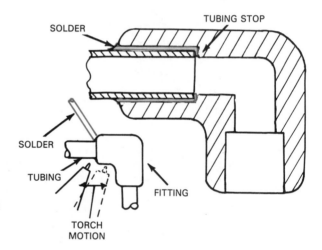

Fig. 7-20. Soldered tubing or pipe joint. Solder is drawn into the joint by capillary action. Note that the solder is applied at a point away from the torch flame.

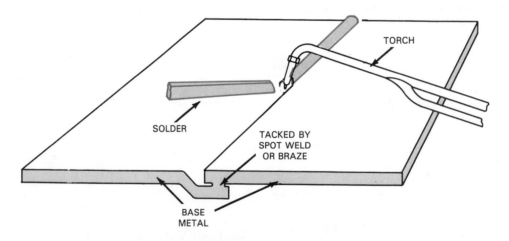

Fig. 7-19. This illustration shows the cavity of an offset lap joint being filled by using the torch soldering method to form a smooth upper surface for final metal finishing.

To solder cast iron, it must first be filed, machine blasted, or shot blasted to remove the oxide skin. The mechanically cleaned surface must then be degreased and cleaned in a special molten flux bath that will remove the surface graphite.

Many solder joints are made automatically. Parts to be soldered are mounted in fixtures. Flux and solder are often added to the joint automatically. Heating is automatic and the soldered joint is cooled before the parts are ejected from the fixtures. See Fig. 7-21. Some installations use pre-placed solder rings or solder foil instead of the automatic feeding of a solder-flux mixture or a solder filler wire. Pre-formed and pre-placed solder comes in any designed form such as sheets, rings, or strips. Pre-cleaning and post-cleaning are sometimes a part of the automatic process.

## 7-11. DIP SOLDERING METHOD

This method consists of melting a quantity of solder in a tank or pot. The solder is protected by means of a hood or chemical covering, such as powdered charcoal, to prevent oxidation of the solder. The articles to be soldered are dipped in a flux bath, and then in the solder bath. This method is a labor saving device to either solder coat surfaces to make them rustproof, or to fasten the various parts of an assembly together.

The articles to be soldered together are usually assembled and then acid cleaned (pickled). They are then thoroughly washed and dried before being dipped into the molten solder. The articles are lifted out of the bath and any excess solder is allowed to drain from the surfaces.

Parts to be dipped into molten solder must be dry. A small amount of moisture will produce

Fig. 7-21. A twelve station rotary soldering machine. The flux which contains powdered solder is automatically applied. Part assemblies are progressively heated until the flux and solder flow. The parts are then cooled and automatically ejected from the machine.

instant high-temperature steam. This may cause an eruption of the molten solder with possible injuries to workers and destruction to property.

## 7-12. WAVE SOLDERING

Molten solder in large tanks may be made to create small waves as in the ocean. The wave height is precisely controlled by means of venturis, baffles, and screens to contact the bottom of the part to be soldered.

Parts such as electronic circuit boards may be soldered by passing them over the molten solder waves. See Fig. 7-22. A typical solder wave tank may be up to 24 in. (.61 m) wide. Parts may be

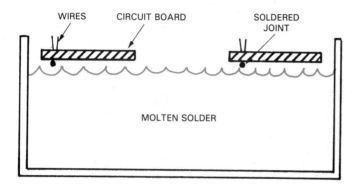

Fig. 7-22. Wave soldering schematic. Electronic circuit boards are passed over the molten solder wave. The crest of the waves just touch the joints. Several thousand joints per minute can be made with this process.

moved over the wave crest at a rate of 12 ft. (3.66 m) per minute. With this speed, up to 1200 electronic joints could be soldered per minute per foot (.3 m) of wave width. See Fig. 7-23.

## 7-13. OVEN AND INFRARED SOLDERING

Ovens heated by fuel gas or electricity are used to solder a large number of joints at one time.

To OVEN SOLDER, parts are assembled with pre-formed and pre-placed solder forms in each joint. A precise amount of flux is placed on each joint. The parts are then placed in the soldering oven. The parts are heated until the solder flows. In large production facilities, the parts are passed through the oven on a conveyor assembly. After the joints are made, the parts are cooled and cleaned.

INFRARED SOLDERING is similar to oven soldering. The difference is that the parts are heated by infrared lamps. This heating usually occurs in an enclosed space so that nobody is burned by the infrared light rays.

## 7-14. RESISTANCE AND INDUCTION SOLDERING

In RESISTANCE SOLDERING parts are assembled with pre-placed and pre-formed solder forms and with flux applied to each joint. Electricity is passed through the base metal which causes it to

Fig. 7-23. Electronic circuit boards are moving through a wave soldering tank. At "1" the solder wave may be seen. A solid solder ingot is constantly fed into the tank at "2" to maintain the solder level.

heat because of its resistance to electrical flow. The metal is heated until the solder begins to flow at which time the electric current is stopped. After soldering, the parts are cooled and cleaned.

In INDUCTION SOLDERING the parts are prepared as above. The heating of the base metal is caused by induction heating. In this process, a current is passed through an insulated copper wire coil. The part to be soldered is placed inside the coil of copper. As the current passes through the coil a current is caused to flow in the part also. This current in the part is an induced current. It occurs without actual electric contacts to the part. The part is thus heated by induction.

When the part is hot enough to cause the solder to flow the current to the coil is turned off. There is no longer any induction heating and the part begins to cool.

## 7-15. STAINLESS STEEL SOLDERING

Many kinds of stainless steels are now being used. Some of the more common stainless steels are:

1. 200 series (approximately 16 - 19 percent chromium, 3.5 - 6 percent nickel).
2. 301 to 308 series (approximately 16 - 24 percent chromium, 6 - 12 percent nickel).
3. 309 to 314 series (approximately 22 - 26 percent chromium, 12 - 22 percent nickel).
4. 316 to 347 series (approximately 16 - 19 percent chromium, 10 - 13 percent nickel).

Properties of stainless steels are described in Chapter 19.

Stainless steel joints can be either soldered, brazed, or welded. Soldering is used often to make low strength, leakproof joints, and to eliminate cracks and crevices.

A corrosive flux is needed to promote adhesion of the solders as the chromium surface film on the stainless steel resists ordinary fluxes. A highly active, inorganic flux is generally used. It is very important that these corrosive fluxes be completely removed after soldering. A solution of warm water with 2 percent hydrochloric acid is used to clean the parts after soldering. This cleaning is followed by a washing with detergent and warm water. Follow all safety precautions carefully.

A 50-50 solder makes a good joint but its color does not match the color of stainless steel. Special solders which match stainless steel's color (usually with no lead content) are available.

Solder for stainless steel comes in rods 1/8 in. (3.2 mm) in diameter by 15 in. (.38 m) long.

## 7-16. STAINLESS STEEL SOLDERING FLUXES

The surface of stainless steel must be cleaned prior to soldering. This can be done by mechanical means such as rubbing with a stainless steel wire brush. Stainless steel may also be pre-cleaned by chemical pickling. Pickling is accomplished with a nitric-hydrofluoric acid mixture. This is a mixture of 15 - 20 percent nitric acid ($HNO_3$) and 2 - 5 percent hydrofluoric acid (HF) at 140° - 180°F (60° - 82°C). Follow all safety precautions.

Only experienced persons should mix the acid (always add the acid SLOWLY to the water—NOT water to acid). Wear goggles, face guards, rubber lined clothing, and rubber gloves when preparing this cleaning mixture.

After the acid has been on the stainless steel surface for approximately four minutes, rinse with warm water. After drying, apply a zinc chloride flux and carefully heat the metal and apply solder as usual. Great care is needed as stainless steel is difficult to solder. One will find that the solder will adhere easier and better if the stainless steel surface is scratched, using a small scrubbing motion with the solder filler rod. This action seems to mechanically remove any residual film which prevents the solder from adhering to the surface.

## 7-17. SOLDERING ALUMINUM ALLOYS

Aluminum and many of its alloys can be soldered in the same manner as other metal. The most easily soldered alloys are those which contain no more than 5 percent silicon and 1 percent magnesium. Other alloys are more difficult to solder because of a poorer flux wetting action.

Wrought alloy products are easiest to solder. Castings are harder to solder because of their varied composition and surface irregularities.

Pre-cleaning is necessary. It may be done by solvent degreasing only. Occasionally mechanical cleaning with a stainless steel wire brush is necessary.

A modified, chemical flux such as zinc chloride and ammonia compounds may be used.

Zinc chloride is the main ingredient in the reaction flux used on aluminum.

When heated, the flux mixes with the aluminum oxides on the surface and cleans the surface. Both the chemical and reactive types of fluxes are highly corrosive and should be removed after soldering.

There are three soldering alloys generally recommended for aluminum. These are:

1. Tin-lead plus zinc and cadmium.
2. Zinc-cadmium or zinc-tin base.
3. Zinc base with aluminum and copper.

The zinc based alloys with additions of aluminum and copper have the best wetting action, strength, and corrosion resistance. However, it is the hardest to apply.

The same basic procedure is necessary for soldering aluminum as for any other metal. See

Heading 7-5. The metal, the flux, and the filler rod must be clean. Good support for the parts is important while they are being joined.

Usually, the joint is heated with a carburizing flame. A rather large torch movement is used. The solder should be melted only by the heat from the metal to be joined. The surface of the metal is often rubbed with the soldering wire as it is applied. This action helps to remove any remaining aluminum oxides from the surface.

## 7-18. SOLDERING DIE CASTINGS

Some zinc die castings can be soldered. However, it is a difficult operation because of the varying composition of zinc die casting metal. The solder used is usually an alloy of 82.5 percent cadmium and 17.5 percent zinc. This alloy melts (flows) at 508 °F (264 °C). Any surface coatings or plating must be removed prior to soldering.

Die castings can sometimes be soldered without flux. Die castings may be successfully soldered by first coating the die casting with nickel and then soldering with a tin-lead solder.

Cadmium fumes are toxic! Soldering on or with cadmium or cadmium alloys should only be done under extremely good ventilation conditions.

## 7-19. TESTING AND INSPECTING SOLDERED JOINTS

Soldered joints are usually tested in two ways.
1. Test for being leakproof (where it is applicable).
2. Test for being immune to humidity.

The tightness of the joint can be tested hydrostatically (under pressure) using water.

The joint is tested for its corrosion resistance powers by putting the joint in a humidity cabinet where it is kept at a temperature of 100 °F (38 °C) and at approximately 100 percent humidity for at least 72 hours. A good joint will reveal no evidence of corrosion under low magnification inspection.

Visual inspection of the joint will usually detect such defects as poor adhesion, incomplete soldering, too much solder, overheating, dirt inclusion, etc.

## 7-20. REVIEW OF SOLDERING SAFETY PRACTICE

Soldering, brazing or welding with or on alloys containing cadmium or beryllium can be extremely hazardous.

Fumes from cadmium or beryllium compounds are extremely toxic. Several deaths have been reported from inhaling cadmium oxide fumes.

Skin contact with cadmium and beryllium should also be avoided.

An expert in industrial hygiene should be consulted whenever cadmium or beryllium compounds are to be used or when repairs are to be made on parts containing these metals.

Fluxes containing fluoride compound are also toxic.

Good ventilation is essential when soldering or brazing and the operator should always observe good safety practices.

A common hazard when soldering is exposure of the skin, eyes, and clothing to acid fluxes. Always work in a way that flux will not be spilled on the skin or clothing. Always wear chemical type eye goggles, rubber gloves, and long sleeves when using cleaning solutions, pickling solutions, or acids. Chemical type eye goggles do not have holes in the shield area above the eyes.

If any of these chemical solutions, acids, or fluxes come in contact with the skin, wash the area thoroughly under running water. If such chemicals contact the eyes, wash them thoroughly under running water and have a doctor check the eyes immediately.

Heating soldering coppers sometimes presents a fire or heating hazard if an open flame is used. Be sure flammable material is kept away from the heating flames.

Be sure that there are no flammable fumes such as gasoline, acetylene or other flammable gases present where soldering is being performed.

Perhaps the greatest hazard one meets in soldering is the attempt to solder gasoline or other fuel tanks. The job should never be attempted before all safety references have been read and a fire marshal consulted. The fire marshal will normally advise on the safe and proper procedures for soldering or welding on containers which have held combustible liquids or gases.

## 7-21. TEST YOUR KNOWLEDGE

Answer these questions on separate sheets of paper. Do not write in this book.
1. Can soldering be used to join two different base metals together?
2. How does soldering differ from welding?
3. How does soldering differ from brazing?
4. A soldered joint typically has the same strength as the base metal. True or False?
5. What does pickling do to a metal?
6. List three common solder alloys.
7. What element which is used in some solders is a health hazard?
8. There are three classifications of fluxes. List all three.
9. What is the purpose of a soldering flux?

10. Which classification of fluxes will best clean a surface?
11. What type soldering flux is recommended in Fig. 7-6 for use on brass?
12. What method is used after soldering to clean a solder joint when an inorganic flux has been used?
13. Name five (5) soldering methods.
14. Should the clearance between two metals to be soldered be large or small?
15. When torch soldering, where does the heat required to melt the solder come from?
16. When the dip soldering method is used, the molten solder is in a tank or pot. What is used to prevent the solder from oxidizing?
17. What type flux should be used when soldering stainless steel?
18. How is the surface of aluminum prepared for soldering?
19. What type solder is used on aluminum?
20. What is the chemical composition of Rose's fusible alloy solder?

Industry photo of brazing operation. The brazing process will not affect the heat treatment of the original metals as much as the welding process. Note the safety equipment of the operator.   (Handy and Harman)

# 8 BRAZING AND BRAZE WELDING

Both brazing and braze welding are metal joining processes which are performed at temperatures above 840 °F (450 °C), as compared to soldering which is performed at temperatures below 840 °F (450 °C). Refer to Heading 1-4 and to Fig. 1-8.

The American Welding Society defines these processes as follows:

BRAZING — "A group of welding processes which produces coalescence of materials by heating them to a suitable temperature and by using a filler metal having a liquidus above 840 °F (450 °C) and below the solidus of the base metal. The filler metal is distributed between the closely fitted surfaces of the joint by capillary action." Coalescence is a joining or uniting of materials.

BRAZE WELDING — "A welding process variation in which a filler metal, having a liquidus above 840 °F (450 °C) and below the solidus of the base metal, is used. Unlike brazing, in braze welding the filler metal is not distributed in the joint by capillary action."

Brazing has been used for centuries. Blacksmiths, jewelers, armorers, and other crafters used the process on large and small articles before recorded history. This joining method has grown steadily both in volume and popularity. It is an important industrial process, as well as a jewelry making and repair process. The art of brazing has become more of a science as the knowledge of chemistry, physics, and metallurgy has increased. Refer to Fig. 1-2 which shows the oxyacetylene flames used for brazing and braze welding. Refer to Fig. 1-9 which shows the oxygen-LP gas flames used for brazing and braze welding.

The usual terms Brazing and Braze Welding imply the use of a nonferrous alloy. These nonferrous alloys consist of alloys of copper, tin, zinc, aluminum, beryllium, magnesium, silver, gold, and others.

BRASS is an alloy consisting chiefly of copper and zinc. BRONZE is an alloy consisting chiefly of copper and tin. Most rods used in both brazing and braze welding on ferrous metals are brass alloys rather than bronze. The brands which are called bronze usually contain a small percent (about one percent) of tin.

## 8-1. BRAZING AND BRAZE WELDING PRINCIPLES

Brazing is an adhesion process in which the metals being joined are heated but not melted; the brazing filler metal melts and flows at temperatures above 840 °F (450 °C). Adhesion is the molecular attraction exerted between surfaces.

A brazed joint is stronger than a soldered joint because of the strength of the alloys used. In some instances it is as strong as a welded joint. It is used where mechanical strength and leakproof joints are desired. Brazing and braze welding are superior to welding in some applications, since they do not affect the heat treatment of the original metals as much as welding.

Brazing and braze welding warp the original metals less, and it is possible to join dissimilar metals. For example, steel tubing may be brazed to cast iron, copper tubing brazed to steel, and tool steel brazed to low carbon steel.

Brazing is done on metals which fit together tightly. The metal is drawn into the joint by capillary action. (A liquid will be drawn between two tightly fitted surfaces. This drawing action is known as CAPILLARY ACTION.) Very thin layers of filler metal are used when brazing. The joints and the material being brazed must be specially designed for the purpose. When brazing, poor fit and alignment result in poor joints and in inefficient use of brazing filler metal.

In braze welding, joint designs used for oxyfuel gas or arc welding are satisfactory. Such joint designs are discussed in Chapters 4 and 10. When braze welding, thick layers of the brazing filler metal are used.

When brazing or braze welding the metals must be cleaned thoroughly prior to adding the filler metal. The proper flux and filler metal must be used and post-cleaning is sometimes required. It is important to provide good ventilation when brazing or braze welding. Fumes from the heated fluxes and vaporized filler metals (such as zinc and cadmium) may affect the respiratory system, eyes, or skin.

## 8-2. JOINT DESIGNS FOR BRAZING AND BRAZE WELDING

The brazing process is similar to the soldering process because it uses a thin layer of filler metal. It also requires specially designed, tightly fitting joints. Brazing also requires that the joint be well supported until it has cooled. Examples of well designed brazing joints may be seen in Fig. 7-11. A few examples of joints designed for brazing, both good and bad, are shown in Figs. 8-1 and 8-2.

Braze welding is a process where a thick layer of filler metal is used. A convex or concave bead is usually used. The fit of the joint is not as critical as

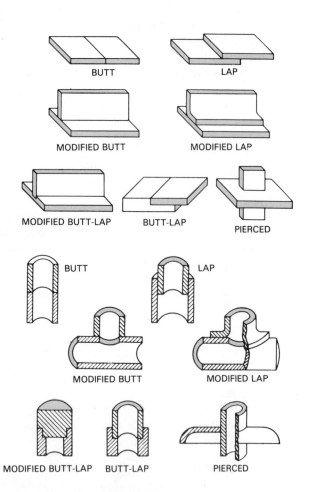

Fig. 8-1. Joints designed to produce good brazing results. (Handy and Harman)

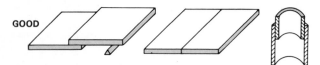

MATING SURFACES SHOULD BE ASSEMBLED AND SUPPORTED PARALLEL THROUGHOUT THE JOINT AREA TO OBTAIN UNIFORM FLOW OF ALLOY AND MAXIMUM STRENGTH.

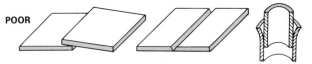

MISMATED LAP JOINTS, V-TYPE JOINTS, AND FLARED TUBULAR JOINTS WASTE BRAZING ALLOY AND MAY REDUCE THE JOINT STRENGTH.

PROPER FIT AND ALIGNMENT IN TUBULAR JOINTS IS YOUR ASSURANCE OF HIGH STRENGTH JOINTS.

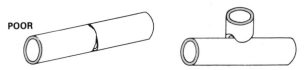

POOR FIT CAN INTERRUPT THE PULL OF CAPILLARY ACTION, REDUCE STRENGTH AND MAKE IT DIFFICULT TO GET LEAK—TIGHT JOINTS.

Fig. 8-2. Well-designed joints which have been prepared for brazing. Some poorly-designed joints shown for comparison.

that required for brazing. Figs. 8-3, 8-4, and 8-5 illustrate several well designed braze welding joints.

The proper clearance between the parts in a brazing joint is critical. If the spacing is too large, capillary action may not take place. If the space is too small, an insufficient amount of filler metal may be deposited in the joint. See Fig. 8-6 for recommended brazing joint clearances.

Brazing is often done on dissimilar metals. Dissimilar metals have different expansion rates. Special attention to the clearance in these joints is required. The varying expansion rates must be considered if the parts are clamped, restrained in jigs, or fit into one another.

Brass expands more than steel at a given temperature. If a brass part is fitted into a steel part, the clearance between them, at room temperature, must be larger than normal. If a steel part fits inside a brass part, the clearance should be closer than normal. See Fig. 8-7.

## 8-3. CLEANING BASE METALS PRIOR TO BRAZING OR BRAZE WELDING

In order to braze, capillary action must occur. It will occur properly only when the parts to be joined

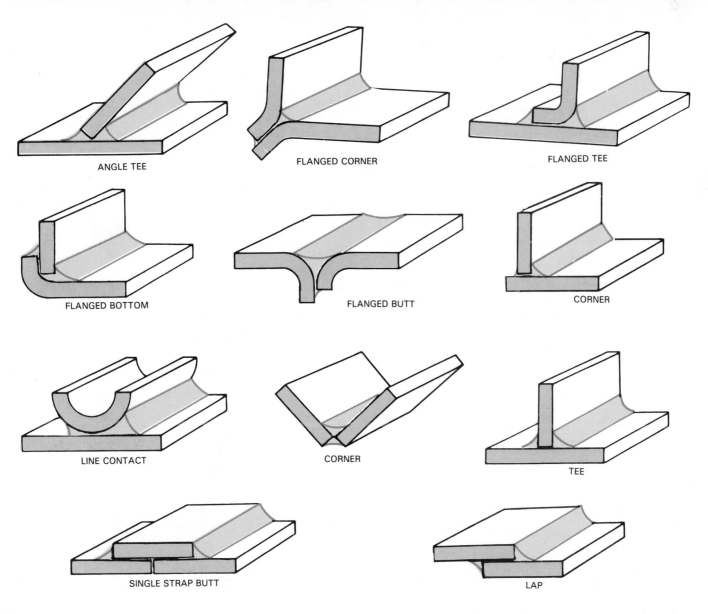

ANGLE TEE

FLANGED CORNER

FLANGED TEE

FLANGED BOTTOM

FLANGED BUTT

CORNER

LINE CONTACT

CORNER

TEE

SINGLE STRAP BUTT

LAP

Fig. 8-3. Well designed braze welding joints. Note the thickness of the filler metal applied. (Aluminum Co. of America)

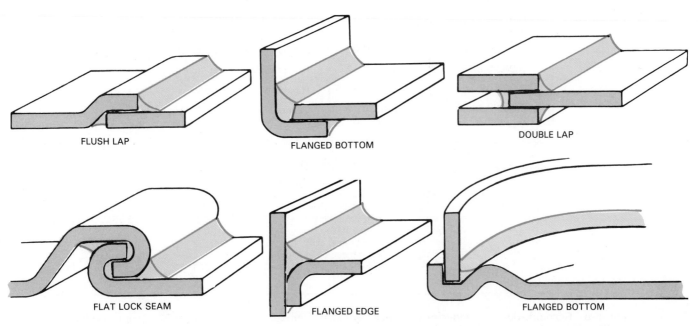

FLUSH LAP

FLANGED BOTTOM

DOUBLE LAP

FLAT LOCK SEAM

FLANGED EDGE

FLANGED BOTTOM

Fig. 8-4. Typical braze welded joint designs for use on containers. (Aluminum Co. of America)

**Brazing and Braze Welding / 197**

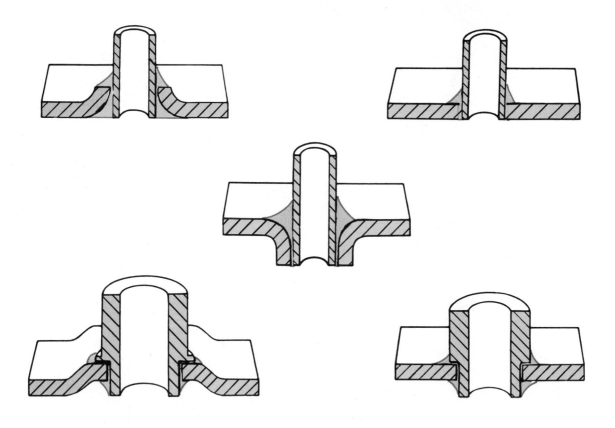

Fig. 8-5. Designs for braze welding sheet and tube joints. (Aluminum Co. of America)

| Filler Metal AWS Classification | Joint Clearance | | Notes |
|---|---|---|---|
| | in. | mm | |
| BAlSi group | 0.006-0.010 | 0.15-0.25 | For length of overlap less than 1/4 in. (6.35 mm) |
| | 0.010-0.025 | 0.25-0.61 | For length of overlap greater than 1/4 in. (6.35 mm) |
| BCuP group | 0.001-0.005 | 0.03-0.12 | |
| BAg group | 0.002-0.005 | 0.05-0.12 | Flux brazing (mineral fluxes) |
| | 0.001-0.002 | 0.03-0.05 | Atmosphere brazing (gas phase fluxes) |
| BAu group | 0.002-0.005 | 0.05-0.12 | Flux brazing (mineral fluxes) |
| | 0.000-0.002 | 0.00-0.05 | Atmosphere brazing (gas phase fluxes) |
| BCu group | 0.000-0.002 | 0.00-0.05 | Atmosphere brazing (gas phase fluxes) |
| BCuZn group | 0.002-0.005 | 0.05-0.12 | Flux brazing (mineral fluxes) |
| BMg group | 0.004-0.010 | 0.10-0.25 | Flux brazing (mineral fluxes) |
| BNi group | 0.002-0.005 | 0.05-0.12 | General applications (flux or atmosphere) |
| | 0.000-0.002 | 0.00-0.05 | Free flowing types, atmosphere brazing |

### Chemical Abbreviations

BAlSi  = Aluminum silicon alloy
BCuP   = Copper phosphorus alloy
BAg    = Silver base alloy
BAu    = Gold base alloy
BCu    = Copper base alloy
BCuZn  = Copper zinc alloy
BMg    = Manganese base alloy
BNi    = Nickel base alloy

Fig. 8-6. Recommended brazing joint clearances at brazing temperatures for various filler metal groups. The "B" in the filler metal classification stands for brazing. Chemical abbreviations are also given.

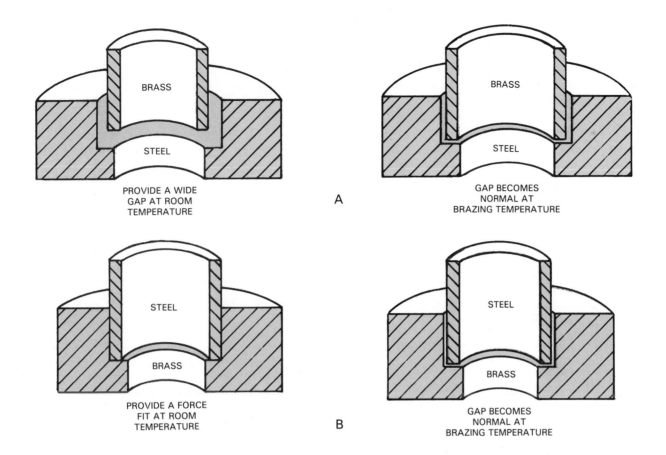

Fig. 8-7. The fitting of dissimilar metals must allow for the different rates of expansion which occur as they are heated. Brass expands at a faster rate than steel. A—Loosely fitted brass in steel fits properly at the brazing temperature. B—Steel force fitted into brass at room temperature has the correct clearance at the brazing temperature.

are clean. Clean parts are also required when braze welding. Part cleaning prior to brazing or braze welding may be accomplished by chemical and/or mechanical means.

Oil and grease must be removed using a proper solvent. Attempting to mechanically clean oily or greasy surfaces will usually result in scrubbing the oil and grease more deeply into the surfaces of the parts.

Following the removal of the oil and grease, a chemical pickling solution is used. This pickling solution removes the scale and rust from the surfaces of the parts. The chemicals used on various metals are the same as those described in Heading 7-2. Be sure to follow all safety procedures.

Mechanical cleaning is done with a grinding wheel, grit blasts, emery cloth, and/or a wire brush. Mechanical cleaning should be followed with a rinsing and drying operation. This rinsing will wash away small particles removed during the mechanical cleaning.

## 8-4. BRAZING AND BRAZE WELDING FLUXES

The American Welding Society defines a FLUX as, "Material used to prevent, dissolve, or facilitate removal of oxides and other undersirable surface substances."

A brazing or braze welding flux must be of a composition that keeps both the brazing filler metal and the metals being joined clean during the joining operation. The fluxes must be chemically pure. Generally, the manufacturers mark their fluxes C.P. (chemically pure).

The ingredients of a brazing flux are generally chlorides, fluorides, borax, borates, fluroborates, boric acid, wetting agents, and water.

These ingredients are mixed in varying amounts by manufacturers of brazing fluxes. The American Welding Society (AWS) has classified brazing fluxes into six categories. These fluxes are designated for use on various metals. Fig. 8-8 lists the AWS flux categories and their suggested uses. The base metal combinations used in brazements often vary greatly. It is often necessary to experiment with fluxes to determine the best one to use for a given combination of metals.

Fluxes used for braze welding must withstand higher temperatures and for longer periods than are required of brazing fluxes.

Borax works well for brazing or braze welding steel and iron. Borax or boric acid (borax plus

| AWS brazing flux type no. | Base Metals | Recommended filler metals | Recommended useful temperature range | | Flux Ingredients |
|---|---|---|---|---|---|
| | | | °F | °C | |
| 1 | All brazable aluminum alloys | BAlSi | 700-1190 | 371-643 | Chlorides, Fluorides |
| 2 | All brazable magnesium alloys | BMg | 900-1200 | 482-649 | Chlorides, Fluorides |
| 3A | All except those listed under 1, 2, and 4 | BCuP, BAg | 1050-1600 | 566-871 | Boric acid, Borates, Fluoborates, Wetting agent |
| 3B | All except those listed under 1, 2, and 4 | BCu, BCuP, BAg, BAu, RBCuZn, BNi | 1350-2100 | 732-1149 | Boric acid, Borates, Fluoborates, Wetting agent |
| 4 | Aluminum bronze, aluminum brass, and iron or nickel base alloys containing minor amounts of Al or Ti, or both | BAg (all) BCuP (copper base alloys only) | 1050-1600 | 566-871 | Chlorides, Fluorides, Borates, Wetting agent |
| 5 | All except those listed under 1, 2, and 4 | Same as 3B (excluding BAg-1 through -7) | 1400-2200 | 760-1204 | Borax, Boric acid, Borates, Wetting agent |

Fig. 8-8. AWS flux classifications for brazing. For specific metals and applications, suppliers of commercial fluxes should be consulted. "RB" means resistance brazing. See Fig. 8-6 for abbreviation meanings.

water) is a common base for brazing and braze welding fluxes for use with steel and iron. A popular mixture is 75 percent borax (powdered form) and 25 percent boric acid (liquid form) mixed to form a paste. Other ratios of these two chemicals are also used to form more solid or more fluid fluxes as may be required. These mixtures range from 75 percent to 25 percent borax with the remainder being boric acid. Some of the commercial fluxes also contain small amounts of phosphorus and halogen salts (halogen means any one of the iodines, bromine, fluorine, chlorine, and astatine chemical elements).

Alkaline bifluoride is used as a flux for brazing or braze welding stainless steel, silicon bronzes, aluminum or beryllium copper alloys. As most of the fumes from these fluxes are harmful to health, good ventilation is very necessary.

A special flux made of sodium cyanide salts is excellent when silver brazing tungsten to copper. THE FUMES ARE VERY DANGEROUS. AVOID BREATHING THE FUMES AND DO NOT LET THE FLUX CONTACT THE SKIN. This flux should only be handled by people specially trained in its use as it must be kept dry, away from acids, nitrates, and other oxidizing agents.

The following criteria should be considered when choosing a brazing flux:
1. Base metal or metals used.
2. Filler metal used.
3. Heat source; oxyfuel gas, resistance, carbon arc, etc.
4. Ease of flux residue removal.

5. Possible corrosive action on the metals used.
6. In resistance brazing the flux should be a conductor of electricity.
7. Fluxes containing water must not be used when dip brazing. The water may turn to steam and cause the liquid solder to erupt from the dip bath.
8. Health hazards in using toxic fluxes.

Fluxes may be purchased as powders, pastes, or liquids.

## 8-5. METHODS OF APPLYING BRAZING AND BRAZE WELDING FLUX

Fluxes are applied in whatever way is most efficient. Flux may be sprayed, brushed, or applied with a pressurized applicator. The amount used should be sufficient to last throughout the brazing or braze welding operation.

When braze welding, the filler metal is usually coated. The brazing rod may be purchased with the flux applied. To apply the flux to the rod on the job, heat the rod and dip it or roll it in the proper flux. The flux will adhere to the rod. This process must be repeated as the rod and flux is consumed. See Fig. 8-9.

A method which eliminates the extra handling of separate flux is to feed the flux into the fuel gas, and have the flux brought to the joint being brazed or braze welded along with the flame gases, as shown in Fig. 8-10. The flux is usually dissolved in alcohol. The installation requires two separate lengths of fuel gas hose with fittings.

The fluxing equipment, complete with the reserve flux tank, is shown in Fig. 8-11.

The fuel gas is fed through a container of liquid flux, and a controlled amount of the flux mixes

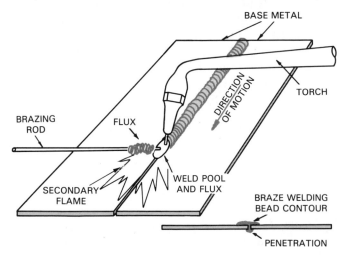

Fig. 8-9. A butt joint being braze welded using the oxyfuel gas flame as the heat source. Flux has been applied to the brazing rod. The torch is being moved in a forehand direction.

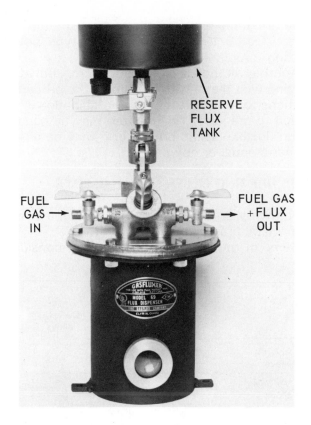

Fig. 8-11. A gas fluxing unit. (Gasflux Co.)

with the fuel gas and is fed to the operation through the torch tip. This method not only eliminates the separate operation of adding flux, but assures a continuous flow of flux of the correct amount, and results in an excellent and clean joint.

Powdered braze metals can also be injected into the welding flame. Using this method, the heated powder particles are protected from oxidation as

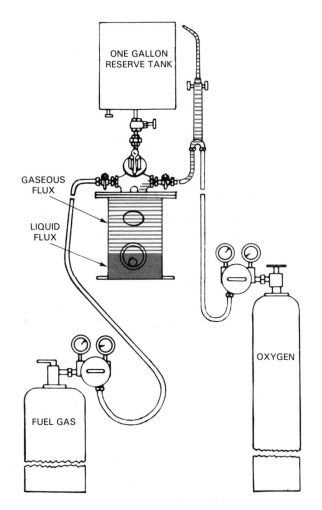

Fig. 8-10. An oxyfuel gas brazing station with a gas fluxing unit in the fuel gas supply line. (Gasflux Co.)

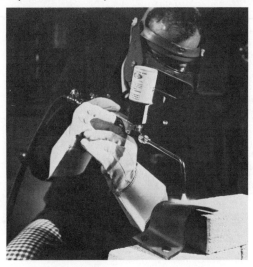

Fig. 8-12. A special oxyfuel gas torch with a hopper feed for ultra fine powder fluxes and/or metals. The torch can be used for brazing, braze welding, metal surfacing, and welding. (Wear Technology Div., Cabot Corp.)

they are transferred to the surface of the base metal. Fig. 8-12 shows a hopper type powder brazing torch.

Another method is to thoroughly mix powdered brazing filler metal with the flux in the proper proportions to form a paste. This combination is then added to the joint. This mixture may be hand fed using flexible plastic bottles or it may be gun fed (either manual or power).

## 8-6. BRAZING FILLER METAL ALLOYS

For every brazable metal there are one or more filler metal brazing rods manufactured.

Some of the factors which must be considered when selecting a brazing filler metal are:
1. Brazing and service temperature required.
2. Compatibility with the base metal(s).
3. Method of heating.

Fig. 8-13 lists a number of brazable metals and suggested metal combinations. The composition of a number of brazing filler metal alloys is shown in Fig. 8-14. Aluminum-silicon filler metals are used on the following aluminum and aluminum alloys: 1060, 1100, 1350, 3003, 3004, 5005, 5050, 6053, 6061, 6062, 6063, 6951, and A712.0 and C712.0 cast alloys. These brazed assemblies will generally withstand a constant temperature of 300 °F (149 °C). Fig. 26-16 lists the different aluminum alloy designations.

| | Al & Al alloys | Mg & Mg alloys | Cu & Cu alloys | Carbon & low alloy steels | Cast Iron | Stainless steels | Ni & Ni alloys |
|---|---|---|---|---|---|---|---|
| Al & Al alloys | BAlSi | | | | | | |
| Mg & Mg alloys | X | BMg | | | | | |
| Cu & Cu alloys | X | X | BAg, BAu, BCuP, RBCuZn | | | | |
| Carbon & low alloy steels | BAlSi | X | BAg, BAu, RBCuZn | BAg, BAu, BCu, RBCuZn, BNi | | | |
| Cast Iron | X | X | BAg, BAu, RBCuZn | BAg, RBCuZn | BAg, RBCuZn, BNi | | |
| Stainless steel | BAlSi | X | BAg, BAu | BAg, BAu, BCu, BNi | BAg, BAu, BCu, BNi | BAg, BAu, BCu, BNi | |
| Ni & Ni alloys | X | X | BAg, BAu, RBCuZn | BAg, BAu, BCu, RBCuZn, BNi | BAg, BCu, RBCuZn | BAg, BAu, BCu, BNi | BAg, BAu, BCu, BNi |
| Ti & Ti alloys | BAlSi | X | BAg | BAg | BAg | BAg | |
| Be, Zr, & alloys (reactive metals) | X BAlSi (Be) | X | BAg | BAg, BNi | BAg, BNi | BAg, BNi | BAg, BNi |
| W, Mo, Ta, Cb & alloys (refractory metals) | X | X | BAg | BAg, BCu, BNi | BAg, BCu, BNi | BAg, BCu, BNi | BAg, BCu BNi |
| Tool steels | X | X | BAg, BAu, RBCuZn, BNi | BAg, BAu, BCu, RBCuZn, BNi | BAg, BAu, RBCuZn, BNi | BAg, BAu, BCu, BNi | BAg, BAu, BCu, RBCuZn, BNi |

*Filler metals:  BAlSi - Aluminum silicon      BCuP - Copper phosphorus
BAg - Silver base       RBCuZn - Copper zinc
BAu - Gold base        BMg - Magnesium base
BCu - Copper          BNi - Nickel base

Fig. 8-13. Brazeable metal combinations and suggested filler metals. This table may be used if you are brazing two different metals. Find one metal in the left column and the other on the top line of the chart. Where the vertical and horizontal lines from these two metals intersect, the suggested filler metal may be found. ''X'' means brazing this combination is not recommended.

Fig. 8-14. A table of brazing filler metal alloys. The nominal chemical composition and melting and flowing temperatures are shown. NOTE: The cadmium (Cd) alloys are printed in red. Cadmium fumes are very toxic. See Fig. 8-6 or Chapter 31 for chemical abbreviations.

| AWS Filler Metal Classification | Ag | Al | Au | Cd | Cr | Cu | Ni | Si | Zn | B | C | Fe | Li | Mg | Mn | Sn | Ti | P | Pd | Zr | Others | Solidus[1] °F | Solidus[1] °C | Liquidus[2] °F | Liquidus[2] °C |
|---|---|---|---|---|---|---|---|---|---|---|---|---|---|---|---|---|---|---|---|---|---|---|---|---|---|
| BAlSi-2 | | 91.0 | | | | .25 | | 7.5 | .20 | | | .8 | | — | .10 | | | | | | .15 | 1070 | 577 | 1135 | 613 |
| BAlSi-3 | | 84.4 | | | .15 | 4.0 | | 10.0 | .20 | | | .8 | | .15 | .15 | | | | | | .15 | 970 | 521 | 1085 | 585 |
| BAlSi-4 | | 86.3 | | | | .30 | | 12.0 | .20 | | | .8 | | .10 | .15 | | | | | | .15 | 1070 | 577 | 1080 | 582 |
| BAlSi-5 | | 88.4 | | | | .30 | | 10.0 | .10 | | | .8 | | .05 | .05 | | | | | | .15 | 1070 | 577 | 1095 | 591 |
| BAlSi-6 | | 88.5 | | | | .25 | | 7.5 | .20 | | | .8 | | 2.5 | .10 | | | | | | .15 | 1038 | 559 | 1125 | 607 |
| BAlSi-7 | | 87.0 | | | | | | 10.0 | | | | .8 | | 1.5 | .10 | | | | | | .15 | 1038 | 559 | 1105 | 596 |
| BAlSi-8 | | 85.9 | | | | .25 | | 12.0 | .20 | | | .8 | | 1.5 | .10 | | | | | | .15 | 1038 | 559 | 1075 | 579 |
| BCuP-1 | | | | | | 94.9 | | | | | | | | | | | | 5.0 | | | .15 | 1310 | 710 | 1695 | 924 |
| BCuP-2 | | | | | | 92.6 | | | | | | | | | | | | 7.3 | | | .15 | 1310 | 710 | 1460 | 793 |
| BCuP-3 | 5.0 | | | | | 88.9 | | | | | | | | | | | | 6.0 | | | .15 | 1190 | 643 | 1495 | 813 |
| BCuP-4 | 6.0 | | | | | 86.6 | | | | | | | | | | | | 7.3 | | | .15 | 1190 | 643 | 1325 | 718 |
| BCuP-5 | 15.0 | | | | | 79.9 | | | | | | | | | | | | 5.0 | | | .15 | 1190 | 643 | 1475 | 802 |
| BCuP-6 | 2.0 | | | | | 90.9 | | | | | | | | | | | | 7.0 | | | .15 | 1190 | 643 | 1450 | 788 |
| BCuP-7 | 5.0 | | | | | 88.1 | | | | | | | | | | | | 6.8 | | | .15 | 1190 | 643 | 1420 | 771 |
| BAu-1 | | | 37.5 | | | 62.4 | | | | | | | | | | | | | | | .15 | 1815 | 991 | 1860 | 1016 |
| BAu-2 | | | 80.0 | | | 19.9 | | | | | | | | | | | | | | | .15 | 1635 | 891 | 1635 | 891 |
| BAu-3 | | | 35.0 | | | 64.9 | | | | | | | | | | | | | | | .15 | 1785 | 974 | 1885 | 1029 |
| BAu-4 | | | 82.0 | | | 17.9 | | | | | | | | | | | | | | | .15 | 1740 | 949 | 1740 | 949 |
| BAu-5 | | | 30.0 | | | .05 | 36.0 | | | | | | | | | | | | 34.0 | | .15 | 2135 | 1166 | 2130 | 1166 |
| BMg-2a | | 9.0 | | | | | .005 | .05 | 2.0 | | | .005 | | 88.4 | .15 | | | | | | .30 | 830 | 443 | 1110 | 599 |
| BMg-1 | | 12.0 | | | | | | | 5.0 | | | | | 82.7 | | | | | | | .30 | 770 | 410 | 1050 | 566 |
| BCu-1 | | | | | | 99.9* | | | | | | | | | | | | | | | .1 | 1980 | 1082 | 1980 | 1082 |
| BCu-1a | | | | | | 99.0* | | | | | | | | | | | | | | | .3 | 1980 | 1082 | 1980 | 1082 |
| BCu-2 | | | | | | 86.5* | | | | | | | | | | | | | | | .5 | 1980 | 1082 | 1980 | 1082 |
| BAg-1 | 45.0 | | | 24.0 | | 15.0 | | | 16.0 | | | | | | | | | | | | .15 | 1125 | 607 | 1145 | 618 |
| BAg-1a | 50.0 | | | 18.0 | | 15.5 | | | 16.5 | | | | | | | | | | | | .15 | 1160 | 627 | 1175 | 635 |
| BAg-2 | 35.0 | | | 18.0 | | 26.0 | | | 21.0 | | | | | | | | | | | | .15 | 1125 | 607 | 1295 | 702 |
| BAg-2a | 30.0 | | | 20.0 | | 27.0 | | | 23.0 | | | | | | | | | | | | .15 | 1125 | 607 | 1310 | 710 |
| BAg-3 | 50.0 | | | 16.0 | | 15.5 | 3.0 | | 15.5 | | | | | | | | | | | | .15 | 1170 | 632 | 1270 | 688 |
| BAg-4 | 40.0 | | | | | 30.0 | 2.0 | | 28.0 | | | | | | | | | | | | .15 | 1240 | 671 | 1435 | 779 |
| BAg-5 | 45.0 | | | | | 30.0 | | | 25.0 | | | | | | | | | | | | .15 | 1250 | 677 | 1370 | 743 |
| BAg-6 | 50.0 | | | | | 34.0 | | | 16.0 | | | | | | | | | | | | .15 | 1270 | 688 | 1425 | 774 |
| BAg-7 | 56.0 | | | | | 22.0 | | | 17.0 | | | | | | | 5.0 | | | | | .15 | 1145 | 618 | 1205 | 652 |
| BAg-8 | 72.0 | | | | | 28.0 | | | | | | | | | | | | | | | .15 | 1435 | 779 | 1435 | 779 |
| BAg-8a | 72.0 | | | | | 27.6 | | | | | | | .38 | | | | | | | | .15 | 1410 | 766 | 1410 | 766 |
| BAg-13 | 54.0 | | | | | 40.0 | 1.0 | | 5.0 | | | | | | | | | | | | .15 | 1325 | 718 | 1575 | 857 |
| BAg-13a | 56.0 | | | | | 42.0 | 2.0 | | | | | | | | | | | | | | .15 | 1420 | 771 | 1640 | 893 |
| BAg-18 | 60.0 | | | | | 29.8 | | | | | | | | | | 10.0 | | .025 | | | .15 | 1115 | 602 | 1325 | 718 |
| BAg-19 | 92.5 | | | | | 7.2 | | | | | | | .23 | | | | | | | | .15 | 1435 | 779 | 1635 | 891 |
| BAg-20 | 30.0 | | | | | 38.0 | | | 32.0 | | | | | | | | | | | | .15 | 1250 | 677 | 1410 | 766 |
| BAg-21 | 63.0 | | | | | 28.5 | 2.5 | | | | | | | | | 6.0 | | | | | .15 | 1275 | 691 | 1475 | 802 |
| BNi-1 | | | | | 14.0 | | 72.4 | 4.5 | | 3.2 | .75 | 4.5 | | | | | | .02 | | | .50 | 1790 | 977 | 1900 | 1038 |
| BNi-1a | | .05 | | | 14.0 | | 73.1 | 4.5 | | 3.2 | .06 | 4.5 | | | | | .05 | .02 | | .05 | .50 | 1790 | 977 | 1970 | 1077 |
| BNi-2 | | | | | 7.0 | | 81.3 | 4.5 | | 3.2 | .06 | 3.0 | | | | | | .02 | | | .50 | 1780 | 971 | 1830 | 999 |
| BNi-3 | | | | | | | 91.5 | 4.5 | | 3.2 | .06 | .5 | | | | | | .02 | | | .50 | 1800 | 982 | 1900 | 1038 |
| BNi-4 | | | | | | | 92.5 | 3.5 | | 1.8 | .06 | 1.5 | | | | | | | | | .50 | 1800 | 982 | 1950 | 1066 |
| BNi-5 | | | | | 19.0 | | 70.0 | 10.2 | | .03 | .10 | | | | | | | | | | .50 | 1975 | 1079 | 2075 | 1135 |
| BNi-6 | | | | | | | 88.3 | | | | .10 | | | | | | | 11.0 | | | .50 | 1610 | 877 | 1610 | 877 |
| BNi-7 | | | | | 14.0 | | 74.9 | .10 | | .01 | .08 | .2 | | | | | | 10.1 | | | .50 | 1630 | 888 | 1630 | 888 |
| BNi-8 | | | | | | 4.5 | 64.7 | 7.0 | | | .10 | | | | 23.0 | | | .02 | | | .50 | 1800 | 982 | 1850 | 1010 |

1-Solidus - Melting temperature    2-Liquidus - Flow temperature    * - Minimum

(AWS A5.8-89 Specification for Filler Metals for Brazing)

There are two magnesium filler metals in general use. The BMg-1 is used to join AZ10A, K1A, M1A magnesium alloys. Filler metal BMg-2a has a lower melting temperature and is used with magnesium alloys, AZ31B and ZE10A. Service temperature for these alloys is generally 250°F (121°C).

Copper and copper-zinc filler metals are generally used on ferrous metals. They may also be used on nonferrous metals. The continuous operating temperature for these alloys is 400°F (204°C).

Copper-phosphorus alloys are used to join copper and copper alloys. They are also used for silver, tungsten, and molybdenum. They should not be used on ferrous, nickel based, or copper-nickel (with more than 10 percent nickel) alloys. The service temperature is 300°F (149°C).

Silver based filler metal alloys may be used to join ferrous or nonferrous metals. Silver alloy filler metals are not recommended for use on aluminum or magnesium. Cadmium added to the silver-zinc-copper alloy makes the alloy more fluid. It also improves the wetting qualities. Cadmium also lowers the melting and flow temperatures of silver based filler metals. The service temperature is 400°F (204°C).

In the fabrication of jewelry, small precision articles and instruments which require strong joints, various silver alloys are used as the joining metal. The original use of silver alloys was in jewelry manufacturing. At present, most industries have some industrial applications for this method of brazing.

SILVER BRAZING is often called silver soldering. However, the correct term is silver brazing because the temperature of the process is above 840°F (450°C).

Many alloys of silver have been developed. Each has a different melting point, flow characteristics, strength, and color. Some silver alloys melt at relatively low temperatures while others melt at fairly high temperatures.

Due to the cost of silver alloys, they are most frequently used for brazing operations rather than for braze welding. The application or the method of performing the brazing varies little from the brazing previously described.

The metals which form the alloys of silver are gold, copper, cadmium, and zinc. The best grades of silver alloys are the ones which are formed partly of gold. This type of silver alloy is used mostly in jewelry work and precision instrument work. The silver alloy compositions are shown in Fig. 8-14.

Many silver brazing alloys are on the market. Each alloy has its own particular fields of application. These alloys vary in melting temperature and in flow temperature. The term MELTING TEMPERATURE means the temperature at which the alloy starts to melt. FLOW TEMPERATURE means the temperature which the alloy must reach so all of the metal alloy is liquid. These temperatures may vary with some alloys from 1125°F (607°C) melting to 1145°F (618°C) flow.

Wide temperature difference between melting temperature and flow temperature presents some difficulties. While applying one of these alloys, the lower melting temperature metals in the alloy may flow into the cracks and leave the higher temperature alloy metals behind.

This action causes a change in color and strength. It also causes difficulty in flowing the remaining alloy onto the base metal. It is best, therefore, to heat the alloy quickly, first to minimize oxidation, and second to prevent alloy separation. The separation feature of the wide temperature range alloys is an advantage in poor fit-up joints. The higher temperature metals in the alloy will help to bridge the gaps.

Cadmium and zinc are used as alloying metals in silver brazing alloys, because they have the peculiar ability to "wet" or flow and alloy with iron. They also lower the alloy melting and flowing temperatures.

There is some danger of producing harmful fumes if alloys containing zinc and cadmium are overheated and the metals vaporize. The work area should be well-ventilated.

Silver brazed joints are very strong if properly made. The strength of any brazed joint varies with the thickness of the filler metal thickness. When joining stainless steel butt joints by silver brazing, the tensile strength varies as shown in Fig. 8-15.

| STRENGTH OF SILVER BRAZED BUTT JOINTS | | | |
|---|---|---|---|
| Thickness of the brazing filler metal | | Tensil Strength | |
| inches | millimeters | (psi) | Megapascals |
| .002 | .05 | 133,000 | 917 |
| .003 | .08 | 115,000 | 793 |
| .006 | .15 | 90,000 | 621 |
| .009 | .23 | 83,000 | 572 |
| .012 | .30 | 76,000 | 524 |
| .015 app. 1/64 | .38 | 70,000 | 483 |

Fig. 8-15. A table showing how the strength of a silver brazed joint is affected by varying thicknesses of silver brazing filler metal in the joint.

Silver brazing is one of the best methods used to connect parts in a leakproof manner and to provide maximum strength. These joints are strong and will stand up under severe conditions. The excellent adhesion qualities are shown in Fig. 8-16.

When joining copper, brass and bronze parts, a copper, silver, and phosphorus alloy may be used. This alloy is less expensive than most silver brazing alloys. No flux is necessary on copper, but

Fig. 8-16. A microphotograph of a silver alloy brazed steel joint. The brazing filler metal in the joint is Easy-Flo which is a 45 percent Ag, 15 percent Cu, 16 percent Zn, and 24 percent Cd alloy.

brass (copper-zinc) is usually brazed using a flux. This type of alloy is not used on steel or iron alloys. The silver content is approximately 15 percent. The alloy flows at 1300 °F (704 °C). A microphotograph of this joint is shown in Fig. 8-17. The absence of flux when brazing copper improves the visibility of the joint during the brazing operation.

Gold filler metals are used in electronics and many space technology applications. It is used to braze ferrous, nickel, and cobalt base metals. Gold alloys are used when volatile filler metal components are undesirable. They are also used when oxidation and corrosion resistance is required.

Fig. 8-17. A microphotograph of a copper and brass part joined by a 15 percent silver plus copper and phosphorus brazing alloy.

Gold filler metal alloys have a service temperature of 800 °F (427 °C).

Nickel based filler metals work well in cryogenic (very low temperatures) applications. They also have a high service temperature of 1800 °F (982 °C). They are used on carbon and stainless steels, and cobalt or copper based alloys.

Brazing filler metals may be applied by means of an applicator. These applicators are generally pneumatically (air) powered. Filler metals are applied in this manner for automatic multistation brazing machines or for oven heated processes. Such a filler metal applicator is shown in Fig. 8-18.

Cobalt based filler metals have the highest service temperature at 1900 °F (1038 °C). It is generally used with cobalt alloys.

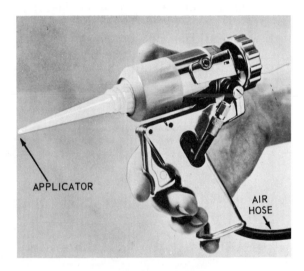

Fig. 8-18. A pneumatically powered filler metal applicator. This Nicrobraz applicator gun is used to apply filler metal paste prior to oven brazing.    (Wall Colmonoy Corp.)

## 8-7. BRAZE WELDING FILLER METALS

There are six AWS classifications for copper or copper-zinc based brazing and braze welding filler metals. The chemical composition and solidus and liquidus temperatures may be seen in Fig. 8-19. Tin, iron, silicon, and magnanese are added to some filler rods. These alloying elements improve the flow characteristics, scavenge oxygen, and increase strength and hardness. These elements also decrease the tendency of zinc to vaporize. The addition of 10 percent nickel also improves the strength of brass rods. The service temperature is 500 °F (260 °C). Brass (copper-zinc) brazing rods have a tensile strength of between 40,000 - 60,000 psi (276 - 414 MPa).

It should be remembered that a braze welded joint generally involves two dissimilar metals.

| AWS Classification | Nominal Chemical Composition (Percentage) | | | | | | | | | | | Temperature | | | |
|---|---|---|---|---|---|---|---|---|---|---|---|---|---|---|---|
| | Cu | Zn | Sn | Fe | Mn | Ni | P | PB | Al | Si | Others (max.) | Solidus °F | °C | Liquidus °F | °C |
| BCu-1 | 99.9 min. | | | | | | .08 | .02 | .01 | | .10 | 1980 | 1082 | 1980 | 1082 |
| BCu-1a | 99.0 min. | | | | | | | | | | .30 | 1980 | 1082 | 1980 | 1082 |
| BCu-2 | 86.5 min. | | | | | | | | | | .50 | 1980 | 1082 | 1980 | 1082 |
| RBCuZu-A | 59.0 | 39.8 | .63 | | | | | .05 | .01 | | .50 | 1630 | 888 | 1650 | 899 |
| RBCuZn-C | 58.0 | 39.4 | .95 | .72 | .26 | | | .05 | .01 | .10 | .50 | 1590 | 866 | 1630 | 888 |
| RBCuZn-D | 48.0 | 41.0 | | | | 10.0 | .25 | .05 | .01 | .15 | .50 | 1690 | 921 | 1715 | 935 |

B - brazing          RB - welding or brazing rod

Fig. 8-19. Copper and copper-zinc based brazing filler metals. See also AWS A5.8-89 entitled "Specification for Filler Metals for Brazing" for more information.

Whenever two dissimilar metals are in contact a small electrical flow is created under certain conditions. This also causes corrosion called galvanic corrosion.

Braze welding filler metal may be less resistant to chemical solutions than the base metal. This lessened corrosion resistance and the possibility of galvanic corrosion must be considered in design applications.

## 8-8. BRAZING AND BRAZE WELDING PROCESSES

Heat sources for brazing may be:
1. A molten bath of brazing metal alloy similar to the dip solder method. See Heading 7-11.
2. Torch heating with an oxyfuel gas torch. Fig. 8-20 illustrates a multistation torch brazing machine.

Fig. 8-20. A multistation torch brazing machine. Parts are manually loaded and unloaded. The flux is applied and the part is heated and cooled automatically as the table rotates.    (Handy and Harman)

3. Controlled atmosphere furnaces.
4. Electric resistance heating.
5. Carbon arc. (Refer to Heading 10-22.)
6. Induction heating.

Braze welding usually requires more heat than brazing. Since braze welding is similar to welding, the operation is usually done with an oxyfuel gas torch, carbon arc torch, or gas tungsten arc.

Brazing and braze welding procedures are quite similar, the main difference being in the joint design and the quantity of brazing rod applied to the joint.

Some good pointers applicable to both brazing and braze welding follow:

1. Metals to be joined must be mechanically and chemically clean.
2. Two pieces to be brazed together must be fitted properly: that is, the metals should not be spaced too far apart or forced together. The braze welded joint is prepared in the same manner as any oxyfuel gas or arc welded joint.
3. The two pieces must be firmly supported during the brazing or braze welding and cooling operations. Any movement of either part while the joining metal is molten or plastic will weaken the joint.
4. The metals to be joined must be heated to a temperature slightly above the melting temperature of the brazing filler metal, but below the melting temperature of the metal being brazed or braze welded.
5. Clean, fresh flux must be used as the operation proceeds to reduce oxidation and to float the oxides to the surface.
6. A heat source, as listed previously, is required to obtain a high enough temperature to obtain a good joint.

The oxyfuel gas torch flame is usually adjusted to a neutral flame. Refer to Fig. 1-2. A carburizing (reducing) flame will produce an exceptionally neat looking joint, but strength will be sacrificed. An oxidizing flame will produce a strong joint, but with a rough looking surface. A neutral flame will give the best results under ordinary conditions.

Automobile body panels are often soldered or leaded after a repair. The solder smooths out any flaws in the panel prior to painting. See Fig. 8-21.

7. As in steel welding, the brazing filler metal must penetrate to the other side of the joint. This penetration must be such that the joining metal covers 100 percent of the joint surfaces and adheres to these two surfaces. A minimum amount of joining material should be used.
8. The metal must be cooled properly to obtain the desired properties in both the original metal and the joining metal. Heat treatment of metals is covered in Chapter 27.

Fig. 8-21. This autobody repair person is soft soldering or leading the deck lid of an automobile. A propane torch heats the solder and a wooden stick smooths the molten solder.

## 8-9. BRAZING OR BRAZE WELDING STEEL WITH COPPER AND ZINC ALLOYS

Clean the surfaces as described in Heading 8-3. An emery wheel or emery cloth is not recommended due to the possibility of imbedding abrasive particles and oil in the metal. When braze welding, provide for contraction and expansion as in the welding process. If the metal is thicker than 8 gauge, (.128 in. or 3.25 mm) it must be grooved or chamfered to permit adequate penetration.

The joint to be brazed must be designed and fitted so that a close fit of the two base metals is obtained.

Adjust the torch for the flame desired, using the same size tip as would be used for welding the same thickness of the base metal. When brazing, the flux is usually applied to the joint, while in braze welding the flux is often applied to the brazing rod. Apply the torch to the metals to be brazed or braze welded, heating the metals at the joint (each one equally) to a dull cherry red. The width of the bead when braze welding will be determined by how wide a portion of the metal is heated to cherry red. The brazing filler metal will not flow over the surface unless the surface is at the brazing filler metal flow temperature.

The width of the braze weld bead should be a little wider than a steel weld on the same thickness of metal. While heating the metal, the brazing filler metal rod should be kept near the torch flame to maintain a fairly high rod temperature as shown in Fig. 8-9. After the metals have been heated, bring the brazing rod (flux-coated portion) into contact with the cherry red metals, meanwhile maintaining the torch motion. The brazing rod will quickly melt and flow over or between the parent metals. Do not overheat the brazing rod. Keep it away from the inner cone.

When braze welding, the bead should proceed along the joint just as in the welding process except the procedure should be faster. The torch flame is not held as close to the metal as in steel welding (approximately double the distance). The width of the bead may be controlled by raising and lowering the torch flame. When the flame is held close to the metal a wide bead will result. By drawing the torch away, the metal cools slightly and the bead will not be so wide.

The finished braze or braze weld should have the appearance of adequate fusion with the base metal. The brazing filler metal should penetrate through the joint and appear on the other side. A white deposit on the outside of the brazed or braze welded joint indicates an overheated joint. The color of the braze filler metal in the joint will indicate if it has been overheated. The best looking brazed or braze welded joint will show a color exactly similar to the brazing filler metal used. If the brazing filler metal is heated to an excessive temperature, some of the zinc will be burned out leaving a coppery appearance. If an oxidizing flame is used, the brazing filler metal will have a red color due to the oxidation of the copper.

## 8-10. BRAZING WITH SILVER ALLOYS

Silver brazing is similar in all ways to other soldering and brazing processes. The points to be remembered are:
A. Clean the joints mechanically and chemically (use clean materials and tools).
B. Fit the joint closely and support the joint.
C. Apply the proper flux.
D. Heat to the correct temperature.
E. Apply the silver brazing material.
F. Cool the joint.
G. Clean the joint thoroughly.

An oxyacetylene torch is an excellent heat source for silver brazing, as shown in Fig. 8-22. Liquefied petroleum fuel in conjunction with oxygen is acceptable.

Fluxes used for silver brazing must be clean, chemically pure and fresh. Chlorides are popularly used as silver brazing fluxes. Borax made into a

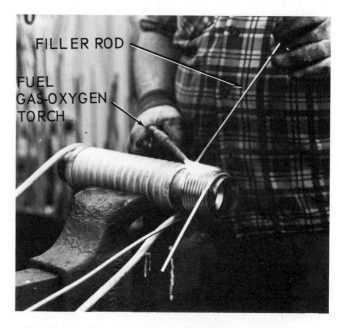

Fig. 8-22. Silver brazing small tubing to large pipe. (Handy and Harmon)

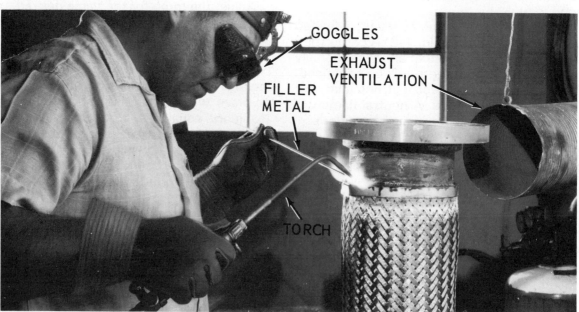

Fig. 8-23. Silver brazing a flexible tube to a flange, using an oxyacetylene flame as the source of heat. (Handy and Harman)

paste with water may also be used successfully as a silver brazing flux. See Fig. 8-8 for suggested fluxes to use.

The behavior of the flux is most helpful in silver brazing. It will indicate the temperature changes in the joint as follows:

1. The flux will dry out as moisture (water) boils away at 212 °F (100 °C).
2. Then the flux will turn milky and start to bubble at about 600 °F (316 °C). Finally, it will turn into a clear liquid at about 1100 °F (593 °C). This is just short of the brazing temperature. THE CLEAR APPEARANCE OF THE FLUX WILL INDICATE THE TIME TO START ADDING THE FILLER METAL.

Fig. 8-23 shows a flexible tube being brazed to a flange. The BAg-5 classification alloy melts at 1230 °F (662 °C) and flows at 1370 °F (743 °C). See Fig. 8-14 for other silver brazing filler metal alloy melting and flow temperatures.

A large tip is recommended for heating the joint, as the extra heat permits a shorter brazing time, thus reducing the time for oxides to form. The basics of a brazing operation when using a flame as the source of heat are shown in Fig. 8-24. The joint should be kept covered with the flame during the whole operation to prevent air from getting to the joint. The flame should not blow either the flux or the molten metal. Use a slight feather (reducing flame) on the inner cone for best joint appearance. Fig. 8-25 shows a large assembly being silver brazed using an oxyacetylene flame as the source of heat.

It is necessary to heat both pieces evenly. If a thick piece is being joined to a thin piece, the heat must be applied more to the thick piece. Keep the torch in motion while the alloy is being added or local "hot" spots may develop and result in a poor joint.

Fig. 8-25. A large assembly being silver brazed using the oxyacetylene source of heat.    (Handy and Harman)

A popular way to apply silver brazing filler metal when making a silver braze on a tubing joint is to use silver alloy rings as shown in Fig. 8-26. This is a practical way to add silver alloy. Preplaced brazing and soldering forms are economical when used on a production basis. See Fig. 8-27.

A method of brazing using preplaced brazing shims is shown in Fig. 8-28.

It is necessary to thoroughly clean the completed silver brazed joint by washing in water or scrubbing. Any flux left on the metals will tend to corrode them and also hinder any painting or plating operations.

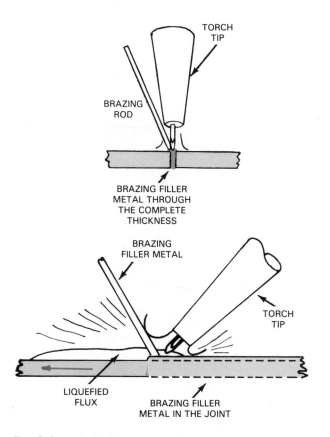

Fig. 8-24. Basic flame method of brazing a butt joint.

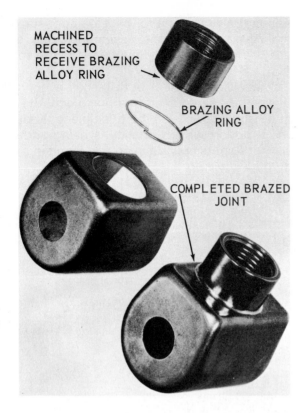

Fig. 8-26. Silver brazed joints designed to use preplaced silver alloy rings. The alloy forms almost perfect fillets and no further finishing is necessary. (Handy and Harman)

Fig. 8-27. Brazing filler metal may be obtained in sheet form, as wire, or in preshaped forms as shown. The preshaped forms are placed on or in the brazement before it is heated. (Handy and Harman)

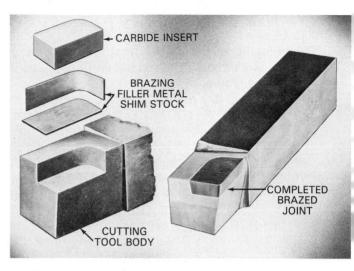

Fig. 8-28. A machining tool bit showing how the carbide insert is brazed to the tool bit body using preplaced brazing filler metal shims.

The joint may be cooled quickly or slowly. Cooling with water is permissible. Water quenching may also serve to wash the flux from the joint.

Visual inspection of the joint will quickly reveal any places where the braze metal did not adhere. It is best to watch for this adherence and make any corrections during the brazing operation.

## 8-11. BRAZING STAINLESS STEEL

Stainless steel may be silver brazed as easily as it is welded. A silver brazed joint is strong if properly designed and made. The metal does not warp

Fig. 8-29. A microphotograph of a silver brazed stainless steel and copper joint.

to any great extent because of the relatively low silver brazing temperatures. A good color match can usually be obtained. With stainless steel, a 45 percent silver alloy filler rod produces a good silver brazed joint.

Fig. 8-29 shows adhesion properties of a silver brazing alloy between stainless steel and copper.

## 8-12. BRAZING AND BRAZE WELDING ALUMINUM ALLOYS

Aluminum alloys can be brazed or braze welded using a number of the methods of heating. Oxy-fuel gas is a popular means of heating aluminum for brazing. Dip brazing and furnace brazing are used when enough production warrants the cost. The carbon arc and electric resistance heating are also used.

Practically all of the common joints have been successfully brazed or braze welded with aluminum. See Fig. 8-3, 8-4, and 8-5.

In the torch method of aluminum brazing and braze welding, both the oxyhydrogen and the oxy-acetylene flames may be used. In either case, a reducing flame with a 1 to 2 in. (25 - 50 mm) intermediate cone should be used. The following procedure produces good results:

1. Clean and degrease the surface. Braze or braze weld the joint as quickly as possible after cleaning the metal, while the surface is free of oxides.
2. Apply a fresh, chemically clean flux, specially compounded for aluminum brazing or braze welding, along the joint and on the brazing rod. The flux may be in either powder or paste form.
3. Heat the joint and as the flux chemicals turn liquid, start to apply the brazing rod. The rod will melt and flow into the joint and through the crevices to produce a heat brazed or braze welded joint. Hold the torch so the inner flame cone is 1 to 2 in. (25 - 50 mm) away from the joint.
4. Cool the finished braze or braze welded joint. Do not disturb the parts until the temperature has dropped to below 900 °F (482 °C).
5. Clean the flux from the joint. Use hot water, then a concentrated nitric acid solution. Follow all safety procedures. After this, wash again in boiling water. Any flux left on the metal will seriously corrode it.

Danger - The handling and use of concentrated acids should only be undertaken by an experienced person who has been given thorough training in the use of dangerous acids.

The finished weld should show 100 percent adhesion and the surface should be smooth and clean. The brazing filler metal should penetrate, with good adhesion, the full thickness of the joint. See Fig. 8-30.

Fig. 8-30. A 15x microphotograph of an aluminum brazed inside corner joint. Note that little or no parent metal has melted.

## 8-13. BRAZING AND BRAZE WELDING MAGNESIUM ALLOYS

Magnesium and magnesium alloys may be brazed or braze welded using the same procedures as used on other metals.

Pre-cleaning methods vary with the magnesium alloy being brazed. See Fig. 8-31. Mechanical cleaning with steel wool or abrasive cloth works very well. Chemical cleaning includes degreasing, bright pickling, chrome pickling, and modified chrome pickling. Degreasing is normally done with an alkaline solution. Refer to Chapter 31 for the

| Magnesium Alloy | Filler Metal Used | | |
|---|---|---|---|
| | AZ125Xa | GA432 | AZ92A |
| AZ10A | 1,2 | 1 | 1 |
| AZ31B | 1,2 | 1,2 | * |
| AZ61A | * | 1 | * |
| K1A | 1 | * | 1 |
| M1A | 1 | 1 | 1 |
| ZE10A | 1, 2, 3 | 1, 3 | * |
| ZK21A | 1 | * | 1 |
| ZK60A | * | 1, 4 | * |

1 - Mechanical cleaning
2 - Bright pickle (2 min.)
3 - Chrome pickle (2 min.)
4 - Modified chrome pickle followed by a 20% hydrofluoric acid dip (2 min.)

Fig. 8-31. Magnesium pre-cleaning methods.

chemical compositions of the various pickling solutions used on magnesium.

The fluxes used in magnesium brazing and braze welding are shown in Fig. 8-32.

| Brazing Process | Flux Composition (Percentage) | Approximate Melting Point | |
|---|---|---|---|
| | | °F | °C |
| Torch | KCl - 45<br>NaCl - 26<br>LiCl - 23<br>NaF - 6 | 1000 | 538 |
| Torch, Dip or Furnace | KCl - 42.5<br>NaCl - 10<br>LiCl - 37<br>NaF - 10<br>AlF₃·3NaF - 0.5 | 730 | 388 |

Fig. 8-32. Magnesium brazing flux composition and use.

The brazable magnesium alloys and the suggested filler metals are shown in Fig. 8-33. The composition of the magnesium brazing filler metal is shown in Fig. 8-34.

Post-cleaning of magnesium brazed and braze welded joints is extremely important. The cleaning should be done with hot flowing water. Mechanical scrubbing may also be required. After this initial cleaning, the parts are dipped into a chrome pickle for about two minutes.

Magnesium may be brazed using the torch, dip, or furnace methods. Heating must be very carefully controlled. The brazing temperature varies between 1080° and 1160°F (582° - 627°C).

The procedure will not vary from that used with other metals as indicated below:

1. Insure that the joint is well designed and, if brazing, clearances are proper to permit capillary action to occur.
2. Pre-clean.
3. Apply flux.
4. Use the proper flame if using oxyfuel gas.
5. Heat the joint evenly to the correct temperature.
6. Apply the braze filler rod.
7. Support the metal during the heating and cooling process to avoid stress cracks.
8. Post-clean the metal.

A good quality brazed magnesium joint is shown in Figs. 8-35 and 8-36.

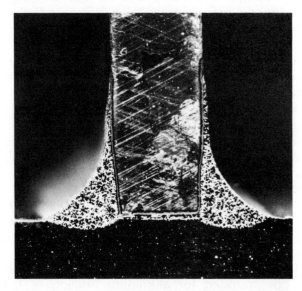

Fig. 8-35. A magnesium brazed joint. The sample has been etched; the macrophotograph is 8x. (Dow Chemical Co.)

| ASTM alloy designation | Base metal | | | | Brazing range | | Suitable filler metal | |
|---|---|---|---|---|---|---|---|---|
| | Solidus | | Liquidus | | | | BMg-1 | BMg-2a |
| | °F | °C | °F | °C | °F | °C | | |
| AZ10A | 1170 | 632 | 1190 | 643 | 1080-1140 | 582-616 | X | X |
| AZ31B | 1050 | 566 | 1160 | 627 | 1080-1100 | 582-593 | | X |
| KIA | 1200 | 649 | 1202 | 650 | 1080-1140 | 582-616 | X | X |
| MIA | 1198 | 648 | 1202 | 650 | 1080-1140 | 582-616 | X | X |
| ZE10A | 1100 | 593 | 1195 | 646 | 1080-1100 | 582-593 | | X |
| ZK21A | 1159 | 626 | 1187 | 642 | 1080-1140 | 582-616 | X | X |

Fig. 8-33. Magnesium brazing alloys.

| AWS A5.8 classification | ASTM Alloy designation | Filler Metal | | | | Brazing range | |
|---|---|---|---|---|---|---|---|
| | | Solidus | | Liquidus | | | |
| | | °F | °C | °F | °C | °F | °C |
| BMg-1 | AZ92A | 830 | 443 | 1110 | 599 | 1120-1140 | 604-616 |
| BMg-2a | AZ125A | 770 | 410 | 1050 | 566 | 1080-1130 | 582-610 |

Fig. 8-34. Magnesium brazing and braze welding filler metals.

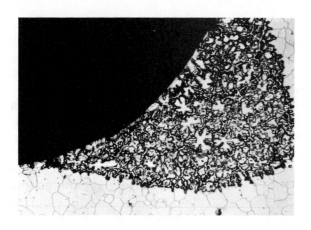

Fig. 8-36. A microphotograph of a magnesium brazed inside corner joint (65x). Note how the brazing alloy has penetrated the grain boundaries of the parent metal.

## 8-14. BRAZING AND BRAZE WELDING CAST IRONS

There are various types of cast iron. These include white cast iron, gray, malleable, and ductile. White cast iron is difficult to braze and is seldom attempted.

Cast iron is pre-cleaned by mechanical and chemical means. Grinding, filing, grit blasting, and chipping are utilized to remove the cast iron surface crust and oxides. Electrochemical cleaning is done to remove graphite and silicon oxides from the surfaces.

Graphite and silicon oxides prevent the filler metal from wetting and smoothly flowing over the surface.

Methods of supporting and reinforcing a cast iron joint are shown in Chapter 18.

Fluxes recommended for cast irons contain the following ingredients: boric acid, borates, fluorides, fluorbates, and agents to improve the wetting action. Common borax makes a good cast iron brazing or braze welding flux. Recommended brazing and braze welding filler metal classifications are BCuZn, BAg, BNi. The BCuZn or brass filler metals adhere better when some nickel is added. See Fig. 8-13 for filler metals recommended when joining various metal combinations.

For best results cast irons should be preheated to 400 - 600 °F (204 - 316 °C). When brazing or braze welding malleable or ductile cast iron the temperature must be kept below 1400 °F (760 °C). Above this temperature their metallurgical structures may be damaged. The procedure for brazing cast irons are the same as for all metals.

Braze welding joints are the same design as those used with other metals. See Fig. 8-3. To produce a good braze weld on cast iron, the sur-

faces must be tinned. Tinning is done by heating the cast iron to a dull cherry red. A thin layer of filler metal is then applied to each surface. If the tinning operation is done improperly, the strength of the joint will be low.

When brazing cast iron, a tip which will provide a high heat output with low gas pressures is advisable. This will provide a ''soft'' flame. A high velocity, high pressure flame tends to blow the flux away from the joint.

Braze welding is the most satisfactory way of repairing malleable iron castings. Welding malleable iron is not recommended. Fig. 8-37 shows cast iron being braze welded. Since cast iron sections are generally quite thick the backhand method is recommended. Multiple passes may also be required.

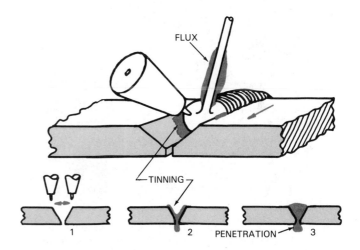

Fig. 8-37. A schematic showing the braze welding of cast iron. The backhand method is being used on this joint.

After brazing or braze welding the metals should be cooled very slowly. Slow cooling will prevent white cast iron from forming. Fig. 8-38 shows a cast iron joint which has been brazed with a silver based filler metal.

## 8-15. CONTROLLED ATMOSPHERE FURNACE

A method of fabricating small articles is the controlled atmosphere brazing furnace. It consists of a typical furnace, usually heated on the inside by electrical resistance coils or partly combusted fuel gases. The inside of the furnace is usually kept as nearly gas-tight as possible, and little or no air is allowed to enter. Instead, the gas fed into the furnace is a low pressure mixture of air and fuel gas. Sufficient temperature is maintained inside the furnace to melt copper, brass, bronze, and silver alloys without melting the base metals.

Fig. 8-38. A microphotograph of steel being joined to cast iron by the silver brazing process. The cast iron was cleaned by a special process prior to the brazing operation.
(Kolene Corp.)

Developments in aluminum alloys enable the brazing of aluminum and its alloys in the reducing atmosphere furnace. The operation of the furnace is as follows.

For example, two articles, such as the two ends of a small steel cylinder, are to be fastened together. They are first cleaned. Then they are assembled in their final form with a small wire of the alloy, which is used to join them, placed inside the assembled joint. The assembly is then placed in the controlled atmosphere furnace.

Upon being subjected to the high temperature in the reducing atmosphere furnace, the brazing filler metal wire melts and joins the two surfaces. The article is then removed from the furnace and cooled slowly in a neutral atmosphere. When finished, the joint is approximately as strong as a weld. The method is well adapted to production work as it insures a high quality joint. The reducing atmosphere in the furnace also simultaneously cleans the complete article, eliminating one manufacturing step. The reducing atmosphere of partly burned fuel gases eliminates any oxidation of the metals during the brazing operation.

Furnaces which are filled with an inert gas have been used for brazing operations. Brazing has also been done in vacuum furnaces.

## 8-16. HEAT RESISTANT BRAZED JOINTS

There are several brazing applications where the brazed joint must retain most of its physical properties when exposed to high temperatures and/or highly corrosive conditions. Brazing filler metal alloys have been developed to withstand temperatures in the range of 2000 °F (1093 °C). These special brazing filler metal alloys must also have good corrosion resistance.

These alloys are usually made of either a nickel-chromium alloy or a silver-manganese alloy, as shown in Fig. 8-39.

The nickel-chromium alloy has good corrosion resistance. It is used on jet engine blades, stainless steel, and on low carbon steel joints. The brazing is usually done in a furnace.

The silver-manganese brazing filler metal is used mainly for joining stainless steel and high nickel alloys. The silver-manganese alloy is also usually applied in a furnace.

## 8-17. SAFETY PRACTICES FOR BRAZING AND BRAZE WELDING

Brazing and braze welding operations are safe to perform if reasonable safety precautions are

| Filler Metal Classification | Composition (Percentage) | | | | | | | | Temperatures | | | | | |
|---|---|---|---|---|---|---|---|---|---|---|---|---|---|---|
| | Ni | Cr | Fe | Si | C | B | Ag | Mn | Solidus | | Liquidus | | Brazing | |
| | | | | | | | | | °F | °C | °F | °C | °F | °C |
| BNiCr | 70 | 16 | * | * | * | 4 | | | 1850 | 1010 | 1950 | 1066 | 2000-2150 | 1093-1177 |
| BAgMn | | | | | | | 85 | 15 | 1760 | 960 | 1780 | 971 | 1780-2100 | 971-1149 |
| *The total of these three elements is approximately 10 percent maximum. | | | | | | | | | | | | | | |

Fig. 8-39. A table of brazing filler metals intended for use in high temperature and/or corrosive conditions. In the American Welding Society Classification BNiCr and BAgMn, the B means brazing; Ni-Nickel; Cr-Chromium; Ag-Silver; Mn-Manganese.

carefully followed. The normal precautions as explained in previous chapters about the safe use of gas welding and arc welding equipment, must also be followed when brazing or braze welding. Refer to Headings 4-35, 4-36, 7-20, and 10-23.

The special precautions which are needed when brazing or braze welding are:

1. Ventilation must be excellent to eliminate the health hazards which are presented by toxic metal and flux fumes so often present when brazing or braze welding.
2. Many of the fluxes used are harmful to the skin and care should be taken in handling them so that no direct contact is made. If the fluxes do come into contact with the skin, the skin should be washed thoroughly with soap and water.
3. When acids are used, these acids should be handled only by persons thoroughly trained for this type work.
4. Some brazing filler metal alloys contain cadmium. When molten, and especially if overheated, these alloys emit cadmium oxide fumes to the atmosphere. Cadmium oxide fumes are very dangerous if inhaled. The limit value for cadmium oxide fumes is 0.1 milligrams per cubic meter of air for daily eight-hour exposures. This value represents the maximum tolerance under which workers may be exposed without adverse effects.

Cadmium fumes have no odor, and a lethal dose need not be sufficiently irritating to cause discomfort until after the worker has absorbed sufficient quantities to be in immediate danger of his life. Symptoms of headache, fever, irritation of the throat, vomiting, nausea, chills, weakness, and diarrhea generally may not appear until some hours after exposure. The primary injury is to the respiratory passages.

Most states have industrial hygiene personnel, usually in the health department or in the department of labor. If there is any question concerning the amount of cadmium oxide fumes in the work area, the industrial hygiene department should be asked to take air samples in the work area. They will evaluate and report on the concentrations of cadmium oxide fumes present. Recommendations may then be made to correct situations that are potentially hazardous.

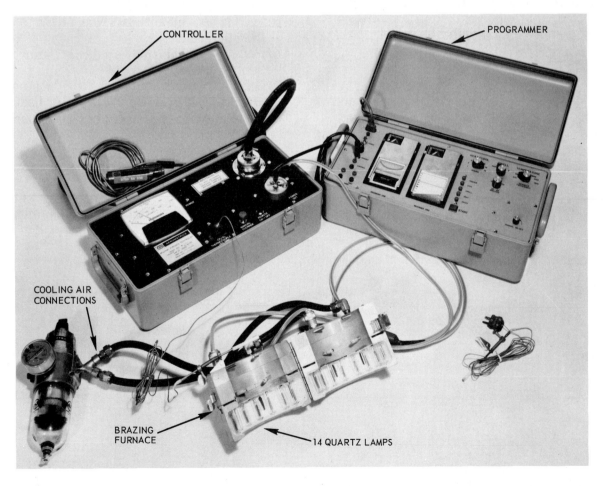

A quartz lamp heating device which may be used for brazing operations. Note the controller which controls the temperatures, the programmer which determines the heating time cycle, and the device for cooling the lamps.

Eating or storing of lunches should not be permitted in the work area. Workers should wash both hands and faces before eating or leaving from work. Workers brazing with cadmium-bearing alloys should be made aware of the hazard involved and trained to take precautionary measures relative to the environment of the particular job.

Alloys free of cadmium are available and should be used wherever possible. Workers should be trained to recognize the brazing alloys which contain cadmium.

An increase in temperature over the molten state accelerates the quantity of the fumes produced. The use of an oxyacetylene flame as the heating source may produce higher concentrations of cadmium oxide fumes due to its higher temperatures. Flames produced by air-acetylene, air-natural gas, or oxygen LP gas are therefore recommended.

Obviously, it is of great importance that the work space used in brazing operations be thoroughly ventilated.

5. Another brazing operation which must be handled carefully is the brazing of beryllium. Oxides from this metal are also very dangerous if inhaled. Thorough ventilation is a must when heating and working with this metal.

## 8-18. TEST YOUR KNOWLEDGE

Write your answers on a separate sheet of paper. Do not write in this book.
1. Explain the difference between brazing and soldering.
2. Explain the difference between brazing and braze welding.
3. Brass is an alloy of copper and _____. Bronze is an alloy of copper and _____.
4. Is the base metal melted when brazing? When braze welding?
5. Can brazing or braze welding be used to join two different base metals? If so, give an example.
6. What is meant by capillary action?
7. Why is some metal "pickled" before it is brazed or braze welded?
8. List four methods of mechanically cleaning a metal prior to brazing.
9. What filler metal and flux ingredients are recommended for brazable aluminum alloys? (See Fig. 8-8.)
10. List at least five criteria which must be considered when selecting a flux for brazing or braze welding.
11. Name three methods for applying a brazing flux.
12. A chloride flux is often used when silver brazing. How can you tell when to add the filler metal? (See Heading 8-10.)
13. What brazing filler metal classifications are recommended for brazing low alloy steel to copper or copper alloy? (Use Fig. 8-13.)
14. What service temperature can most brazing alloys withstand?
15. What is meant by the term "melting temperature"?
16. What is meant by the term "flow temperature"?
17. List two metals or elements which present a health hazard when in a filler or base metal.
18. Which process, brazing or braze welding, requires more heat?
19. When oxyfuel gas welding, what type of flame adjustment produces the strongest joint?
20. What type of flame adjustment should be used when torch brazing aluminum alloys?

part
3

# SHIELDED METAL ARC WELDING

**Shielding Metal Arc Welding / 217**

Industry photo. Shielded Metal Arc Welding is the most popular and useful welding process used on construction sites.    (Miller Electric Mfg. Co.)

# 9 SHIELDED METAL ARC WELDING EQUIPMENT AND SUPPLIES

Shielded metal arc welding (SMAW), sometimes called "stick welding," is the most popular form of electric arc welding. Refer back to Heading 1-5 and to Fig. 1-10 for a review of the principles and processes involved in SMAW. It is important that you have this understanding before you proceed with your study of this chapter. Shielded metal arc welding equipment is used on farms, in service stations, in home shops, on large construction sites, and on pipelines.

The arc welding machines may be of an alternating current (AC) or a direct current (DC) type. This chapter will describe the various machines, accessories, and supplies required for AC or DC, shielded metal arc welding. The next chapter will cover the techniques and skills of shielded metal arc welding.

## 9-1. THE SHIELDED METAL ARC WELDING STATION

A complete shielded metal arc welding (SMAW) station consists of an AC or DC welding machine, an electrode lead, an electrode holder, a work lead, and a work (ground) clamp. The station also includes a welding booth, work bench, stool, and the ventilation system. A typical station is shown in Fig. 9-1.

The arc welding machines, other components of the SMAW station, the accessories, and personal safety items will be discussed in the following headings.

## 9-2. ARC WELDING MACHINE CLASSIFICATIONS

Arc welding machines or welding power sources are either AC, DC, or combination AC/DC machines. They are constructed to produce either a constant current measured in amperes, or a constant voltage (potential) measured in volts.

ALTERNATING CURRENT (AC) MACHINES (see Heading 9-4 for details) may be of the following types:
1. Transformer.
2. Motor or engine driven alternator.

DIRECT CURRENT (DC) ARC WELDERS (see Heading 9-6 for details) are of the following types:
1. Transformer with a DC rectifier.
2. Motor or engine driven generator.
3. Motor or engine driven alternator with a DC rectifier.

COMBINATION AC AND DC ARC WELDING MACHINES are of the following types:
1. Transformer with a DC rectifier.
2. Alternator with a DC rectifier.

To properly describe an arc welding machine, the following minimum information should be given:
1. Type of power source.
2. Whether constant current or constant voltage.
3. Whether AC, DC, or AC/DC.

An example of this description is: "The welding machine is a transformer, constant current, AC machine." Other information, which will be covered later at Heading 9-9, should include: 1. Duty cycle, 2. Welding current rating, and 3. Power requirement.

To meet the variety of demands on different jobs, welding machines are designed to produce AC or DC current. Each machine may have a different design and output capacity. The decision for selecting a machine will also depend upon additional factors such as cost, portability, and the personal preferences of the user.

Arc welding machines also may be classified according to a graph of the electrical output of the particular machine. Such a visual representation is called the STATIC VOLT-AMPERE CURVE. A typical volt-ampere curve will be studied in detail under Heading 9-3 and as shown in Fig. 9-2. The volt-ampere curve plots the electrical current output, measured in amperes, against the voltage

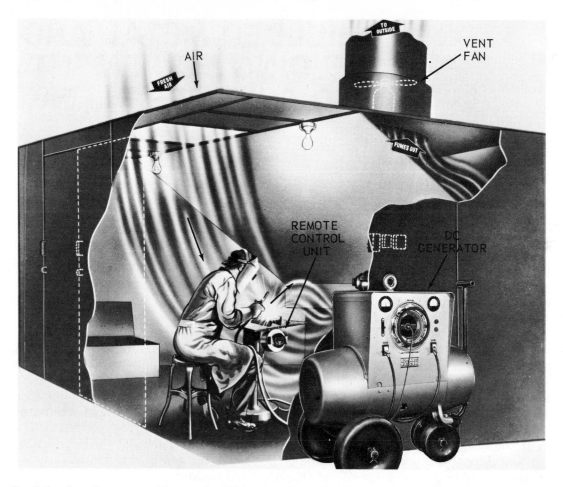

Fig. 9-1. Complete arc welding station. Note that the ventilating air enters above the welder and is directed down and away from the welder's face.    (Hobart Brothers Co.)

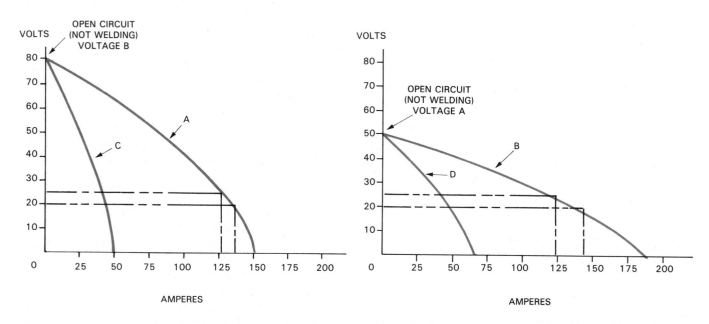

Fig. 9-2. Characteristic volt-ampere curves, or output slopes, for a constant current arc welder. Left view shows steep slope of ''drooper'' machine. Right view shows output slopes of same machine with lower voltage setting. On the curves A and C, notice the relative small percentage change in current for a large percentage change in voltage.

output. Remember that current relates to the welding heat and that voltage relates to the length of the arc.

The general slope of the volt-ampere curve is called the OUTPUT SLOPE of the power source. Some output slopes are very flat while others are very steep. The almost horizontal volt-ampere curves, or output slopes, are produced by constant voltage machines. An arc welding machine that produces a steep output slope, is called a CONSTANT CURRENT MACHINE.

The output slope, or volt-ampere curve, shows the changes in the welding current (heat) as the arc voltage (arc length) is changed. Two types of volt-ampere curves are produced by arc welding machines. The two curves correspond to the constant current arc welding machine and the constant voltage arc welding machine.

## 9-3. CONSTANT CURRENT ARC WELDING MACHINES

Arc welding machines are designed to produce an output which has a nearly constant current or a nearly constant voltage. See Chapter 11 for more information on constant voltage machines. Fig. 9-2 shows a typical volt-ampere curve for a CONSTANT CURRENT WELDING MACHINE. Notice the relatively steep slope or "droop" of this curve. These machines are known as DROOP CURVE MACHINES or DROOPERS.

The arc voltage varies with the size or length of the arc gap. Examining Fig. 9-2, you see a steeply sloping volt-ampere curve A in the left view. The open circuit voltage, or the voltage when not welding, has been selected to be 80 volts. The closed circuit or welding voltage is 20 volts. An increase in arc length causes an increase in the closed circuit voltage.

On the output slope or volt-ampere curve A, a change from 20 volts to 25 volts will result in a decrease in amperage from 135 amps to 126 amps. With a change of 25 percent in voltage, only a 6.7 percent change occurs in the delivered current in curve A. Thus, if the welder varies the length of the arc, causing a change in voltage, the output slope tells that there will be very little change in the current and the weld quality will be maintained. The current in this machine, even though it varies somewhat, is CONSIDERED constant.

The constant current arc welder is the type of machine desired when doing manual arc welding. The two manual arc welding processes are shielded metal arc welding (SMAW) and gas tungsten arc welding (GTAW). Additional information will be covered in Chapter 11.

The open circuit voltage curve for a setting of 50 volts on the machine is shown as curve B in the right view of Fig. 9-2. The same 20 volt to 25 volt (25 percent) change in the welding voltage will result in a drop in current from 143 amps to 124 amps or 13.3 percent. This slower sloping volt-ampere curve output causes a larger change in amperage with the same small change in voltage. A welder may wish to have this slower sloping (flatter) volt-ampere output curve.

With a flatter output slope the welder can control the molten pool and electrode melt rate by making small changes in the arc length. Control of the molten pool and electrode melt rate are most important when welding in the horizontal, vertical, and overhead positions.

## 9-4. ALTERNATING CURRENT ARC WELDERS

Alternating current (AC) arc welders are of either the transformer or alternator type, as classified in Heading 9-2.

An ALTERNATOR creates an AC electrical current from a mechanical source. A ROTOR wrapped with multiple turns of wire is rotated inside a number of magnets. The rotor may be turned by an electric motor or a gasoline or diesel engine. The magnets used in the alternator are made of wire wrapped around a metal core and are known as ELECTRO-MAGNETS. The electro-magnets create a strong magnetic field within the alternator. As the rotor turns within the magnetic field, an ALTERNATING CURRENT is created in the rotor windings. This created (generated) current is the welding current. The amount of welding current produced is in direct relation to the following:

1. The diameter of the wire on the rotor and magnets.
2. The amount of current carried in the rotor and magnets.
3. The number of winding on the rotor and magnet.
4. The speed of rotation of the rotor.

The current output may be adjusted by controlling the amount of current going through the magnet windings (field windings). A bucking (reverse current) field winding may also be used to control the machine's output. Fig. 9-3 illustrates a gasoline engine driven AC alternator type arc welder.

Transformers are constructed of three principal electrical components. They are a primary winding, secondary winding, and an iron core. See Fig. 9-4. The primary windings use thinner wire than the secondary winding. There are also many more windings in the primary coil. Since there are more turns on the primary than on the secondary,

Fig. 9-3. An AC arc welder with the metal outer cover removed. A four cycle engine is used to drive the alternator in this machine. This alternator will produce 150A at a 100% duty cycle for welding. It will also produce 4500 watts at 115V or 230V to power electric appliances and tools.    (The Lincoln Electric Co.)

ARC WELDING MACHINE

Fig. 9-4. The principal electrical components of an AC transformer. $E_1$ is the input or primary voltage; $I_1$ is the primary amperage. $E_2$ and $I_2$ are the secondary voltage and amperage.

welding transformers are STEP-DOWN TRANSFORMERS. They decrease the voltages and increase the amperage from the primary to the secondary circuit.

When current flows in a wire, a magnetic field builds up around the wire. When the current stops flowing in the wire, the magnetic field collapses.

If this current carrying wire is wrapped into a coil with many turns, the coil may be called a

PRIMARY WINDING. When a second coil, usually of larger wire and with fewer turns is placed next to the primary coil, it is called a SECONDARY COIL.

To cause a current to flow in the secondary, an alternating current is made to flow in the primary windings. As it flows, a magnetic field builds up. The current momentarily stops when the alternating current changes direction in the primary circuit. When the current stops, the magnet field collapses and passes across the secondary windings. This collapse of the field INDUCES (creates) a current in the secondary windings in one direction. Current in the primary begins to flow in the opposite direction. It builds a magnetic field and collapses when the current stops to change direction again. The secondary windings are again cut by the magnetic field and a current is induced in the opposite direction. The process of inducing a current in the secondary continues at the rate of 120 times per second and creates an alternating current.

A LAMINATED IRON CORE is placed inside the primary and secondary windings. Its purpose is to keep the magnetic field from wandering too far from the windings.

If the primary and secondary windings are moved away from one another, the amount of current inducted into the secondary decreases. The MOVABLE COIL is one means of adjusting the output of a welding transformer.

## 9-5. METHODS OF CONTROLLING TRANSFORMER OUTPUT

Constant current, AC transformers are controlled by a variety of methods. These control designs are:
1. Movable coil.
2. Movable shunt.
3. Movable core reactor.
4. Tapped secondary.
5. Saturable reactor.
6. Magnetic amplifier.

Current output from a transformer may be controlled by moving the primary and secondary coils together or apart. Either the primary or secondary winding coil may be moved to adjust the secondary output current. Normally the secondary coil is fixed and the primary coil is moved. See Fig. 9-5 for examples of current output with changing coil positions. A screw arrangement is often used to move the primary coil. The adjustment may be on

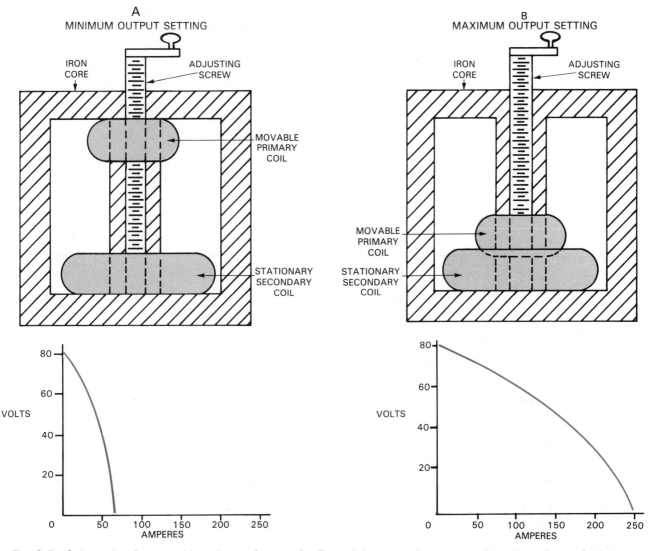

Fig. 9-5. Schematic of a movable coil transformer. A—The minimum setting occurs when the coils are farthest apart. B—The maximum output occurs when the coils are closest.

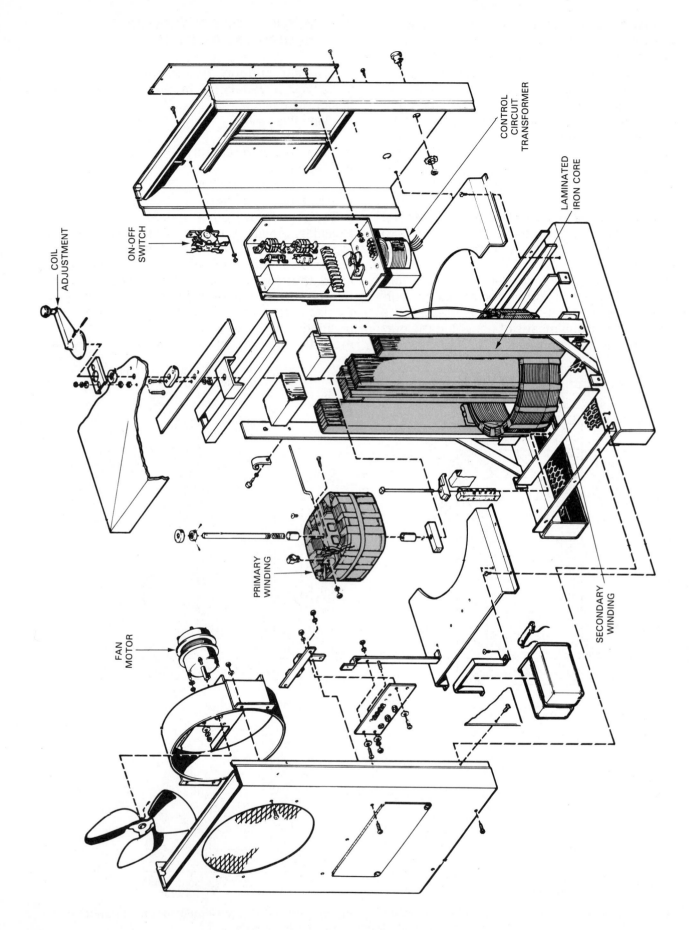

COIL ADJUSTMENT

ON-OFF SWITCH

CONTROL CIRCUIT TRANSFORMER

LAMINATED IRON CORE

SECONDARY WINDING

PRIMARY WINDING

FAN MOTOR

Fig. 9-6. Exploded view of a movable coil transformer type AC arc welder.

the top or front of the machine. See Fig. 9-6.

A schematic of the MOVABLE SHUNT TRANS-FORMER CONTROL is shown in Fig. 9-7. The movable shunt is made of the same material as the main iron core. Both the primary and secondary coils are stationary. With the shunt moved away from the coils, the maximum number of magnetic lines of flux cut the secondary windings when the primary current stops. In this position, the maximum welding current is produced. The primary current stops each time the current changes direction in a 60 Hz (cycle) system. When the shunt is adjusted all the way in, the output current is lowest. See Fig. 9-7.

The MOVABLE CORE REACTOR type transformer uses a primary and secondary coil. In addition, a third coil with a movable core is used to control the current output. See Fig. 9-8. The transformer and reactor produce a constant voltage. The reactor is wound to create a bucking voltage (counter voltage). This bucking voltage or inductive reactance increases the resistance of the secondary circuit. The varying inductive reactance controls the output current. The inductive reactance is least when the movable core is turned

out. In this position, the output current from the transformer is greatest. When the movable core in

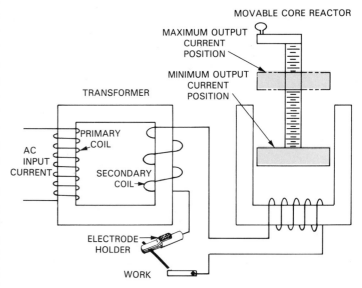

Fig. 9-8. Movable core reactor. The position of the reactor core will cause the inductive reactance (resistance) of the secondary output coil to vary. This will cause the current output to change.

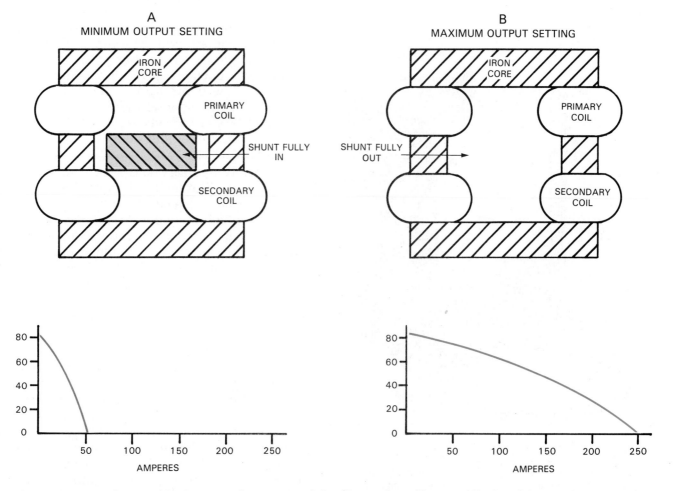

Fig. 9-7. Schematic of a movable shunt transformer control. A—Shunt adjusted in to provide the minimum output current. B—Shunt withdrawn to provide the maximum output current.

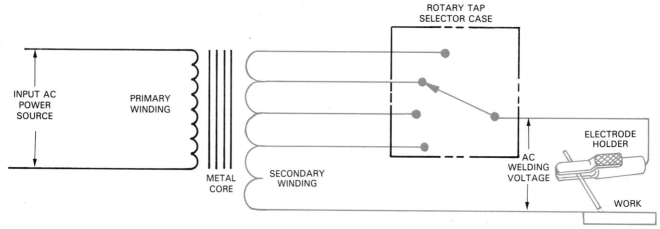

Fig. 9-9. AC welding transformer with tapped secondary winding. The taps may be a plug-unplug or rotary contact type.

the reactor is turned in, the transformer output is lowered.

Some welding transformers use a TAPPED SECONDARY WINDING as shown in Fig. 9-9. In this machine, only a portion of the secondary windings are used. The amount of the secondary winding used depends on where the tap is placed.

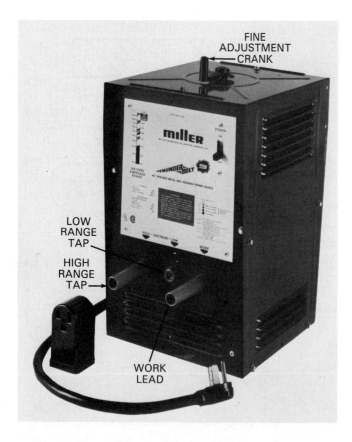

Fig. 9-10. This AC welding machine has a high and low range electrode lead tap. In addition, a crank adjustment is provided with an indicator scale for amperage adjustments. The high range tap provides 40-225 amps. The low range covers 30-150 amps.    (Miller Electric Mfg. Co.)

A small range of current is selected by placement of the tap. See Fig. 9-10.

Finer control within the selected range is provided by another means such as a movable core reactor.

Transformers may use a SATURABLE REACTOR CONTROL. This system of control uses a low amperage, low voltage DC current to create a magnetic field. This magnetic field effects the magnetic field strength of the reactor coils in the secondary circuit. Large changes in the AC output current will occur with small changes in the DC current to the control coils. See Fig. 9-11. This type control requires no mechanical operation. It therefore lends itself well to remote control of the welding machine output current.

The MAGNETIC AMPLIFIER CONTROL uses the welding current coils and diodes in series with the control coils. The load coil is used to assist the control coil to increase the magnetic field of the cores. The high magnetic field in the cores cause an inductive reactance in the secondary welding current. This increased reactance decreases the welding current from the transformer. See Fig. 9-12 for the location of the cores and diodes.

## 9-6. DIRECT CURRENT ARC WELDERS

Direct current (DC) arc welders are of either the transformer-rectifier or generator type, as classified in Heading 9-2.

In the transformer section of the transformer-rectifier welder, AC is changed from line voltage and current to an AC welding voltage and current. Line voltage supplied to the transformer is generally 220 or 440 volts at 60 Hz (cycles). The transformer changes the high voltage to an open circuit (no-load) welding voltage of 60-80 volts. The current (amperage) provided may be as high as 1500

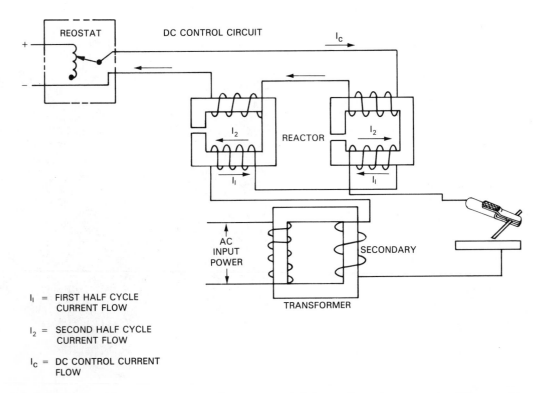

$I_l$ = FIRST HALF CYCLE
        CURRENT FLOW

$I_2$ = SECOND HALF CYCLE
        CURRENT FLOW

$I_c$ = DC CONTROL CURRENT
        FLOW

Fig. 9-11. Saturable reactor type transformer control. A low voltage, low amperage DC current is used in the control circuit. When the DC current is changed, the magnetic field in the reactor is changed. This in turn changes the secondary output current.

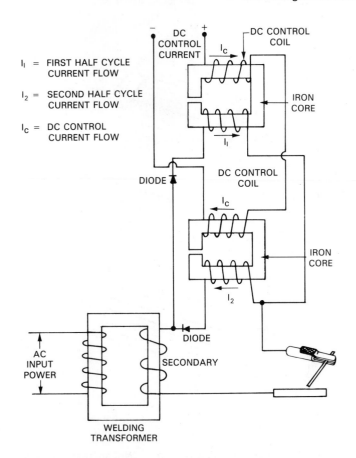

$I_l$ = FIRST HALF CYCLE
        CURRENT FLOW

$I_2$ = SECOND HALF CYCLE
        CURRENT FLOW

$I_c$ = DC CONTROL
        CURRENT FLOW

Fig. 9-12. A schematic of the magnetic amplifier transformer output control. A diode allows current flow in only one direction. This control can be operated from remote locations.

amps, depending on the construction of the welder. After leaving the transformer section, the welding current enters the rectifier. The rectifier changes the AC to DC.

A DC CONSTANT CURRENT TRANSFORMER-RECTIFIER may be either a single-phase or three-phase machine. Fig. 9-13 illustrates both a single-phase rectifier schematic and a three-phase rectifier schematic. These schematics illustrate the function of a diode. Alternating current leaving the transformer can only flow through the diode in one direction. The alternating current flowing into the rectifier is changing direction 120 times per second. The current flows from the rectifier in one direction only and has therefore been converted to direct current (DC). Fig. 9-14 shows an AC transformer which uses rectifiers to convert AC to DC. The control of the current output of a DC transformer-rectifier is accomplished in the same manner as in the AC transformer. Refer to Heading 9-5.

Direct current may also be produced through the use of the DC generator or AC alternator with a rectifier.

The DC GENERATOR is similar in construction to the AC alternator. The generator uses a rotating armature, similar to the stator. The armature is wound with many separately wound coils. Each end of the armature coil winding is soldered to a

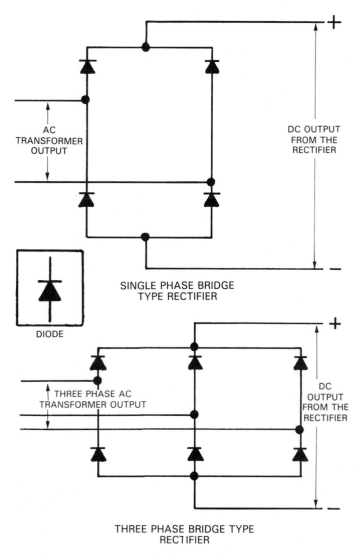

SINGLE PHASE BRIDGE
TYPE RECTIFIER

DIODE

THREE PHASE BRIDGE TYPE
RECTIFIER

Fig. 9-13. Schematic drawings of single and three-phase bridge type rectifiers. Note that diodes allow current to travel in the direction of the arrow only. AC is therefore changed to DC in the rectifier.

Fig. 9-14. This DC arc welder uses a rectifier to change the AC input to DC output. The polarity is changed by reversing the welding leads on the machine.    (Miller Electric Mfg. Co.)

copper terminal called a COMMUTATOR. At least two carbon or copper contacts called BRUSHES touch the commutator. One brush is a positive (+) terminal and one is a negative (−) terminal. See Fig. 9-15 for a schematic of a generator. Two or more stationary, wire wound magnets are used. These are called field windings. The armature windings are rotated by some mechanical means such as a motor or engine. As the armature windings rotate, they pass through (cut) the magnetic fields created by the field windings. This cutting of the magnetic field by the armature windings causes a current to be created (inducted) in the armature coil. In Fig. 9-16A, the armature wire is horizontal and not cutting through the magnetic field so no current is created. This position occurs twice in each revolution. In Fig. 9-16B, the armature is vertical. It cuts through the magnetic

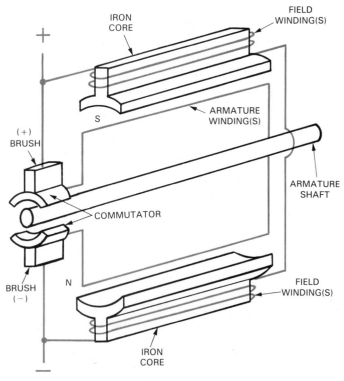

Fig. 9-15. Schematic of a generator. Notice that the field windings are in parallel with the armature windings. The commutator spins with the armature winding(s). The brushes remain stationary.

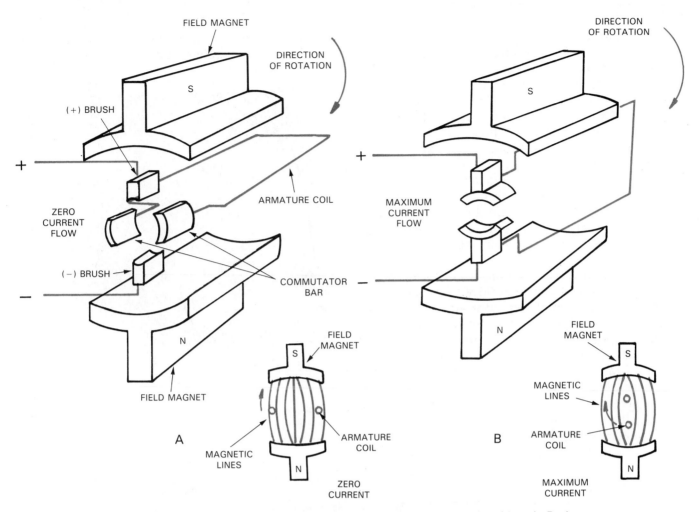

Fig. 9-16. Generator schematic. At A, the armature coil is at the zero current position. At B, the armature is at the maximum current position.

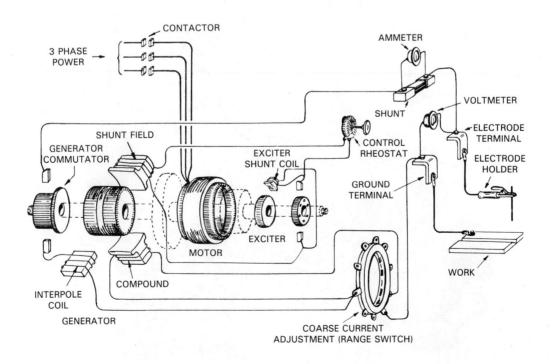

Fig. 9-17. Pictorial wiring diagram of DC motor-generator welding machine, with electric motor drive.

field as it moves up from horizontal to a maximum current position at the top and bottom vertical positions. The induced current is picked up from the commutator by the brushes. A direct current is produced because the induced current is always picked up in one direction. The more armature windings, field windings, and brushes there are, the more evenly the DC will flow.

Welding generators are specially constructed to produce high current flow at low voltage. A variety of generator sizes are available. The current produced by the generator should be steady, and the voltage must not fluctuate during the welding procedure. A steady current is maintained by special devices incorporated in the design of the generator.

Some machines use a separate exciter to maintain good voltage and current flow characteristics. An EXCITER is a small generator electrically connected to the field windings of the large generator. The exciter keeps a constant voltage on the main fields and also prevents them from reversing their polarity. Trace current flow in Fig. 9-17. Fig. 9-18 illustrates a motor driven DC generator. Fig. 9-19 shows an engine driven DC generator.

The alternator produces alternating current (AC). See Heading 9-4 for an explanation of the alternator function. The AC alternator used along with a rectifier will produce direct current. See Fig. 9-13 for a schematic diagram of a single

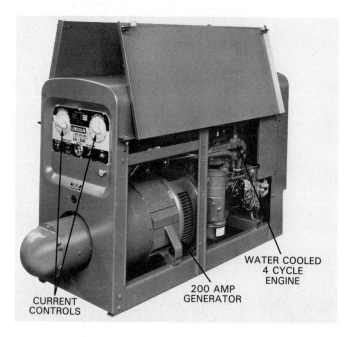

Fig. 9-19. A 200 amp engine driven DC generator. This type unit is often mounted on a pickup truck or two wheel trailer. (The Lincoln Electric Co.)

Fig. 9-18. DC motor-generator welding machine on wheels. (Hobart Brothers Co.)

phase rectifier and a three phase rectifier. SILICON CONTROLLED RECTIFIERS (SCR) are used in solid state rectifier units. These solid state units are compact and have high efficiency. See Figs. 9-20 and 9-21 for examples of small engine driven DC welding machines.

Direct current may flow in one of two directions. It may flow from the welding machine to the electrode holder, across the arc gap, and return to the machine through the work lead. Current flowing in this direction is called DIRECT CURRENT ELECTRODE NEGATIVE (DCEN). The electrode is negative in this case. Direct current electrode negative is also known as DIRECT CURRENT STRAIGHT POLARITY (DCSP).

By reversing the position of the electrode and work lead at the machine, the DC will first flow to the work. From the work, the current crosses the arc to the electrode holder, and returns to the machine through the electrode lead. Current flowing in this direction is called DIRECT CURRENT ELECTRODE POSITIVE (DCEP) or DIRECT CURRENT REVERSE POLARITY (DCRP).

COMBINATION AC/DC ARC WELDING

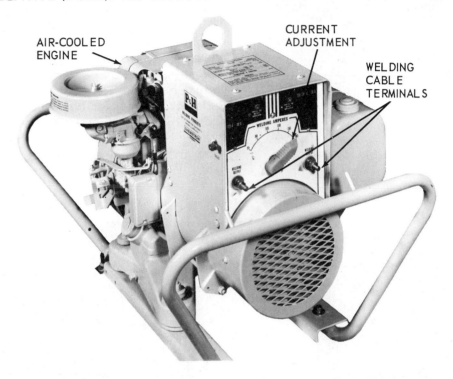

Fig. 9-20. A direct current arc welding generator driven by an air-cooled engine.

Fig. 9-21. An engine driven portable AC/DC welding machine. 115 VAC and 230 VAC outlets are provided for power tools, lights, and accessories.
(The Lincoln Electric Co.)

MACHINES are also produced. An AC transformer is generally used to produce AC welding current. Rectifiers in series with the transformer are switched in when DC is required. Fig. 9-22 illustrates an engine driven AC/DC arc welding machine.

A switch is generally provided on the machine to allow direct current electrode negative (DCEN) or direct current electrode positive (DCEP) to be selected. DCEN polarity is when the work is the positive pole and the electrode is the negative pole. Current, therefore, travels from the electrode to the work. In DCEP, the current travels from the work to the electrode. If the machine is a combination AC/DC welder, a switch is included to provide AC also. Polarity changes may also be made by manually changing the position of the electrode and work leads on a DC arc welding machine.

Fig. 9-22. An engine driven AC/DC arc welding machine with the outer metal shell removed. AC is produced by the alternator. DC is made available by switching in the rectifiers. This unit is mounted on a movable dolly.
(The Lincoln Electric Co.)

## 9-7. COOLING ARC WELDING MACHINES

Most AC welding machines are air cooled. Some of the smaller units have natural or gravity air flow through the mechanisms.

Some machines, particularly the larger sizes, use forced air circulation. An electric motor and fan is connected into the primary circuit and provides forced air circulation automatically when the unit is turned on, and while it is in operation. Fig. 9-23 shows an exposed view of an AC arc welder equipped with a motor-fan, forced air circulation cooling system. Refer also to Fig. 9-6.

Air must get in easily, pass around the mechanism easily, and exit easily. Air passageways must be kept open at all times. The machine should be kept clear of obstructions and should have its ventilation openings uncovered. Periodically, once or twice each year, the power should be disconnected and the casing should be removed for cleaning. Dust should be removed from the internal air passages and from other areas. Wear goggles if compressed air is used. A

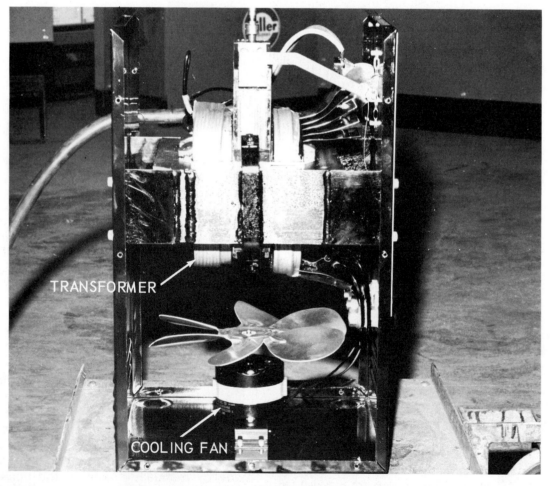

Fig. 9-23. The internal construction of an AC arc welder. Notice the cooling fan which blows air upward to cool the transformer. (Miller Electric Mfg. Co.)

vacuum cleaner in combination with a brush, produces excellent results.

## 9-8. MACHINE INSTALLATION

Since arc-welding machines are single-phase and three-phase machines, they tend to disturb the electrical power circuit. Because of the power factor which will be imposed on the power circuit used, an arc welder will noticeably affect a single-phase or a three-phase circuit to which it may be connected. Power factor correction capacitors are used to improve the power factor of other electrical machinery on the power line. It is best to consult with the electrical utility company and with an electrical contractor before purchasing and installing an arc welding machine.

It is recommended when using an alternating current machine, that the arc welding leads be as short as possible. The electrical resistance of long cables decreases the current available for welding. The cables should also be kept close together to minimize the reactance of the cables, as this reactance will reduce the current output of the machine.

## 9-9. ARC WELDING MACHINE SPECIFICATIONS

An arc welding machine is described and specified by the following:
1. Rated output current rating.
2. Power requirements.
3. Duty cycle.

The RATED OUTPUT CURRENT RATING is a term used to describe the amount of current (amperage) a welder is rated to supply at a given voltage. These ratings are described in the National Electrical Manufacturer's Association (NEMA) publication EW-1. Fig. 9-24 shows the NEMA rated output current for various classes of welding machines. See Heading 9-10 for a description of NEMA classifications.

| RATED OUTPUT CURRENT IN AMPERES | | |
|---|---|---|
| **Class I** | **Class II** | **Class III** |
| 200 | 150 | 180-230 |
| 250 | 175 | 235-295 |
| 300 | 200 | |
| 400 | 225 | |
| 500 | 250 | |
| 600 | 300 | |
| 800 | 350 | |
| 1000 | | |
| 1200 | | |
| 1500 | | |

Fig. 9-24. National Electrical Manufacturer's Association (NEMA) rated current output for various classifications of arc welders.

The RATED LOAD (welding) VOLTAGE for class I and II arc welding machines under a 500 ampere rating is calculated as follows:
Voltage (E) = 20 + 0.04 × rated amperes.

As an example, using a class I machine, as found in the table in Fig. 9-24, with an output current rating of 300 amps, the rated load voltage would be:

$$\text{Voltage} = 20 + 0.04 \times 300$$
$$= 20 + 12$$
$$= 32V.$$

The welding rated output current rating for this machine is 300 amps at 32 volts. For machines with output currents above 600 amps the rated load voltage used is 44 volts.

ELECTRICAL POWER INPUT REQUIREMENTS for NEMA class I and II transformer arc welders are as follows:

for 50 Hz (cycles) – 220, 380, and 440V
60 Hz (cycles) – 200, 230, 460, and 575V.

The electrical power input requirements for NEMA class III transformer is at 60 Hz – 230V.

DUTY CYCLE is: "the length of time that a welding machine can be used continually at its rated output in any ten (10) minute period."

Most welding machines are not required to operate 100 percent of the time. If they are required to weld 100 percent of the time, a machine with a 100 percent duty cycle would be required. Work must be loaded and unloaded and electrodes must be changed. Metal must be chipped, cleaned, and inspected. The duty cycle normally recommended for manual welding is 60 percent. Automatic and semiautomatic welding operations usually require 100 percent duty cycles. Light duty work such as done in the home hobby shop could possibly use a duty cycle of 20 percent.

When purchasing a welder, the maximum duty cycle requirements must be considered.

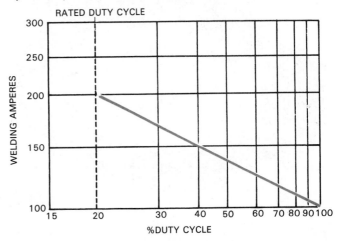

Fig. 9-25. A duty cycle-welding amperage chart. This chart is for the Miller Thunderbolt 225V machine only.
(Miller Electric Mfg. Co.)

See Heading 9-10 for the NEMA arc welding machine classifications by duty cycle.

If an arc machine is used at lower than rated current outputs, the duty cycle may be increased under certain conditions. A graph supplied by the manufacturer will indicate how the duty cycle can vary with a given current setting. Fig. 9-25 is an example of a graph of welding amperage and duty cycle.

## 9-10. NEMA ARC WELDER CLASSIFICATIONS

The National Electrical Manufacturers Association (NEMA) has classified electric arc welders primarily by duty cycle. These classifications are as follows:

NEMA CLASS I.   Machines manufactured to Class I standards deliver rated outputs at a 60, 80, or 100 percent duty cycle.

NEMA CLASS II.   Machines which deliver rated outputs at a 30, 40, 50 percent duty cycle.

NEMA CLASS III.   Machines which deliver rated outputs at a 20 percent duty cycle.

These classes were used in the table in Fig. 9-24.

## 9-11. WELDING MACHINE LEADS

Large diameter, superflexible leads (cables) are used to carry the current from the welding machine to the work and back, Fig. 9-26. The lead from the machine to the electrode holder is known as the ELECTRODE LEAD. The lead from the work to the machine is known as the WORK (GROUND) LEAD. These leads are well insulated with rubber and a woven, fabric reinforcing layer, as shown in Fig. 9-27. The leads are usually subjected to considerable wear and should be checked periodically for breaks in the insulation. The voltage carried by the leads is not excessive, varying between approximately 14 and 80 volts. Leads are produced in several sizes. The smaller the number, the larger the diameter of the lead. Fig. 9-28 is a list of sizes and current capacities for copper leads. The lead must be flexible in order to reduce the strain on the arc welder's hand when welding, and also to permit easy installation of the cable. To produce this flexibility, as many as 800 to 2500 fine wires are used in each cable. The same diameter electric cable must be used on both the electrode and work lead.

The length of the lead has considerable effect on the size to be used for certain capacity machines.

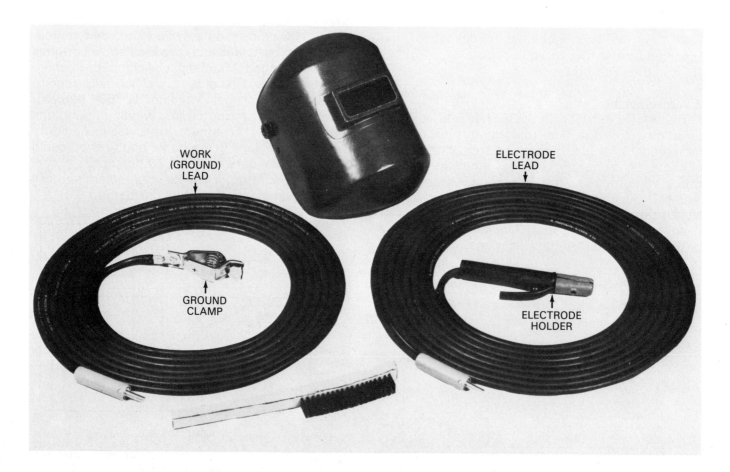

Fig. 9-26. Arc welding leads.   (Miller Electric Mfg. Co.)

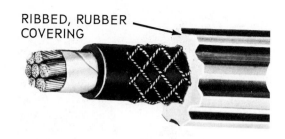

RIBBED, RUBBER COVERING

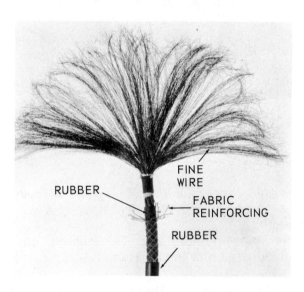

RUBBER

FINE WIRE

FABRIC REINFORCING

RUBBER

Fig. 9-27. Above. Welding lead. Note the heavy ribbed outer rubber covering. Below. The hundreds of individual wires used to form the complete lead (cable). This construction produces the flexibility needed in these leads.

The wiring to the motor, in case a motor driven generator is used, must conform to local, state, and national electrical codes. Copper leads are preferred, however aluminum leads have been used. Aluminum has about 61 percent the current carrying capacity of copper. For a given current capacity, an aluminum lead will be larger in diameter, but only about one-half as heavy.

Aluminum leads are pure, semi-annealed, electrolytic aluminum.

## 9-12. CONNECTIONS FOR LEADS

To consistently carry the large currents used in welding, all parts of the welding circuit must be of heavy duty design and construction including all terminals.

The copper or aluminum leads are fastened to the welding machine and work by means of insulated or uninsulated terminals. The uninsulated terminals are called lugs and are shown in Fig. 9-29. These lugs are soldered or mechanically attached to the leads, as shown in Fig. 9-30. The

Fig. 9-29. Lugs for welding leads. The three sizes will fit cables from No. 6 to 4/0. They may be connected to the lead by soldering or mechanical crimping.   (Tweco Products, Inc.)

lugs provide a firm means of attaching the electrode lead and the work lead to the machine or work table. These connections must be durable and must have low resistance or the joint will overheat during welding. Unsatisfactory current flow will result if the connection is loose. Connections are also available for connecting one lead to another, as shown in Fig. 9-30.

Insulated lead terminals are available as shown in Fig. 9-31. Fig. 9-32 illustrates a quick disconnecting terminal. Terminals are connected to the leads by the following methods:

| | | | CURRENT CAPACITY IN AMPERES LENGTH IN FEET AND METERS | | |
|---|---|---|---|---|---|
| **Lead No.** | **Lead Dia.** | | **Length 0-50 ft. 0-15.2 m** | **Length 50-100 ft. 15.4-30.5 m** | **Length 100-250 ft. 30.5-76.2 m** |
| | in. | mm | Amperage | Amperage | Amperage |
| 4/0 | .959 | 24.4 | 600 | 600 | 400 |
| 3/0 | .827 | 21.0 | 500 | 400 | 300 |
| 2/0 | .754 | 19.2 | 400 | 350 | 300 |
| 1/0 | .720 | 18.3 | 300 | 300 | 200 |
| 1 | .644 | 16.4 | 250 | 200 | 175 |
| 2 | .604 | 15.3 | 200 | 195 | 150 |
| 3 | .568 | 14.4 | 150 | 150 | 100 |
| 4 | .531 | 13.5 | 125 | 100 | 75 |
| Note: Lengths given are for the total combined length of the electrode and work leads. | | | | | |

Fig. 9-28. Arc welding lead recommendations. The voltage drop in these copper leads will be approximately 4 volts with all connections clean and tight.

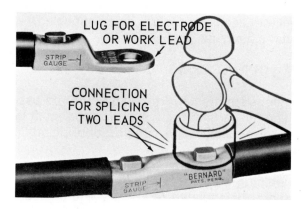

Fig. 9-30. Welding lead connections. The connections are made mechanically in this type connection by striking the malleable punch area with a hammer. (Bernard Welding Equipment Co.)

1. Mechanical.
2. Soldering.
3. Brazing.
4. Welding.

In the case of aluminum cables, it is claimed that it is best to clamp the aluminum to the electrode holder and to the other terminals. However, the lead can be successfully aluminum brazed to either aluminum connections or copper connections. It is recommended that twisting of the lead, especially at the connections, be avoided. The lead may tend to separate from its terminals if this is done. Mechanical connections must be tight and clean.

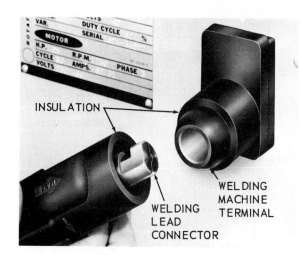

Fig. 9-32. An insulated quick-connect and disconnect arc welding machine terminal. (Cam-Lok Div., Empire Products, Inc.)

Soldered connections must be done skillfully or only a portion of the electrical flow area will be connected.

Silver brazing must also be done by an experienced person to insure enough electrical flow area.

A method of connecting copper welding cables is to copper weld the leads to the terminals or to each other. One process used is a method of welding copper in which no outside source of heat is required. Powdered copper oxide and powdered aluminum are placed in a small graphite crucible

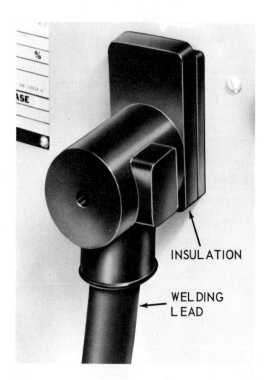

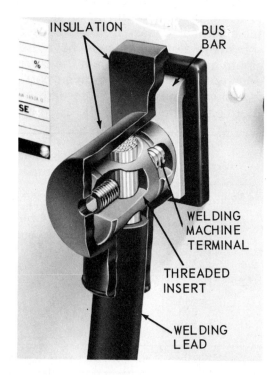

Fig. 9-31. An insulated welding lead terminal. This terminal uses screws to insure a good electrical connection. Left. The complete insulated terminal. Right. Cutaway of the insulated terminal. (Cam-Lok Div., Empire Products, Inc.)

Fig. 9-33. Tools needed to make a fused copper cable splice or cable-to-lug weld. The cable is clamped in the graphite mold. The welding and starting materials are placed in the top of the mold and ignited with a spark lighter. Melting of the charge and the weld that results is similar to the reaction in thermit welding. (Erico Products, Inc.)

and ignited by means of a spark. The rapid oxidation of the powder creates enough heat to weld the copper lead and terminal and produce a sound weld. The tools required and two completed thermite welds are shown in Figs. 9-33, 9-34, and 9-35.

The cable is prepared by stripping approximately one inch of insulation from each end to be

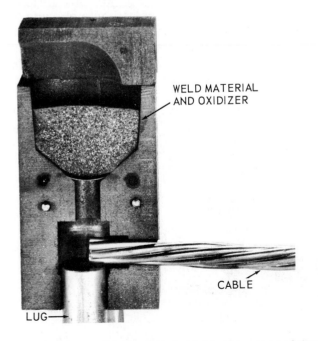

Fig. 9-34. Welding lead parts being joined by means of thermit type welding process. (Erico Products, Inc.)

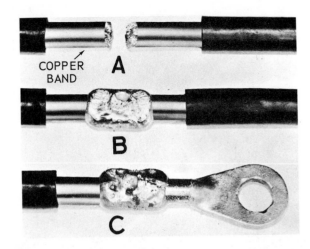

Fig. 9-35. Fused copper lead splices. A—The lead is prepared by removing the insulation, and wrapping a thin band of copper around the strands to keep them in place during the thermit welding operation. B—Fused copper lead splice. C—Fused copper lead and terminal lug. (Erico Products, Inc.)

joined. Both ends are then placed in the welder, butted together under the center of the tap hole, and locked by the clamp type crucible. The flint spark gun ignites the mixture; and in about ten seconds, the weld is completed.

Good work (ground) lead and electrode lead connections are important when welding with either AC or DC. The work lead is usually fastened to the welding bench or table by means of a lug or insulated terminal. This practice is practical when welding can be done on a welding table.

Frequently, however, the ground cable must be fastened to the article being welded, due to its size or location. The spring loaded work lead, as shown in Fig. 9-36, is often used for on-site

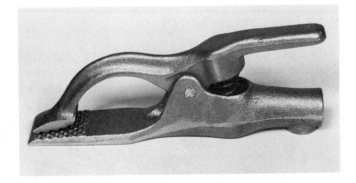

Fig. 9-36. A spring-loaded ground clamp for the work lead (NLC—New Lenco Co., Inc.)

welding. A C-clamp type work lead is shown in Fig. 9-37. It is sometimes difficult to use a clamping device on a metal fabrication. Clamps must be used carefully on finished surfaces to avoid marring the surface. A magnetic work lead terminal is

Fig. 9-37. A work lead held in contact with the work by a C-clamp. (Jackson Products)

available which permits quick and secure fastening of the work lead to the weldment. This makes it easy to change the position of the ground to obtain better arc characteristics, and does not injure or mar the article to be welded. The work lead is either soldered or mechanically fastened to this permanent magnet grounding device, and the operator may easily position it on any ferrous (iron) surface. The magnets are replaceable and are quite powerful.

## 9-13. ELECTRODE HOLDERS

The ELECTRODE HOLDER is the part of the arc welding equipment held by the operator when welding. It is used to hold metallic or carbon elec-

trodes. Many different styles and models have been produced, but they all have similar characteristics. The electrode lead is usually fastened to the electrode holder by means of a mechanical connection inside of the electrode handle. The handle itself is made of an insulating material which has high heat and low electrical resistance qualities. Electrode holders are built to produce a balanced feeling when held in the operator's hand. There should be a good balance with the cable draped over the operator's arm, and with the average length of the metallic electrode in the holder.

There are a number of methods used to clamp the electrode in the holder. One is of pincher construction and has a spring to produce the necessary pressure to obtain a good contact between the holder and the electrode. See Fig. 9-38.

The electrode should be clean at the point where it is to contact the electrode holder. The electrode contact area may be cleaned with a wire brush or steel wool. The electrode holder jaws should also be kept clean by using a file, sandpaper, or other suitable means.

When welding heavy work, the electrode holders are sometimes equipped with SHIELDS (small heat resisting plate). This shield prevents the radiation of heat from the work onto the operator's hand.

## 9-14. ELECTRODE COVERING FUNDAMENTALS

The basic function of the covering (coating) on electrodes is described in this heading. During the arc process the covering changes to NEUTRAL or REDUCING GASES such as carbon monoxide (CO) or hydrogen ($H_2$). These gases, as they surround the arc proper, prevent air from coming in contact with the molten metal. They prevent oxygen from the air which may approach the molten metal, from combining with it. However, the gases usually do not protect the hot metal after the arc leaves that point on the weld. The covering also contains

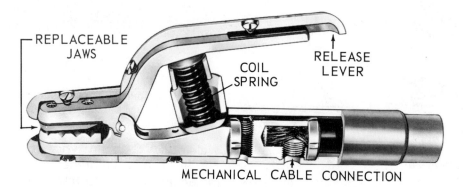

Fig. 9-38. An electrode holder which uses a coil spring for the clamping pressure. (Fibre-Metal Products)

special fluxing ingredients which promote fusion and tend to remove impurities from the molten metal.

As the electrode flux coating residue cools, it forms a coating of material over the weld. This residue coating prevents the air from contacting the hot metal. This coating over the bead is called SLAG.

FLUX which forms the covering on an electrode, commonly consists of feldspar, mica steatite, titanium dioxide, calcium carbonate, magnesium carbonate, and various alumina. The neutral or reducing gas producers are: carbon hydrates such as paper, cotton, wood flour, cellulose, starch, and dextrin. Some special electrodes have metallic salts included in the covering to add alloying elements to the weld metal. A good flux covered electrode can produce a weld that has excellent chemical and physical properties.

## 9-15. SMAW ELECTRODES

Shielded metal arc welding (SMAW) electrodes have a solid metal wire core and a thick flux covering (coating). These electrodes are specified by the wire diameter and by the type of flux covering.

The common SMAW electrode wire diameters are 1/16 (1.6 mm), 3/32 (2.3 mm), 1/8 (3.2 mm), 5/32 (4.0 mm), 3/16 (4.8 mm), 7/32 (5.6 mm), 1/4 (6.4 mm), 5/16 (7.9 mm), and 3/8 in. (9.5 mm). They may be obtained in lengths from 9-18 in. (229-457 mm). The most frequently used length is 14 in. (356 mm).

They are usually purchased in 50 lb. (22.7 kg) packages. Electrodes may be packaged in cardboard cartons or in hermetically (air tight) sealed metal cans. Fig. 9-39 shows one method of storing electrodes.

SMAW electrodes are produced for welding on many metals and alloys, including the following:
1. Carbon steels (AWS specification A5.1).
2. Low alloy steels (AWS A5.5).
3. Corrosion resistant steels (AWS A5.4).
4. Gray and ductile cast irons (AWS A5.15).
5. Aluminum and aluminum alloys (AWS A5.3).
6. Copper and copper alloys (AWS A5.6).
7. Nickel and nickel alloys (AWS A5.11).

Covered electrodes serve many purposes in addition to adding filler metal to the molten weld pool. Some or all of these functions are served by the covered electrode. These functions are:
1. Produce a GAS WHICH SHIELDS the base metal, molten pool, and weld bead from oxidation during welding.
2. Provide FLUXING AGENTS, IMPURITY SCAVENGERS, and DEOXIDIZERS to clean the weld pool.
3. Produce a SOLIDIFIED COVERING over the hot weld area to prevent oxidation. This solidified covering is called SLAG.
4. The SLAG COVERING permits the weld metal to cool slowly. This helps prevent a hard, brittle weld.
5. ESTABLISH THE POLARITY and electrical characteristics of the electrode.
6. Adds alloying ingredients to the weld metal and weld area to change the physical properties.

Fig. 9-39. Storing electrodes. Each bin is marked with AWS number and electrode diameter.

TEMPERATURE CONTROL

Fig. 9-40. Electrode drying oven.   (Despatch Industries, Inc.)

**Shielded Metal Arc Welding Equipment and Supplies / 239**

Dampness destroys the effectiveness of most electrode coverings. Many welding procedures require that electrodes be thoroughly dried prior to use. This is done in specially built drying ovens. See Fig. 9-40. The time it takes for an electrode to pick up moisture from the air varies from thirty minutes to four hours, depending on the electrode. This time period is known for each electrode in use. Welders, therefore, take only enough electrodes from the oven to last for this moisture pick-up time period. Electrodes must be handled carefully to prevent breaking the flux coating. Many welding procedures will not allow an electrode to be used if the flux coating is chipped. Fig. 9-41 illustrates an electrode dispenser which is carried by the welder. This dispenser is sealed to keep the electrodes dry. It is tough enough to prevent damage to the electrodes.

E60XX and E70XX electrodes is exactly the same. The composition of the coatings on the electrodes vary considerably, according to the planned use of the electrode.

The American Welding Society has developed a series of IDENTIFYING NUMBER CLASSIFICATIONS. See Figs. 9-42 and 9-43.

The CARBON AND LOW ALLOY STEEL ELECTRODE CLASSIFICATION NUMBER uses four or five digits. For carbon steels the electrodes are either in the E60XX or E70XX series. The minimum allowable tensile strength, for a weld made with an electrode in the 60 series, is 60,000 psi (414 MPa). For the 70 series, the minimum tensile strength as welded is 70,000 psi (480 MPa). The metals are similar in composition for each classification number. Each manufacturer has its own compounds for the coverings.

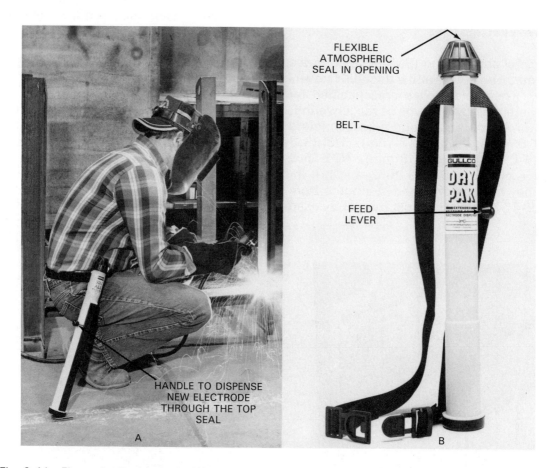

Fig. 9-41. Electrode dispenser. A—Welder wearing a loaded dispenser. B—With top closed, dispenser is sealed. Moving lever feeds a new electrode through the top seal.   (Gullco)

## 9-16. CARBON AND LOW ALLOY STEEL COVERED ELECTRODE CLASSIFICATION

There are a large variety of CARBON AND LOW ALLOY STEEL ELECTRODES on the market. The chemical composition of the steel wire used in the

Therefore, very few of the electrodes behave exactly the same even though the classification number may be identical. The letter E preceding the four or five digit number (EXXXX) indicates a welding electrode used in arc welding. See Fig. 9-42. This is contrasted with the letters RG,

| AWS Classification | Type of covering | Capable of producing satisfactory welds in position shown[a] | Type of current[b] |
|---|---|---|---|
| E60 series electrodes | | | |
| E6010 | High cellulose sodium | F, V, OH, H | DCEP |
| E6011 | High cellulose potassium | F, V, OH, H | AC or DCEP |
| E6012 | High titania sodium | F, V, OH, H | AC or DCEN |
| E6013 | High titania potassium | F, V, OH, H | AC or DC, either polarity |
| E6020 | High iron oxide | H-fillets | AC or DCEN |
| E6022[c] | High iron oxide | F | AC or DC, either polarity |
| E6027 | High iron oxide, iron powder | H-fillets, F | AC or DCEN |
| E70 series electrodes | | | |
| E7014 | Iron powder, titania | F, V, OH, H | AC or DC, either polarity |
| E7015 | Low hydrogen sodium | F, V, OH, H | DCEP |
| E7016 | Low hydrogen potassium | F, V, OH, H | AC or DCEP |
| E7018 | Low hydrogen potassium, iron powder | F, V, OH, H | AC or DCEP |
| E7024 | Iron powder, titania | H-fillets, F | AC or DC, either polarity |
| E7027 | High iron oxide, iron powder | H-fillets, F | AC or DCEN |
| E7028 | Low hydrogen potassium, iron powder | H-fillets, F | AC or DCEP |
| E7048 | Low hydrogen potassium, iron powder | F, OH, H, V-down | AC or DCEP |

a. The abbreviations, F, V, V-down, OH, H, and H-fillets indicate the welding positions as follows:
    F = Flat
    H = Horizontal
    H-fillets = Horizontal fillets
    V-down = Vertical down
    V = Vertical } For electrodes 3/16 in. (4.8 mm) and under,
    H = Overhead } except 5/32 in. (4.0 mm) and under for classifications E7014, E7015, E7016, and E7018.

b. The term DCEP refers to direct current, electrode positive (DC reverse polarity). The term DCEN refers to direct current, electrode negative (DC straight polarity).

c. Electrodes of the E6022 classification are for single-pass welds.

Fig. 9-42. AWS Carbon Steel Covered Arc Welding Electrodes. This information is from AWS A5.1-91. (American Welding Society)

which indicate a welding rod used for gas welding.

The meaning of digit numbers in the AWS designation follows: The first two or three digits of the four or five digit number E60XX or E100XX represents the TENSILE STRENGTH. This is, 60 means 60,000 psi (414 MPa) and 100 means 100,000 psi (689 MPa). This 60,000 pounds per square inch (psi) may be shown as 60 ksi. The letter "k" represents 1000 lbs. or kilopounds. The tensile strength may be given in the "as welded" or "stress relieved" condition. See the electrode

manufacturer's specification to determine under what condition the indicated tensile strength occurs. "As welded" means without post heating. "Stress relieved" means the welding is given a heat treatment after welding to relieve stress caused while welding. See Chapter 27 for an explanation of stress caused by welding. The second digit from the right indicates the RECOMMENDED POSITION of the joint that the electrode is designed to weld. For example, EXX1X: this electrode will weld in all positions; EXX2X electrodes are used for welds in the flat or horizontal position. The EXX4X is recommended for flat, horizontal, overhead, and vertical down welding.

The right hand digit indicates the POWER SUPPLY (AC, DCEN, or DCEP), TYPE OF COVERING, and PRESENCE OF IRON POWDER or LOW HYDROGEN CHARACTERISTICS, or both.

The last two digits should be looked at together to determine the proper application and the covering composition for an electrode. For example:

| ELECTRODE NUMBER | COVERING COMPOSITION |
|---|---|
| EXX10 | High cellulose, sodium |
| EXX11 | High cellulose, potassium |
| EXX12 | High titania, sodium |
| EXX13 | High titania, potassium |
| EXX14 | Iron powder, titania |
| EXX15 | Low hydrogen, sodium |
| EXX16 | Low hydrogen, potassium |
| EXX18 | Iron powder, low hydrogen, potassium |
| EXX20 | High iron oxide |
| EXX22 | High iron oxide |
| EXX24 | Iron powder, titania |
| EXX27 | Iron powder, high iron oxide |
| EXX28 | Iron powder, low hydrogen, potassium |
| EXX48 | Iron powder, low hydrogen, potassium |

See Fig. 9-42 and Fig. 9-43 which show the recomended position and polarity for various carbon and low alloy steel electrodes.

Occasionally an electrode number may have a letter and number after the normal four or five numbers, such as E7010-A1 or E8016-B2. This letter and number combination or suffix is used with low alloy steel electrodes. The suffix indicates the chemical composition of the deposited weld metal. See Fig. 9-44. The letter "A" indicates a carbon molybdenum steel electrode. The letter "B" stands for chromium-molybdenum steel electrode. The letter "C" is a nickel steel electrode and the letter "D", a manganese molybdenum steel electrode. The final digit in the suffix indicates the chemical composition under

| AWS Classification[a] | Type of covering | Capable of producing satisfactory welds in positions shown[b] | Type of current[c] |
|---|---|---|---|
| E70 series — Minimum tensile strength of deposited metal, 70,000 psi (480 MPa) | | | |
| E7010-X | High cellulose sodium | F, V, OH, H | DCEP |
| E7011-X | High cellulose potassium | F, V, OH, H | AC or DCEP |
| E7015-X | Low hydrogen sodium | F, V, OH, H | DCEP |
| E7016-X | Low hydrogen potassium | F, V, OH, H | AC or DCEP |
| E7018-X | Iron powder, low hydrogen | F, V, OH, H | AC or DCEP |
| E7020-X | High iron oxide | H-fillets / F | AC or DCEN / AC or DC, either polarity |
| E7027-X | Iron powder, iron oxide | H-fillets / F | AC or DCEN / AC or DC, either polarity |
| E80 series — Minimum tensile strength of deposited metal, 80,000 psi (550 MPa) | | | |
| E8010-X | High cellulose sodium | F, V, OH, H | DCEP |
| E8011-X | High cellulose potassium | F, V, OH, H | AC or DCEP |
| E8013-X | High titania potassium | F, V, OH, H | AC or DC, either polarity |
| E8015-X | Low hydrogen sodium | F, V, OH, H | DCEP |
| E8016-X | Low hydrogen potassium | F, V, OH, H | AC or DCEP |
| E8018-X | Iron powder, low hydrogen | F, V, OH, H | AC or DCEP |
| E90 series — Minimum tensile strength of deposited metal, 90,000 psi (620 MPa) | | | |
| E9010-X | High cellulose sodium | F, V, OH, H | DCEP |
| E9011-X | High cellulose potassium | F, V, OH, H | AC or DCEP |
| E9013-X | High titania potassium | F, V, OH, H | AC or DC, either polarity |
| E9015-X | Low hydrogen sodium | F, V, OH, H | DCEP |
| E9016-X | Low hydrogen potassium | F, V, OH, H | AC or DCEP |
| E9018-X | Iron powder, low hydrogen | F, V, OH, H | AC or DCEP |
| E100 series — Minimum tensile strength of deposited metal, 100,000 psi (690 MPa) | | | |
| E10010-X | High cellulose sodium | F, V, OH, H | DCEP |
| E10011-X | High cellulose potassium | F, V, OH, H | AC or DCEP |
| E10013-X | High titania potassium | F, V, OH, H | AC or DC, either polarity |
| E10015-X | Low hydrogen sodium | F, V, OH, H | DCEP |
| E10016-X | Low hydrogen potassium | F, V, OH, H | AC or DCEP |
| E10018-X | Iron powder, low hydrogen | F, V, OH, H | AC or DCEP |
| E110 series — Minimum tensile strength of deposited metal, 110,000 psi (760 MPa) | | | |
| E11015-X | Low hydrogen sodium | F, V. OH, H | DCEP |
| E11016-X | Low hydrogen potassium | F, V, OH, H | AC or DCEP |
| E11018-X | Iron powder, low hydrogen | F, V, OH, H | AC or DCEP |
| E120 series — Minimum tensile strength of deposited metal, 120,000 psi (830 MPa) | | | |
| E12015-X | Low hydrogen sodium | F, V, OH, H | DCEP |
| E12016-X | Low hydrogen potassium | F, V, OH, H | AC or DCEP |
| E12018-X | Iron powder, low hydrogen | F, V, OH, H | AC or DCEP |

a. The letter suffix 'X' as used in this table stands for the suffices A1, B1, B2, etc. (see Fig. 9-44) and designates the chemical composition of the deposited weld metal.
b. The abbreviations F, V, OH, H, and H-fillets indicate welding positions as follows: F = Flat; H = Horizontal; H-fillets = Horizontal fillets. V = Vertical } For electrodes 3/16 in. (4.8 mm) and under, except 5/32 in. (4.0 mm) and under for classifications OH = Overhead } EXX15-X, EXX16-X, and EXX18-X.
c. DCEP means electrode positive (reverse polarity). DCEN means electrode negative (straight polarity).

Fig. 9-43. AWS Low Alloy Steel Covered Arc Welding Electrodes. This information is from AWS 5.5-81.
See Fig. 9-44 for the letter suffixes used in place of the "X" with these electrode numbers.
(American Welding Society)

one of these broad chemical classifications. The exact chemical composition may be obtained from the electrode manufacturer.

The letter "G" is used for all other low alloy electrodes with minimum values of: molybdenum (0.20 percent minimum); chromium (0.30 percent minimum); manganese (1.00 percent minimum); silicon (0.80 percent minimum); nickel (0.50 percent minimum); and vanadium (0.10 percent minimum) specified. Only one of these elements is required to meet the alloy requirements of the "G" classification.

An example of a complete electrode classification is the E8016-B2:

| | |
|---|---|
| -A1 | 1/2% Mo |
| -B1 | 1/2% Cr, 1/2% Mo |
| -B2 | 1 1/4% Cr, 1/2% Mo |
| -B3 | 2 1/4% Cr, 1% Mo |
| -C1 | 2 1/2% Ni |
| -C2 | 3 1/4% Ni |
| -C3 | 1% Ni, .35% Mo, .15% Cr |
| -D1 and D2 | .25-.45% Mo, 1.25-2.00% Mn |
| -G | .50 min Ni, .30 min Cr, .20 min. Mo, .10 min V, 1.00 min Mn, .80 min Si (Only one of the listed elements is required for the G classification). |

Fig. 9-44. Approximate chemical composition of suffix numbers of the AWS Electrode Numbering System.

1. E indicates electrode.
2. 80 indicates tensile strength, 80,000 psi or 80 ksi.
3. 16 indicates a low hydrogen, potassium covering used with AC or DCEP (reverse polarity). DCEP means direct current electrode positive. See Heading 10-2 for more information on electrode polarity.
4. The 1 indicates it is an all position electrode.
5. The suffix B2 indicates that the deposited metal chemical composition is a low alloy chromium-molybdenum steel with 1 1/4 percent chromium and 1/2 percent molybdenum.

Manufacturers imprint AWS numbers on the covering material near the grip end for identification. Fig. 9-45 indicates how metallic arc welding electrodes are marked with AWS numbers.

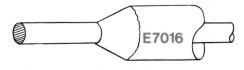

Fig. 9-45. The American Welding Society has standardized a numbering system for identifying welding electrodes. This electrode number is placed on the covering near the end of the electrode. This number may be repeated at intervals along the length of the electrode.

## 9-17. LOW HYDROGEN ELECTRODES

Hydrogen has harmful effects on alloy steels. It causes a low ductility weld and underbead cracking. This is called HYDROGEN ENBRITTLEMENT or HYDROGEN CRACKING.

LOW HYDROGEN ELECTRODES deposit a minimum of hydrogen in the weldment. The low hydrogen condition is obtained by using a special covering in the EXXX5, 6, or 8 category. See Fig. 9-42. Sodium, potassium, and iron powder coverings are used. These electrodes conform to AWS E7015, E7016, E7018, E7028, and E7048 specifications. They are used on hard to weld free machining steels, low carbon, low alloy, and hardenable steels. The slag is very fluid, but good flat or convex beads are easily obtained. Fig. 9-46 shows the composition and application of some low hydrogen electrode coverings. These special coverings contain practically no organic material. Recommended arc welding machine settings for these electrodes are shown in Fig. 9-47.

| CURRENT SETTINGS FOR LOW HYDROGEN ELECTRODES | | | |
|---|---|---|---|
| ELECTRODE DIAMETER | AMPS (FLAT) | AMPS (VERTICAL AND OVERHEAD) | VOLTS |
| 1/8 | 140-150 | 120-140 | 22-26 |
| 5/32 | 170-190 | 160-180 | 22-26 |
| 3/16 | 190-250 | 200-220 | 22-26 |
| 7/32 | 260-320 | | 24-27 |
| 1/4 | 280-350 | | 24-27 |
| 5/16 | 360-450 | | 26-29 |

Fig. 9-47. Recommended arc machine settings when using low hydrogen electrodes in flat, vertical, and overhead positions.

The deposited metal has excellent tensile and ductile qualities and is exceptionally clean as may be seen by X-ray inspection. Fig. 9-48 illustrates the physical properties of a weld made with a low hydrogen E7016 electrode.

| RIGHT HAND DIGIT | COVERING COMPOSITIONS | APPLICATION (USE) |
|---|---|---|
| 5 E-7015 | Low hydrogen sodium type. | This is a low hydrogen electrode for welding low carbon, alloy steels. Power shovels and other earth moving machinery require this rod. The weld files or machines easily. Use DCEP (DCRP) only. |
| 6 E-7016 | Same as "5" but with potassium salts used for arc stabilizing. | It has the same general application as (5) above except it can be used on either DCEP (DCRP) or AC. |
| 8 E-7028 | Iron powder (Low Hydrogen) Flat position only. | For low carbon alloy steels, use DC or AC. |
| E-8018 | Iron powder plus low-hydrogen sodium covering. | Similar to (5) and (6) DCEP (DCRP) or AC. Heavy covering allows the use of high speed drag welding. AC or DCEP (DCRP may be used. |

Fig. 9-46. Low hydrogen electrode covering compositions and applications. These coverings will withstand a high temperature and therefore high currents (amperages) may be used.

| | AS WELDED |
|---|---|
| Yield Point | 60,000 psi (414 MPa) |
| Tensile (minimum) | 70,000 psi (480 MPa) |
| Elongation, % in 2 inches (50.8 mm) | 22% |
| Charpy V Notch at −20 °F (−29 °C) | 20 ft. lbs. (27.1 Nm) |

Fig. 9-48. AWS specifications for welds made with E7016, low hydrogen electrodes.

The E7016 electrodes may be used with either AC or DCEP (DCRP). Low hydrogen electrodes should be dried by baking at 250 °F (121 °C) before using. If they have been exposed to the atmosphere for an appreciable period they should be rebaked at 500 to 700 °F (260 to 371 °C) for one hour. This baking will remove any moisture which may be in the coating. They are then stored at 250 °F (121 °C) until they are used. Never exceed a 1/2 in. (12.7 mm) motion, and practice considerable care during vertical and overhead passes to avoid molten metal flow.

## 9-18. CARBON AND LOW ALLOY STEEL IRON POWDER ELECTRODES

The addition of iron powder to the covering of shielded arc electrodes changes the arc behavior. It also greatly increases the amount of filler metal deposited. Much higher currents can be used to produce welds faster. The addition of iron powder also produces more easily cleaned welds, less spatter, and better bead shapes. Fig. 9-49 shows a change in filler metal deposit from 4.5 pounds (2.0 kg) per hour with the E6012 electrode to 9 pounds (4.1 kg) per hour with the E7028 electrode.

| | ELECTRODE CLASSIFICATION | | | |
|---|---|---|---|---|
| | E7028 | E7024 | E7018 | E6012 |
| Iron powder content | 45% | 39% | 24% | 0% |
| Amount Deposited | 9 lbs/hr | 8 lbs/hr | 5.5 lbs/hr | 4.5 lbs/hr |
| Deposit Efficiency Increase over E6012 | 100% | 78% | 22% | 0% |

Fig. 9-49. Effect of using iron powder in electrode coatings. Note the change in the amount of filler metal deposited per hour. The deposit efficiency increase is based on the E6012 electrode.

The arc obtained is smooth and steady. One may use about 25 percent more current with the EXX18 iron powder electrode, and as much as 50 percent more current with the EXX24 and EXX27 iron powder electrodes, because of their heavy coatings.

The weld puddle is so fluid when the 39 percent iron powder covering is used that this electrode is recommended for downhand (or flat position) welding only.

## 9-19. CHROMIUM AND CHROMIUM-NICKEL STEEL COVERED WELDING ELECTRODES

CHROMIUM and CHROMIUM-NICKEL STEEL ELECTRODES are corrosion resisting. They are used on weldments which are subjected to chemical corrosion or other highly corrosive atmospheres.

Below is a listing of all the AWS classification numbers:

| | | |
|---|---|---|
| E209 | E310 | E330 |
| E219 | E310H | E330H |
| E240 | E310Cb | E347 |
| E307 | E310Mo | E349 |
| E308 | E312 | E410 |
| E308H | E316 | E410NiMo |
| E308L | E316H | E430 |
| E308Mo | E316L | E502 |
| E308MoL | E317 | E505 |
| E309 | E317L | E630 |
| E309L | E318 | E7Cr |
| E309Cb | E320 | E16-8-2 |
| E309Mo | E320LR | |

Occasionally these electrode numbers may be followed by a two digit suffix.

Examples: E320-15.
E309-16.

The -15 electrode is used with direct current electrode positive (DCRP). The suffix -16 is used with AC or direct current electrode positive (DCRP).

Electrodes up to 5/32 in. (3.97 mm) in size may be used in all positions. Electrodes 3/16 in. (4.76 mm) and larger are used only in the flat or horizontal-fillet positions.

The numbering system in this classification has no meaning as do the carbon and low alloy steel electrode numbers. The 'L' which follows some electrode designations means the electrode has a low carbon content.

The letter E means electrode. To find the chemical composition of these corrosion resisting electrodes, see the AWS 5.4-81 publication.

## 9-20. COVERED ELECTRODES FOR CAST IRON

Cast iron may be arc welded with cast iron, copper based, nickel based, or mild steel arc welding electrodes.

The CAST IRON ELECTRODE is designated ECI. It contains 3.25-3.50 percent carbon and 2.75-3.00 percent silicon. It also contains less than 1.00 percent each of manganese, phosphorus, and .10 percent sulphur. The remainder, or approximately 92 percent, of the metal composition is pure iron.

COPPER BASED ELECTRODES for cast iron are designated: ECuSn-A, ECuSn-C, and ECuAl-A₂. The ECuSn-A electrode contains between 4.0 and 6.0 percent tin (Sn), .1 to .35 percent phosphorus, and traces of aluminum and lead. The remainder, or about 94 percent of the metal, is copper (Cu). The ECuSn-C electrode contains 7.0 to 9.0 percent tin (Sn), and traces of aluminum and lead. It also contains between .05 and .35 percent phosphorus. The remainder, or about 91 percent, is copper (Cu). The ECuAl-A₂ classification contains 9.0 to 11.0 percent aluminum and 1.5 percent iron. It also contains a small amount of silicon, zinc, and lead. The remainder, or about 87 percent, is copper. The four NICKEL BASED ELECTRODE classifications are: ENi-Cl, ENiFe-Cl, ENiCu-A, ENiCu-B. The nickel based electrodes contain from 45.0 to 85 percent nickel depending on their designation. They also contain small percentages of carbon, silicon, manganese, sulphur, and copper. ECuFe-Cl may contain as much as 44 percent iron. The single mild steel electrode for cast iron is classified ESt. ESt contains .15 percent carbon, .30 to .60 percent manganese, .03 percent silicon, and .04 percent phosphorus and sulphur. The remainder or about 99 percent of the electrode is pure iron.

Exact chemical compositions for the electrodes may be found in the AWS 5.15-90 publication.

## 9-21. NON-FERROUS ELECTRODE CLASSIFICATIONS

The American Welding Society has produced the following booklets on non-ferrous electrodes:
AWS A5.3-88, "Specification for Aluminum and Aluminum Alloy Electrodes for Shielded Arc Welding."
AWS A5.6-84, "Specification for Covered Copper and Copper Alloy Arc Welding Electrode."
AWS A5.11-90, "Specification for Nickel and Nickel Alloy Welding Electrodes for Shielded Metal Arc Welding."
All of the non-ferrous electrode designations convey an idea of the major alloying ingredients in the letters which make up the designation. The letters are abbreviations for chemical elements as follows:

Al - aluminum   Fe - iron        Ni - nickel
Cr - chromium   Mn - manganese   Si - silicon
Cu - copper     Mo - molybdenum  Sn - tin

There are three ALUMINUM ELECTRODES in the AWS A5.3-88 specifications. They are the E1100 with an aluminum content of 99 percent;

the E3003 with an aluminum content of about 96.7 percent; and the E4043 with about a 92.3 percent aluminum content.

The COPPER BASED ELECTRODES range in copper content from about 68.9 to 99.0 percent. ECuNi is the exception. It contains up to 33 percent nickel. The copper content is about 62 percent with the remainder made up of other elements. The AWS classifications are shown below:

ECu       ECuSn-C    ECuAl-B
ECuSi     ECuNi      ECuNiAl
ECuSn-A   ECuAl-A2   ECuMnNiAl

The NICKEL BASED ELECTRODES contain from 59 to 92 percent nickel. The exact chemical composition may be found in AWS A5.11-90. The eight AWS classifications are:

ENi-1      ENiCrFe-3
ENiCu-7    ENiCrFe-4
ENiCrFe-1  ENiMo-1
ENiCrFe-2  ENiMo-3

## 9-22. CARBON ELECTRODES

CARBON ELECTRODES are used for carbon arc welding, twin carbon arc welding, and carbon arc cutting. Since there are many more efficient methods of cutting, very little carbon arc cutting is done. These electrodes come in sizes ranging from 1/16 up to 1 in. diameter (1.59-25.4 mm). Rods may be obtained in 12, 18, and 24 in. (305, 457, and 610 mm) lengths. The quality of the rod must be extremely high and the structure of the carbon must be uniform. The two types of electrodes obtainable are the CARBON ELECTRODE and the GRAPHITE ELECTRODE. Graphite has better conductivity and is usually of more uniform quality.

The rod should be inserted in the holder with the end, or the point, of the carbon approximately 10 times the diameter of the rod away from the electrode holder. For example, a 1/2 in. (12.7 mm) rod will have its end not over 5 in. (127 mm) away from the holder. As the rod is being used, it tends to slowly burn back toward the holder. The rod should be pointed with the taper of the point approximately 6 to 8 times the diameter of the electrode, as shown in Fig. 10-63.

Currents required for different sizes of carbon electrodes are shown in Fig. 9-50.

More information on carbon electrodes will be found in Chapter 15.

## 9-23. ELECTRODE CARE

It is important that the user follow the electrode manufacturer's recommendations as to ampere settings, base metal preparation, welding technique, welding position, and the like. Electrodes

| ELECTRODE DIAMETER INCHES | WELDING CURRENT | | MAXIMUM CURRENT DENSITY AMPS. PER SQ. IN. | POUNDS PER HOUR DEPOSITED |
|---|---|---|---|---|
| | MIN. | MAX. | | |
| 1/8 | 0 | 35 | 2890 | . . . |
| 3/16 | 25 | 60 | 2200 | . . . |
| 1/4 | 50 | 90 | 1855 | . . . |
| 5/16 | 80 | 125 | 1650 | . . . |
| 3/8 | 110 | 165 | 1510 | . . . |
| 7/16 | 140 | 210 | 1420 | 1.5 |
| 1/2 | 170 | 260 | 1340 | 2.5 |
| 5/8 | 230 | 370 | 1220 | 4.5 |
| 3/4 | 290 | 490 | 1125 | 6.0 |
| 7/8 | 350 | 615 | 1035 | . . . |
| 1 | 400 | 750 | 965 | . . . |

Fig. 9-50. Carbon electrode current requirements. (National Carbon Co.)

must not be used after being exposed to dampness because the steam generated by the heat of the arc may cause the covering to be "blown" away, and also cause hydrogen inclusions. Questionable electrodes should be "baked" at 250 °F (121 °C) for several hours.

Because of the similarity of appearance of electrodes that are much different in welding properties, it is important to label and store them in carefully marked bins. Use masking tape to bind electrodes and label carefully when putting them in storage. Electrodes are costly, and loss of identification may mean loss in time and money.

If an electrode is used beyond its ampere rating, the electrode will overheat, and the covering will crack, thus spoiling the rod. The excess current will also cause considerable splattering of the molten metal.

## 9-24. ARC WELDER REMOTE CONTROLS

Welders work on a variety of joints and metal thicknesses. They use different electrodes and weld in various positions. This all requires changes in polarity and welding machine adjustments.

To eliminate the time required to travel back and forth to the welding machine, several manufacturers provide remote control devices which may be kept near the operator for convenient control of the machine.

The small portable remote control panel, shown in Fig. 9-51, provides for voltage and current adjustment. A switch to allow remote hand or foot switch current control is also included. Using a panel of this type, the operator may climb into a restricted place, turn on the machine, and adjust it to any current setting without returning to the machine. The saving in time is important, and better quality welds are produced because the

Fig. 9-51. Remote control panel. This type of control panel provides adjustment of arc voltage and current. The machine may also be turned on and off. In addition, a foot control may be used to control the current (amperage) from the machine. This control may be mounted on the machine or near the operator. (Vickers, Inc.)

machine is more accurately adjusted to the job requirements at all times. Fig. 9-52 illustrates a remote control used to adjust welding current.

A pen sized cordless control is also available for use on a specially constructed welding machine. See Fig. 9-53. The operator sets the machine for the maximum current anticipated. At the weld site, if the welder wants to change the current, the electrode is replaced by the pen sized control.

Fig. 9-52. Remote current setting device. The connecting cable is not shown. This device will vary the current within the coarse range set on the machine. (Miller Electric Mfg. Co.)

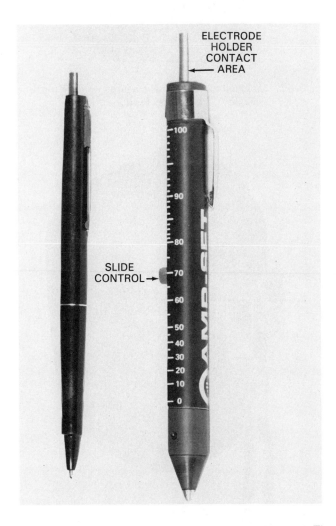

Fig. 9-53. A pocket size cordless remote current control. The control is placed into the electrode holder, set, and touched to the work for 2 - 5 seconds. Circuitry on the arc welder senses the change and resets the output current. (Miller Electric Mfg. Co.)

The control is moved to a desired setting and touched to the work for from 2 to 5 seconds. Special circuitry built into the arc welder senses the change and automatically readjusts the output current.

A foot switch, Fig. 9-54, is used for fine adjustment. The machine is set for a certain range of power, then the foot switch provides variations within that range.

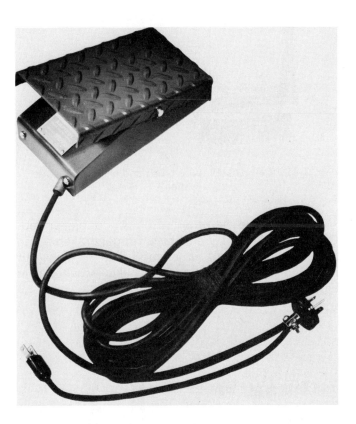

Fig. 9-54. Foot operated remote control for changing machine settings.

## 9-25. WELD CLEANING EQUIPMENT

It is very important that the base metals in weldments be cleaned prior to welding. It is difficult to weld dirty or corroded surfaces. If attempted, the resulting welds will normally be of poor quality. Many types of equipment and tools have been developed for the purpose of cleaning joints and welds. Cleaning may be done by using sand blasting machinery, rotary wire wheels, and tools such as chipping chisels, hammers, and wire brushes. Nonferrous metals may be chemically cleaned, especially in production welding situations. The amount and size of the welding done usually determines the kind of cleaning apparatus needed.

Slag which covers the finished weld must be removed before the next weld bead is laid, to prevent inclusions in the finished weld. The slag on the final bead must also be removed before the weld can be inspected or painted. This coating may be removed by a rotary wire wheel, or by tapping the scale with a CHIPPING HAMMER. In either case, suitable eye protection must be provided. Fig. 9-55 shows the welder's chipping hammer.

Fig. 9-55. Combination wire brush and chipping hammer. (Atlas)

The chipping hammer often is double ended. One end is shaped like a chisel for general chipping and the other end is shaped like a pick, for reaching into corners, and the like, as shown in Fig. 9-56.

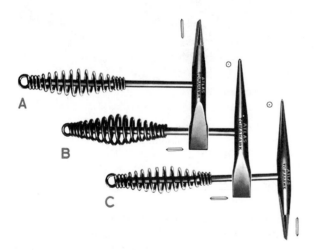

Fig. 9-56. Variety of chipping hammers. Note that in A the blades are turned 90 deg. to each other. In B and C the chipping hammers have a blade at one end, and a pick at the other end. (Atlas)

## 9-26. SHIELDS AND HELMETS

Electric arc welding necessitates the use of special protective devices for skin surfaces, such as the hands, face, and eyes. The arc welding shield or helmet is used to protect the face and eyes. It is mounted and supported on the head,

Fig. 9-57. The hand held welding shield is sometimes preferred, Fig. 9-58.

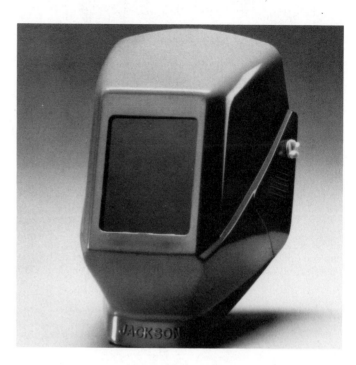

Fig. 9-57. Molded helmet with extra large lens for maximum visibility. (Jackson Products, Inc.)

Fig. 9-58. Arc welding hand shield used for welding inspection work and instructional purposes.

The Occupational Safety and Health Act (OSHA) requires the use of a hard hat with the arc welding helmet on construction work. A combination helmet and hard hat is shown in Fig. 9-59. The face shield is usually made of fiber, plastic or fiber glass, and formed in a shape which covers

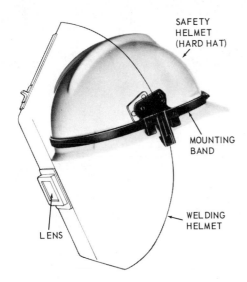

SAFETY HELMET (HARD HAT)

MOUNTING BAND

WELDING HELMET

LENS

Fig. 9-59. An arc welder's helmet used in connection with a hard hat. (Kedman Co.)

the front half of the head. An aperture or opening at the level of the eyes provides visibility. This aperture is approximately 4 1/4 x 2 in. (108 x 50.8 mm) and is provided with at least two glass lenses. Larger square lenses are available in a 4 1/2 x 5 1/4 in. (114 x 133 mm) size. The larger lens is good for use with bi-focal glasses. The outer lens, which is of double strength glass or plastic, is used to protect the inner and more expensive welding lens from metal spatter and abuse. Many operators also put another clear lens on the face side of the helmet to protect the colored lens from that side. Fig. 9-60 shows recommended shade numbers for various arc welding applications.

| | Lens Shade Number |
|---|---|
| SMAW (Shielded metal arc welding) 1/16 - 5/32 in. electrodes | 10 |
| 3/16 - 1/4 in. electrodes | 12 |
| 5/16 - 3/8 in. electrodes | 14 |
| GMAW (Gas metal arc welding) (nonferrous) 1/16 - 5/32 in. electrodes | 11 |
| GMAW (Gas metal arc welding) (ferrous) 1/16 - 5/32 in. electrodes | 12 |
| GTAW (Gas tungsten arc welding) | 10 to 14 |
| CAW (Carbon arc welding) | 14 |

Fig. 9-60. Suggested lens shade numbers for various arc welding applications.

A good grade of colored arc welding filter lens will remove approximately 99.5 percent of the infrared rays and 99.75 percent of the ultraviolet rays. These figures have been developed by U.S. Bureau of Standards. Shade numbers 10, 12, and 14 are the common shades used for shielded metal arc welding (SMAW). Use shade number 14 for carbon arc welding. Numbers 11 or 12 are used for gas metal arc welding (GMAW). When gas tungsten arc welding (GTAW) shade numbers 10-14 should be used. However, other shade numbers may be obtained and used, following the manufacturer's recommendations. The higher the shade number, the lower the transmission of infrared or ultraviolet rays. A gold plating is being applied to some filter lenses. This gold plating reflects most of the harmful rays. The plating greatly improves the filtering capabilities of the regular filter lens. The same lens numbering system is used for the gold plated lenses.

Excess ultraviolet rays may cause eye pain for 8 to 18 hours after exposure. Infrared light rays tend to injure the sight and every precaution should be taken to shield the eyes from these rays.

Filter lenses are of such density or shade that the operator cannot see through them until the arc is struck. A helmet has been developed which has a battery powered, photoelectric cell built into it. The lens is clear until an arc is struck. The circuitry of the photoelectric cell darkens the lens instantly when the arc is struck. This feature will protect the welder from an arc flash from a nearby worker. The weld may be started and restarted with great precision, see Fig. 9-61.

Some outer cover lenses are especially treated to resist the adhesion of metallic particles spattering from the weld. Spring clips are generally used to provide a snug fitting of the lenses in the helmet. Head bands are adjustable.

The helmet, or head shield, has a swing mounting which permits the forward part of the helmet to be lifted above the operator's face, without removing the head band from the head. Some welders who work continuously find that wearing a pair of ordinary welding goggles under the helmet helps to reduce eye strain. These goggles eliminate reflected glare around the back of the helmet. Also available are light weight goggles with a #1 or #2 filter lens. These goggles are called FLASH GOGGLES. They enable the operator to set up work, chip welds (if the lens is tempered), peen, etc., and still have eye protection from flying particles and adjacent arc rays. A welder must wear either a pair of welding goggles or a pair of safety glasses at all times.

Some helmets are available into which fresh air is fed by means of a hose to increase the comfort of the operator. Helmets are also available which

PHOTO SENSITIVE LENS

BATTERY POWERED ACTUATOR

Fig. 9-61. The Winkin® arc welding helmet. A special optical lens remains clear until the arc is struck. The built in actuator contains a 9 volt battery, photo electric cell, and integrated circuitry. (Gor-Vue Corp.)

MOLDED HARD HAT AND AIR SUPPLY CONNECTION

FACE SHIELD AND LENS

BATTERY POWERED AIR FILTER

Fig. 9-62. Arc welding helmet with built-in hard hat and battery powered positive pressure air filter. Under the arc shield is a clear face shield. (Racal Health & Safety, Inc.)

filter the air which the welder breathes. Refer to Fig. 9-62.

## 9-27. SPECIAL ARC WELDING CLOTHING

While an arc weld is in progress, the molten flux and the metal itself sometimes spatter for a considerable distance around the joint being welded. The operator must, therefore, be protected from the danger of being burned by these hot particles. Such clothing as gloves, gauntlet sleeves, aprons, and leggings are sometimes necessary, depending upon the type of welding being performed. When performing welds in the overhead position, it is recommended that one wear a jacket or cape to protect the shoulders and arms. A cap or a special hooded arc helmet is required to protect the head and hair. All these clothing articles should be made of leather. They are referred to as LEATHERS. Chrome leather is usually used for arc welding "leathers." Leggings and gloves are sometimes made of a combination of cloth and asbestos.

It is further recommended that the operator use high-top shoes or boots. Trousers worn by the welder should not have cuffs. Cuffs may catch burning particles as they fall. Gloves should be worn to cover the hands and wrists and to prevent "sunburn."

All clothing worn should be carefully inspected to eliminate any place where the metal may catch and burn. Open pockets are especially dangerous.

Clothing worn, other than the "leathers" mentioned, should be of heavy material. Thin clothing will permit infrared and ultraviolet rays to penetrate to the skin. If the skin is not properly protected, the operator will become "sunburned." Such burns, if they do occur, should be treated as a severe sunburn. If the burns are severe, a physician should be consulted. Easily ignited material such as flammable combs, pens, butane cigarette lighters, and the like should not be on the operator while welding.

## 9-28. TEST YOUR KNOWLEDGE

Write your answers on a separate sheet of paper. Do not write in this book.

1. List the equipment required to outfit a SMAW station.
2. What welding processes use constant cur-

rent arc welders?

3. What minimum information should be provided when describing an arc welding machine?

4. What is a typical open circuit (no load) voltage for an arc welding machine?

5. What are the advantages of using a constant current arc welding machine?

6. In an alternator, what four factors will cause the current output to change?

7. How is DC created in an AC alternator or transformer?

8. Using a movable coil transformer, how is the current changed?

9. What type of welding machine control works best with remote control devices?

10. In the DC generator, how is current inducted into the armature windings?

11. What direct current polarity flows from the electrode, across the arc, to the work?

12. How is an arc welding machine cooled?

13. What size welding leads are required to carry 300 A a distance of 90 ft. (27.4 m)?

14. What are five (5) functions of the flux covering on the SMAW electrode?

15. What information is obtained from the first and second (possibly third) number in the AWS electrode Classification?

16. The last two digits together tell the composition of the electrode covering. What is in a EXX24 electrode covering?

17. a. List four electrode numbers which contain low hydrogen.

    b. Which three last digits designate a low hydrogen electrode?

18. List three AWS electrodes which contain iron powder.

19. What does the electrode classification E316-16 tell the welder?

20. What number filter lens is recommended for SMAW?

A variety of inverter type arc welding machines, both constant current (CC) and constant voltage (CC/VC). (Miller Electric Mfg. Co.) Below. Welding student practices SMAW in the flat position.

# 10 SHIELDED METAL ARC WELDING (AC AND DC)

The most frequently used welding process is shielded metal arc welding (SMAW). Weldments in the shop and in the field are made with this process. Industrial machines, farm equipment, and vehicles are all repaired most frequently using SMAW.

Both alternating current (AC) and direct current (DC) arc welding are done. Each has its own advantages and disadvantages.

The welding technique is similar for both AC and DC. Electricity, as it arcs across the gap between the metal electrode and the work, creates a temperature of approximately 6500-7000 °F (3593-3871 °C). With the correct size electrode this temperature is sufficient to melt any weldable metal.

## 10-1. DIRECT CURRENT (DC) ARC WELDING FUNDAMENTALS

ELECTRIC ARC WELDING is defined by the American Welding Society as: ''A group of welding processes which produce coalescence of metals by heating them with an arc, with or without the application of pressure, and with or without filler metal.''

Shielded metal arc welding (SMAW) is done by producing an arc between the base metal and a consumable, flux covered, metal electrode. The electrode acts as an electrical conductor and filler metal. Fig. 10-1 shows the electrical circuit for shielded metal arc welding.

DC arc welding machines are of the following types:
1. Motor or engine driven generator.
2. Motor or engine driven AC alternator with DC rectifiers.
3. AC transformer with DC rectifiers.

See Heading 9-6 for more information on DC arc welding machines.

It is important to understand the voltage-current (amperes) characteristics of the DC arc welding

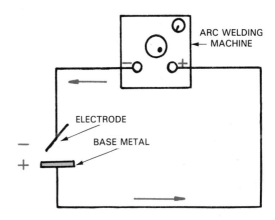

Fig. 10-1. A DC shielded metal arc welding circuit.

machine. Under a no-load (or open circuit) condition, when not welding, the voltage of the machine is about 60-80 volts. When the arc is struck, the current (measured in amperes) will go up and the voltage will come down to about 15-40 volts. See Fig. 10-2 for typical voltage and amperage meter readings on an arc welding machine.

OHM'S LAW for electricity states that voltage in a closed circuit has a constant relationship to the current and the resistance of the circuit. The Ohm's Law formula is:

$$V = I \times R$$

V is the voltage.
I is the intensity of current in amperes.
R is the resistance in the circuit.

On a constant current arc welding machine the only adjustment on the welding machine is for amperage. If the electrode gap or the resistance in the circuit is held constant, when the voltage is increased, the amperage increases. This may be seen in the formula $V = I \times R$ (constant).

The amperage output is determined by the

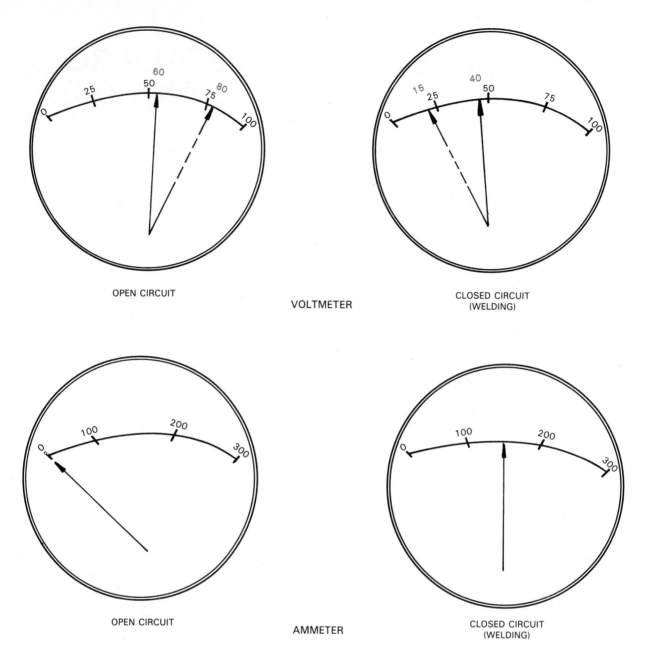

Fig. 10-2. Typical arc welding machine voltage and current readings. Open circuit voltage varies from about 60 - 80 volts. Closed circuit amperage will correspond to the machine setting.

voltage from the main power source. Amperage is also limited by the wiring of the arc welding machine circuits.

In a constant current machine, if the arc gap distance is increased, the resistance will increase in the circuit. Therefore, to maintain a constant amperage, the voltage output of the machine must increase.

Using Ohm's Law, if the arc distance decreases, the voltage must decrease to maintain a constant amperage (current).

Therefore, with a constant current (amperage) arc welding machine, the most important control of the heat generated in the electric arc is the length of the arc gap. The arc gap must be held

constant once the amperage is set on the machine. If the arc gap changes, the amperage across the gap will vary and so will the welding heat developed.

To make a good weld, the operator must consider the following:
1. The current (amperage) output of the welding machine.
2. The diameter, polarity, and type of electrode.
3. The arc and its manipulation.
4. The preparation of the base metal.
5. The type of base metal.

The ELECTRODE wire may be a ferrous or non-ferrous metal. See Heading 9-15 and Chapter 9 for more information on electrodes.

The ARC, when viewed through the helmet lens, is seen to be divided into two separate parts: the STREAM and the FLAME, as shown in Fig. 10-3. The ARC FLAME consists of neutral gases which appear to be pale red. The VAPORIZED METAL in the stream appears yellow. LIQUID METAL in the stream appears green. If the arc is longer than normal, the flame gases can no longer protect the arc stream from oxidation. With a long arc the weld will form oxides and nitrides, resulting in a very weak and brittle weld.

If the correct current flow and arc length are maintained, a good weld should result with direct current. The voltage and amperage required for any particular weld may be easily obtained from established tables. The correct arc length is entirely the operator's responsibility.

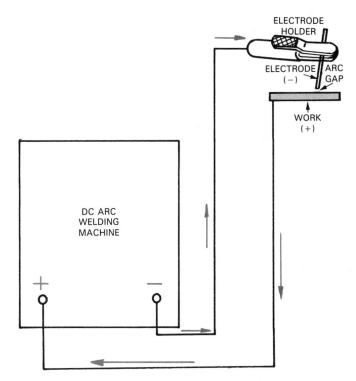

Fig. 10-4. A wiring diagram for a direct current electrode negative (DCEN) arc welding circuit. This circuit is also known as a direct current straight polarity (DCSP) circuit.

It is possible, and sometimes desirable, to reverse the direction of electron flow or polarity in the arc welding circuit. This may be done by disconnecting the electrode and work leads and reversing their positions. Some machines have a switch which will change the circuit polarity. When electrons flow from the negative terminal or pole of the arc welder to the base metal, this circuit is known as DIRECT CURRENT ELECTRODE POSITIVE (DCEP). It is also known as a direct current reverse polarity (DCRP) circuit. In this circuit, the electrons flow from the negative pole of the welding machine to the work. Electrons travel across the arc to the electrode and then return to the positive terminal (pole) of the machine from the electrode side of the arc, as shown in Fig. 10-5.

The choice of when to use direct current electrode negative (DCEN) or direct current electrode positive (DCEP) is primarily determined by the electrode being used. Some SMAW electrodes are designed to use only DCEN, or only DCEP. Other electrodes can be used with either DCEN or DCEP. See Heading 10-8 for information on electrodes.

The choice of the electrode is often not made by the welder in industrial practice. The electrode which must be used for a given job is generally determined by the welding procedure specification (WPS), and/or codes used. See Chapter 29. When the electrode to be used is known, the polarity may be determined by referring to the

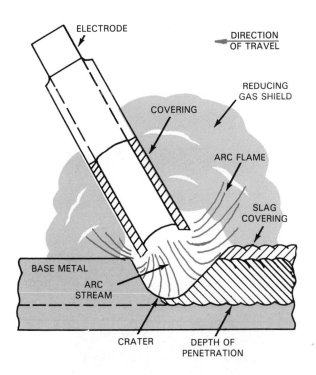

Fig. 10-3. A covered electrode arc weld in progress.

## 10-2. DIRECT CURRENT ELECTRODE NEGATIVE AND ELECTRODE POSITIVE FUNDAMENTALS

The welding circuit shown in Fig. 10-4 is known as a DIRECT CURRENT ELECTRODE NEGATIVE (DCEN) circuit. This circuit used to be defined by the American Welding Society as direct current straight polarity (DCSP). In this circuit the electrons are flowing from the negative terminal (or pole) of the machine to the electrode. The electrons continue to travel across the arc into the base metal and to the positive terminal or pole of the machine.

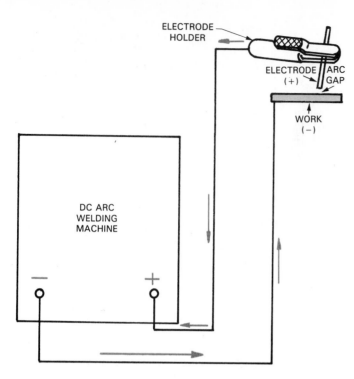

Fig. 10-5. A wiring diagram for a direct current electrode positive (DCEP) arc welding circuit. This circuit is also known as direct current reverse polarity (DCRP) circuit.

manufacturer's recommendations. See Figs. 10-17 and 10-18.

The decision to use DCEN (DCSP) or DCEP (DCRP) often depends on such variables as:
1. The depth of penetration desired.
2. The rate at which filler metal is deposited.
3. The position of the joint.
4. The thickness of the base metal.
5. The type of base metal.

DCEP (DCRP) produces better penetration than DCEN (DCSP). The SMAW electrodes that have the best penetrating abilities are E6010, E6011, and E7010. These electrodes use DCEP. There is a theory that with a DCEP covered electrode there is a jet action and/or expansion of gases in the arc at the electrode tip. This expansion causes the molten metal to be propelled with great speed across the arc. The molten metal impacts the base metal with greater force. This heavy impact on the base metal helps to produce deep, penetrating welds.

When a high rate of filler metal deposit is required, an EXX2X electrode is recommended. DCEN (DCSP) is usually recommended for the EXX2X electrodes. Examples of the EXX2X electrodes which deposit a high rate of filler metal are E6020, E6027, E7027, or E7028. See Fig. 10-17.

To weld out of position, an electrode intended for all positions must be used. Either DCEN or DCEP can be used. The electrode used will deter-

mine which polarity to use.

Base metal thickness will affect which polarity is required. On thick material, a welder must obtain good penetration. However, on thin material, excessive penetration should be avoided.

Shielded metal arc welding (SMAW) can be used to weld nickel, aluminum, and copper. Electrodes designed to weld these metals are generally used with DCEP (DCRP).

These considerations and others discussed in Heading 10-8 will determine what electrode and which current, DCEN or DCEP, should be used on a particular welding job.

## 10-3. ALTERNATING CURRENT (AC) ARC WELDING FUNDAMENTALS

Several types of AC arc welding machines have been produced and used. The two basic types are:
1. Motor and engine driven alternator.
2. Transformer type.

A transformer type machine is shown in Fig. 10-6. Fig. 10-7 illustrates an engine driven alternator. Also see Heading 9-4.

Fig. 10-6. An AC transformer arc welding machine. The output current maximum is 295 amps at a 20% duty cycle or 110 amps at a 100% duty cycle.   (Applied Power, Inc.)

Fig. 10-7. This engine driven, tap type, AC alternator develops 180 amps for welding. It also has 120 volt outlets for auxiliary power. (Air Products & Chemicals, Inc.)

In alternating current (AC), the current reverses its direction of flow 120 times per second. As shown in Fig. 10-8, it requires 1/60 of a second to complete a cycle or hertz (Hz). The current flow completes 60 Hz (cycles) per second and it is called 60 Hz (cycle) current.

Most AC arc welding machines have transformers which step down the voltage and increase the current (amperes) for welding purposes. Headings 9-4 and 9-5 describe how a transformer works.

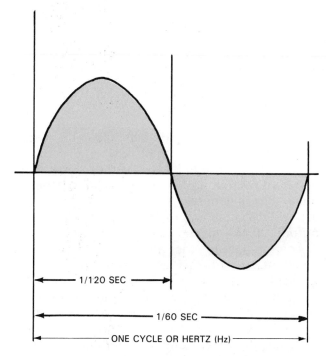

Fig. 10-8. Sine wave form of single-phase alternating current.

Fig. 10-9 shows what happens at the arc in one cycle of a typical AC transformer type arc welder. The voltage at A and B is zero. Beginning at the left side of the graph, the voltage builds up to a maximum in one direction to point C, and then back to zero at point A. The voltage then builds up to maximum in the other direction to point D, then back to zero again at point B. This action is repeated at the rate of 60 Hz (cycles) per second.

When AC welding with 60 Hz current, the voltage and current are zero 120 times per second as shown in Fig. 10-9. Each time the current crosses the zero point, the welding arc stops. To reestablish the arc, the voltage must increase enough to enable the current to jump the arc gap and maintain the arc. It is important that the voltage leads the current as each passes through zero. This will help make the AC arc stable. The AC arc welding machine must be designed to have the voltage lead the current.

Another method used to stabilize the AC arc is to increase the ionization of the material in the arc. IONIZATION is a physical phonomenon in which a particle obtains an electrical charge. These ionized or charged particles in the space between the electrode and the work make it easy for the arc to jump the gap. Electrodes which are intended for AC welding have ionizing agents in the electrode covering. These agents help to ionize the materials in the arc gap and help to stabilize the AC arc.

AC welding can only be done with electrodes that are intended for use with alternating current. If an electrode is not intended for use with AC, the arc will be very unstable.

Welds performed with AC electrodes show good penetration. Larger diameter electrodes are used to increase the metal deposition rates. Higher travel speeds can be obtained when large AC currents and large electrodes are used. There is no arc blow when AC welding. See Heading 10-13 for a description of arc blow.

Fig. 10-9 shows the current flow measured in amperes, in red, as related to the voltage (potential). It should be noted that the voltage leads the current, or the current lags the voltage slightly in time. This slight difference between the voltage and current must be designed into an AC arc welding machine.

## 10-4. SELECTING EITHER AC OR DC ARC WELDING MACHINES

Whether to buy or use an AC or DC arc welding machine is dependent on several factors. Selection of the type of current to use should be made after considering their individual advantages and disadvantages.

**Shielded Metal Arc Welding (AC and DC) / 257**

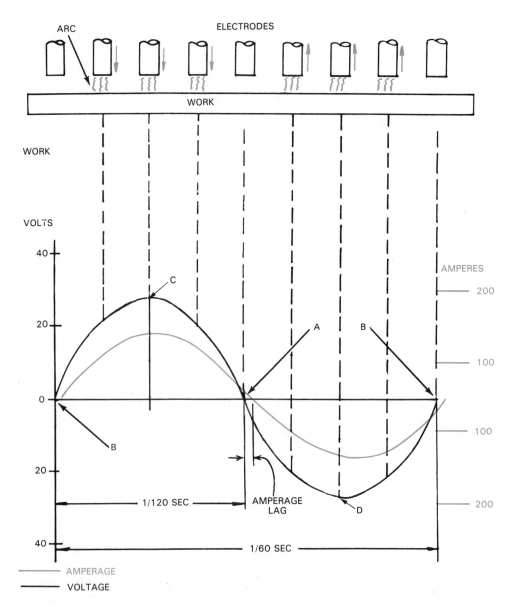

Fig. 10-9. The sine wave curve of alternating current at 60 Hz (cycles). At points A and B the voltage value is zero. At C and D, voltage is at a maximum. These two zero values which occur in each cycle (every 1/60 of a second) may make it difficult to strike and maintain an AC arc at small current values. Because a certain voltage is required to overcome electron inertia in a circuit there is usually a small lag or lead between the voltage and amperage (current). The voltage usually leads the amperage.

Advantages of the direct current (DC), constant current, arc welding machine:
1. The ability to choose direct current electrode positive (DCEP) or DCRP.
   a. DCEP or DCRP produces deeper penetrating welds than DCEN.
   b. DCEP or DCRP can be used in positions other than flat or downhand welding.
   c. Electrodes designed to weld nickel, aluminum, and copper generally use DCEP.
2. The ability to choose direct current electrode negative (DCEN) or DCSP.
   a. DCEN or DCSP is recommended for EXX2X electrodes which have high metal deposition rates.

   b. DCEN or DCSP can also be used in positions other than flat.
   The disadvantage of the direct current (DC), constant current arc welding machine is that a DC arc welder is generally more expensive than an AC arc welding machine of the same quality, current output, NEMA classification, and duty cycle.
   Advantages of the alternating current (AC), constant current, arc welding machine:
1. Welds made with AC arc welding machines and electrodes have moderate penetration.
2. Large diameter electrodes can be used with high AC currents to obtain greater filler metal deposition rates and faster welding speeds.
3. AC arc welding machines are generally less

expensive than DC arc welding machines of equal quality, current output, NEMA classification, and duty cycle rating.

The disadvantage of AC arc welding machines is that not all SMAW electrodes can be used with alternating current.

The choice of which arc welding machine to use or buy must be decided on the basis of what type welds are to be made, the economics of the welding machine purchase, and personal preference.

Welding machines may also be purchased with a selection of AC or DC. Combination AC/DC welding machines are more expensive. However, they offer the welder a better opportunity of matching the current output to the welding requirements of the job.

## 10-5. INSPECTING AN ARC WELDING STATION

The ARC WELDING STATION, as you may recall from Heading 9-1, includes:
1. Arc welding machine (AC or DC).
2. Electrode lead and terminals.
3. Work lead and terminals.
4. Electrode holder.
5. Workbench.
6. Ventilation.
7. Stool.
8. Booth.

Before beginning to weld it is advisable to check all parts of the arc welding station. This should be done to insure the welder's safety and the efficiency of the station.

Before the inspection is made, the arc welding machine should be turned OFF. The arc welding machine should be as close to the booth as possible to eliminate long leads. Check that the electrode and work leads are tightly attached to the machine. Inspect each lead, checking for damage to the covering which may have occurred. Damage to the leads, particularly the work lead, can occur from rolling over them with lift trucks, pallet movers, and other wheeled vehicles. If the electrode and work lead must temporarily run across an aisleway, it is advisable to cover them with something like channel iron.

Inspect the electrode holder to make certain that the handle insulation is not cracked. Check also that the electrode lead is tightly fastened into the holder. The electrode holder jaws should be clean for good electrode contact.

Make certain that the work lead is making good contact with a cleaned area on the workbench. Any loose connections will cause an increase in resistance in the welding circuit. The booth curtains or walls should not have holes in them. Holes in the booth could present arc flash dangers to persons outside the booth.

Check the ventilation pickup hose or duct for holes which would lower the efficiency. Turn on the ventilation system and check to see that it is working.

If the ventilation system efficiency is suspected of being inadequate, it should be checked and repaired.

The inlet to the ventilation pickup duct should be placed so that fumes are removed before they can reach the welder's face.

Each booth must have an insulated hook or hanger on which the electrode holder is hung when not in use.

If a portable arc welding machine is used it is advisable to set up a portable booth to protect others from arc flash.

## 10-6. SAFETY, PROTECTIVE CLOTHING AND SHIELDING

Before proceeding further with the study of arc welding operation, this safety lesson should be studied and observed.

Arc welding performed with proper safety equipment presents no great safety hazards. The beginner should be made aware of the hazards. They should learn the correct procedures for arc welding in order that the hazards which exist may

Fig. 10-10. An arc welding helmet. It is used to protect the eyes and face while arc welding. The lens reduces the amount of harmful rays which would reach the welder's eyes, but will still allow the welder to see the weld and arc crater while welding.   (Fibre Metal Products)

be properly observed and injury avoided.

The chief hazards to be avoided in arc welding are:

1. Radiation from the arc, ultraviolet and infrared rays.
2. Flying sparks, globules of molten metal.
3. Electric shock.
4. Fumes.
5. Burns.

RADIATION from the arc presents some dangers. Eyes must be protected from radiation from the arc by the use of an arc welders helmet or face shields with approved lenses. NEVER LOOK AT AN ARC FROM ANY DISTANCE, UNLESS EYES ARE PROTECTED BY APPROVED LENSES, see Fig. 10-10.

Face, hands, arms, and other skin surfaces must be covered. Gloves should be worn and other parts of the body covered by clothing of sufficient weight to shut out the rays from the arc. Without

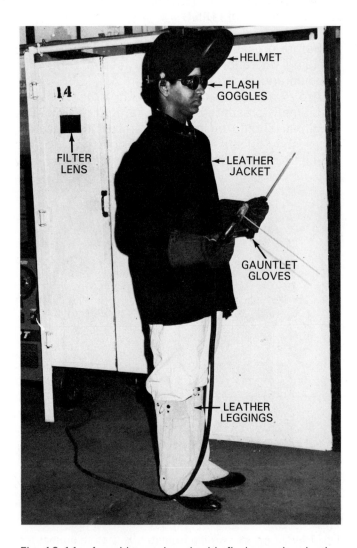

Fig. 10-11. A welder equipped with flash goggles, leather jacket, gauntlet gloves, and leather leggings. Note that the arc welding booth is fitted with a filter lens in order that the instructor may observe the arc from outside the booth.

proper clothing, burns comparable to sunburn will result.

The arc welding operation should be shielded so that no one may accidentally look directly at the arc or have it shine or reflect into their eyes. An arc "flash" may cause a person to be temporarily blinded. The effect is to see a white spot similar to the effect of a photographer's flash. The severity of an arc flash and the time it will take to recover varies with the length of time a person was exposed to the arc. A long exposure has been known to cause permanent damage to the retina of the eye. If someone is severely "flashed," special treatment should be administered at once by a physician.

Arc welding is usually accompanied by FLYING SPARKS. These present a hazard if they strike unprotected skin, lodge on flammable clothing, or hit other flammable material. It is advisable to wear suitable weight clothing and cuffless trousers. See Fig. 10-11.

Pockets should be covered so they will not collect sparks. REMOVE FLAMMABLE MATERIALS, SUCH AS MATCHES, PLASTIC COMBS, OR FOUNTAIN PENS. HIGH SHOES WITH SAFETY TOES SHOULD BE WORN.

The possibility of dangerous ELECTRIC SHOCK can be avoided by working on a dry floor, using insulated electrode holders and wearing dry welding gloves. AVOID USING ARC WELDING EQUIPMENT IN WET OR DAMP AREAS.

The health hazard of FUMES, developed by the electrode covering and molten metal, may be avoided by the use of proper ventilating equipment. Certain special jobs require forced ventilation into the welder's helmet. The fumes generated in the welding arc may contain poisonous metal oxides. ARC WELDING SHOULD NEVER BE DONE IN AN AREA WHICH IS NOT WELL VENTILATED.

HOT METAL will cause severe burns. Use leather gloves with tight fitting cuffs which fit over the sleeves of the jacket. Many welders wear an apron of leather or other heavy material for protection. Hot metal should be handled with tongs or pliers. In a welding shop all metal should be first cooled in the quenching tank before it is handled with bare hands.

## 10-7. STARTING, STOPPING, AND ADJUSTING THE ARC WELDING MACHINE FOR SMAW

Before starting an arc welding machine, inspect the complete arc welding station to make certain it is safe for use. See Heading 10-5. An arc welding machine should never be started or stopped under load (with a closed circuit). Make certain that the electrode holder is hung on an insulated hanger before starting or stopping the machine. The electrode holder should never be left on the workbench.

AC and DC electrically powered arc welding machines are easy to start and stop. An off and on switch is provided on the machine. The engine must first be started on an engine driven AC or DC arc welding machine. Once the engine is running and up to its operating speed, the welding current may be turned on using an off and on switch.

The constant current type arc welding machine is the type recommended for manual arc welding processes. The desired current is set on the machine. The voltage on a constant current machine is not set. It varies as the welding circuit resistance changes to maintain a constant or relatively constant current output. All electrical connections must be tight. The length of the arc gap then controls the welding circuit resistance and therefore the circuit voltage.

Amperage (current) controls vary in appearance, location, and operation on various manufacturer's machines.

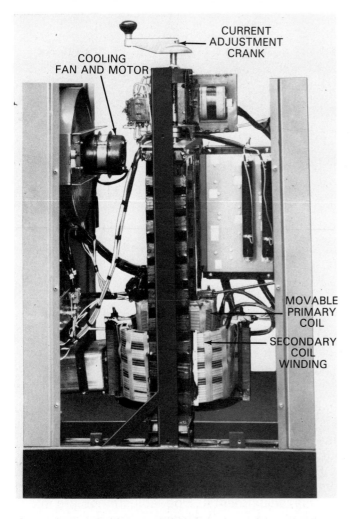

Fig. 10-12. Internal construction of a movable coil AC arc welder. (Applied Power, Inc.)

Fig. 10-12 shows an AC transformer with a hand crank on the top of the machine. As this hand crank is rotated, the primary coil is moved to vary the current output. A pointer also moves with the primary coil and indicates the current setting on a scale on the outside of the cabinet. Fig. 10-13 illustrates a tap type control on an AC transformer. Coarse current adjustments are made by placing one lead into the high, medium, or low current tap. Fine adjustments are made by placing the other welding lead into one of the fine

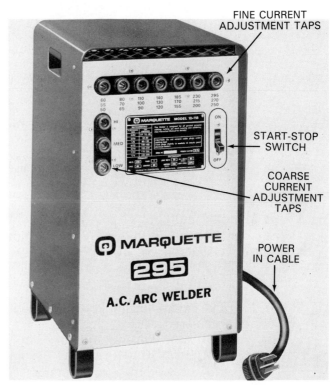

Fig. 10-13. The external appearance of an AC arc welder. The electrode lead may be plugged into one of the fine current tap sockets to obtain the desired amperage. The work lead may be connected to one of the three coarse adjustments. (Applied Power, Inc.)

adjustment tap holes. Fig. 10-14 illustrates an arc welding machine which uses several taps for coarse current range selections. The fine adjustment within a selected range is made on a rotary, resistant type control. The fine adjustment knob is on the face of the machine. Fine adjustment controls are often marked with numbers from 0 - 100. These numbers do not represent current, but rather percentages.

Once a coarse current range tap is selected, the fine adjustment will adjust the current within percentages of the coarse range.

The machine will deliver the amperage set on the low end of the coarse setting plus the percentage of the range set on the fine adjustment.

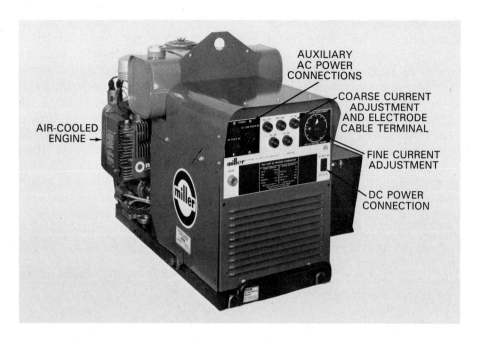

Fig. 10-14. A portable, engine driven arc welder and electrical power supply machine. It can supply either AC or DC arc welding current and AC auxiliary power. (Miller Electric Mfg. Co.)

Example #1:

If the coarse range setting is 90-110 amps the current delivered = 90 amps:

Plus

If the fine adjustment setting is 50 percent, the current delivered is:

$$50\% \times \text{current range or}$$
$$.50 \times (110-90) =$$
$$.50 \times 20 \text{ amps} = 10 \text{ amps}$$

Then

total current delivered = 100 amps

Example #2:

If the coarse range setting is 200-250 amps the current delivered = 200 amps:

Plus

If the fine adjustment is set on 70 percent, the current delivered is:

$$70\% \times \text{current range or}$$
$$.70 \times (250-200) =$$
$$.70 \times 50 \text{ amps} = 35 \text{ amps}$$

Then

total current delivered = 235 amps

Fig. 10-15 illustrates an arc welding machine which uses a large handwheel that is rotated to

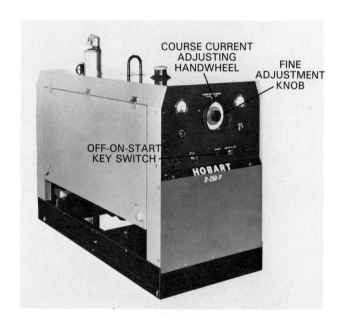

Fig. 10-15. An engine driven DC arc welding machine. This machine uses a large handwheel for coarse current settings. It uses a small knob for fine adjustments. (Hobart Brothers Co.)

make coarse current adjustments. Fine adjustments are made using the smaller knob inside the larger handwheel. The scale for fine adjustments is in percentages (0% - 100%).

Fig. 10-16 illustrates an engine driven arc welding machine which uses a handle to select coarse adjustments. Fine adjustment are made in a range of 0-100 percent of the coarse adjustment range.

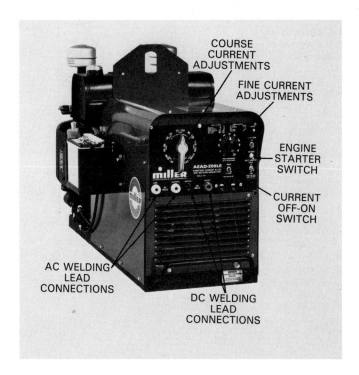

COURSE CURRENT ADJUSTMENTS

FINE CURRENT ADJUSTMENTS

ENGINE STARTER SWITCH

CURRENT OFF-ON SWITCH

AC WELDING LEAD CONNECTIONS

DC WELDING LEAD CONNECTIONS

Fig. 10-16. An engine driven AC/DC arc welding machine. The fine adjustment is set as a percentage (0 - 100) of the coarse range. (Miller Electric Mfg. Co.)

## 10-8. SELECTING THE PROPER ELECTRODE

When attempting to select an electrode for shielded metal arc welding (SMAW) the following must be considered:
1. The weld groove design.
2. Tensile strength of the required weld.
3. The base metal composition.
4. The position of the weld joint.
5. The rate at which you want to deposit the weld metal.
6. The type of arc welding current used.
7. The penetration required.
8. The metal thickness.
9. The experience of the welder.
10. The specifications for the weld to be made.

When a GROOVE WELD is made the electrode must be small enough in diameter to easily manipulate it in the root of the weld. A small diameter electrode is used for the root pass to insure full penetration. After the root pass is made and cleaned, larger electrodes may be used to finish the weld. E60XX, 70XX, 80XX, 90XX, 100XX, and higher may be used depending on the weld strength required. Refer to Figs. 10-17 and 10-18 as you study this material.

The METAL COMPOSITION of the base metal will determine the metal composition of the electrode wire used, refer to Headings 9-15 to 9-20.

The WELD JOINT POSITION will determine the electrode used. If the joint is in the flat position, a

| AWS Classification | Type of covering | Capable of producing satisfactory welds in position shown[a] | Type of current[b] |
|---|---|---|---|
| E60 series electrodes | | | |
| E6010 | High cellulose sodium | F, V, OH, H | DCEP |
| E6011 | High cellulose potassium | F, V, OH, H | AC or DCEP |
| E6012 | High titania sodium | F, V, OH, H | AC or DCEN |
| E6013 | High titania potassium | F, V, OH, H | AC or DC, either polarity |
| E6020 | High iron oxide | H-fillets | AC or DCEN |
| E6022[c] | High iron oxide | F | AC or DC, either polarity |
| E6027 | High iron oxide, iron powder | H-fillets, F | AC or DCEN |
| E70 series electrodes | | | |
| E7014 | Iron powder, titania | F, V, OH, H | AC or DC, either polarity |
| E7015 | Low hydrogen sodium | F, V, OH, H | DCEP |
| E7016 | Low hydrogen potassium | F, V, OH, H | AC or DCEP |
| E7018 | Low hydrogen potassium, iron powder | F, V, OH, H | AC or DCEP |
| E7024 | Iron powder, titania | H-fillets, F | AC or DC, either polarity |
| E7027 | High iron oxide, iron powder | H-fillets, F | AC or DCEN |
| E7028 | Low hydrogen potassium, iron powder | H-fillets, F | AC or DCEP |
| E7048 | Low hydrogen potassium, iron powder | F, OH, H, V-down | AC or DCEP |

a. The abbreviations, F, V, V-down, OH, H, and H-fillets indicate the welding positions as follows:
   F = Flat
   H = Horizontal
   H-fillets = Horizontal fillets
   V-down = Vertical down
   V = Vertical } For electrodes 3/16 in. (4.8 mm) and under, except 5/32 in. (4.0 mm) and under for classifications E7014, E7015, E7016, and E7018.
   H = Overhead }

b. The term DCEP refers to direct current, electrode positive (DC reverse polarity). The term DCEN refers to direct current, electrode negative (DC straight polarity).

c. Electrodes of the E6022 classification are for single-pass welds.

Fig. 10-17. AWS Carbon Steel Covered Arc Welding Electrodes. This information is from AWS A5.1-91. (American Welding Society)

larger diameter electrode may be used. An iron powder electrode with high metal deposition rates may be chosen. If the weld is in the vertical, horizontal, or overhead positions, a smaller diameter electrode may be selected. A smaller diameter electrode will form a smaller weld pool and will be easier to control. The metal in a larger weld pool tends to run out due to gravitational force.

DCEP (DCRP) is generally used for vertical, horizontal, and overhead welding. The covering

| AWS Classification[a] | Type of covering | Capable of producing satisfactory welds in positions shown[b] | Type of current[c] |
|---|---|---|---|
| | E70 series — Minimum tensile strength of deposited metal, 70,000 psi (480 MPa) |||
| E7010-X | High cellulose sodium | F, V, OH, H | DCEP |
| E7011-X | High cellulose potassium | F, V, OH, H | AC or DCEP |
| E7015-X | Low hydrogen sodium | F, V, OH, H | DCEP |
| E7016-X | Low hydrogen potassium | F, V, OH, H | AC or DCEP |
| E7018-X | Iron powder, low hydrogen | F, V, OH, H | AC or DCEP |
| E7020-X | High iron oxide | H-fillets<br>F | AC or DCEN<br>AC or DC, either polarity |
| E7027-X | Iron powder, iron oxide | H-fillets<br>F | AC or DCEN<br>AC or DC, either polarity |
| | E80 series — Minimum tensile strength of deposited metal, 80,000 psi (550 MPa) |||
| E8010-X | High cellulose sodium | F, V, OH, H | DCEP |
| E8011-X | High cellulose potassium | F, V, OH, H | AC or DCEP |
| E8013-X | High titania potassium | F, V, OH, H | AC or DC, either polarity |
| E8015-X | Low hydrogen sodium | F, V, OH, H | DCEP |
| E8016-X | Low hydrogen potassium | F, V, OH, H | AC or DCEP |
| E8018-X | Iron powder, low hydrogen | F, V, OH, H | AC or DCEP |
| | E90 series — Minimum tensile strength of deposited metal, 90,000 psi (620 MPa) |||
| E9010-X | High cellulose sodium | F, V, OH, H | DCEP |
| E9011-X | High cellulose potassium | F, V, OH, H | AC or DCEP |
| E9013-X | High titania potassium | F, V, OH, H | AC or DC, either polarity |
| E9015-X | Low hydrogen sodium | F, V, OH, H | DCEP |
| E9016-X | Low hydrogen potassium | F, V, OH, H | AC or DCEP |
| E9018-X | Iron powder, low hydrogen | F, V, OH, H | AC or DCEP |
| | E100 series — Minimum tensile strength of deposited metal, 100,000 psi (690 MPa) |||
| E10010-X | High cellulose sodium | F, V, OH, H | DCEP |
| E10011-X | High cellulose potassium | F, V, OH, H | AC or DCEP |
| E10013-X | High titania potassium | F, V, OH, H | AC or DC, either polarity |
| E10015-X | Low hydrogen sodium | F, V, OH, H | DCEP |
| E10016-X | Low hydrogen potassium | F, V, OH, H | AC or DCEP |
| E10018-X | Iron powder, low hydrogen | F, V, OH, H | AC or DCEP |
| | E110 series — Minimum tensile strength of deposited metal, 110,000 psi (760 MPa) |||
| E11015-X | Low hydrogen sodium | F, V. OH, H | DCEP |
| E11016-X | Low hydrogen potassium | F, V, OH, H | AC or DCEP |
| E11018-X | Iron powder, low hydrogen | F, V, OH, H | AC or DCEP |
| | E120 series — Minimum tensile strength of deposited metal, 120,000 psi (830 MPa) |||
| E12015-X | Low hydrogen sodium | F, V, OH, H | DCEP |
| E12016-X | Low hydrogen potassium | F, V, OH, H | AC or DCEP |
| E12018-X | Iron powder, low hydrogen | F, V, OH, H | AC or DCEP |

a. The letter suffix 'X' as used in this table stands for the suffixes A1, B1, B2, etc. (see Fig. 9-44) and designates the chemical composition of the deposited weld metal.

b. The abbreviations F, V, OH, H, and H-fillets indicate welding positions as follows: F = Flat; H = Horizontal; H-fillets = Horizontal fillets. V = Vertical / OH = Overhead } For electodes 3/16 in. (4.8 mm) and under, except 5/32 in. (4.0 mm) and under for classifications EXX15-X, EXX16-X, and EXX18-X.

c. DCEP means electrode positive (reverse polarity). DCEN means electrode negative (straight polarity).

Fig. 10-18. AWS Low Alloy Steel Covered Arc Welding Electrodes. This information is from AWS 5.5-81.
See Fig. 9-44 for the letter suffixes used in place of the "X" with these electrode numbers.
(American Welding Society)

on a DCEP (DCRP) electrode melts slower than the electrode wire. This forms a deep cup shape at the end of the electrode. Because of this cup shape, the molten metal travels across the arc in a jet-like action. Thus the metal transfers from the electrode to the arc crater rather than drop due to gravity.

Electrodes are made to be most effective with AC, DCEN (DCSP), or DCEP (DCRP). When selecting an electrode, the type of current produced by the arc welding machine must be known.

AC and some DCEP (DCRP) electrodes produce deeper penetrating welds than other electrodes. Refer to Heading 9-16 and Figs. 10-17 and 10-18.

When welding on THIN METAL, DCEN (DCSP) may be used. Small DCEN (DCSP) electrodes used with a low current setting produce a soft arc action with small penetrating abilities. See electrodes E6012 and E6013 in Fig. 10-19.

The largest electrode diameter possible should be used. However, it must not create too large a weld or overheat the base metal.

The experience of the welder is a large factor in choosing an electrode. An experienced welder can produce a sound weld with a much larger diameter electrode than a beginner can.

Very often no choice is allowed in the selection of an electrode. Whenever welding specifications are used, the diameter and type of electrode are specified and must be used. Figs. 10-19 and 10-20 provide information regarding electrode diameters, current ranges used, and suggested metal thickness applications for E60XX and E70XX series electrodes.

## 10-9. STRIKING AN ARC

One of the first lessons to be mastered when learning to arc weld, is to produce an arc between the metal electrode and the base metal. To strike an arc, the electrode must first touch the base metal. The end of the electrode must then be withdrawn to the correct arc distance or length.

At the first attempt to strike the arc, the electrode may tend to stick. That is to say, the electrode may weld itself to the base metal. The beginner may also tend to withdraw the electrode too far after touching the metal. This will cause the voltage requirement to be too great to maintain the arc. The arc will therefore break and go out. Only experience and practice will overcome these difficulties.

There are two common methods of striking (producing) the arc. The welder may use a GLANCING OR SCRATCHING MOTION with the end of the electrode or a STRAIGHT DOWN-AND-UP MOTION with the electrode. Fig. 10-21 illustrates both methods of striking an arc.

If the arc breaks continually, regardless of how careful the operator may be, it is probably due to too-low a current adjustment on the machine. If the electrode spatters excessively, and if it becomes overheated while welding, the current setting is too high.

As soon as the arc is struck, and the arc

| Suggested Metal Thickness | | Electrode size | | E6010 and E6011 | E6012 | E6013 | E6020 | E6022 | E6027 |
|---|---|---|---|---|---|---|---|---|---|
| in. | mm | in. | mm | | | | | | |
| 1/16 & less | 1.6 & less | 1/16 | 1.6 | | 20-40 | 20-40 | | | |
| 1/16-5/64 | 1.6-2.0 | 5/64 | 2.0 | | 25-60 | 25-60 | | | |
| 5/64-1/8 | 2.0-3.2 | 3/32 | 2.4 | 40-80 | 35-85 | 45-90 | | | |
| 1/8-1/4 | 3.2-6.4 | 1/8 | 3.2 | 75-125 | 80-140 | 80-130 | 100-150 | 110-160 | 125-185 |
| 1/4-3/8 | 6.4-9.5 | 5/32 | 4.0 | 110-170 | 110-190 | 105-180 | 130-190 | 140-190 | 169-240 |
| 3/8-1/2 | 9.5-12.7 | 3/16 | 4.8 | 140-215 | 140-240 | 150-230 | 175-250 | 170-400 | 210-300 |
| 1/2-3/4 | 12.7-19.1 | 7/32 | 5.6 | 170-250 | 200-320 | 210-300 | 225-310 | 370-520 | 250-350 |
| 3/4-1 | 19.1-25.4 | 1/4 | 6.4 | 210-320 | 250-400 | 250-350 | 275-375 | | 300-420 |
| 1 - up | 25.4 - up | 5/16 | 8.0 | 275-425 | 300-500 | 320-430 | 340-450 | | 375-475 |

Fig. 10-19. A table of E60XX series electrodes with suggested metal thickness applications, and amperage ranges. These values are suggested and may be varied as required.

| Suggested Metal Thickness | | Electrode size | | E7014 | E7015 and E7016 | E7018 | E7024 and E7028 | E7027 | E7048 |
|---|---|---|---|---|---|---|---|---|---|
| in. | mm | in. | mm | | | | | | |
| 5/64-1/8 | 2.0-3.2 | 3/32* | 2.4* | 80-125 | 65-110 | 70-100 | 100-145 | | |
| 1/8-1/4 | 3.2-6.4 | 1/8 | 3.2 | 110-160 | 100-150 | 115-165 | 140-190 | 125-185 | 80-140 |
| 1/4-3/8 | 6.4-9.5 | 5/32 | 4.0 | 150-210 | 140-200 | 150-220 | 180-250 | 160-240 | 150-220 |
| 3/8-1/2 | 9.5-12.7 | 3/16 | 4.8 | 200-275 | 180-255 | 200-275 | 230-305 | 210-300 | 210-270 |
| 1/2-3/4 | 12.7-19.1 | 7/32 | 5.6 | 260-340 | 240-320 | 260-340 | 275-365 | 250-350 | |
| 3/4-1 | 19.1-25.4 | 1/4 | 6.4 | 330-415 | 300-390 | 315-400 | 335-430 | 300-420 | |
| 1 - up | 25.4 - up | 5/16* | 8.0* | 390-500 | 375-475 | 375-470 | 400-525 | 375-475 | |

Note: When welding vertically up, currents near the lower limit of the range are generally used.
*: These diameters are not manufactured in the E7028 classification.

Fig. 10-20. A table of E70XX series electrodes with suggested metal thickness applications, and amperage ranges. These values are suggested and may be varied as required.

**Shielded Metal Arc Welding (AC and DC) / 265**

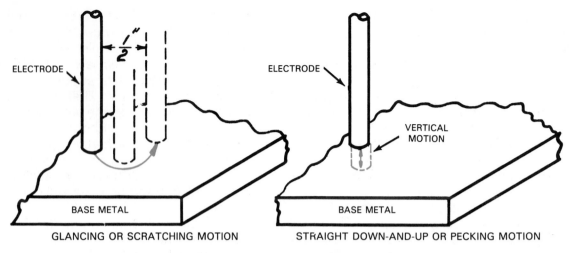

ELECTRODE

BASE METAL

GLANCING OR SCRATCHING MOTION

ELECTRODE

VERTICAL MOTION

BASE METAL

STRAIGHT DOWN-AND-UP OR PECKING MOTION

Fig. 10-21. Methods of striking an arc. The vertical motion method is most often used, however it takes some practice to be skillful in its use.

becomes stabilized, the base metal begins melting and the filler metal deposit begins. It is therefore important to strike the arc in exactly the right spot or the metal may be marred. Most welders position the arc end of the electrode just above the exact spot where the weld is to start. They then lower the helmet in front of their eyes before actually contacting the metal with the electrode. A photoelectric lens remains clear until the arc is struck. This helps the welder to strike the arc in precisely the right spot without raising or lowering the helmet. See Fig. 9-61.

## 10-10. RUNNING A BEAD

Once the arc is struck and the arc stabilizes, the ARC CRATER will begin to form. As the welder moves the electrode forward, the BEAD forms. To make a good weld on any joint in any position, the first skill which must be mastered is to run (form) a bead.

To run a good bead in SMAW, the following factors must be controlled manually by the welder:
1. Arc gap distance or arc length.
2. Speed of forward motion.
3. Bead width.
4. Electrode angle or position.

The ARC LENGTH must be varied slightly as various electrode diameters are used. However, for covered electrodes the arc length will be about 3/16 in. (about 5 mm) to 1/4 in. (about 6 mm). Arc welding should be done with one hand. A welder must use one hand at times to hold a part while tacking it in place.

One way of checking if the arc length is proper is to listen to the sound of the arc. A proper arc length will produce a crankling or hissing sound. Too short an arc length may short out while

welding. When small solidified metal drops are seen on the base metal surface, SPATTERING is occurring. Too long an arc length will cause a great deal of filler metal spattering.

To create a bead which has a uniform width and height and uniformly spaced ripples, a UNIFORM FORWARD SPEED must be maintained.

There are two types of beads used in arc welding called STRINGER BEADS and WEAVING BEADS.

To make a STRINGER BEAD, no motion of the electrode is made except forward. Its width should be 2 to 3 times the diameter of the electrode. For example, with a 1/8 in. (3.2) mm electrode, the normal stringer bead width should be: 1/8 in. × 2 = 1/4 in. wide (6.4 mm) to 1/8 in. × 3 = 3/8 in. (9.6 mm). See Fig. 10-22, view A.

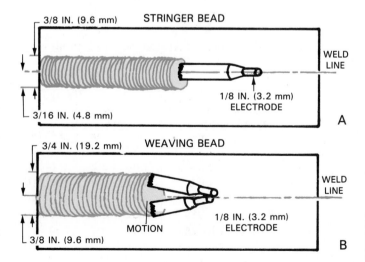

3/8 IN. (9.6 mm)    STRINGER BEAD

WELD LINE

1/8 IN. (3.2 mm) ELECTRODE

3/16 IN. (4.8 mm)

A

3/4 IN. (19.2 mm)    WEAVING BEAD

WELD LINE

MOTION

1/8 IN. (3.2 mm) ELECTRODE

3/8 IN. (9.6 mm)

B

Fig. 10-22. Suggested dimensions for a stringer bead and a weaving bead. The stringer bead is the electrode diameter times three wide. The weaving bead is the electrode diameter times six wide.

When a WEAVING BEAD is made, the electrode is moved uniformly back and forth across the weld line. With such a motion the bead may be made as wide as desired. However, for best bead control it is recommended that a weaving bead be no wider than six times the electrode diameter. For example, using a 3/16 in. (4.8 mm) electrode, the weaving bead should be: 3/16 in. × 6 = 1 1/8 in. wide (4.8 mm × 6 = 28.8 mm). See Fig. 10-22, view B.

The ELECTRODE ANGLE or position is tipped forward 20 degrees in the direction of travel. It is kept in line with the weld line for a stringer bead. See Fig. 10-23.

To practice the stringer bead, use a piece of mild steel about 1/4 in. (6.4 mm) thick, 2 in. (50.8 mm) wide, and 6 in. (152.4 mm) long. Select an AC or DC electrode 1/8 in. (3.2 mm) in diameter, see Figs. 10-19 or 10-20.

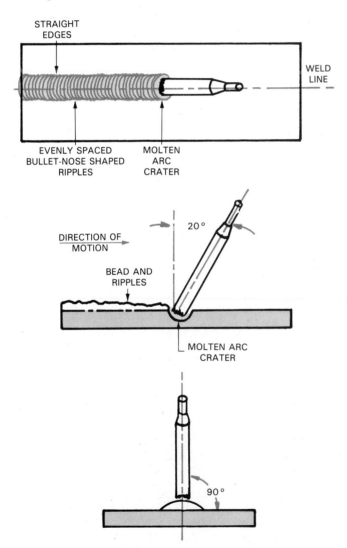

Fig. 10-23. Three views of an arc bead in progress. Note that the electrode is inclined 20 degrees in the direction of travel. The completed arc bead should have straight edges, evenly spaced ripples, and uniform height.

Clean the bare end of the electrode if necessary and place it into the electrode holder. The arc should always be struck about 3/8 in. (9.5 mm) ahead of where the bead should begin. The electrode and arc must then be moved rapidly to the spot where the weld is to begin. During this brief period of time the arc has a chance to stabilize. The bead can then begin with a steady arc. While holding a uniform arc length move slowly in the direction of motion. Right handers usually weld best from left to right. Left handers usually weld best from right to left. With the electrode tipped 20 degrees in the direction of travel a good view of the arc pool is possible. This slight angle also permits the force of the arc to push the molten metal up the rear of the arc pool to form the bead ripples.

When the width of the stringer bead reaches the desired size, move slightly forward. Watch the weld pool grow in size and move slightly forward again. This action of watching and moving is continued until it appears to be a uniform forward motion.

The SPEED OF FORWARD MOTION is judged by two factors:

1. The bead width.
2. The bullet-nose shaped appearance of the ripples at the rear of the molten pool in the arc crater.

If the forward speed is proper, the back of the crater will have a bullet-nose shaped ripple. This will also indicate from above that the WELD BEAD HEIGHT is proper. When the forward speed is too slow, the crater shape will be less curved. This will also indicate that the bead is becoming too high. If the crater shape becomes more pointed, the speed is too fast and the build up is too low.

The completed stringer bead will be even in width, with evenly spaced, bullet-shaped ripples. It will have the proper width and bead height. The weld height is normally about one quarter of the bead width.

The proper current setting is very important to make a quality weld or bead. A bead that is made with the correct current, arc length, and forward speed will have a cross section like the one shown in Fig. 10-24A.

If the current is too low, the bead will be high with poor penetration, see Fig. 10-24B.

If the current is too high the electrode will overheat and also spatter excessively. The bead will also be low. See Fig. 10-24.C.

A very porous bead or weld will result. Gas pockets and impurities will be trapped within the bead. With a short arc, the bead will be high, with poor penetration and overlap, see Fig. 10-24D.

If the arc length is too long, the bead will be too

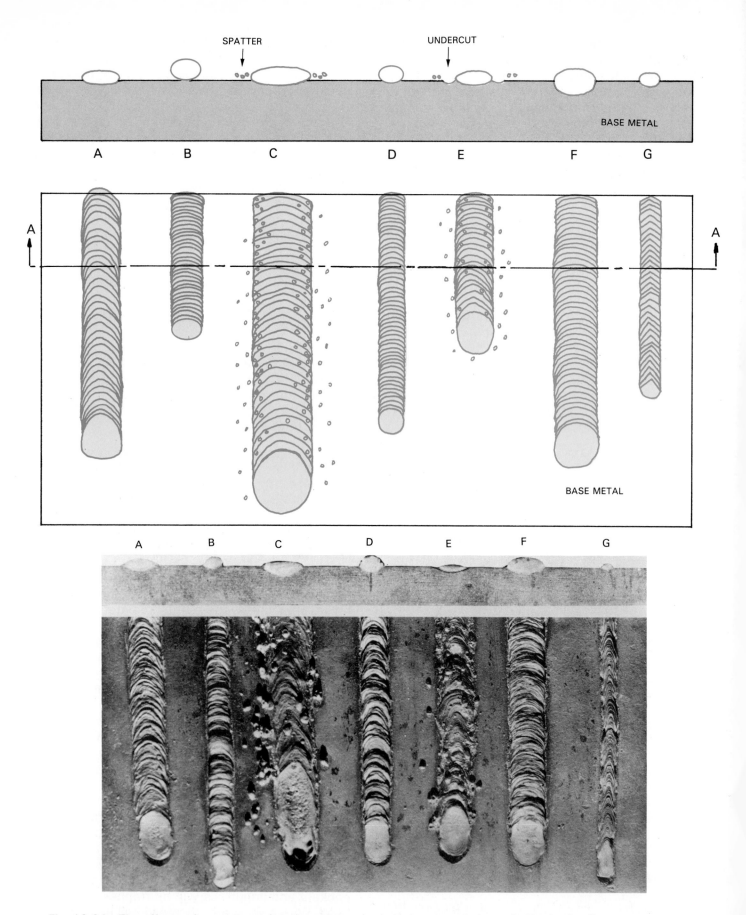

Fig. 10-24. The effects of current, arc length, and travel speed on covered electrode beads. A—Correct current, arc length, and travel speed; B—Amperage too low; C—Amperage too high; D—Too short an arc length; E—Arc length too long; F—Travel speed too slow; G—Travel speed too fast.   (American Welding Society)

low with poor penetration and undercut, see Fig. 10-24E.

Moving too slowly will create a bead which is wide and high, see Fig. 10-24F.

A rapid forward motion will result in a low, narrow bead, see Fig. 10-24G.

Fig. 10-25 labels effects of arc bead created with less than ideal conditions.

A practical application of the arc bead is the rebuilding of worn surfaces in welding maintenance work. Shafting, excavation implements, gear teeth, wheels of various kinds, journals, etc., frequently are worn to the extent that they must either be discarded or rebuilt. Rebuilding these surfaces by laying arc beads side by side over the worn surface in one or more layers, then refinishing, has become an important arc welding maintenance operation.

Another application of bead work is hard surfacing or wear-resistant surfacing. Laying beads of special metallic alloys side by side on a soft steel surface provides a surface that is extremely hard and resistant to abrasion. See Chapter 26.

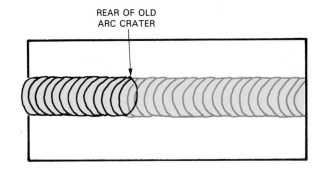

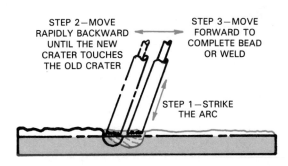

Fig. 10-26. Steps to restarting an arc bead.

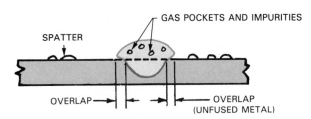

Fig. 10-25. Results of improperly made arc bead.

## 10-11. RESTARTING AND FINISHING AN ARC WELDING BEAD

When a SMAW bead is stopped prior to completion, a deep crater is left in the base metal. Restarting the arc and completing the bead must be done with care. If the restart is done correctly the bead ripples will be uniform. It will be difficult to see where the bead was stopped and restarted.

Before restriking the arc, the previous bead must be cleaned as described in Heading 10-12. To restrike the arc, the arc is struck about 3/8 in. (about 10 mm) ahead of the forward edge of the crater. It is then moved backward rapidly until the new crater just touches the rear edge of the previous crater. As soon as the edges of the two craters touch, the electrode is moved forward to complete the weld. If this is done correctly the ripples of the old and the new bead will match. See Fig. 10-26.

There are two ways to finish a bead or weld without leaving a crater. One method is to use a RUN-OFF TAB. A piece of metal of the same type

and thickness is tack welded to the end of the base metal being welded. The arc bead or weld is completed on the base metal and continued on the run-off tab. When the weld is stopped the crater is on the run-off tab. The run-off strip is cut off leaving a full thickness bead at the end of the base metal.

A similar procedure is used to start a weld. A RUN-ON TAB is tacked to the base metal at the starting end. The arc is struck on the run-on tab. When the run-on tab is cut off it leaves a well defined, full thickness bead at the beginning of the base metal.

Another method used to finish a weld without leaving a crater is to REVERSE ELECTRODE MOTION as the end of the weld is reached. The electrode is moved to the trailing edge of the crater. When the crater is filled the electrode is then lifted until the arc is broken.

## 10-12. CLEANING THE ARC BEAD

When shielded metal electrodes are used, a brittle SLAG COATING is left covering the weld bead. This slag covering must be removed prior to restarting a bead. It must also be removed prior to welding over a bead, and prior to painting.

If the slag is not removed prior to restarting or welding over a bead, many slag inclusions will result. SLAG INCLUSIONS are pieces of slag trapped or included in the weld.

Slag is generally removed manually with a chipping hammer and a wire brush. The slag may also be removed mechanically by shot peening, wire brushing, or chipping.

## 10-13. DC ARC BLOW

As explained earlier, once started, the AC arc is quite stable. The DC arc, however, at times may have a tendency to wander from the weld line. This wandering is usually caused by the forces of the magnetic field around the DC electrode called ARC BLOW. All electrical conductors are surrounded by a magnetic field when current is flowing. If the current travels continually in one direction the magnetism can become quite strong. AC electrodes are not affected because of the constantly changing direction of the current. These reversals virtually cancel the magnetic blow effects in the AC circuit.

The magnetic field or lines of flux travel easily in metal. They travel with greater difficulty in air. Fig. 10-27 shows a butt weld with the work lead connected near the beginning of the weld. When the arc is struck, a magnetic field is created

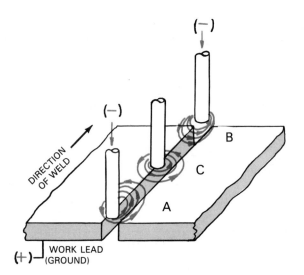

Fig. 10-27. The magnetic field around the electrode is deflected at the ends of joint A and B. It attempts to flow in the metal and not through the air. This concentration of the magnetic flux at the ends of the metal forces the arc toward the center of the base metal. Thus the arc is seen to "blow" away from the area directly under the electrode. Notice that the magnetic field is not distorted in the center areas of the weld at C.

around the DC electrode. The magnetic field prefers to travel in the base metal, not in the air. Therefore the magnetic field forces the molten filler metal to blow in from the end of the weld joint toward the center of the work. This is called FORWARD ARC BLOW. See Figs. 10-27 and

10-28. In the center area of the weld joint the arc and molten filler metal act normally.

As the electrode nears the end of the joint, the magnetic flux intensifies ahead of the electrode. This happens as the magnetic flux tries to stay in the metal rather than travel out into the air. The arc and molten metal are now blown back toward the beginning of the weld. This action is known as BACKWARD ARC BLOW. See Fig. 10-27 and 10-28. Very seldom does arc blow occur across the weld axis (sideways).

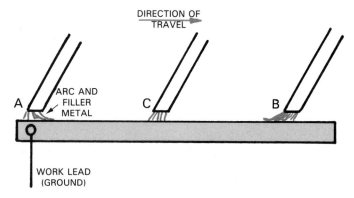

Fig. 10-28. The effects of DC arc blow on the arc and filler metal. As the arc is started at A, the arc is blown toward the right. In the center, C, the arc travels straight down. As the arc approaches the end of the weld at B, the arc and filler metal are blown toward the ground end of the weld.

If the arc blow is extremely strong, certain preventive or corrective measures can be taken. One or more of the following may be used to correct magnetic arc blow:

1. Place the ground connections as far from the weld joint as possible.
2. If forward arc blow is a problem, connect the work lead (ground) near the end of the weld joint.
3. If back blow is a problem, place the work lead (ground) near the start of the weld. It will also help to weld toward a large tack weld. The large tack weld will give the magnetic field a place to flow. This will prevent a crowding of the magnetic field which causes arc blow.
4. Reduce the welding current. This will reduce the strength of the magnetic field.
5. Position the electrode so that the arc force counteracts the arc blow force.
6. Use the shortest arc that will produce a good bead. The short arc will permit the filler metal to enter the arc pool before it is blown away. A short arc will also permit the arc force to overcome the arc blow force.
7. Weld toward a run-off tab or heavy tack weld.

8. Wrap the electrode lead around the base metal in the direction which will counteract the arc blow force.
9. Change to an AC machine and electrodes.
10. Use the backstep method of welding.

The BACKSTEP METHOD uses a number of short welds. The weld bead is divided into several sections. The first segment is started away from the beginning of the joint. The weld is made toward the beginning. Each section is welded back toward the previous section. See Fig. 10-29.

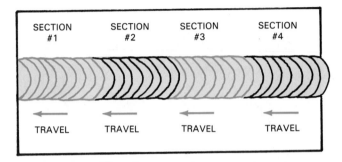

Fig. 10-29. An example of the backstep method of welding. Each section is welded back toward the previous section.

## 10-14. ARC WELDED JOINT DESIGNS

Arc welded joint designs are similar to the joints used in oxyfuel gas welding in Chapter 4 and braze welding in Chapter 8.

The five basic joint designs are:
1. Butt.
2. Lap.
3. Corner.
4. T-Joint.
5. Edge.

See Fig. 10-30. Refer also to Heading 2-1 for a complete discussion of joint designs and edge preparation.

The lap joint is made with square edges on the base metal. The butt joint, corner joint, T-joint, and edge joint weld may require edge preparation. The edges may be flame cut, ground, or machined to the required shape or angle. Edge preparation is done to insure complete penetration of the weld.

It is necessary to know the proper names for the various parts of a weld in order to discuss welds and their quality. Refer to Fig. 10-31 for the proper names used in joints and welds.

The weld joint may be in any position in the job of welding. The weld must be made in any of the five joints in every conceivable position. These welding positions are:
1. FLAT (downhand) — When the weld axis and weld face are horizontal, See Fig. 10-32.

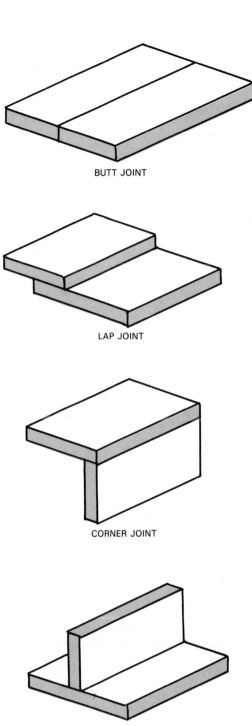

BUTT JOINT

LAP JOINT

CORNER JOINT

T-JOINT

EDGE JOINT

Fig. 10-30. Common arc welding joint designs. See also Heading 2-1.

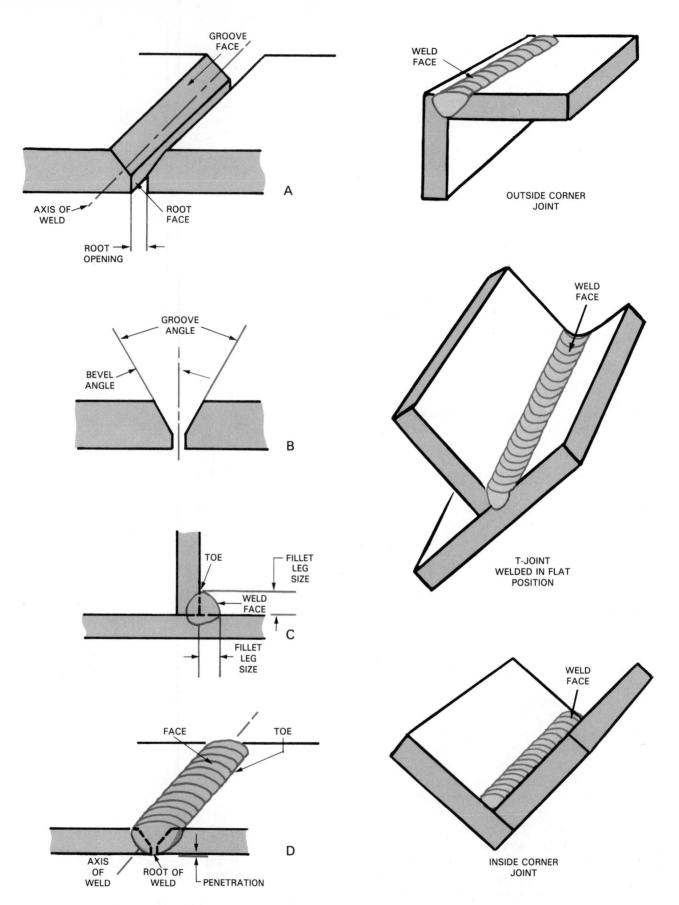

Fig. 10-31. Proper names for parts of a weld joint and weld. A and B illustrate parts of the joint. C and D illustrate parts of the weld.

Fig. 10-32. Examples of welds made in the flat (downhand) position. Note that the face of each weld is in a horizontal position.

2. HORIZONTAL — When the weld axis is horizontal and the weld face is vertical or near vertical, see Fig. 10-33, view A.
3. VERTICAL — When the weld axis is vertical, see Fig. 10-33, view B.
4. OVERHEAD — When the weld axis is horizontal, but the weld is made from the underside, see Fig. 10-33, view C.

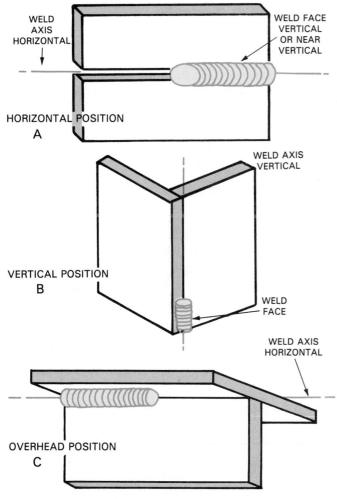

Fig. 10-33. The horizontal, vertical, and overhead welding positions.

## 10-15. WELD DEFECTS

Complete welds may have a variety of defects. Many of them can be seen with the naked eye. Other defects may only be found by means of destructive or nondestructive testing. Refer to Chapter 28.

By means of a visual inspection, a welder may find the following weld defects:
1. Poor weld proportions.
2. Undercutting.
3. Lack of penetration.
4. Surface defects.

The weld face should have relatively small, evenly spaced ripples. The weld face on a groove joint should be wide enough to span the complete groove. A groove weld should have complete penetration.

On the lap joint and corner joint the weld normally does not penetrate to the other side. On a groove type corner joint the weld may penetrate to the metal face opposite the bevel. Refer to Fig. 10-34.

A weld with a properly CONTOURED (shaped) face is shown in Fig. 10-35, view A.

The weld illustrated in Fig. 10-35, view B shows an undercut condition. Note that the toes of the weld are cut deep into the base metal. This weakens the base metal and is a defect that must be repaired. Undercutting is caused by improper welding technique. Often a long arc or too high a current is the cause. In Fig. 10-35, view B, the cause was high current. The excessive penetration indicates a high current setting.

In Fig. 10-35, view C, the weld has been made with a low current setting, or a short arc length. The weld is overlapped at the toe of the weld. The weld also has poor penetration.

Fig. 10-36 illustrates the cross section of three well-formed fillet welds. The weld at A has a flat face contour. At B the weld face is convexed (curved outward) and at C the weld face is concaved (curved inward). Note that there is no

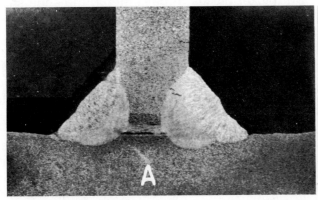
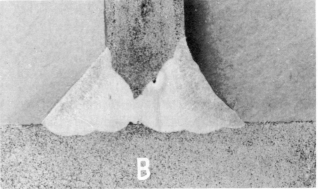

Fig. 10-34. Conventional DC fillet weld on the left at A compared to the deep penetration fillet weld on the right at B made with an AC arc.

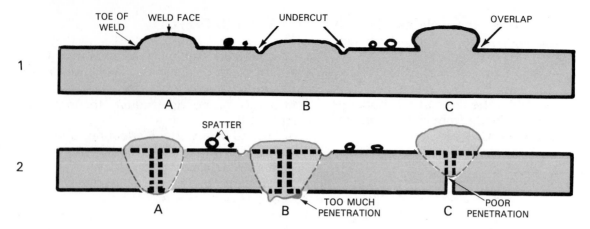

Fig. 10-35. A properly made bead is shown at A; it has good contour and penetration. At B, the bead is undercut and the base metal is thus weakened. At C, the bead was made with insufficient heat and the bead is overlapped with poor fusion and penetration.

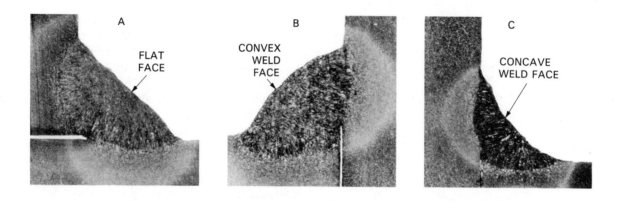

Fig. 10-36. Cross sections of acceptable fillet welds (etched and magnified four times).    (U.S. Steel)

undercut at the toe of the weld. Fig. 10-37 is a magnified cross section of a fillet weld. Note the undercut at the toe of the weld at the vertical metal surface.

The three fillet welds shown in cross section in Fig. 10-38 are unacceptable welds. At A the weld contour is poorly shaped, at B the weld is undercut, and at C the weld is overlapped.

The metal surfaces of the finished weld should be free from excessive spatter. HIGH AND LOW SPOTS in the weld bead are not acceptable. This generally indicates a weld speed which was not uniform. Highs and lows may also indicate a poor restarting technique. If the bead has a number of small pit holes this indicates porosity. POROSITY is an indication of gases trapped inside the weld. The cause may be a welding speed which is too fast. The weld crater must remain liquid long enough to allow all the gases to reach the surface before the metal becomes solid.

If a defect is found, the weld must be repaired. The poor section is ground down or cut out with a gouging electrode and rewelded.

## 10-16. EDGE JOINT SMAW

The EDGE WELD is defined by AWS as: "A joint between the edges of two or more parallel or nearly parallel members." See Fig. 10-39. The edge weld may be made in any position. This type of joint is the easiest of the various joints to arc weld.

On thin metal no edge preparation is necessary. On thicker pieces of base metal, the edge should be ground, gouged, or machined to provide a bevel, V, U, or J-groove. See Fig. 2-4.

To gain experience in joining two pieces together by arc welding, the welder should obtain two pieces of metal approximately 1/4 in. (about 6.5 mm) thick. A 1/8 in. (3.2 mm) E60XX electrode should be used. The current setting on the machine should be about 100 amps. Prior to beginning this weld or any weld, it is advisable to run a few practice beads. A piece of the same metal being welded should be used for practice. Using the suggested E60XX electrode and about 100 amps, run a bead. If the bead is narrower than 3/8 in. or about 10 mm (3 × electrode

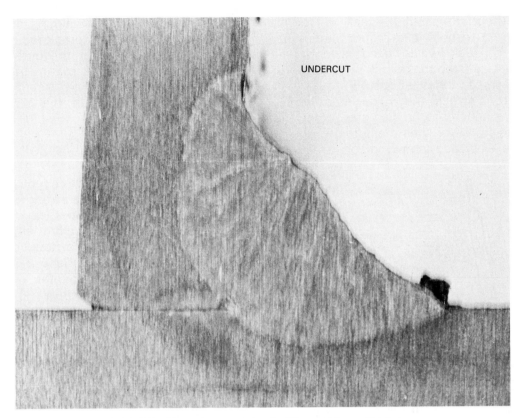

UNDERCUT

Fig. 10-37. Photomacrograph of a fillet weld on a T-joint using DC. Note the undercut on the vertical piece of metal.

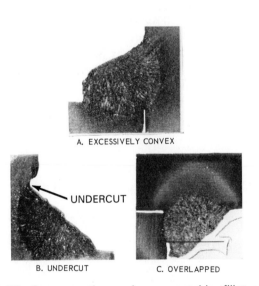

A. EXCESSIVELY CONVEX

UNDERCUT

B. UNDERCUT

C. OVERLAPPED

Fig. 10-38. Cross sections of unacceptable fillet welds (etched and magnified four times). These are unacceptable because of excessive contour in A, undercutting in B, and overlapping in C. (This weld was made in the overhead position.)

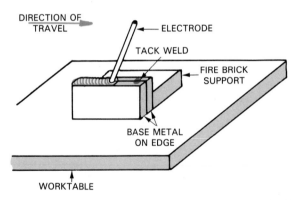

DIRECTION OF TRAVEL

ELECTRODE

TACK WELD

FIRE BRICK SUPPORT

BASE METAL ON EDGE

WORKTABLE

Fig. 10-39. A suggested setup for arc welding an edge joint in the flat position.

diameter), the current is probably too low. If the bead is wider than 3/8 in. (about 10 mm), the current may be set too high. Run test beads and continue to reset the arc welding machine current until a bead three times the electrode diameter (3 × Dia) is achieved. The bead should be run at a forward speed which is comfortable and easy to control. If the bead width and forward speed are proper, the amperage is right for the electrode and the welder (operator). Using test beads may insure a better bead on the finished part.

This thickness of metal (about 1/4 in. or 6.4 mm) should require no edge preparation. The pieces should be clamped together and grounded to the table. Firebricks may be used to prop the pieces up so that a flat weld position is possible.

A TACK WELD is a small, well-fused weld in one spot used to hold parts in proper alignment while welding. A weak tack weld may break during the welding operation and allow the metal to shift its position. Before making the edge joint or any welded joint, the metal should be tack welded as often as required to hold it in proper alignment. The number of tack welds used is optional. However, it is suggested that a tack weld be made at intervals of three inches (about 75 mm).

When the arc is struck, the welder must watch the arc crater as it increases in diameter. When the edge of the arc crater touches the outside edge of the metal, the electrode is moved forward. This process is repeated in an almost continuous process until the weld is completed. On extremely thick pieces, the groove in the joint may not touch the outside edge of the metal. In this case the electrode is moved forward when the arc crater just touches the edge of the groove. Refer to Fig. 10-40.

## 10-17. FLAT POSITION LAP JOINT SMAW

The flat position LAP JOINT type of joint is common, although it is not the strongest joint design. The weld normally used on a lap joint is the fillet weld. A fillet weld is basically triangular in shape. The weld face may have a flat, convex, or concave shape. The desired shape is normally given on the weld symbol, or by the designer of the joint. If not specified, a convex shape is slightly stronger and normally preferred. See Fig. 10-41. The electrode should be aimed into the bottom of the joint. The electrode should point more toward the metal surface than the edge. Remember, it takes more heat to melt the surface. On thin metal no sideways motion is needed to form a good fillet. When larger fillets are required a weaving motion may be necessary.

Fig. 10-42 illustrates the position of the base metal while making a fillet weld on a lap joint in the flat position. The fillet weld on a lap joint is made on the edge of one piece and the surface of the other piece. More heat is required to melt the surface than is required to melt the edge. To distribute the heat, a weaving motion must be used with most of the motion taking place on the surface, not the edge.

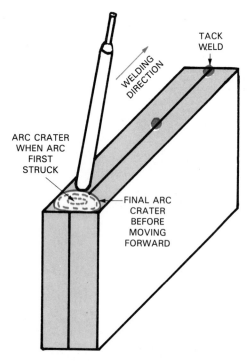

Fig. 10-40. An edge weld as the weld bead is begun. Note that the arc crater must increase in size until it touches the edges of the metal before the electrode is moved forward.

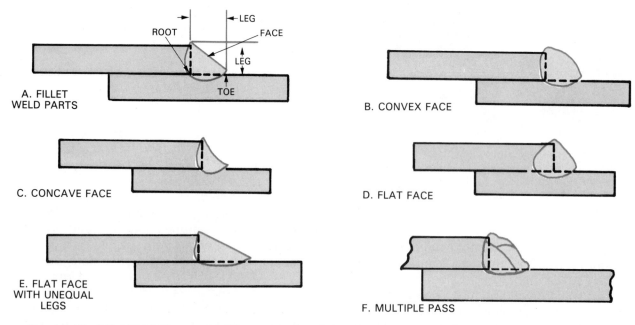

Fig. 10-41. Fillet Weld Shapes. A—Fillet weld parts. The leg size is given on the weld symbol. B—Convex face. C—Concave face. D—Flat face. E—Fillet with unequal legs. F—Multiple pass weld lap joint.

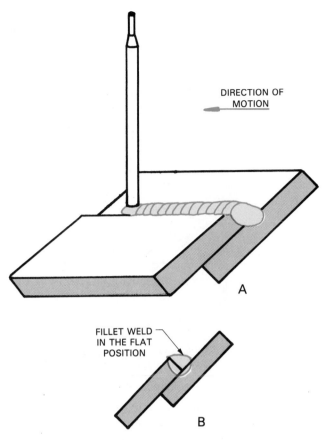

DIRECTION OF MOTION

A

FILLET WELD IN THE FLAT POSITION

B

Fig. 10-42. Fillet weld on a lap joint done in the flat position. A—The weld in progress. B—The base metal position for welding a lap joint in the flat position.

The finished bead should have the proper contour, must be straight, even in width, smooth, and clean. It should show good fusion between the bead and the parent metal. The bead should blend smoothly with the base metal. If a distinct line is seen, the fusion is probably poor. There should be no overlap or undercutting in the fillet. The weld must penetrate through the root of the joint. It should not penetrate through the base metal.

## 10-18. CORNER OR T-JOINT SMAW IN THE FLAT POSITION

The CORNER JOINT may be made as an inside or outside corner joint.

The inside corner joint may be made as a bevel or J-groove or without edge preparation as a square groove. A fillet weld is used on the inside corner joint. See Fig. 10-43.

The outside corner joint is prepared as a butt joint. The joint may be cut as a bevel, V, J, or U-groove or without edge preparation. Refer to Fig. 10-44.

The weld on either the inside or outside corner should penetrate through the root. The weld should not penetrate through the base metal. Both

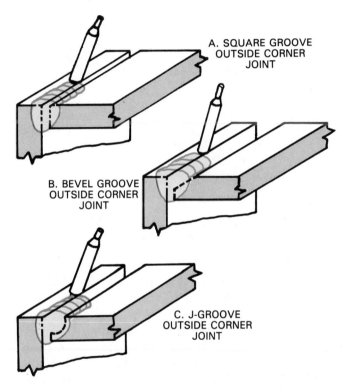

A. SQUARE GROOVE OUTSIDE CORNER JOINT

B. BEVEL GROOVE OUTSIDE CORNER JOINT

C. J-GROOVE OUTSIDE CORNER JOINT

Fig. 10-44. Outside corner joints welded in the flat position.

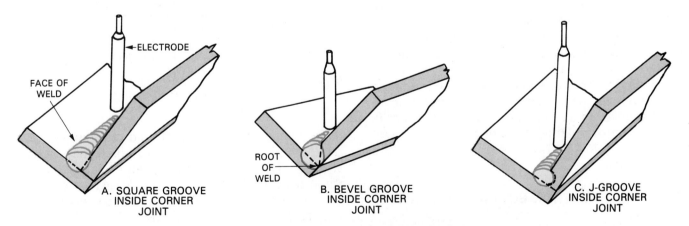

ELECTRODE

FACE OF WELD

A. SQUARE GROOVE INSIDE CORNER JOINT

ROOT OF WELD

B. BEVEL GROOVE INSIDE CORNER JOINT

C. J-GROOVE INSIDE CORNER JOINT

Fig. 10-43. Inside corner joints welded in the flat position. The fillet weld is used in each case. Note that the weld face is horizontal in the flat position.

edges of the bead should blend smoothly with the metal surfaces. There should be no overlapping or undercutting. The bead must be straight and have a uniform width and contour, see Fig. 10-45.

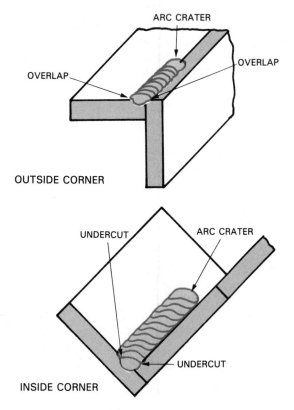

Fig. 10-45. Examples of overlap and undercut on an outside and inside corner joint.

The T-JOINT is formed by placing one piece of base metal on the other to form a T-shape. This joint may be welded from one or both sides. The joint may be in the form of a square, V, or J-groove. If the T-joint is to be welded from both sides, it is usually easier to complete the welds if the entire work piece is repositioned after one weld is completed. See Fig. 10-46.

The T-joint fillet weld is made in the same manner as the weld on an inside corner joint.

These welds, as all welds, are easier to make in the flat (downhand) position. A weaving motion is required only on large, wide joints. The torch angles are the same as those used when running a bead. However, the welder should be certain to melt the surfaces as the inside corner or T-joint is welded. Too often the beginner makes an inside corner or T-joint without proper fusion of the metal surfaces.

## 10-19. FLAT POSITION BUTT JOINT SMAW

It is suggested that the butt weld be practiced on 3/8 to 1/2 in. (9.5 - 12.7 mm) thick low car-

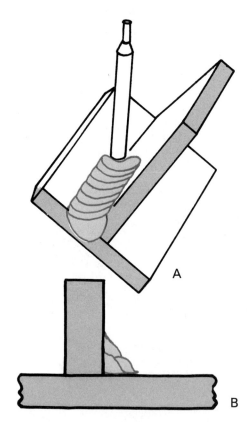

Fig. 10-46. A—Fillet weld in progress on a flat position T-joint. B—T-joint welded with multiple passes.

bon steel. Steel thicker than 1/4 in. (6.4 mm) should have the edge prepared prior to welding. This will insure complete penetration of the base metal. A V-groove joint is suggested. Both metal edges should be ground to about a 45 degree angle. The angle should be cut to within 1/16 in. (1.6 mm) of the bottom. See Fig. 10-47.

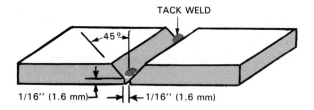

Fig. 10-47. Preparing a V-groove butt joint for welding. The edges have been beveled and the metal tack welded to hold it in proper alignment. Note that the root opening and root face are 1/16 in. (1.6 mm).

The base metal joint should be tapered, or tacked, to prevent or control distortion while welding. This welding may be done in one, two, or three passes, as shown in Fig. 10-48. If done in one operation, the welder must use a weaving motion to distribute the filler metal in the V-groove and to secure adequate fusion. If two or more beads are used to complete the weld, the operator

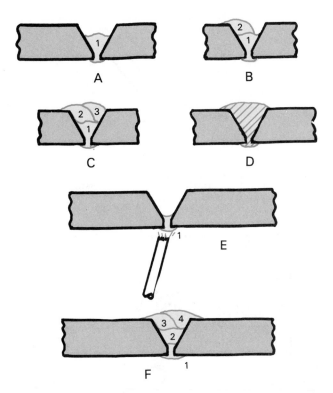

Fig. 10-48. Arc welding joints welded with multiple passes. A—The first or root pass in a butt joint. B—Second pass. C—Third pass. D—Finished weld. E and F—Show the butt joint where the root pass is welded from the penetration side of the joint.

should clean the first bead before attempting to make the next bead. This operation prevents including slag in the next bead or pass. This weld should be penetrated slightly, and it should also be built above the original metal.

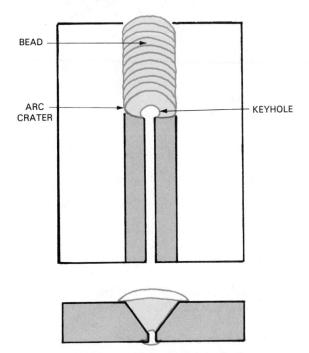

Fig. 10-49. Two views of a V-groove butt weld in progress. Note the keyhole near the leading edge of the arc crater.

To insure complete penetration in a groove weld, the KEYHOLE METHOD should be used. The keyhole is seen only on the first or root pass. As the arc crater forms and melts through the thinner metal at the root face, a small hole is formed. The keyhole is continually filled in as the weld moves ahead. The hole looks like the old-fashioned keyhole. See Fig. 10-49. If the diameter of the keyhole is kept constant, the amount of penetration will be uniform also.

In the case of a long weld where more than one electrode is used, the arc should be restarted about 3/8 in. (about 10 mm) ahead of the crater. The electrode is moved back over the crater to properly fill it and keep a uniform contour. Starting the arc in this way allows time for the arc to stabilize. It also insures proper preheating of the base metal.

When the end of the seam is reached, move the arc to the back of the crater. This motion will fill the trailing crater. Then lift the electrode to break the arc.

Cool the weld, remove the slag, if any, and wire brush the weld surface. Guard your eyes against flying particles. ALWAYS USE SAFETY GOGGLES WHEN CLEANING METAL.

Inspect the weld for straightness, constant width, smoothness, penetration, gas bubbles, fusion, spatter, and buildup. It should have a clean looking bead with straight edges and constant width of bead. The height of the bead should be constant. The ripples should be evenly spaced. The penetration should just show through the under part of the metal joint. The weld should have no small cavities which would indicate too long an arc. It should have good fusion. FUSION means a good bond between the filler metal and the base metal. Fusion is indicated by a smooth blending of the filler and base metals at the edges of the weld. There should not be a distinct edge or line between the filler metal and the base metal. There should be little or no spatter. Spattering is the result of too long an arc or too high a current. If the metal is loosely connected to the table, the resulting wandering arc may give poor fusion and weld appearance.

## 10-20. ARC WELDING IN THE HORIZONTAL POSITION

When welding a butt, edge, or outside corner joint in the HORIZONTAL POSITION, the electrode should be pointed upward at about 20 degrees. This is to counteract the sag of the molten metal in the crater. The electrode is also inclined about 20 degrees in the direction of travel of the weld as shown in Fig. 10-50. Gravity tends to work against the welder in all positions other than the

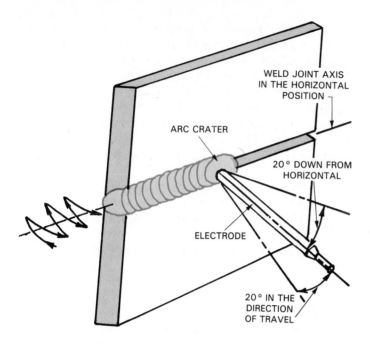

Fig. 10-50. Suggested electrode position for welding the butt joint in the horizontal position. The electrode is tipped 20° forward and 20° downward. Use the suggested electrode motion or one similar to it. The upward and backward motion uses the force of the arc to push the filler metal into position. Stopping momentarily at the end of each motion will help to eliminate undercutting.

flat position. Therefore, a short arc gap should be used to insure the filler metal will travel across the horizontally positioned arc.

Be sure to eliminate undercutting at the edge of the bead. Undercutting is usually the result of excess current for the size of the electrode used or poor electrode motion. Stop momentarily at the end of each motion or swing of the electrode. This will deposit additional metal and will help to eliminate undercutting.

When welding a T or inside corner joint with one piece of the base metal in a near vertical position, and one in a near horizontal position, the electrode is inclined 20 degrees in the direction of travel. The electrode is also positioned at about a 40 - 50 degree angle to the horizontal piece of metal as shown in Fig. 10-51. The motion used is usually some type of backward slanting weave motion. The forward motion may take place on the vertical piece. Fig. 10-52 shows a T-joint ready for welding.

The suggested electrode angles for making a fillet weld on a horizontal lap joint are shown in Fig. 10-53. The electrode should point more toward the surface than toward the edge. It takes more heat to melt the surface of the metal than the edge of the metal.

Whatever electrode motion is used, remember that it should have a slight backward slant. This

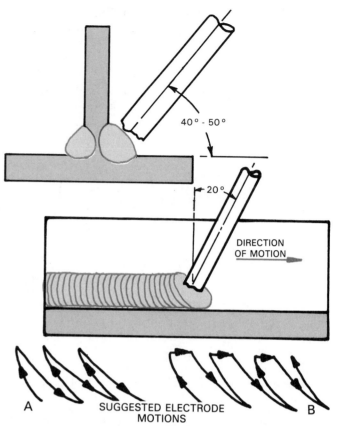

Fig. 10-51. Suggested electrode position for fillet welding on an inside corner or T-joint. The two suggested electrode motions have a backward slant. In B the forward motion takes place on the vertical piece. Both motions use the arc force to overcome gravity.

Fig. 10-52. A welder about to tack weld a T-joint prior to making a fillet weld in the horizontal position. Note the position of the electrode.    (The Lincoln Electric Co.)

backward slant lets the arc force push the filler metal up as it attempts to sag down with gravity.

## 10-21. ARC WELDING IN THE VERTICAL POSITION

Whenever possible, welds should be made with the seams in the flat position. In some industries, special turntables are used to rotate the work so

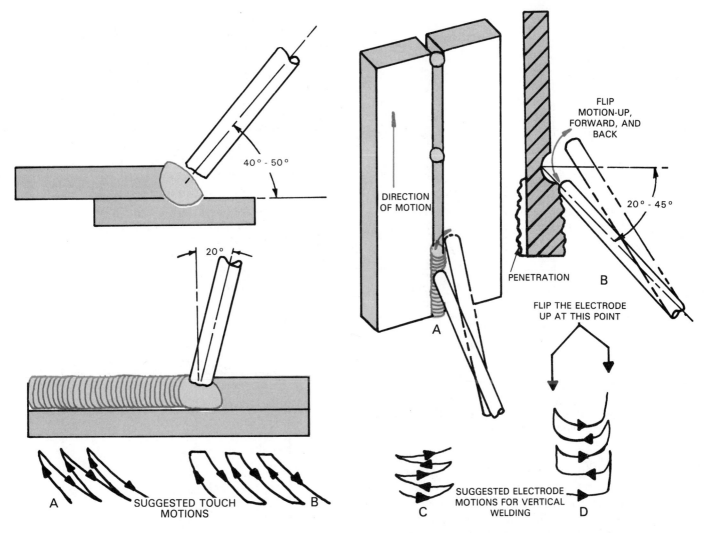

Fig. 10-53. Suggested electrode position for fillet welding on a horizontal lap joint. Note the two suggested backward slanting electrode motions.

Fig. 10-54. A—The flip motion used to control and cool the arc crater. This motion may be used on any type of joint. B—The electrode should point upward at 20°- 45°. The angle may vary as required. Note the motions shown in C and D.

that this position may be obtained. However, many welds have to be done in a vertical, horizontal, or in an overhead position because of the size of the project. These welds must be of the same quality and strength as welds done in a flat position. Fig. 10-54 illustrates the metal and electrode positions for VERTICAL butt joint welding.

Welding in the vertical position may be done in either of two directions. The welder may weld vertically up or vertically down. Using either direction, the weld must be made so that the electrode flux or slag is not entrapped or included in the weld metal. The weld must also be made so that it does not run or drip. This is caused by allowing the filler metal in the crater to stay molten too long.

The following actions may be taken to prevent or control filler metal sagging:

1. Hold a short arc. This will permit more filler metal to transfer from the electrode to the crater.

2. Use as low a current setting as possible. But one which will produce well fused welds.
3. Make multiple pass welds. Metal in a narrow arc crater will cool more rapidly and tend to sag less.
4. Use a flip motion to allow the metal in the crater to cool.
5. Use an electrode motion which will allow the metal in the crater to cool. See Fig. 10-54, C and D.

The vertical up method of welding is generally preferred. When welding vertically down the electrode slag or flux has a tendency to run into the molten weld metal. This is prevented by welding at a speed fast enough to stay ahead of the molten slag. Welding at such a rapid speed generally does not produce adequate penetration. The vertical down method may work best on thin metal. See Fig. 10-55.

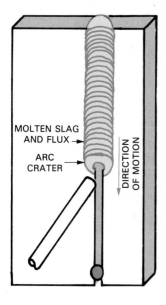
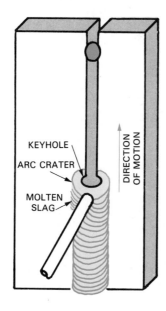

WELDING VERTICALLY DOWN      WELDING VERTICALLY UP

Fig. 10-55. The vertical down and vertical up welding methods. Vertical up is generally preferred. When welding vertically down there is a danger of having the falling slag or flux mix with the molten metal.

When welding vertically up the greatest problem is controlling the metal in the arc crater. If this metal becomes too hot, or the crater becomes too wide, the metal may drip down and off the weld. To control the arc crater heat and allow the metal time to cool, a flip motion is used. With a flip motion, the electrode end is momentarily moved forward and raised slightly. It cannot be raised too far or the arc will stop. The electrode is then brought back to the rear of the arc crater to continue the

weld. During the time that the electrode and arc are moved up, forward, and back again, the arc crater cools slightly. This flip action is continued throughout the weld. If more time is required for cooling, the electrode is moved farther forward before returning to the rear of the crater. Fig. 10-56 shows a vertical weld in progress.

The vertical butt, edge, and outside corner joints may be a square groove, V-groove, bevel, J or U-groove. Practice pieces may be positioned and tack welded in the flat position. These pieces must then be held in the vertical welding position. This may be done by clamping them in a welding fixture such as shown in Figs. 4-46 and 4-47.

The motion used on the vertical lap is shown in Fig. 10-57. A small weaving motion is often used. Forward movement and any flip motion required is done on the surface, not on the edge. The arc crater should just touch the edge of the one piece momentarily. The electrode and arc crater are then moved away to the surface.

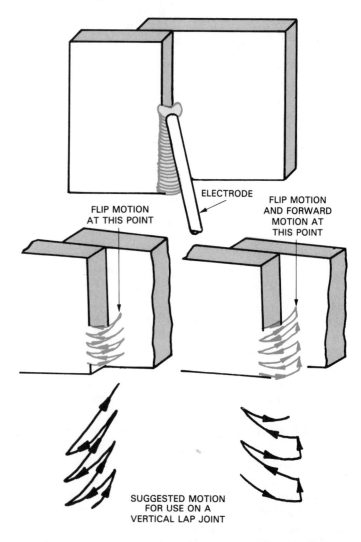

ELECTRODE

FLIP MOTION AT THIS POINT

FLIP MOTION AND FORWARD MOTION AT THIS POINT

SUGGESTED MOTION FOR USE ON A VERTICAL LAP JOINT

Fig. 10-57. A view of a vertical fillet weld in progress on a lap joint. Note the suggested motions. Both use a flip motion on the surface piece.

Fig. 10-56. A welder shown making a fillet weld in the vertical position on a large field welded structure.
(Hobart Brothers Company)

The electrode angles used for the vertical inside corner and T-joint are shown in Fig. 10-58. This figure also shows the electrode motion suggested for the filler weld used on this joint.

All vertical welds should have the same appearance as a weld done in the flat position. The bead must be straight with a uniform width. The weld bead must be properly fused, with no overlap or undercutting. The ripples of the bead should be uniform. The face of the weld should have the required contour.

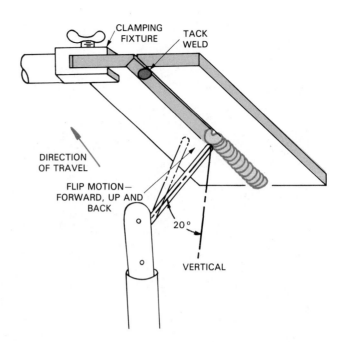

Fig. 10-59. A V-groove butt weld in progress. This is the overhead position. Note the position of the electrode in the electrode holder.

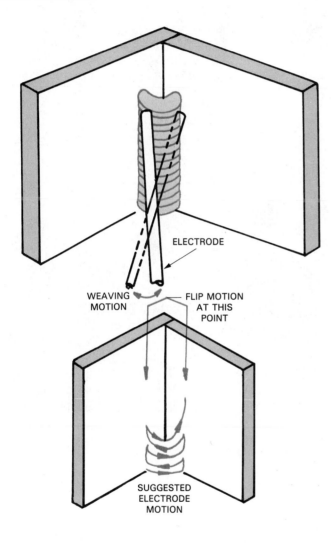

Fig. 10-58. A fillet weld on a vertical inside corner joint. A weaving motion is often used with a flip motion to keep the filler metal from sagging.

## 10-22. ARC WELDING OVERHEAD

OVERHEAD ARC WELDING is generally the most difficult. It can also be dangerous for any operator who is not wearing the correct protective clothing. The beginner should practice making beads in the overhead position before attempting to weld seams. Fig. 10-59 illustrates a V-groove

butt joint being welded in the overhead position. On practice pieces the joint is tack welded in the flat position. It is then placed in a fixture to hold it during overhead welding. See Figs. 4-46 and 4-47. The angles of the electrode are much the same as for flat or horizontal welding. To keep the metal in the arc crater from overheating a flip motion is used.

Most electrode holders have jaws designed to allow the electrode to be held in a variety of positions. The electrode should be placed into the electrode holder at an angle which is comfortable for the welder to use. See Fig. 10-60. A covered electrode should never be bent to change its angle for welding. Bending a covered electrode will crack the covering and cause it to fall off. Lack of a flux covering on the electrode will result in a poor weld. It is most important to keep metal in the molten crater from falling due to gravity. To keep it cool and to control the molten crater, a flip motion is used. See Heading 10-19.

A cap, leather cape or coat, and leather quality gloves should be worn when overhead welding. Coveralls must be buttoned at the collar and all pockets should have closed flaps on them. Fig. 10-60 shows a welder practicing an overhead T-joint. Notice the position of the electrode in the electrode holder.

## 10-23. CARBON ARC WELDING (CAW) PRINCIPLES

The CARBON ARC WELDING or CAW produces an arc between the metal and a carbon electrode

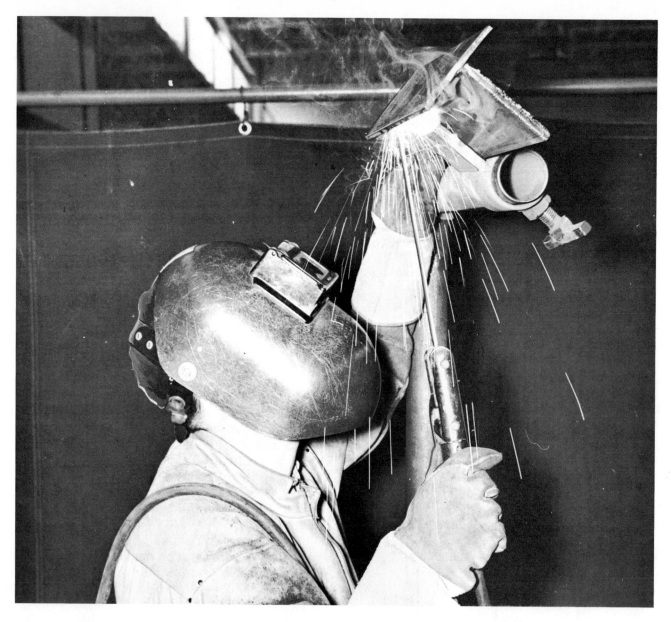

Fig. 10-60. Making a practice overhead arc weld. Note the position of the electrode in the electrode holder. This welder is wearing a cape to protect his arms and shoulders, a cap, and gauntlet gloves.

or between two carbon electrodes. The arc produced will create enough heat to melt the base metal. The operator may add metal to the molten puddle by means of a welding rod, as shown in Fig. 10-61.

In the twin carbon arc welding process, two carbon electrodes are used. An arc is struck between the two electrodes. The heat from this arc is high enough for heat treating, soldering, brazing, and light gauge metal welding. A double carbon arc welding electrode holder is shown in Fig. 10-62.

The carbon electrode is prepared for use by grinding the electrode to the shape of a cone. The taper of the cone should be 6 - 8 times the diameter of the electrode. When placing the carbon electrode in the holder, the electrode should

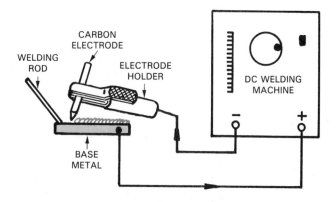

Fig. 10-61. A single carbon electrode arc welding circuit. The arc between the electrode and work creates the heat. The welder must add filler metal as in oxyfuel gas welding.

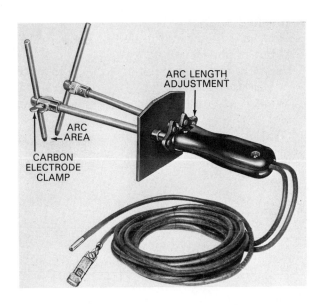

Fig. 10-62. A twin carbon electrode holder with carbon electrodes in place. Two leads are required because the arc is created between the two electrodes.

stick out from the holder a distance equal to 10 times the electrode diameter as shown in Fig. 10-63.

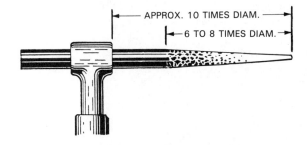

Fig. 10-63. The method of preparing a carbon electrode for use in carbon arc welding. The carbon should taper a distance equal to 6-8 times its diameter, and should stick out from the holder a distance equal to 10 times its diameter.
(National Carbon)

## 10-24. REVIEW OF SAFETY IN SMAW

The following are some safety rules that must be carefully observed if accidents, when welding, are to be prevented:

1. The eyes and face must be protected from harmful rays and sparks by using an approved type of helmet. See Figs. 9-57, 9-60, and 9-61.
2. Recommended clothing and shoes must always be worn. Leathers should be worn especially during overhead welding.
3. Open pockets and cuffs are not recommended because they may catch hot sparks and the clothing may be ignited.
4. The floor on which the operator stands should be kept dry to eliminate the chance of an electrical shock. The welder may choose to stand on a wooden platform for comfort and insulation.
5. Only an experienced electrician should work on electrical power connections used in the electric arc welding machine.
6. The operator should wear heavy, gauntlet-type gloves.
7. When arc welding, all skin should be covered to prevent burns from the arc rays.
8. The operator should have adequate ventilation to protect the nose, throat, and lungs from harmful and irritating fumes generated in the electric arc.
9. Always use National Electrical Manufacturers Association (NEMA) approved equipment.
10. Never use homemade or non-approved transformer equipment. It may be dangerous particularly if the primary and secondary winding should become electrically shorted and/or connected.
11. Never operate an AC welding machine with the welding cables wrapped around the welding machine. The magnetic field produced by the welding cables may interfere with the magnetic circuit inside the welder.
12. The primary circuit should always be installed according to the prevailing electrical code.

## 10-25. TEST YOUR KNOWLEDGE

Write your answers on a separate sheet of paper. Do not write in this book.

1. What is the approximate temperature of the arc in SMAW?
2. List three types of DC arc welding machines.
3. List two types of AC arc welding machines.
4. What is the average no-load amperage and voltage in a DC arc welding circuit?
5. How is the AC arc stabilized?
6. What is arc blow?
7. Using the electron theory of current flow, in what direction does the current flow in a DCEN (DCSP) circuit?
8. List three variables which may determine the choice of which DC polarity to use.
9. Why is it important that the electrode and work lead connections are tight?
10. What precautions should be taken when an electrode or work lead must run across an aisle way temporarily?
11. What is the danger of arc flash?
12. When is a small diameter electrode used (two reasons)?
13. To run a good bead, what factors must the

welder control in manual SMAW?

14. How wide is a stringer bead: a weaving bead?

15. What does a bead made with too long an arc gap look like?

16. How is an arc restarted on a previous, un-finished bead?

17. How is an arc bead ended properly?

18. What is undercut?

19. What is the best position in which to weld any shape joint?

20. What is the flip motion and why is it used?

A welder properly dressed for out-of-position arc welding. The welder is wearing a leather jacket, leather trousers, leather gauntlet gloves, hard-toed shoes, a cap, and an arc helmet.

# part 4

# GAS TUNGSTEN AND GAS METAL ARC WELDING (TIG & MIG)

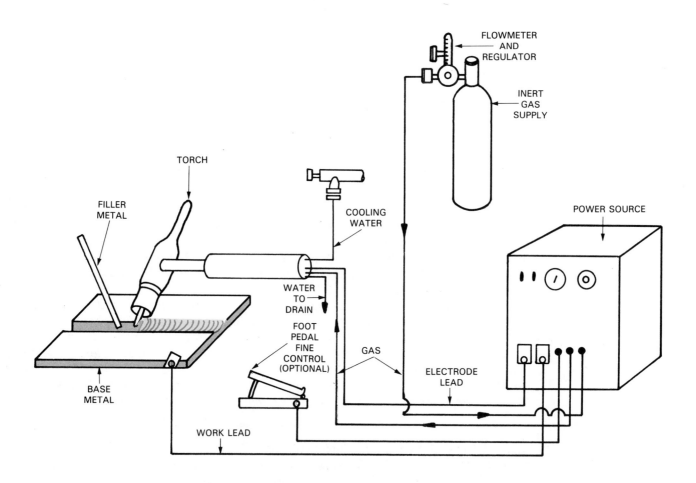

Fig. 11-1. A diagrammatic drawing of a complete gas tungsten arc welding outfit.

**Authors' Note:** MODERN WELDING follows the American Welding Society's approved welding process terminology. Gas Tungsten Arc Welding or GTAW and Gas Metal Arc Welding or GMAW are used in this text as the professional terms for the processes. On the job, the informal terms such as MIG for Metal Inert Gas welding and TIG for Tungsten Inert Gas welding will be encountered. These convenient conversational terms have the same meaning to the welder as the professional terms from the text.

# 11 GTAW and GMAW EQUIPMENT AND SUPPLIES

GAS TUNGSTEN ARC WELDING (GTAW) is a welding process where an arc is struck between a nonconsumable tungsten electrode and the metal workpiece. The weld area is shielded by inert (chemically inactive) gas to prevent contamination. Filler metal may or may not be added to the weld. It is possible to weld most weldable metals with GTAW. This process is excellent for welding root passes in heavy metal sections. This process is also known as tungsten inert gas or TIG welding in shop terms. Heading 1-6 describes the basics of GTAW.

GAS METAL ARC WELDING (GMAW) is a welding process where an arc is struck between a consumable metal electrode and the metal workpiece. The consumable electrode wire is fed to the welding torch from large wire spools which hold several hundred feet (meters) of wire. The consumable electrode is the filler metal. The weld area is protected by a shielding gas. This process is used in production, in welding shops, and in automobile body repair shops. It is capable of making excellent welds, almost continuously. The manual welding skills required for this process are not as great as those required for some other manual welding processes. This process is also known as metal inert gas or MIG welding in shop terms. Refer to Heading 1-7 for a basic description of the GMAW process.

FLUX CORED ARC WELDING (FCAW) is similar in most respects to the Gas Metal Arc Welding (GMAW) process. The difference is in the electrode wire used. FCAW uses a hollow cored electrode which contains flux or alloying materials. The flux core provides a gaseous shield around the arc. FCAW may also use a shielding gas provided through the torch similar to the GMAW process. Refer to Heading 1-8 for a basic description of the FCAW process.

The equipment and supplies used for each process will be explained in this chapter. Chapter 12 will cover Gas Tungsten Arc Welding techniques

and principles, while Chapter 13 will cover Gas Metal Arc Welding and Flux Cored Arc Welding principles and techniques. Studying these chapters coupled with actual welding practice will build the skills and techniques needed to master these welding processes.

## 11-1. THE GAS TUNGSTEN ARC WELDING STATION

The typical gas tungsten arc welding (GTAW) outfit will contain the following equipment and supplies:
1. An AC or DC arc welding machine.
2. Shielding gas cylinders or facilities to handle liquid gases.
3. A shielding gas regulator.
4. A gas flowmeter.
5. Shielding gas hoses and fittings.
6. Electrode lead and hoses.
7. A welding torch (electrode holder).
8. Tungsten electrodes.
9. Welding rods.
10. Optional accessories.
   a. A water cooling system with hoses for heavy duty welding operations.
   b. Foot rheostat.
   c. Arc timers.

Fig. 11-1 shows a schematic drawing of a gas tungsten arc welding outfit. The booth and exhaust system are not shown in this illustration. Refer also to Fig. 1-11.

## 11-2. ARC WELDING MACHINES FOR GTAW

The power source for gas tungsten arc welding (GTAW) can be an alternating current (AC) or direct current (DC) arc welding machine. These machines may be either transformers, generators, alternators, or transformer-rectifier type machines. Gas tungsten arc welding machines must produce or supply a CONSTANT CURRENT. In these constant current machines, the volt-ampere curve is relatively steep. Because of the shape of this curve, the machine is known as a DROOPING VOLTAGE TYPE MACHINE.

Fig. 11-2 shows a drooping voltage curve. Two curves are shown in this figure. Curve 1 is for an 80 volt open circuit voltage machine set for a 150 amps welding current. A five volt change in the closed circuit (welding) voltage from 20 volts to 25 volts will result in an 8 amp change in the amperage. The 20 volts to 25 volts change represents a 25 percent change in voltage with only a change of about 5 percent in the current flowing. Curve 2 represents a curve for an 80 volt open circuit voltage set for 50 amps of welding current. A 25 percent change in voltage from 20 volts to 25 volts will only change the amperage about 2 amps or 4 percent. The amperage is therefore considered to be relatively constant with large changes in voltage.

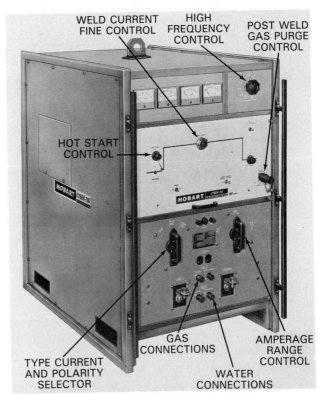

Fig. 11-3. A transformer-rectifier type AC/DC welder designed for gas tungsten arc welding (GTAW) or TIG. This machine has a high frequency control, hot start, post weld purge control, and gas and water connections in addition to the normal controls. (Hobart Bros. Co.)

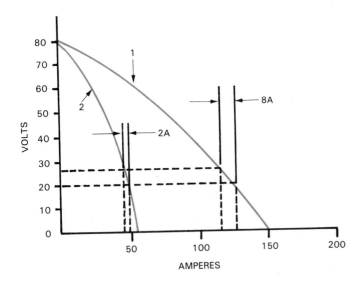

Fig. 11-2. A volt-ampere curve for a drooping voltage or constant current arc welding machine. Slope 1—The curve for a machine set at 150 amps with an 80 volt open circuit voltage. Slope 2—The curve for a machine set at 50 amps with an 80 volt open circuit voltage.

AC transformers, generators, or alternators may be used with GTAW. Gas tungsten arc welding may also be done with alternators or transformers with DC rectifiers built in. Figs. 11-3 and 11-4 are AC or DC transformer-rectifier arc welders for gas tungsten arc welding. The direct current arc welding machine must be equipped with a device which produces a high frequency voltage and feeds it into the welding circuit. High frequency voltage is produced by an oscillator or a similar device. Several thousand volts are produced at a frequency of several million cycles or megahertz

per second. The current is only a fraction of an ampere in this high frequency circuit. High frequency voltage, when used in a DC circuit, is used only for starting the arc. Once the DC welding arc is stabilized, the superimposed high frequency starting voltage is automatically stopped.

A high frequency voltage producer, such as an oscillator, is also required on all AC welding machines used for GTAW. This high frequency voltage is fed into the AC welding circuit and is used to start the arc. In alternating current GTAW, the high frequency is used to stabilize the arc. It remains on constantly until the AC machine is turned off. High frequency voltage stabilizes the flow of welding current. It keeps the gases in the arc ionized so that the arc is more easily maintained during the zero points of each AC cycle.

The alternating current arc is most difficult to maintain during the half of the AC cycle when the electrode is positive. The methods used to produce the required high frequency voltage are shown below. These methods are used to stabilize the AC arc:

1. A 3000-5000 volts superimposed high frequency voltage with low amperage.
2. A transformer with a 150-200 volt open cir-

AC/DC SWITCH

DC POLARITY SWITCH

CURRENT CONTROL

GAS, COOLANT, AND POWER CONNECTION UNDER THIS COVER

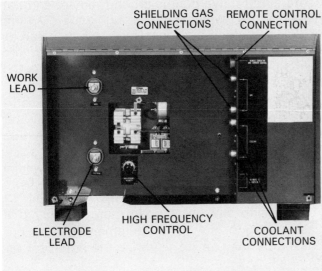

SHIELDING GAS CONNECTIONS

REMOTE CONTROL CONNECTION

WORK LEAD

ELECTRODE LEAD

HIGH FREQUENCY CONTROL

COOLANT CONNECTIONS

Fig. 11-4. An AC/DC gas tungsten arc welding machine. This is a constant current transformer-rectifier type machine with a high frequency switch. The connections for gas coolant and power are under the lower cover as shown. This machine may also be used for shielded metal arc welding. (Miller Electric Mfg. Co.)

cuit voltage. This method will require protection for the operator.

3. A 200-400 volts capacitor charge to provide a surge voltage. This surge voltage is injected in-to the welding circuit at the start of each reverse polarity (electrode positive) half of a cycle.

4. An oscillator or "ringing" circuit connected across the power and supply terminals. When the arc is extinguished, a resonant (vibrating) frequency circuit which contains a capacitor and inductance creates a high frequency voltage. This high frequency voltage reignites the arc.

The construction, operation, and control of AC and DC arc welding transformers, transformer-rectifiers, generators, and alternators are covered in Heading 9-3 to 9-6. Review this basic information for a more complete understanding of welding equipment.

## 11-3. BALANCED AND UNBALANCED AC WAVES

Each alternating current cycle when plotted against time on a graph has a shape like a wave, as shown in C, Fig. 11-5. Theoretically, alternating current has equal current flowing during both halves of each wave, or cycle. In actual practice, the current is less during the half of the AC cycle when the electrode is positive. This may be explained as follows.

During one half of AC cycle the electrode is negative and work positive. After going through the zero current point, the cycle reverses. The electrode then becomes positive and the work negative.

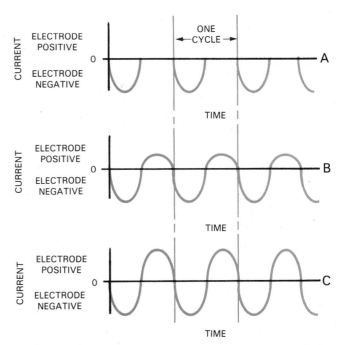

Fig. 11-5. Alternating current plotted against time. A—Completely rectified AC wave with no current flow during electrode positive. B—Unbalanced wave. C—Balanced wave.

Electrons flow from negative to positive. During the electrode negative half of the cycle, the hotter electrode gives off electrons readily. Since the electrode tip has a small surface, the electrons are concentrated and flow easily toward the large surface of the work.

When the work is negative, the electrons must flow from a large surface to a small point (the electrode). The electrons on the surface are not concentrated at any one point. Electrons, therefore, cannot easily travel from the surface to the electrode. This resistance of electron flow from a surface to a point is called RECTIFICATION.

Rectification may be so severe that the entire flow of current in the electrode positive half of the cycle is stopped as shown in A, Fig. 11-5.

The question may be asked by the welder: ''What happens to the current during the electrode positive half cycle? No current flows during the electrode positive half cycle. This lack of current flow makes the AC arc very unstable. Thus rectification of the current is undesirable for the welding operation.

To overcome this lack of flow during one half of the alternating current cycle, a high frequency voltage is created and wired into the welding circuit. Several thousand volts, with only a fraction of an ampere, are created. This high voltage has a frequency of several megahertz (million cycles per second).

The high frequency unit is normally built into the gas tungsten arc welding machine. It consists of a step-up transformer, capacitors, a control rheostat, a set of spark gap points, and a coil to couple the high frequency unit to the welding circuit.

High frequency voltage flows continually in the welding circuit. This high frequency voltage keeps the shielding gas in the arc area ionized. When the gas is ionized, the arc is maintained during the half of the cycle when the electrode is positive. While the arc is maintained some current will flow across to the electrode. However, because the work surface does not emit (give off) electrons as easily, less current flows during this half of the AC cycle, or wave, as shown in B, in Fig. 11-5. The wave shown in B, Fig. 11-5 is an unbalanced wave.

A completely balanced wave, as shown in C, Fig. 11-5, may be obtained by the use of a large number of capacitors in series. A capacitor is a device used to store electricity. When the capacitor is discharged a surge of current flows from it. With capacitors in series, a flow can be maintained for a longer period. The capacitors are recharged during the electrode negative half of the AC cycle.

The balance wave AC power source is recommended where the utmost in weld quality is desired. Balanced wave GTAW machines are usually more expensive than an unbalanced wave arc welder.

The advantages of a balanced wave AC arc welding circuit are:
1. Excellent oxide cleaning is provided because during the electrode positive half of the cycle more current flows.
2. A more stable arc.
3. A reliable reignition of the arc during the electrode positive half of the cycle.
4. Excellent for high quality production welding.

The advantages of the unbalanced wave are:
1. Higher currents can be used with a given electrode since the electrode positive half of the cycle has less current flowing than with a balanced wave.
2. The arc welding machine is less expensive than a balanced wave arc welder.
3. Better penetration is possible with an unbalanced wave than with a balanced wave. This is possible because penetration takes place during the electrode negative half of the cycle.

## 11-4. SHIELDING GASES USED WITH GTAW

INERT GASES are gases which will not react with metals or other gases. Inert gases are used to create a protective bubble around the arc and hot metals while welding is being done. Argon, helium, and argon-hydrogen mixtures are used as the inert gases when doing gas tungsten arc welding. This process is also known as tungsten inert gas or TIG welding.

ARGON (Ar) has an atomic weight of 40 and is therefore a relatively heavy gas. It is usually furnished in heavy steel cylinders similar in construction to oxygen cylinders. The usual cylinder size is 330 cu. ft. (9.34 m³). Argon as used for welding is 99.99 percent pure. It is capable of being shipped in the liquid form at $-300\,°F$ ($-184\,°C$). Liquid argon can be shipped more cheaply than gaseous argon. However, equipment must be purchased by the user to vaporize the argon for use in welding.

Argon is used more often than helium for the following reasons:
1. It provides a smoother, quieter arc action.
2. It requires a lower arc voltage than other shielding gases for a given arc length and current used.
3. It provides easier arc starting.
4. It has a lower overall cost and is more readily available.
5. It provides a greater metal cleaning action on aluminum and magnesium with alternating current.

6. It is heavier than other inert gases and requires a lower flow rate for good shielding.
7. Because it is heavy, it provides for better cross-draft shielding.

Because it requires lower arc voltages to maintain the arc, argon has great advantages for use with thin metal and in out-of-position welding.

HELIUM (He) is the lightest inert gas. It has an atomic weight of 4. Argon therefore is ten times heavier than helium. Helium is 99.99 percent pure as used in welding. It is normally shipped and used as a gas. The cylinders are similar to those used for oxygen. Helium is usually used in cylinders of the 330 cu. ft. (9.34 m³) size. Because helium is so light, it requires about two to three times as much helium as it would argon to shield a weld area.

The chief advantage of helium over argon is that helium can be used with greater arc voltages. Helium also yields a much higher available heat on the metal than is possible with argon. Helium is therefore used to weld thick sections of metal or metals with a high heat conductivity like copper and aluminum.

Higher currents must be used with argon than with helium to produce the same available heat at the metal. Studies have shown that undercutting will occur at the same current levels with either gas. Helium therefore will produce better weld results at higher speeds without undercutting than is possible with argon.

Both helium and argon provide good cleaning action with direct current. With AC which is used on aluminum and magnesium, argon provides a better cleaning action. Argon also provides better arc stability with AC than helium.

ARGON-HYDROGEN MIXTURES are used to produce higher welding speeds. The welding speed possible is in direct proportion to the amount of hydrogen added to argon gas. The addition of hydrogen to argon permits the mixture to carry higher arc voltages. Too much hydrogen will cause POROSITY (gas pockets) in the weld metal. Porosity weakens a weld. Mixtures of 65 percent argon and 35 percent hydrogen have been used on stainless steel with a .010-.020 in. (approximately .250-.500 mm) root opening. The most common argon-hydrogen ratio is 85 percent argon and 15 percent hydrogen. Argon-hydrogen mixtures are used with stainless steels, nickel-copper, and nickel based alloys. Hydrogen produces negative welding effects with most other metals.

Argon, helium, or argon and helium mixtures can be used for most GTAW welding jobs. Manual welding on thin metals is done best with argon because of the low arc voltages and welding currents required.

Whenever there is a cross draft, it is advisable to construct a wind break. This will reduce the possibility of the shielding gas being blown away from the weld area.

## 11-5. SHIELDING GAS CYLINDERS

The gases described in the previous paragraph can be obtained in cylinders of various sizes. These cylinders are similar to the oxygen cylinders described in Chapter 3. They are manufactured to Interstate Commerce Commission (ICC) specifications.

Some gases are stored in cylinders as a gas. Some gases may also be stored as liquids in thermos-like tanks similar to the Dewar flask shown in Fig. 3-3. The quantity of gas in the cylinder is determined by:
1. The high pressure gauge or volume scale, if stored as a gas.
2. The weight, if stored as a liquid.

The gases, with the exception of hydrogen, are not flammable and therefore combustion and/or support of combustion is not a problem. The high pressures in the full cylinders make it necessary to handle the cylinders with care.

The cap should be securely threaded over the cylinder valve whenever the cylinders are being moved, or when they are in storage. The cylinder should be fastened to a wall or very stable object when in use. Cylinders should be placed where it is virtually impossible to accidentally injure the cylinder with an arc or cutting torch. Gas cylinders should be stored and used in an upright position.

## 11-6. INERT GAS REGULATORS

SHIELDING GAS REGULATORS are designed to perform in the same way as the oxygen, acetylene, and hydrogen regulators described in Chapter 3. Review this material for additional understanding.

Inert gas regulators have either a gauge or a pressure indicator to show the cylinder pressure. Some of them have only a flow meter gauge on the gas delivery side. Fig. 11-6 shows a regulator with a high pressure gauge on the inlet side. On the outlet side, a gauge indicates the volume of gas flowing. This regulator has a constant outlet pressure to the flowmeter of approximately 50 psig (344.7 kPa).

Fig. 11-7 shows a flowmeter and regulator mounted on a cylinder. The fitting used to connect the regulator to the cylinder varies with the kind of gas. These regulators will deliver gas flows up to 60 cu. ft./hr. (28.3 L/min). The flowmeter scales are accurate only if the gas entering them is at

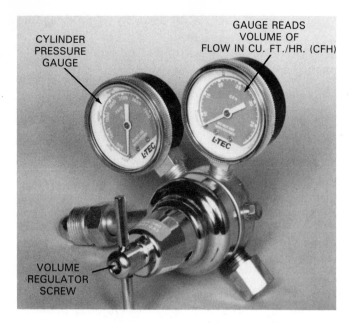

Fig. 11-6. A regulator with a flow rate gauge on the outlet side. The flow rate gauge reads volume of flow in cu. ft./hr. (L-TEC Welding & Cutting Systems)

Fig. 11-7. An argon gas cylinder with the pressure regulator and flowmeter attached. Note that the regulator pressure is preset and seldom needs adjustment. (National Welding Equip. Co.)

approximately 50 psig (344.7 kPa). If higher inlet pressures are used, the gas flow rate will be higher than the actual reading, and the reverse is true if the inlet pressure is lower than 50 psig (344.7 kPa). It is therefore important to use accurately adjusted regulators. Fig. 11-8 illustrates a two stage regulator for argon gas. The gauge is a high pressure gauge and is used to indicate the pressure in the cylinder.

## 11-7. FLOWMETERS AND GAS MIXERS

The amount of gas around the arc can best be measured by the VOLUME of gas coming out of the

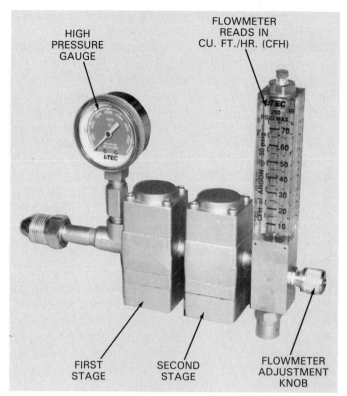

Fig. 11-8. A two-stage regulator with a flowmeter. The gas is fed to the flowmeter at 50 psig (344.7 kPa). This flowmeter has four separate calibrated gauges to accurately show the flow of argon, helium, carbon dioxide, or nitrogen. (L-TEC Welding & Cutting Systems)

nozzle, rather than the pressure of the gas. Therefore the shielding gas system is usually equipped with a flowmeter that is calibrated in cubic feet per hour (cfh) or liters per minute (L/min).

In one type of flowmeter, a tapered plastic or glass tube contains a loosely fitted ball. As the gas flows up the tube, it passes around the ball and lifts the ball. The more gas that moves up the tube, the higher the ball is lifted. Fig. 11-9 shows a cross section of a tapered tube flowmeter.

The tube and return gas housing are either clear plastic or glass. Some have a metal protecting cover. The joint between plastic tubes and body must be gastight. The scale on the inner tube is usually calibrated from 0 cu. ft./hr. (cfh) to 60 cfh (0-28.3 L/min). The flowmeter scale is usually read by aligning the top of the ball with the cfh or L/min reading desired.

For an accurate reading, it is important that this type instrument be mounted in a vertical position. Any slant will cause an inaccurate reading.

Because gas densities vary, it is necessary to use different flowmeters for different gases.

Some universal type flowmeters are available with a 0-100 scale on the flow tube. A chart accompanying the flowmeter will indicate the flow

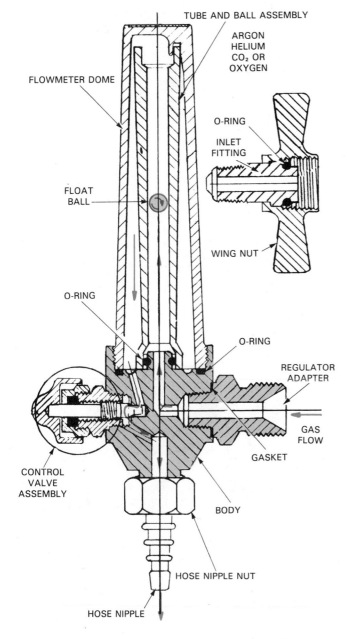

Fig. 11-9. A cross section of a floating ball type flowmeter.

of gases in cfh or L/min for various density gases.

Gas volume is related to pressure if the orifices and gas passages remain constant in size and shape. Therefore a gas pressure gauge can be used as a volume gauge when the scale is calibrated as shown in Fig. 11-6.

Most flowmeters have a needle valve to turn the gas flow on and off and to control the volume of the gas flow. Because this valve controls the volume, and because the pressure to the valve is constant, the valve orifice and the needle must be accurate and in good condition at all times. Any abuse of the needle and/or seat will result in erratic volume feeds. Also, if the packing around the needle or the needle threads are abused, the nee-

dle will not stay in adjustment. It is necessary that this valve be handled carefully. It should not be forced. FINGERTIP HANDLING ONLY is recommended.

It is sometimes necessary to use mixtures of inert gases in GTAW or GMAW. Such mixing must be done with great accuracy. Gases are mixed by percentages of volume. Each gas to be mixed is fed to a separate flowmeter where the desired volume of each gas is set. These gases are then mixed by volume in a mixing chamber and fed to the welding torch through a final flowmeter.

## 11-8. ELECTRODE LEADS AND HOSES USED FOR GTAW

A water cooled gas tungsten arc welding (GTAW) torch generally has two hoses and one combination hose and electrode lead connected to it. These hoses may be made of rubber or plastic materials. One hose carries the shielding gas to the torch. The second is a combination water hose and electrode lead. The electrode lead is a woven metal tube with excellent current carrying capabilities. This electrode lead is surrounded by a rubber or plastic tube. Welding current travels through the woven metal tube while water travels through the center of this tube. The third hose carries the water from the torch to the water drain. In a closed water system the water is returned to the water cooling device and recirculated, Fig. 11-10.

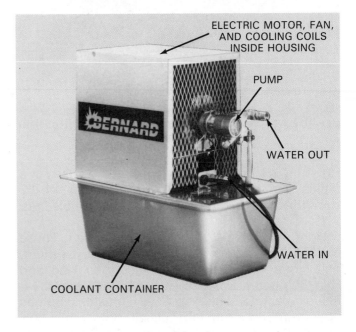

Fig. 11-10. A water circulating and cooling unit used with closed circuit, water cooled systems. The coolant container comes in larger sizes. This unit can be used with GTAW, GMAW, PAW, and resistance and electroslag welding. (Bernard Welding Equipment Co.)

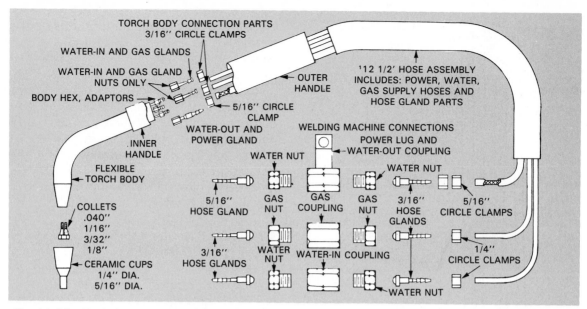

Fig. 11-11. Exploded view of a gas tungsten arc welding torch and hose assembly. This torch has a flexible body which can be bent to suit the job. (Falstrom Co.)

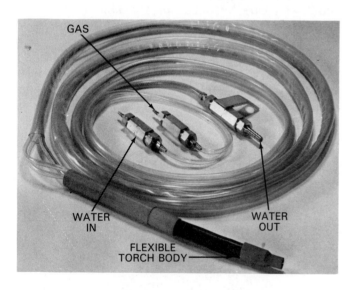

Fig. 11-12. Complete torch, hose, and electrode lead assembly. This is a flexible body torch which is water cooled. (Falstrom Co.)

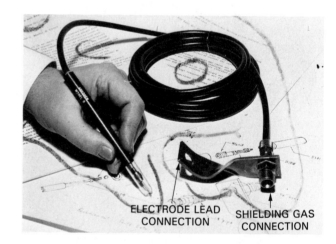

Fig. 11-13. A pen size gas tungsten arc welding torch. In use, the operator would wear gloves. (Argopen)

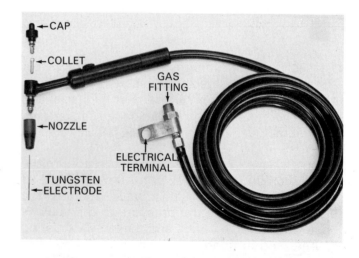

Fig. 11-14. Air cooled, gas tungsten arc welding torch. A short cap and electrode permit use of this torch in difficult to reach places. (Airco Welding Products, Inc.)

Fig. 11-11 shows the hoses and fittings used in a typical water cooled GTAW torch. Fig. 11-12 shows a similar torch completely assembled for use.

Light duty GTAW torches are air cooled. These torches have only one hose connected to them. This hose is a combination electrode lead and shielding gas hose. The electrode lead may be a flexible cable or a woven tube. The shielding gas flows through a rubber or plastic tube which surrounds the electrode lead. Fig. 11-13 shows an extremely light duty GTAW torch which is air cooled. Fig. 11-14 illustrates a light duty air cooled GTAW torch. Fig. 11-15 shows an air cooled torch with a separate gas hose and electrode lead.

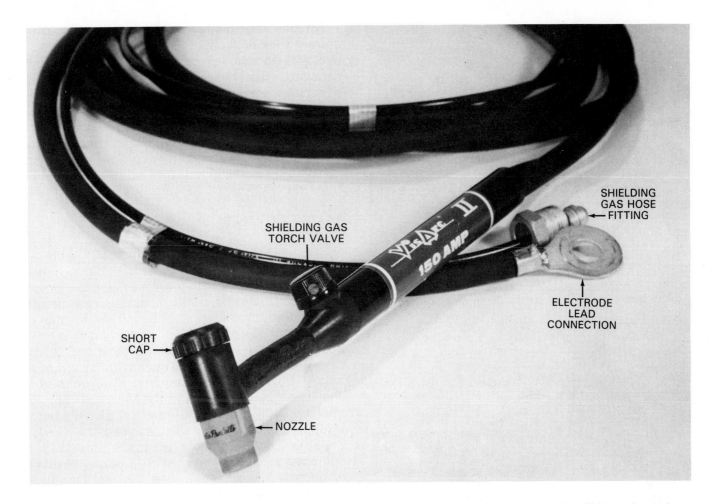

Fig. 11-15. An air cooled GTAW torch which uses a short cap and stub length tungsten electrodes. This torch can be used in confined spaces. The valve allows the welder to control the shielding gas flow at the torch. Notice the separate gas hose and electrode lead. (Air Products & Chemicals, Inc.)

## 11-9. GTAW TORCHES

GAS TUNGSTEN ARC WELDING ELECTRODE TORCHES come in many sizes, amperages, capacities, and shapes. Two basic models are available:

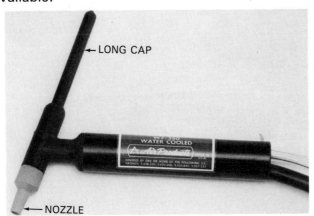

Fig. 11-16. This GTAW torch has a long cap and uses 6 or 7 in. (152.4 or 177.8 mm) long tungsten electrodes. (Air Products & Chemicals, Inc.)

1. For short length tungsten electrodes (stub) as shown in Fig. 11-15.
2. For tungsten electrodes of 6 to 7 in. (152.4 to 177.8 mm) length as shown in Fig. 11-16.

Fig. 11-14 shows the construction of a typical GTAW torch. Tungsten electrodes are held in the collet or chuck. Electrode collets are made with various inside diameters. This is necessary to accommodate the various electrode diameters produced. Most collets are tightened by turning in the electrode cap. The electrode collet makes electrical contact within the torch. Shielding gas flows around the collet and into the nozzle. The nozzle is attached to the torch by a press fit or threads.

The collet used to hold the tungsten electrode is usually copper or stainless steel. Fig. 11-17 shows a copper collet being installed in a light duty GTAW torch.

Some of the torches have a manually operated switch on the torch to operate the gas solenoid valve, as shown in Fig. 11-18. The valve should be opened before the arc is struck, and it should be kept open for a few moments after the arc is

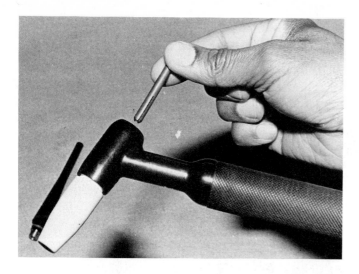

Fig. 11-17. A copper electrode collet being installed in a GTAW torch. Notice the long electrode cap in the background.

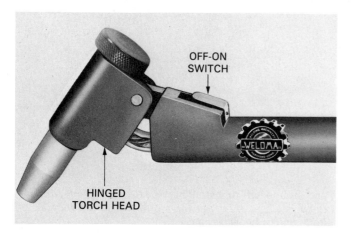

Fig. 11-18. A gas tungsten arc welding torch which has a solenoid valve switch that controls the flow of the shielding gas.

broken (until the tungsten electrode cools) to prevent oxidation of the tungsten electrode. The gas, electrical, and water connections must be clean and tight.

Torches vary in size according to the amount of current they are to handle. Air cooled torches are available in capacities up to about 150 amp capacity. Torches of the higher amperage capacities are usually water cooled.

Gas tungsten arc welding torches are also made for automatic welding. This type torch, shown in Fig. 11-19, has an appearance similar to an automatic gas cutting torch. Fig. 11-20 is a GTAW torch designed for automatic welding. Most production torches are water cooled.

## 11-10. NOZZLES

GAS NOZZLES used for gas tungsten arc welding torches are made in a variety of heat resistant materials. They are available in various diameters, lengths, and shapes. The length, shape, and diameter are chosen to fit the requirements of each job. If the job is hard to reach because of close fitting parts, a short nozzle may be used. If the weld joint is deep within a structure a long nozzle may be required. The volume of shielding gas required and the width of the joint opening will determine the nozzle diameter. Various shapes have been designed for nozzles. These shapes are in most cases designed to decrease gas turbulence. With a nozzle designed to decrease turbulence, a welder may be able to extend the electrode up to 1 in. (approximately 25 mm) from the nozzle without gas turbulence. With the electrode extended that far from the end of the nozzle, the weld may be more easily seen. With a large electrode extension, more shielding gas is required to shield the weld properly.

Nozzles used on gas tungsten arc welding torches are usually made of ceramic materials, chrome plated steel, plastic, or pyrex glass.

Nozzles must be fastened to the torch body with a leakproof joint. They must have the correct gas flow opening. The nozzles must also be centered

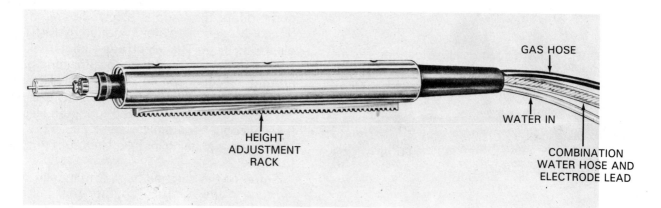

Fig. 11-19. A gas tungsten arc welding torch designed for automatic welding. It can be mounted on motorized trackers or climbers. (Tec Torch Co., Inc.)

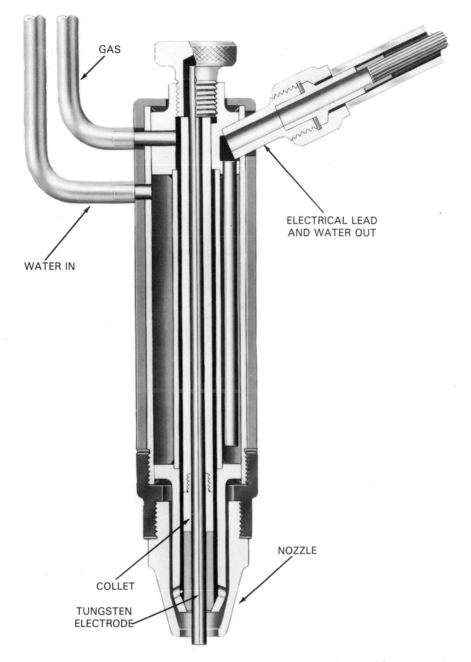

Fig. 11-20. A cross section of a water cooled gas tungsten arc welding torch designed for automatic welding. This torch is designed with a capacity of up to 500 amps.　(Weldma Co.)

around the electrode to provide a concentric gas flow. Nozzles are also called "cups," however, "nozzle" is preferred by the American Welding Society.

Ceramic nozzles are usable up to approximately 250-300 amps. Fig. 11-21 illustrates several different ceramic nozzles. Above this current level, water cooled metal or metal coated ceramic nozzles are needed. Metal nozzles are sometimes made from powdered (sintered) metals.

Plastic type nozzles are made of a transparent plastic with high temperature stability. Pyrex glass nozzles have been used to some extent.

## 11-11. TUNGSTEN ELECTRODES

The ELECTRODES used in gas tungsten arc welding may be one of the following types:
1. Pure tungsten.
2. Tungsten with one or two percent thoria (thorium dioxide).
3. Tungsten with 0.15 to 0.40 percent zirconia (zirconium oxide).
4. Pure tungsten using a core of tungsten with one to two percent thoria. The overall percentage of thoria is .35-.55 percent when the pure tungsten and the core combine.

Fig. 11-21. Ceramic nozzles for gas tungsten arc welding torches. (Diamonite Products)

Tungsten electrodes are supplied in diameters of .010, .020, .040, 1/16, 3/32, 1/8, 5/32, 3/16, and 1/4 in. (.254, .508, 1.02, 1.59, 2.38, 3.18, 3.97, 4.76, and 6.35 mm). These electrodes come in lengths of 3, 6, 7, 12, 18, or 24 in. (76.2, 152.4, 177.8, 304.8, 457.2, 609.6 mm). The surface of tungsten electrodes when purchased is either ground or chemically cleaned.

The chemical composition of tungsten electrodes is shown in Fig. 11-22 from AWS A5.12. In Figs. 11-22 and 11-23 the letters and numbers used may be interpreted as follows:

E - electrode
W - tungsten
P - pure
Th-1 - 1% thoria
Th-2 - 2% thoria
Th-3 - tungsten using a core of tungsten with one to two percent thoria
Zr - 0.15 to 0.40 percent zirconia.

Pure tungsten electrodes are less expensive than those that contain zirconia or thoria. The lower current carrying capacity and ease of contamination of pure tungsten makes it the least desirable for critical jobs.

Tungsten electrodes with one or two percent thoria added are most desirable. These thoriated electrodes carry a higher current and have a greater resistance to contamination. It is easier to strike the arc and maintain a stable arc with thoriated tungsten electrodes.

| AWS Classification | Defined | Color |
|---|---|---|
| EWP | pure tungsten | green |
| EWTh-1 | 1% thoria | yellow |
| EWTh-2 | 2% thoria | red |
| EWTh-3 | 0.35-0.55% thoria | blue |
| EWZr | 0.15-0.40% zirconia | brown |

Fig. 11-23. The AWS color code applied to various types of tungsten electrodes.

Zirconia added to tungsten electrodes gives the electrode qualities which fall somewhere between both pure tungsten and tungsten with thoria added.

The AWS color codes for tungsten electrodes are shown in Fig. 11-23.

See Chapter 12 for a table of suggested currents to use with various types and diameters of tungsten electrodes.

## 11-12. CARE OF TUNGSTEN ELECTRODES

The electrode must be straight. If it is off center in the nozzle, the weld may become contaminated. As the electrode is used, it may become brittle up to 3/8 in. (approximately 10 mm) back from the arc end. If the electrode becomes brittle or contaminated (dirty), that part of the electrode should be broken off. A pair of pliers may be used to break the electrode off easily

| AWS Classification | Tungsten, min, percent (by difference) | Thoria, percent | Zirconia, percent | Total other elements max, percent |
|---|---|---|---|---|
| EWP | 99.5 | — | — | 0.5 |
| EWTh-1 | 98.5 | 0.8 to 1.2 | — | 0.5 |
| EWTh-2 | 97.5 | 1.7 to 2.2 | — | 0.5 |
| EWTh-3[a] | 98.95 | 0.35 to 0.55 | — | 0.5 |
| EWZr | 99.2 | — | 0.15 to 0.40 | 0.5 |

a. A tungsten electrode with an integral lateral segment throughout its length which contains 1.0 to 2.0 percent thoria. The average thoria content of the electrode shall be as specified in this table.

Fig. 11-22. The chemical composition of various AWS classifications for tungsten electrodes. (AWS A5.12-80)

to form a new clean arc end.

Tungsten electrodes must be clean and must have good electrical contact with the collet. The tungsten should be adjusted to extend about 1/8 in. (about 1.6 mm) beyond the end of the cup. It is extremely important that the shielding gas hose connections be tight to prevent air or moisture from mixing with the shielding gas. Such contamination of the shielding gas would be harmful to the weld and to the electrode.

Fig. 11-24 shows the proper way to break and retip a tungsten electrode for best results. If the retipping is not properly done, the tungsten electrode may shatter and/or split.

Conditioning the electrode end for DC welding is very important. Refer to Heading 12-7 on selecting and preparing a tungsten electrode for additional information.

## 11-14. GAS METAL ARC WELDING STATION

The typical GAS METAL ARC WELDING (GMAW) outfit contains the following equipment and supplies:
1. A constant voltage AC or DC arc welding machine.
2. An electrode wire feed mechanism.
3. Shielding gas cylinders. See Heading 11-5.
4. A gas regulator. See Heading 11-6.
5. A shielding gas flowmeter. See Heading 11-7.
6. Shielding gas, coolant hoses and fittings, and the electrode lead.
7. A GMAW torch.
8. Metal electrode wire.
9. Optional equipment.
   a. A coolant system. See Fig. 11-10.
   b. Remote controls.

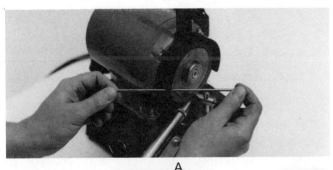
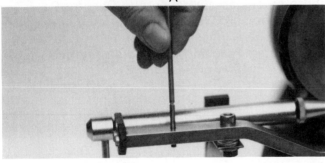

Fig. 11-24. Three steps required to prepare a tungsten electrode for welding. A—Notching the electrode prior to breaking. B—Breaking the electrode in special fixture. C—Grinding the electrode in a special grinder.
(Ind. Schweisstechnik E. Jankus)

## 11-13. FILLER METALS USED WITH GTAW

Gas tungsten arc welding is used to weld almost all weldable metals. The choice of filler metals therefore is very large.

Fig. 11-25 lists the American Welding Society (AWS) filler metal specifications booklets for each type of base metal welded. Each ''A5'' specification booklet contains a list of filler metals, their suggested uses, and current settings in addition to their chemical compositions.

Fig. 11-26 is a schematic drawing of a GMAW outfit. A complete station would include the arc welding booth, a ventilation system, and a welding bench. Refer also to Fig. 1-12.

## 11-15. GAS METAL ARC WELDING MACHINES

Gas metal arc welding (GMAW) is performed with rectified AC or DC machines. Direct current electrode positive (DCEP) or (DCRP) is the most popular and practical current used with GMAW.

| AWS SPECIFICATION NUMBER | BOOKLET TITLE |
|---|---|
| A5.2 | Iron and Steel Gas-Welding Rods. |
| A5.7 | Copper and Copper Alloy Bare Welding Rods and Electrodes. |
| A5.9 | Corrosion-Resisting Chromium and Chromium-Nickel Steel Bare and Composite Metal Cored and Stranded Arc Welding Electrodes and Welding Rods. |
| A5.10 | Aluminum and Aluminum Alloy Welding Rods and Bare Electrodes. |
| A5.13 | Surfacing Welding Rods and Electrodes. |
| A5.14 | Nickel and Nickel Alloy Bare Welding Rods and Electrodes. |
| A5.16 | Titanium and Titanium Alloy Bare Welding Rods and Electrodes. |
| A5.18 | Mild Steel Electrodes for Gas Shielded Arc Welding. |
| A5.19 | Magnesium-Alloy Welding Rods and Bare Electrodes. |
| A5.21 | Composite Surfacing Welding Rods and Electrodes. |
| A5.24 | Zirconium and Zirconium Alloy Bare Welding Rods and Electrodes. |

Fig. 11-25. A list of AWS specifications and booklets for filler metals which may be used in gas tungsten arc welding (GTAW).

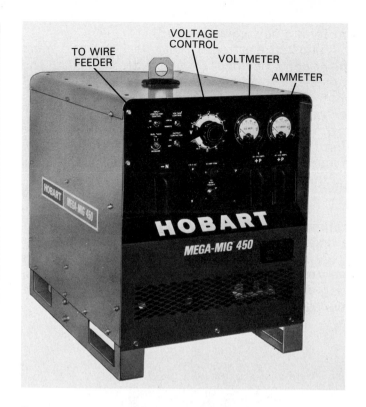

Fig. 11-27. A constant voltage DC arc welder designed for GMAW. (Hobart Bros. Co.)

Regardless of which type current is used, the machine must be manufactured to produce a CONSTANT VOLTAGE.

GMAW may be accomplished with the following types of machines:
1. AC generator or alternator with a DC rectifier.
2. AC transformer with a DC rectifier, as shown in Fig. 11-27.

For gas metal arc welding, AC has not been used successfully.

The construction, operation, and control of these machines are covered in Chapter 9.

Gas metal arc welding transformers are normally three-phase machines. Small single-phase constant voltage machines are produced in capacities of 200 amperes or below.

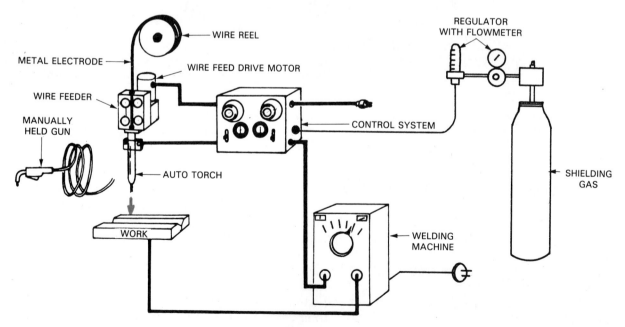

Fig. 11-26. A schematic drawing of a combination manual and automatic gas metal arc welding outfit.

The characteristics of the volt-ampere curve for a constant voltage machine is seen in Fig. 11-28. This type of curve results in a relatively constant voltage with large changes in current. This curve illustrates that a small change of only 6 volts from 18 volts to 24 volts causes a decrease in amperage from 300 amps to 200 amps. This type volt-ampere curve is most desirable when a constant feed electrode process is used.

Constant feed electrode processes include gas metal arc welding (GMAW), flux cored arc welding (FCAW), and submerged arc welding (SAW). Fig. 11-29 lists the types of constant voltage machines used with various welding processes. A constant voltage power supply used with a constant speed wire feed produces a system with a self correcting arc length.

The greatest voltage drop or circuit resistance occurs as the arc length changes. Referring to Fig. 11-28 again, it may be seen that a small change in voltage will result in a large current change. The volt-ampere constant voltage curve for machines used for GMAW of nonferrous metals is shown as curve A on Fig. 11-30. This same curve is used for

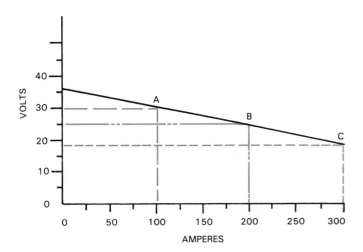

Fig. 11-28. Characteristic volt-ampere curve for a constant voltage arc welder. A 100 amp change from 200 amperes to 300 amperes results from a voltage change of only 6 volts from 24 volts to 18 volts.

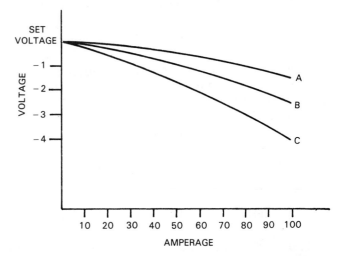

Fig. 11-30. Volt-amperage curves for various welding machine slope settings. Curve A—1 1/2 to 2 volts per 100 amperes for GMAW of nonferrous metals. Curve B—2 to 3 volts per 100 amperes for GMAW with $CO_2$. Curve C—3 to 4 volts per 100 amperes for GMAW short circuiting arc transfer.

| WELDING PROCESS | DC Constant Voltage | AC Constant Voltage |
|---|---|---|
| Carbon arc welding, cutting and gouging (AAC) | yes | yes |
| Electrogas (EGW) | yes | no |
| Electroslag welding (EW) | yes | 1 |
| Flux cored arc welding (FCAW) | — | — |
|   with external shielding gas | yes | no |
|   without shielding gas (self shield) | yes | no |
| Gas metal arc welding (GMAW): | — | — |
|   Inert gas (MIG)-nonferrous | yes | no |
|   Spray arc (argon-oxygen) | yes | no |
|   $CO_2$ (big wire) | yes | no |
|   Shorting arc (fine wire) | yes | no |
| Gas tungsten arc welding (GTAW) or TIG | no | no |
| Plasma arc welding (PAW) | no | no |
| Shielded metal arc welding (SMAW) stick | no | no |
| Stud arc welding (SW) | yes | no |
| Submerged arc welding (SAW) | yes | yes |

1. Used in some applications.

Fig. 11-29. A list of welding processes indicating which use direct current constant voltage (DC-CV) or constant voltage alternating current (AC) power sources.

submerged arc, and flux cored welding with large electrodes. The slope is 1 1/2 to 2 volts per hundred amperes. A medium slope of 2 to 3 volts per hundred amperes is used for GMAW with carbon dioxide ($CO_2$) gas and small flux cored wires as shown as curve B in Fig. 11-30. A steeper slope, curve C, in Fig. 11-30, has a slope of 3 to 4 volts per hundred amperes. This steeper slope is recommended for short circuiting arc transfer.

The constant voltage arc welding machine is wired so that as the arc voltage changes the current output of the machine changes.

If the arc length shortens slightly, and decreases the voltage by 1 1/2 volts, the amperage may increase 100 amps. See Fig. 11-30, curve A. The change in arc length will cause an increase in amperage and an increase in the electrode burn-off rate. The increase in burn-off rate will quickly bring the arc length back to normal.

If the arc length increases, the opposite actions occur. The arc voltage will increase, the current output will decrease, and the burn-off rate will decrease. The arc length will shorten and return to normal.

## 11-16. WIRE FEEDERS USED WITH GMAW

The wire electrode used in gas metal arc welding comes in large coils (spools) or drums. These wire coils or drums contain hundreds of feet (meters) of wire.

Small coils of wire are mounted on free turning axles near the wire feed mechanism. If the larger drums of wire are used, the drum is placed near the wire feeder. The wire is smoothly drawn from either the coil or drum.

Wire feeders consist of a coil mounting device, a set of drive wheels for the wire, and an adjustable, constant speed motor to turn the wire drive wheels. Fig. 11-31 shows a wire feeder with the drive wheels exposed. Both drive wheels have gear teeth on the outside circumference. The driven wheel moves the second wheel by means

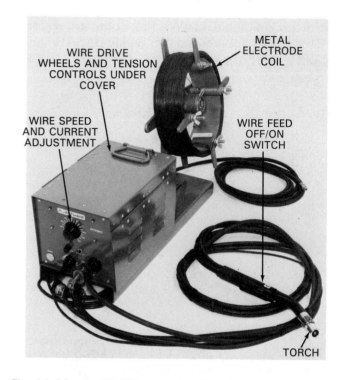

Fig. 11-32. A GMAW wire feed mechanism. The drive motor, drive wheels, and wire tensioning device are within the control box. (Air Products & Chemicals, Inc.)

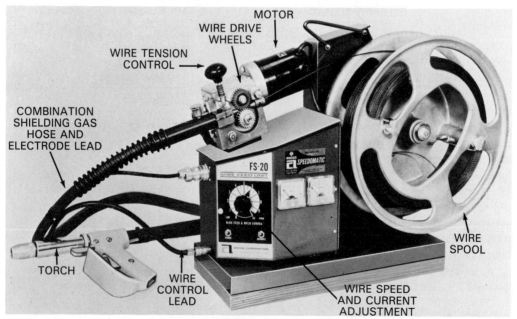

Fig. 11-31. A gas metal arc welding wire drive unit. This photo shows the wire spool, wire feed rollers, electrode lead, control lead, controls, and torch. (Arcos Corp.)

of the gear teeth on each. One of the drive wheels is driven by the adjustable, DC type, constant speed motor. The tension on the wire passing between these wheels is adjusted by moving the wheels closer together. This is done by turning down on a tensioning device.

Fig. 11-32 illustrates a wire drive mechanism with the GMAW torch attached. Fig. 11-33 shows a wire drive which has a digital read-out for the wire feed speed and voltage.

A switch on the wire drive control panel is used to rotate the drive motor slowly. This is done to move the wire slowly through the electrode lead to the torch. The wire is moved slowly so that it does not bend or kink within the electrode cable. The switch on the wire drive control panel is called an "inch switch." Most wire feeders have an inch switch as shown in Fig. 11-33.

The shielding gas hose may have air in it prior to its first use after a long shut-down period. To clear or purge the hose and torch, the shielding gas is turned on for a short period prior to welding. A purge switch is provided for this purpose on the wire feed control box.

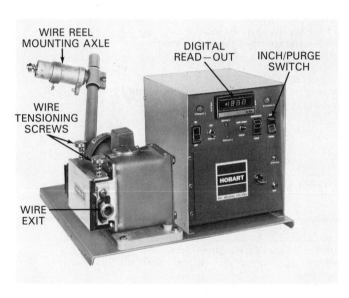

Fig. 11-33. A wire drive unit with a digital read-out for wire speed and voltage. Notice the inch/purge switch on this control. (Hobart Bros. Co.)

## 11-17. GMAW SHIELDING GASES

The shielding gases used with gas metal arc welding (GMAW) may be inert, reactive, or mixtures of both types of gases.

Inert gases used are argon (Ar) and helium (He). These gases, being INERT, will not react with other chemical elements. REACTIVE GASES will react with other chemical elements. Used properly, reactive gases will not cause defects in welds.

The reactive gases used in GMAW are carbon dioxide ($CO_2$) and oxygen ($O_2$). Hydrogen (H) and nitrogen ($N_2$) are used, but only in highly specialized applications. Their use results in better control of penetration. Hydrogen and nitrogen will cause embrittlement and porosity in the welds on most metal. These gases have been used, however, in a few highly specialized applications.

Mixtures of argon and helium; argon and oxygen; argon and carbon dioxide; and helium, argon and carbon dioxide are used. The gas or gas mixture used will be determined by the metal being welded and the type of arc transfer method desired. See Headings 13-3, 13-4, 13-5, and 13-6 for an explanation of metal transfer methods. Figs. 11-34 and 11-35 list metals and suggested gas mixtures for use with GMAW spray and short circuiting transfer.

## 11-18. ARGON AND HELIUM GASES USED IN GMAW

Pure argon (Ar) and helium (He) gas are excellent for protecting the arc, metal electrode, and weld metal from contamination. They are not, however, as suitable for some GMAW processes as mixtures of gases. Gas mixtures seem to create arc stability, reduce spatter, and improve the bead contour.

Reactive gases like carbon dioxide, oxygen, and nitrogen are not practical to use alone as shielding gases. Carbon dioxide is the exception. It is inexpensive and works well on carbon and low alloy steels. Carbon dioxide generally costs about one tenth as much as pure argon gas.

Argon and helium are the gases generally used with nonferrous metal as seen in Figs. 11-34 and 11-35. Helium conducts heat better in the arc than argon. Helium is used when high heat input is required in a welding application. Helium gas, therefore, is chosen for use on thick metals. It is also used on metals like copper and aluminum which conduct heat away from the weld area rapidly. When welding thin metal and metals which conduct heat poorly, argon is a good choice. Argon is often used to weld out-of-position because of its lower heat conductivity. Because argon is ten times heavier than helium, it shields better. Less gas is required to provide a good shield; therefore, argon is cheaper to use.

The weld bead contour and penetration are also affected by the gas used. Welds made with argon generally have deeper penetration. They also have a tendency to undercut at the edges. Welds made with helium generally have wider and thicker beads. Fig. 11-36 shows the shape of welds made with various gases and gas mixtures.

| METAL | SHIELDING GAS | ADVANTAGES |
|---|---|---|
| Aluminum | Argon | 0.1 in. (0.25 mm) thick; best metal transfer and arc stability; least spatter. |
| | 75% Helium-25% argon | 1-3 in. (25-76 mm) thick; higher heat input than argon. |
| | 90% Helium-10% argon | 3 in. (76 mm) thick; highest heat input; minimizes porosity. |
| Copper, nickel, & their alloys | Argon | Provides good wetting; good control of weld pool for thickness up to 1/8 in. (3.2 mm) |
| | Helium-argon | Higher heat inputs of 50 and 75% helium mixtures offset high heat conductivity of heavier gages. |
| Magnesium | Argon | Excellent cleaning action. |
| Reactive metals (Titanium, Zirconium, Tantalum) | Argon | Good arc stability; minimum weld contamination. Inert gas backing is required to prevent air contamination on back of weld area. |
| Steel, carbon | Argon-3-5% oxygen | Good arc stability; produces a more fluid and controllable weld pool; good coalescence and bead contour, minimizes undercutting; permits higher speeds, compared with argon. |
| Steel, low alloy | Argon-2% oxygen | Minimizes undercutting; provides good toughness. |
| Steel, stainless | Argon—1% oxygen | Good arc stability; produces a more fluid and controllable weld pool, good coalescence and bead contour, minimizes undercutting on heavier stainless steels. |
| | Argon-2% oxygen | Provides better arc stability, coalescence, and welding speed than 1% oxygen mixture for thinner stainless steel materials. |

Fig. 11-34. Suggested gases and gas mixtures for use in GMAW spray transfer.

| METAL | SHIELDING GAS | ADVANTAGES |
|---|---|---|
| Aluminum, copper, magnesium, nickel, and their alloys | Argon and argon-helium | Argon satisfactory on sheet metal; argon-helium preferred on thicker sheet metal. |
| Steel, carbon | Argon-20-25% $CO_2$ | Less than 1/8 in. (3.2 mm) thick; high welding speeds without melt-through; minimum distortion and spatter; good penetration. |
| | Argon-50% $CO_2$ | Greater than 1/8 in. (3.2 mm) thick; minimum spatter; clean weld appearances; good weld pool control in vertical and overhead positions. |
| | $CO_2$[a] | Deeper penetration; faster welding speeds; minimum cost. |
| Steel, low alloy | 60-70% Helium-25-35% argon-4-5% $CO_2$ | Minimum reactivity; good toughness; excellent arc stability, wetting characteristics, and bead contour; little spatter. |
| | Argon-20-25% $CO_2$ | Fair toughness; excellent arc stability; wetting characteristics, and bead contour; little spatter. |
| Steel, stainless | 90% Helium-7.5% argon-2.5% $CO_2$ | No effect on corrosion resistance; small heat-affected zone; no undercutting; minimum distortion; good arc stability. |

a - $CO_2$ is used with globular transfer also.

Fig. 11-35. Suggested gases and gas mixtures for use in GMAW short circuiting transfer.

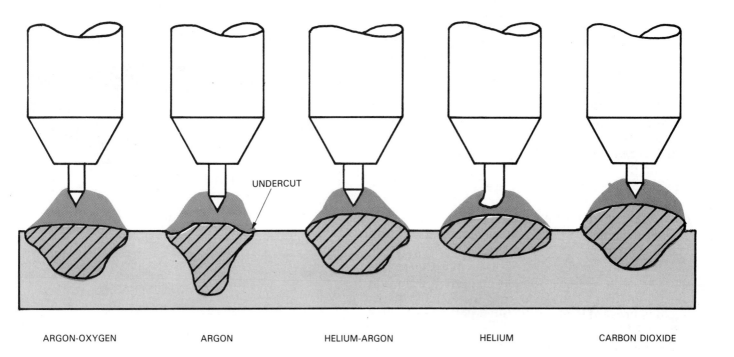

ARGON-OXYGEN     ARGON     HELIUM-ARGON     HELIUM     CARBON DIOXIDE

Fig. 11-36. GMAW bead shapes and depth of penetration for various gases and gas combinations. These shapes are typical for DCEP (DCRP). Note deep penetration and undercut with pure argon.

Argon used with the gas metal arc spray transfer process tends to produce deeper penetration through the centerline of the bead. GMAW spray transfer occurs more easily in argon gas than in helium. The GMAW globular and short circuiting arc metal transfer methods produce wider beads with shallower penetration.

## 11-19. GAS MIXTURES USED IN GMAW

Welds made with pure argon have deep penetration. On carbon and low alloy steels the filler metal tends to draw in from the toe of the weld (weld edge) when argon is used as the shielding gas. This can cause undercutting. To overcome this tendency of the filler metal not to flow out to the edges of the weld, a gas mixture is used. For steels, oxygen ($O_2$) or carbon dioxide ($CO_2$) are added to argon. This addition of $O_2$ or $CO_2$ to argon tends to cause better metal transfer and flow. It also reduces metal spatter and stabilizes the arc.

As little as 0.5 percent oxygen will cause noticeable improvements. Additions of 1 to 5 percent oxygen or 3 to 10 percent carbon dioxide are usually used.

Adding oxygen or carbon dioxide to argon or helium causes the shielding gas to become slightly oxidizing. This may cause porosity in some ferrous metals. To offset this oxidizing tendency, a deoxidizer is added to the electrode wire.

## 11-20. CARBON DIOXIDE GAS

Carbon dioxide ($CO_2$) reduces to carbon monoxide (CO) and oxygen ($O_2$) in the arc. However, these gases return to $CO_2$ as they cool. A 25 percent higher current is used with $CO_2$ than with argon or helium. This causes more agitation of the weld pool. This agitation makes it easier for the entrapped gases to rise to the surface of the metal, thus reducing weld porosity.

Direct current electrode positive (DCEP) or DCRP is generally used with $CO_2$. Carbon dioxide gas used with direct current electrode negative (DCEN), or DCSP, causes larger drops in the metal transfer, more electrode spatter, and a more unstable arc. Carbon dioxide may be used with DCEN when the electrode wire is treated with cesium and sodium. Such wire treatment results in a more stable arc and permits a spray transfer with DCEN. Cesium and sodium treated wire is expensive and, therefore, seldom used.

Moisture free carbon dioxide must be used or the hydrogen generated while welding will cause weld porosity and brittleness.

Carbon dioxide is furnished in liquid form in 50 lb. (22.7 kg) cylinders. These cylinders are approximately 9 in. (228.6 mm) in diameter and 51 in. (1.30 m) high and weigh 105 lbs. (47.6 kg), when empty. Each pound (.45 kg) of liquid $CO_2$ will furnish 8.7 cu. ft. (.25 m³) of gas which is equal to 435 cu. ft. (12.3 m³) per cylinder. It must change from a liquid to a gas as it is being used. This change from liquid to gaseous $CO_2$ is dependent on the room temperature. When the delivery line from the liquid tank is opened, carbon dioxide gas "boils" or bubbles out of the liquid. The expansion of the gas as it leaves the liquid and passes through the regulator causes the $CO_2$ gas to cool. If moisture is present, it may condense and freeze in the regulator, causing a blocking of the gas passage. Excessive moisture may also be indicated by erratic flowmeter operation. It is recommended that $CO_2$ with a $-20\,°F$ ($-29\,°C$) dew point or lower be used.

One cylinder can only furnish about 35 cfh (16.5 L/min) when the cylinder is in a 70 °F (21 °C) room. Sometimes two or more cylinders must be connected in parallel to furnish enough gas for welding. The pressure in the cylinder when liquid is present is about 835 psig at 70 °F (5757 kPa at 21 °C).

Carbon dioxide is also obtainable in the gaseous form in cylinders. These cylinders are similar to the oxygen cylinders described in Chapter 3.

Because carbon dioxide is 50 percent heavier than air, its ability to shield the arc is quite satisfactory.

Carbon dioxide has a rather high electrical resistance and it, therefore, has a rather critical arc length. Even small changes in the arc length will produce a wild arc and spattering. A very short and constant arc length must be maintained. Gas metal arc welding with $CO_2$ lends itself particularly well to automatic machine welding. However, many successful manual applications are also in regular use.

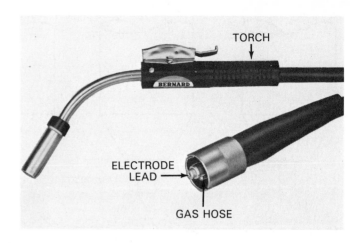

Fig. 11-37. An air cooled GMAW torch. The gas hose and electrode lead are built into a combination hose and lead. (Bernard Welding Equipment Co.)

## 11-21. THE GMAW TORCH

A typical air cooled gas metal arc welding torch and its electrode lead and gas hose are shown in Fig. 11-37. The gas hose, the electrode lead, and the electrode wire carrier are all built into one large hose. The torch has a fully insulated handle, a trigger switch, a current contact tube, and a nozzle. The trigger switch turns on the wire feed and starts the shielding gas flowing. The nozzle controls the flow of shielding gas around the electrode wire and the weld area. Various sizes and shapes of nozzles are available. At the end of the torch and underneath the nozzle is the current contact tube. Current flows from the current contact tube to the electrode wire as the wire passes through

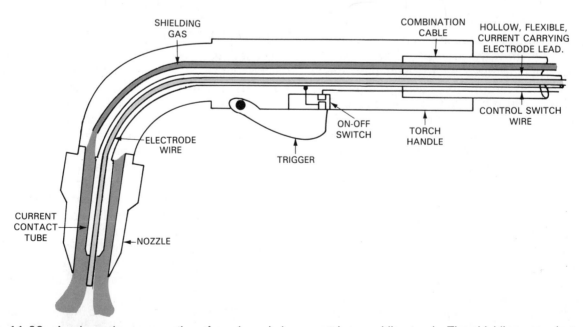

Fig. 11-38. A schematic cross section of an air cooled gas metal arc welding torch. The shielding gas, electrode wire, and control switch are carried in a combination cable. The electrode wire runs through the hollow electrode lead.

the tube. The inside diameter of the current contact tube must match the diameter of the electrode wire being used. Several replaceable sizes are available for use with each wire size used in the torch. Fig. 11-38 shows the main parts of an air cooled GMAW torch.

Gas metal arc welding torches, used with argon or helium, are generally air cooled up to 200 amps. Carbon dioxide GMAW torches may be used for intermittent duty up to 600 amps. Figs. 11-39 and 11-40 show air cooled GMAW torches and their combination electrode lead and shielding gas hose.

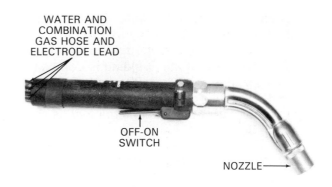

Fig. 11-41. A water cooled GMAW torch. Note the water in, water out, and combination gas and electrode lead used on the water cooled torch. (L-TEC Welding & Cutting Systems)

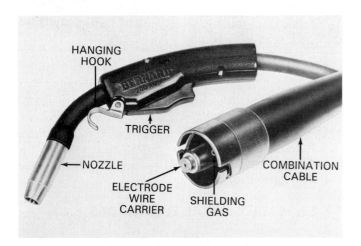

Fig. 11-39. An air cooled GMAW torch. Note how the electrode wire travels through the center or the combination cable. (Bernard Welding Equipment Co.)

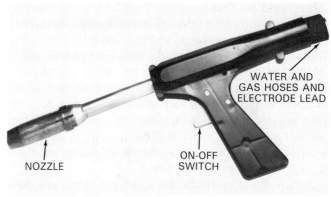

Fig. 11-42. A water cooled GMAW torch. This torch has a pistol grip for added control. (Miller Electric Mfg. Co.)

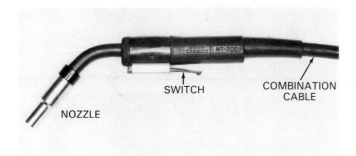

Fig. 11-40. An air cooled GMAW torch. (L-TEC Welding & Cutting Systems)

Torches used with argon and helium above 200 amps and torches used with $CO_2$ above 600 amps or for continuous duty are water cooled. Figs. 11-41 and 11-42 illustrate two different types of water cooled GMAW torches. With a water cooled torch, three hoses are used. A combination tube carries the outgoing cooling water and the electrode lead and carrier. A second hose carries the cooling incoming water to the torch. The third hose carries the shielding gas to the torch.

Gas metal arc welding torches are available which have self contained wire drive units. This torch has an electric motor to drive the wire feed unit and a small coil of wire enclosed in the torch.

There are other torches called pull type welding torches. These torches may be used when welding at distances of over 50 feet (approximately 15 m) from a regular wire drive unit. At such distances the wire in a flexible conduit may kink (bend). To prevent bending the wire, a wire drive is built into the torch. It pulls the wire while the regular wire drive pulls or pushes it.

Gas metal arc welding torches are usually used with higher amperages and temperatures than are used in gas tungsten arc welding. The nozzles used with GMAW torches are usually made from metal. Powdered metal is sometimes used. The exterior of each GMAW nozzle is generally brightly chromed. This is done to reflect heat and keep the nozzle cooler. The shape and diameter of the nozzle vary with the joint being welded and the type of gas being used.

Torches used for automatic GMAW have a straight body and nozzle. They do not have a handle. They are firmly attached to the automatic machine. Since automatic welding is done almost continuously, this torch is water cooled. Electrode wire and shielding gas is fed through the torch body to the arc area in the same manner as shown in Fig. 11-38.

## 11-22. SMOKE EXTRACTING TORCH SYSTEMS

The GMAW process often generates a great deal of smoke. When doing GMAW with carbon dioxide ($CO_2$) there is some carbon monoxide (CO) created. In addition to carbon monoxide, some ozone is generated. Both of these gases are toxic. It is especially important to prevent breathing these gases.

Removal of these gases from the weld area is done by means of a GMAW torch equipped with a fume extractor. Fume extractors may be built into the torch as shown in Fig. 11-43. Gas fumes, contaminated air, and air near the weld are drawn away by the torch, filtered, and released to the atmosphere. Fig. 11-44 illustrates a torch with the fume extractor and a smoke collector and filter cabinet system. Fig. 11-45 illustrates a SMA weld being made with a portable fume extractor located to pick up fumes at the weld.

Conventional ventilation systems also are ad-

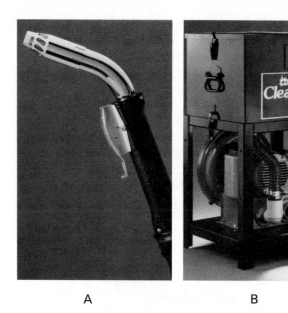

A       B

Fig. 11-44. A GMAW torch with a fume extractor built in. The torch extractor, A, is connected to the smoke collector and filter system shown in B. (Bernard Welding Equipment Co.)

vised to remove contaminated air from any welding operation.

## 11-23. GMAW METAL ELECTRODES

Bare metal electrodes used for gas metal arc welding can be very small in diameter. The

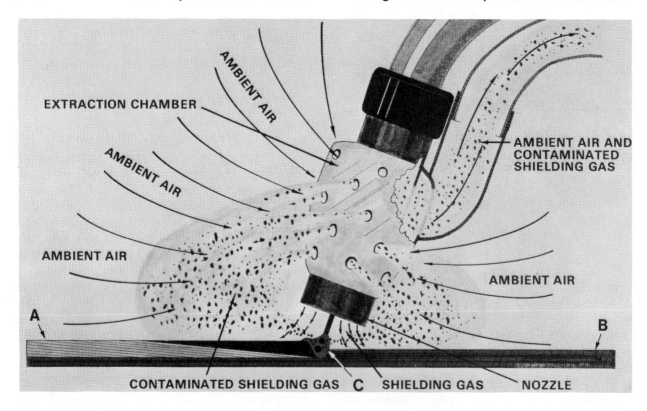

Fig. 11-43. A schematic of a gas metal welding torch equipped with a smoke extracting device. A—Weld metal. B—Base metal. C—Arc crater. (Bernard Welding Equipment Co.)

MOVABLE PICKUP

PORTABLE VENTILATOR AND FILTER

Fig. 11-45. A portable fume extractor mounted on a dolly. This unit can remove fumes from the weld area during on-site repairs. (Nederman, Inc.)

diameters used generally range from .020-.125 in. (.51-3.18 mm). They come in spools of several hundred feet or meters or in drums with up to several thousand feet or meters of wire coiled inside.

A variety of wire compositions are made available. They are packaged in coils and drums so that a large variety of metals can be welded using the GMAW process.

Deoxidizers are added to most electrode wires. This is done to reduce the porosity in the weld. During the welding process, the deoxidizer added to the wire will react with oxygen, nitrogen, or hydrogen which may be in the shielding gases. By reacting with the shielding gases, the deoxidizers reduce the possiblity of these gases producing porosity in the finished weld which would lower the mechanical strength of the weld.

The American Welding Society chemical composition specifications for carbon steel filler metal used with GMAW are shown in Fig. 11-46. Fig. 11-47 lists the AWS filler wire specifications for GMAW low alloy steels.

The numbers and letters used in the AWS classifications in Figs. 11-46 and 11-47 have the following meanings:

Example:  ER70S-2

E - electrode
R - rod

| CHEMICAL COMPOSITION, WEIGHT PERCENT | | | | | | | | | | | | | |
| --- | --- | --- | --- | --- | --- | --- | --- | --- | --- | --- | --- | --- | --- |
| AWS Classification | C | Mn | Si | P | S | Ni[a] | Cr[a] | Mo[a] | V[a] | Cu[b] | Ti | Zr | Al |
| ER70S-2 | 0.07 | 0.90 to 1.40 | 0.40 to 0.70 | 0.025 | 0.035 | | | | | 0.50 | 0.05 to 0.15 | 0.02 to 0.12 | 0.05 to 0.15 |
| ER70S-3 | 0.06 to 0.15 | 0.90 to 1.40 | 0.45 to 0.70 | | | | | | | | — | — | — |
| ER70S-4 | 0.07 to 0.15 | 1.00 to 1.50 | 0.65 to 0.85 | | | | | | | | — | — | — |
| ER70S-5 | 0.07 to 0.19 | 0.90 to 1.40 | 0.30 to 0.60 | | | | | | | | — | — | 0.50 to 0.90 |
| ER70S-6 | 0.07 to 0.15 | 1.40 to 1.85 | 0.80 to 1.15 | | | | | | | | — | — | — |
| ER70S-7 | 0.07 to 0.15 | 1.50 to 2.00 | 0.50 to 0.80 | | | | | | | | — | — | — |
| ER70S-G | No chemical requirements[c] | | | | | | | | | | | | |

a. These elements may be present but are not intentionally added.
b. The maximum weight percent of copper in the rod or electrode due to any coating plus the residual copper content in the steel shall be 0.50.
c. For this classification, there are no chemical requirements for the elements listed, with the exception that there shall be no intentional addition of Ni, Cr, Mo, or V.

Fig. 11-46. Chemical composition specifications for carbon steel filler metal used with GMAW. For exact limitations and more information, see AWS specification A5.18-79.

## CHEMICAL COMPOSITION

| AWS Classification | Carbon | Manganese | Silicon | Phosphorus | Sulfur | Nickel | Chromium | Molybdenum | Vanadium | Titanium | Zirconium | Aluminum | Copper[a] | Total other elements |
|---|---|---|---|---|---|---|---|---|---|---|---|---|---|---|
| Chromium-molybdenum steel electrodes and rods | | | | | | | | | | | | | | |
| ER80S-B2 | 0.07-0.12 | 0.40-0.70 | 0.40-0.70 | 0.025 | 0.025 | 0.20 | 1.20-1.50 | 0.40-0.65 | — | — | — | — | 0.35 | 0.50 |
| ER80S-B2L | 0.05 | 0.40-0.70 | 0.40-0.70 | 0.025 | 0.025 | 0.20 | 1.20-1.50 | 0.40-0.65 | — | — | — | — | 0.35 | 0.50 |
| ER90S-B3 | 0.07-0.12 | 0.40-0.70 | 0.40-0.70 | 0.025 | 0.025 | 0.20 | 2.30-2.70 | 0.90-1.20 | — | — | — | — | 0.35 | 0.50 |
| ER90S-B3L | 0.05 | 0.40-0.70 | 0.40-0.70 | 0.025 | 0.025 | 0.20 | 2.30-2.70 | 0.90-1.20 | — | — | — | — | 0.35 | 0.50 |
| Nickel steel electrodes and rods | | | | | | | | | | | | | | |
| ER80S-Ni1 | 0.12 | 1.25 | 0.40-0.80 | 0.025 | 0.025 | 0.80-1.10 | 0.15 | 0.35 | 0.05 | — | — | — | 0.35 | 0.50 |
| ER80S-Ni2 | 0.12 | 1.25 | 0.40-0.80 | 0.025 | 0.025 | 2.00-2.75 | — | — | — | — | — | — | 0.35 | 0.50 |
| ER80S-Ni3 | 0.12 | 1.25 | 0.40-0.80 | 0.025 | 0.025 | 3.00-3.75 | — | — | — | — | — | — | 0.35 | 0.50 |
| Manganese-Molybdenum steel electrodes and rods | | | | | | | | | | | | | | |
| ER80S-D2 | 0.07-0.12 | 1.60-2.10 | 0.50-0.80 | 0.025 | 0.025 | 0.15 | — | 0.40-0.60 | — | — | — | — | 0.50 | 0.50 |
| Other low alloy steel electrodes and rods | | | | | | | | | | | | | | |
| ER100S-1 | 0.08 | 1.25-1.80 | 0.20-0.50 | 0.010 | 0.010 | 1.40-2.10 | 0.30 | 0.25-0.55 | 0.05 | 0.10 | 0.10 | 0.10 | 0.25 | 0.50 |
| ER100S-2 | 0.12 | 1.25-1.80 | 0.20-0.60 | 0.010 | 0.010 | 0.80-1.25 | 0.30 | 0.20-0.55 | 0.05 | 0.10 | 0.10 | 0.10 | 0.35-0.65 | 0.50 |
| ER110S-1 | 0.09 | 1.40-1.80 | 0.20-0.55 | 0.010 | 0.010 | 1.90-2.60 | 0.50 | 0.25-0.55 | 0.04 | 0.10 | 0.10 | 0.10 | 0.25 | 0.50 |
| ER120S-1 | 0.10 | 1.40-1.80 | 0.25-0.60 | 0.010 | 0.010 | 2.00-2.80 | 0.60 | 0.30-0.65 | 0.03 | 0.10 | 0.10 | 0.10 | 0.25 | 0.50 |
| ERXXS-G | Subject to agreement between supplier and purchaser.[b] | | | | | | | | | | | | | |

a. The maximum weight percent of copper in the rod or electrode due to any coating plus the residual copper content in the steel shall comply with the stated value.
b. In order to meet the requirements of the G classification, the electrode must have as a minimum either 0.50 percent nickel, 0.30 percent chromium, or 0.20 percent molybdenum.

Fig. 11-47. The chemical composition specifications for low alloy steel filler metals used with GMAW. For exact limitations and more information, see AWS specification A5.28-79.

70 - tensile strength in ksi
      (1000 psi)
 S - solid rod
 2 - variations in chemical
      compositions
Examples of dash numbers:
      B2 - chrome-molybdenum steel
     B3L - chrome-molybdenum steel
            with a lower carbon content
   Ni(1-3) - Nickel steel
      D2 - manganese-molybdenum
           steel.

## 11-24. FLUX CORED ARC WELDING (FCAW)

Ferrous metals are being welded with increasing frequency with the FLUX CORED ARC WELDING (FCAW) process. Flux cored arc welding is very similar to gas metal arc welding. The difference is in the metal electrode used. Flux cored electrode wire is hollow or tubular. Fluxing ingredients are contained within the tubular electrode. The shielding of the weld from atmospheric contamination is accomplished by two methods. The self shielding FCAW method uses the gases formed when the fluxing agents within the hollow electrode vaporize in the heat of the arc. See Fig. 11-48. A second method is called gas shielding flux cored arc welding. In this method a shielding gas such as argon or carbon dioxide is used in addition to the vaporizing flux in the electrode core. See Fig. 11-49. Refer to Heading 1-8 for an additional description of FCAW.

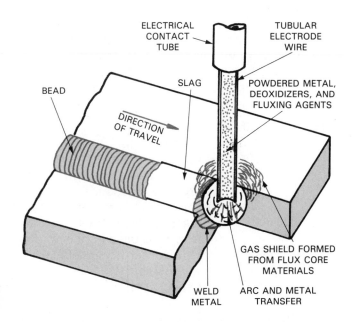

Fig. 11-48. A sketch showing a self shielding flux cored arc weld in progress. Note that there is no nozzle or shielding gas required.

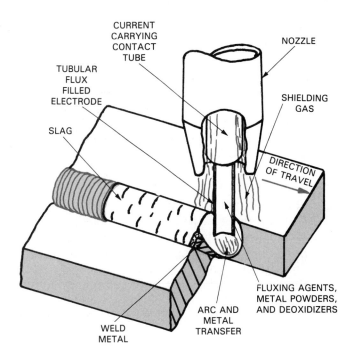

Fig. 11-49. A diagrammatic sketch of a gas shielded flux cored arc weld in progress.

Flux cored arc welding and gas shielded FCAW combine the benefits of shielded metal arc welding (SMAW), submerged arc welding (SAW), and gas metal arc welding (GMAW). These include:
1. The use of fluxing agents to dissolve and remove oxides and undesirable substances.
2. A thin slag layer to shield the hot weld bead.
3. The ability to weld continuously for long periods.
4. The metallurgical effects which can be controlled by using various fluxing and alloying elements.

Flux cored arc welding is generally done semiautomatically. The welding torch or gun is held and manipulated by the welder when doing semiautomatic welding. Flux cored arc welding may be used on fully automatic machines with excellent results.

Self shielding FCAW gives better results than gas shielded FCAW where cross winds are present. The shielding gas for gas shielded FCAW may be disturbed or removed by a cross wind. This would result in poor welds. The self shielding FCAW process also is preferred for welding in hard to reach or hard to see places. Since there is no shielding gas to worry about, the electrode extension may be greater. Moving the electrode out and away from the electrode contact tube permits the welder to see the joint more easily.

## 11-25. THE FLUX CORED ARC WELDING STATION

The equipment required in a flux cored arc welding station is similar in most respects to the equipment used in gas metal arc welding. The equipment required for FCAW is as follows:

1. A constant voltage DC or rectified AC arc welding machine. See Heading 11-15.
2. An electrode wire feed mechanism. See Heading 11-16.
3. Shielding gas cylinders. See Heading 11-5.
4. A gas regulator. See Heading 11-6.
5. A shielding gas flowmeter. See Heading 11-7.
6. Shielding gas and coolant hoses and fittings and the electrode lead. See Heading 11-21. Shielding gas is used only for gas shielded FCAW.
7. A FCAW torch. See GMAW torch, Heading II-21.
8. Flux cored metal electrode wire.
9. Optional equipment:
   a. A coolant system. See Fig. 11-10.
   b. Remote controls.
10. A booth, table, and a ventilation system.

The differences occur in the wire drive and shielding gas equipment.

Shielding gas is used with the gas shielded FCAW process. Therefore, shielding gas cylinders, regulators, flowmeters, gauges, and hoses are required. When self shielding FCAW wire electrodes are used, none of the shielding gas equipment listed above is needed. Also when self shielding FCAW is done there is no need for a gas nozzle. See Fig. 11-48.

The torch or gun for FCAW may be air cooled or water cooled. Air cooled torches are generally used up to 200 amperes. Water cooled torches are used for torches using more than 200 amperes. Water cooling is also used on torches which operate on a 100 percent duty cycle.

Flux cored arc welding electrodes are tubular and easily flattened. The wire drive wheels used for FCAW must be specially designed to prevent crushing the tubular electrodes. The adjustment of these drive wheels must be carefully done to permit the wire to be driven, but not flattened.

## 11-26. FLUX CORED ARC WELDING ELECTRODE WIRE

The tubular, flux cored filler wire used in flux cored arc welding is made as follows:
1. Flat strip steel is formed by rolls into "U" shape.
2. The "U" shape area is then filled with a carefully prepared granular flux and/or alloying material.

3. After filling, the metal is closed and rolled into a round shape. This closing and rolling compresses the flux material.
4. The tubular wire is then passed through drawing (forming) dies. This further compresses the granular flux and forms a perfectly round form of an exact diameter.
5. In a continuous operation, the completed flux cored wire is then wrapped on spools or into coil drums.

Flux cored wire has the advantage of varying the core ingredients to match any weld requirements. Such welding requirements may include:

1. Adding deoxidizers such as silicon, manganese, or aluminum to reduce weld porosity.
2. Adding denitrifiers such as aluminum to reduce nitrogen by forming stable nitrides.
3. Providing mechanical, metallurgical, and corrosion resistant properties to the weld metal by adding alloying elements.
4. Forming gases to shield the weld area from the oxygen and nitrogen in the atmosphere.
5. Creating a slag covering over the weld bead to shield it while it cools.
6. Stabilizing the arc by providing for better ionization of the arc.
7. Scavenging the impurities in the molten weld metal and floating them to the top of the weld to form slag.

Fig. 11-50 gives the chemical composition for flux cored arc welding electrode wire used on carbon steels. These electrodes may be used for self shielding or gas shielded FCAW. Dash numbers are used at the end of the carbon steel electrode classifications. They are used to indicate electrode polarity, the shielding gas used, or the number of weld passes recommended. See Fig. 11-51 for the meaning of these dash numbers.

The electrode number and letter classifications are explained as follows:

Example: EXXT-1

E - electrode

XX - two digit number that specifies minimum tensile strength of the electrode in thousands of pounds per sq. in (ksi). The second digit represents the position the electrode is to be used in. 0 is for flat and horizontal fillet welds. 1 is for all-position welding.

T - indicates the electrode is tubular

−1 to −11 - a grouping of chemical composition, method of shielding, or its suitability to make single or multiple pass welds.

| CHEMICAL COMPOSITION, PERCENT[a] | | | | | | | | | | |
|---|---|---|---|---|---|---|---|---|---|---|
| AWS Classification | Carbon | Phos-phorus | Sulfur | Vana-dium | Sili-con | Nickel | Chro-mium | Molyb-denum | Manga-nese | Alumi-num[b] |
| EXXT-1<br>EXXT-4<br>EXXT-5<br>EXXT-6<br>EXXT-7<br>EXXT-8<br>EXXT-11<br>EXXT-G | — | 0.04 | 0.03 | 0.08 | 0.09 | 0.50 | 0.20 | 0.30 | 1.75 | 1.8 |
| EXXT-GS<br>EXXT-2<br>EXXT-3<br>EXXT-10 | No chemical requirements | | | | | | | | | |

a. Values are maximums. Composition limits are intended to insure a plain carbon steel deposit.
b. For self-shielded electrodes only.

Fig. 11-50. The chemical composition of various carbon steel electrodes for flux cored arc welding. Note that aluminum is used as a denitrifier only for self shielding electrodes. See Fig. 11-51 for the meanings of the dash numbers.    (American Welding Society)

| AWS CLASSIFICATION | EXTERNAL SHIELDING MEDIUM | CURRENT AND POLARITY |
|---|---|---|
| EXXT-1 (Multiple-pass) | $CO_2$ | dc, electrode positive |
| EXXT-2 (Single-pass) | $CO_2$ | dc, electrode positive |
| EXXT-3 (Single-pass) | None | dc, electrode positive |
| EXXT-4 (Multiple-pass) | None | dc, electrode positive |
| EXXT-5 (Multiple-pass) | $CO_2$ | dc, electrode positive |
| EXXT-6 (Multiple-pass) | None | dc, electrode positive |
| EXXT-7 (Multiple-pass) | None | dc, electrode negative |
| EXXT-8 (Multiple-pass) | None | dc, electrode negative |
| EXXT-10 (Single-pass) | None | dc, electrode negative |
| EXXT-11 (Multiple-pass) | None | dc, electrode negative |
| EXXT-G (Multiple-pass) | a | a |
| EXXT-GS (Single-pass) | a | a |

a. As agreed upon between supplier and user.

Fig. 11-51. The meanings and uses of the dash numbers at the end of carbon steel FCAW electrode designations. The letter "G" is used when a manufacturer and purchaser develop a special electrode.

Example: E70T-3

This is a tubular electrode with 70 ksi (482.6 MPa) tensile strength. The dash three ( − 3) in this case means it is self shielding and intended for single pass welds. For a complete description of these FCAW electrodes refer to the AWS 5.20 electrode specification manual.

Chemical compositions and AWS classification numbers for low alloy FCAW electrodes are shown in Fig. 11-52.

The meaning of electrode numbers and letters used in Fig. 11-52 are shown below:

Example: E80T1-B2H

    E - electrode
    80 - 80 ksi (80,000 psi) tensile strength

T1 - intended usage
B - major alloying ingredients as follows:
    A - carbon-molybdenum;
    B - chromium-molybdenum;
    Ni - nickel;
    D - manganese-molybdenum;
    K - all other low alloy electrodes
2 - Chemical composition group. See brackets on Fig. 11-52.
H - comparative carbon content
    H - higher carbon;
    L - lower carbon.

Fig. 11-53 lists the chemical composition of the deposited weld metal when using chromium and chromium-nickel flux cored electrodes.

**CHEMICAL COMPOSITION, PERCENT**

| AWS Classification | C | Mn | P | S | Si | Ni | Cr | Mo | V | Al[a] | Cu |
|---|---|---|---|---|---|---|---|---|---|---|---|
| Carbon-molybdenum steel electrodes | | | | | | | | | | | |
| E70T5-A1 / E80T1-A1 / E81T1-A1 | 0.12 | 1.25 | 0.03 | 0.03 | 0.80 | — | — | 0.40/0.65 | — | — | — |
| Chromium-molybdenum steel electrodes | | | | | | | | | | | |
| E81T1-B1 | 0.12 | 1.25 | 0.03 | 0.03 | 0.80 | — | 0.40/0.65 | 0.40/0.65 | — | — | — |
| E80T5-B2L | 0.05 | 1.25 | 0.03 | 0.03 | 0.80 | — | 1.00/1.50 | 0.40/0.65 | — | — | — |
| E80T1-B2 / E81T1-B2 / E80T5-B2 | 0.12 | 1.25 | 0.03 | 0.03 | 0.80 | — | 1.00/1.50 | 0.40/0.65 | — | — | — |
| E80T1-B2H | 0.10/0.15 | 1.25 | 0.03 | 0.03 | 0.80 | — | 1.00/1.50 | 0.40/0.65 | — | — | — |
| E90T1-B3L | 0.05 | 1.25 | 0.03 | 0.03 | 0.80 | — | 2.00/2.50 | 0.90/1.20 | — | — | — |
| E90T1-B3 / E91T1-B3 / E90T5-B3 | 0.12 | 1.25 | 0.03 | 0.03 | 0.80 | — | 2.00/2.50 | 0.90/1.20 | — | — | — |
| E100T1-B3 / E90T1-B3H | 0.10/0.15 | 1.25 | 0.03 | 0.03 | 0.80 | — | 2.00/2.50 | 0.90/1.20 | — | — | — |
| Nickel-steel electrodes | | | | | | | | | | | |
| E71T8-Ni1 / E80T1-Ni1 / E81T1-Ni1 / E80T5-Ni1 | 0.12 | 1.50 | 0.03 | 0.03 | 0.80 | 0.80/1.10 | 0.15 | 0.35 | 0.05 | 1.8 | — |
| E71T8-Ni2 / E80T1-Ni2 / E81T1-Ni2 / E80T5-Ni2 / E90T1-Ni2 / E91T1-Ni2 | 0.12 | 1.50 | 0.03 | 0.03 | 0.80 | 1.75/2.75 | — | — | — | 1.8 | — |
| E80T5-Ni3 / E90T5-Ni3 | 0.12 | 1.50 | 0.03 | 0.03 | 0.80 | 2.75/3.75 | — | — | — | — | — |
| Manganese-molybdenum steel electrodes | | | | | | | | | | | |
| E91T1-D1 | 0.12 | 1.25/2.00 | 0.03 | 0.03 | 0.80 | — | — | 0.25/0.55 | — | — | — |
| E90T5-D2 / E100T5-D2 | 0.15 | 1.65/2.25 | 0.03 | 0.03 | 0.80 | — | — | 0.25/0.55 | — | — | — |
| E90T1-D3 | 0.12 | 1.00/1.75 | 0.03 | 0.03 | 0.80 | — | — | 0.40/0.65 | — | — | — |
| All other low alloy steel electrodes | | | | | | | | | | | |
| E80T5-K1 | 0.15 | 0.80/1.40 | 0.03 | 0.03 | 0.80 | 0.80/1.10 | 0.15 | 0.20/0.65 | 0.05 | — | — |
| E70T4-K2 / E71T8-K2 / E80T1-K2 / E90T1-K2 / E91T1-K2 / E80T5-K2 / E90T5-K2 | 0.15 | 0.50/1.75 | 0.03 | 0.03 | 0.80 | 1.00/2.00 | 0.15 | 0.35 | 0.05 | 1.8 | — |
| E100T1-K3 / E110T1-K3 / E100T5-K3 / E110T5-K3 | 0.15 | 0.75/2.25 | 0.03 | 0.03 | 0.80 | 1.25/2.60 | 0.15 | 0.25/0.65 | 0.05 | — | — |
| E110T5-K4 / E111T1-K4 / E120T5-K4 | 0.15 | 1.20/2.25 | 0.03 | 0.03 | 0.80 | 1.75/2.60 | 0.20/0.60 | 0.30/0.65 | 0.05 | — | — |
| E120T1-K5 | 0.10/0.25 | 0.60/1.60 | 0.03 | 0.03 | 0.80 | 0.75/2.00 | 0.20/0.70 | 0.15/0.55 | 0.05 | — | — |
| E61T8-K6 / E71T8-K6 | 0.15 | 0.50/1.50 | 0.03 | 0.03 | 0.80 | 0.40/1.10 | 0.15 | 0.15 | 0.05 | 1.8 | — |
| E101T1-K7 | 0.15 | 1.00/1.75 | 0.03 | 0.03 | 0.80 | 2.00/2.75 | — | — | — | — | — |
| EXXXTX-G | — | 1.00 min* | 0.03 | 0.03 | 0.80 min* | 0.50 min* | 0.30 min* | 0.20 min* | 0.10 min* | 1.8 | — |
| E80T1-W | 0.12 | 0.50/1.30 | 0.03 | 0.03 | 0.35/0.80 | 0.40/0.80 | 0.45/0.70 | — | — | — | 0.30/0.75 |

a. For self-shielded electrodes only.
* All values are maximum except where min (minimum) is indicated.

Fig. 11-52. The chemical composition and AWS classification numbers for low alloy FCAW electrodes. Numbers and letters used are explained in Heading 11-26.

| AWS Classification | C | Cr | Ni | Mo | Cb + Ta | Mn | Si | P | S | Fe | Cu |
|---|---|---|---|---|---|---|---|---|---|---|---|
| E307T-X | 0.13 | 18.0-20.5 | 9.0-10.5 | 0.5-1.5 | — | 3.3-4.75 | 1.0 | 0.04 | 0.03 | Rem | 0.5 |
| E308T-X | 0.08 | 18.0-21.0 | 9.0-11.0 | 0.5 | — | 0.5-2.5 | | | | | |
| E308LT-X | a | 18.0-21.0 | 9.0-11.0 | 0.5 | — | | | | | | |
| E308MoT-X | 0.08 | 18.0-21.0 | 9.0-12.0 | 2.0-3.0 | — | | | | | | |
| E308MoLT-X | a | 18.0-21.0 | 9.0-12.0 | 2.0-3.0 | — | | | | | | |
| E309T-X | 0.10 | 22.0-25.0 | 12.0-14.0 | 0.5 | — | | | | | | |
| E309CbLT-X | a | 22.0-25.0 | 12.0-14.0 | | 0.70-1.00 | | | | | | |
| E309LT-X | a | 22.0-25.0 | 12.0-14.0 | | | | | | | | |
| E310T-X | 0.20 | 25.0-28.0 | 20.0-22.5 | | — | 1.0-2.5 | | 0.03 | | | |
| E312T-X | 0.15 | 28.0-32.0 | 8.0-10.5 | | — | 0.5-2.5 | | 0.04 | | | |
| E316T-X | 0.08 | 17.0-20.0 | 11.0-14.0 | 2.0-3.0 | — | | | | | | |
| E316LT-X | a | 17.0-20.0 | 11.0-14.0 | 2.0-3.0 | — | | | | | | |
| E317LT-X | a | 18.0-21.0 | 12.0-14.0 | 3.0-4.0 | — | | | | | | |
| E347T-X | 0.08 | 18.0-21.0 | 9.0-11.0 | 0.5 | 8 × C min to 1.0 max | | | | | | |
| E409T-X | 0.10 | 10.5-13.0 | 0.60 | 0.5 | — | 0.8 | | | | | |
| E410T-X | 0.12 | 11.0-13.5 | 0.60 | 0.5 | — | 1.2 | | | | | |
| E410NiMoT-X | 0.06 | 11.0-12.5 | 4.0-5.0 | 0.40-0.70 | — | 1.0 | | | | | |
| E410NiTiT-X | a | 11.0-12.0 | 3.6-4.5 | 0.05 | — | 0.70 | 0.60 | 0.03 | | | |
| E430T-X | 0.10 | 15.0-18.0 | 0.60 | 0.5 | — | 1.2 | 1.0 | 0.04 | | | |
| E502T-X | 0.10 | 4.0-6.0 | 0.40 | 0.45-0.65 | — | 1.2 | | | | | |
| E505T-X | 0.10 | 8.0-10.5 | 0.40 | 0.85-1.20 | — | 1.2 | | | | | |
| E307T-3 | 0.13 | 19.5-22.0 | 9.0-10.5 | 0.5-1.5 | — | 3.3-4.75 | | | | | |
| E308T-3 | 0.08 | 19.5-22.0 | 9.0-11.0 | 0.5 | — | 0.5-2.5 | | | | | |
| E308LT-3 | 0.03 | 19.5-22.0 | 9.0-11.0 | 0.5 | — | | | | | | |
| E308MoT-3 | 0.08 | 18.0-21.0 | 9.0-12.0 | 2.0-3.0 | — | | | | | | |
| E308MoLT-3 | 0.03 | 18.0-21.0 | 9.0-12.0 | 2.0-3.0 | — | | | | | | |
| E309T-3 | 0.10 | 23.0-25.5 | 12.0-14.0 | 0.5 | — | | | | | | |
| E309LT-3 | 0.03 | 23.0-25.5 | 12.0-14.0 | | — | | | | | | |
| E309CbLT-3 | 0.03 | 23.0-25.5 | 12.0-14.0 | | 0.70-1.00 | | | | | | |
| E310T-3 | 0.20 | 25.0-28.0 | 20.0-22.5 | | — | 1.0-2.5 | | 0.03 | | | |
| E312T-3 | 0.15 | 28.0-32.0 | 8.0-10.5 | | — | 0.5-2.5 | | 0.04 | | | |
| E316T-3 | 0.08 | 18.0-20.5 | 11.0-14.0 | 2.0-3.0 | — | | | | | | |
| E316LT-3 | 0.03 | 18.0-20.5 | 11.0-14.0 | 2.0-3.0 | — | | | | | | |
| E317LT-3 | 0.03 | 18.5-21.0 | 13.0-15.0 | 3.0-4.0 | — | | | | | | |
| E347T-3 | 0.08 | 19.0-21.5 | 9.0-11.0 | 0.5 | 8 × C min to 1.0 max | | | | | | |
| E409T-3 | 0.10 | 10.5-13.0 | 0.60 | 0.5 | — | 0.80 | | | | | |
| E410T-3 | 0.12 | 11.0-13.5 | 0.60 | 0.5 | — | 1.0 | | | | | |
| E410NiMoT-3 | 0.06 | 11.0-12.5 | 4.0-5.0 | 0.40-0.70 | — | 1.0 | | | | | |
| E410NiTiT-3 | 0.04 | 11.0-12.0 | 2.6-4.5 | 0.5 | — | 0.70 | 0.50 | 0.03 | | | |
| E430T-3 | 0.10 | 15.0-18.0 | 0.60 | 0.5 | — | 1.0 | 1.0 | 0.04 | | | |
| EXXXT-G | As agreed upon between supplier and purchaser. | | | | | | | | | | |

a. The carbon content shall be 0.04% maximum when the suffix "X" is "1"; it shall be 0.03% when the suffix "X" is "2".

Fig. 11-53. The chemical composition of the deposited weld metal for flux cored chromium and chromium-nickel steel electrodes.

The electrode classification numbers are explained below:

Example: E308T-X and E308LT-X

    E - electrode
    308 - the stainless steel classification
    T - tubular electrode
    LT - low carbon, tubular electrode
    -X - numbers from 1 to 3 plus the letter "G"
        1 - used with carbon dioxide ($CO_2$)
        2 - used with argon plus 2% oxygen
        3 - self shielding. No external shielding gas used.
        G - not specified. Used for special compositions.

Example: E316T-3

A tubular (T) stainless steel (316) electrode (E) used without any external shielding gas (−3).

For complete specifications and information on flux cored arc welding electrodes refer to one of the following AWS electrode specifications:

A5.20 - Specifications for Carbon Steel Electrodes for Flux Cored Arc Welding.

A5.22 - Specifications for Flux Cored Corrosion-Resisting Chromium and Chromium-Nickel Steel Electrodes.

A5.29 - Specifications for Low-alloy Steel Flux Cored Arc Welding Electrodes.

## 11-27. ACCESSORIES

When water cooled torches are used, it is important that water flow be maintained. If the water flow stops or slows, the torch and the electrode lead may quickly overheat and be damaged. To protect the equipment, a safety switch is sometimes placed in the water circuit. If the water flow decreases below a set limit, an electrical switch opens and shuts off the electrical power source. Fig. 11-54 shows a flow safety device.

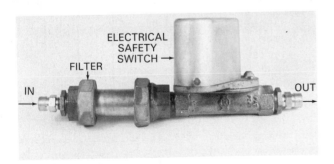

Fig. 11-54. Water flow safety control. If the water flow decreases below a set limit, the current is shut off until the flow of water resumes.

Some systems use a heat fuse to protect the water cooled torch. The fuse will interrupt the current flow when and if the water flow is stopped, or decreases to a dangerous minimum. This may cause the fuse link to overheat to interrupt the electrical flow. See Fig. 11-55.

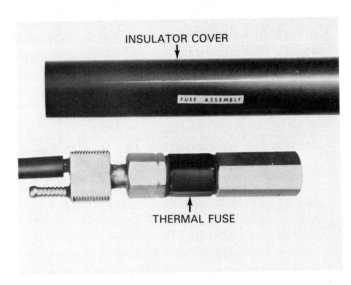

Fig. 11-55. Fuse and hose assembly with insulator cover removed. If the thermal fuse overheats, it will shut off the current flow to the torch. (L-TEC Welding & Cutting Systems)

To economize on water flow and/or gas flow, a lever operated shutoff valve is sometimes used. Fig. 11-56 shows this shutoff valve called an economizer. The valve operating lever also acts as the torch holder. When the torch is not in use, it is hung on this lever and its weight closes the valves. When the torch is removed, the valves automatically open, allowing the water and gas flow to start.

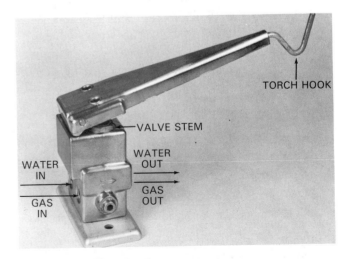

Fig. 11-56. A shielding gas and water flow economizer. When the torch is hung on the hook, the shielding gas and water flow are shut off.

Most arc welding machines are equipped with a remote control circuit switch and external plug. The remote control is used to vary the voltage or amperage within a rough setting range. The welder at the job site can adjust the arc welding machine at a distance with this control. Fig. 11-57 shows one type of remote control device. Several types of remote controls and their uses are described in Heading 9-24.

Fig. 11-57. A remote current adjuster. (Airco Welding Products, Div. of Airco, Inc.)

## 11-28. FILTER LENSES FOR USE WHEN GAS SHIELDED ARC WELDING

Because of the clearer atmosphere around the arc, the operator must use arc welding lenses in a darker shade to reduce eye fatigue and possible eye damage. Most helmets for gas arc welding use a clear cover lens and a filter lens. Sometimes a clear cover lens is used on the inside of the helmet also. It is very important that all these lenses be clean. It is also recommended that flash goggles of approximately #2 shade be worn under the helmet when welding for extended periods.

The recommended lens shade for use with various processes with electrodes up to 5/32 in. (4.0 mm) in diameter is shown below:

| Process | Metal | Shade Number |
|---|---|---|
| GTAW | all | 12 |
| GMAW & FCAW | nonferrous | 11 |
| GMAW & FCAW | ferrous | 12 |

For electrodes larger than 5/32 in. (4.0 mm), use a darker lens.

## 11-29. PROTECTIVE CLOTHING

Fire resistant cloth and leather clothing and accessories are recommended. Wool is satisfactory. Cotton does not provide sufficient protection and deteriorates rapidly under infrared and ultraviolet rays. Always wear dark clothing to reduce reflection of light behind the helmet.

The clothing should be without cuffs or open pockets as these will collect sparks.

Leather or leather palm gloves should be worn. See Heading 9-27 for a more in-depth description of clothing recommended for arc welding.

## 11-30. SAFETY REVIEW

When using gas metal, gas tungsten, or flux cored arc welding equipment, use all the safety precautions normally used with arc welding equipment.

1. Electrical connection leads should be in good condition and tight. They should be protected from accidental damage from shop traffic.
2. Adequate ventilation and filtration equipment must be used.
3. When arc welding, it is advisable to draw the fumes and contaminated air away before they reach the welder's face.
4. Proper clothing should be worn to prevent burns from hot metal and ultraviolet and infrared rays.
5. No flammable materials should be carried in the pockets of clothing. Pockets should be closed and cuffs rolled down to prevent hot metal from going into them.
6. A welding helmet with the proper number filtering lens must be used for the type of welding being done.
7. Shielding curtains should be placed around all jobs so that workers in the area are not bothered by the arc flashes.

## 11-31. TEST YOUR KNOWLEDGE

Write your answers on a separate sheet of paper. Do not write in this book.
1. Name the welding processes covered in this chapter which may use a shielding gas to shield the weld area.
2. What type welding current may be used with GTAW?
3. The GTAW and GMAW electrode holder is more correctly called a _____.
4. GTAW welding machines are constant _____ machines.
5. What is the advantage of using a constant voltage machine?

6. Why is a high frequency voltage used continually in an AC circuit used for GTAW?
7. Name five reasons why argon is more frequently used for GTAW than helium.
8. Which shielding gas: Is best for use in cross drafts? Gives the best cleaning action? Provides the greatest heat conductivity?
9. A regulator controls and measures _____. A flowmeter measures and controls the _____ of gas flow.
10. The maximum amperage used on an air cooled GTAW torch is generally _____ amperes.
11. What determines the nozzle diameter when selecting a nozzle for GTAW or GMAW?
12. What is the advantage of using thoriated tungsten electrodes?
13. The color of a 2% thoriated tungsten electrode is _____.
14. The arc welding machine used for GMAW and FCAW produce a constant _____.
15. GMAW machines have a sloping voltage curve. How much slope should the curve have for use with carbon dioxide ($CO_2$) gas?
16. Increasing the rate of speed on the wire feed in GMAW or FCAW increases the _____ in the circuit.
17. The shielding gas mixture suggested in Fig. 11-34 for use with low alloy steel is a mixture of _____ and 2% _____.
18. Which gas produces the deepest penetration when used in GMAW?
19. To produce better metal transfer through the arc, reduce spatter, and stabilize the arc, _____ and _____ are added to argon or helium.
20. Describe the following GMAW electrode completely: ER90S-B3L
21. What two methods are used in FCAW to shield the weld from atmospheric contamination?
22. When feeding a FCAW electrode through the wire feed mechanism care must be taken not to _____ the flux cored wire.
23. List five ways that FCAW wires are altered to meet the requirements of a weld.
24. Describe the following FCAW electrode completely: E80T-2.
25. What number filter lens is recommended for welding steel with GMAW? FCAW? GTAW?

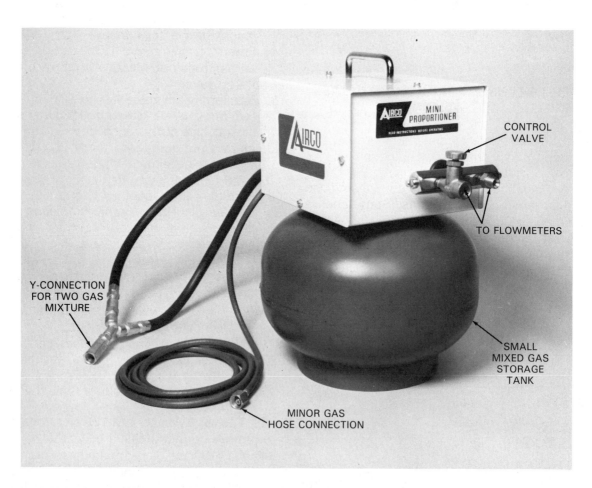

This demand gas mixing system will mix two or three gases at factory set ratios and store them in a small tank until used in GTAW or GMAW processes.

# 12 GAS TUNGSTEN ARC WELDING

GAS TUNGSTEN ARC WELDING (GTAW) is a process used to produce high quality welds in virtually all weldable metals. See Fig. 12-1. It is done manually or automatically. Gas tungsten arc welding (GTAW) is also known as TIG (tungsten inert gas) welding. It was originally called Heliarc welding when it was first developed.

Welding is done normally with one tungsten electrode, but multiple electrodes have been used in special applications. Gas tungsten arc welding may be done in any welding position. It is used to weld thin walled pipe and tube. The process is also used almost exclusively to weld the root bead in heavy walled pipe in petroleum, chemical, and power generating applications. Filler metal may or may not be required with GTAW. Flange joints on thin metal may be designed for welding without filler metal.

Inert gases are used to shield the GTAW from atmospheric contamination.

| METAL | WELDABLE |
|---|---|
| Aluminum and Aluminum Alloys | Yes |
| Bronze and Brass | Yes |
| Copper and Copper Alloys | Yes |
| Cast Iron | |
|   Malleable | Possible, but not popular |
|   Nodular | Possible, but not popular |
| Inconel | Yes |
| Lead | Possible, but not popular |
| Magnesium | Yes |
| Monel | Yes |
| Nickel and Nickel Alloys | Yes |
| Precious Metals (gold, silver, platinum, etc.) | Yes |
| Steel, alloy | Yes |
| Steel, stainless | Yes |
| Titanium | Yes |
| Tungsten | Possible, but not popular |

Fig. 12-1. A list of metals showing their weldability.

The welder must manipulate the gas tungsten arc welding torch to control the arc length. The welder must also carefully add the filler metal when doing manual GTAW. Manual gas tungsten arc welding therefore requires more welder skill than gas metal arc welding.

## 12-1. GAS TUNGSTEN ARC WELDING PRINCIPLES

Gas tungsten arc welding requires the use of a torch, an inert gas, gas regulating equipment, a constant current power supply, and filler metal when required. Refer to Heading 1-6 and Fig. 1-11.

Direct current (DC) or alternating current (AC) power supplies may be used. Either direct current electrode negative (DCEN) (DCSP) or direct current electrode positive (DCEP) (DCRP) may be used. When doing GTAW with DCEN, two thirds of the heat generated in the arc is released on the work piece and one third is released at the electrode. In DCEP, two thirds of the heat generated is released at the electrode and one third on the work. When DCEP is used for GTAW, a larger diameter electrode must be used than when DCEN is used. Fig. 12-2 shows how the electrons flow and how the heat is distributed when DCEN, DCEP, or AC is used with GTAW. The current carrying capacity of an electrode using DCEP is only about one tenth that of an electrode using DCEN. DCEN is therefore used most often.

Filler metal is added to the arc pool either manually or automatically. When doing manual GTAW, the filler metal is added in much the same way as when doing oxyfuel gas welding. A flange may be bent up on thin metal and used as the filler metal.

Using automatic GTAW, metal as thin as .003 in. (.076 mm) may be welded using a flanged joint.

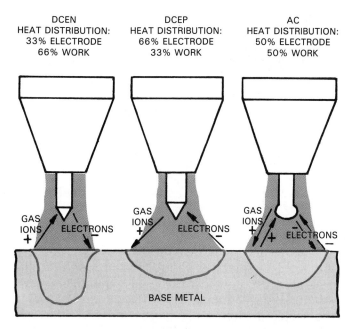

DCEN
HEAT DISTRIBUTION:
33% ELECTRODE
66% WORK

DCEP
HEAT DISTRIBUTION:
66% ELECTRODE
33% WORK

AC
HEAT DISTRIBUTION:
50% ELECTRODE
50% WORK

Fig. 12-2. The approximate heat distribution for GTAW using DCEN, DCEP, and AC. Note the larger electrode diameter required for DCEP and the depth of penetration for DCEN.

The filler metal used for automatic GTAW is usually in the form of spooled wire. The wire is fed into the arc pool as shown in Fig. 12-3. Thicknesses above .02 in. (.51 mm) are generally joined using an added filler metal.

Alternating current or direct current electrode positive (DCEP) or DCRP is used when surface ox-

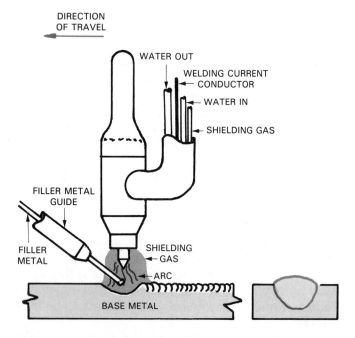

Fig. 12-3. Automatic GTAW. The filler metal guide feeds the filler metal to the desired position in the arc at a preset feed rate.

ides must be removed. Surface oxides occur on aluminum, magnesium, and some other nonferrous metals. These metal oxides melt at a higher temperature than the base metal. The oxides therefore make it hard to weld the base metal.

Electrons flow from the work to the electrode when DCEP is used and during one half of the AC cycle. However, positively charged shielding gas ions travel from the nozzle to the negatively charged work; see Fig. 12-2. These shielding gas ions strike the work surface with sufficient force to break up the oxides. DCEP and AC both work well in breaking up the surface oxides on aluminum and magnesium. See Fig. 12-4. AC gives better penetration. AC can also be used with a smaller electrode diameter for a given current flow.

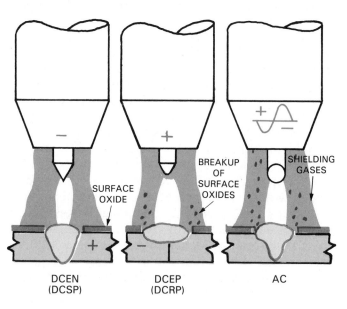

DCEN
(DCSP)

DCEP
(DCRP)

AC

Fig. 12-4. Examples of how the surface oxides are broken up when DCEP and AC current is used. The heavy gas ions striking the surface is what causes the oxides to break up near the weld bead area. The gas ions travel from the work to the electrode in DCEN. The gas ions in DCEN therefore do not strike the surface. No surface oxides are removed when DCEN is used.

Gas tungsten arc welding is done with AC, a steady flowing direct current, or direct current that is pulsing. Fig. 12-5 graphically shows the current flow used in STEP-PULSED GTAW. The amperage is high and then low. Special wiring within the welding machine creates the square wave form shown in Fig. 12-5. Welding is done during the high current interval. This high current is usually referred to as the peak current.

During the low amperage period the arc is maintained, but the heat output of the arc is reduced. The low amperage period is called the BACKGROUND CURRENT. During a welding

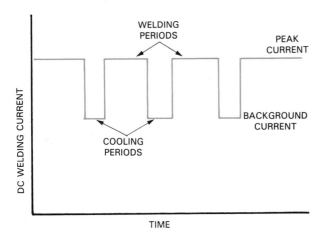

Fig. 12-5. The high and low amperage in a step-pulsed GTAW sequence.

operation the arc crater cools slightly during the low amperage period. PULSED ARC GTAW is ideal for welding out of position. It allows for controlled heating and cooling periods. A FLIP MOTION is used in shielded metal arc welding (SMAW) to allow a period for cooling. With pulsed arc GTAW, the welder does not have to manipulate the torch to accomplish a cooling period. Out of position GTAW therefore can be done with less welder skill. A pulsed current gas tungsten arc welded seam is a series of overlapping arc spot welds. The torch is moved from one arc spot location to the next during the cooling period. Each arc spot weld overlaps the previous one as shown in Fig. 12-6. The physical characteristics and appearance of the finished weld is excellent.

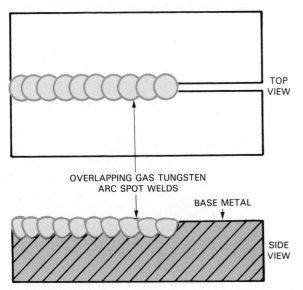

Fig. 12-6. Two views of a pulsed current GTAW. A series of overlapping gas tungsten arc spot welds make up the welded seam.

Not all GTA welding machines can be programmed to provide a pulsed current as shown in Fig. 12-7. Some machines, in addition to providing a pulsed current, can also provide upslope and downslope currents. UPSLOPE is used to gradually increase the current from zero to the welding current. DOWNSLOPE is a gradual decrease in current. Downslope allows the final weld crater to be completely filled and slowly cooled. When automatic GTAW, the welder must set the machine to control the arc voltage and welding current. The welder must also set the wire feed rate and the torch travel speed.

Pulsed current is used with excellent results where the metals are not well fitted. It is also used when parts of unequal thicknesses are welded. Gas tungsten arc welding, like other welding processes, works best when accurately fitted metals of equal thickness are welded.

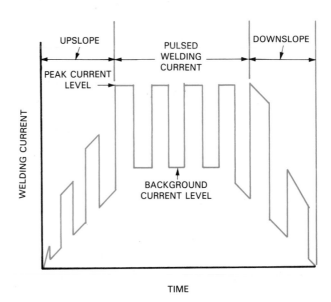

Fig. 12-7. A programmed upslope, pulsed welding current, and downslope for GTAW on tubing or piping.

## 12-2. GTAW POWER SOURCES

Direct current electrode negative (DCEN) (DCSP) is used when the greatest amount of heat is to be on the base metal. DCEN is also used when welding thicker sections and for deepest penetration. See Fig. 12-2. Direct current electrode positive (DCEP) (DCRP) is used to weld thin metal sections. DCEP or AC is used for the best surface cleaning action when welding aluminum or magnesium. An alternating current source is chosen when equal heating is preferred. AC is also used for medium penetration welds.

The power source may be able to furnish steady or pulsing current. Pulsing current is the best choice when welding out of position.

During half of the alternating current (AC) cycle, the electrode is positive. Electrons do not travel easily from the flat work surface to the relatively small tungsten electrode tip. This may cause a blocking (rectification) or unbalancing of the current flow during the electrode positive half of the AC cycle. See A, Fig. 12-8. Rectification can be avoided or reduced by increasing the open circuit voltage of the welding machine. See B, Fig. 12-8. The alternating current wave form is said to be stabilized when some current flows during the electrode positive half of the cycle. It may also be stabilized by adding a high frequency voltage circuit in series with the welding circuit. This added high frequency circuit provides several thousand volts with an extremely low current or amperage. A high frequency voltage is continually applied in the AC welding circuit for GTAW.

High frequency or a higher open circuit voltage will stabilize the AC wave form, but the wave form is still unbalanced. To balance the AC wave form, capacitors are used to increase the current flow during the EP half cycle. Capacitors store electricity during the EN half cycle and release the electricity during the EP half cycle. Through the use of capacitors, a balanced AC wave form is obtained. See C, Fig. 12-8.

Gas tungsten arc welding is always done with a power source which furnishes a CONSTANT CURRENT supply. The current setting will be determined by the size of the electrode, metal thickness, shielding gas used, and the type of current supplied.

## 12-3. SETTING UP GTAW POWER SOURCE

The arc welding machine should be given a brief safety inspection prior to its use. With the main power switch off, check the ground and electrode lead connections for a tight connection on the machine. Check the entire length of each lead for evidence of wear or cuts. This could indicate internal damage to the conducting cables. Keep the leads protected with a steel channel whenever they must run across an aisleway. The ground lead clamp or connection should be checked for a good connection. The contact area of the ground clamp or connection must be clean for the best current flow conditions. All shielding gas connections must be tight. This will prevent expensive gas leaks. Loose connections may also allow air to enter the shielding gas lines. This will cause contamination of the electrode and the weld.

A remote control foot switch, shown in Fig. 12-9, performs many functions. When the pedal

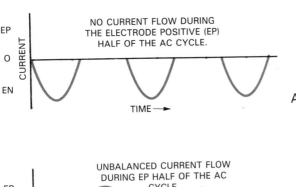

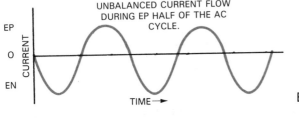

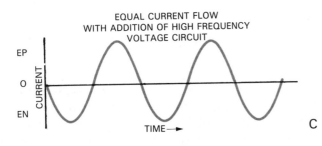

Fig. 12-8. Alternating current wave forms as current is plotted against time. A—Curve where the electrode positive half of each cycle is rectified (stopped). B—Unbalanced curve where partial rectification is occuring. C—Balanced curve which results from the use of capacitors in the welding circuit.

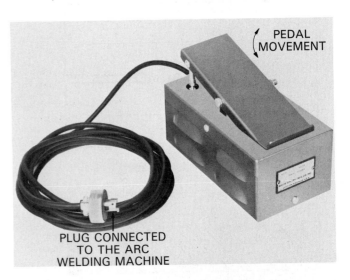

Fig. 12-9. Foot switch which controls the arc current. The foot switch also turns the shielding gas and cooling water on or off.   (Miller Electric Mfg. Co.)

is pressed, the shielding gas and cooling water begin to flow. The switch is used to control the welding current. As the pedal is pressed, more current is supplied to the torch. The pedal can also be used as a current on-off switch without varying the current. If a remote control foot switch is used, the plug must be firmly installed in the arc welder receptacle. A torch mounted off-on control (contactor) is shown in Fig. 12-10. This thumb operated button is clamped onto the GTAW torch.

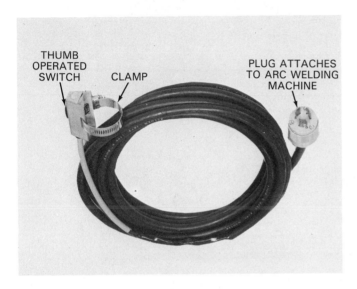

Fig. 12-10. A torch mounted on-off switch. This thumb operated switch is clamped on to the GTAW torch. (Miller Electric Mfg. Co.)

Fig. 12-11 illustrates one type of GTAW (TIG) arc welding machine and its control panel. The ground and electrode lead connections, the 115 volt receptacles, and the gas and water connection fittings are under the lower panel on this machine.

Fig. 12-12 is a close up of the control panel for the arc welding machine in Fig. 12-11. This control panel contains the following controls which are explained below:

1. The RANGE SELECTOR is used to select a high or low current range.
2. The CURRENT CONTROL DIAL AND CONTROL KNOB is used to set the desired current. This dial is marked in white for the high range settings. It is marked in black for the low range settings.
3. Alternating or direct current may be selected by the setting of the AC/DC SELECTOR SWITCH.
4. Direct current electrode negative (DCEN) (DCSP) or direct current electrode positive (DCEP) (DCRP) may be selected by the setting of the POLARITY SWITCH.

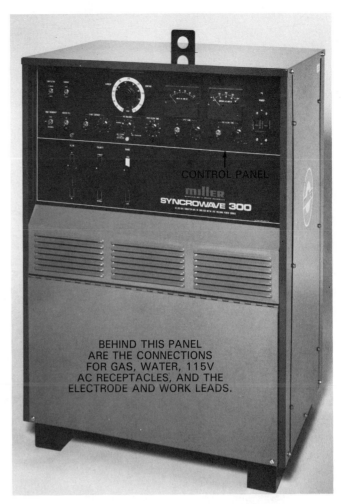

CONTROL PANEL

BEHIND THIS PANEL ARE THE CONNECTIONS FOR GAS, WATER, 115V AC RECEPTACLES, AND THE ELECTRODE AND WORK LEADS.

Fig. 12-11. An arc welding machine designed for gas tungsten arc welding (GTAW) or TIG welding. It may also be used for SMAW. See Fig. 12-12 for a close up of the control panel. (Miller Electric Mfg. Co.)

5. The POWER SWITCH is used to turn the arc welding machine primary circuit on or off. The machine should not be turned off while the arc is struck. The machine should only be turned on or off while the GTAW torch is hung on an insulated hook.
6. The POST-FLOW TIMER or after flow timer is used to control how long in seconds the shielding gas will flow after the arc is stopped. The gas should flow for 10-15 seconds to keep the hot electrode from becoming contaminated. A good rule to use is to continue the gas flow one second for each 10 amperes of current used.
7. The VOLTMETER will show open circuit voltage and it will show the closed circuit voltage while welding.
8. The AMMETER will indicate the current flow while welding.
9. The CONTACTOR SWITCH has two positions, standard and remote. In the standard

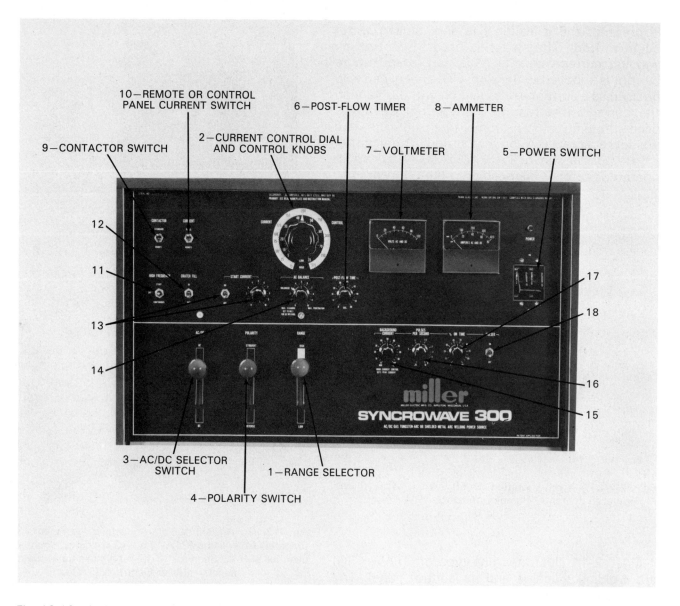

Fig. 12-12. A close up view of the control panel of one type of gas tungsten arc welding machine. See also Fig. 12-11. The function of each number and callout is explained in the text. (Miller Electric Mfg. Co.)

position, a switch on the control panel is used to turn the machine's secondary circuit on or off. The standard position is generally used for SMAW. In this position the electrode has open circuit voltage to it. When the switch is in the remote position, a thumb switch on the torch or a remote foot switch is used to turn the secondary welding circuit on or off. In the remote position the secondary current, and the water, and gas flow are started when the thumb switch or foot switch is depressed slightly.

10. The REMOTE OR CONTROL PANEL CURRENT SWITCH has two positions. In the remote position, the current may be varied at the welding site. In the panel position, the current is changed on the machine panel only.

11. The HIGH FREQUENCY SWITCH has two positions. In the start position, high frequency is applied to the welding circuit only until the arc is struck. This position is used when DC is used. In the continuous position, high frequency is applied to the welding circuit constantly. This position is used when AC is used for GTAW.

12. The CRATER FILL feature provides a 5-6 second current taper for uniform weld finishes. This feature permits welder to fill the ending crater with a gradually lowering arc current.

13. The START CURRENT SWITCH AND CONTROL work together. The control is marked from 0 - 10. This control will set a starting current from low (1) to high (10). After the arc is stabilized, the regular welding current will automatically come into use.

14. The AC BALANCE DIAL provides the means for controlling the penetration and cleaning action; see Fig. 12-13. Refer to Heading 11-3 for an explanation of balanced and un-balanced wave forms.

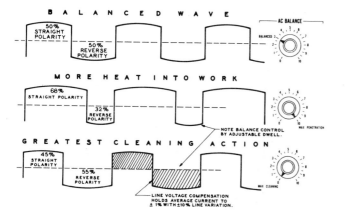

Fig. 12-13. Variations in the square wave patterns which occur with the setting of the AC balance control. (Miller Electric Mfg. Co.)

15. Some DC gas tungsten arc welding machines have the ability to pulse the welding current. The welding current can be pulsed from one controllable current level to another, as shown in Fig. 12-14.

   During the peak current, the welding is done at a high heat input. This peak current is held at the set level for the duration of the pulse time. The arc is maintained during the BACKGROUND CURRENT pulse time but at a lower current level. During the duration of the background pulse time, the weld cools slightly. This feature is excellent for out of position welding.

16. The duration of the pulse cycle when used may be varied. Typical pulse frequencies will vary from ten per second to one in two seconds. The number of pulses per second are set on the PULSES PER SECOND CONTROL.

17. Control of the duration of the peak current pulse is set on the PERCENT ON TIME CONTROL. Even with ten pulse cycles per second, the percent on time control may be set for a relatively large percentage. The duration of the background current in this case would be relatively short.

18. The PULSER ON-OFF SWITCH activates the pulsed arc circuits in the arc welding machine. If the pulser switch is off, the background current, pulses per second, and percent on time controls are not operating.

   Fig. 12-15 illustrates another type of GTAW (TIG) arc welding machine and its control panel. This arc welding machine contains many of the same controls shown in Fig. 12-12. Controls and switches may appear in a different form or location on various machines. These controls however, serve the same purposes as those described for the machine shown in Fig. 12-11.

   The suggested type of welding current and polarity to use when welding various base metals is shown in Fig. 12-16.

   Suggested current settings, gas types and flow rates, and electrode and filler rod sizes for various base metals are shown in the following figures: Fig. 12-17—mild steel; Fig. 12-18—aluminum; Fig. 12-19—stainless steel; Fig. 12-20—magnesium; and Fig. 12-21—copper.

## 12-4. SELECTING PROPER SHIELDING GAS FOR GTAW

High quality welds using the gas tungsten arc can only be made using either argon (Ar), helium (He), or argon-helium mixtures. Both argon and helium are inert gases. See Heading 11-4 for more information on argon and helium.

Argon (Ar) provides a smoother, quieter arc. A lower arc voltage is required and it provides a better cleaning action than helium. Argon is ten (10) times heavier than helium. Argon provides better shielding than helium with less gas.

Helium (He) however, provides a higher available heat at the workpiece than does argon. GTAW done with helium gas produces deeper penetration than does argon gas.

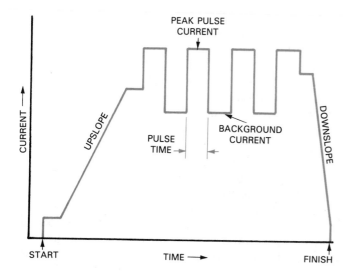

Fig. 12-14. A pulsed GTAW program. The peak current is for welding. The background current maintains the arc but allows the weld to cool.

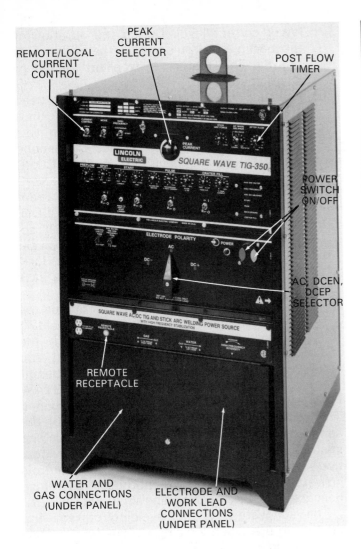

PEAK CURRENT SELECTOR

REMOTE/LOCAL CURRENT CONTROL

POST FLOW TIMER

POWER SWITCH ON/OFF

AC, DCEN, DCEP SELECTOR

REMOTE RECEPTACLE

WATER AND GAS CONNECTIONS (UNDER PANEL)

ELECTRODE AND WORK LEAD CONNECTIONS (UNDER PANEL)

Fig. 12-15. A GTAW machine with square wave output. An explanation of the switches and dials is given in the text. This machine also has controls for pulse setting, crater fill, and GTA spot welding. (The Lincoln Electric Co.)

| Base Material | Direct Current DCEN | DCEP | Alternating Current |
|---|---|---|---|
| | DCSP | DCRP | AC |
| Aluminum up to 3/32 | P | G | E |
| Aluminum over 3/32 | P | P | E |
| Aluminum bronze | P | G | E |
| Aluminum castings | P | P | E |
| Beryllium copper | P | G | E |
| Brass alloys | E | P | G |
| Copper base alloys | E | P | G |
| Cast Iron | E | P | G |
| Deoxidized copper | E | P | P |
| Dissimilar metals | E | P | G |
| Hard facing | G | P | E |
| High alloy steels | E | P | G |
| High carbon steels | E | P | G |
| Low alloy steels | E | P | G |
| Low carbon steels | E | P | G |
| Magnesium up to 1/8'' | P | G | E |
| Magnesium over 1/8'' | P | P | E |
| Magnesium castings | P | G | E |
| Nickel & Ni-alloys | E | P | G |
| Stainless steel | E | P | G |
| Silicon bronze | E | P | P |
| Titanium | E | P | G |
| E-Excellent    G-Good    P-Poor | | | |

Fig. 12-16. Suggested current and polarity for use with GTAW on various base metals. (Welding & Fabrication Data Book)

Both argon and helium produce good welds with direct current. Helium is best for use on thicker metal sections than argon because of its higher available heat.

Alternating current GTAW cannot be done acceptably with helium gas. AC with argon is suggested only for use on aluminum and its alloys. Argon and helium gas mixtures are used in some applications. These mixtures contain up to 75 percent helium. They produce a weld with deeper penetration and they have a good cleaning action.

Hydrogen (H₂) is added to argon when welding stainless steel, nickel-copper, and nickel based alloys. The addition of hydrogen to argon permits increased welding speeds. Hydrogen is not recommended for use on other metals because it produces hydrogen cracks in the welds. The argon-hydrogen gas mixture contains up to 15 percent hydrogen.

Fig. 12-22 lists the type of shielding gases and power sources suggested for various metals.

## 12-5. SELECTING CORRECT SHIELDING GAS FLOW RATE FOR GTAW

The FLOW RATE is the volume of gas flowing. The rate of flow is measured in cubic feet per hour or liters per minute. This flow rate varies with the base metal being welded, the thickness of the base metal, and the position of the welded joint. A higher gas flow rate is required when welding overhead. This is due to the fact that argon, being heavier than air, tends to fall away from the overhead joint.

Fig. 12-23 lists suggested flow rates for GTA welding various base metals, thickness and welding positions.

After determining the correct gas and flow rate, the flow rate must be properly set on the flowmeter. Refer to Heading 11-7 for an explanation of flowmeters and flow gauges. The vertical tube gas flowmeter is most common. See Fig. 12-24 for a schematic of a gas flowmeter.

Before setting the flowmeter, the shielding gas cylinder must be opened. The procedure for open-

| Metal Thickness | Joint Type | Tungsten Electrode Diameter | Filler Rod Diameter (If req'd.) | Amperage | Gas | | |
|---|---|---|---|---|---|---|---|
| | | | | | Type | Flow CFH | L/Min* |
| 1/16 in. (1.59 mm) | Butt | 1/16 in. (1.59 mm) | 1/16 in. (1.59 mm) | 60-70 | Argon | 15 | 7.08 |
| | Lap | 1/16 in. | 1/16 in. | 70-90 | Argon | 15 | |
| | Corner | 1/16 in. | 1/16 in. | 60-70 | Argon | 15 | |
| | Fillet | 1/16 in. | 1/16 in. | 70-90 | Argon | 15 | |
| 1/8 in. (3.18 mm) | Butt | 1/16-3/32 in. (1.59-2.38 mm) | 3/32 in. (2.38 mm) | 80-100 | Argon | 15 | 7.08 |
| | Lap | 1/16-3/32 in. | 3/32 in. | 90-115 | Argon | 15 | |
| | Corner | 1/16-3/32 in. | 3/32 in. | 80-100 | Argon | 15 | |
| | Fillet | 1/16-3/32 in. | 3/32 in. | 90-115 | Argon | 15 | |
| 3/16 in. (4.76 mm) | Butt | 3/32 in. (2.38 mm) | 1/8 in. (3.18 mm) | 115-135 | Argon | 20 | 9.44 |
| | Lap | 3/32 in. | 1/8 in. | 140-165 | Argon | 20 | |
| | Corner | 3/32 in. | 1/8 in. | 115-135 | Argon | 20 | |
| | Fillet | 3/32 in. | 1/8 in. | 140-170 | Argon | 20 | |
| 1/4 in. (6.35 mm) | Butt | 1/8 in. (3.18 mm) | 5/32 in. (4.0 mm) | 160-175 | Argon | 20 | 9.44 |
| | Lap | 1/8 in. | 5/32 in. | 170-200 | Argon | 20 | |
| | Corner | 1/8 in. | 5/32 in. | 160-175 | Argon | 20 | |
| | Fillet | 1/8 in. | 5/32 in. | 175-210 | Argon | 20 | |

*Liters per minute

Fig. 12-17. Variables for manual gas tungsten arc welding mild steel using DCEN (DCSP).

| Metal Thickness | Joint Type | Tungsten Electrode Diameter | Filler Rod Diameter (If req'd.) | Amperage | Gas | | |
|---|---|---|---|---|---|---|---|
| | | | | | Type | Flow CFH | L/Min |
| 1/16 in. (1.59 mm) | Butt | 1/16 in. (1.59 mm) | 1/16 in. (1.59 mm) | 60-85 | Argon | 15 | 7.08 |
| | Lap | 1/16 in. | 1/16 in. | 70-90 | Argon | 15 | |
| | Corner | 1/16 in. | 1/16 in. | 60-85 | Argon | 15 | |
| | Fillet | 1/16 in. | 1/16 in. | 75-100 | Argon | 15 | |
| 1/8 in. (3.18 mm) | Butt | 3/32-1/8 in. (2.38-3.18 mm) | 3/32 in. (2.38 mm) | 125-150 | Argon | 20 | 9.44 |
| | Lap | 3/32-1/8 in. | 3/32 in. | 130-160 | Argon | 20 | |
| | Corner | 3/32-1/8 in. | 3/32 in. | 120-140 | Argon | 20 | |
| | Fillet | 3/32-1/8 in. | 3/32 in. | 130-160 | Argon | 20 | |
| 3/16 in. (4.76 mm) | Butt | 1/8-5/32 in. (3.18-4.0 mm) | 1/8 in. (3.18 mm) | 180-225 | Argon | 20 | 11.80 |
| | Lap | 1/8-5/32 in. | 1/8 in. | 190-240 | Argon | 20 | |
| | Corner | 1/8-5/32 in. | 1/8 in. | 180-225 | Argon | 20 | |
| | Fillet | 1/8-5/32 in. | 1/8 in. | 190-240 | Argon | 20 | |
| 1/4 in. (6.35 mm) | Butt | 5/32-3/16 in. (4.0-4.76 mm) | 3/16 in. (4.76 mm) | 240-280 | Argon | 25 | 14.16 |
| | Lap | 5/32-3/16 in. | 3/16 in. | 250-320 | Argon | 25 | |
| | Corner | 5/32-3/16 in. | 3/16 in. | 240-280 | Argon | 25 | |
| | Fillet | 5/32-3/16 in. | 3/16 in. | 250-320 | Argon | 25 | |

Fig. 12-18. A table of the variables used when manually welding aluminum with the gas tungsten arc using AC and high frequency.

ing the shielding gas cylinder is as follows:

1. Turn the regulator adjusting screw outward in a counter clockwise direction. This insures that the regulator is closed. Note: When a preset pressure is set on the regulator, no ad-justing handle is used. Once the pressure is set, an acorn nut is placed over the adjusting screw. In this case, the regulator is always open.

2. Open the cylinder valve SLOWLY. Continue to

| Metal Thickness | Joint Type | Tungsten Electrode Diameter | Filler Rod Diameter (If req'd.) | Amperage | Gas Type | Flow CFH | L/Min |
|---|---|---|---|---|---|---|---|
| 1/16 in. (1.59) | Butt | 1/16 in. (1.59 mm) | 1/16 in. (1.59 mm) | 40-60 | Argon | 15 | 7.08 |
| | Lap | 1/16 in. | 1/16 in. | 50-70 | Argon | 15 | |
| | Corner | 1/16 in. | 1/16 in. | 40-60 | Argon | 15 | |
| | Fillet | 1/16 in. | 1/16 in. | 50-70 | Argon | 15 | |
| 1/8 in. (3.18 mm) | Butt | 3/32 in. (2.38 mm) | 3/32 in. (2.38 mm) | 65-85 | Argon | 15 | 7.08 |
| | Lap | 3/32 in. | 3/32 in. | 90-110 | Argon | 15 | |
| | Corner | 3/32 in. | 3/32 in. | 65-85 | Argon | 15 | |
| | Fillet | 3/32 in. | 3/32 in. | 90-110 | Argon | 15 | |
| 3/16 in. (4.76 mm) | Butt | 3/32 in. (2.38 mm) | 1/8 in. (3.18 mm) | 100-125 | Argon | 20 | 9.44 |
| | Lap | 3/32 in. | 1/8 in. | 125-150 | Argon | 20 | |
| | Corner | 3/32 in. | 1/8 in. | 100-125 | Argon | 20 | |
| | Fillet | 3/32 in. | 1/8 in. | 125-150 | Argon | 20 | |
| 1/4 in. (6.35 mm) | Butt | 1/8 in. (3.18 mm) | 5/32 in. (4.0 mm) | 135-160 | Argon | 20 | 9.44 |
| | Lap | 1/8 in. | 5/32 in. | 160-180 | Argon | 20 | |
| | Corner | 1/8 in. | 5/32 in. | 135-160 | Argon | 20 | |
| | Fillet | 1/8 in. | 5/32 in. | 160-180 | Argon | 20 | |

Fig. 12-19. A table of the variables used when welding stainless steel manually with the gas tungsten arc and DCEN or DCSP.

| Metal Thickness | Joint Type | Tungsten Electrode Diameter | Filler Rod Diameter (If req'd. Red) | Amperage[1] With Backup | W/O Backup | Gas Type | Flow CFH | L/Min |
|---|---|---|---|---|---|---|---|---|
| 1/16 in. (1.59 mm) | All | 1/16 in. (1.59 mm) | 3/32 in. (2.38 mm) | 60 | 35 | Argon | 13 | 6.14 |
| 3/32 in. (2.38 mm) | All | 1/16 in. (1.59 mm) | 1/8 in. (3.18 mm) | 90 | 60 | Argon | 15 | 7.08 |
| 1/8 in. (3.18 mm) | All | 1/16 in. (1.59 mm) | 1/8 in. (3.18 mm) | 115 | 85 | Argon | 20 | 9.44 |
| 3/16 in. (4.76 mm) | All | 1/16 in. (1.59 mm) | 5/32 in. (4.0 mm) | 120 | 75 | Argon | 20 | 9.44 |
| 1/4 in. (6.35 mm) | All | 3/32 in. (2.38 mm) | 5/32 in. (4.0 mm) | 130 | 85 | Argon | 20 | 9.44 |
| 3/8 in. (9.53 mm) | All | 3/32 in. (2.38 mm) | 3/16 in. (4.76 mm) | 180 | 100 | Argon | 25 | 11.80 |
| 1/2 in. (12.7 mm) | All | 5/32 in. (4.0 mm) | 3/16 in. (4.76 mm) | — | 250 | Argon | 25 | 11.80 |
| 3/4 in. (19.05 mm) | All | 3/16 in. (4.76 mm) | 1/4 in. (6.35 mm) | — | 370 | Argon | 35 | 16.52 |

1 - Use alternating current with a constant high frequency (AC-HF)

Fig. 12-20. Variable for welding magnesium manually with the gas tungsten arc using AC-HF.

| Metal Thickness | Joint Type | Tungsten Electrode Diameter | Filler Rod Diameter (If req'd. Red) | Amperage[1] | Gas Type | Flow CFH | L/Min |
|---|---|---|---|---|---|---|---|
| 1/16 in. (1.59 mm) | All | 1/16 in. (1.59 mm) | 1/16 in. (1.59 mm) | 110-150 | Argon | 15 | 7.08 |
| 1/8 in. (3.18 mm) | All | 3/32 in. (2.38 mm) | 3/32 in. (2.38 mm) | 175-250 | Argon | 15 | 7.08 |
| 3/16 in. (4.76 mm) | All | 1/8 in. (3.18 mm) | 1/8 in. (3.18 mm) | 250-325 | Argon | 18 | 9.50 |
| 1/4 in. (6.35 mm) | All | 1/8 in. (3.18 mm) | 1/8 in. (3.18 mm) | 300-375 | Argon | 22 | 10.38 |
| 3/8 in. (9.53 mm) | All | 3/16 in. (4.76 mm) | 3/16 in. (4.76 mm) | 375-450 | Argon | 25 | 11.80 |
| 1/2 in. (12.7 mm) | All | 3/16 in. (4.76 mm) | 1/4 in. (6.35 mm) | 525-700 | Argon | 30 | 14.16 |

1 - Use DCEN (DCSP)

Fig. 12-21. Variables for welding deoxidized copper using the gas tungsten arc and DCEN or DCSP.

| METAL | THICKNESS | MANUAL | AUTOMATIC (MACHINE) |
|---|---|---|---|
| Aluminum and aluminum alloys | Under 1/8 in. (3.2 mm)<br>Over 1/8 in. (3.2 mm) | Ar[1] (AC-HF)<br>Ar (AC-HF)[3] | Ar (AC-HF) or He[2] (DCEN)<br>Ar-He (AC-HF) or He (DCEN)[4] |
| Copper | Under 1/8 in. (3.2 mm)<br>Over 1/8 in. (3.2 mm) | Ar-He (DCEN)<br>He (DCEN) | Ar-He (DCEN)<br>He (DCEN) |
| Nickel alloys | Under 1/8 in. (3.2 mm)<br>Over 1/8 in. (3.2 mm) | Ar (DCEN)<br>Ar-He (DCEN) | Ar-He (DCEN) or He (DCEN)<br>He (DCEN) |
| Steel, carbon | Under 1/8 in. (3.2 mm)<br>Over 1/8 in. (3.2 mm) | Ar (DCEN)<br>Ar (DCEN) | Ar (DCEN)<br>Ar-He (DCEN) or He (DCEN) |
| Steel, stainless | Under 1/8 in. (3.2 mm)<br>Over 1/8 in. (3.2 mm) | Ar (DCEN)<br>Ar-He (DCEN) | Ar-He (DCEN) or Ar-H₂[5] (DCEN)<br>He (DCEN) |
| Titanium and its alloys | Under 1/8 in. (3.2 mm)<br>Over 1/8 in. (3.2 mm) | Ar (DCEN)<br>Ar-He (DCEN) | Ar (DCEN) or Ar-He (DCEN)<br>He (DCEN) |

1 - Ar (argon)
2 - He (helium)
3 - AC-HF (alternating current, high frequency)
4 - DCEN (direct current electrode negative, also DCSP)
5 - H₂ (hydrogen)

Fig. 12-22. Suggested choices of shielding gases and power sources for welding various metals.

| Base Metal | Joint | Thickness | Weld* Position | Flow (CFH) Argon | Flow (CFH) Helium | Flow (L/Min) Argon | Flow (L/Min) Helium |
|---|---|---|---|---|---|---|---|
| Aluminum and Aluminum Alloys | Fillet, lap, edge, and corner | 1/16 | FVH<br>O | 16<br>20 | | 7.55<br>9.44 | |
| | | 3/32 | FVH<br>O | 18<br>20 | | 8.50<br>9.44 | |
| | Butt | 1/16, 3/32, 1/8 | FVH<br>O | 20<br>25 | | 9.44<br>11.80 | |
| | Fillet, lap, edge, and corner | 1/8 | FVH<br>O | 20<br>25 | | 9.44<br>11.80 | |
| | Butt | 3/16 | FVH<br>O | 25<br>30 | | 11.80<br>14.16 | |
| | Butt | 1/4 | FVH<br>O | 30<br>35 | | 14.16<br>16.52 | |
| | Fillet, lap, edge, and corner | 3/16, 1/4 | FVH<br>O | 30<br>35 | | 14.16<br>16.52 | |
| | Butt, fillet, lap, edge, and corner | 3/8 | FVH<br>O | 35<br>40 | | 16.52<br>18.88 | |
| Nickel and Nickel Alloys | | to max. for GTAW (3/8 approx.) | | 10-20 | 1½ to 3 times the argon flow | 4.72- 9.44 | |
| Copper and Copper Alloys | | 1/16<br>1/8<br>3/16<br>1/4<br>1/2 | FVH<br>O | 10-15<br>14-20<br>16-22<br>20-30<br>25-35 | 8-12<br>10-15<br>12-18<br>16-25<br>20-30 | 4.72- 7.08<br>6.61- 9.44<br>3.56-10.38<br>9.44-14.16<br>11.80-16.52 | 3.78- 5.66<br>4.72- 7.08<br>5.66- 8.50<br>3.56-11.80<br>9.44-14.16 |

* Positions — F-Flat. H-Horizontal. V-Vertical. O-Overhead.

Fig. 12-23. Suggested shielding gas flow rates for GTAW various metals, thicknesses, and positions.

open it until it is fully opened. This is necessary because a back seating valve is used to seal the valve stem from leakage. A back seating valve is also used in an oxygen cylinder. Refer to Heading 3-3 and Fig. 3-5.

3. If a regulator adjusting screw is used, turn it in to the pressure at which the flowmeter is calibrated. Most flowmeters are calibrated at 50 psig (344.7 kPa). The calibrating pressure should be indicated somewhere on the flowmeter.

4. Remove the GTAW torch from the gas economizer (if used). A gas economizer is a valve to turn off the flow of the shielding gas. See Fig. 11-56. This is done to insure a flow of gas when the cylinder is opened. Or, activate

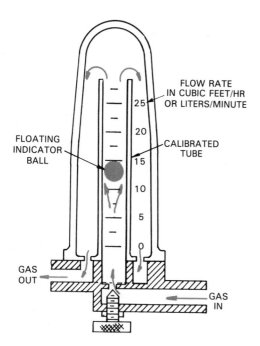

Fig. 12-24. A schematic of a gas flowmeter used on GTAW or GMAW outfits. When adjusting this gauge, the top of the ball indicates the gas flow through the meter. The rate of flow is measured in cubic feet per hour (ft³/hr) or liters per minute (L/min). This flowmeter is set for 15 cubic feet per hour (ft³/hr).

the contractor switch to cause the gas to flow through the torch.

5. Turn the adjusting knob out on the flowmeter until the top of the floating indicator ball is at the flow rate line desired.

6. Place the GTAW torch back on to the gas economizer (if used) to stop the flow of the shielding gas. Or, close the contactor switch to stop the gas flow.

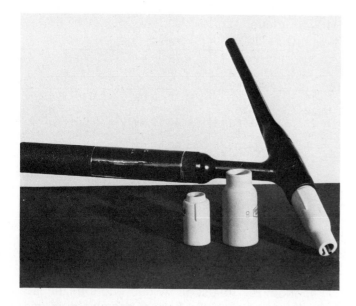

Fig. 12-25. This GTAW torch is shown with three ceramic nozzles of different lengths and discharge diameters. (Diamonite Products)

## 12-6. SELECTING CORRECT GTAW TORCH NOZZLE

Nozzles used on gas tungsten arc welding torches vary in size and method of attachment, Fig. 12-25. The end of the nozzle which attaches to the torch varies in design. This design variation is necessary to permit attachment to different manufacturers' torches.

Most nozzles used for GTAW are manufactured from ceramic materials.

The diameter of the nozzle closest to the arc or the exit diameter is manufactured in a variety of sizes; see Fig. 12-26. GTAW nozzles are also made in various lengths as shown in Fig. 12-27.

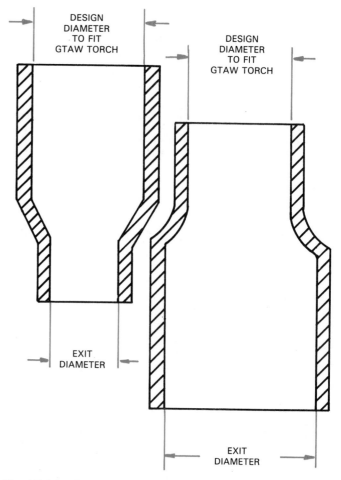

Fig. 12-26. Two different GTAW torch nozzles. These nozzles show the design diameter which permits a nozzle to fit a certain torch. The variation in exit diameter is also shown.

Manufacturers may put their own distinct part number on each nozzle. In addition, a single digit number is used to identify the exit diameter. This number designates the exit diameter of the nozzle in 1/16 in. (1.6 mm). A number 6 nozzle is:

6 × 1/16 in. = 3/8 in. in diameter or
6 × 1.6 mm = 9.6 mm in diameter

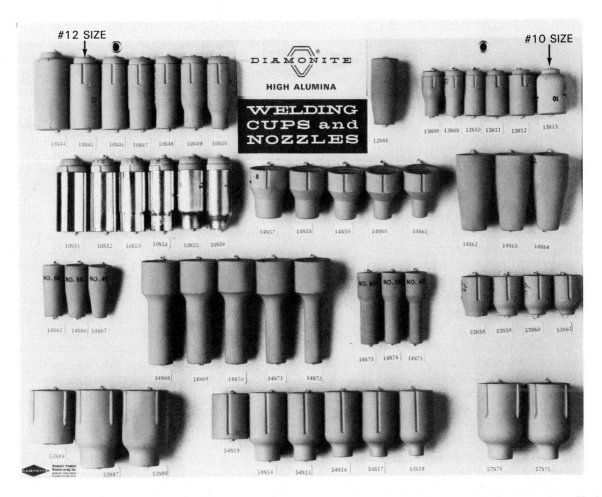

Fig. 12-27. A variety of ceramic GTAW torch nozzles. Notice the various lengths and exit diameter sizes. Notice also the design ends which will fit various torches. Six of the nozzles on line two have a metallic coating to reflect heat. Notice that a few of the nozzles are turned to show their number size.    (Diamonite Products)

A number 8 nozzle is:

$$8 \times 1/16 \text{ in.} = 1/2 \text{ in. in diameter or}$$
$$8 \times 1.6 \text{ mm} = 12.8 \text{ mm}$$

The diameter of the nozzle must be large enough to allow the entire weld area to be covered by the shielding gas. The action of the shielding gas will vary with a given gas flow rate as the diameter of the nozzle changes. With a gas flow rate of 14 cubic feet/hr (cfh) or 6.61 liters/minute (L/min) and a number 8 nozzle, the gas will flow out of the nozzle slowly and gently. With the same flow rate and a number 4 nozzle, the gas will flow out more rapidly and may be blown away from the joint more quickly. However, a low number nozzle will have to be used in a narrow groove in order to reach the bottom of the joint.

The choice of the nozzle size is often a compromise which is necessary to meet the requirements of the job. Small diameter nozzles are often used to permit a constant arc length. As an example, see Fig. 12-28. In this case, the important root pass is to be welded. The nozzle is constantly touching the sides of the groove as it is rocked back and forth across the root opening. It is also kept in contact with the groove opening as the weld moves forward slowly. This choice of a small nozzle diameter allows the welder to reach

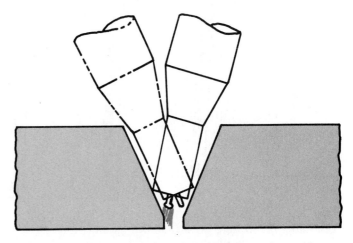

Fig. 12-28. Using a small exit diameter allows the welder to reach the bottom of a groove weld. It also allows the welder to hold a constant arc length if the nozzle is held against the sides of the groove.

the bottom of the groove. It also allows the welder to keep a constant arc length. A higher gas flow may be needed in this case to compensate for the smaller nozzle diameter.

## 12-7. SELECTING AND PREPARING A TUNGSTEN ELECTRODE

The selection of the correct type and diameter of tungsten electrode is extremely important to performing a successful gas tungsten arc weld. Heading 11-11 describes the various types of tungsten electrodes used. It also lists the diameters and lengths of electrodes available.

The types of tungsten electrodes are:
1. Pure tungsten (painted with a green color band).
2. Tungsten with one or two percent thoria added. These are called thoriated tungsten electrodes (1%-yellow band, 2%-red band).
3. Tungsten with from 0.15 to 0.40 percent zirconia added (brown band).
4. Pure tungsten with a core of tungsten which contains one to two percent thoria. When the core and pure tungsten combine, the percentage of thoria is from 0.35-0.55 (blue band).

PURE TUNGSTEN ELECTRODES are preferred for AC welding on aluminum or magnesium. They are preferred because they form a ball or hemisphere at the tip when they are heated. This shape is preferred because it reduces current rectification and allows the AC to flow more easily. See A, Fig. 12-29.

THORIATED TUNGSTEN ELECTRODES are preferred for DC applications. See B, Fig. 12-29. Thoria increases the electron emissions from a tungsten electrode. The addition of one percent to two percent thoria also keeps the electrode tip from forming a ball. Thoriated electrodes are usually ground to a point or to a near point when used with direct current (DC). Refer to B, Fig. 12-29.

If a thoriated electrode is used with AC, it will not form a ball. It will form a number of small projections at the tip. These projections will cause a wild, wandering of the arc. Therefore thoriated electrodes should only be used for direct current GTAW. See C, Fig. 12-29.

A pure tungsten electrode used with AC is shown in D, Fig. 12-29. The amperage in this case was too high. The ball on the end has begun to melt and bend to one side. With continued use, the ball may have fallen into the weld and contaminated it. Excessive current may also cause tungsten electrodes to split near the end.

A pure tungsten electrode will always form a ball at the tip when used with AC. When used with AC, the pure tungsten electrode is not ground to a point. A ball the same size as the electrode diameter forms when the arc is struck. This ball may be up to one and one half times the size of the electrode diameter, but should never exceed that

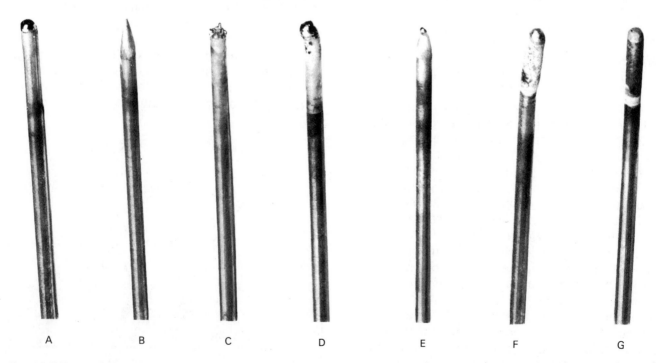

A    B    C    D    E    F    G

Fig. 12-29. The appearance of several properly and improperly used tungsten electrodes. A—Pure tungsten electrode with ball or hemisphere. B—Thoriated electrode ground to point. C—Thoriated electrode used with AC. D—Pure tungsten electrode used with excessive amperage AC. E—Pure tungsten electrode ground to a point. F—Tungsten electrode contaminated by filler rod. G—Tungsten electrode contaminated by air.    (Miller Electric Mfg. Co.)

limit. Above the one and one half times the electrode diameter limit, the tungsten in the ball may melt off and fall into the weld. If the ball on the end of the tip is much larger than the electrode diameter, the current may be set too high.

A pure tungsten electrode which was ground to a point and used with direct current electrode negative (DCEN or DCSP) is illustrated at E, Fig. 12-29. Notice the small ball formed at the tip. Pointing of pure tungsten is not recommended. The small ball formed at the tip may melt off and fall into the weld.

Tungsten electrodes of any type must be protected from contamination. The electrode should not be touched to the base metal, weld metal, or filler metal while it is hot. View F, Fig. 12-29 il-

lustrates a tungsten electrode which was touched by the filler rod. Such contamination prevents the electrode from emitting or receiving electrons effectively. The end of this electrode must be broken off and reshaped. See Fig. 11-24 for the proper ways of retipping a contaminated or split electrode.

The hot tungsten electrode and metal in the weld area may be contaminated by oxygen and nitrogen in the air and by airborne dirt. To prevent this contamination, a shielding gas is used. The shielding gas is allowed to flow over the electrode and the weld area after the arc is broken. The timing of this shielding gas AFTER FLOW is set on the arc welding machine panel. See Fig. 12-12 and Heading 12-3.

| PURE TUNGSTEN | CURRENT RANGE—AMPERES | | | |
|---|---|---|---|---|
| Diameter (mm) | DCEN or DCSP Argon | DCEP or DCRP Argon | AC-HF Argon | AC Balanced Wave-Argon |
| .010" (.254) | Up to 15 | * | Up to 15 | Up to 10 |
| .020" (.508) | 5-20 | * | 5-20 | 10-20 |
| .040" (1.02) | 15-80 | * | 10-60 | 20-30 |
| 1/16" (1.59) | 70-150 | 10-20 | 50-100 | 30-80 |
| 3/32" (2.38) | 125-225 | 15-30 | 100-160 | 60-130 |
| 1/8" (3.18) | 225-360 | 25-40 | 150-210 | 100-180 |
| 5/32" (3.97) | 360-450 | 40-55 | 200-275 | 160-240 |
| 3/16" (4.76) | 450-720 | 55-80 | 250-350 | 190-300 |
| 1/4" (6.35) | 720-950 | 80-125 | 325-450 | 250-400 |
| 2% Thorium Alloyed Tungsten Diameter (mm) | | | | |
| .010" (.254) | Up to 25 | * | Up to 20 | Up to 15 |
| .020" (.508) | 15-40 | * | 15-35 | 5-20 |
| .040" (1.02) | 25-85 | * | 20-80 | 20-60 |
| 1/16" (1.59) | 50-160 | 10-20 | 50-150 | 60-120 |
| 3/32" (2.38) | 135-235 | 15-30 | 130-250 | 100-180 |
| 1/8" (3.18) | 250-400 | 25-40 | 225-360 | 160-250 |
| 5/32" (3.97) | 400-500 | 40-55 | 300-450 | 200-320 |
| 3/16" (4.76) | 500-750 | 55-80 | 400-550 | 290-390 |
| 1/4" (6.35) | 750-1000 | 80-125 | 600-800 | 340-525 |
| Zirconium Alloyed Tungsten Diameter (mm) | | | | |
| .010" (.254) | * | * | Up to 20 | Up to 15 |
| .020" (.508) | * | * | 15-35 | 5-20 |
| .040" (1.02) | * | * | 20-80 | 20-60 |
| 1/16" (1.59) | * | * | 50-150 | 60-120 |
| 3/32" (2.38) | * | * | 130-250 | 100-180 |
| 1/8" (3.18) | * | * | 225-360 | 160-250 |
| 5/32" (3.97) | * | * | 300-450 | 200-320 |
| 3/16" (4.76) | * | * | 400-550 | 290-390 |
| 1/4" (6.35) | * | * | 600-800 | 340-525 |
| *NOT RECOMMENDED | | | | |
| The figures listed are intended as a guide, and are a composite of recommendations from American Welding Society and electrode manufacturers. | | | | |

Fig. 12-30. The suggested current range for various types and sizes of tungsten electrodes.

In G, Fig. 12-29, this electrode was contaminated by the air. The black surface of the electrode indicates that the shielding gas did not have sufficient after flow time to protect it. An electrode contaminated by oxidation must not be reused. If it is reused as is, the oxides will fall from the electrode into the weld. The electrode must be broken off behind the contamination. Such an electrode must then be retipped for use.

Tungsten electrodes come in sizes from .010 in. to 1/4 in. (.25-6.4 mm). The most frequently used lengths are 6 and 7 in. (152.4 and 177.8 mm).

Each diameter and type of electrode has a maximum amperage which it can carry. This amperage limit varies with the current, polarity, and shielding gas used.

Fig. 12-30 lists the operating current range for each type of tungsten electrode. Pure tungsten electrodes used with AC are sometimes ground to a blunt point. The arc is then struck to form a small ball at the tip. This point must not be sharp or the tip may fall into the crater when the ball forms. Thoriated tungsten electrodes are preferred for DC welding applications. They should be ground to a point or to a near point for use with high currents.

Special grinding wheels should be used for pointing tungsten electrodes. These wheels should be used only for grinding tungsten electrodes. Freedom from contamination is essential.

Silicon carbide or alumina oxide grinding wheels are preferred. Alumina oxide cuts more slowly, but lasts longer than silicon carbide wheels. Electrodes should be rough ground on an 80 grit grinding wheel. The finish grinding should be done on a 120 grit wheel. A grinding wheel with an open structure is best because it will run cooler and pick up less contamination.

Electrodes should be ground in a lengthwise direction. The grinding marks on the tapered area must run in a lengthwise direction. This method of grinding insures the best current carrying characteristics. Figs. 12-31 and 12-32 illustrate the suggested method for grinding tungsten electrodes. Always wear safety goggles when grinding.

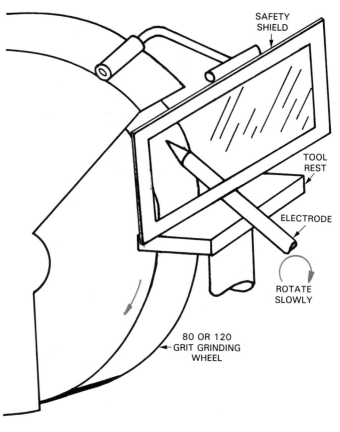

Fig. 12-32. The correct position for grinding a tungsten electrode. The grinding marks on the taper must run lengthwise on the electrode. Use an 80 grit wheel for rough grinding and a 120 grit wheel for finish grinding.

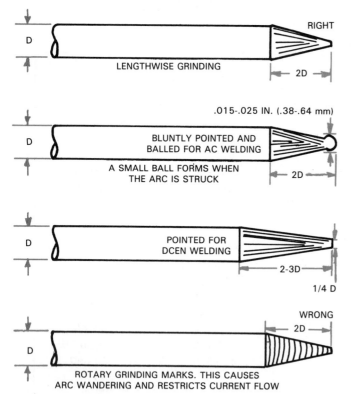

Fig. 12-31. Methods of grinding tungsten electrodes. Note the lengthwise direction of the grinding marks on the properly ground electrodes.

After the electrode diameter has been selected, a collet of the same inside diameter must be placed into the torch. The electrode is then placed into the torch collet. A torch collet or chuck is shown in Fig. 11-14. An electrode may be adjusted even with the end of the nozzle. It may extend up to 1/2 in. (approximately 13 mm) beyond the nozzle. However, as a general rule, an extension distance equal to one electrode diameter is an average setting. The extension beyond the nozzle is determined by the shape of the joint. A longer extension permits the welder to see the arc crater better. Higher gas flow rates are required with longer extensions in order to protect the electrode from contamination. The extension distance of the electrode generally should not be greater than the exit diameter of the nozzle.

## 12-8. SELECTING CORRECT FILLER METAL FOR USE WITH GTAW

Filler metal used for gas tungsten arc welding (GTAW) is generally bare wire. This filler metal is produced in a solid wire form. Corrosion resisting chromium and chromium-nickel steel filler metal comes as a solid, composite cored, or stranded rod. The filler metal is purchased in coils of several hundred feet or in precut lengths. The coiled wire is cut to any desired length by the welder. The precut wire is usually purchased in lengths of 24 or 36 in. (610 or 914 mm).

Steel welding rods should not be copper coated as in oxyfuel gas welding. The copper coating will cause spatter which may contaminate the tungsten electrode.

The diameters of the filler wires most frequently used varies from 1/16 in.-1/4 in. (1.59-6.35 mm). Smaller diameter wire is readily available in coils to .015 in. (.38 mm). Precut filler rods are readily available up to 1/4 in. (6.35 mm). See Fig. 11-25 for a listing of the AWS specifications for filler metals used with GTAW.

The proper filler rod diameters to use when welding various thicknesses of metals are shown in Figs. 12-17 to 12-21.

## 12-9. PREPARING METAL FOR WELDING

GTAW is generally chosen as a welding method because it produces very high quality welds. Therefore, before welding is attempted, the surface of the base metal must be cleaned. All rust, dirt, oil, grease, and other materials which may be on the metal surface must be removed. Any such materials left on the base metal will cause contamination of the weld metal. This will undoubtedly cause weak and defective welds.

Aluminum and magnesium oxides on the surface of aluminum and magnesium are removed while welding with AC or DCEP (DCRP). No such surface cleaning occurs when DCEN (DCSP) is used. If it is necessary to use DCEN when welding aluminum or magnesium, the surface oxides must be removed prior to welding. This generally is done chemically or with abrasive cloth, a stainless steel brush, a grinding wheel, or steel wool.

The joint designs used for GTAW are the same as those used for SMAW. On thin metal sections, the flange type joint is used without filler metal. When gas tungsten arc welding metal over 3/16 in. (5 mm), the edge is generally beveled or machined to the desired groove shape.

If parts must be beveled or cut to shape or size, they may be cut by thermal or mechanical means. Plasma arc cutting produces a clean, fairly smooth and accurate cut. Machining, sawing, shearing, and grinding are the mechanical processes usually used.

Caustic soda and other suitable cleaning solutions may be used to clean the surface. Refer to the AWS, ''Welding Handbook, Volume 4'' for specific chemical cleaning solutions for use with each type of metal. Always wear goggles to protect the eyes when using chemical solutions. Phosgene gas is formed when arc welding on parts that have been cleaned with chlorinated hydrocarbon solvents if a film of the solvent is present. Follow all safety rules!

Metal parts of a weldment must be held in alignment while welding. This may be done by using jigs and fixtures. It is often done by tack welding the parts into place.

To control the depth of penetration, a backing strip or ring is used. The backing ring or strip is also used when welding metals that may experience HOT SHORTNESS. Hot shortness occurs on metals like aluminum, magnesium, and copper. These metals, when they reach the melting temperature, become very weak. As metal in the weld area reaches the melting temperature, it may fall away. This would leave a large hole in the base metal. A backing strip or ring helps to prevent this falling away of the metal.

It is suggested that shielding gas be used on the back or root side of the weld when welding metals like magnesium, titanium, or zirconium. This backing gas is applied by various means. It may be applied through a metal channel attached to the root side of the weld. In pipes or closed containers, the inside of these weldments may be filled with the shielding gas. A vent must be provided when gas is used in a closed container. This vent allows the gas to escape, thus avoiding a pressure buildup which may prevent good penetration.

CAUTION: Never enter an area filled with a shielding gas unless supplied air breathing equipment is worn. A person can lose consciousness in as little as seven seconds in spaces filled with shielding gases like argon. Death from suffocation has resulted. This is because shielding gases displace oxygen.

## 12-10. METHODS OF STARTING THE ARC

The gas tungsten arc may be started in one of three ways. These are by touch starting, by the application of a superimposed high frequency, and high voltage starting.

To start the arc, the remote finger or foot operated contactor switch must be depressed. See Heading 12-3 and Figs. 12-9 and 12-10. This switch also causes the shielding gas and cooling water to flow prior to starting the arc.

When TOUCH STARTING is used, the electrode is touched to the base metal and withdrawn about 1/8 in. (about 3 mm). After a few seconds when the arc is stabilized (running smoothly), the arc may be brought down to a short arc length of about 1/32-3/32 in. (.8-2.4 mm). By touching the tungsten electrode to the base metal it may become contaminated.

To start the arc using the SUPERIMPOSED HIGH FREQUENCY, place the nozzle on the metal as shown in Fig. 12-33. With the electrode and nozzle in this position, the contactor switch is turned on to start the high frequency current. The machine contactor switch may be operated by a foot or finger operated remote switch. Another method when using high frequency start is to hold the electrode horizontally about 1 in. (about 25 mm) above the metal. The electrode is then rotated toward a vertical position. As the electrode comes near the base metal, the high frequency will jump the gap to start the arc. When using direct current, the high frequency will turn off automatically when the arc is stabilized. The high frequency should remain on constantly when using alternating current.

HIGH VOLTAGE STARTING is done with a high voltage surge. The electrode is brought close to the base metal as in high frequency starting. When the contactor switch is turned on, a high voltage surge causes the arc to jump and start the arc. After the arc is stabilized, the voltage surge stops automatically.

When welding with alternating current, the electrode forms a ball or spherical shape on the end. This ball end can be formed before the actual weld begins. To form the ball on the electrode tip, strike the arc on a clean piece of copper. Copper will not melt easily and will not contaminate the electrode readily if a touch occurs. This piece of copper may be 2 in. × 2 in. (roughly 50 × 50 mm) and 1/16 in. (1.59 mm) thick. It should be kept by the welder as a part of the station equipment.

## 12-11. GAS TUNGSTEN ARC WELDING TECHNIQUES

One advantage of GTAW is a weld may be made with a small heat affected zone around the weld. Oxyfuel gas and SMAW heat a large area while the metal is raised to the melting temperature. This causes a large heat affected zone and a potentially weaker metal area around the weld.

Another advantage of GTAW is that there is no metal transfer through the arc. There is no spattering of metal globules from the arc or crater. The arc action is very quiet and the completed weld is of high quality.

GTAW should be done with the lowest current necessary to melt the metal. The highest welding speed possible, which will insure a sound weld, should be used.

Once the arc is struck, it is directed to the area to be melted. A molten weld pool is formed under the arc and the filler rod is added to fill the pool. The width of the pool, when making stringer beads, should be about 2-3 times the diameter of the electrode used. If the bead must be wider, a weaving bead is used. Several stringer beads may also be used to fill a wide groove joint. Sufficient shielding gas must flow to protect the molten metal in the weld area from becoming contaminated.

The filler rod must not be withdrawn from the area protected by the shielding gas. If the filler rod is withdrawn while it is molten, it will become contaminated. If it is then melted into the weld, the weld will be contaminated.

After the arc is struck, heat a spot until a molten pool forms. The electrode should be held at about a 60-75 degree angle from the work. Hold the filler rod at about a 15-20 degree angle to the work. See Fig. 12-34. When the molten pool

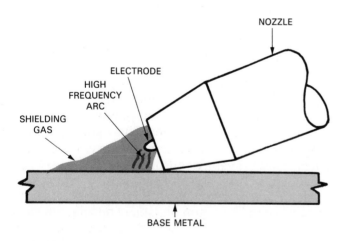

Fig. 12-33. The position of the electrode and nozzle for high frequency arc starting.

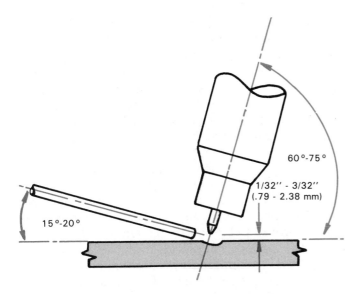

Fig. 12-34. The relative positions of the electrode and filler rod when GTAW.

reaches the desired size, add the filler rod to the pool. When the filler rod is to be added, the electrode should be moved to the back of the pool. The filler rod is then added to the forward part of the molten pool. Refer to Fig. 12-35. This technique of adding the filler metal to the pool may be used for all weld joints in all positions.

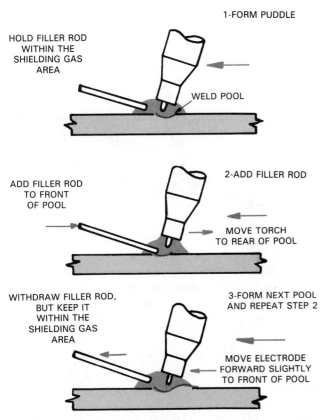

Fig. 12-35. The steps required to add filler metal during GTAW. This process is continuous until the weld is completed.

## 12-12. SHUTTING DOWN THE GTAW STATION

Each time the arc is broken, the shielding gas continues to flow for a few seconds. This is done to protect the weld metal, electrode, and filler metal from becoming contaminated by the surrounding atmosphere. The gas also continues to flow after the torch or foot operated contactor (off-on) switch is turned off.

When welding is stopped for a short time, the GTAW torch is hung on an insulated hook. It may be hung on the hook of a gas economizer valve.

If welding is to be stopped for a long period of time, the station should be shut down. After the gas post-flow period, hang up the torch. Shut off the shielding gas cylinder. Turn on the torch or foot operated contactor switch to start the gas flow. Lift the torch from the economizer, if used. This is done to drain the complete shielding gas system of gas. Hang the torch up again. Turn out the regulating screw on the regulator to turn it off. Screw in the flowmeter to shut it off. If the flowmeter is not turned off, the float ball will hit the top of the flowmeter very hard when the regulator is opened again. Turn off the arc welding machine power switch.

## 12-13. WELDING JOINTS IN FLAT POSITION

Welding joint designs for GTAW are the same as those used for oxyfuel gas and shielded metal arc welding. The flat or downhand position is generally the easiest position in which to weld.

### SQUARE GROOVE WELD ON A FLANGE JOINT

The metal should be bent up as shown in Fig. 12-36. This bent up edge is melted down to form a weld. No filler rod is required. The electrode should be held at about 60-75 degrees from the base metal. Heat is applied to the flanged metal to form a weld pool on both edges. As the arc crater increases in size, it will touch the outer edges of the flange. When this occurs, the electrode must be moved ahead to the front edge of the crater. When the crater touches the outer edges of the flange, the electrode is again moved ahead. This is repeated until the weld is completed. When the end of the weld is reached, the electrode should be moved backward slightly and raised slowly until the arc stops. Hold the electrode near the end of the weld for a short time. This is done so that the post-flow of the shielding gas protects the weld and electrode as they cool. The post-flow of the shielding gas is set on the arc welder. This flow is necessary to prevent the hot metal and electrode from becoming contaminated.

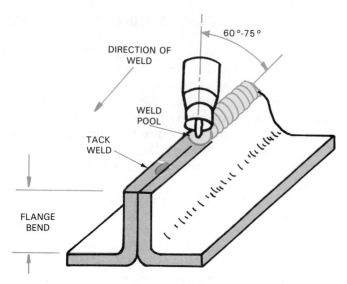

Fig. 12-36. A flange weld in progress. Note the torch angle, puddle size, and the tack weld.

## FILLET WELD ON A LAP JOINT

The fillet weld on a lap weld should be set up as shown in Fig. 12-37 and tacked about every 3 in. (about 75 mm). An edge and a surface are being

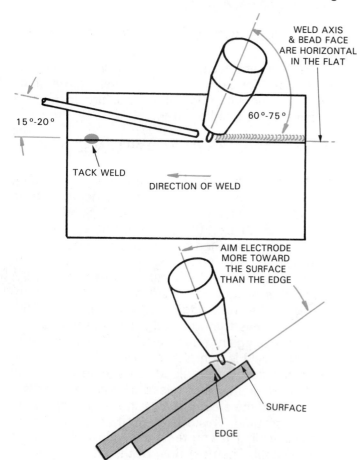

Fig. 12-37. A lap weld being welded with GTAW in the flat (downhand) position. Notice the suggested angles for the electrode and filler metal.

heated on a lap joint. The electrode should be aimed more toward the surface, since it will take more heat to melt. The metal should be heated to the melting point. When the crater forms a "C" shape as shown in Fig. 12-38, the electrode is moved to the rear of the crater. At the same time the filler rod is added to the front of the crater. The "C" shape crater indicates that both pieces are molten and flowing together. Enough filler rod is melted into the crater to form the desired bead contour. A flat or convexed bead is generally desired. The filler rod is then withdrawn slightly, but should remain within the shielding gas area. As the filler rod is withdrawn, the electrode is moved to the front of the crater. The "C" shaped crater is again formed and the filler rod added. This procedure is continued to the end of the weld. When the end of the weld is reached, the electrode is slowly raised and moved to the rear of the crater. Filler metal is added to fill the crater as the electrode is withdrawn. The electrode is continually withdrawn until the arc stops. If a wide joint is welded, several passes may be required to fill the joint.

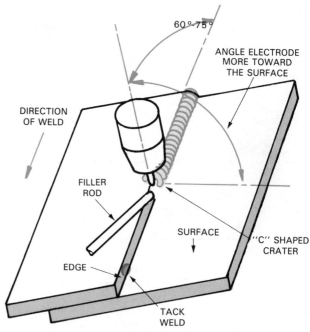

Fig. 12-38. A fillet weld in progress on a lap joint. Notice that the electrode is aimed more toward the surface than toward the edge. The edge will melt more easily.

## FILLET WELD ON AN INSIDE CORNER JOINT

The inside corner joint should be set up as shown in Fig. 12-39. It should be tack welded about every 3 in. (about 75 mm). The fillet weld is being made on two surfaces in this case. Each piece must be heated evenly. Therefore, the electrode should be kept about a 45 degree angle to each piece as

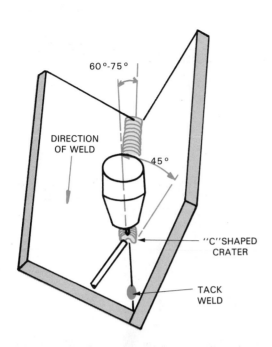

Fig. 12-39. An inside corner joint. A fillet weld is in progress in the flat position using GTAW. In the flat position, the face of the weld should be near horizontal.

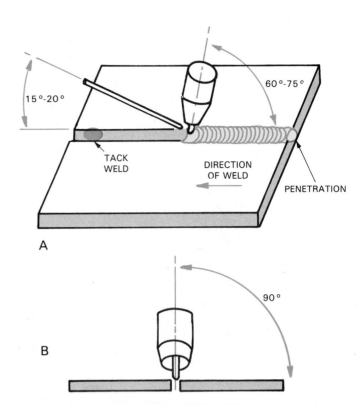

Fig. 12-40. A square groove butt weld in progress in the flat position. Note the penetration is 100%. The torch angles are the same as for other weld joints. The electrode should point straight down the weld line as shown in view B.

shown in Fig. 12-39. When both pieces are melted sufficiently to run together, the crater will appear "C" shaped. When the "C" shaped crater forms, the electrode is moved to the rear of the crater. At the same time the filler rod is added to the front of the crater. Continue to melt the filler rod until a flat or convex bead forms. Withdraw the filler rod and move the electrode to the front of the crater. Reform the "C" shaped crater. Add filler rod to the front of the crater while the electrode is moved to the rear of the crater. Repeat this action to the end of the weld. When the filler rod is withdrawn, always keep it within the area protected by the shielding gases. Finish off the weld by adding filler rod to the end of the crater as the arc is withdrawn and stops. If a wide joint is welded, several passes will be required to fill the joint.

## SQUARE GROOVE AND V-GROOVE WELDS ON A BUTT JOINT

Metal under 3/16 in. (4.8 mm) in thickness may not require edge preparation. If full penetration is difficult to obtain from one side, the edges should be machined or flame cut. When metals are over 3/16 in. (4.8 mm) in thickness, the edges must be machined or flame cut to obtain full penetration.

A square groove weld on a butt joint is shown in Fig. 12-40. Note the electrode and filler rod position in relation to the weld line. The electrode should

point straight down the weld line as shown in view B, Fig. 12-40. If the electrode is pointed toward one piece or the other, the weld bead may pile up on one side. Notice also the suggested electrode and filler rod angles from the base metal. The weld joint designs used for GTAW are the same as those used for SMAW. See Figs. 2-1 and 10-30 for additional information on basic welding designs.

GTAW is seldom used to weld metals over 1/4 in. (6.4 mm) thick except in very critical applications. On carbon steel, alloy steel, stainless steel, steel, copper, and other metals over 1/4 in. (6.4 mm), GTAW is used only for the root pass. After the root pass is laid, other types of welding processes are generally used. Other processes used may be SMAW, SAW, and GMAW.

The groove design shapes may be square, U, J, V, or bevel. These design shapes may be cut on both sides of a thick metal butt weld.

Fig. 12-41 shows a single sided, V-groove butt weld in progress. The torch and filler rod angles are the same as those used on other GTAW butt joints. Complete penetration is required in a butt weld. To obtain complete penetration, the KEYHOLE method of welding is used. See Fig. 4-21 and refer to Heading 4-21 for an explanation of keyhole welding.

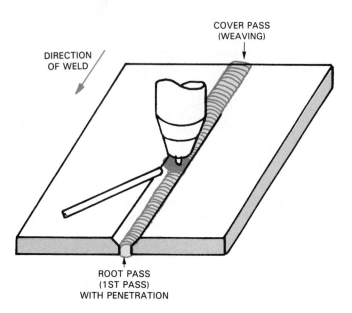

Fig. 12-41. A V-groove butt weld, welded in the flat position with the gas tungsten arc. Two or more weld passes may be required to weld a thick joint.

The width of a stringer bead, like the root pass, should only be 2-3 times as wide as the electrode diameter. To complete a V-groove or other wide mouthed joint, a weaving bead might be used. A weaving bead done with the gas tungsten arc should only be about 6 times the electrode diameter in width. If a wider area must be welded, then several narrower passes must be used.

## SQUARE OR V-GROOVED OUTSIDE CORNER JOINT

The square or V-grooved outside corner joints are set up as shown in Fig. 12-42. They are welded in the same manner as a square groove or V-grooved butt weld.

### 12-14. WELDING JOINTS IN HORIZONTAL POSITION

The welding procedures for GTAW explained in Heading 12-13 are used when welding in the horizontal position.

When horizontal welding is done, the weld line is horizontal and the face of the bead is in a near vertical position. Molten weld metal has a tendency to move downward. To prevent this downward movement of the weld metal, the following actions can be taken:

1. Do not create a large diameter weld crater.
2. Add the filler rod at the upper edge of the weld crater.
3. Point the electrode slightly upward. This upward angle will use the force of the arc to reduce sagging.

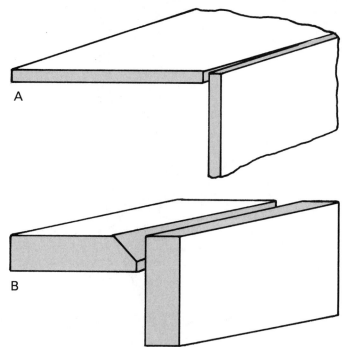

Fig. 12-42. Outside Corner Joints. A—Square groove butt joint. B—Bevel groove butt joint. These joints are welded in the same manner as the square groove and V-groove butt joints.

Figs. 12-43 and 12-44 illustrate a lap joint being welded in the horizontal position. Notice that the angle of the electrode and filler rod from the base metal surface are similar to the angles used when welding in the flat position. The electrode may be pointed upward slightly to reduce sagging of the molten metal.

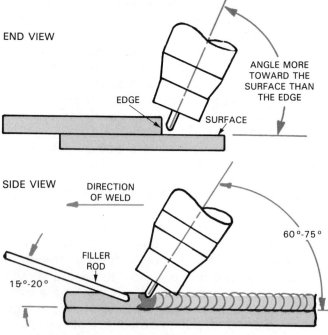

Fig. 12-43. A lap joint being welded with GTAW in the horizontal position. Notice the suggested angles for the electrode and filler rod.

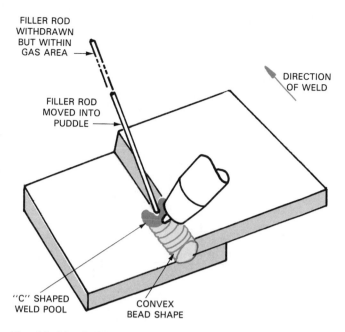

Fig. 12-44. A fillet weld on a lap joint in the horizontal position.

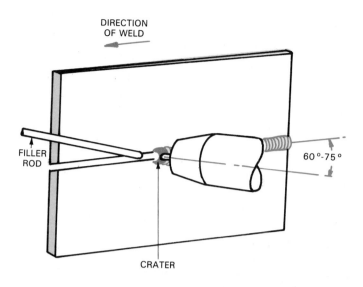

Fig. 12-46. Horizontal square groove butt weld in the horizontal position. The filler rod is held about 15°-20° from the base metal. The outer end of the filler rod is held slightly above the horizontal weld line.

An inside corner joint is shown being welded in the horizontal position in Fig. 12-45. A square groove weld on a butt joint is shown in Fig. 12-46. Fig. 12-47 illustrates the suggested electrode and filler rod angles for use on a V-groove butt joint.

The outside corner joints can be welded horizontally in the same manner as a square groove or V-groove butt joint.

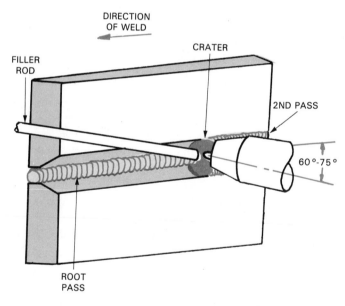

Fig. 12-47. Horizontal V-groove butt weld in the horizontal position. The filler rod is held about 15°-20° from the base metal. The outer end of the filler rod is held slightly above the horizontal weld line.

## 12-15. WELDING JOINTS IN VERTICAL POSITION

When welding vertically, it is important to use the lowest current possible to make a good weld. Lower current will help to keep the crater from becoming too large. If the crater is too large, the molten metal may flow away from the weld.

A pulsed arc may be used to provide a good cooling period which will help control the molten

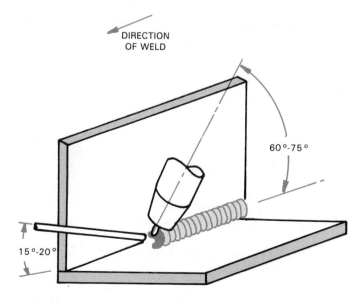

Fig. 12-45. GTAW used to make a fillet weld on an inside corner joint. This weld is being done in the horizontal position.

crater. Refer to Heading 12-1 for an explanation of the pulsed arc.

Electrode and filler rod angles are the same as for welding in the flat position. The electrode should be held about 60-75 degrees from the base metal surface. The filler rod is held about 15-20 degrees from the base metal.

Vertical welds may be made from the bottom up (vertically up) or from the top down (vertically down). Metal thicknesses over 1/2 in. (approximately 13 mm) are seldom welded vertically down.

There is no flux to run into the molten pool when doing GTAW. Therefore, both vertically up and vertically down produce welds of excellent quality.

Fig. 12-48 illustrates a fillet weld being made on a lap joint in the vertical position. The electrode should be aimed more toward the surface. This prevents the edge of the lapped metal from melting too quickly. Wait for the crater to form a ''C'' shape before adding the filler rod to the crater. This crescent shape indicates that both

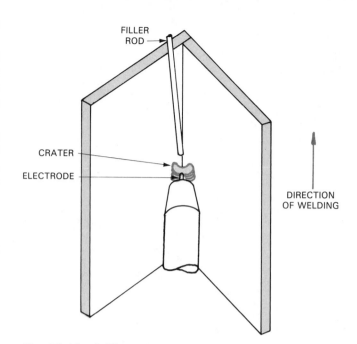

Fig. 12-49. A fillet weld being made on an inside corner joint in the vertical position. Note that the electrode is distributing the heat evenly to both surfaces. This weld is being made vertically up.

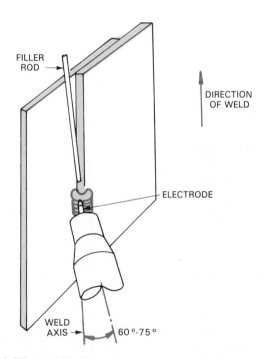

Fig. 12-48. A fillet weld being made on a lap joint in the vertical position.

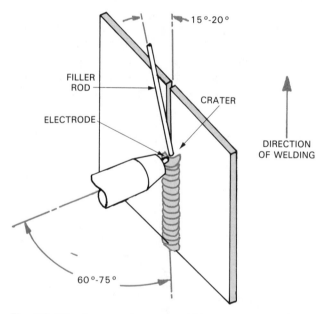

Fig. 12-50. A square groove weld in progress on a butt joint in the vertical position. The weld is being made in the vertical up direction.

pieces are molten and are running together. If the filler rod is added before the ''C'' shaped crater forms, the filler rod may not fuse with both pieces of the joint. A fillet weld on an inside corner joint in the vertical position is shown in Fig. 12-49. Like the fillet weld on a lap joint, do not add the filler rod until the ''C'' shaped crater forms.

Fig. 12-50 illustrates a square groove weld being made on a butt joint.

The outside corner joint shown in Fig. 12-51 is being joined with a beveled groove weld.

## 12-16. WELDING JOINTS IN OVERHEAD POSITION

Welds in the overhead position are relatively easy to make using the gas tungsten arc. The temperature and size of the crater must be con-

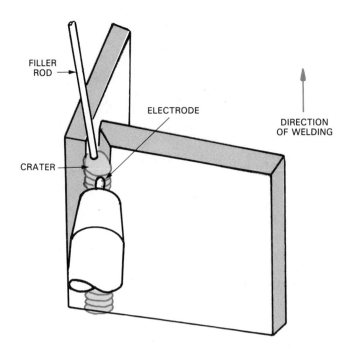

Fig. 12-51. A bevel groove weld being welded vertically up on an outside corner joint.

trolled. This may be done by selecting the lowest possible, effective current. Cooling of the crater is also possible by using a pulsed arc. Refer to Heading 12-1 for an explanation of the pulsed arc. Small beads may be used to keep the crater small and controlled. Several passes will be necessary in this case to completely fill the weld groove.

The angle of the electrode is kept at about 60-75 degrees from the surface of the base metal. An angle of 15-20 degrees is considered correct for the filler rod.

Fig. 12-52 illustrates a fillet weld being made on a lap joint in the overhead position.

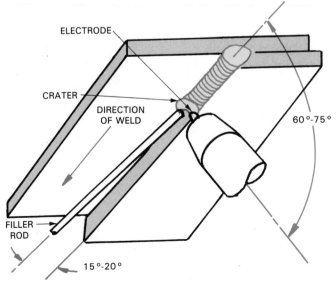

Fig. 12-52. A fillet weld on a lap joint in an overhead position. Note that the angles of the electrode and filler rod are the same as in the flat position.

An overhead fillet weld is being made on an inside corner joint in Fig. 12-53.

The overhead weld shown in Fig. 12-54 is a U-groove on a butt joint.

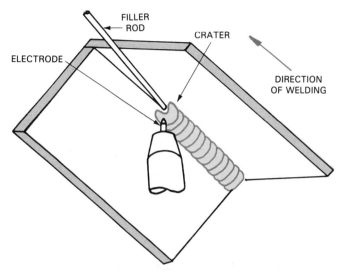

Fig. 12-53. A fillet weld on an inside corner joint in the overhead position.

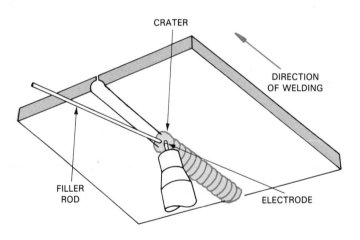

Fig. 12-54. A U-groove weld on a butt joint in the overhead position. The angle of the electrode and filler rod from the base metal is the same as for flat position welding.

Fig. 12-55 shows a square groove weld being made overhead on outside corner joint.

## 12-17. AUTOMATIC GTAW

Gas tungsten arc welding is also done in the fully automatic manner. The GTAW torch is moved over or around the weld. The torch may be mounted on an electrically driven tracker, on a rigid

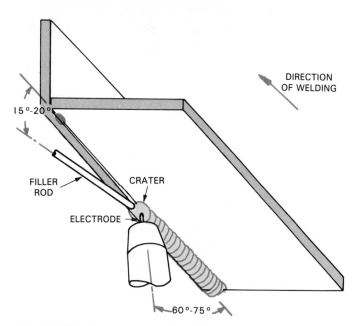

15°-20°

FILLER ROD

CRATER

ELECTRODE

DIRECTION OF WELDING

60°-75°

Fig. 12-55. A square groove weld on an outside corner joint in the overhead position. Note the tack weld.

frame as when automatic flame cutting, or on a robot arm. A filler metal wire feed motor and drive mechanism is used. The filler wire is brought to the weld pool or crater at a constant rate by the feed mechanism. See Fig. 12-56.

Extremely accurate and sound welds are possible using automatic GTAW. Automatic GTAW can

be used on metal as thin as .003 in. (.076 mm). Generally metals up to .02 in. (.51 mm) are welded without filler rod. This is possible by using a flange joint design and melting the flange which serves as the filler metal. Above .02 in. (.51 mm), filler wire or rod is added. Fig. 12-57 illustrates an automatic gas tungsten arc welding machine. The torch and wire feeder are mounted on a horizontal beam along with various controls. Torch motion is possible in two directional axes. The torch and weld can be made to follow a pattern in the same manner as when automatic flame cutting. See Heading 5-9.

The filler wire feed rate must be carefully set and the wire directed into the crater. The wire feed rate will determine the size of the bead buildup. A pulsed arc may be used to produce a strong, attractive, continuous, overlapping bead of gas tungsten arc spot welds.

Automatic GTAW may be done in all positions using the pulsed arc.

In heavy duty automatic welding applications, the wire guide may be water cooled.

To speed the rate of welding, a hot filler wire process is sometimes used. The filler wire is fed, controlled, and directed in the same manner as cold filler wire. The hot wire process heats the wire before it enters the arc crater. The filler wire is heated as it travels through the electrically

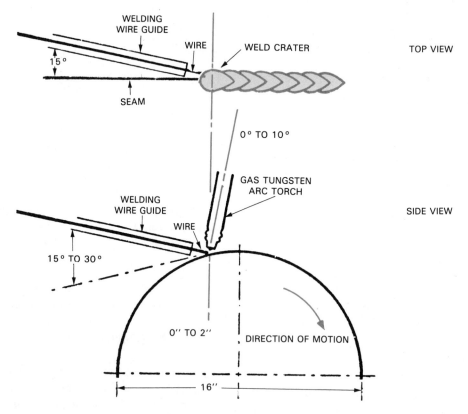

WELDING WIRE GUIDE

WIRE

WELD CRATER

TOP VIEW

15°

SEAM

0° TO 10°

GAS TUNGSTEN ARC TORCH

SIDE VIEW

WELDING WIRE GUIDE

WIRE

15° TO 30°

0″ TO 2″

DIRECTION OF MOTION

16″

Fig. 12-56. A schematic of an automatic GTAW torch and wire feed used to weld a circular bead.

Fig. 12-57. An automatic gas tungsten arc welding arrangement. The torch, wire feed and welding control panel are mounted on a horizontal beam. The carriage moves along the beam. The beam can also be moved.
(L-TEC Welding & Cutting Systems)

heated wire guide. See Fig. 12-58. This saves much of the heat of the arc for melting the base metal. Since the filler metal is preheated, the welding speed is much faster.

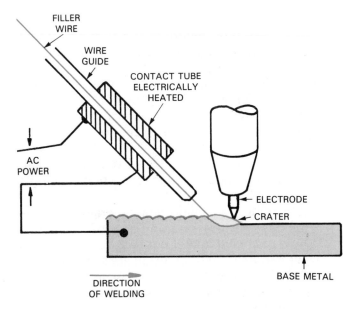

Fig. 12-58. A schematic of the hot wire, filler wire process in use. The wire is heated and fed at a constant rate directly into the crater.

## 12-18. GTAW TROUBLESHOOTING GUIDE

Occasionally a problem such as those listed below will occur which will affect the quality of the completed weld:
1. An unstable arc.
2. Rapid electrode consumption.
3. Tungsten in the weld (inclusions).
4. Porous welds.

Fig. 12-59 lists these problems, their possible cause, and how to correct the problem.

## 12-19. GTAW SAFETY

ALL safety precautions required for other electric arc welding processes apply to GTAW also. See Heading 10-24 for a list of safety precautions. When welding out of position and particularly overhead, be certain to wear a cap and leathers. This is done as protection against burns from falling molten metal.

The ultra-violet and infrared rays given off in GTAW may be more intense than those given off in SMAW. Therefore, the welder must be more careful to keep all exposed skin covered. The arc welding lens used for GTAW, according to American National Standard ANSI Z49.1-1973, is

| TROUBLE | POSSIBLE CAUSES | HOW TO CORRECT |
|---|---|---|
| Arc Instability | 1. Dirty, contaminated base material. | 1. Use chemical cleaners, wire brush, abrasives as appropriate to clean base material. |
| | 2. Joint is too narrow. | 2. Make groove wider; bring electrode closer to the work; decrease voltage. |
| | 3. Contaminated electrode. | 3. Cut off end of electrode tip, dress tip. |
| | 4. Electrode diameter too large. | 4. Use smaller electrode, smallest diameter that will handle required current. |
| | 5. Arc too long. | 5. Bring electrode closer to work. |
| Rapid Electrode Consumption (Use) | 1. Inert shielding is inadequate, allowing oxidation of the electrode. | 1. Clean nozzle; bring nozzle closer to work; increase gas flow. |
| | 2. Operating on reverse polarity. | 2. Change to straight polarity or use larger electrode. |
| | 3. Electrode too small for required current. | 3. Use larger electrode - See Fig. 12-17 to 12-21. |
| | 4. Electrode holder is too hot. | 4. Change collet; use ground finish electrodes; check for proper collet contact. |
| | 5. Electrode contamination. | 5. Remove contaminated section of electrode. Electrode will continue to degrade as long as contaminants are present. |
| | 6. Oxidation of electrode during cooling. | 6. Continue gas flow for 10-15 seconds after arc stops. Rule: 1 second for each 10 amps. |
| Tungsten Inclusions in Work | 1. Touch starting with electrode. | 1. Use high-frequency starting device; use a copper striking plate. |
| | 2. Electrode melts and alloys with base plate. | 2. Use lower current or larger electrode; use thoriated or zirconiated tungsten electrode (they run cooler) |
| | 3. Fragmentation of electrode by thermal shock. | 3. Be sure electrode ends are not cracked, especially when using high currents. Use embrittled tungsten for a clean easy break. |
| Porosity | 1. Gas impurities present; hydrogen, nitrogen, air, water vapor. | 1. Use welding grade inert gas (99.995 percent pure), purge all lines before striking arc. |
| | 2. Use of old acetylene hose. | 2. Use new hose only. Acetylene impregnates a hose and makes it unsuitable with an inert gas. |
| | 3. Gas and water hoses interchanged. | 3. Never interchange gas and water hoses. Use hoses of different colors. |
| | 4. Oil film on base material. | 4. Clean base material with a chemical cleaner that does not dissociate in the arc. Do not weld while material is wet. |

Fig. 12-59. A GTAW Troubleshooting Guide.

a number 12. With electrodes larger than 5/32 in. (4.0 mm), a number 14 lens should be considered.

There is little danger from molten metal spatter when using the GTAW process.

Wear supplied air breathing equipment when entering a space which is filled with an inert gas, CO or $CO_2$.

Phosgene gas is formed when arc welding on parts that have been cleaned with chlorinated hydrogen solvents, if a film of the solvent is present. Therefore, adequate ventilation must be provided to remove this gas.

## 12-20. TEST YOUR KNOWLEDGE

Write your answers on a separate sheet of paper. Do not write in this book.

1. Which DC electrical polarity, DCEN or DCEP, requires a larger electrode diameter? Why?

2. The best direct current polarity to use to clean off surface oxides on aluminum or magnesium is DCE__ __?

3. Why do the electrons not flow easily during the DCEP half of the AC cycle?

4. How many seconds of gas after flow should be set if the machine is set to deliver 130 amps?

5. When GTAW 1/8 in. (3.18 mm) mild steel, what amperage is suggested when welding a lap joint?

6. Which inert shielding gas generally provides the best surface cleaning action?

7. How many cubic feet per hour (cfh) and liters per minute (L/min) of argon are suggested for GTAW a 3/16 in. (4.76 mm) aluminum butt joint? What diameter electrode is suggested for this weld?

8. When adjusting a flowmeter, what part of the floating ball is aligned with the desired

flow rate line on the flowmeter?

9. What diameter is the exit diameter of a gas tungsten arc welding nozzle with the number 4 on it?
10. List two reasons why a small diameter nozzle is often used.
11. What kind of a tungsten electrode has a yellow painted band on it?
12. Pure tungsten electrodes are used to form a ball at the tip when using _____ current.
13. What will cause a tungsten electrode to split at the tip?
14. When grinding a point on a thoriated tungsten electrode, in what direction would the grinding marks run?
15. As a general rule, how far should the electrode extend beyond the nozzle?
16. Why is it extremely dangerous to enter a space which is filled with an inert gas without breathing equipment?
17. List and describe three ways of starting the GTAW arc.
18. When gas tungsten arc welding, how is the electrode moved as the filler rod is added?
19. When GTAW, the electrode is held at _____ - _____ degrees from the base metal and the filler rod is held at _____ - _____ degrees from the base metal.
20. List two causes for rapid tungsten electrode consumption (use).

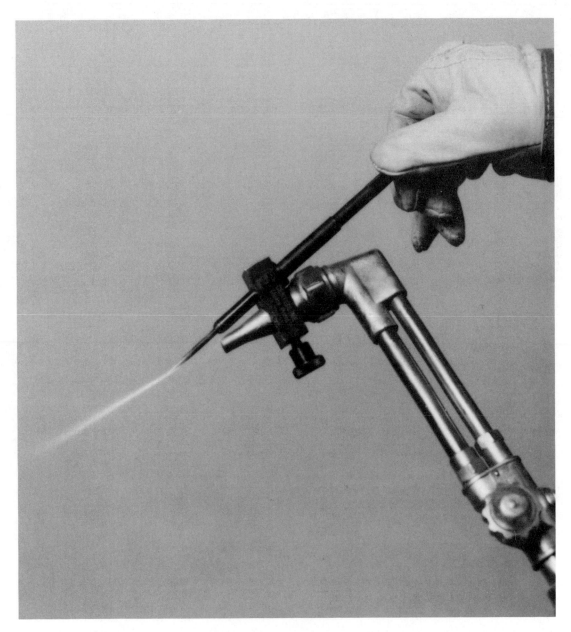

A new, patented flame sharpening device for accurately pointing tungsten electrodes.
(Swanson Tools USA)

DEMAND GAS MIXER

Industry photo. A single demand gas mixture system combines two shielding gases while using four flowmeters to supply four different welders.

# 13 GAS METAL ARC WELDING

GAS METAL ARC WELDING (GMAW) is also known as metal inert gas or MIG welding.

The GMAW process uses a continuously fed wire electrode which is fed into the arc crater. The wire electrode is consumed and also becomes the filler metal. Refer to Headings 1-7 and 1-8, also to Figs. 1-12 and 1-13.

The tremendous growth of the use of GMAW is the result of several events. The development and perfection of constant voltage arc welding machines has made GMAW more useful. The low cost of GMAW makes it an attractive choice for welding. GMAW can be used to produce high quality welds on all commercially important metals such as aluminum, magnesium, stainless, carbon, and alloy steels, copper, and others. GMAW may also be done easily in all welding positions.

## 13-1. GMAW PRINCIPLES

Gas metal arc welding (GMAW) is generally used because of its high productivity. It can be used in all positions and on all commercially important metals. GMAW is also easy to use and it creates high quality welds at a low overall cost.

GMAW can be done using solid wire, flux cored, or a specially coated solid wire electrode. See Fig. 13-1. A shielding gas or gas mixture must be used with GMAW.

GMAW is generally done using DCEP (DCRP). Alternating current is never used. DCEN (DCSP) is used with only one special electrode, called an EMISSIVE ELECTRODE. (AWS designation E70U-1).

For every pound of solid electrode wire used, 92-98 percent becomes deposited weld metal.

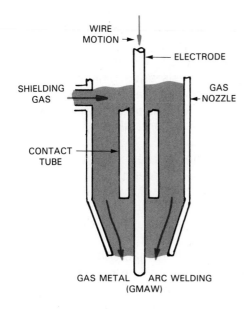

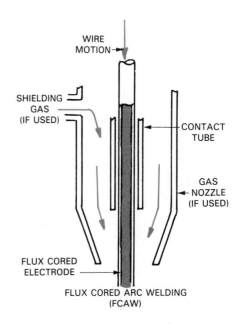

Fig. 13-1. A GMAW and FCAW gas nozzle and electrode shown schematically. Shielding gas is not always used with FCAW. If shielding gas is not used, no nozzle is required.

Flux cored arc welding (FCAW) wire is deposited with a wire efficiency of 82-92 percent. As a comparison, shielded metal arc welding (SMAW) deposits 60-70 percent of the electrode wire as weld metal. Some spatter does occur in the GMAW and FCAW processes. Very little stub loss occurs when continuously fed wire is used.

There is a very thin glass-like coating over the weld bead after GMA welding. No heavy slag is required because the weld area is shielded by a gas. When FCAW, a slag covering is present. Some of the flux in the FCAW forms a gas around the weld area. Some of the flux forms a slag, covering the weld. Shielding gas may or may not be used when FCAW. More welder time can be spent on the welding task with a continuously fed wire process. This improves the cost efficiency of GMA and FCA welding.

The GMAW process can be suited to a variety of job requirements by choosing the correct shielding gas, electrode size, and welding parameters. Welding parameters include the voltage, travel speed, and wire feed rate. The arc voltage and wire feed rate will determine the filler metal transfer method.

Metal transfer occurs in two ways. One is by the short circuiting method. The second way is to transfer metal across the arc. Methods of transferring metal across the arc include:
1. Globular transfer.
2. Spray transfer.
3. Pulsed spray transfer.
4. Rotating spray transfer.

## 13-2. SHORT CIRCUIT GMAW

SHORT CIRCUIT gas metal arc welding is used with relatively low welding currents. It also uses electrode wire sizes under .045 in. (1.14 mm). This process is particularly useful on thin metal sections in all positions. All position welds are made easily because there is no metal transfer across the arc. The weld pool or crater cools or solidifies rapidly using the short circuit arc.

Short circuit gas metal arc welding is also used to weld thick sections in the overhead or vertical position. It is very effective in filling the large gaps of poorly fitted parts.

Refer to Fig. 13-2 to see how the short circuiting arc method deposits metal. When the electrode touches the molten pool, the arc is no longer present. The surface tension of the pool pulls the molten metal, on the end of the electrode, into the pool. The PINCH FORCE around the electrode squeezes the molten end of the electrode. The surface tension and the pinch force separate the molten metal and the electrode. The arc is then

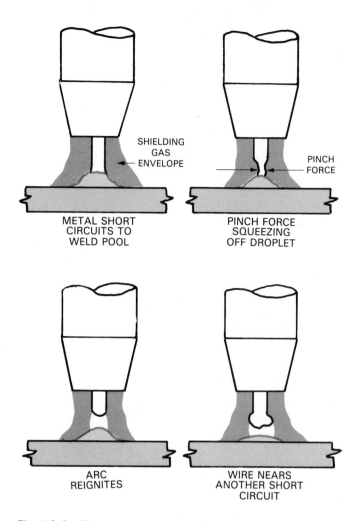

Fig. 13-2. The sequence of metal transfer during the short circuiting, GMAW method.

reestablished. The continuously fed electrode again touches the molten pool and the process repeats itself. The droplet transfer or short circuit process repeats itself about 20-200 times per second. The strength of the pinch force depends on the arc voltage, the slope of the power supply, and the circuit resistance. These factors, voltage, slope, and resistance affect the welding current. The frequency of the pinch force and the formation of droplets is controlled by the inductance of the power supply.

If 150 amperes are set on the arc welding machine, when the electrode short circuits, the amperage may rise rapidly to the maximum output of the machine. This could be 500 amperes or more. To control and slow down this possible rapid rise in current, an inductance circuit is built into the arc welding machine.

INDUCTANCE is the property in an electric circuit that slows down the rate of the current change. Some arc welding machines have an electric coil built in near the welding current transformer coils. See Heading 9-5 for a discussion of inductance coils. The current traveling through the inductance coil creates a magnetic field. This magnet field creates a current in the welding circuit which is in opposition to the welding current. Increasing inductance in a welding machine will slow down the increase of the welding current.

Decreasing the inductance will increase the rate of change of the welding current.

When too much welding current is used, the pinch force is so great that the molten metal at the end of the electrode literally explodes. A great deal of spattering occurs in this case. When too little inductance is used, the current rises too rapidly. The metal droplets will then form at a rapid rate and will squirt off the electrode in an uncontrolled manner.

By properly balancing the inductance and slope, an ideal droplet transfer rate and pinch force can be obtained. See Fig. 13-3 for the metal deposition rate for the short circuit transfer method.

| GMAW METHOD | METAL DEPOSITED | |
|---|---|---|
| | lbs/hr | kg/hr |
| Short circuiting | 2-6 | 0.9-2.7 |
| Globular | 4-7 | 1.8-3.2 |
| Spray | 6-12 | 2.7-5.4 |
| Pulsed spray | 2-6 | 0.9-2.7 |
| Rotating spray | 14-30 | 6.3-13.5 |

Fig. 13-3. The approximate rate at which filler metal is deposited with various GMAW methods. (American Welding Society)

## 13-3. GLOBULAR TRANSFER

GLOBULAR METAL TRANSFER GAS METAL ARC WELDING occurs when the welding current is set above the range for short circuit metal transfer. This range will vary with the electrode diameter. In the globular metal transfer process, the metal transfers across the arc as large, irregularly shaped drops. See Fig. 13-4. The drops are usually larger than the electrode diameter. The drop forms on the end of the electrode. It grows so large that it falls from the electrode due to its own weight. Magnetism in the arc causes the drops to travel across the arc in random patterns. This nondirectional flow of the drops results in increased amounts of spatter. Globular transfer generally occurs when carbon dioxide ($CO_2$) or $CO_2$ and low percentages of argon are used as the shielding gas.

To reduce spatter, a short arc length is used. When using a short arc, the arc may actually occur deep within the crater and below the metal surface. This is referred to as a BURIED or SUB-

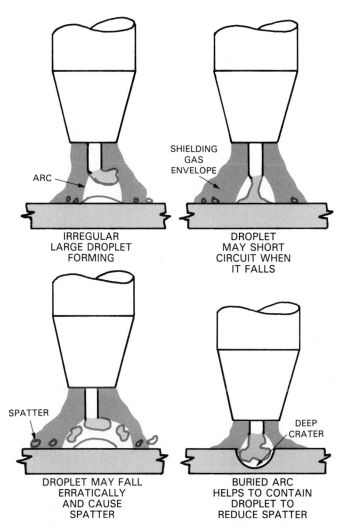

Fig. 13-4. GMAW globular metal transfer. Drops may fall erratically and cause spatter. Note that the buried arc may help contain the drops to reduce spatter.

MERGED ARC. Using a buried arc, much of the spatter is contained within the crater. With a buried arc, a combination of globular and short circuiting transfer occurs.

Large diameter electrode wire may be used with globular transfer. Currents between 350 to 450 amperes with 30-32 volts may be used. Welds of sufficient quality for many applications can be produced with this process. Welds may be made faster with this process than with the short circuit transfer method.

See Fig. 13-3 for the rate at which metal is deposited with this method.

## 13-4. SPRAY TRANSFER

SPRAY TRANSFER GAS METAL ARC WELDING will occur by increasing the current setting above the current required for globular transfer.

When spray transfer occurs, very fine droplets of metal form. These droplets travel at a high rate

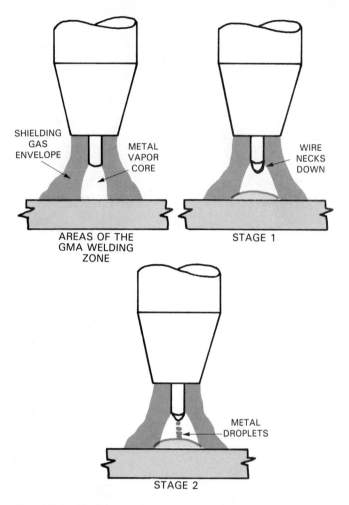

SHIELDING
GAS
ENVELOPE

METAL
VAPOR
CORE

AREAS OF THE
GMA WELDING
ZONE

WIRE
NECKS
DOWN

STAGE 1

METAL
DROPLETS

STAGE 2

Fig. 13-5. Spray transfer method. Note how the droplets are concentrated within the metal vapor core. Spray transfer will only occur when a high percentage of argon gas is used.

of speed directly through the arc stream to the weld pool. Fig. 13-5 illustrates the spray arc metal transfer method.

Before spray transfer can occur, a current setting above the TRANSITION CURRENT level must be set on the welding machine. The transition current varies with the electrode diameter, its composition, and the amount of electrode extension. A higher transition current is required for steel than aluminum. The transition current increases with the electrode diameter. It decreases as the electrode extends farther from the wire contact tube. See Fig. 13-40. Until the transition current is exceeded, the metal transfers as large globules. When the transition current level is exceeded, the pinch force becomes great enough to squeeze the metal off the tip of the electrode as fine droplets. See Fig. 13-6 for various transition current levels.

The droplets are squeezed off cleanly and transferred across the arc gap in a straight path. Spray transfer only occurs when at least 80 percent argon is used as the shielding gas. Common shielding gas mixtures are: 98 percent Ar plus 2 percent $O_2$; 95 percent Ar plus 5 percent $O_2$; 95 percent Ar plus 5 percent $CO_2$; and 90 percent Ar plus 10 percent $CO_2$. The spray transfer method produces deep penetration. The arc can be directed easily by the welder. This is because the arc and metal spray pattern are stable and concentrated. Spray transfer is best done in the flat or horizontal position, and on metal over 1/8 in. (approximately 3.2 mm) thick. See Fig. 13-3 for the metal deposition rate.

| Wire electrode type | Wire electrode diameter | | Shielding gas | Minimum spray arc current, A |
|---|---|---|---|---|
| | in. | mm | | |
| Mild steel | 0.030 | 0.76 | 98% argon-2% oxygen | 150 |
| Mild steel | 0.035 | 0.89 | 98% argon-2% oxygen | 165 |
| Mild steel | 0.045 | 1.14 | 98% argon-2% oxygen | 220 |
| Mild steel | 0.062 | 1.59 | 98% argon-2% oxygen | 275 |
| Stainless steel | 0.035 | 0.89 | 99% argon-1% oxygen | 170 |
| Stainless steel | 0.045 | 1.14 | 99% argon-1% oxygen | 225 |
| Stainless steel | 0.062 | 1.59 | 99% argon-1% oxygen | 285 |
| Aluminum | 0.030 | 0.76 | Argon | 95 |
| Aluminum | 0.045 | 1.14 | Argon | 135 |
| Aluminum | 0.062 | 1.59 | Argon | 180 |
| Deoxidized copper | 0.035 | 0.89 | Argon | 180 |
| Deoxidized copper | 0.045 | 1.14 | Argon | 210 |
| Deoxidized copper | 0.062 | 1.59 | Argon | 310 |
| Silicon bronze | 0.035 | 0.89 | Argon | 165 |
| Silicon bronze | 0.045 | 1.14 | Argon | 205 |
| Silicon bronze | 0.062 | 1.59 | Argon | 270 |

Note: Spray transfer will only occur when high percentages of Argon are used.

Fig. 13-6. Approximate transition current levels for various metals welded with DCEP (DCRP).
(American Welding Society)

## 13-5. PULSED SPRAY TRANSFER

The PULSED SPRAY TRANSFER GAS METAL ARC WELDING method is similar to the spray transfer method. See Fig. 13-7. The current level for pulsed spray must be above the transition current level. Special circuits within the arc welding machine cause the current to pulse. A low level current in the globular transfer range is used to maintain the arc. This current is called the BACKGROUND CURRENT. The current is increased at a regular frequency to the PEAK CURRENT. The peak current is above the transition current level. Since the background current is on for only a short time, no globular transfer actually occurs. During the peak current time period, spray transfer occurs. In this method, no necking down of the wire occurs. The metal leaves the electrode in a spray of small droplets. The spray transfer does not occur continually. Therefore the name pulsed spray transfer. The rate of metal transfer increases and the droplet size decreases as the pulse frequency is increased from 60-120 pulses per second. The coolest spray transfer occurs at 60 pulses per second.

A lower average current level is used in pulsed spray than in spray transfer. This lower average current level makes it possible to weld out of position. Thin metal sections may also be welded more easily with the pulsed spray. This method creates very little metal spatter.

The pulsed spray transfer method can use larger diameter electrode wire. This is an advantage. Larger diameter electrodes are cheaper. Also, nonferrous wires of larger diameter can be fed through the wire drive unit more easily without bending.

See Fig. 13-3 for the metal deposition rate for the pulsed spray transfer method. Pulsed spray is used also to weld parts with silicon bronze filler wire. This process is sometimes called MIG brazing. Light steel parts in auto repair shops can be welded with very low heat inputs. This reduces the problems of distortion and burn through.

## 13-6. ROTATING SPRAY TRANSFER

ROTATING SPRAY TRANSFER GAS METAL ARC WELDING occurs when the amperage is increased above the normal levels required for spray transfer. The end of the electrode becomes molten and rotates in a spiraling or helical pattern. See Fig. 13-8. A constant, controlled stream of metal

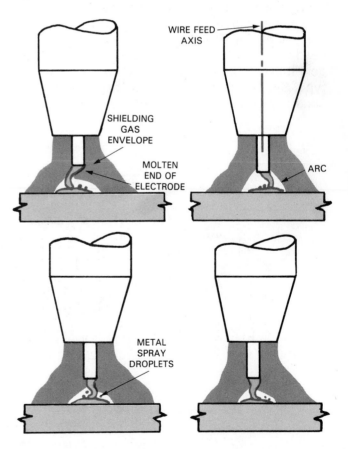

Fig. 13-7. Pulsed spray metal transfer method. Spray transfer only occurs during peak current.

Fig. 13-8. GMAW rotating spray metal transfer. The end of the electrode becomes molten. The electrode tip, arc, and metal spray rotate around the wire feed axis.

leaves the end of the electrode. The weld pool is wide due to the wide, rotating metal spray pattern.

Rotating spray transfer can only occur when argon shielding gas is mixed with 2-5 percent oxygen or 5-10 percent $CO_2$. The arc appears over a wide area as shown in Fig. 13-8. Penetration is less due to the widely spread weld crater (pool).

The current required for rotating spray transfer increases as the electrode diameter increases. The current required decreases as the electrode extension increases. ELECTRODE EXTENSION is the distance that the electrode extends beyond the electrode contact tube.

Rotating spray transfer requires wire feed equipment with rates up to 2000 inches per minute (847 mm/second). Electrodes with low copper coatings and high tensile strength are also used.

Rotating spray transfer can deposit metal up to five times faster than with short circuiting transfer. See Fig. 13-3.

It is used best in the flat position because of the large crater formed.

## 13-7. GMAW POWER SOURCES

Fig. 13-9 illustrates a complete GMAW outfit. The same equipment may be used with flux cored electrodes for FCAW. The arc welding machine used for gas metal arc welding (GMAW) is generally a constant voltage (constant potential) type. See Heading 11-15 for an explanation of the constant voltage arc welding machine.

Direct current electrode positive (DCEP), also called direct current reverse polarity (DCRP), is usually used. DCEP (DCRP) provides a stable arc and smooth electrode metal transfer. The type of metal transfer determines the amount of penetration and the amount of metal spatter. DCEP (DCRP) is used for all types of metal transfer.

Direct current electrode negative (DCEN) or DCSP is seldom used. Only one electrode, a special emissive electrode (AWS E70U-1), is produced for DCEN. Alternating current (AC) is not used for GMAW.

The GMAW machine will have a control to set the voltage. If the wire feed is built into the arc welding machine, the machine may also have a wire speed control. Setting the wire speed also sets the amperage of the welding circuit. Fig. 13-10 illustrates an arc welding machine with controls for setting arc voltage and wire speed.

## 13-8. SETTING UP THE ARC WELDING MACHINE

The GMAW gun electrode lead and ground lead must be tightly connected to the machine terminals. In DCEP, the gun electrode lead is connected to the positive terminal. The ground lead is connected to the negative terminal. All the gas and water hose connections must be checked for tightness.

Connect the plug for the remote contactor switch into the remote contactor receptacle on the machine. Connect the plug for the remote voltage control into the remote control receptacle.

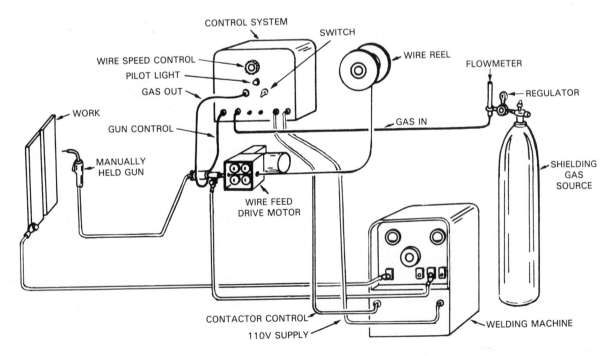

Fig. 13-9. Diagrammatic view of a complete gas metal arc welding (GMAW) outfit.

Fig. 13-10. A constant voltage (CV) arc welding machine. The wire speed is set on the wire drive panel. (The Lincoln Electric Co.)

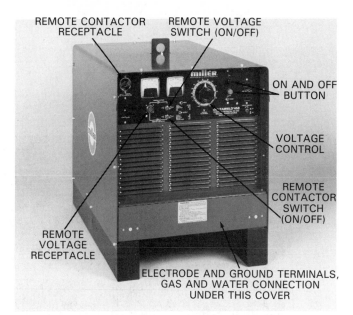

Fig. 13-11. A GMAW machine with a voltage adjustment only. (Miller Electric Mfg. Co.)

Fig. 13-12. A GMAW machine used for pulsed spray arc transfer. See Fig. 13-13 for a close up view of the control panel. (Miller Electric Mfg. Co.)

Move the remote control switch and the remote contactor switch on the machine into the remote or standard position.

If the gas metal arc welding machine is to be used for short circuit, globular, spray, or rotating transfer, only the voltage needs to be set. Fig. 13-11 illustrates an arc welding machine which has a voltage control only. Some machines provide a slope control to vary the slope. Most machines have a preset slope for constant voltage (constant potential) use. The same machine can be used for GMAW and FCAW.

When pulsed spray transfer GMAW is used, a special arc welding machine is needed. This machine must have controls to regulate the voltage and the background and peak current levels. If the pulse frequency is variable, the arc welding machine should have a means of controlling the frequency of the pulses. Fig. 13-12 shows a machine designed for pulsed spray metal transfer. Fig. 13-13 is a close up view of the control panel shown in Fig. 13-12.

To set the voltage adjustments, refer to the following figure numbers for arc or load voltage ranges:

| Metal | Metal Transfer Method | Figure No. |
|---|---|---|
| Mild and low alloy steel | Short circuit | 13-14 |
| | Spray transfer | 13-15 |
| Stainless steel (300 series) | Short circuit | 13-16 |
| | Spray transfer | 13-17 |
| Aluminum and aluminum alloys | Short circuit | 13-18 |
| | Spray transfer | 13-19 |

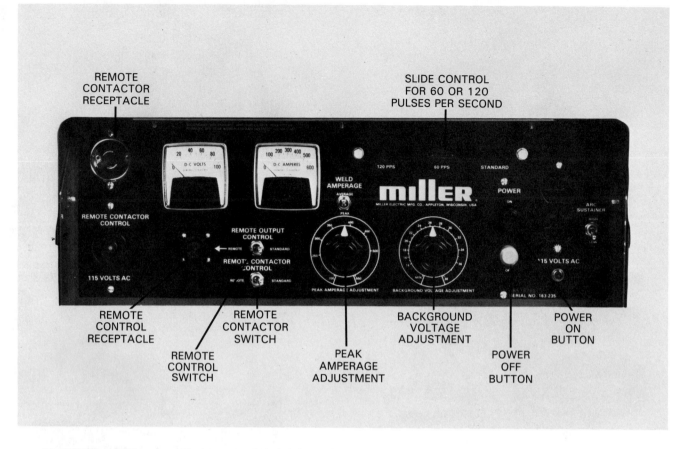

Fig. 13-13. A close up of the control panel for the machine shown in Fig. 13-12. The peak current and background voltage must be set to do pulsed spray welding. (Miller Electric Mfg. Co.)

| ELECTRODE DIAMETER | | ARC VOLTAGE | AMPERAGE RANGE |
|---|---|---|---|
| in. | mm | | |
| .030 | .76 | 15-21 | 70-130 |
| .035 | .89 | 16-22 | 80-190 |
| .045 | 1.14 | 17-22 | 100-225 |

Note: The values shown are based on the use of CO₂ for mild steel and Argon-CO₂ for low alloy steel.

Fig. 13-14. Approximate machine settings for short circuit metal transfer on mild and low alloy steel.

| ELECTRODE DIAMETER | | ARC VOLTAGE | AMPERAGE RANGE |
|---|---|---|---|
| in. | mm | | |
| .030 | .76 | 17-22 | 50-145 |
| .035 | .89 | 17-22 | 65-175 |
| .045 | 1.14 | 17-22 | 100-210 |

Note: The values shown are based on a mixture of 90% Helium; 7 1/2% Argon; 2 1/2% CO₂. The flow rates were about 20 cfh (9.44 L/min).

Fig. 13-16. Approximate machine settings for short circuit transfer on series 300 stainless steel.

| ELECTRODE DIAMETER | | ARC VOLTAGE | AMPERAGE RANGE |
|---|---|---|---|
| in. | mm | | |
| .030 | .76 | 24-28 | 150-265 |
| .035 | .89 | 24-28 | 175-290 |
| .045 | 1.14 | 24-30 | 200-315 |
| 1/16 | 1.59 | 24-32 | 275-500 |
| 3/32 | 2.38 | 24-33 | 350-600 |

Note: The values shown are based on the use of Argon-5% Oxygen for mild and low alloy steel.

Fig. 13-15. Approximate machine settings for spray arc transfer on mild or low alloy steel.

| ELECTRODE DIAMETER | | ARC VOLTAGE | AMPERAGE RANGE |
|---|---|---|---|
| in. | mm | | |
| .030 | .76 | 24-28 | 160-210 |
| .035 | .89 | 24-29 | 180-255 |
| .045 | 1.14 | 24-30 | 200-300 |
| 1/16 | 1.59 | 24-32 | 215-325 |
| 3/32 | 2.38 | 24-32 | 225-375 |

Note: The values shown are based on the use of Argon-Oxygen shielding gas. The oxygen percentage varies from 1-5%.

Fig. 13-17. Approximate machine settings for spray transfer on series 300 stainless steel.

| ELECTRODE DIAMETER | | ARC VOLTAGE | AMPERAGE RANGE |
|---|---|---|---|
| in. | mm | | |
| .030 | .76 | 15-18 | 45-120 |
| .035 | .89 | 17-19 | 50-150 |
| (3/64) .047 | 1.19 | 16-20 | 60-175 |

Note: The values shown are based on the use of Argon shielding gas.

Fig. 13-18. Approximate machine settings for short circuit transfer on aluminum and aluminum alloys.

| ELECTRODE DIAMETER | | ARC VOLTAGE | AMPERAGE RANGE |
|---|---|---|---|
| in. | mm | | |
| .030 | .76 | 22-28 | 90-150 |
| .035 | .89 | 22-28 | 100-175 |
| (3/64) .047 | 1.19 | 22-28 | 120-210 |
| 1/16 | 1.59 | 24-30 | 160-300 |
| 3/32 | 2.38 | 24-32 | 220-450 |

Note: The values shown are based on the use of Argon as the shielding gas.

Fig. 13-19. Approximate machine settings for spray transfer on aluminum and aluminum alloys.

Globular transfer voltages will fall in the range between those shown for short circuit and spray transfer. Pulsed spray transfer settings will use a setting below the transition current level for the background current. The peak current must be equal or above the spray transfer value.

Rotating spray voltage and currrent values are above those for spray transfer. They should not exceed 320-380 amps. This will make it possible to maintain a controllable arc crater and bead shape.

## 13-9. SETTING UP THE WIRE DRIVE UNIT

Two mated gears are located in the wire drive unit. One gear is driven by an electric variable speed motor. A drive roll is screwed onto each gear. Fig. 13-20 illustrates a two drive roll wire drive system. The upper drive roll and gear are adjustable by means of a spring loaded thumb screw. Adjustment is required to obtain the correct amount of force on the electrode wire. Only enough force should be applied to drive the wire without slippage. Too much force on the rolls and wire may cause the solid wire to flatten. Flux cored electrodes may be crushed. If the wire is damaged, it will not feed through the wire cable and torch properly.

The lower roll on the wire drive unit shown in Fig. 13-20 is adjustable in and out. The lower drive gear has spring washers behind it. By turning the adjustment bolt, in the center of this gear, the gear and drive roll can be moved in or out. This adjustment is provided to align the groove in the wire drive roll with the center of the wire. Fig. 13-21 illustrates the adjustment of the wire drive rolls.

Fig. 13-22 shows a wire drive unit with four drive rolls. This drive unit is similar to the two drive roll unit shown in Fig. 13-20. The unit

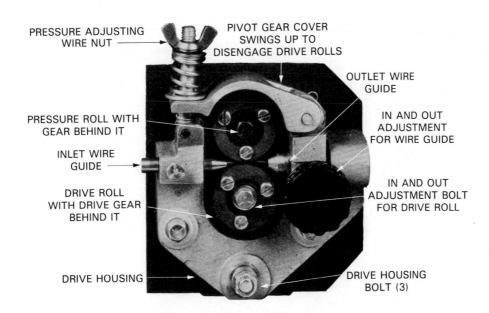

Fig. 13-20. A two drive roll wire drive system. The upper pressure roll is pivoted out of the way when the wing nut is loosened and the gear cover lifted up.   (Miller Electric Mfg. Co.)

Gas Metal Arc Welding / 359

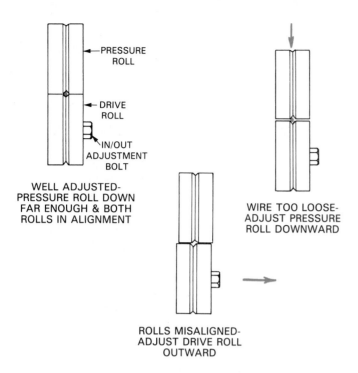

WELL ADJUSTED-
PRESSURE ROLL DOWN
FAR ENOUGH & BOTH
ROLLS IN ALIGNMENT

WIRE TOO LOOSE-
ADJUST PRESSURE
ROLL DOWNWARD

ROLLS MISALIGNED-
ADJUST DRIVE ROLL
OUTWARD

Fig. 13-21. Adjustment of drive rolls. The pressure (upper) roll is adjusted up and down by means of the pressure adjusting wing nut as in Fig. 13-20. The lower drive roll is adjusted in and out by means of an adjustment bolt.

shown in Fig. 13-22 has three wire guides, four drive rolls, and five gears. The gear in the center is connected to the electric drive motor and drives the two lower wire drive gears. Both pressure gears and rolls are adjustable with wing nuts. The wire guides must also be in alignment with each other and with the center of the drive rolls. Fig. 13-23 illustrates properly and improperly adjusted wire guides. The alignment of the wire guides is made at the factory. In time, a readjustment may be necessary. Each wire guide must be in perfect alignment with the other as shown on the drive units in Figs. 13-20 and 13-22. The wire guides are mounted on the drive housing. This drive housing may move up or down. Such movement will cause the guides to be misaligned with the drive rolls. Adjustment is made by loosening the drive housing mounting bolts. The guides and drive rolls are realigned and the bolts retightened. The inner end of each wire guide should be adjusted as close to the drive rolls as possible without touching them. After the wire guide is set, the securing bolt is tightened to hold the guide in place.

To load the wire into the wire feed unit, the spool of wire is first placed on the wire spool hub. The wire spool is secured to the hub. See Fig. 13-24. The pressure gear and drive roll are loosened and swung out of the way. Wire is fed from the spool through the inlet and outlet guide and 2-3 in. (50-75 mm) into the gun electrode cable.

Reposition the pressure gear and upper drive roll and lightly tighten the pressure adjusting wing nut.

The machine controls shown in Fig. 13-24 are explained below:

INCH SWITCH: (See Fig. 13-24.) Press the inch button (switch) on the wire drive control panel. This will cause the wire drive motor to slowly feed the wire inch by inch through the electrode cable to the gun. Continue to press the inch button until

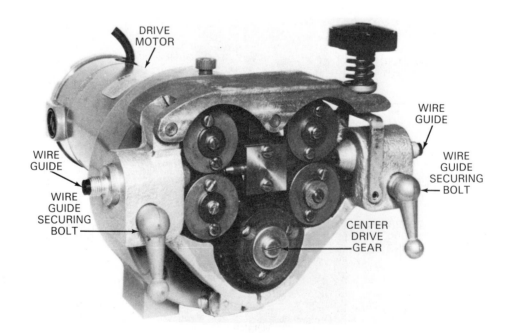

Fig. 13-22. A four drive roll wire drive system. The parts are similar to those shown in Fig. 13-21. The center drive gear drives the gears behind the two lower rolls. (L-TEC Welding & Cutting Systems)

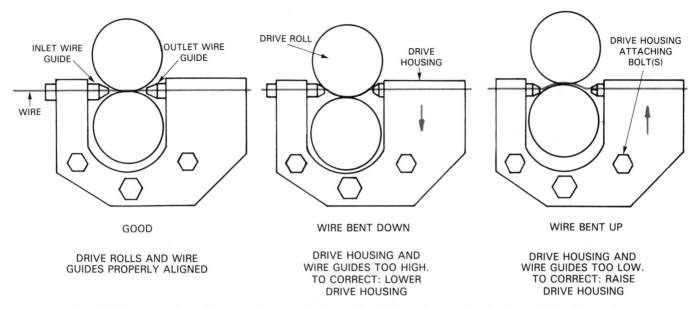

GOOD

DRIVE ROLLS AND WIRE
GUIDES PROPERLY ALIGNED

WIRE BENT DOWN

DRIVE HOUSING AND
WIRE GUIDES TOO HIGH.
TO CORRECT: LOWER
DRIVE HOUSING

WIRE BENT UP

DRIVE HOUSING AND
WIRE GUIDES TOO LOW.
TO CORRECT: RAISE
DRIVE HOUSING

Fig. 13-23. Properly and improperly aligned wire guides. If the wire bends going through the drive rolls, adjust the drive housing up or down. Loosen the drive housing bolts, align, and retighten the bolts.

the wire comes out of the contact tube 2-3 in. (50-75 mm). The contact tube is in the center of the gas nozzle.

FEED AND RETRACT SWITCH: Most wire feed units have a switch to reverse the direction of the wire feed. This switch in Fig. 13-24 is called the feed or retract switch. The feed/retract switch is moved to change the direction of the wire movement. The inch button is pressed to feed or retract the electrode wire.

SHIELDING GAS PURGE BUTTON: This shielding gas purge button causes the shielding gas to

Fig. 13-24. A wire drive mechanism for GMAW. This unit has a 4 drive roll system. The two lower rolls are driven by the drive gear between them. (Miller Electric Mfg. Co.)

flow through the system while the button is depressed. This is done to clear the system of any contaminating gases or moisture.

WIRE SPEED CONTROL: Adjustment of the wire feed speed is made on the wire feed unit control panel. Changing the drive motor speed changes the wire feed speed. The wire drive motor is on an electrical circuit which is separate from the welding circuit. Changing the wire speed also changes the amperage delivered by the constant potential arc welding machine. Set the wire speed to deliver the amperage required for the job being done. The amperage suggested may be supplied by the machine manufacturer or the tables in Figs. 13-14 to 13-19 may be used.

If the wire feed speed is too high, the wire will not be able to melt off fast enough to maintain the desired arc length. When the wire speed is too high, the wire may stub or snake on the surface. See Fig. 13-25.

REMOTE CONTROL: A remote contactor switch is often used by the welder to turn the wire feeder and welding current on and off. If a remote unit is used, the plug for the unit is plugged into the wire feed control panel receptacle. To activate the remote circuits, the remote/standard switch is placed on remote.

Fig. 13-26 illustrates a wire feed unit which uses a digital readout to display the voltage and

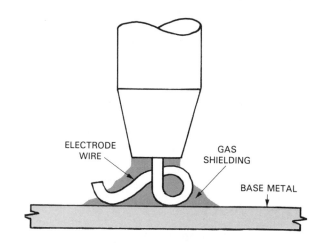

Fig. 13-25. The result of a wire speed that is too fast. The electrode will stub or snake on the metal surface.

wire speed. The values set and the values actually occurring can be read separately by pressing a button.

## 13-10. INERT GASES AND GAS MIXTURES USED FOR GMAW

The inert shielding gases and other gases used in shielding gas mixtures for GMAW are:
Argon (Ar), helium (He), nitrogen ($N_2$), oxygen ($O_2$), and carbon dioxide ($CO_2$).

Fig. 13-26. A wire drive unit with a digital readout of both voltage and wire speed. The values that are set and actually delivered can be read on this unit. (The Lincoln Electric Co.)

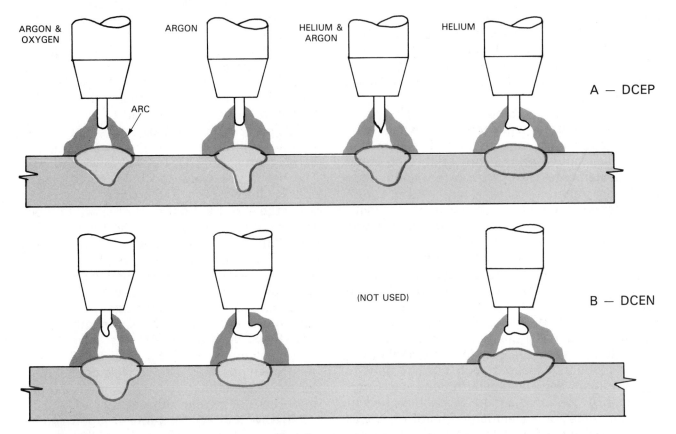

ARGON & OXYGEN    ARGON    HELIUM & ARGON    HELIUM

ARC

A — DCEP

(NOT USED)

B — DCEN

Fig. 13-27. Bead contours and penetration shapes which occur with various gases and electrode polarities. A—Electrode positive. B—Electrode negative. Notice the difference in penetration with pure argon and helium in DCEP.

Inert gases used should be at least 99.9 percent pure. Carbon dioxide gas is generally supplied 100 percent pure. Each shielding gas and mixture has a different effect on the shape of the bead and the penetration. See Fig. 13-27.

Some of the gases and gas mixtures used on various metals are shown in Fig. 13-28.

The factors which must be considered when

| Shielding Gas or Mixture | Chemical Behavior | Metals and Applications |
|---|---|---|
| Argon | Inert | Virtually all metals except steels. |
| Helium | Inert | Aluminum, magnesium, and copper alloys for greater heat input and to minimize porosity. |
| Ar + He (20-80% to 50-50%) | Inert | Aluminum, magnesium, and copper alloys for greater heat input to minimize porosity (better arc action than 100% helium). |
| Nitrogen | | Greater heat input on copper (Europe). |
| Ar + 25-30% $N_2$ | | Greater heat input on copper (Europe); better arc action than 100% nitrogen. |
| Ar + 1-2% $O_2$ | Slightly oxidizing | Stainless and alloy steels; some deoxidized copper alloys. |
| Ar + 3-5% $O_2$ | Oxidizing | Carbon and some low alloy steels. |
| $CO_2$ | Oxidizing | Carbon and some low alloy steels. |
| Ar + 20-50% $CO_2$ | Oxidizing | Various steels, chiefly short circuiting arc. |
| Ar + 10% $CO_2$ + 5% $O_2$ | Oxidizing | Various steels (Europe). |
| $CO_2$ + 20% $O_2$ | Oxidizing | Various steels (Japan). |
| 90% He + 7.5% Ar + 2.5% $CO_2$ | Slightly oxidizing | Stainless steels for good corrosion resistance, short circuiting arc. |
| 60-70% He + 25-35% Ar + 4-5% $CO_2$ | Oxidizing | Low alloy steels for toughness, short circuiting arc. |

Fig. 13-28. Shielding gases used with various metals.    (American Welding Society)

choosing a shielding gas are:
1. The type of metal transfer desired: short circuit, spray transfer, globular transfer, pulsed spray transfer, or rotating spray transfer.
2. The bead shape, width, and weld penetration desired.
3. The welding speed required.
4. The undercutting tendencies of the gas.

Inert gases such as argon and helium are chemically inactive and do not unite with other chemical elements. Nitrogen, oxygen, and carbon dioxide are reactive gases. They will mix with other chemicals such as the metal in a weld joint.

Reactive gases are not used alone as shielding gases with the exception of $CO_2$.

Each gas has an effect on the bead size and penetration, welding speed, type of metal transfer, and undercutting tendencies. The characteristics of each gas are discussed in the following paragraphs. Refer also to Fig. 13-28.

ARGON: Argon (Ar) quiets the arc and reduces spatter. It causes a squeezing (constricting) of the arc. This action causes a high current density (concentration) arc, deep penetration, and a narrow bead. Argon ionizes more easily than helium and it conducts some electricity. Therefore, lower arc voltages are required for a given arc length. Argon conducts heat through the arc more slowly than helium. Argon has a lower thermal (heat) conductivity. It is an excellent choice for use on thin metal. It is also good for out of position welds because of the low voltages required.

Argon is better than helium for use with the spray transfer method. High percentages of argon, 90 percent or more, must be used to obtain spray transfer when welding steel.

Pure argon used on carbon steel will cause undercutting, poor penetration, and a poor bead contour using the spray transfer method. Because of this, argon is usually mixed with small amounts of oxygen or carbon dioxide. Argon is heavier than helium and therefore less gas is needed to protect a weld.

HELIUM: Helium (He) has a high heat conducting ability. It transfers heat through the arc better than argon. Helium is used to weld thick metal sections. This gas is also used to weld metals which conduct heat well. Such metals as aluminum, magnesium, and copper will conduct heat away from the weld zone rapidly. More heat must be put into the metal and, therefore, helium gas is the best choice. The arc voltages required for helium are higher and spatter is greater. Helium will allow filler metal to be deposited at a faster rate than is possible with argon. This gas is often used on nonferrous metals. It produces welds with wider bead reinforcements. Helium is lighter than argon and will require a greater gas flow to protect a weld as well as argon.

CARBON DIOXIDE: Carbon dioxide ($CO_2$) has a higher thermal (heat) conductivity than argon. This gas requires a higher voltage than argon. Since it is heavy, it covers the weld well. Therefore, less gas is needed.

$CO_2$ costs about 80 percent as much as argon. This price difference will vary in various locations. Beads made with $CO_2$ have a very good contour. The beads are wide and have deep penetration and no undercutting. The arc in a $CO_2$ atmosphere is unstable and a great deal of spattering occurs. This is reduced by holding a short arc. Deoxidizers like aluminum, manganese, or silicon are often used. The deoxidizers remove the oxygen from the weld metal. Good ventilation is required when using pure $CO_2$. About 7-12 percent of the $CO_2$ becomes CO (carbon monoxide) in the arc. The amount increases with the arc length.

NITROGEN: Nitrogen ($N_2$) is used in Europe where helium is not readily available. The addition of nitrogen has been used to weld copper and copper alloys. One mixture used contains 70 percent argon and 30 percent nitrogen.

ARGON-HELIUM: Mixtures of argon and helium help to produce welds and welding conditions which are a balance between deep penetration and a stable arc. A mixture of 25 percent argon and 75 percent helium will give deeper penetration with the arc stability of a 100 percent argon gas. Spatter is almost zero when a 75 percent helium mixture is used. Argon-helium mixtures are used on thick nonferrous sections.

ARGON-CARBON DIOXIDE: $CO_2$ in argon gas makes the molten metal in the arc crater more fluid. This helps to eliminate undercutting when GMA welding carbon steels.

$CO_2$ also stabilizes the arc, reduces spatter, and promotes a straight line (axial) metal transfer through the arc.

ARGON-OXYGEN: Argon-oxygen gas mixtures are used on low alloy carbon, and stainless steels. A 1-5 percent oxygen mixture will produce beads with wider, less finger shaped, penetration. Oxygen also improves the weld contour, makes the weld pool more fluid, and eliminates undercutting. Oxygen seems to stabilize the arc and reduce spatter. The use of oxygen will cause the metal surface to oxidize slightly. This oxidization will generally not reduce the strength or appearance of the weld to an unacceptable level. If more than 2 percent oxygen is used with low alloy steel, a more expensive electrode wire with additional deoxidizers must be used.

HELIUM-ARGON-CARBON DIOXIDE: This shielding gas mixture is used to weld austenitic stainless

| Metals | Gases % | Uses and Results |
|---|---|---|
| Aluminum | Ar | Good transfer, stable arc, little spatter. Removes oxides on DCRP - to 1/2 inch thickness |
| | 50Ar-50He | Hot arc - 3/8 to 3/4 inch thickness |
| | 25Ar-75He | Hot arc, less porosity, removes oxides - 1/2 to 1 inch |
| | He | Hotter, more gas; 1/2 inch and up |
| Magnesium | Ar | Good cleaning DCRP |
| | 75He-25Ar | Hotter, less porosity, removes oxides |
| Copper (deox.) | 75He-25Ar | Preferred. Good wetting, hot |
| | Ar | For lighter gages |
| Carbon Steel | Ar-2CO₂ | Fast, cheap, spattery |
| | CO₂ | Fast, cheap, spattery, deep penetration. Short circuiting arc: high quality, low current, out-of-position, low spatter |
| Carbon Steel | Ar-5 O₂ | Fast, stable, good bead shape, little undercut, fluid puddle |
| | 75Ar-25CO₂ | *Fast, no burnthrough, little distortion and spatter |
| | 50Ar-50CO₂ | *Deep penetration, low spatter |
| Low Alloy Steel | Ar-2 O₂ | Removes oxides, eliminates undercut, good properties |
| High Strength Steels | 60He-35Ar-5CO₂ | *Stable arc, good wetting and bead contour, little spatter. Good impacts |
| | 75Ar-25CO₂ | *Same except low impacts |
| Stainless Steel | Ar-1 O₂ | No undercutting on DCRP. Stable arc, fluid weld, good shape |
| | Ar-5 O₂ | More stable arc DCRP. |
| | 90He-7 1/2Ar-2 1/2 CO₂ | *Small heat affected zone, no undercut, little warping |
| Nickel, Monel | Ar | Good wetting - decreases fluidity |
| Inconel | Ar-He | Stable arc on light gage<br>Hotter arc |

*Short circuiting arc

Fig. 13-29. Shielding Gas Selections for GMAW Various Metals. (Welding & Fabrication Data Book)

steel using the short circuit transfer method. The following mixture is often used and produces a low bead: 90 percent He; 7 1/2 percent Ar; 2 1/2 percent $CO_2$.

The various GMAW metal transfer methods and the gases suggested with their use are discussed below:

SHORT CIRCUITING: Pure argon or helium, or argon and helium mixtures are used on aluminum. For carbon steels, pure $CO_2$ or a mixture of 75 percent argon and 25 percent $CO_2$ is often used. A mixture of helium, argon, and $CO_2$ is used to weld stainless steel.

GLOBULAR: $CO_2$ and argon with high percentages of $CO_2$ are used with globular transfer. When $CO_2$ is used, the electrode tip is not surrounded by the arc plasma. With $CO_2$ the globules leave the wire in a random way and spatter is high. When argon or an argon based gas is used the electrode tip is surrounded by the arc plasma. When argon is used, the metal is squeezed off the wire and travels in a straighter line to the metal.

SPRAY: The spray transfer method will only occur in an atmosphere which has a high argon percentage. Pure argon may be used. The following argon mixtures are also used: Argon with 2-5 percent Oxygen ($O_2$) and also Argon with 5-10 percent $CO_2$.

Small amounts of oxygen lower the transition current. Oxygen appears to decrease the surface tension of the molten metal on the wire. This allows the molten metal droplets to leave the electrode more easily. Oxygen also makes the puddle more fluid and reduces undercutting. It also acts to stabilize the arc.

PULSED SPRAY: Argon with 2-5 percent $O_2$ or Argon with 5-10 percent $CO_2$ is used with the pulsed spray transfer method.

ROTATING SPRAY: Only argon based gases will permit rotating spray to occur. Arc occurs over a long length of wire. Argon mixtures with less than 10 percent $CO_2$ or 2-5 percent $O_2$ may be used.

Fig. 13-29 lists shielding gas selections for GMAW on a number of metals.

## 13-11. SHIELDING GAS FLOW RATES

Enough gas must flow to create a straight line (laminar) flow. If too much gas comes out of the nozzle, the gas may become turbulent. See Fig. 13-30. If it becomes turbulent, the shielding gas will mix with the atmosphere around the nozzle area. This will cause the weld to become contaminated.

When too little gas flows, the weld area is not properly protected. The weld will become contaminated and a porous weld will occur.

The recommended rate of flow for a given nozzle is generally provided by the manufacturer. Once the correct flow rate is known, it can be used at all wire speeds. When welding with the correct amount of shielding gas flowing, a rapid crackling and hissing sound will be heard.

Too little gas will give a popping sound. Spatter will also occur, the weld will have porosity showing, and the bead will be discolored. Refer to Fig. 13-31 for some suggested gas flow rates for use with various metals and thicknesses.

The heavier shielding gases like $CO_2$ and argon will tend to drop away from the weld area when welding out of position. Therefore, the gas flow rates must be increased as the position moves from the flat to the horizontal, vertical, and overhead positions.

When a gas mixture is used, it may be necessary to use a double or triple unit gas mixer. Such units

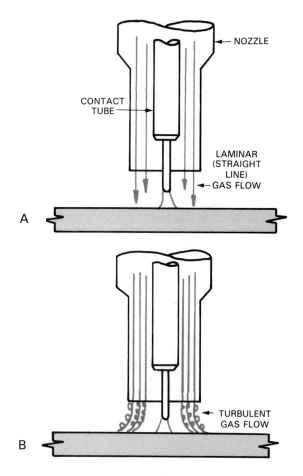

Fig. 13-30. A—Laminar gas flow is the result of the proper gas flow rate. B—Turbulence occurs when too much gas is used.

| METAL | TYPE JOINT | THICKNESS | | WELD POSITION | ARGON FLOW | |
|---|---|---|---|---|---|---|
| | | in. | mm | | ft³/h | L/min |
| Aluminum and Aluminum Alloys | All | 1/16 | 1.59 | F | 25 | 11.80 |
| | | 3/32 | 2.38 | F,H,V,0 | 30 | 14.16 |
| | | 1/8 | 3.18 | F,H,V,0 | 30 | 14.16 |
| | | 3/16 | 4.76 | F,H,V,0 | 23-27 | 10.85-12.74 |
| | | 1/4 | 6.35 | F | 40 | 18.88 |
| | | | | H,V | 45 | 21.24 |
| | | | | 0 | 60 | 28.32 |
| | | 3/8 | 9.53 | F | 50 | 23.60 |
| | | | | H,V | 55 | 25.96 |
| | | | | 0 | 80 | 37.76 |
| | | 3/4 | 19.05 | F | 60 | 28.32 |
| | | | | H,V,0 | 80 | 37.76 |
| Stainless Steel | Butt | 1/16 | 1.59 | | 30 | 14.16 |
| | Butt | 1/8-3/16 | 3.18-4.76 | | (98Ar-20₂) 35 | 16.52 |
| | 60° Bevel | 1/4-1/2 | 6.35-12.7 | | 35 | 16.52 |
| | 60° Double Bevel | 1/2-5/8 | 12.7-15.88 | | 35 | 16.52 |
| | Lap, 90° Fillet | 1/8-5/16 | 3.18-7.94 | | 35 | 16.52 |
| Nickel and Nickel Alloys | All | Up to 3/8 | Up to 9.53 | | 25 | 11.80 |
| Magnesium | Butt | .025-.190 | .64-4.83 | | 40-60 | 18.88-28.32 |
| | | .250-1.000 | 6.35-25.4 | | 50-80 | 23.60-37.76 |

Fig. 13-31. Suggested gas flow rates for various metals and thickness.

have a separate pressure regulator and flowmeter for each gas. See Fig. 13-32. Premixed gas mixtures can be purchased from welding gas suppliers in cylinders, just like pure argon or oxygen.

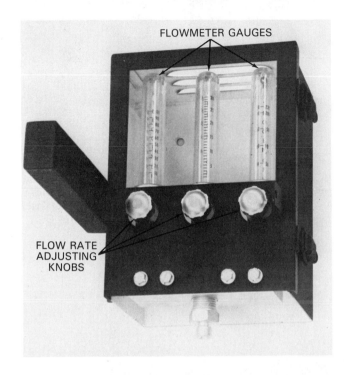

Fig. 13-32. A gas proportioner designed to mix industrial gas like argon, helium, and carbon dioxide. Each gas is adjusted separately, mixed and delivered to the GMAW or FCAW torch through the gas hose.

## 13-12. GAS NOZZLES AND CONTACT TUBES

The GAS NOZZLE is located at the end of the GMAW gun. See Figs. 13-33 and 13-34. It is designed to deliver the shielding gas to the weld area in a smooth, unrestricted manner. The gas

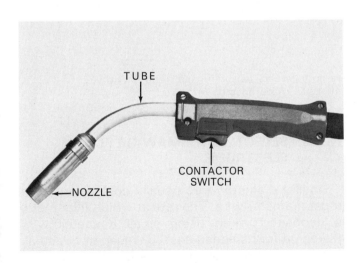

Fig. 13-33. An air cooled GMAW gun. (Miller Electric Mfg. Co.)

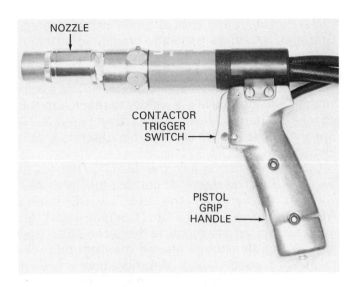

Fig. 13-34. A water cooled GMAW gun. This gun has a pistol grip. (L-TEC Welding & Cutting Systems)

nozzle is usually made of copper. Copper is a very good heat conductor. A copper nozzle will resist melting when exposed to the heat generated in the welding operation. GMAW nozzles and FCAW nozzles (if used) are the same. The construction of the nozzle end of a GMAW torch is shown in Fig. 13-35.

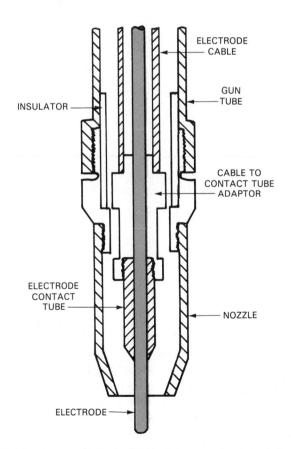

Fig. 13-35. A schematic drawing of the nozzle end of the GMAW or FCAW torch.

Nozzles are made with different exit diameters. Gun manufacturers generally provide information on the correct nozzle to use for various applications. A general purpose nozzle is often used and will work well for most applications. Nozzle extensions are made to allow a welder to reach into thin spaces or hard-to-reach areas. See Fig. 13-36. Special nozzle shapes are also manufactured, as illustrated in Fig. 13-37.

Under the nozzle lies the ELECTRODE CONTACT TUBE. The electrode contact tube is threaded into the end of the ELECTRODE CABLE ADAPTER. See Fig. 13-35. Contact tubes are made with a variety of inside diameters (ID). They must fit tightly enough around the electrode wire to make a good sliding, electrical contact. Each time the wire diameter is changed, the contact tube must be changed also.

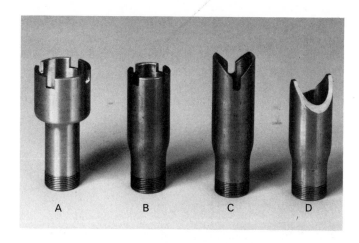

Fig. 13-37. Special GMAW nozzles for spot or tack welding. A and B—Standard spot or tack welding nozzle. C—Nozzle for outside corner spot or tack welds. D—Nozzle for inside corner spot or tack welds. (Miller Electric Mfg. Co.)

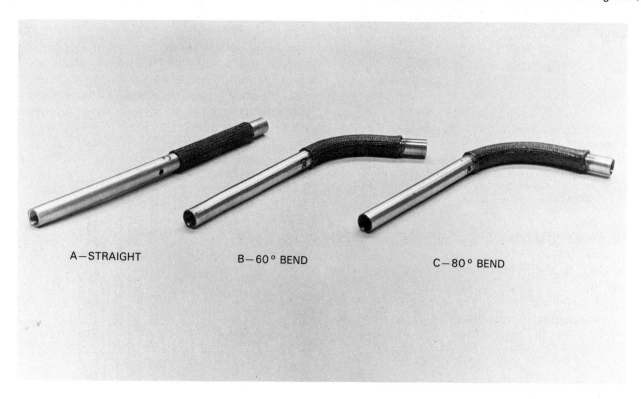

A—STRAIGHT          B—60° BEND          C—80° BEND

Fig. 13-36. A, B, and C—Straight and bend head tubes go between the torch handle and the nozzle. They permit welding in hard to reach locations. (Miller Electric Mfg. Co.)

The inside and outside of the nozzle become spattered during the welding operation. This spatter can be kept from sticking by spraying the nozzle with a special proprietary anti-stick compound. If the inside of the nozzle becomes spattered, the flow of shielding gas will become turbulent. Gas turbulence may cause weld contamination. To remove the spatter from the nozzle a special cleaning reamer is used. One type of nozzle reamer is shown in Fig. 13-38.

## 13-13. SELECTING A GMAW OR FCAW ELECTRODE

Smaller diameter wire usually costs more than larger diameter wire. The rate at which filler metal is deposited when using small diameter wire makes up for its added cost. Because of the small diameter and the high currents generally used in GMAW and FCAW, small diameter filler wire melts more rapidly than larger diameter wire.

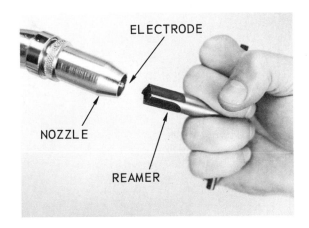

Fig. 13-38. A nozzle reamer used to keep the nozzle clean and free from metal splatter.

in. (50-75 mm) extend beyond the nozzle.

Whenever the wire is changed, the inside of the electrode cable should be cleaned. This may be done by blowing compressed air through the empty wire feed tube. The compressed air will blow out any metal flakes which may have rubbed off the wire. Caution: Be certain that the exit end of the wire feed tube is pointed away from any other persons when blowing through it with compressed air. A small amount of graphite powder is sometimes placed into the tube to lubricate it when each new wire roll is added.

The wire contact tube will wear and must be changed at times. Regular replacement of the contact tube will insure a continuous good electrical contact with the electrode wire.

## 13-14. PREPARING THE METAL

Metal surfaces generally may be cleaned mechanically or chemically. Abrasive cloth or wire brushing may be used. On severely corroded areas grinding may be used. Welding may be done on an oxidized (rusted) carbon or low alloy steel surface without cleaning. However, if the surface is rusty, a deoxidizing electrode wire should be used. This reduces oxidation and weld porosity.

The groove angle used when GMAW or FCAW may be smaller than the angle used when SMAW. See Fig. 13-39. This narrower angle is possible for two reasons. The wire diameters used are smaller and GMAW penetrates better than SMAW. It will take less filler metal to fill a 45 degree groove than a 75 degree groove. Welding time will also be less. Therefore, savings in filler metal and welder's time are possible.

The filler wire or electrode used must match or be compatible with the base metal. When $CO_2$ is used it causes oxidation of the weld metal. Deoxidizer types of electrode wires must be used to neutralize this oxidation. Manganese, silicon, and aluminum are used as deoxidizers in steel electrode wires. Titanium, silicon, and phosphorus are the deoxidizers in copper electrodes.

For more information regarding GMAW electrodes see Heading 11-23. Also refer to Fig. 11-46 for carbon steel electrodes and Fig. 11-47 for low alloy electrodes. More information about FCAW electrodes may be found in Heading 11-26 and Fig. 11-50.

Once the correct electrode is selected it should be loaded in the wire feeder as stated in Heading 13-9. The wire should be fed through the electrode cable using the inch switch until about 2-3

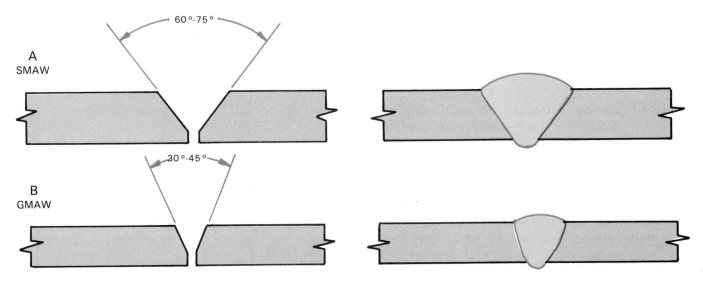

Fig. 13-39. A—Typical groove angle and weld bead for SMAW. B—Typical groove angle and weld bead for GMAW and FCAW. Notice that less filler metal is required to fill the groove at B. Welding time will also be less.

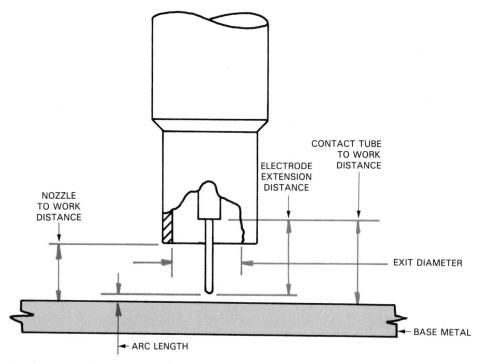

Fig. 13-40. Electrode extension distance. Other distances important in GMAW and FCAW are also shown.

## 13-15. ELECTRODE EXTENSION

ELECTRODE EXTENSION is the amount the end of the electrode wire sticks out beyond the end of the contact tube. See Fig. 13-40. This distance is sometimes referred to as stickout.

A good extension for use with the GMAW short circuit transfer method is about 1/4-1/2 in. (approximately 6-13 mm). The correct electrode extension for all other GMA metal transfer methods varies between 1/2-1 in. (approximately 13-25 mm). Similar electrode extensions may be used with gas shielded flux cored arc welding. The suggested electrode extension for use with self-shielding flux cored arc welding is between 3/4-3 3/4 in. (approximately 19-95 mm).

As the electrode extension increases, the resistance of the electrodes increases. Any increase in resistance will cause the current to heat the wire along the length of extended wire. A long extension may cause too much filler metal to be deposited with a low arc heating. This may cause spatter, shallow penetration, and a low weld bead shape.

After loading the electrode spool on the wire feeder, the electrode wire is fed through the electrode cable. Feed the wire until about 2-3 in. (approximately 50-75 mm) extend beyond the noz-

zle. Cut the wire off with a wire cutter until the correct wire extension is obtained.

## 13-16. WELDING PROCEDURES

Before beginning to weld, the welding station should be checked for safety. All electrical, gas and water connections must be checked for tightness.

Weldments should be tack welded or placed into well designed fixtures prior to welding. When complete joint penetration is required, backing strips are recommended.

Most arc welding processes require the welder to control the arc length, welding speed, and torch or gun angle to obtain a good weld. In GMAW and FCAW the arc length will remain constant and is determined by the arc voltage. The welder in GMAW must watch and control the distance from the nozzle or wire contact tube to the work. See Fig. 13-40. By controlling the nozzle to work distance, the welder will control the electrode extension distance. Heading 13-15 explains the importance of electrode extension.

The welding speed will be determined by the appearance of the bead width and penetration. Torch angle will also affect the bead width and penetration. The terms forehand, backhand, and perpendicular welding are used.

In FOREHAND WELDING, the tip of the electrode points in the direction of travel. When BACKHAND WELDING, the electrode tip points

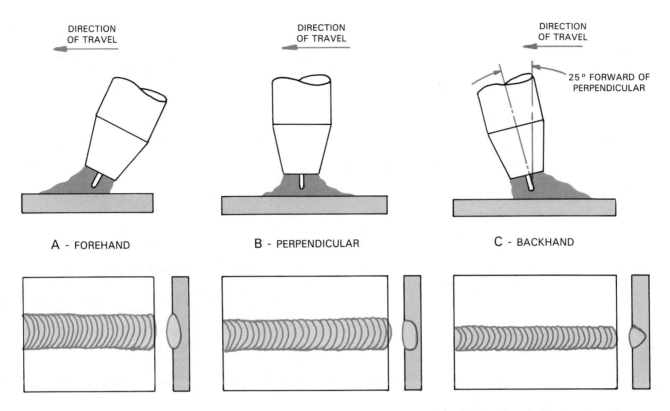

Fig. 13-41. Effects on the weld of welding. A—Forehand. B—Perpendicular. C—Backhand. Notice that the backhand method gives the deepest penetration.

away from the direction of travel. PERPEN-DICULAR WELDING is done with the electrode at 90 degrees to the base metal. Fig. 13-41 shows the effects of these various methods.

The backhand method will give the best penetration. A 25 degree angle forward of perpendicular will give the best penetration in the flat welding position as shown in view C, Fig. 13-41. For the best control of the molten pool, an angle between 5-15 degrees forward of perpendicular is preferred for all positions.

To start welding, the welder tips the top of the gun 5-15 degrees in the direction of travel and places the helmet down over his eyes. To start the arc and the wire, gas, and water (if used), the welder squeezes the trigger on the gun. The wire will arc as soon as it feeds out far enough to touch the metal. No striking or up and down motion is required to start the arc as required with SMAW.

As the arc pool reaches the proper width, the welder moves the electrode forward. The welder continues to move the torch along the weld, watching the width of the weld pool to maintain a uniform size. This procedure is continued until the end of the weld is reached. A run-off tab may be used to insure a full width bead to the end of the weld. Without a run-off tab the end of the weld may have a crater depression. This depression can be reduced by moving the electrode to the end of

the weld and then back over the completed bead about 1/2 in. (approximately 13 mm). At the end of this reverse travel the contractor switch is released. To shield the end of the weld, hold the gun in position to allow the gas post flow to protect the weld until it cools.

More than one pass may be required to fill a weld groove. Each pass should be cleaned before the next pass is laid. This is generally done with a wire brush or wheel. The glass-like coating on some gas metal arc welds is easily removed. The slag layer on a flux cored arc weld is heavier and requires more effort to remove.

Welds made out-of-position require that leathers be worn. Molten base metal, filler metal, and spatter may fall on the welder. Therefore, a cap, coat, cape, and chaps should be worn to protect against burns.

## 13-17. SHUTTING DOWN THE STATION

When welding is stopped for an extended period, the station should be shut down. To shut down the station proceed as follows:
1. Return the wire speed to zero.
2. Turn off the wire drive unit.
3. Turn off the shielding gas cylinder(s).
4. Squeeze the gun trigger and hold it in for a few seconds to bleed the gas lines.

5. Turn the flowmeter adjusting knob(s) in to close it.
6. Turn off the power switch on the arc welding machine.
7. Hang the gun on an insulated hook.
8. Turn out the pressure adjusting knob on the flowmeter regulator if an adjustment knob is provided.

## 13-18. WELDING JOINTS IN THE FLAT POSITION

The face of a weld made in the flat position should be horizontal or nearly horizontal. The weld axis is also horizontal. See Fig. 13-42. Any of the metal transfer methods may be used in the flat position. The method used will depend on the metal thickness and other factors.

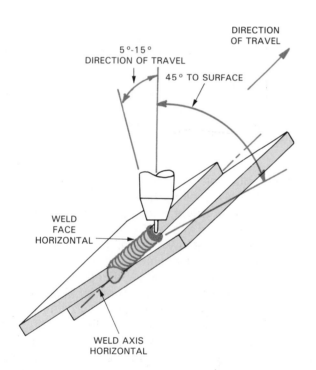

Fig. 13-42. A fillet weld on a lap joint in the flat position. Note the angles used and the deep penetration of the weld. Also, the weld face and axis are horizontal or near horizontal.

### FILLET WELD ON A LAP JOINT

The metal should be set up as shown in Fig. 13-42. It should be tack welded about every 3 in. (approximately 75 mm). This will hold it in position while the weld is made.

To make the fillet weld, the centerline of the electrode should be held at about 45 degrees to each edge and metal surface. The electrode should point more toward the surface if the edge begins to melt too quickly. The electrode and gun should be held between 5-15 degrees forward

from a vertical line to the metal surface.

A C-shaped metal pool will form as when GTAW. When the end of the weld is reached, reverse the direction for about 1/2 in. (approximately 13 mm). This movement will help reduce the crater which occurs if the weld is stopped at the end of the joint. No matter what type weld is made, this same finish movement can be made. A run-off strip will totally eliminate the crater at the end of a weld.

### FILLET WELD ON AN INSIDE CORNER JOINT

Fillet welds may be made on metal up to 3/8 in. (approximately 10 mm) thick without edge groove preparation. This is possible because of the deep penetration possible with the spray transfer method. The centerline of the electrode should be held at 45 degrees to each metal surface. If the backhand welding procedure is used, the electrode and gun are held between 5-15 degrees forward of vertical. See Fig. 13-43.

GMAW can generally weld 1/4 in. (approximately 6 mm) beads in each pass. If the weld is to be over 1/4 in. (6 mm) thick, two or more weld passes will be required.

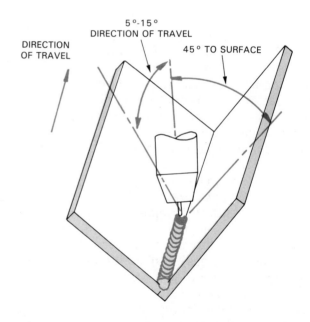

Fig. 13-43. A fillet weld on an inside corner joint in the flat position. The electrode is 45 degrees from each metal surface. It is also tipped 5-15 degrees forward in the direction of travel.

### GROOVE WELD ON A BUTT JOINT

Square groove welds can be made on metal up to 3/8 in. (approximately 10 mm) without edge shaping. Groove welds with shaped edges of any thickness can be made with the GMAW process. The groove angle on a V-groove butt weld can be narrower than is used with SMAW. Because of the penetration possible with the spray transfer methods, the root face or thickness can be larger. The root opening can be smaller with GMAW than the opening used for SMAW.

The centerline of the electrode should be directly over the axis of the weld. An angle between 5-15 degrees forward of vertical is correct for the backhand welding method. See Fig. 13-44.

A keyhole in the weld crater will indicate that complete penetration is occurring. One problem which may occur in a groove weld made with GMAW is whiskers. WHISKERS are lengths of electrode wire which stick through the root side of a groove weld. Whiskers occur when the electrode wire is advanced ahead of the weld pool.

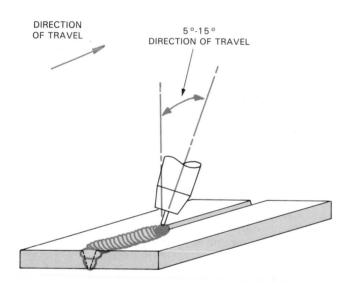

Fig. 13-44. A V-groove weld on a butt joint in the flat position. Note the narrow (45 degree) groove possible with GMAW.

The wire goes through the root, burns off, and starts up again. The burned off length is left stuck in the weld. Whiskers can be prevented by slowing the welding speed. They may also be prevented by reducing the wire feed speed. A small weaving motion may be used to keep the wire from getting ahead of the weld pool.

GROOVE WELD ON AN OUTSIDE CORNER JOINT

The outside corner joint is set up as shown in Fig. 13-45. A square or contoured groove weld may be used. The electrode angles are the same as those used for welds made on a butt joint.

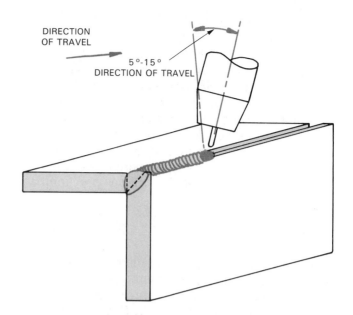

Fig. 13-45. A bevel groove weld on an outside corner joint in the flat position.

Since groove welds are made on the outside corner joint, whiskers can occur.

## 13-19. WELDING JOINTS IN THE HORIZONTAL POSITION

The face of a weld made in the horizontal position is in the vertical or near vertical position. In the horizontal position, the centerline of weld axis runs in a horizontal or near horizontal position. See Fig. 13-46.

Short circuit, globular, spray, or pulsed spray transfer methods may be used in the horizontal position.

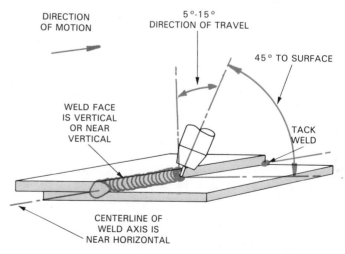

Fig. 13-46. A fillet weld on a lap joint in the horizontal position. In the horizontal position, the weld axis is near horizontal and the face of the weld near vertical.

Gas Metal Arc Welding / 373

## FILLET WELD ON A LAP JOINT

For practice welds the metal should be set up and tack welded as shown in Fig. 13-46. The centerline of the electrode should be about 45 degrees to the edge and metal surface. It may point more toward the surface if the edge melts too quickly. The electrode or gun should tip about 5-15 degrees forward of vertical in the direction of travel. The typical C-shaped puddle will indicate that both the edge and surface are melting properly.

## FILLET WELD ON AN INSIDE CORNER JOINT

Square or prepared groove welds may be made. The use of a V, bevel, U, J type prepared groove will depend on the metal thickness and joint design. The bead width used in GMAW does not have to be as wide for the same thickness as when doing SMAW. This is because the gas metal arc weld penetrates more. It does not need bead width and reinforcement to strengthen the weld.

The electrode should be held at 45 degrees to each metal surface. Aiming the wire more toward the vertical surface may improve the bead shape. This will help compensate for the molten metal sag. Incline the gun and electrode about 5-15 degrees forward of vertical. See Fig. 13-47.

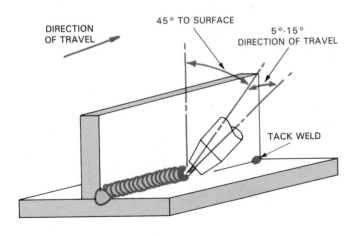

Fig. 13-47. A fillet weld on an inside corner joint in the horizontal position. Note the angles from the metal and in the direction of travel.

## GROOVE WELD ON A BUTT JOINT

A square or prepared groove weld may be used. Fig. 13-48 shows a U-groove weld in progress. The electrode centerline should be directly over

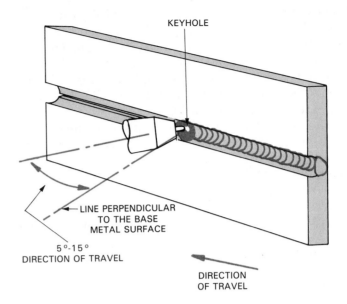

Fig. 13-48. A U-groove weld on a butt joint in the horizontal position. Notice the keyhole at the root of the weld.

the weld line. For best weld pool control the electrode should tip 5-15 degrees in the direction of travel. To insure complete penetration watch for a continuous keyhole through the root pass. More than one pass may be necessary on thicknesses above 1/4 in. (approximately 6 mm). To completely fill the groove an electrode weaving motion may be required.

## GROOVE WELD ON AN OUTSIDE CORNER JOINT

A groove weld on an outside corner is made in the same manner as a butt joint.

### 13-20. WELDING JOINTS IN THE VERTICAL POSITION

GMAW in the vertical position is done using the short circuit or pulsed spray transfer method. Spray transfer may also be used, but only with small diameter wire and a small molten puddle.

In the vertical welding position the weld axis and the weld face are both vertical. Fig. 13-49 shows a vertical outside corner joint being tack welded. The GMAW may be made from the bottom up (uphill) or top down (downhill).

Downhill welding is difficult with FCAW. The flux material might flow into the weld. This can be avoided if the welder can keep the crater ahead of the molten flux.

The centerline of the electrode should be tipped 5-15 degrees in the direction of travel as in other position welds. This angle will permit the easiest weld pool control. Using a weaving motion with the spray metal transfer method will help keep the

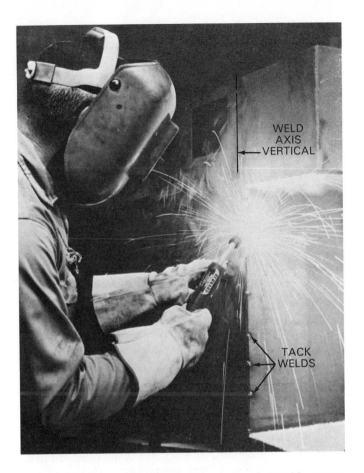

Fig. 13-49. Tack welds being placed on an outside corner joint prior to making a vertical weld.

crater cooled. See Fig. 13-50.The weld pool is relatively cool when the short circuit method of metal transfer is used. A properly adjusted pulsed arc will allow time between pulses for the molten crater to cool.

## FILLET WELD ON A LAP JOINT

Fig. 13-51 illustrates a fillet weld being made in the vertical position. The angles of the electrode and torch are the same as for other positions. The electrode should tip about 5-15 degrees in the direction of motion. The centerline of the electrode should be at about 45 degrees to the edge and the flat surface. If the edge of the metal melts too rapidly, point the electrode more toward the flat surface. Be certain that the edge and surface are melting completely as the filler metal is added. The appearance of a C-shaped molten pool will indicate good fusion.

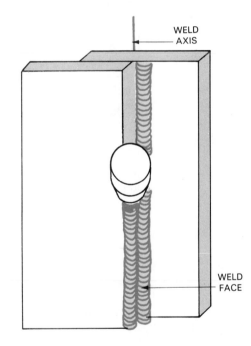

Fig. 13-51. A fillet weld on a lap joint in the vertical position. Two passes are being used on the weld. Note the weld axis and bead face are vertical.

## FILLET WELD ON AN INSIDE CORNER

The centerline of the electrode should be held at 45 degrees to each surface. It should be tipped at 5-15 degrees in the direction of motion. A C-shaped pool will indicate good fusion is occurring. A weaving motion may be necessary with the spray arc process. The weaving will help the crater to cool between the swings of the weaving motion. The welder should hesitate very briefly at the end of each swing. See Fig. 13-50.

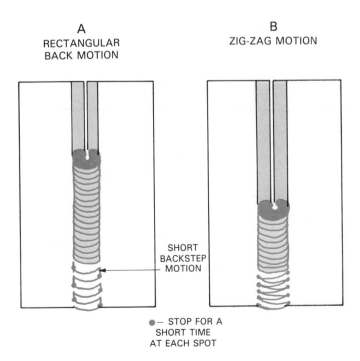

Fig. 13-50. Two weaving motions used with spray arc transfer to control the temperature of the molten pool. Notice the short backstep motion used in the motion at A. Stopping for an instant at each side allows time for the pool to cool slightly.

**Gas Metal Arc Welding / 375**

## GROOVE WELD ON A BUTT JOINT

A V-groove butt weld in progress is shown in Fig. 13-52. The electrode centerline should be directly above the weld line. The electrode and torch should be inclined (tipped) 5-15 degrees in the direction of travel. A keyhole at the root of the weld will indicate complete penetration. A weaving bead may be necessary with the spray arc process.

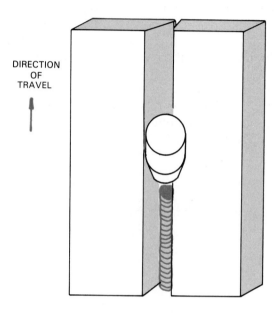

Fig. 13-52. A V-groove weld on a butt joint. The root pass is in progress.

## GROOVE WELD ON AN OUTSIDE CORNER

This weld is made in the same manner as a groove weld on a butt joint.

## 13-21. WELDING JOINTS IN THE OVERHEAD POSITION

The short circuiting and pulsed spray metal transfer methods are recommended for overhead welding. When overhead welding, it is strongly suggested that a cap, coat, cape, and possibly chaps be worn. This is necessary to protect the welder from falling, molten metal.

The angle of the electrode from the joint surfaces is the same as for other welding positions. The electrode should be held more vertically when overhead welding. An angle of between 5-10 degrees is suggested. The weld pool in short circuit and pulsed arc transfer is relatively cool. A weaving motion is not required for the purpose of cooling the pool. As the weld pool increases in size, the possibility of the metal falling out or sag-

ging increases. Several narrower beads are recommended rather than a weaving motion.

## FILLET WELD ON A LAP JOINT

See Fig. 13-53 for an example of angles used.

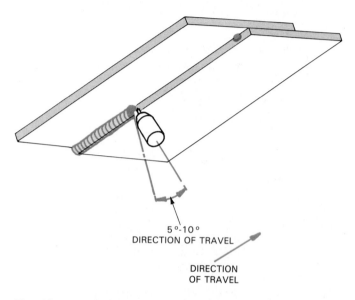

Fig. 13-53. A fillet weld on a lap joint in the overhead position. In the joint, two passes will be made. This is done to keep the pool size smaller.

## FILLET WELD ON AN INSIDE CORNER

Fig. 13-54 shows a fillet weld in progress on an inside corner. The centerline of the electrode should be 45 degrees from each metal surface. It should be tipped about 5-10 degrees in the direction of travel.

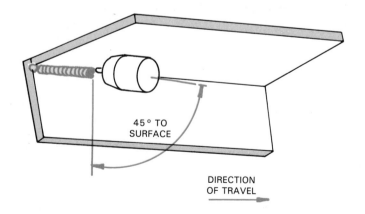

Fig. 13-54. A fillet weld on an inside corner joint. The electrode and gun are tipped 5-10 degrees in the direction of travel.

## GROOVE WELD ON A BUTT JOINT

A square groove butt weld is shown being welded in Fig. 13-55.

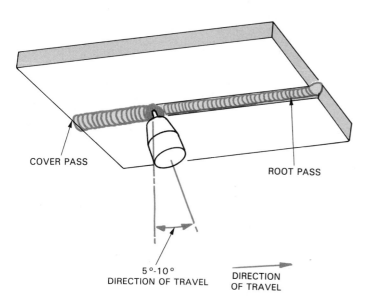

Fig. 13-55. Bevel groove weld on a butt joint in an overhead position.

## GROOVE WELD ON AN OUTSIDE CORNER

Fig. 13-56 shows a bevel groove weld being made on an outside corner joint.

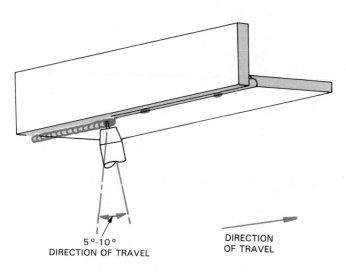

Fig. 13-56. A J-groove weld in an outside corner joint in an overhead position.

## 13-22. AUTOMATIC GMAW AND FCAW

Gas metal arc welding and flux cored arc welding may be semi-automatic or fully automatic processes. Used as a semi-automatic process, the welder must direct and move the arc welding gun while the electrode wire feeds automatically into the weld crater.

Both GMAW and FCAW guns may be mounted on a motor driven carriage or robot. When the arc welding gun is directed and moved by a machine, the process becomes fully automatic. Refer to Chapter 23 for information on robots and other automatic welding equipment.

## 13-23. GAS METAL ARC SPOT WELDING

The gas metal arc welding machine and arc welding gun can be used to produce a weld in one small spot.

Metals commonly welded with the gas metal arc spot welding process are low carbon steel, stainless steel, and aluminum. Gas metal arc spot welding is generally done on metals under 1/16 in. (1.6 mm) thick.

Small tack welds can be made on lap and inside corner joints. See Fig. 13-57. A spot weld is also shown in Fig. 13-58. A spot weld is a weld which joins two metal surfaces at points away from the edges. See Fig. 13-57. The gas metal arc welding machine must be equipped with special controls in order to do spot welding. The arc welding gun must be fitted with a special nozzle for spot welding. See Fig. 13-37.

Several welding variables must be controlled in order to make gas metal arc spot welds. These variables are:
1. Welding current.
2. Welding time.
3. Arc voltage.
4. Electrode size and composition.
5. Electrode extension.
6. Shielding gas.

The type of welding current used is generally DCEP (DCRP). The current is controlled by varying the wire feed speed. This is done in the usual way on the wire feeder.

Penetration increases as the welding time is lengthened. The diameter of the weld area between the two pieces also increases as the welding time is increased. Some timers begin timing when the arc welding gun switch is pressed. However, an arc does not always occur as the wire touches the metal. When the weld finally does start, the time is shorter than required. As a result a poor weld will occur. The best timers do not begin timing until the arc is struck. This timing system gives more uniform welds.

Voltage settings are made on the arc welding machine in the same way as when GMAW. If the voltage is increased, the arc length will increase.

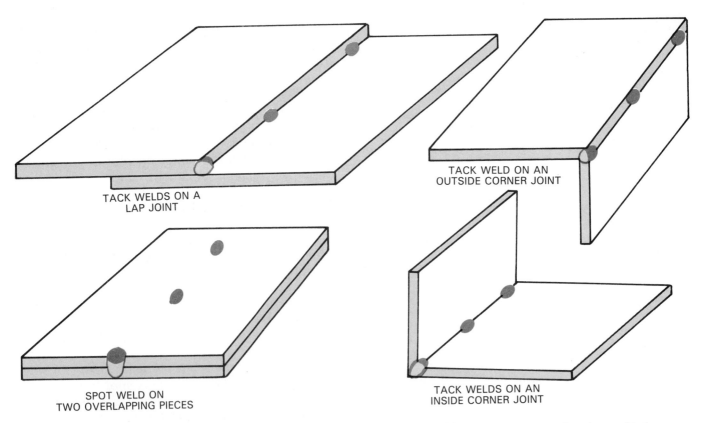

TACK WELDS ON A
LAP JOINT

SPOT WELD ON
TWO OVERLAPPING PIECES

TACK WELD ON AN
OUTSIDE CORNER JOINT

TACK WELDS ON AN
INSIDE CORNER JOINT

Fig. 13-57. Tack welds on lap, inside corner, and outside corner joints. Several spot welds are also shown. Notice the depth of penetration shown in section.

Fig. 13-58. A completed gas metal arc spot weld. (L-TEC Welding & Cutting Systems)

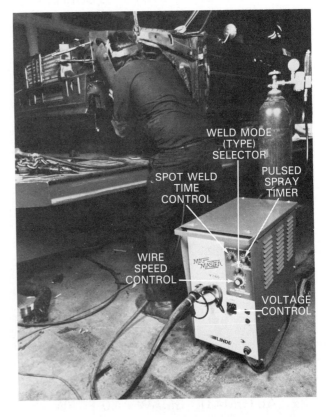

Fig. 13-59. A GMA spot welder being used in automobile body repair. Note the types of controls for regular and spot welding. (L-TEC Welding & Cutting Systems)

As the voltage is decreased, the arc length will decrease. As the voltage is increased, the weld diameter will increase. At the same time the penetration will decrease and the weld reinforcement (buildup) will increase.

The same size and type of solid wire used for welding may be used for spot welding a particular metal.

Electrode extension, see Fig. 13-40, must remain constant during the GMA spot welding process. The extension distance is kept constant by using a special nozzle. Several GMA spot welding nozzle designs are shown in Fig. 13-37. The end of the electrode contact tube is set back from the end of the nozzle. This is done to keep the contact tube out of the weld. This set-back wire will also reduce the possibility of the electrode melting up into the contact tube at the end of the weld cycle.

The shielding gas used may be the same gas or mixture used for welding beads.

GMA spot welds may be made in any position.

Weld quality and uniformity is not as good as that possible with resistance spot welding.

The gas metal arc spot welding controls found on various gas metal arc welding machines differ. Fig. 13-59 shows a GMA spot welder being used in automobile body repair. The appearance and the names used on various control panels may be different. Some controls typically found on the GMA spot welding control panel are:

CONTROL SWITCH: A control switch is used to switch the gas metal arc welding machine from a regular welder to a spot welder.

WELD TIMER. This control is for setting the welding time. The entire spot welding operation takes place in one or two seconds.

BURN BACK ADJUSTMENT: Some machines have a burn back adjustment. This control allows the current to flow for a short time after the wire feed stops. The continued current flow prevents the wire from sticking in the weld pool. If the burn back time is set too high the electrode wire may burn back into the contact tube. If it is not set high enough, the wire will stick in the weld pool at the end of the welding time. Fig. 13-60 shows the spot welding controls for another GMA welding machine.

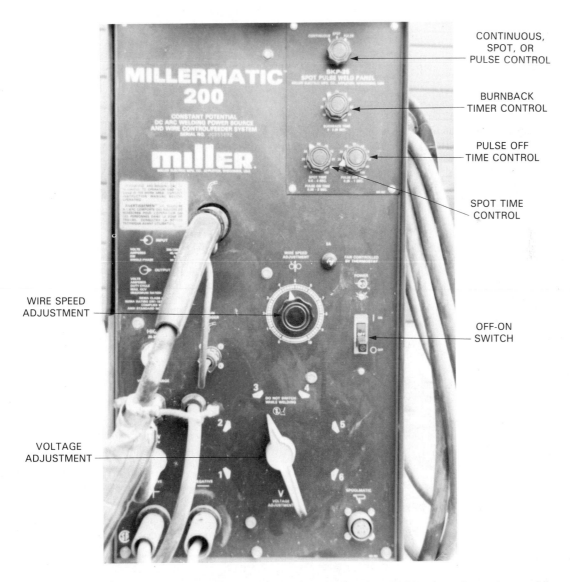

Fig. 13-60. The control panel for a GMA welding machine. This machine is capable of continuous arc welding, spot welding, and pulsed arc welding. The uppermost control must be set for the type of welding to be done.

| Trouble | Possible Causes | How to Correct |
|---|---|---|
| Difficult arc start | Polarity wrong<br>Insufficient shielding gas<br>Poor ground<br>Open circuit to start switch | Check polarity, try reversing<br>Check valves, increase flow<br>Check ground—return circuit<br>Repair |
| Irregular wire feed, burn back | Insufficient drive roll pressure<br>Wire feed too slow<br>Contact tube plugged<br>Arcing in contact tube<br>Power circuit fluctuations<br>Polarity wrong<br>Torch overheating<br>Kinked electrode wire<br>Conduit liner dirty or worn<br>Drive rolls jammed<br>Conduit too long | Increase drive roll pressure<br>Check, adjust wire feed speed<br>Clean, replace contact tube<br>Clean, replace contact tube<br>Check line voltage<br>Check polarity, try reversing<br>Replace with higher amp gun<br>Cut out, replace spool<br>Clean, replace<br>Clean drive case, clean electrode wire<br>Shorten, install push-pull drive |
| Welding cables overheating | Cables too small<br>Cable connections loose<br>Cables too long | Check current requirements, replace<br>Check, tighten<br>Check current carrying capacity |
| Unstable arc | Cable connections loose<br>Weld joint area dirty | Check, tighten<br>Clean chemically or mechanically |
| Arc blow | Magnetic field in DC causes arc to wander | Rearrange or split ground connection<br>Use brass or copper backup bars<br>Counteract "blow" by direction of weld<br>Replace magnetic work bench |
| Undercut | Current too high<br>Welding speed too high<br>Improper manipulation of gun<br>Arc length too long | Use lower current setting<br>Slow down<br>Change angle to fill undercut<br>Shorten arc length |
| Excessively wide bead | Current too high<br>Welding speed too slow<br>Arc length too long | Use lower current setting<br>Speed up<br>Shorten arc length |
| Incomplete penetration | Faulty joint design<br><br>Welding speed too rapid<br>Welding current too low<br>Arc length too long<br>Improper welding angle | Check root opening, root face dimensions, included angle<br>Slow down welding speed<br>Increase welding current<br>Shorten arc length<br>Correct faults, change gun angle |
| Incomplete fusion | Faulty joint preparation<br><br>Arc length too long<br>Dirty joint | Check root, opening, root face dimensions, included angle<br>Shorten arc length<br>Clean chemically or mechanically |
| Dirty welds | Inadequate gas shielding<br><br><br><br><br><br><br>Dirty electrode wire<br><br><br>Dirty base metal | Hold gas cup closer to work<br>Increase gas flow<br>Decrease gun angle<br>Check gun and cables for air and water leaks<br>Shield arc from drafts<br>Center contact tube in gas cup<br>Replace damaged gas cup<br>Keep wire spool on welder covered<br>Keep unused wire in shipping containers<br>Clean wire as it enters wire drive<br>Clean chemically or mechanically |
| Porosity<br><br>See above, *Dirty welds* | Dirty electrode wire<br>Dirty base metal<br>Inadequate gas shielding<br>Improper technique | See above, *Dirty welds*<br><br>See above, *Dirty welds*<br>Change angle of gun to improve shielding |
| Cracked welds | Faulty design<br>Faulty electrode<br><br>Shape of bead<br><br>Travel speed too fast<br>Improper technique<br>Rigidity of joint | Check edge preparation and root spacing<br>Check electrode wire for compatibility with base metal<br>Change travel speed or shielding gas to obtain more convex bead<br>Slow down<br>Change angle of gun to improve deposition<br>Redesign joint, preheat and postheat, weave bead |

Fig. 13-61. A troubleshooting guide for problems which may occur when GMAW.
(Welding and Fabricating Data Book)

## 13-24. GMAW TROUBLESHOOTING

Fig. 13-61 is a chart which describes many typical troubles which may occur when making a gas metal arc weld. The causes of the problem are shown along with methods for correcting each problem.

## 13-25. GMAW AND FCAW SAFETY

The safety precautions for arc welding covered in previous chapters also apply to GMAW and FCAW.

Adequate eye protection should always be worn. If welding for long periods, flash goggles with a #2 lens shade should be worn under the arc helmet. A #11 lens is recommended for nonferrous GMAW and a #12 for ferrous GMAW. Lens shades up to #14 may be worn as required for comfort. All welding should be done in booths or in areas protected by curtains. This is done to protect others in the weld area from arc flashes.

Suitable dark clothing must be worn. This is done to protect all parts of the body from radiation or hot metal burns. Leather clothing offers the best protection from burns.

Ventilation should be provided. This ventilation and/or filtering equipment is necessary to keep the atmosphere around the welder clean. Carbon monoxide is generated when doing GMAW and FCAW using $CO_2$ as a shielding gas. It is suggested that all welding be done in well ventilated areas. Ozone is also produced when doing GMAW and FCAW. Ozone is a highly toxic gas. Metals still covered with chlorinated hydrocarbon solvents will form poisonous, toxic phosgene gas when welded.

Protect arc cables from damage. Do not touch uninsulated electrode holders with bare skin or wet gloves. A fatal shock could result. Welding in wet or damp areas is not recommended.

Shielding gas cylinders must be handled with great caution. Refer to Chapters 3 and 4 for a review of how to handle high pressure cylinders. These chapters should also be referred to for instructions on how to attach regulators and other gas equipment.

## 13-26. TEST YOUR KNOWLEGE

Write your answers on a separate sheet of paper. Do not write in this book.
1. Name three benefits of using GMAW.
2. The polarity used for almost all GMAW and FCAW is DC_ or DC_ _.
3. Name three metal transfer methods.
4. _____ is the property in an electric cir-

cuit that slows down the rate of the current change.
5. Spray transfer will only occur when the current is set above the _____ current.
6. Spray transfer will only occur when at least _____% argon is used.
7. What is MIG brazing?
8. How many pounds and kilograms of filler metal can be deposited per hour with the rotating spray transfer method?
9. Is it possible to do GMAW with DCEN? If so, what electrode is used?
10. A GMAW machine used for pulsed spray transfer must have what additional controls?
11. Using spray arc transfer _____ volts and _____ amperes are used with a .045 in. (1.14 mm) electrode to weld stainless steel.
12. On the wire drive unit shown in Fig. 13-20, how is the lower drive wheel adjusted in and out?
13. On the wire drive unit shown in Fig. 13-20, wire guides and drive rolls are aligned by loosening the _____ _____ securing bolts and moving the _____ _____ up or down.
14. To feed the electrode wire through the electrode cable to the arc welding gun, the _____ switch is operated.
15. What factors must be considered when choosing a shielding gas?
16. Argon has a _____ thermal conductivity then helium so _____ is used to weld thick aluminum or copper sections.
17. Why is good ventilation important when using $CO_2$ gas?
18. What effect does oxygen ($O_2$) have on the arc when mixed with argon?
19. Which gases are suggested for use with pulsed spray transfer?
20. What argon flow rate in ft³/hr and L/min should be used to weld .150 in. (3.81 mm) thick magnesium in a butt joint?
21. What part of the gas metal arc welding gun contacts the electrode wire and passes electricity to the electrode?
22. How can metal spatter be kept from sticking to the nozzle?
23. Electrode extension is the distance from the end of the _____ _____ and the end of the _____.
24. The suggested angle for the electrode and gun for best puddle control with backhand welding in most positions is _____° to _____° forward of vertical.
25. Metals still covered with chlorinated hydrocarbon solvents will form poisonous toxic _____ gas when welded.

This welder is GMA welding a plate onto a pillar using an inverter type arc welding machine and a separate wire drive. (Miller Electric Mfg. Co.)

# part 5

# ARC CUTTING

Pipe cutting operation using one cutting torch mounted on a pipe cutting machine.
(H & M Pipe Beveling Machine Co., Inc.)

# 14 ARC AND OXYGEN ARC CUTTING EQUIPMENT AND SUPPLIES

The electric arc method of metal cutting is a popular cutting process. Oxygen, air, or inert gases along with an electric arc are used to cut and gouge metals. The arc cutting processes permit an operator to make high quality cuts. The cutting speed of the arc process is also very high. Improved equipment and supplies have contributed to the usefulness and efficiency of arc cutting. The arc cutting processes are:

1. Carbon arc cutting (CAC).
2. Air carbon arc cutting (AAC).
3. Oxygen arc cutting (AOC). Also see Chapter 22.
4. Gas metal arc cutting (GMAC).
5. Gas tungsten arc cutting (GTAC).
6. Shielded metal arc cutting (SMAC).
7. Plasma arc cutting (PAC).

Follow all safety precautions listed for both oxy-fuel and arc welding procedures whenever cutting. Refer to Heading 15-7 for a review of arc cutting safety.

## 14-1. CARBON ARC CUTTING (CAC) EQUIPMENT

The CARBON ARC CUTTING process was the original method used to cut or melt away metals. The base metal is melted by the heat of the arc. Metal is removed from the kerf by the force of the arc and gravity. The quality of the cut is generally poor. This method is often used in small shops where more efficient and more expensive equipment is generally not available. The AWS abbreviation for carbon arc cutting is CAC.

The complete cutting outfit consists of:

1. An AC, DC, or an AC/DC Arc Welding Machine.
2. A booth with better than normal ventilation (1,000 CFM or more).
3. An electrode lead.
4. A work lead.
5. A special carbon electrode holder, as shown in Fig. 14-1.
6. Carbon or graphite electrodes.

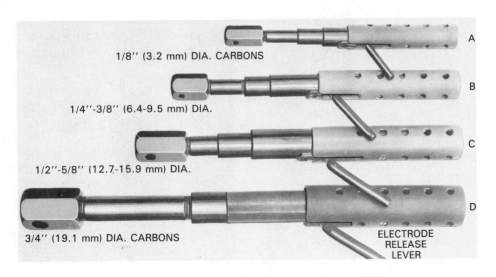

1/8'' (3.2 mm) DIA. CARBONS — A

1/4''-3/8'' (6.4-9.5 mm) DIA. — B

1/2''-5/8'' (12.7-15.9 mm) DIA. — C

3/4'' (19.1 mm) DIA. CARBONS — D

ELECTRODE RELEASE LEVER

Fig. 14-1. Carbon electrode holders. These holders are designed for: A—1/8 in. (3.2 mm) carbon. B—1/4 in.-3/8 in. (6.4 mm-9.5 mm) carbon. C—1/2 in.-5/8 in. (12.7 mm-15.9 mm) carbon. D—3/4 in. (19.1 mm) carbon. (Tweco Products)

7. Heavy-duty gloves.
8. Heavy-duty clothes.
9. A helmet with a lens of #12 to #14 shade.

The high amperage flow at the arc releases a considerable amount of heat. Heavy gloves and clothing must be worn to protect the operator from the intense heat of the carbon arc. The electrode holder often has a shield around the handle to protect the operator's hand from radiant heat. Some carbon electrode holders are water cooled. When using water cooled equipment, less protection from heat is required.

It is very important to have a good ventilation system.

See Chapter 9 for more information about the machines, cables, carbon electodes, and electrode holders used in this process.

## 14-2. AIR CARBON ARC CUTTING (AAC) EQUIPMENT

The equipment used for air carbon arc cutting (AAC) is as follows:
1. An AC, DC or AC/DC are welding machine.
2. An air compressor or compressed air in cylinders (for small applications).
3. An air carbon arc torch equipped with an air jet device.
4. An electrode lead.
5. A compressed air hose. This hose is usually combined with the electrode lead, as shown in Fig. 14-2.

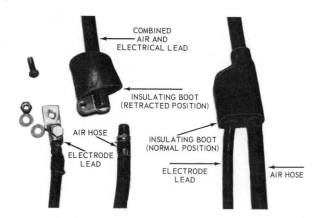

Fig. 14-2. A combination electrode lead and air hose assembly for air carbon arc cutting. (Arcair Co.)

6. A work lead.
7. Carbon or graphite electrodes.

The usual gloves, special clothing, booths, ventilation devices, benches, and the like as used in SMAW are required. See Chapter 9 and Headings 9-26 and 9-27. A #12 - #14 arc welding filter lens is required in the welding helmet. The DC power source used should be direct current electrode

positive, DCEP (DCRP). It must be a constant current machine with a drooping curve. See Heading 9-3. The machine must also be capable of delivering the required current at a 100 percent duty cycle. The normal arc voltage is 40 volts. AC arc welding machines may also be used with special AC electrodes. Refer to Heading 1-9.

The air carbon arc process may be used in conjunction with a motorized carriage. Fig. 14-3 shows an air carbon arc holder mounted on a

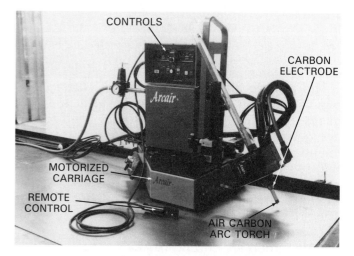

Fig. 14-3. An air carbon arc electrode mounted in a large motor driven carriage. This carriage may be controlled remotely. (Thermadyne Industries, Inc.)

motorized carriage or tractor. The motorized carriage or tractor may be used in the horizontal, vertical, and overhead positions. In these positions, a special track is used and the tractor is specially equipped and known as a "climber." The track is held to magnetic materials by magnets. On non-magnetic materials, the track is held to the metal by suction cups or clamps. Specially equipped tractors may be operated by remote control. Fig. 14-4 shows an air carbon arc machine in use.

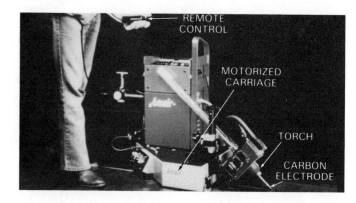

Fig. 14-4. Gouging a steel plate using an air carbon arc torch mounted in a motorized carriage. The operator is using a remote control to steer the carriage. (Thermadyne Industries, Inc.)

## 14-3. AIR CARBON ARC COMPRESSED AIR SUPPLY

Air is fed to the air carbon arc torch jet at 80 to 100 psig (551.6 - 689.5 kPa). The air source must be capable of supplying 50 cu. ft. per minute (1416 L/min) at the torch. For a light duty electrode holder, the pressure may be 40 psig (275.8 kPa) at 3 cu. ft/min (85 L/min). This air should contain a minimum of abrasives and moisture. A sufficient amount of air must flow through the jet to remove the molten metal, see Fig. 14-5. Large air compressors are used for this purpose. The size of the air compressor and hose varies with the size of the cutting job. The air compressor is automatically controlled to turn on and off by an electrical air pressure sensing switch. The air flow to the hose is kept at a constant pressure by means of a pressure regulator. On the hand-held

The air carbon arc torch has an air passageway, an air on-and-off valve, and an air orifice (hole) for the air jet; Fig. 14-5.

The electrode lead and air hose connections are usually made outside the handle of the torch. See Fig. 14-2. The air valve is mounted in the handle and is either lever or button operated. Most of these valves have a lock-open feature. This enables the operator to keep the air flowing, and still have a comfortable hand position on the handle.

It is very important that the air jet be directed as close as possible to the arc crater. Any misalignment or abuse of the air orifice may produce below average results.

## 14-5. AIR CARBON ARC ELECTRODES

AAC electrodes are obtainable in the carbon

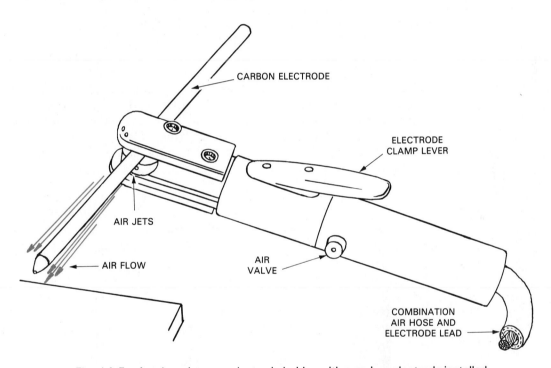

Fig. 14-5. An air carbon arc electrode holder with a carbon electrode installed.
Note the air jet orifices behind the electrode.

air carbon arc torch, a hand valve controls the on and off operation of the air flow. The air pressure on an automatic machine is turned on and off by means of a switch or valve.

Compressed air in cylinders may be used when portability is desired. It may also be used when the cutting project is of small size.

## 14-4. AIR CARBON ARC ELECTRODE HOLDER

The hand-held air carbon arc electrode holder or torch is similar to the standard electrode holder.

form, graphite form or as a mixture of carbon and graphite. Most commercially available electrodes are made of a mixture of carbon and carbon in the graphite form.

There are three basic types of air carbon arc cutting (AAC) carbon electrodes. They are:
1. DCEP (DCRP) plain.
2. DCEP (DCRP) copper coated.
3. AC copper coated.

The copper coating helps to reduce the oxidation of the electrode body. It also helps to keep the electrode cool. The copper coating does not, however, add much to the conductivity of the electrode.

Carbon electrodes come in sizes from 5/32 in. - 1 in. (7.9 - 25.4 mm). Fig. 14-6 shows a table of carbon electrode sizes and the approximate cur-

AC carbon electrodes are available in 3/16, 1/4, 3/8, and 1/2 in. diameters (4.8, 6.4, 9.5, 12.7 mm). Most carbon electrodes are available in standard 12 in. (305 mm) lengths. Carbon electrodes are available which are jointed at the ends. This

| ELECTRODE DIAMETER | | AMPERAGE WITH DCEP (DCRP) ELECTRODE | | AMPERAGE WITH AC ELECTRODE | |
|---|---|---|---|---|---|
| in. | mm | Min. | Max. | Min. | Max. |
| 5/32 | 4.0 | 90 | 150 | -- | -- |
| 3/16 | 4.8 | 150 | 200 | 150 | 200 |
| 1/4 | 6.4 | 200 | 400 | 200 | 300 |
| 5/16 | 7.9 | 250 | 450 | -- | -- |
| 3/8 | 9.5 | 350 | 600 | 300 | 500 |
| 1/2 | 12.7 | 600 | 1000 | 400 | 600 |
| 5/8 | 15.9 | 800 | 1200 | -- | -- |
| 3/4 | 19.1 | 1200 | 1600 | -- | -- |
| 1 | 25.4 | 1800 | 2200 | -- | -- |

Fig. 14-6. A table of suggested current settings for various diameters and types of air carbon arc electrodes.

rent settings for each. The table in Fig. 14-7 shows cutting speeds and electrode sizes recommended for various groove diameters and depths.

Standard carbon electrodes are best used with direct current electrode positive (DCEP) polarity. Electrodes are also available for use with AC. The

permits them to be connected together continually. This eliminates electrode waste and permits continuous cutting. See Fig. 14-8.

Air carbon arc electrodes are not always round in cross section. Flat, half round, and special shapes are also available. These shapes are used

| ELECTRODE SIZE FOR SINGLE GROOVE DIAMETER | | DESIRED DEPTH | | AMPERAGE | TRAVEL SPEED | |
|---|---|---|---|---|---|---|
| in. | mm | in. | mm | DCEP (DCRP) | ipm | mm/min |
| 5/16 | 7.9 | 1/4 | 6.4 | 425 | 36 | 914 |
| 5/16 | 7.9 | 5/16 | 7.9 | 450 | 33 | 838 |
| 3/8 | 9.5 | 1/4 | 6.4 | 500 | 46 | 1168 |
| 3/8 | 9.5 | 3/8 | 9.5 | 500 | 25 | 635 |
| 3/8 | 9.5 | 1/2 | 12.7 | 500 | 17 | 432 |
| 1/2 | 12.7 | 3/8 | 9.5 | 850 | 35 | 889 |
| 1/2 | 12.7 | 1/2 | 12.7 | 850 | 24 | 610 |
| 1/2 | 12.7 | 5/8 | 15.9 | 850 | 20 | 508 |
| 5/8 | 15.9 | 1/2 | 12.7 | 1250 | 28 | 711 |
| 5/8 | 15.9 | 5/8 | 15.9 | 1250 | 22 | 559 |
| 5/8 | 15.9 | 3/4 | 19.0 | 1250 | 17 | 432 |
| 3/4 | 19.0 | 3/4 | 19.0 | 1600 | 20 | 508 |
| 3/4 | 19.0 | 1 | 25.4 | 1600 | 15 | 381 |

Fig. 14-7. A table of operating conditions for N-Series, Arcair air carbon arc torches. (Arcair Co.)

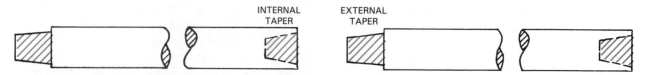

INTERNAL TAPER       EXTERNAL TAPER

Fig. 14-8. Two air carbon arc electrodes joined by matching tapers. New electrodes can be added to old ones to permit a continuous gouging operation.

to produce special groove shapes.

As shown in Fig. 14-6, some current ratings are quite high and the arc welding machine must be of ample capacity. Two or more standard machines may be connected in parallel to give the current capacity needed as shown in Fig. 14-9.

## 14-6. OXYGEN ARC CUTTING (AOC) EQUIPMENT

OXYGEN ARC CUTTING is done by striking an arc on the metal to be cut with a hollow (tubular) electrode. Oxygen under pressure is passed through the hollow electrode. The oxygen flow oxidizes the metal and blows the molten metal in the arc crater away as in oxyfuel gas cutting. This action continues to form a kerf or cut in the base metal. Refer to Heading 1-10.

Oxygen arc cutting equipment is a combination of an oxyfuel gas and arc welding station. It uses the usual oxyfuel gas station equipment as follows:

1. Oxygen cylinder.
2. Oxygen regulator.
3. Oxygen hose.

It uses the following parts of an arc welding station:

1. Arc welding machine.
2. Work lead (ground cable).
3. Electrode lead (cable).
4. Booth.
5. Ventilation system.
6. Bench.

The special equipment needed is:

1. Special oxygen arc torch.
2. Oxygen arc cutting electrodes.

## 14-7. OXYGEN ARC CUTTING ELECTRODE HOLDER

The electrode holder or torch is designed to hold a hollow (tubular) iron or steel electrode, see Fig. 14-10. On some torches, a special cap is designed

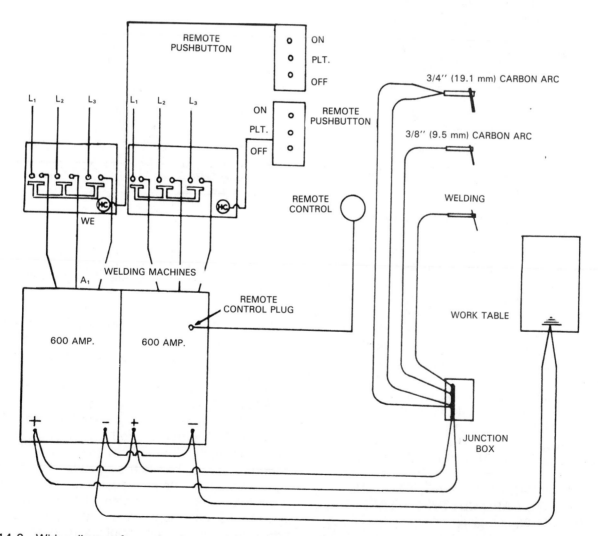

Fig. 14-9. Wiring diagram for connecting two DC arc welders in parallel. An operator can gouge or cut with 3/4 in. (19.1 mm) diameter electrodes, reduce the amperage and remove fins and defects with 3/8 in. (9.5 mm) electrodes, then reduce the amperage further and repair welds and defects which were removed.

to fit over the end of the electrode. This special cap connects the electrode to the oxygen supply. A soft special gasket seals the opening where the electrode contacts the oxygen orifice cap. Neoprene or silicon rubber are common gasket materials. A clamp or collet in the electrode holder holds the electrode and carries the current into the electrode.

Some torches use a collet system rather than a clamp to hold the electrode. The collet provides the joint to seal against oxygen leakage. It also provides an area of electrical contact with the electrodes. See Fig. 14-10.

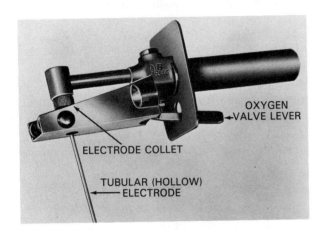

Fig. 14-10. Oxygen arc cutting electrode and holder. (Cut-Mark, Inc.)

A hand valve is mounted on the electrode holder to control the oxygen flow.

## 14-8. OXYGEN ARC CUTTING ELECTRODES

The electrode consists of a steel tube covered with a flux coating. The tube carries the arc current, the inside of the tube is the oxygen passageway, and the covering serves as an insulator. With the heavy flux covering the AOC electrode is dragged across the metal. The covering provides a means of maintaining the correct arc length. It also provides fluxing agents to improve the cutting action.

AOC electrodes are available in 3/16 and 5/16 in. (4.8 and 7.9 mm) outside tube diameters and 1/16 and 1/10 in. (1.6 and 2.5 mm) inside diameters respectively. These electrodes are made in 14 and 18 in. (356 and 457 mm) lengths.

## 14-9. GAS METAL ARC (GMAC) AND GAS TUNGSTEN ARC CUTTING (GTAC) PRINCIPLES

Aluminum, stainless steel, chrome steel, nickel, copper, monel, and inconel have been sucessfully

cut by means of the GAS METAL ARC CUTTING (GMAC) and the GAS TUNGSTEN ARC CUTTING (GTAC) process. Satisfactory severing cuts also have been made on titanium, magnesiuim, and most of the exotic metals. Refer to Heading 1-11.

The principle of cutting is the same for both the GMAC and GTAC processes. An arc is struck between the wire or tungsten electrode and the work. As the base metal melts, it is blown away by the flow of shielding gas. The shielding gas flow must extend through the full thickness of the metal being cut to keep the surface of the cut from oxidizing. Fig. 14-11 shows this cutting process.

The effectiveness of this cutting process is due to the action and velocity of the inert gas. The inert gas keeps the molten metal from forming oxides. These oxides are what make aluminum, stainless steel, and the others listed above, hard to cut. Sufficient gas must flow to prevent oxidation. The use of too much inert gas is wasteful and will not speed the cutting process.

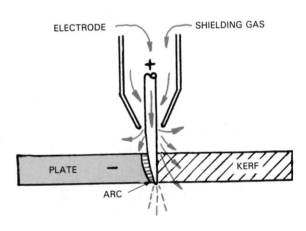

Fig. 14-11. The basic operation of the GMAC process. Notice the use of DCEP (DCRP) and the extension of the steel electrode through the thickness of the metal being cut.

Gas metal arc cutting (GMAC) uses direct current electrode positive (DCEP). Gas tungsten arc cutting (GTAC) uses direct current electrode negative (DCEN).

Argon and helium mixtures are usually used when GTAC. Nitrogen has also been used. These gases along with $CO_2$ have been used for GMAC.

The GTAC process produces a smooth cut in most materials. One side of the kerf or cut is acceptable as cut. This is the side closer to the work lead. The other side of the cut required cleanup to remove the slag on the bottom edge. This process is used to cut aluminum, copper, brass, other nonferrous metals and stainless steel up to 1/2 in. (13 mm) in thickness. Standard GTAW torches can be used for GTAC. Standard GMAW equipment can be used for GMAC. GMAC does require a lot of current and uses a lot of electrode wire.

Manual cuts may be made, but at a slower rate and the results are usually of poorer quality. These processes are not widely used commercially. Manual cutting is used for special applications and where larger, more expensive processes are not available.

## 14-10. GAS METAL ARC AND GAS TUNGSTEN ARC CUTTING EQUIPMENT

Equipment required for the cutting of metals using the gas tungsten arc (GTAC) or gas metal arc (GMAC) process is the same as that required for welding with these same processes. Refer to Chapter 13. However, in all cases the equipment should be heavy duty to make it completely versatile. Fig. 14-12 shows a complete gas metal arc cutting station. The DC power source must be of higher capacity than normal, since the required amperage for some metals, such as 1/4 in. (6.4 mm) monel, may be as high as 1000 amps.

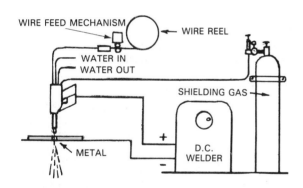

Fig. 14-12. Schematic of a gas metal arc cutting outfit.

## 14-11. SHIELDED METAL ARC CUTTING (SMAC) ELECTRODES

Shielded metal arc cutting electrodes are designed to produce a high velocity gas and particle stream. This action is created by a deep cavity in the electrode end. The resulting jet action of the gases and metal particles produce a driving arc which cuts the base metal. Using special coatings on the electrodes produces this forceful jet action. Somewhat the same result may be obtained with ordinary flux covered electrodes by using a higher than normal current for the size of the electode. Special shielded metal arc cutting (SMAC) electrodes are available in standard diameters and lengths. Common sizes which come in 14 in. (356 mm) length, are 3/32, 1/8, 5/32, 3/16, and 1/4 in. (2.4, 3.2, 4.0, 4.8, and 6.4 mm).

Fig. 14-13 shows the approximate electrode sizes and current settings when using special shielded metal arc cutting electrodes.

When cutting mild steel, regular shielded metal arc welding electrodes may be used. However, a much higher welding current range will be required. Fig. 14-14 lists the current values to use when doing various cutting operations with standard covered steel welding electrodes.

If regular mild steel shielded metal arc welding electrodes are used for cutting, they will last a little longer if they are soaked in water for a period not to exceed 10 minutes before being used. The moisture in the coating slows the vaporizing of the coating material thus producing a deeper cup (cavity) at the arc end of the electrode. This deeper cavity creates a stronger jet action which improves the cutting. Water soaked electrodes

| METAL THICKNESS | | ELECTRODE DIAMETER | | DCEN (DCSP) CURRENT RANGE AMPS |
|---|---|---|---|---|
| in. | mm | in. | mm | |
| 1/8 | 3.2 | 3/32 | 2.4 | 100 - 160 |
| 1/8 - 1 | 3.2 - 25.4 | 1/8 | 3.2 | 160 - 350 |
| 3/4 - 2 | 19.1 - 50.8 | 5/32 | 4.0 | 250 - 400 |
| 1 - 3 | 25.4 - 76.2 | 3/16 | 4.8 | 350 - 500 |
| 3 and over | 76.2 and over | 1/4 | 6.4 | 400 - 650 |

Fig. 14-14. Approximate current settings and electrode sizes for shielded metal arc cutting, when using standard welding electrodes.

| METAL THICKNESS | | ELECTRODE DIAMETER | | AC CURRENT RANGE AMPS | DCEN (DCSP) AMPS |
|---|---|---|---|---|---|
| in. | mm | in. | mm | | |
| 1/8 | 3.2 | 3/32 | 2.4 | 40 - 150 | 75 - 115 |
| 1/8 - 1 | 3.2 - 25.4 | 1/8 | 3.2 | 125 - 300 | 150 - 175 |
| 3/4 - 2 | 19.1 - 50.8 | 5/32 | 4.0 | 250 - 375 | 170 - 500 |
| 1 - 3 | 25.8 - 76.2 | 3/16 | 4.8 | 300 - 450 | |
| 3 and over | 76.2 and over | 1/4 | 6.4 | 400 - 650 | |

Fig. 14-13. A table of suggested uses and approximate current settings for cutting with shielded metal arc cutting electrodes.

should be used immediately. Water soaked electrodes should NOT be used for welding because the moisture produces hydrogen embrittlement in the weld.

## 14-12. PLASMA ARC CUTTING (PAC) PRINCIPLES

PLASMA ARC CUTTING uses the principle of passing an electric arc through a quantity of gas traveling through a restricted outlet. The electric arc heats the gas as it travels through the arc. The gas is heated to such a high temperature that it turns into what physicists call PLASMA. Plasma is the fourth state of matter—not a gas, liquid, or solid. A great deal of heat energy is required to change the gas to a plasma. The heat required is furnished by the electric arc. It is this heat in the plasma which heats the base metal. The heat of the plasma is released on the metal as the plasma turns back to a gas. Refer to Heading 1-13.

Plasma arc cutting (PAC) was originally developed to cut nonferrous metals using inert gases. Developments have made it possible to use oxygen as the plasma gas for cutting steel. Using the plasma process, temperatures as high as 25,200 °F (14 000 °C) have been reached. DCEN (DCSP) is used when cutting. Fig. 14-15 shows a complete plasma arc cutting station. Clean, narrow kerfs are possible at high speeds. See Fig. 14-16.

Several ports are sometimes used in the torch. Only the center port has the arc plasma, while the

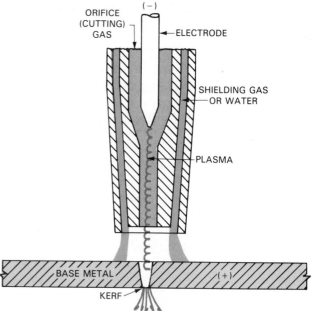

Fig. 14-17. A dual flow plasma arc cutting torch nozzle. The orifice (cutting) gas becomes a plasma in the arc stream. The shielding gas or water flows out around the plasma orifice through a number of holes.

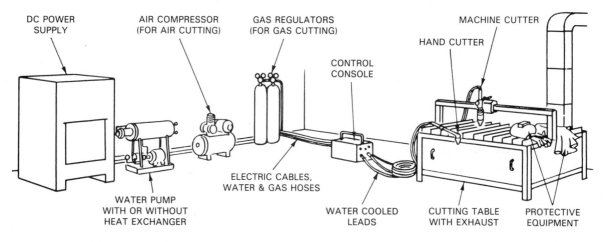

Fig. 14-15. Diagrammatic sketch of the equipment required for a plasma arc cutting station. (Thermal Dynamics Corp.)

Fig. 14-16. Mild steel plate 1 in. (25.4 mm) thick, cut with a plasma arc torch using compressed air as the gas.

surrounding ports provide a shielding gas protection. Gas flow may be as high as 250 cu. ft./hr. (118 L/min). The dual flow cutting process uses one gas for the plasma and one for a shielding gas. See Fig. 14-17. Nitrogen is often used as the plasma gas. The shielding gas for mild steel is $CO_2$ or air. An argon and helium mixture is used as the shielding gas for aluminum.

Water may be used in specially built plasma torches in a similar manner as the shielding gas.

See Fig. 14-17. This process is known as WATER SHIELD PLASMA CUTTING.

A water spray may be used to constrict the plasma gas flow as shown in Fig. 14-18. Narrow cuts can be produced using this method. Water shield and water injection plasma cutting may be done on a cutting table covered with a shallow depth of water. The metal to be cut is placed under water on the cutting table. The water

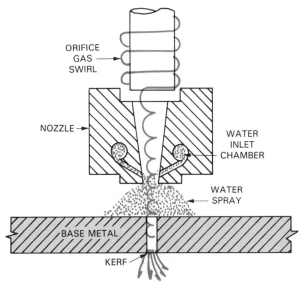

Fig. 14-18. A water injection plasma arc cutting torch nozzle. Water is injected into the plasma stream to constrict it (reduce its diameter).

Fig. 14-19. Plasma arc cutting torch cutting steel plate under water. Cutting under water makes this process much quieter. (L-TEC Welding & Cutting Systems)

shielded or water injection plasma arc torch is started and the cutting is done under water. The water acts as a shielding medium and also quiets the sound of the plasma cutting operation. Water from the table may be filtered and recirculated for reuse in the torch, see Fig. 14-19.

As the gas is heated in the nozzle, it tries to expand but it is restricted by the nozzle. The nozzle acts like a venturi causing the plasma to increase in speed. As the plasma passes through the nozzle, its speed is increased to several thousand feet per second (meters per second). It is this high velocity of gas which removes the molten metal from the work and creates the kerf. The plasma formed from a shielding gas, and the shielding gas or auxiliary shielding gas shields the sides of the kerf from contamination and oxidation.

Two types of plasma arc cutting processes are shown in Fig. 14-20. They are:
1. Transfer arc.
2. Nontransfer arc.

In some plasma arc cutting processes, the TRANSFERRED ARC principle is used. The work is attached so that it becomes a part of the electrical circuit, and an arc is created between the electrode and the work in addition to the arc between the electrode and the holder. Figs. 14-21 and 14-22 show schematics of the transferred plasma arc processes. Direct current electrode negative (DCEN) or (DCSP) is used to create most of the heat on the work.

In the NONTRANSFER ARC process, the electrical circuitry is different from that of the transfer arc process. The arc is struck between the electrode and the torch nozzle. The work is not part of the electrical circuit. A schematic of this system is shown in Fig. 14-20 at the left.

Of the two methods, the transferred arc creates the greatest amount of heat and is used when cutting. The plasma arc cutting method may be used to cut almost all types of metals with great success and economy because of the extreme temperatures and the shielding qualities of the gases used. Speeds of 300 inches per minute (IPM) (7.62 m/min) have been obtained on 1/4 in. (6.4 mm) aluminum plate and 50 IPM (1.27 m/min) on 1 in. (25.4 mm) aluminum plate. Speeds of 100

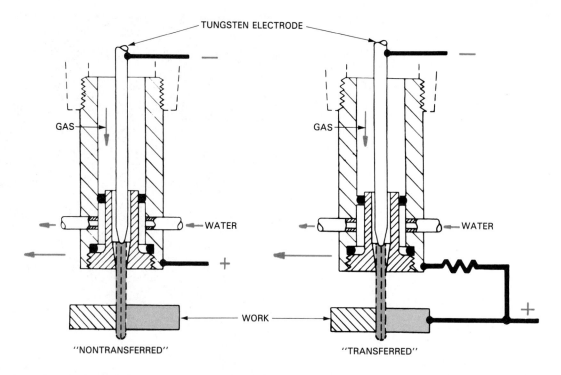

Fig. 14-20. Schematic showing the principle of operation of the nontransferred arc and the transferred arc plasma cutting processes.

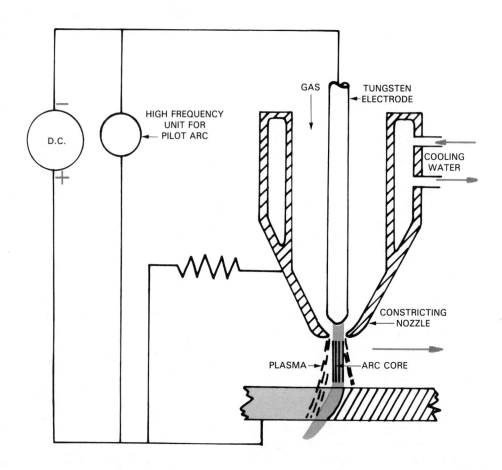

Fig. 14-21. A cross section schematic of a torch nozzle used in the transfer arc plasma cutting process.

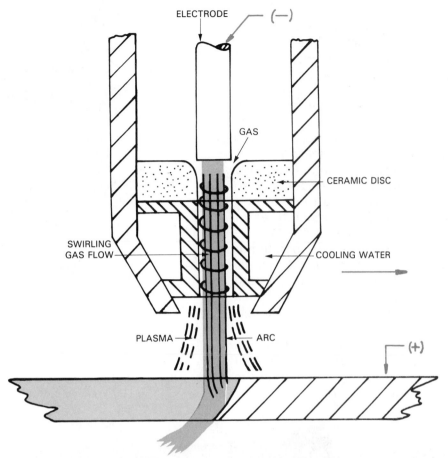

ELECTRODE — (—)

GAS

CERAMIC DISC

SWIRLING GAS FLOW

COOLING WATER

PLASMA — — ARC

(+)

Fig. 14-22. A schematic drawing of the transferred arc plasma cutting process.

IPM (2.54 m/min) have been made on 1/4 in. (6.4 mm) stainless steel plate and 30 IPM (762 mm/min) on 1 in. (25.4 mm) stainless steel plate. All of these cuts may be made dross (slag) free. Underwater cuts have also been made by some operators using this process.

The plasma arc process has also been used to "machine" inconel and stainless steel metals which rapidly wear down ordinary cutting tools. The metal is held and rotated in a lathe and the plasma arc torch is held in a special fixture on the movable tool holder. In this way, accurate cuts may be made over the entire length of the rotating metal.

The plasma transfer arc process is more often used for surfacing operations. See Chapter 26.

## 14-13. PLASMA ARC CUTTING EQUIPMENT AND SUPPLIES

Plasma arc cutting may be done with automatic equipment or manually as shown in Fig. 14-23. The equipment, with the exception of the power source, is the same as that used for plasma arc

Fig. 14-23. A welder cutting with a manual plasma arc cutting torch. (Thermal Dynamics Corp.)

welding. See Chapter 18. The manual type torch for heavy duty cutting is sometimes equipped with a heat shield to protect the operator's hands from the intense heat. It is necessary for heavy duty cutting up to 5 in. (127 mm) to have a power source capable of supplying DC power of 700 amps and at 170 volts. A power source of 400 amps and 80 volts under load may be used to cut up to 1 1/2 in. (38.1 mm) aluminum and 1 in. (25.4 mm) stainless steel. A high frequency unit is required to start the pilot arc at the beginning of a cut. Cooling water pumps and a shielding gas or gases under pressure are also required. For heavy duty cutting, cooling water pumps are required to circulate enough water through the torch to prevent damage from the extreme temperatures of the plasma arc as shown in Fig. 14-24.

The torches may be either manually operated or machine operated. Fig. 14-25 shows a manual and automatic type torch. The machine operated torches are usually designed with a gear and rack adjustment similar to a oxyacetylene machine cutting torch.

Common shielding gases used for plasma arc cutting are nitrogen, $CO_2$, and varying mixtures of argon and helium. Air has also been used as the plasma gas with acceptable results.

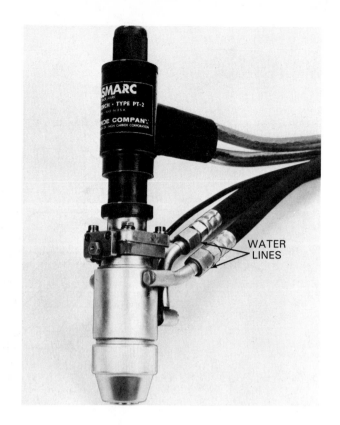

Fig. 14-24. A water cooled plasma arc torch. (L-TEC Welding & Cutting Systems)

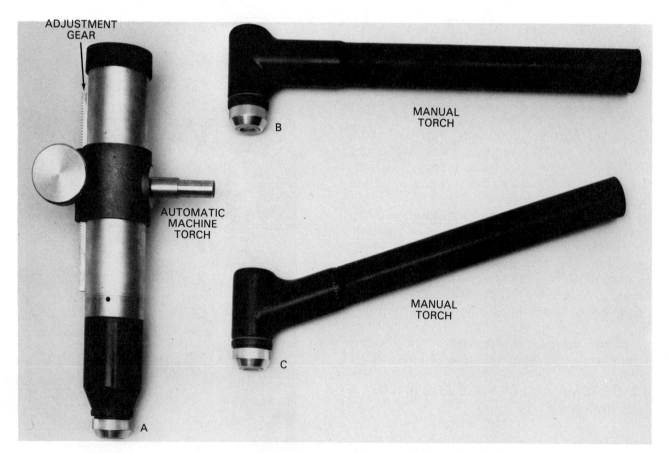

Fig. 14-25. A—Torch used on an automatic plasma arc cutting machine. B and C—Manual type torches. (Thermal Dynamics Corp.)

## 14-14. TEST YOUR KNOWLEDGE

Write your answers on a separate sheet of paper. Do not write in this book.

1. List six arc cutting processes with their AWS letter abbreviations.
2. When carbon arc cutting, what shade filter lens should be used?
3. What DC polarity is used when air carbon arc cutting?
4. When doing air carbon arc cutting a carbon or _____ electrode can be used.
5. On non-magnetic metals how may the motorized carriage or climber be attached to the metal?
6. How much air pressure is supplied to a heavy duty air carbon arc electrode holder?
7. The electrode used for carbon arc cutting may be plain or _____ coated.
8. What current is suggested for a 3/8 in. (9.5 mm) air carbon arc electrode with DCEP (DCRP)?
9. How does the electrode used in oxygen arc cutting differ from other cutting electrodes?
10. When doing GMAC and GTAC what DC polarity is suggested?
11. Name four (4) types of shielding gases which are used with GMAC and GTAC.
12. In shielded metal arc cutting a high velocity gas and particle stream is caused by a _____ _____ in the end of the electrode.
13. What amperage is suggested for use with a 1/4 in. (6.4 mm) gas metal arc cutting electrode?
14. Regular arc welding electrodes will last longer when used for cutting if they are soaked in _____.
15. What is a plasma?
16. The temperature of a plasma arc may reach _____ °F or _____ °C.
17. In a water injection plasma arc cutting torch nozzle, what is the purpose of the water besides cooling?
18. What is a "dual flow" plasma cutting process?
19. Using the PAC process, what cutting speed has been reached when cutting 1/4 in. (6.4 mm) aluminum?
20. When plasma arc cutting, list all materials which may be used as a shielding medium.

Hand held plasma arc cutting torch. This torch is being used to cut 1/4 in. stainless steel at high speed.   (Thermal Dynamics Corp.)

Fig. 15-1. Air carbon arc gouging U-grooves in I-beams prior to welding. (Thermadyne Industries, Inc.)

# 15 ARC AND OXYGEN ARC CUTTING

The intense heat of the electric arc makes it possible to melt and make fluid very small areas of a metal. If the fluid metal can be made to flow away either by gravity or by gas pressure, the metal will be removed leaving a cavity or cut.

In the arc cutting processes, either a carbon or metal electrode is used. Many devices and processes have been designed to make this type of metal cutting faster and more accurate. Great progress has been made in the development of electrodes for shielded metal arc cutting. Arc cutting using air, oxygen, and inert gases is perfected in gas metal arc and gas tungsten arc cutting. Plasma arc cutting is an accurate, easily used process. It is a very popular arc cutting process used in industry today. The air carbon arc process is used widely for gouging.

## 15-1. CARBON ARC CUTTING (CAC)

Metals may be successfully cut by using the carbon electrode arc. See Fig. 15-1. The CARBON ARC CUTTING process is generally used only when arc or oxyfuel gas cutting is not available.

Since no foreign metals are introduced at the arc, the cut is clean. The manual carbon arc cut is generally very ragged and of poor appearance. The amperages used for carbon arc cutting are generally higher than for welding on the same metal thickness. Manufacturers or suppliers of carbon electrodes will normally supply amperage ranges for use with various electrode diameters. See Fig. 15-2 for suggested electrode diameters and current settings.

Prior to use, the carbon electrode should be ground to a very sharp point. The length of the taper should be 6 - 8 times the electrode diameter. See Fig. 10-63. The electrode should stick out from the electrode holder a distance equal to 10 times the electrode diameter. This is necessary to reduce electrical resistance and to reduce the heating effect on the electrode. If the carbon wears away too fast, shorten the electrode extension out of the electrode holder to as little as 3 in.

During the actual cutting, the carbon electrode should be manipulated in a vertical elliptical movement to undercut the metal. This motion facilitates the removal of the molten metal. In addition

| THICKNESS OF PLATE | | CURRENT SETTING AND CARBON ELECTRODE DIAMETER | | | |
|---|---|---|---|---|---|
| | | 300 AMPS 1/2 in. Dia. (12.7 mm) | 500 AMPS 5/8 in. Dia. (15.9 mm) | 700 AMPS 3/4 in. Dia. (19.1 mm) | 1000 AMPS 1 in. Dia. (25.4 mm) |
| in. | mm | SPEED OF CUTTING IN MINUTE PER FOOT | | | |
| 1/2 | 12.7 | 3.5 | 2.0 | 1.5 | 1.0 |
| 3/4 | 19.1 | 4.7 | 3.0 | 2.0 | 1.4 |
| 1 | 25.4 | 6.8 | 4.1 | 2.9 | 2.0 |
| 1 1/4 | 31.8 | 9.8 | 5.6 | 4.0 | 2.9 |
| 1 1/2 | 38.1 | — | 8.0 | 5.8 | 4.0 |
| 1 3/4 | 44.5 | — | — | 8.0 | 5.3 |
| 2 | 50.8 | — | — | — | 7.0 |

Fig. 15-2. A table of recommended electrode sizes, current settings, and speeds for carbon arc cutting various thicknesses of steel.

to the vertical motion, a side to side crescent motion is recommended along the line of the cut. The electrode angle is about 20 degrees from vertical. Fig. 15-3 illustrates the relative position of the electrode and the work when cutting cast iron.

a vertical position. The gouge should be made from top to bottom, or vertically down. This will keep the molten metal flowing out of the gouge area. See Fig. 15-4.

The graphite form of carbon electrode is pre-

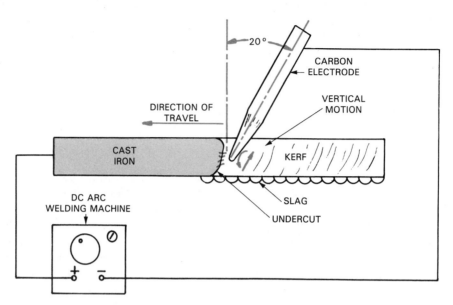

Fig. 15-3. Carbon arc cutting. Note that the carbon electrode is connected to the negative terminal of the welding machine, and the work to the positive terminal.

The carbon arc method of cutting may be used quite successfully on cast iron because the temperature of the arc is sufficient to melt the iron oxides formed. It is important to undercut the cast iron "kerf" if you desire an even cut. Fig. 15-4 shows a cast iron gear hub being gouged using a carbon electrode. The molten metal must flow away from the gouge or cutting area by gravity. Therefore, gouging is done best with the metal in

ferred because high currents are needed.

A constant current arc welding machine and other items of arc welding station equipment may be used for carbon arc cutting. Refer to Heading 14-1. When using a DC arc welding machine with DC electrodes, DCEN (DCSP) is used. When using an AC arc welding machine, special AC cutting electrodes are used.

When carbon arc cutting, the light from the arc is more intense than when shielded metal arc welding. A shade #12 - #14 is recommended for both carbon arc welding and cutting.

Follow all safety precautions listed for arc welding procedures whenever cutting. Refer to Heading 15-7 before attempting carbon arc cutting.

## 15-2. CUTTING WITH THE AIR CARBON ARC (AAC)

AIR CARBON ARC CUTTING and gouging may be done manually or automatically. Study Heading 1-9. The manual torch is shown in Fig. 15-5 removing a weld bead. Fig. 15-6 is a photograph of an air carbon arc torch.

In both cases, the welding machine amperage and electrode diameter are selected according to the width and depth of the desired groove. See Figs. 14-6 and 14-7 for suggested current set-

Fig. 15-4. Carbon arc gouging a cast iron gear hub. Note the type electrode holder, and the position of the gear which permits the molten cast iron to flow down and away from the gouge. (Tweco Products, Inc.)

Fig. 15-5. An air carbon arc electrode holder and electrode being used to remove a weld bead. Notice the flat, rectangular cross section of the electrode, the electrode angle, and the welder's clothing. (Arcair Co.)

Fig. 15-6. An air carbon arc electrode holder. This torch has two air jet heads installed. This will allow the air jet to be behind the electrode regardless of the direction the operator may move. (Arcair Co.)

tings. Refer to Headings 14-3, 14-4, and 14-5 for additional information on equipment.

When gouging manually, the air stream must be turned on prior to striking the arc. The air stream must be directed from behind the carbon electrode. This permits the metal to be blown out of the arc pool, as shown in Fig. 15-7. Fig. 15-8 shows the air jets in the air carbon arc electrode

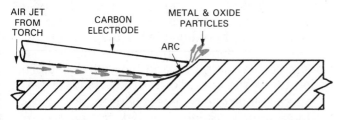

Fig. 15-7. A schematic of an air carbon arc gouging operation in progress. Note how the compressed air blows under the carbon electrode to remove the molten metal.

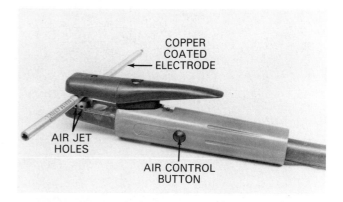

Fig. 15-8. An air carbon arc torch with an electrode in place. The electrode is copper coated and has a tapered hole at one end and an external taper on the other. This permits attaching a new electrode to the old one to reduce electrode waste. (Arcair Co.)

holder. If the air carbon arc is used to cut completely through a part, the electrode angle should be about 20 - 45 degrees from vertical, as shown in Fig. 15-9. Fig. 15-10 shown a V groove gouged into a plate. Fig. 15-11 illustrates the sequence of gouges used to remove a large fillet weld.

In the vertical cutting position, gouging should be done from the top down. This permits gravity to help remove the molten metal from the arc groove. Gouging in the horizontal position may be

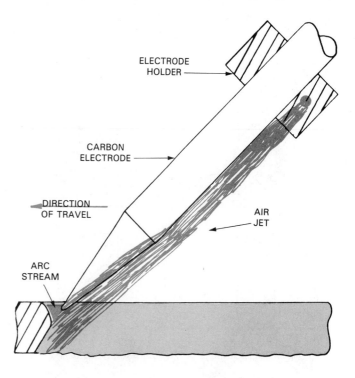

Fig. 15-9. Air carbon-arc cutting schematic. Note the air jet provided to blow away the molten metal.

Fig. 15-10. A V-groove gouged in 2 in. (50.8 mm) thick carbon steel. Note the use of the run-off tab to get a complete gouge to the end of the plate.

done from the right or left. The air stream must be from behind the electrode in either case. When gouging overhead, the electrode should be placed into the electrode holder so that it is nearly parallel to the centerline of the holder. With this electrode position, the gouging action will occur as far away

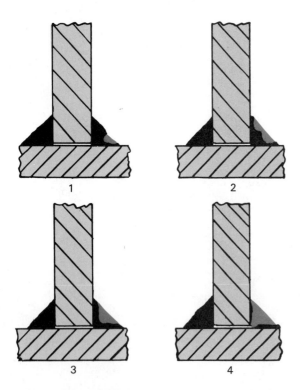

Fig. 15-11. Recommended gouging sequence to be used when removing a fillet weld.

from the welder as possible. This may prevent hot metal from falling on the welder.

When manually gouging on steel, the electrode should extend from the electrode holder about 6 in. (about 150 mm). When gouging on aluminum or aluminum alloys, the electrode extension is about 4 in. (about 100 mm). The welder must reposition the electrode in the holder regularly to maintain the ideal electrode extension. Once the arc is struck, the electrode is withdrawn a short distance. It should remain in close contact with the metal being gouged. The electrode angle is about 20 degrees from the base metal. Forward speed and smoothness of travel will determine the quality of the finished gouge.

When automatic gouging machines are used, the amperage and the air pressure and air volume are set. The travel speed of the motorized tractor is set according to Fig. 14-7. The electrode extension should be approximately the same as for manual gouging. In most automatic machines, the electrode is automatically fed out to maintain a uniform extension at all times. The operator must add new electrodes to permit continuous gouging. The electrode angle is about 45 degrees or less when gouging automatically.

The accuracy of the depth of the gouge may be held to ±0.03 in. (±.76 mm) with the automatic process.

Carbon deposits may indicate an improper air jet flow or excessive travel speed. A rough cut results from a slow travel speed. Poor cuts may also result from a poorly positioned air jet or a jet with low air pressure or volume.

### 15-3. OXYGEN ARC CUTTING (AOC)

The OXYGEN ARC CUTTING process is a combination of an electric arc and a jet of oxygen. The equipment required may be either an AC or DC welding machine, an oxygen arc electrode holder, a source of oxygen, and suitable oxygen arc cutting electrodes. Refer to Headings 14-6, 14-7, and 14-8. Oxygen arc cutting is used to cut alloy steels, aluminum, and cast iron. These metals are difficult to cut using the oxyfuel gas cutting process. This process is also used on carbon steel.

The oxygen arc cutting process usually makes use of a tubular metal covered electrode. An electric arc is struck between the hollow electrode and the work. Oxygen fed through the hollow electrode oxidizes the metal and blows it away. This action forms the cut or kerf. Study Heading 1-10. A covered electrode is generally used, especially if fluxing ingredients are needed for cutting the metal. The covering also enables the operator to hold the electrode end against the base metal and still maintain the arc.

Fig. 15-12 lists suggested amperages and oxygen pressure for use in cutting various metals and metal thicknesses. An oxygen cylinder and a high volume regulator are required. The oxygen arc torch is connected to the regulator by means of a heavy duty oxygen hose. The pressures used for cutting may be as high as 75 psig (517 kPa). The electrical lead is connected to a current source as shown schematically in Fig. 15-13.

To operate the cutting device, the operator first strikes as arc between the electrode and the metal to be cut. The oxygen valve on the holder is then

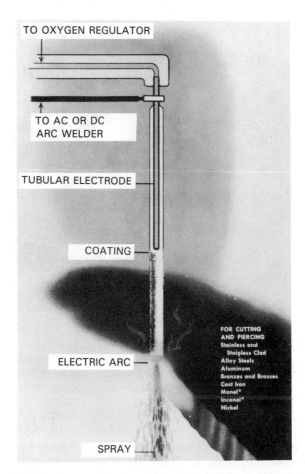

Fig. 15-13. A schematic illustration of the main parts of a tubular electrode and holder as used in oxygen arc cutting. This process may be used on many different metals such as stainless steel, alloy steel, aluminum, and cast iron. (Cut-Mark, Inc.)

| SUGGESTED AMPERAGES AND OXYGEN PRESSURES FOR OXYGEN ARC CUTTING | | | | |
|---|---|---|---|---|
| **CAST IRON** | | | | |
| Thickness | | Amps | Oxygen Pressure | |
| in. | mm | | psig | kPa |
| 1/4 | 6.4 | 180 | 10 | 68.9 |
| 1/2 | 12.7 | 185 | 10-15 | 68.9-103.4 |
| 3/4 | 19.1 | 190 | 15-20 | 103.4-137.9 |
| 1 | 25.4 | 200 | 20-25 | 137.9-172.4 |
| 1 1/4 | 31.2 | 210 | 25-30 | 172.4-206.8 |
| 1 1/2 | 38.1 | 215 | 30-35 | 206.8-241.3 |
| 1 3/4 | 44.5 | 220 | 35-40 | 241.3-275.8 |
| 2 | 50.8 | 225 | 40-45 | 275.8-310.3 |
| 2 1/4 | 57.2 | 225 | 45-50 | 310.3-344.7 |
| 2 1/2 | 63.5 | 230 | 50-55 | 344.7-379.2 |
| 2 3/4 | 69.9 | 230 | 55-60 | 379.2-413.7 |
| 3 | 76.2 | 235 | 60 | 413.7 |
| **ALUMINUM** | | | | |
| Thickness | | Amps | Oxygen Pressure | |
| in. | mm | | psig | kPa |
| 1/4 | 6.4 | 200 | 30 | 206.8 |
| 1/2 | 12.7 | 200 | 30 | 206.8 |
| 3/4 | 19.1 | 200 | 30 | 206.8 |
| 1 | 25.4 | 200 | 30 | 206.8 |
| 1 1/4 | 31.2 | 200 | 35 | 241.3 |
| 1 1/2 | 38.1 | 200 | 35 | 241.3 |
| 1 3/4 | 44.5 | 200 | 40 | 275.8 |
| 2 | 50.8 | 200 | 40 | 275.8 |
| 2 1/4 | 57.2 | 175 | 45 | 310.3 |
| 2 1/2 | 63.5 | 175 | 45 | 310.3 |
| 2 3/4 | 69.9 | 175 | 45 | 310.3 |
| 3 | 76.2 | 175 | 45 | 310.3 |
| **CARBON AND LOW ALLOY HIGH TENSILE STEEL** | | | | |
| Thickness | | Amps | Oxygen Pressure | |
| in. | mm | | psig | kPa |
| 1/4 | 6.4 | 175 | 75 | 517.1 |
| 1/2 | 12.7 | 175 | 75 | 517.1 |
| 3/4 | 19.1 | 175 | 75 | 517.1 |
| 1 | 25.4 | 175 | 75 | 517.1 |
| 1 1/4 | 31.2 | 200 | 75 | 517.1 |
| 1 1/2 | 38.1 | 200 | 75 | 517.1 |
| 1 3/4 | 44.5 | 200 | 75 | 517.1 |
| 2 | 50.8 | 200 | 75 | 517.1 |
| 2 1/4 | 57.2 | 225 | 75 | 517.1 |
| 2 1/2 | 63.5 | 225 | 75 | 517.1 |
| 2 3/4 | 69.9 | 225 | 75 | 517.1 |
| 3 | 76.2 | 225 | 75 | 517.1 |

Fig. 15-12. Amperages and oxygen pressure suggested for various metals and metal thicknesses when oxygen arc cutting.

opened, and the cut begins, as shown in Fig. 15-14. A continuous arc is maintained throughout the cutting operation. While cutting, the electrode is dragged across the metal. The electrode angle used is near vertical, with a slight angle in the direction of travel. See Fig. 15-14. Holes may be pierced easily with this process. Once the arc is struck and the oxygen jet turned on, the electrode is forced through the metal. The electrode is gradually consumed as the cut is produced.

## 15-4. THE GAS METAL ARC (GMAC) AND GAS TUNGSTEN ARC CUTTING (GTAC) PROCEDURE

When cutting with GAS METAL ARC or GAS TUNGSTEN ARC, the amperage and shielding gas volume required should be carefully set in accordance with the equipment manufacturer's manuals. DCEP (DCRP) should be used to obtain the required arc force. If the torch is water cooled, the cooling water flow should be adequate to cool the torch, particularly if high amperage flows are

Fig. 15-14. The oxygen arc process (AOC) being used to cut holes in a large pipe. A hollow, covered electrode is being used. Note the welder's clothing and the electrode position. (Cut-Mark, Inc.)

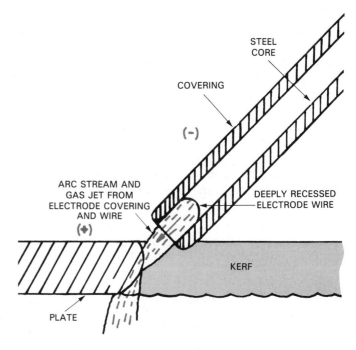

Fig. 15-15. A shielded metal arc cutting electrode being used to cut a metal plate. Note how deeply the electrode wire is recessed within the special slow burning covering.

required. To obtain a good cut using GMAC, the wire must extend just to the bottom surface of the metal. The wire feed rate and the speed of the cut must be carefully set to keep the wire at the correct position during the cutting operation. Study Heading 1-11. Cutting speeds or 140 inches per minute (IPM) (3.56 m/min) on 1/4 in. (6.4 mm) aluminum and 85 IPM (2.16 m/min) on 1/4 in. (6.4 mm) monel have been obtained using automatic gas metal arc and gas tungsten arc cutting equipment. Information concerning equipment may be studied under Headings 14-9 and 14-10.

GMAC and GTAC can be done manually, but the resulting cuts are of generally poor quality.

The electrode angle for both automatic and manual GMAC and GTAC is about 70 - 80 degrees from vertical. The lower end of the electrode should be ahead of the holder as the cut moves across the part.

## 15-5. SHIELDED METAL ARC CUTTING (SMAC)

Metal may be removed with the electric arc using covered metal electrodes.

Most cutting electrodes use a covering that disintegrates at a slower rate than the metal center of the electrode. This action creates a deep recess at the arc end of the electrode and produces a jet action that tends to blow the molten metal away, as shown in Fig. 15-15.

Obviously the electrode metal will melt and add metal to the puddle or crater. This extra metal

must also be removed. Some of the typical operations that may be performed using this cutting method are shown in Figs. 15-16, 15-17, 15-18, and 15-19.

When piercing a hole the welder must be especially careful to protect against burns. The metal being cut will be driven up out of the hole being cut. Only after the hole is cut through will the molten metal go downward.

Fig. 14-13 gives suggested amperages to use with various diameters of electrodes, both AC and

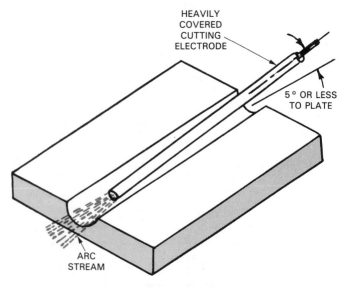

Fig. 15-16. A typical gouge cutting operation, using a shield metal arc cutting (SMAC) electrode.

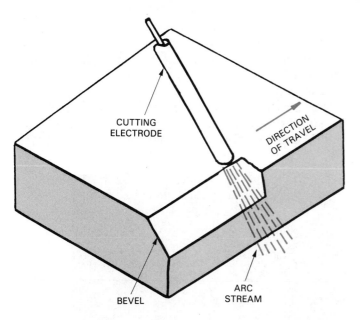

Fig. 15-17. Beveling the edge of a plate using a SMAC electrode.

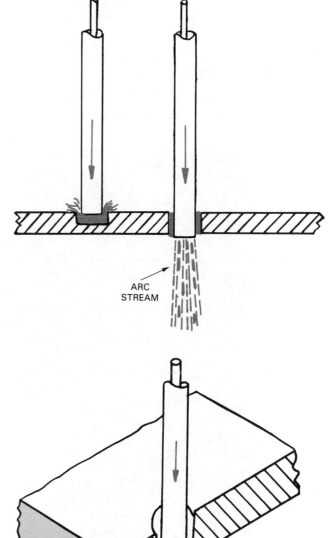

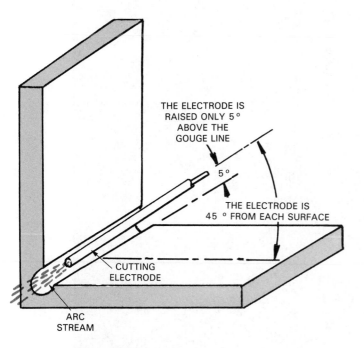

Fig. 15-18. A SMAC electrode being used to gouge an inside corner joint.

Fig. 15-19. Piercing a hole in a plate using a SMAC electrode. The electrode must be pushed downward after the arc is struck.

DCEN (DCSP). Shielded metal arc cutting electrodes come in the following diameters:

3/32'' or 2.4 mm
1/8 '' or 3.2 mm
5/32'' or 4.0 mm
3/16'' or 4.8 mm
1/4 '' or 6.4 mm

This system is easy to set up. A constant current arc welding machine is used. The welder needs to use a shielded metal arc cutting electrode for good gouging or cutting results. Heading 4-11

describes equipment used for SMAC.

Many manufacturers have developed special shielded metal arc cutting electrodes with special coverings which intensify the arc stream for rapid cutting. Shielded metal arc cutting electrodes may be used for cutting stainless steel, copper, aluminum, bronze, nickel, cast iron, manganese, steel or alloy steels.

The process can be used with either AC or with direct current electrode negative (DCEN or DCSP). A very short arc is recommended. The SMAC electrodes may be dragged across the metal without shorting out. This is possible because of the heavy covering on the electrode wire. If SMAC is used to cut under water, the coating must be water-proofed.

### 15-6. PLASMA ARC CUTTING (PAC) PROCEDURE

The first step required in PLASMA ARC CUTTING is to adjust the power supply and gas settings. The cooling water must also be turned on. The water flow safety device will prevent the starter arc circuit from functioning until the water is flowing.

Ear muffs or plugs are recommended since plasma arc cutting is a noisy process. Long exposure to sound may affect hearing. Equipment for PAC is described in Headings 14-12 and 14-13.

Fig. 15-20. A 200 kW plasma arc torch being used to cut a metal plate. Note the smooth appearance of the kerf (cut). (Thermal Dynamics Corp.)

To start the arc, the operator must press the starter button or trigger. The control box will then create a high frequency current to start the pilot arc. At the same time the plasma gas begins to flow, the DC power supply is turned on. After the plasma arc is established through the ionized starter arc, the starter arc is turned off. The cut may proceed as shown in the photograph, Fig. 15-20. Study Heading 1-13.

In the transfer arc process, the arc will go out when the cut is completed. Once the cut runs off the metal there is no longer a complete circuit through the base metal. When the arc goes out, the gas will also stop flowing. If an automatic carriage is being used, it will also stop, due to the wiring of the circuits. In the nontransfer process, the operating switch must be opened when the end of the cut is reached.

Plasma arc cutting may be done in any position, and it will cut virtually any metal. This ability makes it a very useful and versatile process. Plasma arc cutting may be done under several inches (mm) of water as shown in Fig. 15-21. Noise levels are greatly reduced. Fumes and

Fig. 15-21.  Plasma arc cutting being done under several inches (mm) of water. This procedure lowers the noise level. Fumes and ultraviolet radiation are almost eliminated. (L-TEC Welding & Cutting Systems)

ultraviolet radiation are almost eliminated. Even though the cutting is done under water the cutting speed and quality are not affected.

## 15-7. REVIEW OF ARC CUTTING SAFETY

Electric arc cutting does not present any great or new hazards. Some cutting operations may continue for a considerable length of time and at high current rates. Precautions must therefore be taken to avoid skin burns or eye damage. The following precautions are recommended:

1. Be sure that all surfaces of the skin are shielded from the arc rays. Wear approved gloves, helmet, and approved clothing.
2. Use a helmet filter lens a shade or two darker than would be used for welding with the same size electrode. A #12-14 lens is recommended for carbon arc and air carbon arc cutting.
3. Use completely insulated electrode holders.
4. Stand or work only in dry surroundings.
5. A considerable amount of fumes and gases are liberated during all of the arc cutting operations. The ventilation of the working space MUST be such that the operator is working in clean, fresh air at all times.
6. Most of the arc cutting processes are accompanied by a great amount of sparking and showers of molten globules of metal. These hot particles may be thrown some distance from the arc. Wear garments which are fire-resistant. Pockets, cuffs, and other clothing crevices must be covered.
7. Be sure that the working area is fireproof. All flammable objects such as wood benches, floors, or cabinets should be removed from the vicinity or covered with flame resistant materials. Be sure there are no openings in the floor which might allow the sparks to travel the floor below.
8. To reduce the possibility of fire, some electric arc cutting stations use a "wet table." The wet table provides for a thin water film under the flame cutting area. The wet film quenches the sparks.
9. Check local fire codes before cutting a tank which has contained a flammable material.
10. The oxygen arc cutting process uses a higher electrical current than is used for welding the same thickness of metal. Be certain that the welding machine has ample capacity for the current needed. No arc welding machine should be used to exceed its duty cycle rating.
11. Most oxygen arc cutting processes and the plasma arc cutting process are very noisy.

Such stations should be located where the noise will not be objectionable. The operator and those in the vicinity must wear industrial ear muffs to prevent hearing loss.
12. When piercing a hole, precautions must be taken to avoid burns. The molten metal will come up until the hole goes through the metal.

## 15-8. TEST YOUR KNOWLEDGE

Write your answers on a separate sheet of paper. Do not write in this book.

1. Name the arc cutting processes which have the following AWS abbreviations:
   CAC
   PAC
   SMAC
   AAC
   AOC
2. When is the carbon arc process generally used?
3. The manual carbon arc cut is generally very _____ and of _____ appearance.
4. The length of the taper on a carbon electrode used for cutting should be _____ to _____ times the electrode diameter.
5. When cutting with the air carbon arc process, in what position should the air jets be in relation to the electrode?
6. When AAC at which angle to the work should the electrode be?
7. When air carbon arc cutting with the metal in the vertical position, the cutting should be done from the top down or bottom up?
8. The depth of a gouge done with automatic AAC can be controlled to within _____ in. or _____ mm.
9. How does the electrode used in oxygen arc cutting differ from the electrodes used in the other arc cutting processes?
10. What current range and oxygen pressure are suggested for cutting 3/4 in. (19.1 mm) aluminum with the AOC process?
11. What polarity should be used when gas metal arc or gas tungsten arc cutting?
12. To obtain a good cut with GMAC, the electrode speed must be adjusted to keep the electrode at the _____ surface of the metal while cutting.
13. When plasma arc cutting or oxygen arc cutting, how must the hearing be protected?
14. How is the arc in the nontransferred and transferred arc process stopped at the end of the cut?

15. How may the noise, fume, and radiation levels be reduced when doing plasma arc cutting?
16. When piercing a hole with oxyfuel gas or an arc cutting process, what danger is there from molten metal?
17. What amperage should be used with a 1/8 in. (3.2 mm) electrode when shielded metal arc cutting?
18. What DC polarity is recommended for SMAC?
19. What number filter lens is recommended when arc cutting with a given size electrode?
20. Prior to cutting into a tank which may have contained a flammable liquid, what must be done?

Industry photo. Using torch in vertical position. U-grooving seams in off shore drilling rig supports.

Industry photo. Automated station assembles the basic wire components of refrigerator shelving for roller resistance welding. Special fixtures align all of the components for the welding operation.
(Collis Div., Chamberlain Mfg. Corp.)

# part 6

# RESISTANCE WELDING

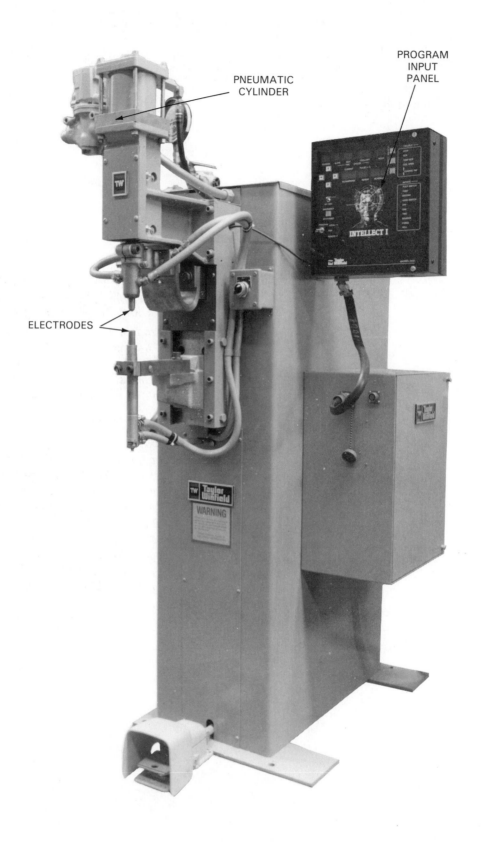

PROGRAM
INPUT
PANEL

PNEUMATIC
CYLINDER

ELECTRODES

INTELLECT I

A pneumatically operated resistance spot welder with a program input panel to accurately set all welding variables.   (Taylor Winfield Corp.)

# 16 RESISTANCE WELDING EQUIPMENT AND SUPPLIES

Designing and constructing resistance welding machines involves a combination of electrical design, machine structures, mechanisms, and controls. The machines vary from simple mechanisms to exceptionally complicated units. Refer to Headings 1-14, 1-15, 1-16, and 1-17.

Accurate control of the resistance welding variables is essential if good resistance welds are to result.

The Resistance Welder Manufacturers' Association (RWMA) and the National Electric Manufacturers' Association (NEMA) have developed standards for resistance welding machines. These standards have enabled the manufacturers to produce machines of known rated capacity and recognized durability.

The parts which make up a resistance welding station include a resistance welding machine, an electronic controller, and resistance welding electrodes.

CAUTION: This chapter is written using the conventional current theory. The CONVENTIONAL CURRENT THEORY states that CURRENT FLOWS FROM POSITIVE TO NEGATIVE.

The American Welding Society (AWS) and the Resistance Welder Manufacturers' Association both use conventional current when explaining resistance welding. In order to conform with the publications of these groups, explanations and diagrams in Chapter 16 and Chapter 17 use the conventional current flow theory.

## 16-1. ELECTRIC RESISTANCE WELDING MACHINES

In design, most electric resistance welding machines are quite similar. However, the manner in which the metal to be welded is held, and the appearance and position of the welding electrodes varies from machine to machine.

The methods used to control the various times, the welding current, and the pressure also vary between machines. Almost all resistance welders use automatic control of the welding time. The machines have hydraulic (fluid), pneumatic (air), or electric solenoids to supply and control electrode forces and electrical flow.

There are two other types of machines available. One type has manual control of the time and electrode force. The second type has manual control of welding time with hydraulic or pneumatic electrode force systems. Electric solenoids are used on these machines to control both the fluid flow, the electrode force, and the electricity flow. These two types are not produced in large quantities.

The parts of a resistance welding machine are:
1. Frame.
2. Transformer.
3. Welding arms and electrodes.
4. Force mechanisms.
5. Controller (not used in a manual control welder).

The TRANSFORMER is of special construction and usually has several tap settings or adjustments. It is usually water cooled, but it may be air cooled. The secondary winding usually consists of one loop or of several parallel loops. The ends of these loops are soldered, bolted, or brazed to the electrode arms of the machine.

The WELDING ARMS and operating mechanisms are different for each type of electric resistance welder. Each one usually uses a foot-operated switch in connection with pneumatic or hydraulic cylinders to press the parts together. The manual machines use foot pressure on a level to force the parts together.

All automatically controlled welders have a controller. The CONTROLLER regulates the timing of the weld cycle. It activates the pressure cylinder, turns the weld current on and off, and releases the force on the electrodes at the proper time. A resistance welding controller can be seen in Fig. 16-1 on the following page.

Fig. 16-1. A resistance welding electrical control cabinet. Most of the electrical connections, switches, and fuses are contained in the control cabinet.
(Robotron Div. Midland-Ross Corp.)

The part of the resistance welding machine that contacts the metal to be welded is called the ELECTRODE. Two electrodes are used and electric current is passed between them. The electrodes are usually copper. The electrodes are made in many different forms, each has its own purpose. The electrodes must be able to carry the large currents used in resistance welding. They must also be able to withstand the high force used.

## 16-2. TRANSFORMERS

A RESISTANCE WELDING TRANSFORMER is designed and built to use the voltage and current provided by the electric utility companies. However, neither the current nor voltage supplied to the machine is suitable for resistance welding.

Resistance welding requires high amperage, low voltage electrical energy. The transformer must lower the supplied voltage and increase the current. This type of transformer is called a STEP-DOWN transformer, since it reduces the supplied voltage.

For example, an ideal step-down transformer will transform 50 amps at 230 volts to 100 amps at 115 volts, or to 1,000 amps at 11.5 volts, or to 10,000 amps at 1.15 volts. In each example, the wattage (volts × amps) equals 11,500 watts. The resistance welding transformer usually delivers secondary welding current to the electrodes in the range of 5,000 to 100,000 amps.

The secondary circuit voltage is usually about 10 volts open circuit and decreases to less than 1 volt when current flows during welding.

Most transformers are intermittent in operation. They only deliver current for short intervals of time.

The amount of time the transformer delivers current, in ratio to the time the current is off, is called the DUTY CYCLE. Thus, if a transformer delivers current for 3 seconds out of each 60 seconds (one minute) the duty cycle is 5 percent:

$$3 \div 60 = .05 \quad \text{or 5 percent.}$$

A transformer has two electrical coils that are wrapped or wound around an iron core. The two coils are not electrically connected. One coil is called the primary coil, the other is called the secondary coil. The PRIMARY COIL is part of the primary circuit; the SECONDARY COIL is part of the secondary circuit. The secondary circuit is known as the secondary.

Alternating current (AC) is made to flow in the primary coil. As the current flows, a magnetic field builds up. The current then stops as the AC changes direction. the magnetic field collapses when the current stops. This occurs 120 times a second. The magnetic field that is created and destroyed passes through the secondary windings. This INDUCES (creates) a current in the secondary. The current in the secondary is also an alternating current.

The resistance welding transformer must be designed to carry the large currents required during welding.

The secondary wiring is usually made of cast copper or rolled copper bars. These bars are usually water-cooled. Copper tubes are welded or

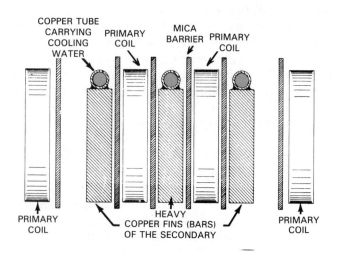

Fig. 16-2. Schematic cross section of a resistance welding transformer. Note that the secondary circuit bars have copper tubes brazed to them.

brazed to the bars to carry the cooling water as shown in Fig. 16-2. Smaller transformers may be air-cooled.

The core of the transformer is laminated. The core is usually made of four percent silicon transformer iron.

The primary windings are insulated so that no accidental grounds can occur. Fig. 16-3 illustrates a typical primary winding. Great care is taken to insulate the ribbon-type primary windings from each other and from the frame. A secondary winding which consists of three parallel loops is shown

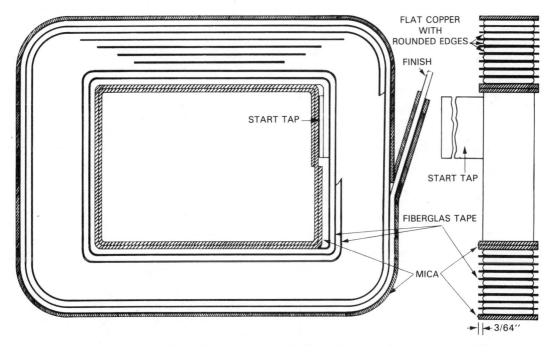

Fig. 16-3. The construction of the primary winding of a typical resistance welding transformer. The conductor is flat or ribbon shaped and the turns are insulated from each other by fiber glass.    (Taylor-Winfield Corp.)

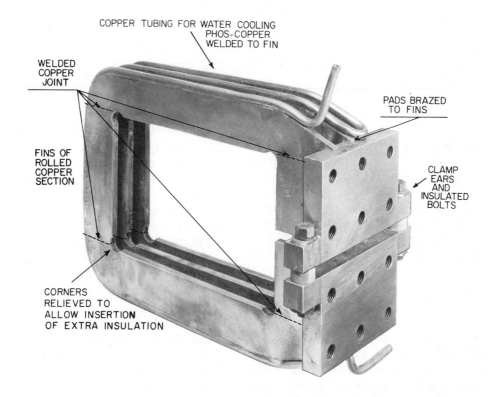

Fig. 16-4. A typical secondary winding design for a resistance welding transformer. In this case three bars are connected in parallel to increase current carrying capacity and still retain the one turn electrical function.

in Fig. 16-4. This design is still considered as one winding. The three parallel paths only make one winding or loop.

Fig. 16-5 shows a schematic wiring diagram for a resistance welding transformer. This machine provides for eight current ranges. The range is dependent on the ratio of the number of turns in

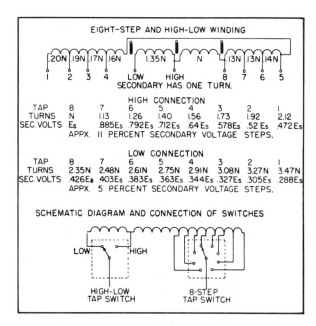

Fig. 16-6. Resistance welding transformer schematic wiring diagram. This unit has a high-low tap plus eight regular taps in the primary winding for purposes of current adjustment.

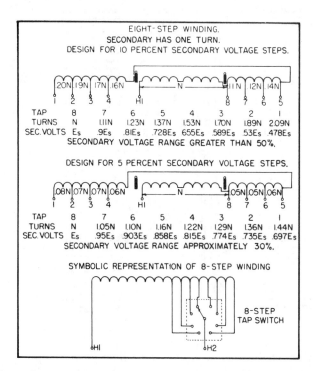

Fig. 16-5. Schematic wiring diagram which shows adjustable primary windings of a resistance welding transformer. This unit has eight taps (adjustments) in the primary circuit.

the primary circuit to the number of turns in the secondary. A wiring diagram for a resistance welder with a high-low tap and an eight-step primary winding is shown in Fig. 16-6.

Transformers vary in capacity and will suit varied welding needs. The capacity is usually listed as a KVA (kilovolt-ampere) rating. Resistance welding transformers are rated on a different basis than most transformers. The KVA is the input or primary circuit rating. For example, with a 230 volt input and 130 amps flowing, the machine would be rated at 30 KVA.
(230V × 130A = 29,900 VA)

$$\frac{29,900}{1,000} = 29.9 \text{ KVA} = 30 \text{ KVA.}$$

A resistance welding transformer is usually rated at 50 PERCENT DUTY CYCLE. This means that it can be used 30 seconds out of each 60 seconds. The duty cycle is based on a one minute period. The transformer cannot be used for three

minutes constantly and then cooled for three minutes. The transformer will overheat.

Duty cycles vary all the way from one percent to 100 percent, with a normal rating of 50 percent. If the duty cycle is high, the KVA setting must be smaller to insure safe transformer temperatures. The manufacturer usually specifies the KVA rating based on the 50 percent duty cycle. A transformer for a seam welding machine should have a 100 percent duty cycle.

The welding transformer is rated on a SPECIFIED INPUT and not on the output like other transformers. The output of the transformer will vary with the conductivity of the metal being welded, the electrode circuit, and other factors. Therefore, the welding manufacturer rates only the primary input capacity of the transformer.

When a machine is used at its rated KVA, the temperatures of the transformer windings will stay within safe limits for the insulation used. When the KVA rating or duty cycle is exceeded, the transformer efficiency decreases rapidly as the temperature rises. There is also the possibility of an insulation breakdown occurring.

Small spot welders may be rated in VA (volt-amperes). This is done when the KVA rating would be a decimal as in the case where 300 VA = .3KVA.

The primary current flow influences the electrical power service. For example, a 100 KVA machine may draw as much as 400 amps. If this load is applied to the power service all at once, a large voltage drop may occur in the line voltage. This drop in line voltage will affect other machines on the same power line. For example, a 10 percent drop in line voltage will cause a 20 to 30 percent reduction in the welding heat available at the electrodes. Thus automatic machines will produce substandard welds if the line voltage varies.

## 16-3. FORCE SYSTEMS

There are five methods of applying force to the electrodes. The five types are magnetic, electric motor, air, hydraulic, and mechanical leverage. The type of force systems depends on the type of welding machine used.

One common spot and projection welding machine is the rocker type. It uses a mechanical leverage, electric motor, or an air force system. A MECHANICAL LEVERAGE SYSTEM is simply a series of levers. When the welding operator presses on the foot pedal, leverage causes more force to be applied to the electrodes. A schematic of a mechanical leverage system is shown in Fig. 16-7.

The most common force system is the air operated or pneumatic system. The AIR OPERATED SYSTEM is simple to use and requires little maintenance. Compressed air is used to supply the force. The desired air pressure is set on a valve. The valve setting is proportional to the force applied at the electrodes. Fig. 16-8 shows an air operated rocker arm welding machine which uses an air cylinder to supply the force. The major parts of a rocker arm spot welding machine are labeled.

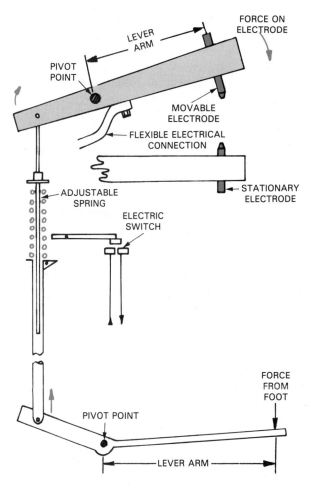

Fig. 16-7. A mechanical leverage system. The electrode force is controlled by adjusting the adjustable spring and by the force applied by the welding operator.

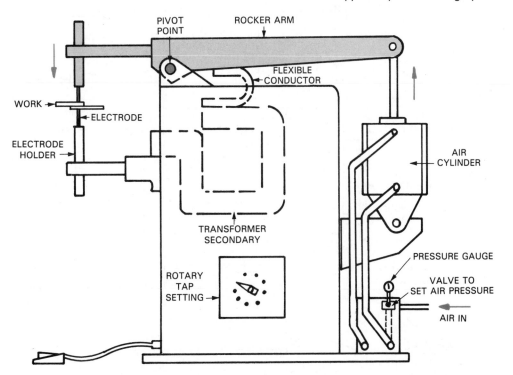

Fig. 16-8. A rocker arm welding machine which uses air pressure to supply the welding force. The air pressure is adjustable. The main parts of a rocker arm machine are labeled.

ELECTRIC MOTOR operated machines are not used much. They are used when compressed air is not available. The electric motor would be mounted on the back of a rocker arm welding machine instead of the air cylinder shown in Fig. 16-8. The motor is used to rotate a cam which applies the correct force.

Hydraulic cylinders can be used to supply the required force for resistance welding. The HYDRAULIC SYSTEM is used on a press type spot and projection welding machine. A press type welding machine uses direct force. The air or hydraulic cylinder and the ram apply the required force directly to the electrodes. The air operated press type welding machine is the more common type. A press type welding machine is shown in Fig. 16-9. Much higher forces are obtainable with a hydraulic force system than an air-operated (pneumatic) force system.

Some force systems can supply a forging force to the movable electrode. The forging force is applied to the electrode after the welding current is turned off and while the weld area is still molten or plastic. It forces the electrodes down and holds them at a set position until the metal cools and solidifies.

The force on the resistance welding electrode may vary from a few pounds or newtons (N) to a few thousand pounds (thousand newtons). Common electrode forces used in industry vary from 300 pounds (1335 N) to 1400 pounds (6228 N). Very thin gauge sheet metal can be welded with 100 pounds (445 N) of force on the electrode.

## 16-4. CONTROLLERS

Resistance welding is a fast operation. A weld is often completed in less than one second. The control of the welding operation must be very accurate. A controller is needed to control the welding operation. There are five VARIABLES in resistance welding. These variables are:
1. Current.
2. Time.
3. Force applied to the metal.
4. Electrode contact area.
5. Welding machine.

The welding controller is used to control the time and the current required to make a weld. There are three different times required to complete one resistance welding schedule. A RESISTANCE WELDING SCHEDULE is the sequence of

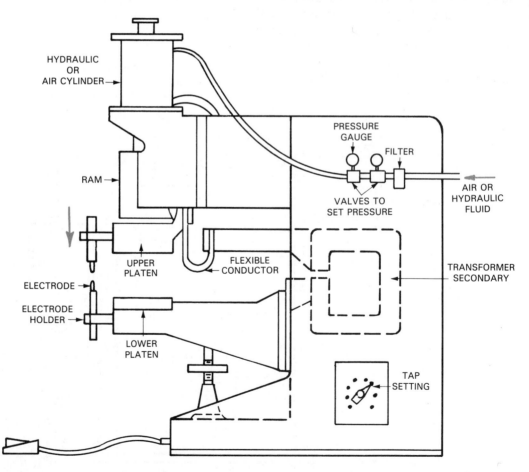

Fig. 16-9. A typical press type welding machine. Hydraulic or air pressure can be used to supply the required force to the electrodes.

events which must occur to complete one weld. These three times are:

1. Squeeze time.
2. Weld time.
3. Hold time.

Each of the times is set on the controller in cycles. One cycle is 1/60 of a second.

The SQUEEZE TIME is the time required for the electrodes to close on the work piece and apply the proper force. This force clamps the pieces and provides for a good electrical contact.

The WELD TIME is the time the current flows through the work piece. As the current flows, it heats the two pieces of metal where they are in contact with one another. The current is stopped by the controller.

The HOLD TIME is required so the molten metal has time to cool. The two pieces of metal are now welded together.

A typical welding controller is shown in Fig. 16-10. The welding operator can set the squeeze time, weld time, and hold time in cycles. An exam-ple of a complete welding schedule is: squeeze for 20 cycles, weld for 10 cycles, and hold for 20 cycles. This weld cycle is shown in Fig. 16-11.

There is also a setting for off time. Off time is used when the welding machine is making spot welds repeatedly. The OFF TIME allows the welding operator to reposition the parts being welded, or to place some new material between the electrodes. The weld schedule is then repeated automatically.

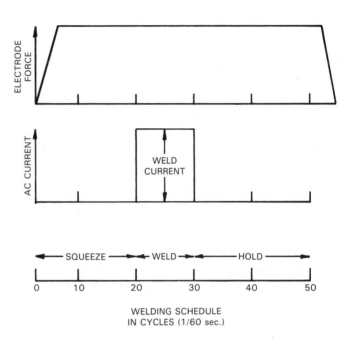

Fig. 16-11. A complete weld schedule. The desired times for each portion of the weld schedule are set in cycles. Notice the time for the pressure system to react.

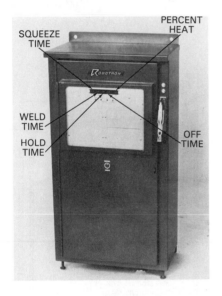

Fig. 16-10. A welding controller. The different times and the percent heat are set on the control panel. Inside of control panel also shown. (Robotron Div. of Midland-Ross Corp.)

The welding controller is also used to adjust the current. Large changes in current are made by changing the TAP SETTING on the transformer. This is similar to changing the range selector on an arc welding machine. Smaller changes within a current range are made by adjusting the PERCENT HEAT CONTROL on the welding control panel. This control is marked from 0 - 100 (percent). See Heading 10-7 for an explanation of setting this type of control.

The welding current does not have to remain a constant. When welding large thicknesses, up slope or an increasing current is often required. The current is gradually increased from 10 percent heat, up to the desired welding percent heat. Down slope or a decreasing current can also be used. It may also be required to use two or more pulses of current. A short cooling time separates the different pulses. See Fig. 16-12 for an example of how these functions are used in a weld schedule.

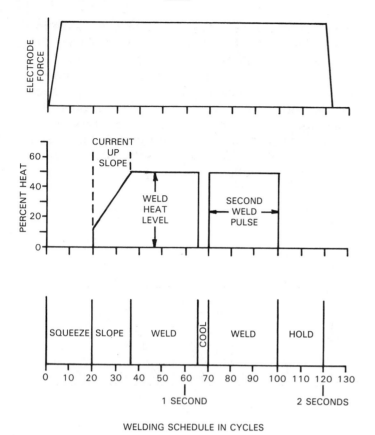

Fig. 16-12. A complex welding schedule which includes up slope, two current pulses, and a cool time.

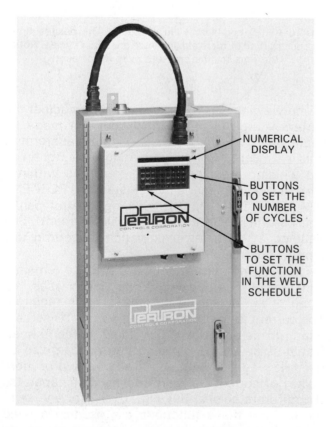

Fig. 16-13. A welding controller. (Pertron Controls Corp.)

An example of a complex weld schedule is as follows: squeeze for 20 cycles, slope the current up (up slope) from 10 percent heat to 50 percent heat in 15 cycles, weld at 50 percent heat for 30 cycles, cool for 5 cycles, weld a second pulse at 50 percent heat for 30 cycles, and hold for 20 cycles. This weld schedule is illustrated in Fig. 16-12.

In order to use these special up slope, down slope, and multiple pulse functions, a more sophisticated controller must be used. An example of a welding controller designed to direct up slope, down slope, and multiple pulse functions is shown in Fig. 16-13.

## 16-5. CURRENT CONTROLS

The primary and secondary circuit of a resistance welding machine must be able to carry

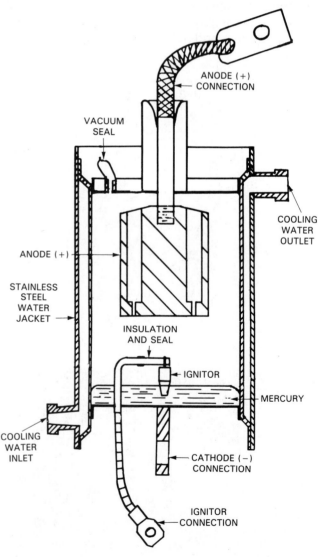

Fig. 16-14. The main parts of an ignitron tube. When the ignitor fires, the mercury vaporizes. The vaporized mercury completes the circuit between the anode and the cathode. (General Electric Co.)

the high currents required during welding. The secondary loop of the transformer and all the secondary current carrying conductors (cables) are of very sturdy construction. See Figs. 16-8 and 16-9.

The IGNITRON TUBES are an important part of the primary circuit. The ignitron tubes are capable of switching the large currents every half cycle. It is almost impossible for any mechanical switch to carry the large currents required in resistance welding and still be able to switch the current on and off.

The operation of the ignitron tube is as follows. The ignitron tube is usually a non-conductor. There is a small pool of liquid mercury at the bottom of the tube as shown in Fig. 16-14. To make the tube conduct, a small current must pass from the ignitor to the cathode through the liquid mercury. This small current is enough to vaporize the liquid mercury. The vaporized mercury acts as a switch as it fills the ignitron tube. The high current may now travel between the anode and the cathodes.

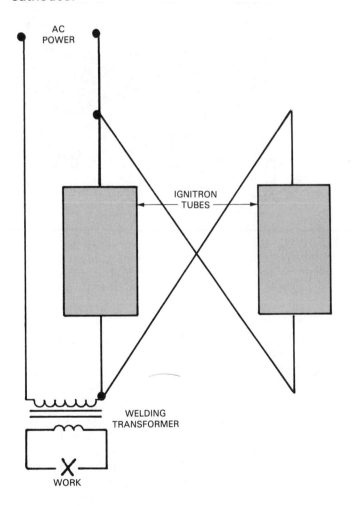

Fig. 16-15. Two ignitron tubes connected in reverse parallel. The anode of one is connected to the cathode of the second. This allows current to flow in both half cycles.

The ignitron tube will only conduct for one half of the AC cycle. A second ignitron tube must be connected in reverse parallel to carry current on the second half cycle. With each ignitron tube carrying current for a half cycle, the welding machine can now carry current continuously. A circuit showing two ignitron tubes connected so current will flow during both half cycles is shown in Fig. 16-15.

There are five different sizes of ignitron tubes: A, B, C, D, and E. The capacities of the different tubes are shown in Fig. 16-16.

| IGNITRON TUBE MODEL | AMPERE CAPACITY BASED ON APPROX. 20% DUTY CYCLE | COOLING WATER GALS/MIN. MINIMUM |
|---|---|---|
| A | 200 | 1 1/2 |
| B | 600 | 1 1/2 |
| C | 1300 | 1 1/2 |
| D | 3000 | 3 |
| E | 8000 | 5 |

Fig. 16-16. Table of ignitron tube properties. Capacities fluctuate depending on many variables. The size tube to be used should be carefully calculated based on Resistance Welder Manufacturers' Association (RWMA) standards.

Since these ignitron tubes carry so much current, they can become very hot. They must be water cooled so they can operate properly. A water jacket is built around each ignitron tube. In addition, some ignitron tubes are protected by a thermostat. An ignitron tube thermostat is shown in Fig. 16-17. The thermostat is shown mounted on an ignitron tube in Fig. 16-18. If the tube

Fig. 16-17. Ignitron tube thermostat.

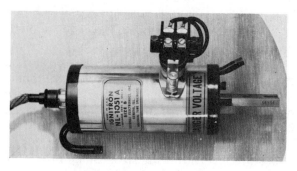

Fig. 16-18. Ignitron tube with a thermostat mounted on the tube.

temperature rises to 125°F (52°C) the thermostat opens and shuts down the machine. The thermostat will close again when the temperature is 105°F (41°C).

Fig. 16-19 shows a complete welding control. The ignitron tubes, the control panel, and the solid state controls can all be seen.

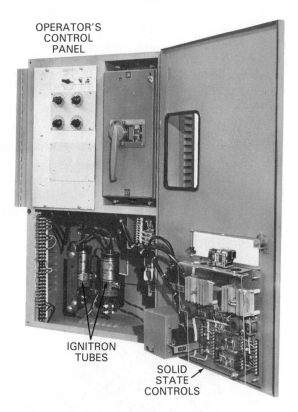

Fig. 16-19. A resistance welding controller. The panel door is open to show the solid state controls and the ignitron tubes.

Some resistance welding machines and controllers do not use ignitron tubes. They are using SILICON CONTROLLED RECTIFIERS (SCR). An SCR performs the same function as the ignitron tubes. Each SCR allows current to flow in one direction. Two SCRs are used and are connected

in reverse parallel. With this type of circuit they can conduct current during both half cycles. The SCRs become hot but cannot be water cooled because the water would cause a short circuit. To cool the SCRs, the frame to which they are attached is water cooled. An SCR controller is shown in Fig. 16-20. The SCRs are much smaller than the ignitron tubes. They cannot carry the large currents that the large ignitron tubes can.

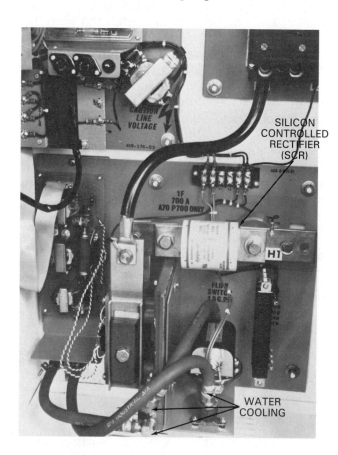

Fig. 16-20. Silicon controlled rectifiers (SCR) are used to control the electric current. Two SCRs are used to control the current in both half cycles. One SCR is hidden from view. (Robotron Div. Midland-Ross Corp.)

The ignitron tubes and the SCRs can be made to begin conducting at various points on the AC cycle. The different times correspond to the heat settings on the control panel. If a high percent heat is desired, the ignitron tube or SCR will begin conducting very early in the AC cycle. As lower percent heat settings are desired, the ignitron tube or SCR will not conduct until later in the AC cycle. Fig. 16-21 shows how the current is delayed as the percent heat setting is lowered.

CAUTION: The following paragraphs use the conventional current flow theory in the explanations. The conventional current flow theory states that current flows from positive to negative.

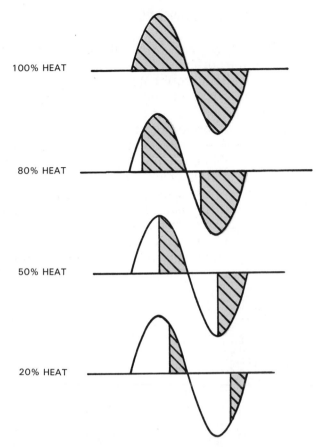

100% HEAT

80% HEAT

50% HEAT

20% HEAT

Fig. 16-21. The percent heat setting changes the point at which the current begins to flow in each cycle. The higher the percent heat setting, the longer the current will flow in each cycle.

The symbol for an SCR is shown in Fig. 16-22. The anode, cathode, and the gate are labeled. The current will only flow in the direction of the arrow.

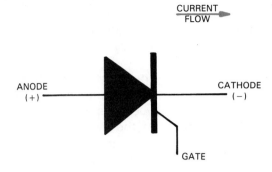

CURRENT FLOW

ANODE (+)

CATHODE (−)

GATE

Fig. 16-22. The symbol for a silicon controlled rectifier (SCR). The SCR will only allow current to flow in one direction. (Conventional current flow.)

The current will not flow through the SCR until a small trigger voltage is applied to the gate. As soon as the proper voltage is applied to the gate, the SCR begins to conduct current. It continues to conduct until the end of the half cycle. When a lower percent heat setting is selected, the trigger voltage is increased. With a higher trigger voltage, the SCR is not activated until later in the AC cycle, thus the current is less.

The ignitron tubes can be delayed from carrying current in a similar fashion. The ignitron tube will not conduct until a small current passes from the ignitor to the cathode. The ignitron tube can be delayed from firing by delaying the small current to the ignitor.

Fig. 16-23 shows one way of delaying the ignitor current. An SCR is placed in the ignitor circuit. As the percent heat setting on the control panel is changed, the trigger voltage on the SCR

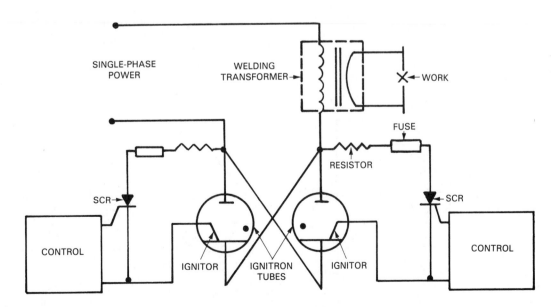

SINGLE-PHASE POWER

WELDING TRANSFORMER

WORK

FUSE

RESISTOR

SCR

SCR

CONTROL

IGNITOR

IGNITRON TUBES

IGNITOR

CONTROL

Fig. 16-23. An SCR is used to control the current to the ignitor. The percent heat controls the trigger voltage of the SCRs.

changes. When the correct trigger voltage is applied to the SCR, it allows current to flow. This current goes to the ignitor and passes through the liquid mercury. The mercury vaporizes, and the ignitron tube conducts.

A second way to adjust the time that the ignitron tube conducts is to use a THYRATRON TUBE. A thyratron tube acts like an SCR. The anode is on one end of the tube, the cathode on the other. A control grid separates the two. The control grid prevents current from flowing. The thyratron tube will conduct when the grid voltage is reduced. A schematic view of two types of thyratron tubes is shown in Fig. 16-24. Thyratron tubes are used in Fig. 16-25 to control the firing of the ignitron tubes.

## 16-6. SPOT WELDING ELECTRODES

Resistance welding ELECTRODES conduct the current to the surfaces of the metals to be welded. There are certain requirements which these electrodes must possess. They must:
1. Be a good conductor of electricity.
2. Be a good conductor of heat.
3. Have good mechanical strength and hardness.
4. Have a minimum tendency to alloy (combine) with the metals being welded.

Pure copper possesses good electrical and thermal properties; however, it is rather soft and does not wear well. Also it tends to soften with heat. Most electrodes are a copper alloy.

The electrode must be a good conductor of elec-

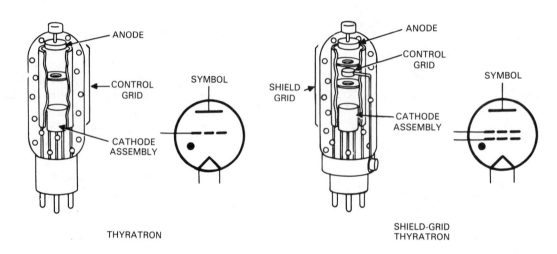

Fig. 16-24.  Schematic of two thyratron control tubes.

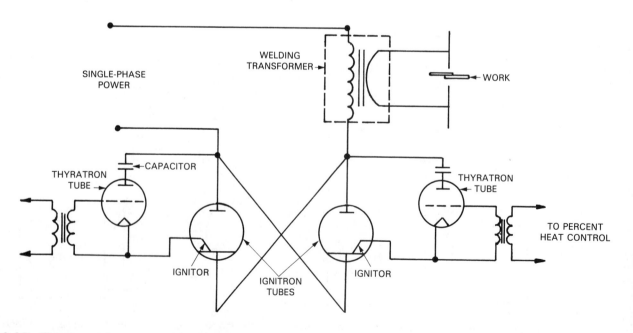

Fig. 16-25.  Thyratron tubes are used in the ignitor circuit. The percent heat control determines the grid voltage.

tricity in order that the current may flow to the work piece without overheating the electrode. It must be a good conductor of heat so that the heat generated at the point of contact with the weldment may be conducted away without causing the electrode to become overheated. Typical spot welding electrodes are shown in Fig. 16-26. Every electrode has a face and a shank. The ELECTRODE FACE is the part of the electrode which contacts the work. The face may have a number of different designs as shown in Fig. 16-27. The ELECTRODE SHANK must be large enough to carry the force on the electrode and the welding current. In order to meet some spot welding

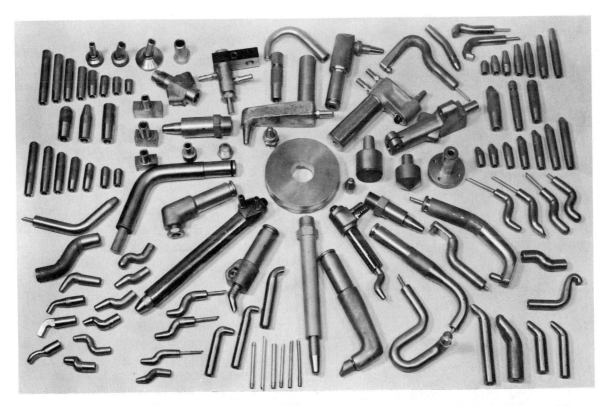

Fig. 16-26. Variety of electrodes, electrode holders, and seam welder wheels used in resistance welding. (Hercules Welding Products DIvision of Obara Corporation)

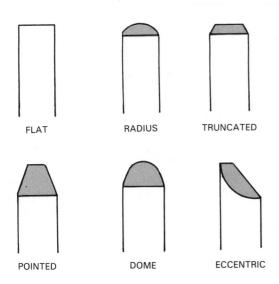

Fig. 16-27. Common electrode face designs.

requirements, some electrode shanks are bent. Examples of bent electrodes can be found in Fig. 16-26.

Some electrodes have two pieces. The two pieces are called the ELECTRODE CAP and the ADAPTOR. The electrode caps can be male or female caps as shown in Fig. 16-28.

Most electrodes and electrode caps are water cooled. There is a hole down the center of the electrode. A flexible water tube is placed inside the electrode. The cooling water flows through the tube and cools the electrode very near the face of the electrode. The water returns around the tube and keeps the entire electrode cool. A section view of a bent electrode is shown in Fig. 16-29. The path of the cooling water is shown.

The Resistance Welder Manufacturers' Association (RWMA) and the Resistance Welding Alloy Association (RWAA) recognize two groups of materials used for resistance welding electrodes.

**Resistance Welding Equipment and Supplies / 425**

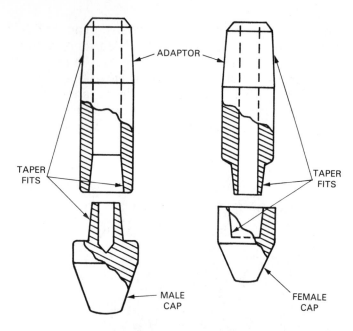

Fig. 16-28. Two piece electrode cap and adaptors. The electrode caps may be male or female and are usually held on the adaptor by a taper fit.

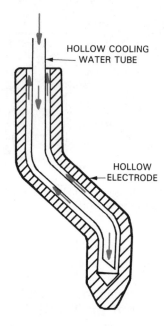

Fig. 16-29. The cooling water path through a bent electrode.

The groups are Group A and Group B.

Group A are copper-base alloys having good electrical and thermal (heat) conductivity and with improved hardness and wear qualities.

Group B are refractory metal alloys. The electrical and thermal properties of this group are not as good as the alloys in Group A. However, they have extremely high melting temperatures, and high compressive strength and wear resistance. These materials are usually mixtures of tungsten and copper.

When ordering or purchasing resistance welder electrodes the following data may be referred to as a guide:

### GROUP A, COPPER-BASE ALLOYS

The most common materials used are Group A, Class 1 and 2. The Class 2 electrodes are the most common. These two types have the best conductivity with good hardness. They are also the least expensive and most available.

Characteristics and uses of various classes of Group A electrodes are:

Class 1. Highest conductivity of 80%; recommended for spot welding aluminum alloys, magnesium alloys, galvanized iron, brass, or bronze. The electrode material is a copper-cadmium alloy.

Class 2. 75% conductivity and good hardness. For high production spot and seam welding, clean mild steel, low alloy steels, stainless steels, nickel alloys, and monel metal. The electrode material is a copper-chromium alloy.

Class 3. 45% conductivity. Higher strength and hardness than Class 2 electrodes. Recommended for projection welding electrodes, flash and butt welding electrodes. Recommended for stainless steel. The electrode material is a copper-zirconium alloy.

Class 4. This is a hard, high strength alloy. This alloy is recommended for use as an electrode material for special application when the forces are extremely high and the wear is severe. The conductivity is only 20%.

Class 5. This alloy is used chiefly as a casting and has high mechanical strength. The electrical conductivity is only 15%.

### GROUP B, REFRACTORY METAL COMPOSITIONS

Characteristics and uses of the refractory alloys in Classes 10-14.

Class 10. This electrode material has the best conductivity of Group B electrodes. The conductivity is 45%. It is recommended for facings for projection welding electrodes and flash and butt welding electrodes.

Class 11, 12. This material is harder than the material in Class 10. It is used when exceptional wear resistance is required.

Class 13, 14. These electrodes are made of commercially pure tungsten and molybdenum, respectively. They are used to weld nonferrous metals that have very high electrical conductivity.

## 16-7. ELECTRODE HOLDER

The resistance welding electrodes and adaptors are held by ELECTRODE HOLDERS. The holders are clamped into the ends of the movable spot welding arm and the stationary arm. Refer to Figs. 16-8 and 16-9. Most of the electrode holders are water-cooled.

The electrode holders hold the electrode in proper position, carry the welding current, and provide the electrode with water-cooling. Fig. 16-30 shows a typical electrode holder and tip. The water tube runs down the center of the electrode holder and continues into the electrode.

copper alloy which provides good current carrying qualities and rigidity. Some examples of electrodes and electrode holders combinations are shown in Fig. 16-31.

The electrode or adaptor can be attached to the electrode holder is three different ways:
1. Taper fit.
2. Threaded.
3. Straight shank.

A tapered fit is the most common, especially for

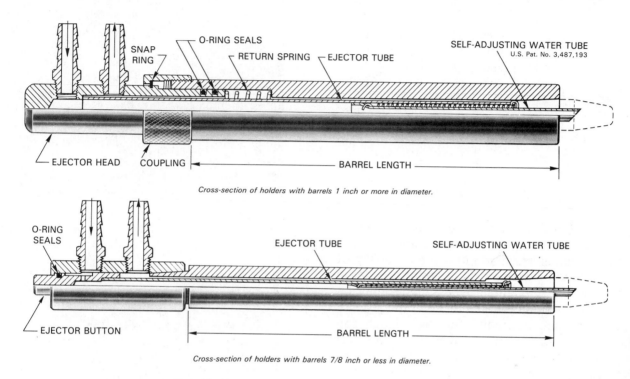

Cross-section of holders with barrels 1 inch or more in diameter.

Cross-section of holders with barrels 7/8 inch or less in diameter.

Fig. 16-30.  Cross section of two electrode holders.
(Tuffaloy Products, Inc.)

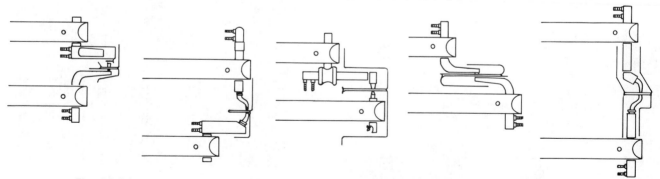

Fig. 16-31.  Electrodes and electrode holders can be combined to meet a job requirement.
(Tuffaloy Products, Inc.)

It should be noted that on most of the spot welding machines the electrode holders are adjustable for length and position. In general, the electrode holders should be adjusted to the shortest length at which the weld metal may be easily inserted. Electrode holders are made of a

high production spot welding. Replacing electrodes and electrode caps is quick and easy when a tapered fit is used. An ejector tube or knock out bar is used to loosen tapered adaptors and electrodes. Two different tapers are available. One taper is called the Standard Morse taper; the other is called the Jarno taper. The Jarno taper is preferred. A table of the Jarno tapers is shown in Fig. 16-32. A threaded attachment is used when high welding forces are used. The straight shank electrode or adaptor is mechanically connected to the holder by a separate coupling or collar.

| JARNO TAPERS | | | | | | |
|---|---|---|---|---|---|---|
| TAPER NUMBER | A | | B | | C | |
| | inch | mm | inch | mm | inch | mm |
| 3 | .3 | 7.62 | .375 | 9.53 | 1.5 | 38.1 |
| 4 | .4 | 10.16 | .500 | 12.70 | 2.0 | 50.8 |
| 5 | .5 | 12.70 | .625 | 15.88 | 2.5 | 63.5 |
| 6 | .6 | 15.24 | .750 | 19.05 | 3.0 | 76.2 |
| 7 | .7 | 17.78 | .875 | 22.23 | 3.5 | 88.9 |
| All tapers are .600 inches/foot (50 mm/meter). | | | | | | |

Fig. 16-32. Dimensions of Jarno tapers.

## 16-8. SPOT WELDING MACHINES

SPOT WELDING MACHINES are the most common of the resistance welding machines. Spot welding machines are made in a great variety of sizes from small bench units to extremely large welding machines. A small bench model may be used to spot weld such items as costume jewelry and electron tube components. Two bench top welding machines are shown in Figs. 16-33 and 16-34. The large welding machines are capable of welding hundreds of spots on large sheet metal productions such as automobile bodies, refrigerator cabinets, and the like as shown in Figs. 16-35 and 16-36. Some resistance welding machines

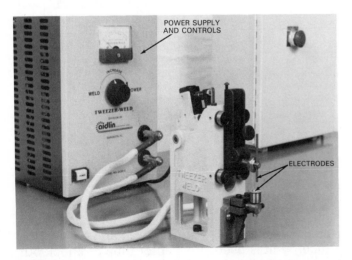

Fig. 16-33. This welder is only 9 in. (23 cm) high and has a 5 KVA capacity. A variety of 1/8 in. (3 mm) diameter electrode shapes may be used. (Aidlin Automation Corp.)

are portable. Portable machines are covered in Heading 16-10. Most machines have only one set of electrodes.

The basic purpose of the spot welder is to make a spot weld (a fused nugget) between two or more lapped pieces of metal. The size of the spot weld is controlled by changing the resistance welding

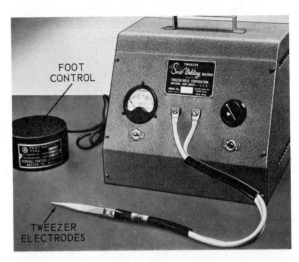

Fig. 16-34. Spot welder for fine precision work. The electrodes are of the tweezer design. (Arvin Automation)

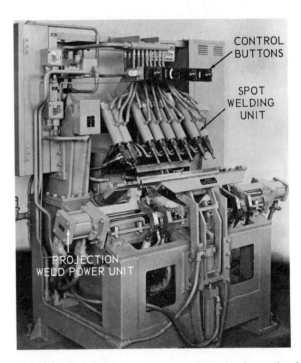

Fig. 16-35. Combination spot welding and projection welding machine with four 55 KVA transformers. The input power is 440V., 60 cycle. The spot welding guns lower, weld, and retract in 1 1/2 seconds. There is one projection welding unit at each end. The units advance, weld, and retract in 1 1/2 seconds. There are dual controls. The operator manually loads and unloads the parts. Note the pneumatic, hydraulic, and electric devices. The necessary parts are water-cooled using water valves and electric timers to control the water flow. (Resistance Welder Corp.)

WELDING CONTROLLERS

WELDING TRANSFORMERS

RH

WELDING ELECTRODES

Fig. 16-36. A large spot welding machine used to weld automobile side panels. Notice the number of welding electrodes, welding transformers, and welding controllers. (Progressive Machinery Corp.)

variables. The variables in resistance welding are:
1. Current.
2. Time.
3. Force.
4. Electrode contact area.
5. Welding machine.

The selection of a resistance welding machine determines the following:
1. The dimensions of the operating area.
2. The KVA rating.
3. The type of pressure system.

The OPERATING AREA has two dimensions. These are called the throat depth and the horn spacing. The THROAT DEPTH is the distance from the center of the electrodes to the frame of the welding machine. The throat depth determines the size (width) of a part which can be welded.

The HORN SPACING is the distance between the electrode arms when the electrodes are closed. The height of a part to be welded is controlled by this dimension. The horn spacing is usually adjustable. The throat depth and horn spacing are shown in Figs. 16-37 and 16-38.

The throat depth should be kept to a minimum.

As the throat depth increases, the KVA required to create a weld increases.

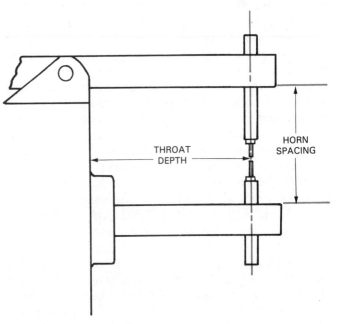

THROAT DEPTH

HORN SPACING

Fig. 16-37. Throat depth and horn spacing on a rocker type machine.

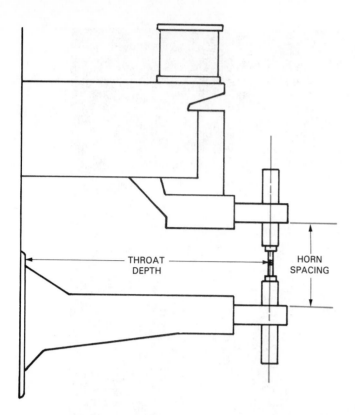

Fig. 16-38. Throat depth and horn spacing on a press type machine. The horn spacing is measured with the electrodes touching one another (power off).

Resistance welding machines obtain their electrical energy in one of three ways:
1. Single-phase machine.
2. Three-phase machine.
3. Stored energy machine.

All resistance welding machines demand a large current for a very short period of time. The SINGLE-PHASE MACHINE is connected to one phase of the electricity supply. The large demand for electrical energy by the single-phase machine can cause a drop in the line voltage of the one phase. If several single-phase resistance welding machines are all connected to the same phase and are all fired at once, a large drop of voltage can affect other machines (welding and non-welding) on the same phase. This decrease in voltage on the one phase causes an imbalance on the power system. To prevent an imbalance, a stored energy or a three-phase machine can be used.

STORED ENERGY WELDING MACHINES are spot welding machines with added electrical characteristics. The stored energy welding machines obtain the energy needed for welding from the service lines at a slow rate. This stored energy, once it has reached the desired energy level, can be released at a high rate for welding. By taking the energy at a slow rate from the service line and storing it, the stored energy machine does

not cause a voltage drop. There are three types of resistance welders developed to do this work:
1. Electromagnetic type.
2. Electrostatic type (capacitor).
3. Electrochemical type (battery).

The ELECTROSTATIC resistance welding machine is the one most often used. Its principle of operation depends upon storing the electrical energy in a capacitor. When the energy is needed to weld, it is released by the capacitors. The current passes through the work piece to form a weld. A spike welding machine uses this principle to obtain extremely high currents for a very short time (three milliseconds or less).

The use of capacitors to store electrical energy enables the welding machine to use a smaller KVA transformer. The off-time part of the cycle of the machine is used to electrically load the capacitors or condensers. An electrostatic welding machine is shown in Fig. 16-39.

Fig. 16-39. An electrostatic (spike) spot welding machine. The electrical energy for welding is stored in capacitors. This machine uses an air cylinder to apply the welding force to the electrodes. (Weldex, Inc.)

Stored energy machines have been largely replaced by three-phase machines. The THREE-PHASE RESISTANCE WELDING MACHINE draws energy from each of the three phases of the electricity supply equally. The three-phase machine does not cause an imbalance in the power supply.

Three-phase machines have three times as much electrical circuitry in the primary as a single-phase machine. A single-phase machine has two ignitron tubes. A three-phase machine has two ignitron tubes for each phase, for a total of six. A three-phase machine also has three primary transformer windings. Fig. 16-40 shows a schematic of a three-phase resistance welding machine. Because of the additional electrical requirements and increased size, a three phase resistance welding machine is more expensive than a single-phase machine.

A press type or rocker type welding machine can be used for projection welding. If there is only one projection weld, flat electrodes can be used in either type machine. When more than one projection weld is made at a time, special dies or electrodes must be used.

Another way to make many projection welds at one time is to place the part between the upper and lower platens on a press welding machine. A platen

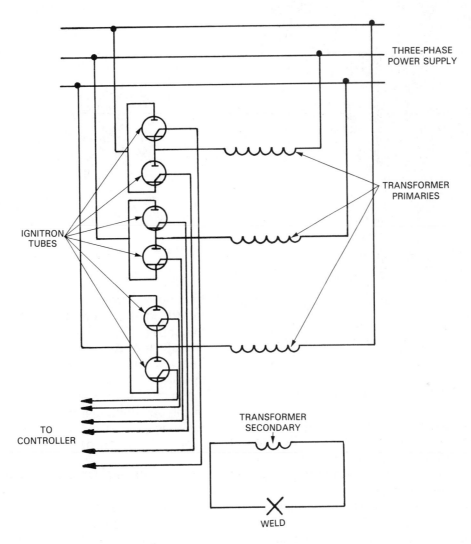

Fig. 16-40. The electrical connections for a three-phase machine. A three-phase machine has six ignitron tubes and three primary transformers.

## 16-9. PROJECTION WELDING EQUIPMENT

Projection welding equipment is similar to the equipment used in spot welding. In PROJECTION WELDING, one of the pieces to be welded has one or more small projections pressed into it. The current passes through the projection and forms a weld between the two metal pieces. Projection welding is used to accurately locate the welds on a part which is in high production.

is a large, generally flat surface through which welding current flows. The upper platen is movable. The lower platen is adjustable, but once adjusted, it remains stationary. An air or hydraulic cylinder forces the upper platen against the lower platen. See Fig. 16-41. The two parts being joined must be flat. If they are not flat, a specially made die can be attached to the platens. Higher forces and currents are required to make multiple projection welds.

Fig. 16-41. A projection welding machine. Special dies can be attached to the upper and lower platens. (Taylor Winfield Corp.)

## 16-10. PORTABLE WELDING MACHINES

One type of spot welding machine is known as a portable spot welding machine. It is usually referred to as a welding gun, or a gun welder. A typical portable spot welding machine has four main parts:
1. A portable welding gun.
2. A welding transformer which may also have a rectifier.
3. An electrical contactor and controller.
4. A cable and hose to bring power and cooling water to the welding gun from the transformer.

A portable spot welding machine is shown in Fig. 16-42. The welding gun can be used to weld within a four foot (1.2 m) area of the transformer. The transformer is supported on a counter balanced arm.

Portable welding guns are designed to do a specific job. The pressure system may use air or hydraulics. Many different electrode shapes are available to make it possible to weld in hard-to-reach locations.

The portable welding guns require a higher

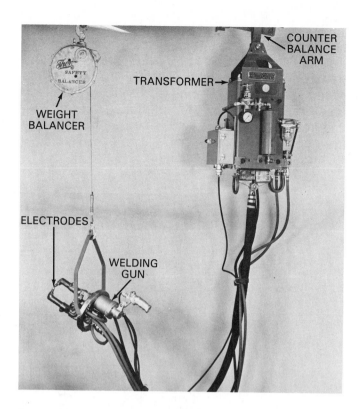

Fig. 16-42. A portable welding gun mounted on a counter balance arm.

secondary voltage because of the increased length in the secondary cables. The transformer secondary may have two or three turns. The transformer and the welding gun develop a lot of heat so they must be water cooled. A portable spot welding gun is shown in Fig. 16-43.

Portable spot welding machines are widely used in the automobile industries. A welder can use the welding gun to make a number of spot welds on irregular shaped parts.

## 16-11. SEAM WELDING MACHINES

A SEAM WELDING MACHINE is a special form of a resistance welder. Seam welding uses two copper alloy wheels as electrodes. These wheels press the two pieces of metal together and roll slowly along the seam. Either a continuous current is passed between electrodes and work or current is passed through the electrodes and work at timed intervals.

A seam welding machine has the following parts:
1. A main frame which contains the transformer and tap switch.
2. Secondary connections to the electrodes.
3. Wheel electrodes.
4. Stationary lower electrode arm.
5. Movable upper electrode arm.
6. Air cylinder and ram to apply required force.

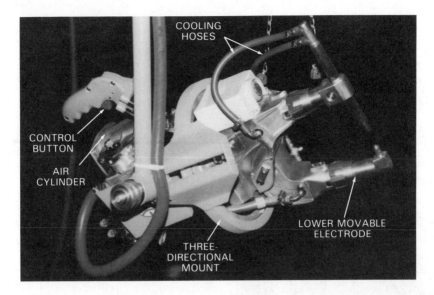

Fig. 16-43. A portable spot welding gun. This unit is supported from overhead to relieve weight from the operator. It has a built-in transformer and electrodes that can be rotated as required. The lower electrode is pneumatically operated.

7. Electrode drive mechanism.

The main frame, air cylinder, and ram are similar to those used on a press type welding machine. The electrode drive mechanism can be one of three types:
1. Gear drive.
2. Knurl drive.
3. Friction drive.

Usually only one electrode wheel is driven; the second rotates freely.

A gear drive turns the center of the wheel at a constant RPM. As the wheel wears down, the welding speed decreases.

A knurl or friction drive uses a small roller which contacts the outer edge of the wheel electrode. The small roller rotates at a constant RPM. The small roller drives the wheel electrode at a constant RPM. With a knurl or friction drive, the welding speed will remain constant as the electrodes wear.

A seam welding machine can be one of three types. The wheels can be parallel to the front of the machine (transverse seam); perpendicular to the front of the machine (longitudinal seam); or a combination of both (universal). See Fig. 16-44. The universal machine can be used to make both longitudinal and circular seams. There are two lower arms and electrode wheels which can be swung in and out of position. There is also one movable upper arm and wheel. See Fig. 16-45.

A seam welding machine is controlled by a controller. Current can be timed to make overlapping spot welds, spots at a given spacing, or current can run constantly to make a continuous seam.

Pneumatic (air) pressure is usually used to con-

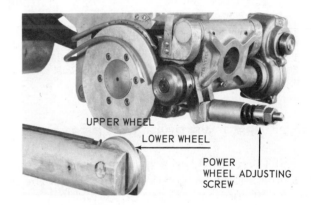

Fig. 16-44. Seam welding electrodes used to make a longitudinal seam. These electrode wheels are perpendicular to the front of the machine.

trol the force applied to the electrodes.

The pressures required are not extremely high. The details of the upper electrode operating mechanism of a universal seam welding machine are shown in Fig. 16-46.

Occasionally seam welds are made on parts that do not allow a continuous seam to be made. A portion of a wheel can be used to get up close to the area which could not be welded by a complete wheel electrode. Fig. 16-47 shows an example of a partial wheel and how it can be used.

The metal part being seam welded and the wheel electrodes get very hot. The electrodes and the work must be water cooled. Flexible copper tubes carry cooling water to the joint. Both electrodes must be cooled by the water spray. A flexible water tube can be seen in Fig. 16-47 in use with a shoe type electrode.

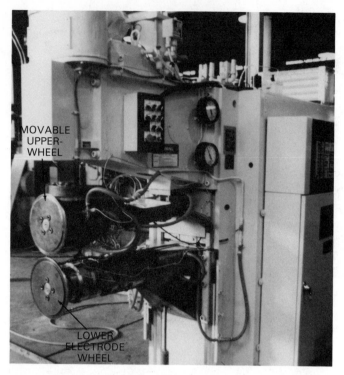

MOVABLE UPPER-WHEEL

LOWER ELECTRODE WHEEL

Fig. 16-45. A universal seam welder. Note that these electrode wheels are set up for transverse welding. The upper wheel may be rotated 90° and the lower wheel may be changed to weld longitudinally. (Ferranti Sciaky, Inc.)

## 16-12. UPSET WELDING MACHINES

UPSET WELDING is a resistance welding process which joins two pieces of metal over the entire area of surface contact. Upset welding is used to join the butt ends of rods and bars. The heat is obtained from resistance to the flow of electric current through the entire area of surface contact. Force is applied before heating is started and is maintained during the heating period. At the end of the heating period, a large force is applied to upset the metal at the joint. The metal cools and forms a weld. The finished weld produces an enlarged metal area at the weld. See Fig. 16-48. This enlargement or upset metal is usually removed so the weld is the same size as the original metal.

The machine uses the same type of transformer as a spot welding machine. The electrodes in this case are clamping dies or vises. One die is movable and one is fixed. The surfaces to be welded must be clean. Fig. 16-49 shows an upset welding machine with the machine opened for loading or unloading. Fig. 16-50 shows a flash butt welding machine. Note the operator controls. Fig. 16-51 shows the power required and the time needed for upset welding 1/2 in. (12.7 mm) square steel bars.

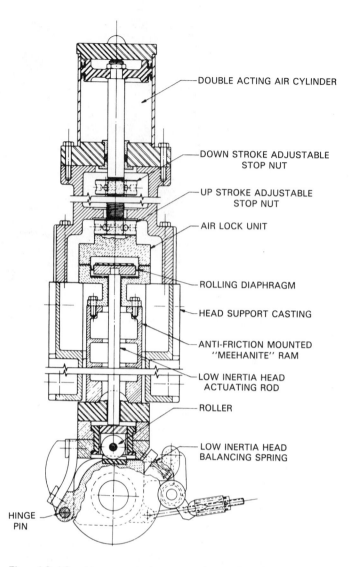

DOUBLE ACTING AIR CYLINDER

DOWN STROKE ADJUSTABLE STOP NUT

UP STROKE ADJUSTABLE STOP NUT

AIR LOCK UNIT

ROLLING DIAPHRAGM

HEAD SUPPORT CASTING

ANTI-FRICTION MOUNTED "MEEHANITE" RAM

LOW INERTIA HEAD ACTUATING ROD

ROLLER

LOW INERTIA HEAD BALANCING SPRING

HINGE PIN

Fig. 16-46. Air operated upper electrode of a universal seam welder. It is used to raise and lower the upper roller and to develop the correct welding pressure. (Taylor-Winfield Corp.)

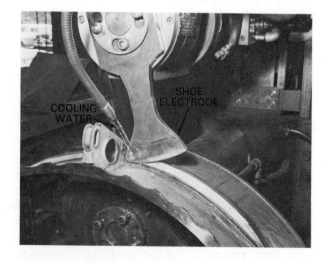

COOLING WATER

SHOE ELECTRODE

Fig. 16-47. Shoe electrode used to weld a seam close to a part in the path of a continuous seam. Note the stream of water which is cooling the electrode and the part.

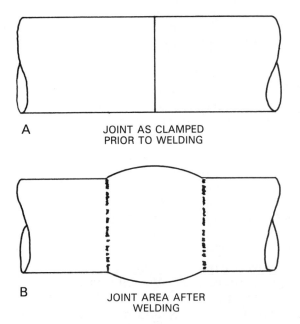

Fig. 16-48. An upset welding joint. At A, the two parts are shown prior to welding. B—shows the metal after welding. Notice the enlarged or upset metal area at the weld joint.

Fig. 16-49. An upset welding machine. The completed weld is shown mounted in the open clamping dies.

## 16-13. SPECIAL RESISTANCE WELDING MACHINES

There are numerous specially designed resistance welding machines. These machines consist of some special application of one or more of the basic type machines.

Two of the more popular models are the cross wire welding machines, shown in Fig. 16-52, and forge welding machines.

CROSS WIRE RESISTANCE WELDING MACHINES use special single electrodes or multiple electrodes which hold the wire or rod in correct alignment. The current passes through the diameters of the wires where they cross. The resistance at the contact spot creates enough heating to cause fusing of the metal. A strong welded joint is produced.

FORGE WELDING is similar to the cross wire and upset welding processes. The current heats the metals to a plastic and/or flow temperature. The metal is upset by a movement of the electrodes (press action) and the parts are welded together.

| UPSET WELDING ENERGY AND TIME CHART | | | |
|---|---|---|---|
| KW | TIME IN SECONDS | DISTANCE BETWEEN GRIPPING DIES | |
| | | in. | mm |
| 19 | 3 | 0.79 | 20.1 |
| 14 | 6 | 1.6 | 40.6 |
| 12 | 9 | 2.4 | 61.0 |
| 10 | 11 | 3.2 | 81.3 |
| 8 | 14 | 3.9 | 99.1 |
| 7.5 | 17 | 4.5 | 114.3 |
| 7 | 20 | 5.5 | 139.7 |

Fig. 16-51. Recommended upset welding energy and time table for 1/2 in. (12.7 mm) square steel bars.

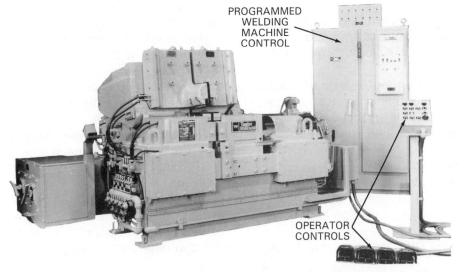

Fig. 16-50. A flash butt welding machine with operator and computer controls.     (Taylor Winfield Corp.)

Fig. 16-52. Welding cross wires, with a resistance welding machine.

## 16-14. CARE OF RESISTANCE WELDING EQUIPMENT

There are several main areas of resistance welder maintenance:
1. Mechanical.
2. Electrical.
3. Hydraulic.
4. Pneumatic.
5. Water-cooling.

The mechanical maintenance consists of:
1. Lubrication.
2. Checking moving parts for wear and alignment.
3. Checking the force applied by the electrodes against the base metal being welded.
4. Checking mechanical safety devices such as guards, machine mounting, etc.

Chapter 30 contains additional information relative to the welding shop.

The maintenance manuals for each machine should be consulted to determine the exact specifications to be used on each particular machine.

Electrical maintenance consists of:
1. Checking all electrical connections.
2. Checking primary voltage and current.
3. Checking secondary voltage and current.
4. Checking, cleaning, installing of fuses, switches, and relays.
5. Loose shunts are a frequent source of trouble. All electrical connections should be checked daily. Either a voltmeter or an ohmmeter may be used. Oscilloscopes and oscillographs are needed to make a complete analysis of the electrical equipment.

CAUTION: Use extreme caution when working with resistance welding machines. The high currents and voltages used in resistance welding can cause death.

Whenever working on the secondary of a welding machine make sure the machine is off. The circuit should also be turned off. If work must be done on the primary, disconnect the machine from the electrical power. Disconnect the circuit breaker and remove the fuses from the machine. If the machine does not have its own circuit breaker, the electrical power must be disconnected at the substation. Whenever a circuit is disconnected, place a large sign on the circuit breaker. The sign should say why the breaker is off and should include the date and the time. This precaution will prevent someone from turning the power back on and possibly killing the person repairing the welding machine. All repair work should be performed by a person knowledgeable in electrical circuits and power.

The hydraulic maintenance consists of:
1. Checking pressure.
2. Checking quantity and condition of hydraulic fluid.
3. Checking hydraulic lines and connections.
4. Checking hydraulic valves for leaks.
5. Checking hydraulic pumps and cylinders for leaks. Do not attempt to repair hydraulic leaks. Turn the equipment off and inform a supervisor or repair person. Hydraulic fluid is under very high pressure and can cause severe harm.

Pneumatic maintenance is similar to the hydraulic maintenance procedures.

The water-cooling circuit is an important component of a resistance welder. Thermometers, pressure gauges, and volume measurements are necessary for checkup purposes.

An adequate supply of cool water should be provided to each welding machine, at a minimum of 30 psi (206.8 kPa) line pressure. Cooling water should be at a temperature less than 85 °F (29 °C).

Be certain that the water hose connections are tight. The welding transformer, the ignitron tubes, the SCRs mounts, the electrode holder, and the electrodes are all water cooled. On small machines some of these may not be water cooled.

Use electrode holders which will pass at least two gallons of water per minute. Make sure the water inlet hose is connected to the electrode holder inlet. Make sure the water tube inside the electrode holder extends to the electrode to keep the electrodes cool.

The seam welding wheels and the work should be sprayed with cooling water. The water should be directed at the point where the wheel contacts the work.

The water hoses should be free of any deposits that might reduce the water flow. Make sure none of the hoses or connections leak because water in the machine could cause an electrical short.

## 16-15. TEST YOUR KNOWLEDGE

Write your answers on a separate sheet of paper. Do not write in this book.

1. Resistance welding requires _____ amperage; _____ voltage electrical energy.
2. What happens to the voltage and the current when it goes through a step-down transformer?
3. The current in the secondary of a resistance welder is (AC or DC).
4. What is a typical duty cycle for a resistance welding machine?
5. On what period of time is the duty cycle based?
6. Name three types of electrode force systems used on a resistance welding machine.
7. What force system can supply the highest force?
8. What are the five variables in resistance welding?
9. List the three times required to make a complete spot weld schedule.
10. Name two ways to increase the current on a resistance welding machine.
11. What is required to make an ignitron tube conduct current?
12. What parts of a resistance welding machine are water cooled?
13. When two piece electrodes are used, what is the name of each piece?
14. The size of the pieces that can be welded by a resistance spot welding machine is limited by two dimensions which are called _____ _____ and _____ _____.
15. What is the advantage of using a stored energy machine?
16. What type of machine is used for projection welding?
17. What are the main parts of a resistance seam welding machine?
18. When seam welding, what is the advantage of using a knurl or friction drive rather than a gear drive?
19. List two processes that use the following principle: Two metals are brought together and heated by passing electrical current through them. When they have been heated, a large force is applied to join the pieces.
20. If a leak in the hydraulic or pneumatic system is discovered, what should be done?

Welding robots used in production. Top photo shows a GMAW on tubing. in the bottom photo, fixtured parts are welded with GMAW as they rotate past the robot. (GMFanuc Robotics Corp.)

# 17 RESISTANCE WELDING

RESISTANCE WELDING is based upon the fundamental principle that when an electrical current is passed through a metal, the resistance of the metal to this electrical flow heats the metal. The majority of the heat is developed where the two pieces of metal to be welded are in contact. By applying sufficient current, very high temperatures can result. When these temperatures reach the fusion (melting) temperature, welding occurs.

The term resistance welding includes a variety of welding applications. The resistance welding processes include spot welding, projection welding, seam welding, upset welding, and others. See Headings 1-14, 1-15, 1-16, and 1-17.

Resistance welding has a number of advantages. It is fast, there is very little warpage of the metal, and the process can be accurately controlled. This type of welding is particularly well suited to all forms of automatic production.

## 17-1. PRINCIPLES OF ELECTRIC RESISTANCE WELDING (RSW)

When two pieces of metal are in contact, the area where they touch has a high resistance to electrical current. Due to surface roughness, the two or more pieces of metal to be welded cannot have their surfaces in perfect or complete contact. The roughness of the metals that form the contacting surfaces offer the highest resistance to current flow. When an electrical current is passed across the surfaces in contact, heat is generated. If enough current is used for a long enough time, the metal surfaces heat until they become molten. If the two pieces are pressed together while their surfaces are molten and allowed to cool, the pieces will fuse into one piece.

A resistance welding machine is basically an electric transformer. It operates from an alternating current circuit. Resistance welding requires very high current (amperes) at a relatively low voltage. This requires the resistance welding machine to have a step down transformer. A step down transformer decreases the voltage supplied and increases the current supplied. This is accomplished by having many turns or windings in the transformer primary. The secondary winding will ordinarily have only one turn. Sometimes two or three turns are used if the secondary leads are long, such as in gun welders. Fig. 17-1 illustrates an elementary electric circuit for a resistance welding machine.

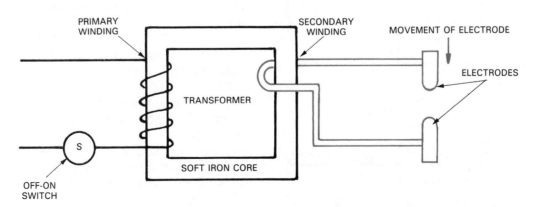

Fig. 17-1. A basic resistance spot welding machine electrical circuit. Note that the primary winding has many more turns than the secondary winding. This is a step down transformer.

## 17-2. TYPES OF ELECTRIC RESISTANCE WELDING

There are several types of welding processes and equipment based on the resistance welding principle. Some of the more common types are:
1. Spot welding.
2. Gun welding.
3. Projection welding.
4. Seam welding.
5. Upset welding.
6. Flash welding.
7. Metal foil welding.
8. Metal fiber welding.
9. High frequency resistance welding.

All of these operations are fundamentally the same, but the preparation of the metal and the construction of the machine may be different.

## 17-3. RESISTANCE SPOT WELDING PRINCIPLES

One way to join sheet metal parts is to drill holes and either rivet or use machine or sheet metal screws to fasten the parts together. Another way to join sheet metal parts is to spot weld them together. Two or more pieces of metal can be spot welded together. Two pieces are most commonly joined together.

Spot welding is the most common resistance welding process. Two pieces of metal are welded together with a small nugget of fused metal. Refer to Heading 1-14. The process consists of lapping two pieces of metal and clamping them between two electrodes. A current is passed between the two electrodes. A small molten spot is formed.

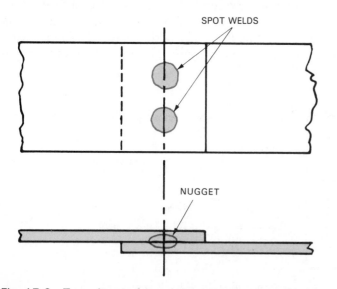

Fig. 17-2. Two pieces of metal, lapped and spot welded in two places. Note the nugget where the fusion takes place between the two pieces.

The current stops, but the electrodes continue to hold the metal to allow the spot to solidify. The two pieces of metal are now fused together by a SPOT WELD or WELD NUGGET. Fig. 17-2 shows a diagram of spot welds. A cross section of a spot weld nugget is shown in Fig. 17-3.

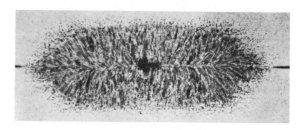

Fig. 17-3. A cross section of a spot weld nugget. The material is .094 in. (2.38 mm) thick mild steel. (Taylor-Winfield Corp.)

In resistance spot welding there are five variables that must be controlled. These variables are:
1. Time.
2. Current.
3. Force.
4. Electrode contact area.
5. Machine selection.

The selection of the type of machine is covered in Heading 17-4. Refer also to Chapter 16. The proper settings for the first four variables are discussed in Heading 17-6.

The time required to make a spot weld has three separate timing periods. These are:
1. Squeeze time.
2. Weld time.
3. Hold time.

The SQUEEZE TIME is the time required for the electrodes to close on the metal and apply the proper pressure. The WELD TIME is the time that the current flows and heats the metal. The HOLD TIME is the time after the current stops when the pressure is still applied to allow the molten metal spot to solidify. These three times make up ONE WELD SEQUENCE or a WELDING SCHEDULE.

One additional time is sometimes used: the off time. The OFF TIME is the time between the end of one weld sequence and the beginning of the next weld sequence.

Each of these times is set on the controller or control panel. The times are set in numbers of cycles or Hertz. One cycle or Hertz is 1/60th of a second. Examples of resistance welding controllers are shown in Figs. 16-10 and 16-12.

The current is an important variable. The amount of current used can be changed in one of two ways:

1. Change the tap setting on the welding machine.
2. Change the percent heat setting on the welding machine controller.

The TAP SETTING is used to make large changes in the current settings. A given tap setting has a range of currents. To vary the current within a given range, the PERCENT HEAT SETTING is changed. The higher the percent heat setting, the larger the current measured in amperes. For example: a given tap setting may be able to supply between 4000 amps and 10,000 amps. If the percent heat setting is set at 75 percent heat, the current will be approximately 4000 amps plus .75 × the current range. The current range in this case is: 10,000 − 4,000 = 6,000. Therefore 4000 amps + (.75 × 6000 amps)

= 4000 amps + 4500 amps

= 8500 amps.

To operate a spot welding machine, the operator must safely clean the electrodes, turn on the cooling water, and set the tap switch on the transformer to the correct current setting. The welding operator must also set the times and percent heat on the controller, and set the desired pressure.

The welding operator places the two pieces of metal to be welded on the lower (stationary) electrode. Each weld sequence is started by pressing the foot pedal or starting switch.

On electronically programmed machines, the weld schedule is completed automatically. At the end of the weld schedule, the two pieces are welded together and the electrodes separate. The part is removed and new pieces are loaded to begin a new weld schedule.

When using a manual spot welding machine, the welding operator controls the times of the weld schedule. The operator first pushes the foot pedal part way. The electrodes close on the metal to be welded and apply the proper force. This is the SQUEEZE TIME. The operator then pushes the pedal a little bit further. This closes an electrical contact switch. The current flows and forms a molten nugget. This is the WELD TIME. The operator then pushes the foot pedal all the way down. This releases the electrical contact, but pressure is still being applied. This is the HOLD TIME when the molten metal becomes solid. Finally, the operator releases the pressure on the foot pedal. The electrodes separate and the welded metal is removed. A schematic of a manual resistance welding machine is shown in Fig. 16-7.

Automatically controlled welding machines are shown in Figs. 17-4 and 17-5. These machines have a pneumatic (air) force system. There is a valve to adjust the air pressure and a valve to turn

the cooling water on. The machine illustrated in Fig. 17-4 has a foot pedal to control the weld operation.

Fig. 17-4. A pneumatic force system on an automatically controlled spot welding machine. Notice the foot control pedal which is used to begin the welding process. (Acro Automation Systems, Inc.)

The machine is Fig. 17-5 has two palm switches which are used to initiate a weld. Both switches on this machine must be pushed in order to make a weld. With this setup, there is no danger of the welding operator getting a hand in between the electrodes. The dual palm switches are used to satisfy the Occupational Safety and Health Administration (OSHA) requirements.

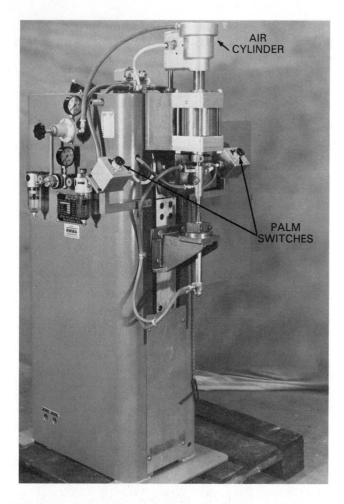

Fig. 17-5. An electric spot welding machine with a pneumatic electrode force system. Dual palm switches are used to initiate the weld schedule. The dual palm switches satisfy OSHA requirements.
(Acro Automation Systems, Inc.)

## 17-4. SPOT WELDING MACHINES

There are many components and systems that make up a resistance spot welding machine. Three basic designs are used. These are the:
1. Rocker arm machine (see Fig. 16-8).
2. Press type machine (see Fig. 16-9).
3. Portable type machine. The parts of a portable type or "gun" welding machine are labeled in Fig. 17-6.

Spot welding machines use different types of force systems. The different types of force systems in use are:
1. Mechanical leverage.
2. An electric motor and rotating threads.
3. Pneumatic (air) pressure.
4. Hydraulic (fluid) pressure.

The pneumatic types are the most common. Refer to these force systems that are described in Heading 16-3.

Spot welding machines can be electronically or

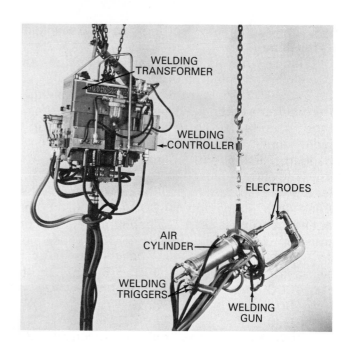

Fig. 17-6. This portable "gun" welding machine uses an air cylinder to apply the electrode force. The machine also has three triggers. Each trigger performs a different weld schedule. (Progressive Machinery Corp.)

manually controlled. The majority are electronically controlled.

Spot welding machines are also classified according to the type of electrical power the machine uses. There are three types:
1. The single-phase machine.
2. The three-phase machine.
3. The stored energy machine.

Refer to Chapter 16 for additional information.

Two other welding machine variables are their KVA rating and their duty cycle rating. Each of these variables is important when identifying or selecting a spot welding machine. There is no machine that is best for all applications. Each machine has advantages and disadvantages. Select the best machine for a specific job. The most common spot welding machine is an automatically controlled, single-phase welding machine which uses a pneumatic force system. Press type and portable type machines are both very common. The KVA ratings vary from 10 KVA to 150 KVA. The most common duty cycle is 50 percent.

## 17-5. PORTABLE SPOT WELDING MACHINES

A PORTABLE SPOT WELDING MACHINE can be moved and positioned on the parts to be welded. Portable spot welding machines are also known as GUN WELDING MACHINES. The gun welding machine is moved to the parts to be welded instead of bringing the parts to the welding machine. Gun welding machines are popular in sheet metal fabrication, such as automobile bodies, washers, refrigerators, and microwave ovens. The electrodes are operated either hydraulically or pneumatically. The timing and current are set on the controller. The pressure is set by adjusting an air or hydraulic pressure valve.

The welding operator selects the place to weld and places the nonmoving electrode against the metal. The operator then pushes the start switch (trigger) and the controller performs the weld schedule. When the weld is completed and the electrodes separate, the gun is moved to the next weld location.

## 17-6. SPOT WELDING SETUP

There are five variables in resistance welding:
1. Time.
2. Current.
3. Electrode force.
4. Electrode contact area.
5. Machine selection.

Machine selection involves choosing the best machine for the given application. The welding operator is often required to set up the weld schedule. The operator must know what type of material is being welded and the thickness of the metal. Refer to Heading 16-8.

When spot welding low carbon steel, the equations in Fig. 17-7 can be used to get an approximate value of the different variables. For example, if two sheets of .060 in. (1.5 mm) thick low carbon steel are to be welded, the contact tip diameter would be $.1 + (0.60 + 0.60) = .220$ in. $(2.54 + [1.5 + 1.5] = 5.54$ mm). The weld time to weld a .040 in. (1.02 mm) and a .060 in. (1.5 mm) piece would be $60 \times (0.40 + .060) = 6.0$ cycles. $(2.36 \times [1.02 + 1.5] = 5.9$ or 6 cycles.) The current to weld a .020 in. (.51 mm) and a .040 in. (1.02 mm) is $100,000 \times (0.20 + .040) = 6,000$ amps or $3937 \times (.51 + 1.02) = 6024$ amps. The elec-

trode force to weld two pieces of .050 in. (1.27 mm) is $5000 \times (0.50 + .050) = 500$ lbs. or $876 \times (1.27 + 1.27) = 2225$ Newtons. Additional information can be obtained from the Resistance Welder Manufacturers' Association (RWMA) Resistance Welding Manual.

Now that the desired values are known, the welding operator must use the values to set the welding machine.

Only certain tip diameters are manufactured. Select a tip diameter that is close to the calculated value.

The welding time is set on the controller in cycles. The operator selects the time by adjusting the thumb wheel controls or by pushing buttons on the controller. The buttons on a controller are used the same way that push buttons are used on a hand held calculator. A push button controller is shown in Fig. 17-8. Three times must be set: the squeeze time, the weld time, and the hold time. The off time is optional.

To set the current requires prior knowledge about the particular welding machine, trial and error, or special equipment. A welding operator who is familiar with a welding machine will know what transformer tap setting and percent heat settings are required to weld two pieces of metal.

The TRIAL AND ERROR METHOD involves welding two pieces together, tearing them apart, and making an adjustment in the current setting. First select a transformer tap setting. Use a higher setting for thicker material and a lower setting for thin material. Set the percent heat control at 70 percent. This is a starting point to make a weld.

Examine the two pieces after they are torn apart. If the weld size (nugget) is too small, more current is required. Increase the transformer tap setting or the percent heat and make two more weld samples. Continue to make test samples and tear them apart until the correct size of weld is obtained.

| RESISTANCE SPOT WELDING LOW CARBON STEEL UP TO 1/8 IN. (3.2 mm) | | | | |
|---|---|---|---|---|
| Variable | Equation (English) | Units of Measure | Equation (Metric) | Units of Measure |
| Contact Tip Dia. | .1 + (total sheet thickness) | inches | 2.54 + (total sheet thickness) | mm |
| Weld Time | 60 × (total sheet thickness) | cycles | 2.36 × (total sheet thickness) | cycles |
| Current | 100,000 × (total sheet thickness) | amps | 3,937 × (total sheet thickness) | amps |
| Electrode Force | 5000 × (total sheet thickness) | lbs. | 876 × (total sheet thickness) | Newton |
| Weld Size | The weld size is the same size or slightly smaller than the contact tip diameter. | | | |

Fig. 17-7. A table of equations used to approximate the values of the variables in resistance spot welding low carbon mild steel. The total sheet thickness is the combination of the two thicknesses being welded.

Fig. 17-8. A controller which uses push buttons to set the time of the weld schedule and the percent heat. (Pertron Controls Corp.)

When molten metal squirts out from a spot weld, it is called an EXPULSION WELD. If when making test welds an expulsion weld occurs, the current is too high. In this case, decrease the tap setting or the percent heat. Weld two more pieces

Fig. 17-9. A current analyser and pickup coil are used to monitor the welding current and weld time. (Duffers Associates, Inc.)

together and pull them apart. Again, make any necessary change in the current to increase or decrease the weld size. When the weld size equals the tip diameter, the proper current is being used.

A third way to set and adjust the current is to use a CURRENT ANALYSER. A coil is placed around the electrode or electrode holder. The coil is connected to the current analyser. When a weld is made, the current analyser will display the current in kiloamps (1 kiloamp = 1000 amps). The tap setting or percent heat is changed to increase or decrease the welding current. A current analyser and a pick up coil are shown in Fig. 17-9. A smaller, portable current analyser is shown in Fig. 17-10.

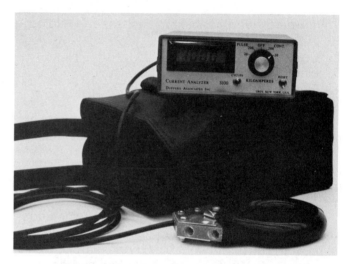

Fig. 17-10. A portable current analyser. This unit can measure currents up to 200,000 amps. (Duffers Associates, Inc.)

Fig. 17-11. Two force gauges used to measure the force applied to resistance welding electrodes. One gauge reads 2000 lbs. maximum (8896 N), the larger gauge reads up to 10,000 lbs. (44482 N). (Tuffaloy Products, Inc.)

The final variable to be set is the electrode force. The force is changed by adjusting a valve. The valve adjusts the hydraulic or pneumatic (air) pressure that is applied to the electrodes. To set the force, a FORCE GAUGE is placed between the electrodes. The electrodes are forced to close over the force gauge. The gauge will read the force applied. Any change required in the force is made and another force reading is taken. This is repeated until the proper force is obtained. Two force gauges are shown in Fig. 17-11.

After setting all of the variables, the spot welding machine is properly set and should make good quality spot welds.

The machine may, however, make poor welds. The metal may be distorted, as shown in Fig. 17-12, or it may have large indentations. This may be caused by improper electrode alignment.

BLOW-HOLES and CAVITIES can occur from improperly set controls, too short a hold time, or improper positioning of parts with reference to the tips. See Fig. 17-13. Other physical conditions such as dirt, burrs, etc., on the surface of the metal result in a quick, excessive concentration of welding current. If the heat developed is not absorbed or conducted to the surrounding metal a blow-hole may form.

Tips which have parallel faces, but whose centers are not aligned, have a reduced effective tip area as shown in view A of Fig. 17-14. When the welding tip surfaces are not parallel, pressure

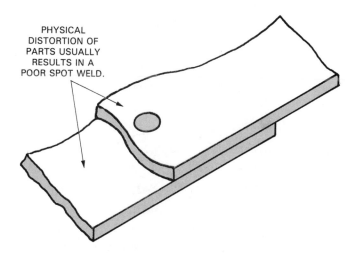

PHYSICAL DISTORTION OF PARTS USUALLY RESULTS IN A POOR SPOT WELD.

Fig. 17-12. Sheet metal warping or distortion will cause improperly made spot welds.

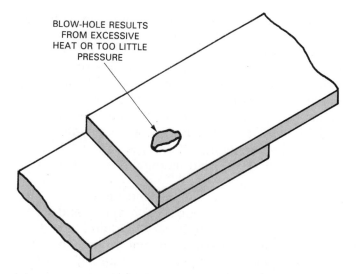

BLOW-HOLE RESULTS FROM EXCESSIVE HEAT OR TOO LITTLE PRESSURE

Fig. 17-13. A faulty spot weld which results in a blow-hole in the sheet metal.

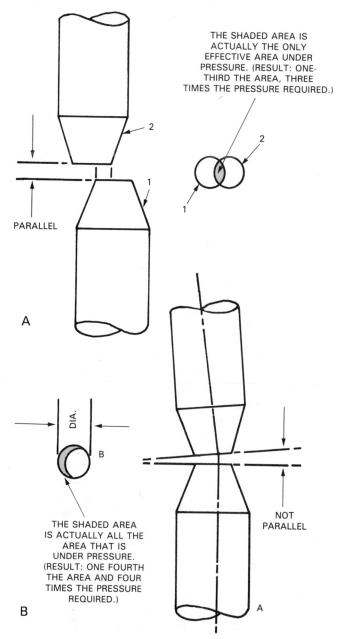

THE SHADED AREA IS ACTUALLY THE ONLY EFFECTIVE AREA UNDER PRESSURE. (RESULT: ONE-THIRD THE AREA, THREE TIMES THE PRESSURE REQUIRED.)

PARALLEL

A

THE SHADED AREA IS ACTUALLY ALL THE AREA THAT IS UNDER PRESSURE. (RESULT: ONE FOURTH THE AREA AND FOUR TIMES THE PRESSURE REQUIRED.)

B

NOT PARALLEL

Fig. 17-14. Effects of using electrodes which are misaligned. A—These electrodes are off center. B—These electrodes are aligned, but not parallel. In either case, the pressure on the contact area is excessive.

and current are confined to a fraction of the proper area as shown in view B of Fig. 17-14. When the tips are misaligned, the pressure (force/area) on the metal between the tips is excessive. The metal surface may be deeply indented by this excessive pressure. The electrodes must be realigned. With the electrodes properly aligned, the welding machine may now be used to make consistent, high quality spot welds.

## 17-7. PROPER RESISTANCE WELDING PRACTICES

The welding machine and the controller must be properly set before any welding is done. The set-up of the resistance welding machine has been described in Heading 17-6.

Material to be welded is placed between the electrodes. Each welding process is initiated by pressing a foot pedal or a palm switch. Squeeze time, weld time, and hold time are preset and provided by the welding controller. After the pieces are welded together, the electrodes separate.

The metal after welding is very hot and should be handled with care. Gloves should be worn to protect the operator from burns. Safety glasses must be worn because an occasional expulsion weld throws a lot of molten metal in every direction.

Some maintenance is required during spot welding due to the heat developed. This heat is necessary to form the weld; however, the heat damages the electrodes. Electrodes can be used to make 2000 spot welds or more on mild steel. After this, the electrode begins to flatten out which increases the electrode contact area. To keep the electrode contact area constant, it is necessary to reshape the electrode. Reshaping is also called DRESSING the electrode. Fig. 17-15 shows hand and power tools used to dress the electrodes. These tools have cutters which clean and shape the electrode.

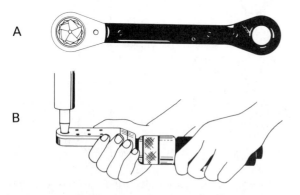

Fig. 17-15. Electrode cleaning and dressing tools. A—Manual dressing tool. B—Power dressing tool. (CMW, Inc.)

It is also necessary to replace the electrodes or electrode tips, especially if they cannot be dressed. A tool for removing tapered electrode tips is shown in Fig. 17-16. Some electrode holders have built in electrode ejectors. See Fig. 16-30. A rubber or lead hammer is used to place new electrode tips in the tapered shank of the electrode holder. A steel hammer should not be used because it could damage the electrode tip or the taper.

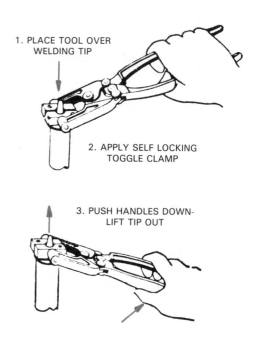

1. PLACE TOOL OVER WELDING TIP

2. APPLY SELF LOCKING TOGGLE CLAMP

3. PUSH HANDLES DOWN-LIFT TIP OUT

Fig. 17-16. A tool for removing tapered electrode tips.

## 17-8. SPOT WELDING ALUMINUM AND STAINLESS STEEL

The spot welding of aluminum and stainless steel differ from the welding of mild steel. Spot welding aluminum requires more current for shorter times than mild steel. Aluminum and its alloys have a tendency to crack while they are cooling. To prevent cracking, additional pressure is applied. The increase in pressure is called a FORGE FORCE. Only certain pressure systems are capable of applying this forge force. A weld schedule used to weld aluminum can be seen in Fig. 17-17.

Stainless steel is also welded faster than mild steel. Stainless steel cools very rapidly and forms martensite. Martensite is very hard and brittle. To improve the properties of the martensite, the finished weld is tempered. Tempering involves heating the metal to a prescribed temperature and then allowing it to cool slowly. (See Heading 27-8

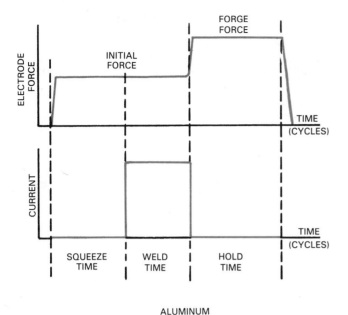

Fig. 17-17. Weld schedule used to weld aluminum. The forge force is applied during the hold time for aluminum.

for more information on martensite and tempering). In spot welding, tempering is performed by passing additional current through the spot weld after it has cooled. This tempering heat lasts for three - six cycles. A complete weld schedule used to weld stainless steel is shown in Fig. 17-18.

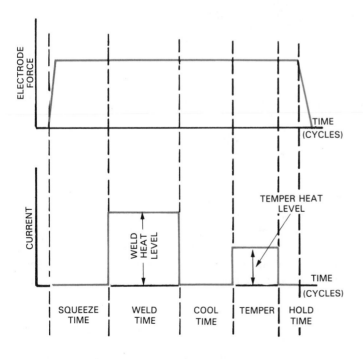

Fig. 17-18. Weld schedule used to weld stainless steel. The stainless steel weld is tempered by applying additional heat after the cool time.

## 17-9. PROJECTION WELDING (PRW)

PROJECTION WELDING requires the forming of projections or bumps on one piece of metal. The projections are accurately formed in precise locations on the metal by a special set of dies. After the projections are formed, the raised portions on one piece are pressed into contact with another piece. Fig. 17-19 shows a projection weld setup. Current is passed through the two pieces. The current is concentrated at the projections. These points heat and form a spot weld. Refer to Heading 16-9.

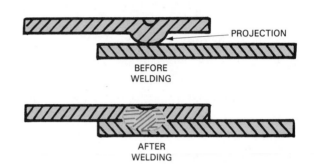

Fig. 17-19. A schematic drawing of metal before and after it is projection welded. Some of the depression formed when the projection is stamped, remains on the top surface of the finished weld.

An advantage of projection welding is that it locates the welds at certain desired points. Also, several spots may be welded at the same time. However, the tooling required makes this method practical only when high production is planned.

Macrographs of a projection weld in various stages of completion are shown in Fig. 17-20. Each successive photograph illustrates the effect of the number of cycles of the electrical current flowing through the weld.

One use of projection welding is to place nuts, screws, or pins on the surface of a part. The nuts, screws, and pins have projections formed in them so they can be projection welded. Some examples are shown in Fig. 17-21. Special electrodes are needed to weld some nuts, screws, and pins.

## 17-10. RESISTANCE SEAM WELDING (RSEW)

SEAM WELDING produces a continuous or intermittent seam. Refer to Heading 1-17. The weld is usually near the edge of two overlapped metals. Two wheel electrodes travel over the metal, and current passes between them. Refer to Heading 16-11. The current heats the two pieces of metal to the fusion point. See Fig. 17-22.

The current can be continuous or intermittent.

COLD SET DOWN— NO CURRENT— ORIGINAL PROJECTION HEIGHT OF .050 DECREASES .010 CAUSED BY .008 INDENTATION AND .002 PROJECTION COLLAPSE. SHEET SEPARATION .040.

1-CYCLE WELD TIME—PRESSURE WELD FORMING WITH TENSILE SHEAR STRENGTH OF 1050 LBS.— SHEET SEPARATION .010.

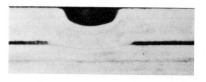

2-CYCLE WELD TIME—STRENGTH 1750 LBS.—SHEET SEPARATION OF .007.

4-CYCLE WELD TIME—STRENGTH 2300 LBS.—SHEET SEPARATION .005.

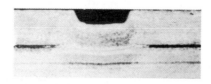

6-CYCLE WELD TIME—STRENGTH 2600 LBS.—SHEET SEPARATION .005.

8-CYCLE WELD TIME—STRENGTH 3000 LBS.—SHEET SEPARATION .002.

10-CYCLE WELD TIME—STRENGTH 3125 LBS.—SHEET SEPARATION .002. THIS IS LAST STAGE OF WELDING WITHOUT NUGGET FORMATION.

12-CYCLE WELD TIME—STRENGTH 3100 LBS.—SHEET SEPARATION .001. NUGGET FORMATION HAS STARTED. NUGGET DIAMETER .150.

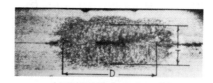

14-CYCLE WELD TIME—STRENGTH 3200 LBS.—SHEET SEPARATION .0. NUGGET DIAMETER .160.

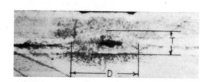

16-CYCLE WELD TIME—STRENGTH 3700 LBS.—SHEET SEPARATION .0. NUGGET DIAMETER .160. THIS IS RECOMMENDED WELD TIME.

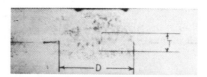

20-CYCLE WELD TIME—STRENGTH 4000 LBS.—NUGGET DIAMETER .300. THIS LAST STAGE INDICATES INCREASE IN NUGGET SIZE WITH INCREASED WELD TIME.

Fig. 17-20. Samples of projection welds as the weld progresses from 0 cycles to 20 cycles weld time. (1 cycle = 1/60 second)   (Taylor-Winfield Corp.)

CONTINUOUS CURRENT produces a continuous seam. INTERMITTENT CURRENT produces spots which can be timed to overlap.

Overlapped spots produce a gas and liquid leakproof lap joint. Examples of continuous and overlapping seam welds are shown in Fig. 17-23.

Another form of seam welding is called BUTT SEAM WELDING. It is used to butt weld a longitudinal seam in pipe. Electric current heats the metal to the molten condition just before it passes between a set of rollers. The two rollers press the pipe edges together, and the metal is welded. The rollers also hold the welded joint in position while it cools to a safe temperature. This resistance welding method uses high frequency current to heat the metal. Fig. 17-24 shows a longitudinal seam weld being made on a pipe.

Seam welding machines must have a higher duty cycle than a spot welding machine. A machine which produces a continuous or overlapping spot weld seam must be rated at a 100 percent duty cycle.

## 17-11. UPSET WELDING (UW) PRINCIPLES

UPSET WELDING (UW) involves clamping two pieces of metal which are to be welded in two separate clamps. One clamp is stationary; the other is movable. Metal to be upset welded must be very clean and aligned properly. The clamps

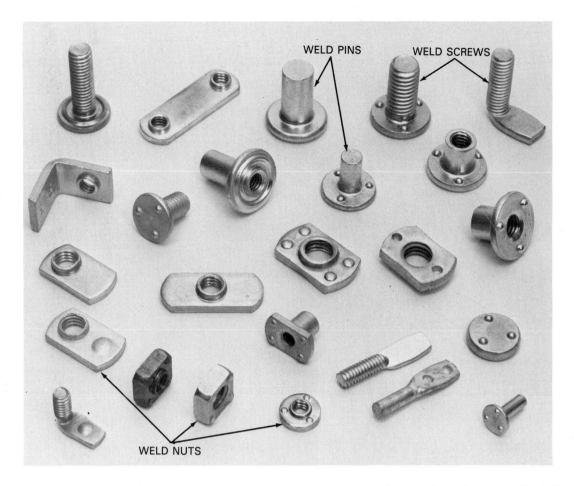

Fig. 17-21. Weld nuts, screws, and pins with projections on them so they can be projection welded. Note the circular projection on some of these pieces. (The Ohio Nut and Bolt Co.)

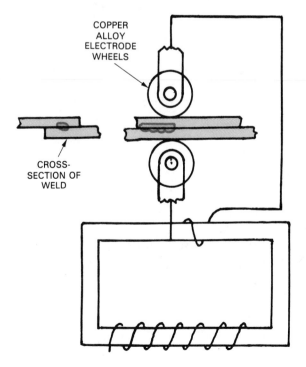

Fig. 17-22. A schematic drawing of metal being seam welded.

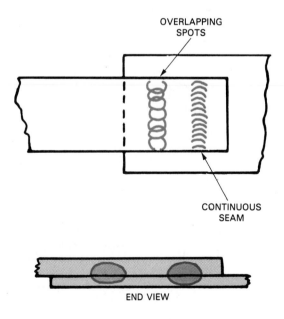

Fig. 17-23. This sketch shows metal that has been seam welded. A seam weld is made with wheel type electrodes. If the current to the electrodes is turned off and on, a series of overlapping spots form the seam. An uninterrupted flow of current to the electrodes will form a continuous seam.

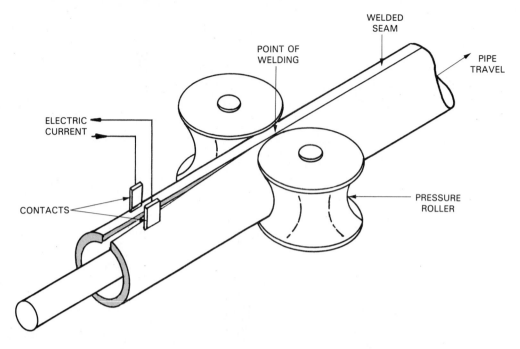

Fig. 17-24. Butt seam welding a longitudinal seam on a pipe. The contacts heat the metal just before it enters the rollers. The rollers squeeze the metal together to fuse the joint. The rollers also hold the metal while it cools.

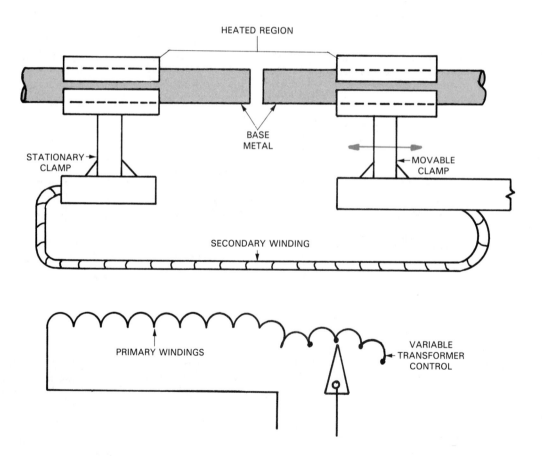

Fig. 17-25. A diagram of an upset welding machine, showing the transformer primary and secondary, and the clamping mechanism. Only the region between the clamps is heated.

and the electrical wiring of an upset welding machine are shown in Fig. 17-25. Refer to Heading 16-12.

In upset welding, the two pieces are forced together, and a large current is passed from one clamp to the other. The metal between the two clamps is heated. A majority of the heat is concentrated at the place where the two metal pieces touch. The ends of the metal are heated to the fusion (melting) temperature. After reaching the fusion temperature, the two pieces are pressed together (upset). This upsetting occurs while the current is flowing and after the current is turned off. The metals join together and upon cooling, become one piece. After a short cooling period, the clamps are released and the part is removed from the welder. Fig. 17-26 shows the final upset weld.

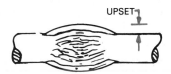

Fig. 17-26. A finish upset weld. Note that the diameter of the rod is increased (upset) in the area of the joint.

## 17-12. FLASH WELDING (FW)

FLASH WELDING is a resistance welding process generally used to weld the butt ends of two pieces. Refer to Heading 1-16. The end surfaces of the two pieces are not cleaned or machined as in upset welding. The parts are held in two clamps. One clamp is stationary, and one clamp is movable. See Fig. 17-27.

When the weld process is started, the butt ends of the two pieces are brought into a light contact. Heat for welding is created when a high current is passed from one piece to the other. The high resistance of the uneven contacting surfaces causes these areas to heat. Because the surfaces are rough and uneven, small arcs occur across the contact surfaces. This action is called FLASHING. These small arcs also heat the contacting surfaces. During the arcing period, the parts are slowly and continuously brought together under a light pressure. When the surfaces are molten, the current is stopped. At this time a very high pressure is applied usually by an air or hydraulic cylinder or by a motor driven cam. The pressure squeezes the parts together and forces some of the molten, unclean metal on the contacting surfaces to be forced outward. Fig. 17-28 shows two pieces of metal before and after they have been flash welded. The parts fuse together, and the parts are held in position for a short time to allow them to cool.

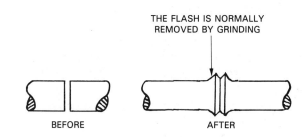

Fig. 17-28. Metal before and after it has been flash welded. Note that the enlarged area is known as "flash."

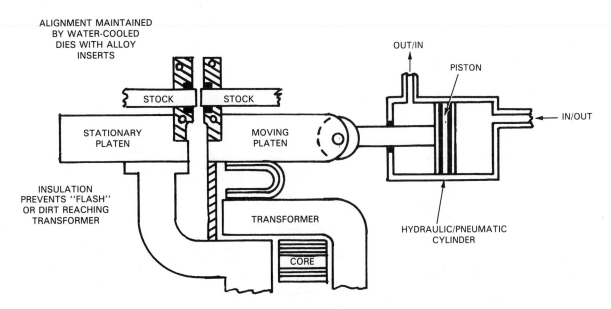

Fig. 17-27. Schematic drawing of a resistance flash welder. Note that the heated stock is forced together rapidly by a hydraulic or pneumatic (air) cylinder.

After the weld is completed the clamps are released and the parts removed. Fig. 17-29 shows a flash welder in operation.

Two advantages of flash welding over upset welding are flash welding is faster, and the metal surfaces do not have to be cleaned before welding.

Fig. 17-29. A flash welder in operation.
(Airmatic-Allied, Inc.)

## 17-13. PERCUSSION WELDING

PERCUSSION RESISTANCE WELDING heats the metals to be welded over their entire surface area by an arc. The arc is a short pulse of electrical energy.

The power supplies for percussion welding are of two types:
1. Capacitor discharge.
2. Magnetic force which uses a transformer.

The parts to be welded are held apart by a small projection, or one part is moved toward the other. The energy in the capacitors is discharged and an arc heats the parts. Pressure is applied very rapidly (percussively) during or after the current flow, thus the name PERCUSSION WELDING.

This process can join parts of equal or unequal cross section. Examples include joining wires end to end, and welding wires to flat surfaces. This process is used in the electronics industry.

## 17-14. METAL FOIL WELDING

Metal foil welding is a patented process. As shown in Fig. 17-30, sheet steel may be butt welded with this process. The sheets to be welded are placed between wheel electrodes, and the joint is covered both top and bottom with a thin metal foil approximately .010 in. (.254 mm) thick. The foil is the same material as the metal being joined.

The foil tends to concentrate the welding current in the immediate area of the weld joint. The weld has a slightly raised bead as compared to a slight indentation from the usual resistance welding operation. Fig. 17-31 shows the weld in its various stages of formation.

## 17-15. METAL FIBER WELDING

A METAL FIBER WELD is usually a lap joint resistance weld. Metal fiber sheets are formed by FELTING very tiny filaments of the metal to be used. The sheets are like a thin sheet of felt cloth.

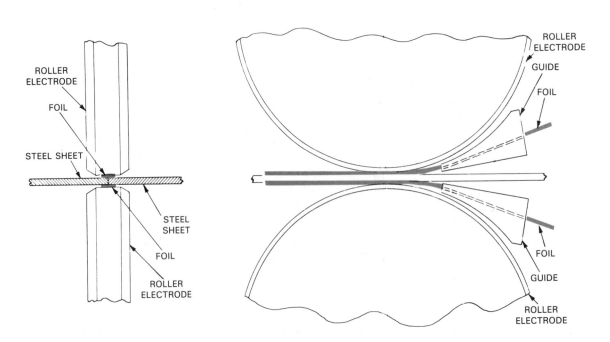

Fig. 17-30. Metal foil welding a butt joint. The foil serves to concentrate the weld current and to add reinforcement.

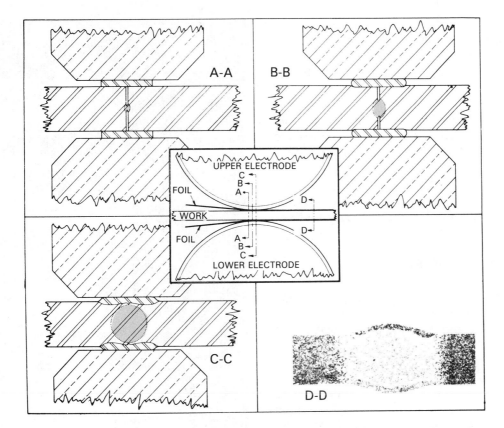

Fig. 17-31. Fusion action during the foil butt welding process; AA—Start of weld. BB—Nugget is approximately one half developed. CC—Weld approximately complete. DD—This is an actual sample cross section of a completed weld etched and macrophotographed.

A strip of this felted fiber is then placed between the two pieces of metal to be joined. Fig. 17-32 shows a cross section of the metal fiber and welding electrodes prepared to make a metal fiber resistance weld.

Metal fiber welding is applicable to a variety of resistance welding jobs. The metal fiber used may be impregnated with various metals such as copper, brazing metals, silver, or an alloy. Such fibers may be used to join two different metals together. Copper may be joined to stainless steel. Other similar joints may be made using metal fiber welding.

When resistance welding using metal fiber, less electrode pressure is required. The fiber offers a greater resistance to the flow of current than the solid metal so it heats first, and to a higher temperature. The high temperatures at the surfaces of the metal to be joined assist in forming a good resistance weld. Also, since a lower electrode pressure is required, less indentation of the metal will be made.

## 17-16. HIGH FREQUENCY RESISTANCE WELDING

When extremely high frequency current flows through a conductor, it flows at or near the surface of the metal and not through the entire thickness.

This phenomenon of high frequency electricity has been used to weld sections as thin as .004 in. (.102 mm).

Low amperage current with frequencies as high as 450,000 cycles per second are used. This high

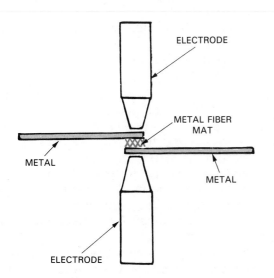

Fig. 17-32. The setup for a metal fiber resistance weld. The metal fiber mat is placed between the base metals before they are welded.

frequency heats the surface of the metal.

When pressure is applied to force the metal parts together, an excellent fusion weld occurs at the surfaces of the parts. This process is used to weld longitudinal seams in pipe, as shown in Fig. 17-24.

Copper has been welded to steel, and alloy steel welded to mild steel by this method. Exotic metals can also be welded in an inert gas atmosphere.

The electronic equipment needed for this high frequency resistance method of welding is complicated, and special training is needed to become a competent technician for maintaining, adjusting, and servicing these machines.

## 17-17. FEEDBACK CONTROL

To improve the quality of spot welds, feedback control and monitoring systems are being used. A monitoring system measures and displays the welding variables. The welding operator must interpret the results and manually make changes to obtain the desired weld quality. A feedback controller also monitors the welding variables. The feedback controller changes the welding variables electronically to keep a constant weld quality.

The variables which can be monitored are the voltage, current, and acoustic emission. Using the voltage and current, the resistance of the weld can be calculated as $R = V/I$. ($R$ = resistance, $V$ = volts, and $I$ = current).

Another method of examining the resistance welding process is to monitor the acoustic (sound) emission during the weld. ACOUSTIC EMISSION is sound that is emitted from the metal during welding. The sound is changed into an electrical signal. Fig. 17-33 shows some of the typical sounds given off during a resistance spot weld.

The information from the acoustic emission and other variables is used to maintain a desired spot weld quality. The weld can be stopped if expulsion occurs. A feedback controller which monitors acoustic emissions is shown in Fig. 17-34.

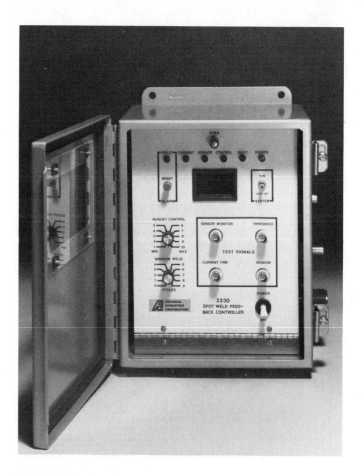

Fig. 17-34. A spot weld feedback controller used to maintain good quality spot welds by monitoring acoustic emissions. (Physical Acoustic Corp.)

## 17-18. REVIEW OF SAFETY IN RESISTANCE WELDING

Resistance welding and resistance welding equipment, if properly handled, are very safe. The greatest dangers are:
1. Flying sparks and molten metal.
2. Electrical shock.
3. Hot or sharp objects.
4. Moving machinery.

It is recommended that all operators of resistance welding equipment wear face shields or flash goggles. There may be some flying sparks or "flash" thrown from the joint being welded. Protective clothing is also necessary.

The welding voltage across the electrodes is very low, but the secondary current is very high. The primary circuit wiring should be handled only by qualified electricians. All resistance welders should be grounded.

Naturally, any surface being welded will be very hot. Operators should always wear protective gloves if it is necessary to handle the materials being welded. Gloves will also protect the hands against sharp or ragged sheet metal edges.

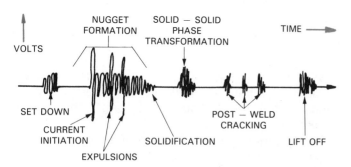

Fig. 17-33. Typical acoustic emissions (sounds) during a resistance spot weld. Not all of these signals are present in every weld. (Physical Acoustics Corp.)

Most modern resistance welding machines have pneumatically operated electrodes. The electrode forces may be quite high. There is always danger of injury if the operator's hand or fingers should accidentally be caught between the electrodes as they come together on the work. Safety devices are available and should always be used to insure that no part of the operator's body can be in a danger area at the time the machine starts to operate.

## 17-19. TEST YOUR KNOWLEDGE

Write your answers on a separate sheet of paper. Do not write in this book.

1. What is the basic principle of all resistance welding?
2. What does a step-down transformer do?
3. Name five variables used in resistance spot welding.
4. A welding schedule is composed of three different times. Name them in order.
5. Name two ways to change the welding current.
6. What is the difference between hold time and squeeze time?
7. Give another word for the word pneumatic?
8. Resistance welding machines are classified by the type of electrical power they use. What are the three types?
9. If two pieces of .040 in. (1.02 mm) mild steel are going to be welded together, what weld time and current should be used?
10. If a piece of .040 in. (1.02 mm) mild steel is to be welded to a piece of .090 in. (2.29 mm) mild steel, what weld time and electrode force should be used?
11. When molten metal squirts out from a spot weld, it is called an _____ _____.
12. How is the electrode force adjusted?
13. What does dressing an electrode mean?
14. What is a forge force? Tempering?
15. Name two advantages of projection welding.
16. Why does flashing occur during flash welding?
17. What are two advantages of flash welding over upset welding?
18. What two types of seams can be welded by resistance seam welding?
19. What variables can be monitored by a feedback control device?
20. What are the dangers associated with resistance welding? How can injury be prevented?

part 7 SPECIAL PROCESSES

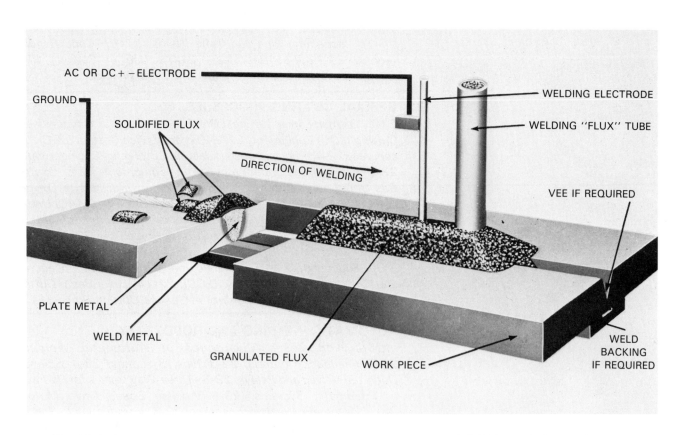

AC OR DC + − ELECTRODE

GROUND

SOLIDIFIED FLUX

DIRECTION OF WELDING

WELDING ELECTRODE

WELDING "FLUX" TUBE

VEE IF REQUIRED

PLATE METAL

WELD METAL

GRANULATED FLUX

WORK PIECE

WELD BACKING IF REQUIRED

Fig. 18-1. Diagrammatic view of a submerged arc weld in progress.   (L-TEC Welding & Cutting Systems)

# 18 SPECIAL WELDING PROCESSES

The general classification of welding includes various methods that have been developed for fusing metals together. Some of these methods are highly specialized. Some are patented, and these may be used only with the permission of the patent owners. Others are special developments of gas flame welding, arc welding, resistance welding, and combinations of these methods.

## 18-1. SPECIAL WELDING PROCESSES

The three main types of welding are OXYFUEL GAS WELDING, ARC WELDING, and RESISTANCE WELDING. Each of these three categories has several related processes. However, any method that can produce intermingling of the molecules (cohesion) is a welding process. Many processes have been developed. Some of the processes are part of the three main categories. Two additional categories contain the rest of the special welding processes. These two categories are SOLID STATE WELDING (SSW) and OTHER WELDING. Some of the welding processes which may be considered special are:

ARC RELATED PROCESSES
Submerged Arc Welding (SAW)
Electroslag (ESW) and Electrogas Welding
 (EGW)
Narrow Gap Welding
Stud Arc Welding (SW)
Arc Spot Welding
Plasma Arc Welding (PAW)
Underwater Shielded Metal Arc Welding

SOLID STATE WELDING (SSW) PROCESSES
Coextrusion Welding (CEW)
Cold Welding (CW)
Diffusion Welding (DFW)
Explosion Welding (EXW)
Forge Welding (FOW)
Friction Welding (FRW)

Hot Pressure Welding (HPW)
Inertia Welding
Roll Welding (ROW)
Ultrasonic Welding (USW)

OTHER WELDING PROCESSES
Thermit Welding (TW)
Electron Beam Welding (EBW)
Laser Beam Welding (LBW)

GAS RELATED PROCESSES
Self-Generating Oxyhydrogen Welding
Solid Pellet Oxyfuel Gas Welding

## 18-2. SUBMERGED ARC WELDING (SAW)

SUBMERGED ARC WELDING is a welding process that has grown rapidly in popularity. It has some outstanding advantages. It is very fast. There is no visible arc, no splatter, and the welds are of high quality. Refer to Heading 1-18 and to Fig. 1-23.

The method involves striking the arc between a consumable electrode and the joint while the arc is buried in a granular flux (such as titanium oxide-silicate). See Fig. 18-1.

Some submerged arc machines are able to produce single pass welds on butt joints up to 3 inches (approximately 75 mm) in thickness, plug welds up to 1 1/2 inches (approximately 40 mm) in thickness, and fillet welds with up to 3/8 inches (approximately 10 mm).

Electrodes up to 1/2 in. (12.7 mm) diameter or multiple electrodes may be used. The current may be as high as 4000 amps, and it may be either AC or DC. Usually, a constant voltage power source is used. When using large diameter electrodes, a constant current power source can be used. The typical submerged arc welding machine, as shown in Fig. 18-2, has a power operated carriage. A universal variable speed motor controls the travel

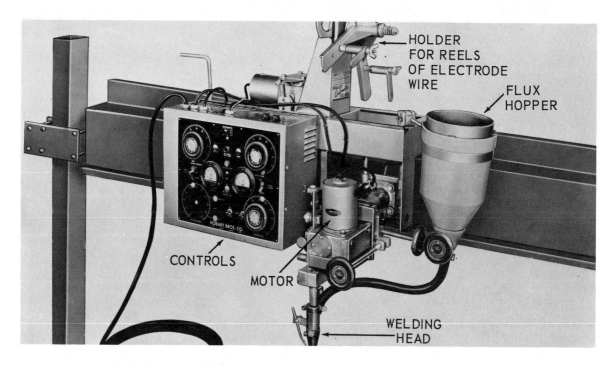

Fig. 18-2. Submerged arc welding outfit. (Hobart Bros. Co.)

speed from 7 to 210 inches per minute (3 - 89 mm/sec). A hopper feeds the granular flux to the joint just ahead of the arc. The heat generated by the arc melts the adjacent flux granules. As the flux solidifies, it covers the weld with an air-tight slag that protects the weld until it cools. This slag may be readily removed. The unused granules can be used again. Special equipment is used on some applications to pick up the unused flux and to feed it back into the hopper; see Fig. 18-3.

Since the arc is submerged and therefore not visible to the operator, the correct setting is indicated by an ammeter reading, and a voltmeter reading.

Submerged arc welding is excellent for production jobs which require welds on large materials.

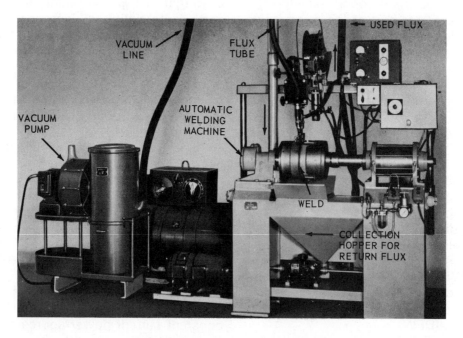

Fig. 18-3. Machine for recovering unused flux from a submerged arc welding machine.
(Invincible Airflow Systems)

The carriage may be used on a standard track, a template track, or the metal being welded may be mounted on a carriage.

The electrode is furnished in coils. A variable speed electric motor controls the electrode feed. The controls for the process are mounted on the machine. Only a few adjustments are required on the machine to handle a variety of jobs.

The welding operator must set the voltage, wire feed speed, travel speed, and electrode stickout. These are the same variables used in gas metal arc welding. The operator must also align the metal to be welded.

Submerged arc welding may be done manually. See Fig. 18-4. Note the small hopper for granular flux built into the electrode holder. A combination electrode and wire feed cable is attached to the electrode holder. The operator guides the electrode holder and is able to make fast welds on either straight or irregular joints.

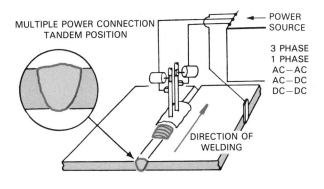

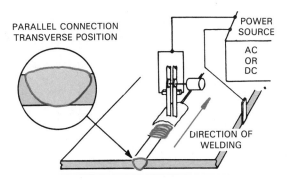

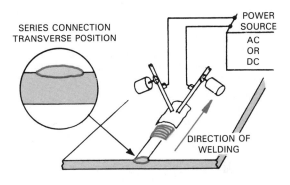

Fig. 18-5. Three different ways to use multiple electrodes in submerged arc welding.

Fig. 18-4. Manually submerged arc welding. The process is very similar to GMAW but the electrode is beneath a layer of flux. (Hobart Bros. Co.)

Submerged arc welding may be done with more than one electrode. The use of more than one electrode may be desirable when the finished weld must be wide or higher welding speeds are desired. See Fig. 18-5 for schematic drawings of three different methods of multiple electrode welding.

The most common use is the single electrode. Two and three electrodes are usually used in a tandem position as shown in Fig. 18-5. Large welds can be completed quickly using this method. Fig. 18-6 shows a comparison of the metal deposited by single and multiple electrodes.

Submerged arc welding is usually done in the flat position. This is because there is a large

amount of molten metal and molten slag. SAW can also be used to make horizontal fillet welds. It cannot be used to weld other joints horizontally, vertically, or overhead. The flux and molten metal would not stay in place.

In submerged arc welding, the flux remaining over the completed weld acts as a heat insulator. As a result, the weld remains very hot for some time after it is completed. Therefore, follow all safety precautions.

## 18-3. ELECTROSLAG (ESW) AND ELECTROGAS (EGW) WELDING

The ELECTROSLAG WELDING (ESW) method has been developed to weld very thick sections or joints. The process eliminates the need for multiple

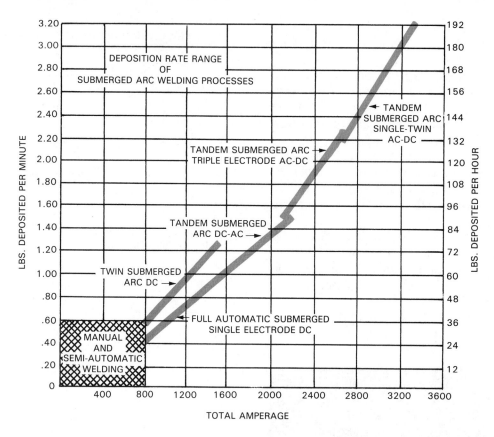

Fig. 18-6. A comparison of weld deposition rates for single and multiple electrode fully automated submerged arc welding. Note that the fully automatic processes deposit more than the manual or semi-automatic processes.

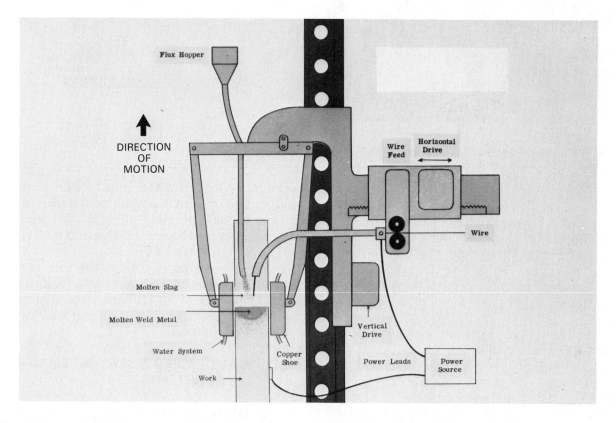

Fig. 18-7. Electroslag welding operation showing the flux hopper, electrode wire feed, shoes, and the rail on which everything moves as the weld progresses.

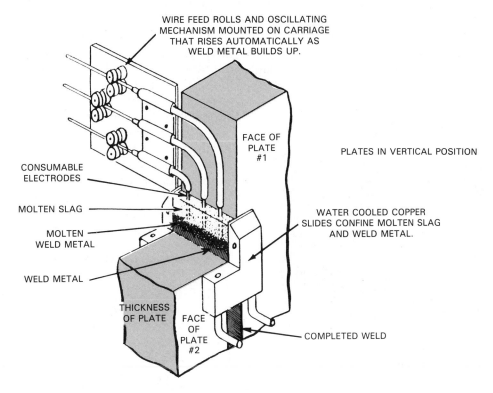

WIRE FEED ROLLS AND OSCILLATING
MECHANISM MOUNTED ON CARRIAGE
THAT RISES AUTOMATICALLY AS
WELD METAL BUILDS UP.

FACE OF
PLATE
#1

PLATES IN VERTICAL POSITION

CONSUMABLE
ELECTRODES

MOLTEN SLAG

MOLTEN
WELD METAL

WELD METAL

WATER COOLED COPPER
SLIDES CONFINE MOLTEN SLAG
AND WELD METAL.

THICKNESS
OF PLATE

FACE
OF
PLATE
#2

COMPLETED WELD

Fig. 18-8. A three wire electroslag welding operation shown diagrammatically. The molten slag floats above the weld metal and prevents oxidation.

passes and for bevel, V, U, or J grooves. Refer to Heading 1-19 and to Fig. 1-24.

The process is used to weld joints in a vertical position. The equipment used for electroslag welding includes:

1. A power source.
2. One or more electrodes and electrode guide tubes.
3. One or more wire feeders and oscillators.
4. Retaining shoes (molds).
5. Flux.

The electroslag process is started by producing an arc between the electrode(s) and the joint bottom. Flux is added and forms a layer of molten slag. When a large layer of slag has formed, the arc is no longer needed. The resistance to electrical current flow through the flux creates the heat necessary to melt the electrode(s) and base metal. The electrodes used are either solid wire or flux cored. The process is fast and requires no edge prepraration of the metal. More than one electrode may be used. This permits a thick joint to be welded faster.

Water cooled, copper shoes are used to contain the molten metal and slag. As the weld is made, the molds move up the joint. The weld is completed in one pass. Butt joints, T joints, corner joints, and others may be made with this process. Butt welds are the most common. Specially

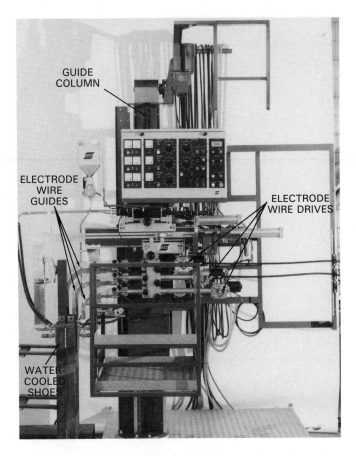

GUIDE
COLUMN

ELECTRODE
WIRE
GUIDES

ELECTRODE
WIRE DRIVES

WATER-
COOLED
SHOES

Fig. 18-9. An electroslag welding station for three electrodes. Each electrode has a separate drive and guide.    (ESAB Group)

shaped shoes are required for T and corner joints. Fig. 18-7 shows an electroslag weld in progress. Figs. 18-8 and 18-9 show three wires used to fill a joint.

The extreme heat produced by the molten slag and metal in the weld causes the base metal to

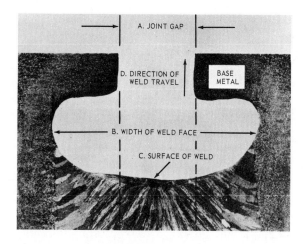

Fig. 18-10. A macrograph (1.5x) of a manganese-molybdenum steel joint welded by the electroslag process. Note the way the base metal melts away from the edges of the original joint gap.

melt away from the original joint gap as shown in Fig. 18-10.

A special application of electroslag welding is to join two pieces of pipe. Fig. 18-11 shows a circumferencial weld schematic as it begins and before it ends. The welding is done at the 3 o'clock position. Fig. 18-12 shows ESW equipment after the weld is completed.

The process of ELECTROGAS WELDING (EGW) is similar to the electroslag process, except in this process, a shielding gas is used. Flux cored wire is automatically fed into the weld joint. An electric arc is continuously maintained between the electrode and the weld puddle. Fig. 18-13 illustrates a typical setup for electrogas welding.

## 18-4. NARROW GAP WELDING

NARROW GAP WELDING is similar to electrogas welding except the weld joint has a very narrow gap. Gas metal arc welding (GMAW) spray transfer is used to fill the weld joint. Since the joint is narrow, less weld metal is required to fill the joint. This process is used to weld thick sections. Fig. 18-14 shows a comparison between the

**GIRTH SEAM WELDING**

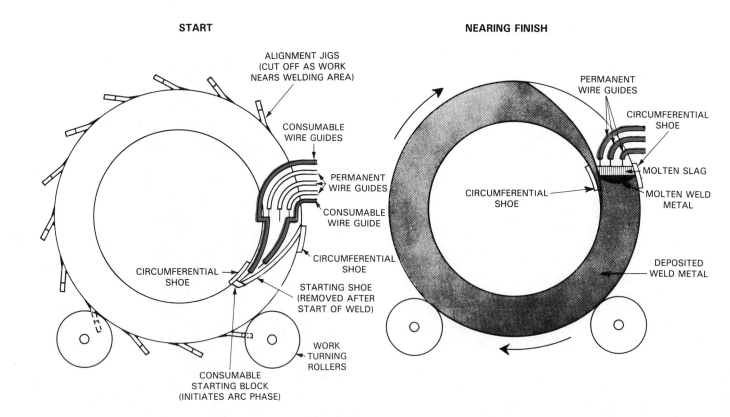

Fig. 18-11. The use of the electroslag process for welding butt joints in cylindrical objects. In this case, two consumable wire guides are used at the beginning of the weld. Three wires are used to fill the joint.

Fig. 18-12. A completed electroslag circumferential weld on a thick walled pipe. Notice the copper shoes. (ESAB Welding Products Div.)

weld joint preparation for shielded metal arc welding (SMAW), submerged arc welding (SAW), and narrow gap welding. The narrow gap joint requires less weld metal to fill the groove.

To feed the electrode to the bottom of the joint, an electrode guide tube is used. One problem with narrow gap welding is directing the arc to both sides of the joint. The arc must be directed to both sides to melt the sides and allow mixing. One method used to direct the arc to both sides of the joint is to make the electrode oscillate between the sides of the joint. To do this, the electrode is bent into a wave form. This is done by bending the wire before it enters the guide tube. When the electrode comes out of the guide tube, it welds first one side of the groove and then the other because of the wave shaped wire. Shielding gas is used to protect the arc and the molten metal. Fig. 18-15 shows a mechanism used to bend the

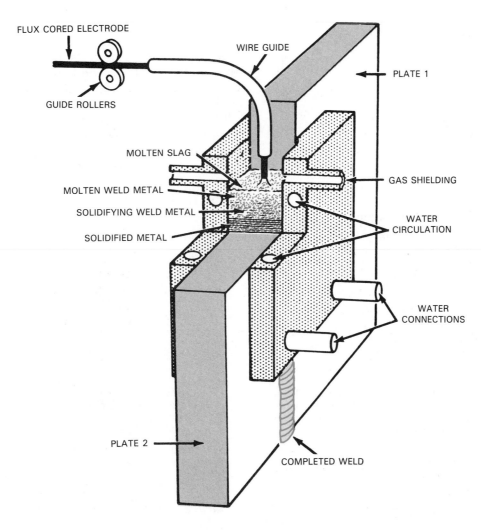

Fig. 18-13. Schematic drawing of an electrogas weld in progress. The shoes are water cooled and are moved up as the weld proceeds. A shielding gas protects the molten weld metal from oxidation.

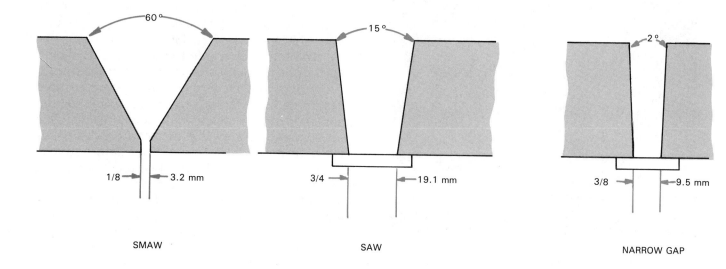

SMAW                    SAW                    NARROW GAP

Fig. 18-14. A comparison between the weld joint preparation for shielded metal arc welding (SMAW), submerged arc welding (SAW), and narrow gap welding. Note that the narrow gap joint requires less weld metal to fill the groove.

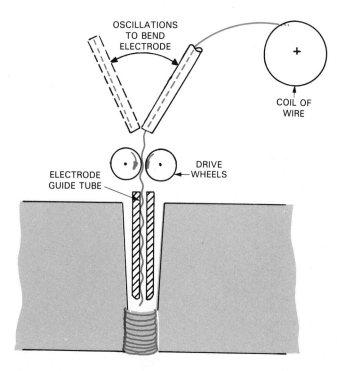

Fig. 18-15. A mechanism used to bend the electrode into a wave form and guide the electrode into the weld joint. The electrode will oscillate between the walls of the weld joint as the weld progresses.

electrode into a wave form and guide the electrode into the welding joint.

## 18-5. STUD ARC WELDING (SW)

Stud arc welding is a special welding development that quickly and efficiently welds studs and other fastening devices to plates and other surfaces. The process permits attaching an assembly device to a structure, and fastening different parts to this structure without piercing the metal; Fig. 18-16. Refer to Heading 1-20 and to Fig. 1-25.

Stud arc welding eliminates drilling or punching holes in the main structure. It saves the work of mechanically fastening an object to the main structure using bolts, rivets, or screws.

A complete set up for stud arc welding is shown in Fig. 18-17. The equipment consists of:

Fig. 18-16. A stud welded to a steel plate. Note the depth of fusion in this application as shown in the stud which has been sectioned.

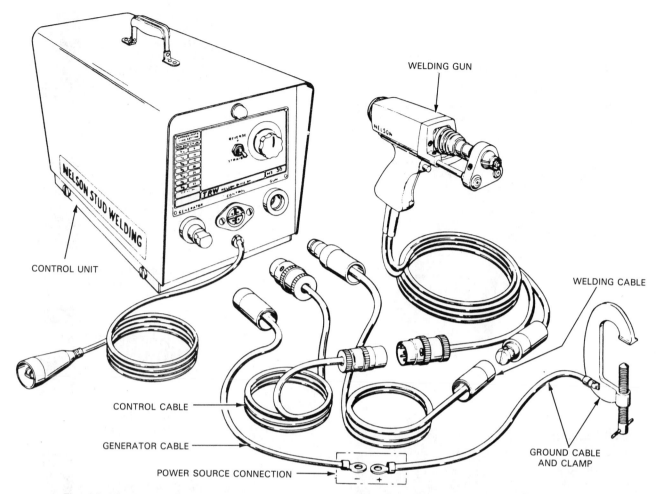

Fig. 18-17. Equipment required for stud arc welding. The generator cable and ground cable are attached to a constant current power source. (Nelson Stud Welding)

CONTROL UNIT

WELDING GUN

WELDING CABLE

CONTROL CABLE

GENERATOR CABLE

POWER SOURCE CONNECTION

GROUND CABLE AND CLAMP

1. A constant current power source or a capacitor discharge power source.
2. Control unit.
3. Stud Arc Welding gun.
4. Cables.
5. Studs.

The control unit serves as a timer which controls the stud arc welding process. The control unit stops the arcing process at the set time. Times may vary from .5 seconds to 2 seconds, depending on the stud size.

A stud arc welding gun has the following main parts:
1. Trigger.
2. Device to hold stud.
3. Solenoid.
4. Springs.

A cross section of a stud arc welding gun is shown in Fig. 18-18.

Stud arc welding is a rapid process. A stud arc weld is being made in Fig. 18-19. The stud arc welding process is illustrated in Fig. 18-20. The stud is placed in a chuck on the gun. A ceramic

ferrule is placed on the end of the stud. The ferrule contains the molten metal during arcing and serves as a mold when metal solidifies. The stud and ferrule are placed in contact with the work, A, Fig. 18-20. When the trigger is pulled, an arc is formed between the stud and the work. The stud is pulled away from the work by solenoid in the gun, B, Fig. 18-20. At the end of the timed interval, the timer stops the welding current and stops the current to the solenoid. Springs force the molten stud into the molten pool, C, Fig. 18-20. The gun is removed after a few seconds and the ceramic ferrule is chipped or broken off. A completed stud arc weld is shown in D, Fig. 18-20.

The studs prior to welding have a small nib or bump on them. This is to help initiate the arc. The stud usually contains a small amount of flux at the arc end under the small bump (nib). Flux helps to stabilize the arc. A variety of studs are shown in Fig. 18-21.

Aluminum can also be stud arc welded. To protect the aluminum from oxidizing, a shielding gas is provided. The process is the same as welding

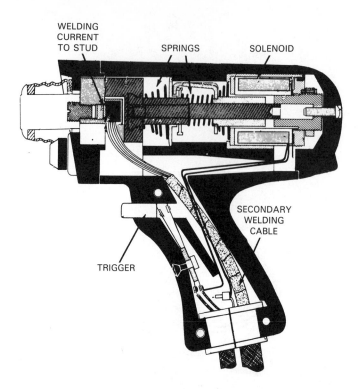

Fig. 18-18. Cross section of a stud arc welding gun.

Fig. 18-19. Stud being welded to a tractor floor pan.
(Nelson Stud Welding)

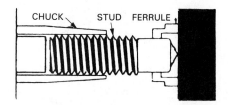

A—FLUXED END OF STUD IS PLACED
IN CONTACT WITH WORK.

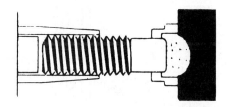

B—STUD IS AUTOMATICALLY RETRACTED
TO PRODUCE AN ARC.

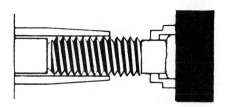

C—STUD IS PLUNGED INTO POOL
OF MOLTEN METAL.

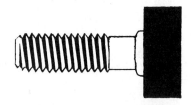

D—OPERATION COMPLETED—STUD
IS WELDED TO WORK.

Fig. 18-20. Steps which occur during the production of a
stud arc weld on steel plate. (Nelson Stud Welding)

steel. Refer to Fig. 18-22. Fig. 18-23 shows a
stud arc welding gun used to weld aluminum.

Some flash and sparking occurs in the stud arc
welding operations. Eyes, face, and hands must
be properly protected from the sparking.

## 18-6. ARC SPOT WELDING

The gas tungsten arc welding (GTAW) process
may be used to produce welds between two
sheets of metal or between a sheet of metal and
a plate or structural metal member.

The process consists of striking an arc and
holding this arc in one place. The top sheet of
metal melts through and fuses with a molten portion of the sheet or structural member underneath.

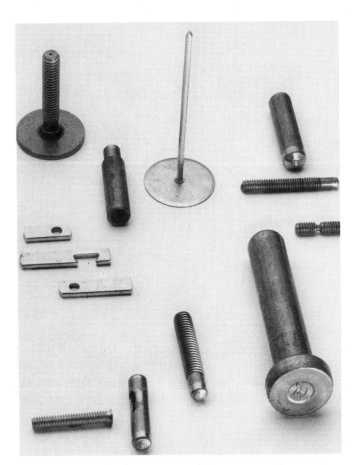

Fig. 18-21. A variety of studs and fastening devices designed for stud welding. (Omark Industries)

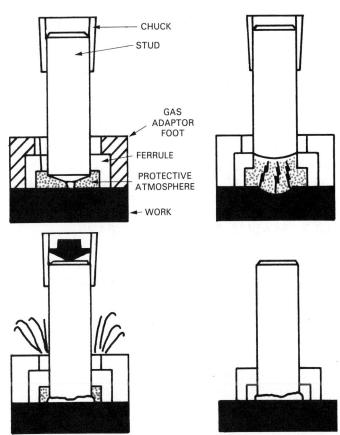

CHUCK
STUD
GAS ADAPTOR FOOT
FERRULE
PROTECTIVE ATMOSPHERE
WORK

Fig. 18-22. Operations required when welding a stud in an aluminum plate. Notice the gas adaptor foot which contains shielding gas. (Nelson Stud Welding)

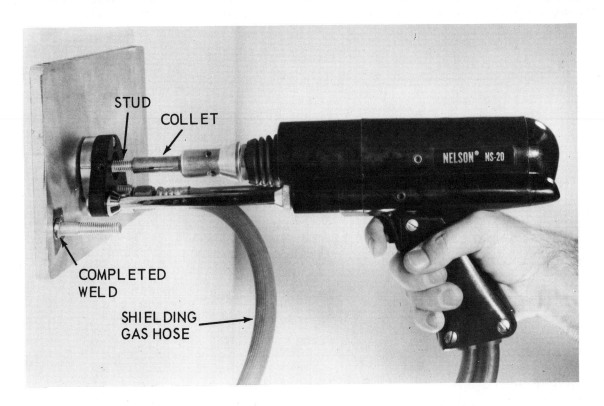

STUD
COLLET
COMPLETED WELD
SHIELDING GAS HOSE
NELSON NS-20

Fig. 18-23. Stud arc welding gun used for welding on aluminum. The weld area is protected by shielding gas fed into a chamber around the stud. (Nelson Stud Welding)

The arc time is usually automatically controlled. The electrode holder usually has an insulator which properly spaces the tungsten electrode from the metal. A trigger moves the electrode up to the sheet metal. When the electrode is withdrawn, the arc is created and the timer is started. The process can only be used on sheet metal less than 1/8 in. (3.2 mm) thick.

Arc spot welding can also be done with shielded metal arc welding (SMAW). A spot is made in the top sheet which passes through the sheet and fuses with the bottom sheet or plate.

## 18-7. PLASMA ARC WELDING (PAW)

The term PLASMA, as used in physics, means an ionized gas (a gas which has lost or gained electrons). When you see a streak of lightning in the sky, what you are actually seeing is the gases of the atmosphere ionized and heated to incandescence. The electric arc is capable of ionizing both solids and gases.

A cross section of a plasma arc torch is shown in Fig. 18-24. The tungsten electrode is surrounded by an orifice gas. This gas is an inert gas like helium or argon, or it could be nitrogen. The orifice gas cannot contain oxygen because it will oxidize the electrode.

This orifice gas becomes a plasma as it passes through the arc. The orifice gas is propelled toward the work, when it passes through the con-

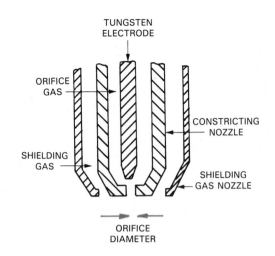

Fig. 18-24. Cross section of a typical plasma arc torch.

stricting nozzle. A shielding gas protects the welding or cutting process.

The advantages of PAW include its concentrated current flow, arc stability with large changes in arc length, and its low heat input and high welding speed.

The arc used in PAW may be a transferred or a nontransferred arc. A TRANSFERRED ARC has a negative electrode and the work is positive (DCEN or DCSP). A NONTRANSFERRED ARC is an arc between the negative electrode and a positive constricting nozzle. In the nontransferred arc, the

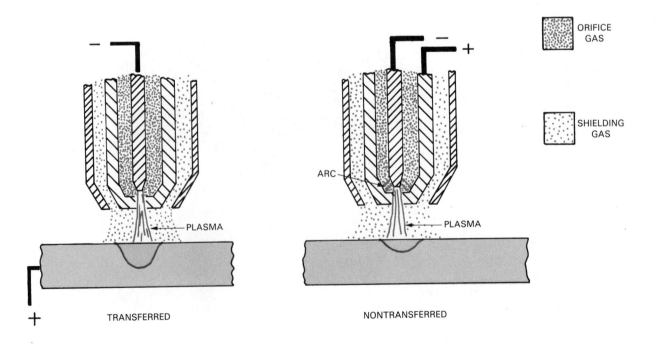

Fig. 18-25. Diagram showing a transferred and nontransferred arc.

work is not a part of the electrical circuit. Even though the arc is nontransferred, the plasma has sufficient energy to melt the work surface. The transferred arc has better penetrating ability. The transferred and nontransferred arc are illustrated in Fig. 18-25. The nontransferred arc can be used to weld or cut materials that do not conduct electricity.

The joint used for plasma arc welding is the square butt joint. A groove weld may also be welded. Welding .25 inch (6.4 mm) stainless steel or .50 inch (12.7 mm) titanium can be completed in one pass. Arc voltage varies from 21 volts for thin material to 38 volts for thicker material. Currents range from 120 amps to 275 amps. Travel speeds can be as high as 30 inches per minute (12.7 mm/second). Fig. 18-26 lists some variables used when welding stainless steel.

Plasma arc welding uses the keyhole method to obtain a full penetration weld. Current and gas flow must be regulated to prevent the creation of too large a keyhole. A plasma arc torch gives the best results when operated automatically. Filler metal is also added automatically. See Fig. 18-27.

| Thickness | | Travel Speed | | Current (DCSP) (DCEN) | Arc Voltage | Nozzle type | Gas flow | | | | Remarks |
| | | | | | | | Orifice | | Shield | | |
| in. | mm | in/min | mm/s | A | V | | ft³/h | L/min | ft³/h | L/min | |
|---|---|---|---|---|---|---|---|---|---|---|---|
| 0.092 | 2.4 | 24 | 10 | 115 | 30 | 111M | 6 | 3 | 35 | 17 | Keyhole, square-groove weld |
| 0.125 | 3.2 | 30 | 13 | 145 | 32 | 111M | 10 | 5 | 35 | 17 | Keyhole, square-groove weld |
| 0.187 | 4.8 | 16 | 7 | 165 | 36 | 136M | 13 | 6 | 45 | 21 | Keyhole, square-groove weld |
| 0.250 | 6.4 | 14 | 6 | 240 | 38 | 136M | 18 | 8 | 50 | 24 | Keyhole, square-groove weld |

Notes:
a. Nozzle type: number designates orifice in thousandths of an inch; "M" designates design.
b. Gas underbead shielding is required for all welds.
c. Gas used: 95% Ar-5% $H_2$.
d. Torch standoff: 3/16 in. (4.8 mm).

Fig. 18-26. Variables to be used when plasma arc welding stainless steel.

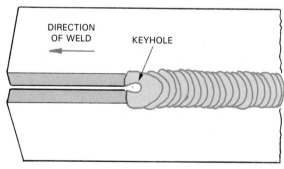

Fig. 18-27. Top. Plasma torch (nontransferred arc) being used to weld a stainless steel joint. Notice how the filler metal is fed into the molten pool. (L-TEC Welding & Cutting Systems) Bottom. Partially completed plasma arc weld showing the "keyhole" at the leading edge of the pool. The size of the keyhole indicates the amount of penetration.

Plasma arc welding can be done manually but at a slower rate. The plasma arc may also be used for cutting.

Plasma arc welding and cutting produce intense sound. It is usually necessary to wear ear protection during the PAW operation.

## 18-8. UNDERWATER SHIELDED METAL ARC WELDING

Shielded metal arc underwater welding is done under water using a well insulated electrode holder and special covered electrodes, Fig. 18-28.

Fig. 18-28. Underwater welding electrode holder. Note that the holder is well insulated with no bare metal areas exposed. (Craftsweld Equipment Corp.)

The electrode covering is protected from water damage by special waterproof outer coatings. About 10 percent higher current settings are used for welding under water than are used when welding the same object in the atmosphere. Because of the problem of rapid heat loss in water, stringer beads should be used rather than wide weaving beads. Also, a short arc length and DCEN (DCSP) is recommended.

Due to the poor visibility under water, a #4 - #8 welding lens is recommended for use on the mask or helmet. When wearing the metal helmet for deep dives, the operator must be careful not to ground the helmet to any part of the welding circuit.

It is recommended that a communication system be used between the diver and the equipment operator who is above water. The arc welding current should be turned on from above water only after receiving the diver's orders when he or she is ready to weld. The current should be turned off as soon as the weld is completed.

In underwater welding, fillet welds are recommended. When properly made, such welds will develop approximately 80 percent of the tensile strength and 50 percent of the ductility of similar welds made above water. Extensive diving and welding training should be completed before attempting to weld under water. Follow all safety precautions for both welding and diving.

## 18-9. SOLID STATE WELDING (SSW)

A number of welding processes are classified as SOLID STATE WELDING processes. These processes do not use an arc, they do not have a beam of energy, they do not use a flame, and they do not use resistance to heat the metal. These processes are done when the metal is cold, warm, or hot. However, the temperature does not exceed the melting point. The metal to be joined is heated (if necessary) to a point where they can be forced together. No filler metal is needed.

The American Welding Society recognizes the following processes as solid state welding processes:
1. Coextrusion welding (CEW).
2. Cold welding (CW).
3. Diffusion welding (DFW).
4. Explosion welding (EXW).
5. Forge welding (FOW).
6. Friction welding (FRW).
7. Hot Pressure welding (HPW).
8. Roll welding (ROW).
9. Ultrasonic welding (USW).
The most common of these processes are covered in the following paragraphs.

## 18-10. COLD WELDING (CW)

The room temperature pressure welding process works best on soft, ductile metals like aluminum and its alloys, copper, alloys of cadmium, nickel, lead, zinc, etc. No heat is needed. The cleaned metal is forced together under considerable pressure, and the ductility of the metals produces a true fusion condition. Refer to Heading 1-21 and Fig. 1-26. When pressed together, enough pressure must be applied to reduce the original thickness of the metal to about 1/4 of the original thickness. Aluminum when welded by this process has a tensile strength of up to 22,000 psi (151.7 MPa). Fig. 18-29 shows three typical cold welds.

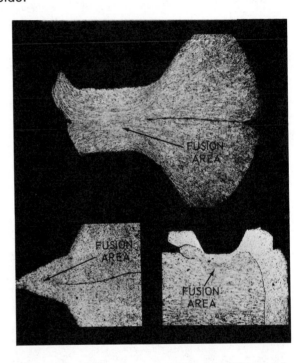

Fig. 18-29. Samples of cold welded parts. The welds are aluminum to aluminum (1003). Notice the large indentation of the surface.    (Kelsey-Hayes Co.)

The metals to be joined by cold welding must be very carefully prepared. The oxides and other contaminants must be completely removed. An abrasive wire wheel, turning at high speed, has been found to be satisfactory. This method is considered to be better than chemical methods because residual solvents often remain on the metal surface.

The theory of the process is that the pressure at the surface causes fusion only a few molecules deep. This fusion is sufficient to hold the material together and provide good strength.

The design of the tool for imposing the pressure on the metals is very important. Dies must be designed to compensate for varying hardness of the metals. The tool must have twice as much contact against the softer aluminum when cold welding aluminum to copper. Fig. 18-30 shows a hand operated welding tool. It can weld up to a combined thickness of .080 in. (2.0 mm) aluminum or .060 in. (1.5 mm) copper.

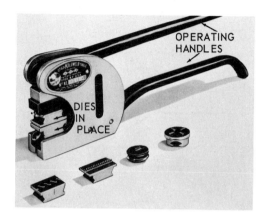

Fig. 18-30. Hand operated cold weld machine.

The tools can be either hand, pneumatically, or hydraulically powered. Fig. 18-31 shows a hydraulically powered unit which can butt weld up to 3/4 in. (19.1 mm) round or strip copper. It can also perform lap welds.

## 18-11. EXPLOSION WELDING (EXW)

This process is very dangerous. It should be performed ONLY by explosive experts, and in specially designed or water filled chambers. Special permits must be obtained from local, state and federal authorities before this type of work can be done.

Metals have been successfully welded using the energy from an explosion. Refer to Heading 1-22

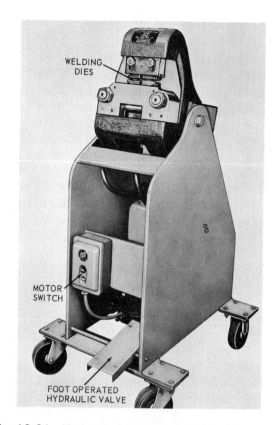

Fig. 18-31. Hydraulically powered cold weld machine.

and to Fig. 1-27. To prepare an explosion weld, the metals are placed at an angle to each other. A protective material is placed over the metal and then the explosive material is placed around the metal. An explosion welding set up is shown in Fig. 18-32. The explosion forces the plates together at high velocity causing surface ripples in the metal. The ripples lock or weld the two metals together. This process has been successfully used to weld steel to steel, aluminum to aluminum, copper to steel, stainless steel to molybdenum, and many other metal combinations.

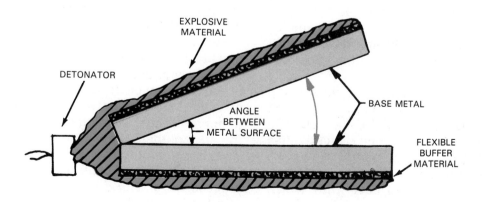

Fig. 18-32. Schematic drawing of a setup for explosion welding. It shows how the explosive material and protective coating are placed on the base metal.

## 18-12. FORGE WELDING (FOW)

The oldest form of fusing two pieces of metal together is FORGE WELDING. Refer to Heading 1-23 and to Fig. 1-28. This type of welding requires considerable skill on the part of the operator. It is usually limited to the joining together of pieces of solid steel stock. The two pieces of metal to be welded together are heated in a blacksmith forge such as shown in Fig. 18-33. This forge uses charcoal or coal and an air blast from below the coals to produce the heat

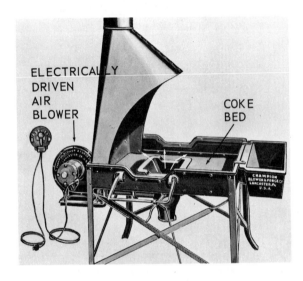

Fig. 18-33. A blacksmith forge used to heat metal for forge welding.

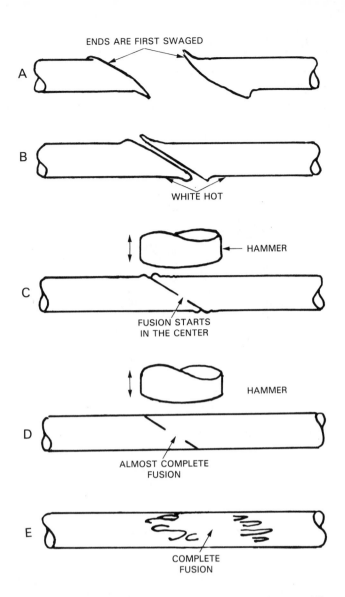

Fig. 18-34. Illustration of the steps used when forge welding.

necessary for forge welding. The metal to be welded is placed in the bed of hot coals. The metals to be joined are heated to a white heat just short of the rapid oxidizing or burning point. They are then placed together so that the surfaces may be forced against one another by pounding with a hammer. A blacksmith usually swages (enlarges) the ends to be joined. The pressure of the hammer blows, and the extra heat produced from the hammering, fuses the two pieces together. See Fig. 18-34.

This same type of welding may be done using power hammers, fixtures, and presses. It is often called FORGE WELDING or HAMMER WELDING.

A weld of this nature, when done correctly, has every quality of the original metal. However, because of the skill necessary to produce a successful joint, and the relative ease with which other processes accomplish the same task, forge welding has virtually been replaced by more modern welding processes.

## 18-13. FRICTION WELDING (FRW)

Friction generates heat. If two surfaces are rubbed together, enough heat can be generated so the parts subjected to the friction may be fused together. Refer to Heading 1-24 and Fig. 1-29.

A friction weld is prepared by mounting the pieces to be welded in a chuck or fixture. One fixture is stationary and the other is revolved. The pieces to be welded are brought together under pressure. Friction between the parts raises the temperature to a welding heat. The parts are then forced together, allowed to cool, and the friction weld is complete. The weld is produced in about 15 seconds. Fig. 18-35 shows the steps in the process.

In friction welding, the spinning part is driven by a motor. It is forced to rotate while the parts are in contact. A small pressure is applied while the

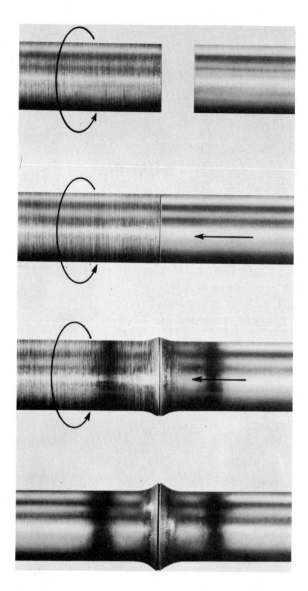

Fig. 18-35. Steps in friction welding two pieces of 1 in. (25.4 mm) diameter carbon steel. The rod at the left is rotated at high speed while the rod on the right is forced against the rotating rod. Friction creates enough heat to perform a sound weld.

metal is being heated. After the heating is completed, great force is applied along the axis to join the surfaces. The process is similar to upset or flash welding but friction is used to generate the heat.

Various metals can be welded using this process, such as:

1. Carbon steel.
2. Stainless steel.
3. Stainless to carbon steel.
4. Tool steel.
5. Copper.
6. Aluminum.
7. Alloy steels.
8. Alloy steels to carbon steel.
9. Titanium.

Most friction welding is used to weld ferrous parts. The electrical, chemical, nuclear, and marine industries use the process to join dissimilar (unlike) metals. The process is well suited for joining dissimilar materials to mild steel in production. No protective atmosphere is required, saving a significant amount of time and materials.

## 18-14. INERTIA WELDING

INERTIA WELDING is similar to friction welding. A movable part revolves against a stationary part to generate heat. The energy for welding comes from a flywheel spinning at high speed. A small amount of energy is required to give the flywheel a high velocity.

When the parts are brought together, the spinning part and flywheel are slowed due to friction. When the flywheel stops spinning, the forging or welding pressure is applied. An inertia welding machine for joining cross sections up to 10 in.² (6451.6 mm²) is shown in Fig. 18-36.

## 18-15. ULTRASONIC WELDING (USW)

In ultrasonic welding, metals are clamped together under pressure and high frequency vibrations are introduced into these metals through a welding tip or sonotrode. The vibrations, at a frequency above those one can hear, break up the surface films and cause the solid metals to bond tightly together. Refer to Heading 1-25 and to Fig. 1-30. This happens without the use of heat and without melting the metal. Careful cleaning of the metals is not necessary. The low clamping pressure reduces deformation.

Ultrasonic welding equipment contains one or more transducers which convert high frequency electrical power into mechanical vibration at the same frequency. (This is usually between 10,000 and 75,000 Hertz or cycles per second.) Beside the transducers, there is a coupling system which transmits the mechanical vibration to the welding tip and thus into the metals being joined.

Ultrasonic welding has several advantages. There is no heat distortion in the parts. No fluxes or filler metals are required. Thin sheets can be welded to thick sections. The pressures used are less and the welding times are shorter than used in resistance welding. Many types of metals can be joined to themselves or to other metals. The equipment can be operated by semi-skilled personnel with minimal training.

For spot welding, the metals are clamped together and transverse or lateral vibration is imposed on the assembly, as shown in Figs. 18-37 and 18-38. An ultrasonic spot welding machine is

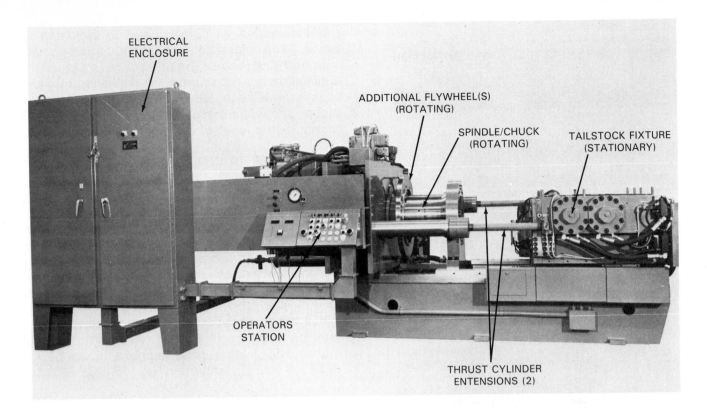

Fig. 18-36. An inertia welder for joining tool steel to oil well drilling pipe up to 10 in.² (6451.6 mm²) cross section. (Manufacturing Technology, Inc.)

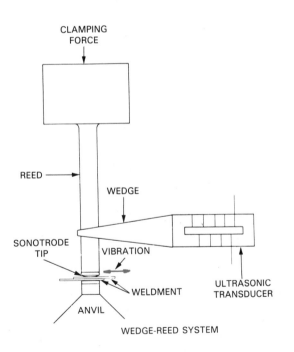

Fig. 18-37. Wedge-reed principle of ultrasonic welding. (Sonobond Corp.)

shown in Fig. 18-39.

Ultrasonic welding can also be used to produce seam welds. A seam welder transducer and wheel are shown in Fig. 18-40. A part to be welded is drawn between two counter rotating rollers.

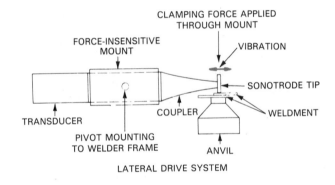

Fig. 18-38. Lateral drive principle of ultrasonic welding.

Vibrations are introduced through the upper roller. A continuous leaktight seam weld is produced between the two metals.

Ultrasonic welding can produce welds in similar or dissimilar metals. The maximum thickness that can be welded is .10 in. (2.54 mm) of aluminum and .040 in. (1.02 mm) of harder materials. This process produces welds very quickly.

## 18-16. THERMIT WELDING (TW)

THERMIT WELDING is based on the fundamental chemical principle that aluminum is a more chemically active metal than iron. The process consists of mixing iron oxide and aluminum, both

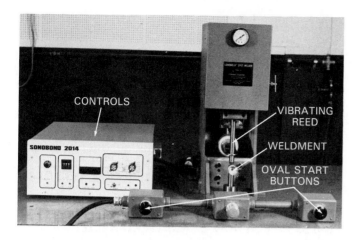

Fig. 18-39. A 20 kilohertz, 1400 watt ultrasonic spot welding machine which uses the wedge-reed coupling system. (Sonobond Corp.)

in a powder form. They are placed in a hopper above a mold which surrounds the area to be welded. This mixture is ignited. The aluminum will combine chemically with the oxygen molecules in the iron oxide producing a temperature of approximately 5000°F (2760°C). An aluminum slag (aluminum oxide) and liquid iron are produced. This high temperature molten iron is heavier than the aluminum oxide and it settles to the bottom of the mold which permits it to come into contact with the steel joint. It will melt the surface of the steel to be welded and fuse with it. This method may be used to weld both large and small steel parts together. It is very popular where large sections are to be welded together. Preheating is often used if the size of the pieces to be welded is large.

Thermit welding may be roughly compared to a foundry casting operation. One difference is that

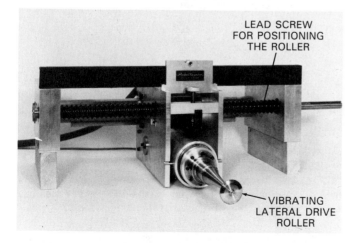

Fig. 18-40. A 500 watt ultrasonic seam welder. The wheel presses against the work and a second wheel (not shown). The parts move through the wheels and a seam weld is produced. (Sonobond Corp.)

the metal being poured is of a considerably higher temperature than metal melted in a furnace.

Welding railroad rails together, welding new teeth on large gears, and welding large fractured crankshafts are common applications of thermit welding. Thermit welding is used to weld sections of castings together when size prevents their being cast in one piece. It is also used to repair large steel structures that are made on special order and would be very costly to replace. Thermit welding has been applied successfully in almost every industry.

Thermit welding is also used for welding pipe. However, in this application, the molten iron does not mix with the pipe metal, but merely furnishes the heat to melt the pipe ends. The pipes are forced together when molten.

## 18-17. PROCEDURE FOR THERMIT WELDING

As thermit welding is a specialized form of casting, molds are necessary to control the flow and shape the liquid iron. The metal parts to be welded together are firmly and accurately set up for welding. A joint is usually machined to provide a V gap all around the joint to enable the molten metal to gain access to all parts. Another method is to use a wax pattern placed in the joint. It has the same shape as the final weld. A sand mold is built around the wax pattern and the work to be welded. Vent holes are necessary on the larger jobs. During the preheating of the metal to be welded, the wax used as a pattern burns away, leaving the correctly shaped cavity to receive the molten metal.

A third process uses a permanent mold. Fig. 18-41 shows a section of a mold assembled around a rail joint to be welded.

On most applications, a funnel shaped container constructed of the same materials as the mold is built above the mold. This funnel contains the aluminum and the iron oxide in their original powder form. Sufficient quantity is required to provide enough iron for the weld. Some molds have a preheating gate, which provides for preheating the metal just prior to the pouring of the iron. This preheat also insures that all moisture is removed. As in all casting work, any moisture present when molten metal is poured may cause a sudden creation of steam and an explosion. An ignition powder which has a low ignition temperature but which burns at a high temperature, must be used in order to start the thermit reaction. The aluminum and iron oxide mixture will ignite only after it has been brought to the temperature of approximately 2000°F (1093°C). Once the mixture starts to burn, it is self propagating. The chemical

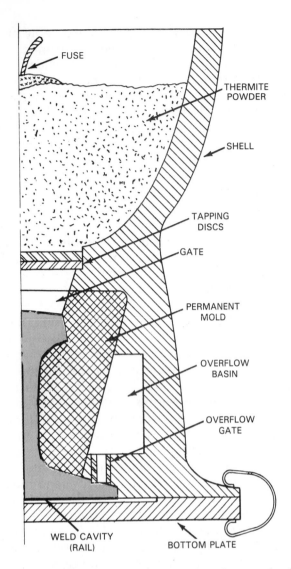

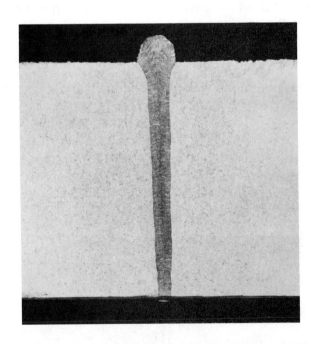

Fig. 18-41. Half a cross sectional drawing of a setup for thermit welding railroad rails together. This setup uses a permanent mold and a shell.

The thermit process is also used for welding cables for electrical conductors such as shown in Fig. 9-34.

## 18-18. HIGH ENERGY DENSITY WELDING

HIGH ENERGY DENSITY WELDING involves concentrating a lot of energy on a small spot. This high energy density can produce deep, narrow welds. Two welding processes have a high energy density. They are ELECTRON BEAM WELDING and LASER BEAM WELDING.

## 18-19. ELECTRON BEAM WELDING (EBW)

Electron beam welding uses a concentrated beam of high velocity electrons to melt the metal to be welded. Refer to Heading 1-27 and to Fig. 1-32. The parts to be welded are placed very close together, almost touching. The edges to be welded must be very straight. An electron beam can make a complete penetration weld in a 10 in. (approximately 254 mm) steel plate. See Fig. 18-42. Electron beam welding is also very fast. Steel .25 inches (approximately 6 mm) thick can be welded at over 200 inches per minute (5.1 m/min).

Fig. 18-42. A cross section of an electron beam weld. Note how narrow the weld is in relation to the thickness of the base metal. (PTR-Precision Technologies, Inc.)

reaction should be allowed to go to completion before the pouring is started. The process is considered safe because of the very high temperature necessary to begin the thermit reaction. This eliminates the chance of accidental ignition of the mixture.

After the weld is completed and the metal has been allowed to cool, the mold is removed. After the removal of the mold, the weld may need trimming because of the excess metal clinging to the weld. Examples are the pouring gate metal, the riser metal, and the metal in the vent holes.

Before the mold is placed around the joint to be welded, it is necessary to clean the surfaces until they are bright and clean. It is especially important to remove any oil, grease, or water from the metal being welded. These materials could vaporize and may build up a dangerously high pressure and cause the mold to burst.

A schematic of the principal parts of an electron beam gun are shown in Fig. 18-43. Current is passed through a tungsten filament which emits electrons. The electrons pass through the bias

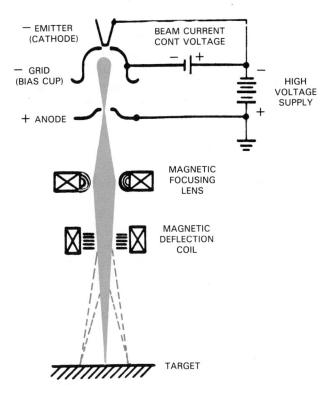

Fig. 18-43. The basic parts of a modern electron beam welder. (PTR-Precision Technologies, Inc.)

electrodes which shape the beam. Electrons are then accelerated toward the anode which has a positive charge. A voltage difference of up to 175,000 volts exists between the emitter and the anode. This is known as the accelerating voltage. An electromagnetic lens is used to focus the electron beam. A cutaway view of an electron beam welder is shown in Fig. 18-44.

Electron beam welding is usually done in a vacuum. In a high vacuum, there are no particles in the air to interfere with the electron beam. In a partial vacuum, there are some particles which cause the electron beam to spread out. In a partial vacuum, the parts to be welded must be closer to the source of the electrons to maintain a high energy density.

Electron beam welding can be done without a vacuum. The parts must be very close to the beam source (1/4 in. [6.2 mm] or less). Fig. 18-45 shows how the beam spreads out as the pressure increases. Fig. 18-46 illustrates the three electron beam systems. The best results are obtained in a high vacuum.

A part being welded must be placed on a carriage that moves the part under the electron beam.

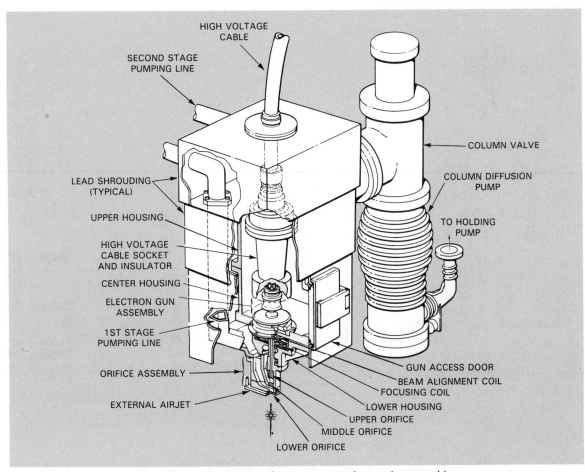

Fig. 18-44. A cutaway of a nonvacuum electron beam welder. (PTR-Precision Technologies, Inc.)

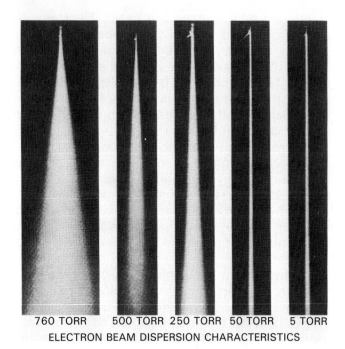

ELECTRON BEAM DISPERSION CHARACTERISTICS

| 760 TORR | 500 TORR | 250 TORR | 50 TORR | 5 TORR |

Fig. 18-45. The electron beam spreads out as the pressure increses. (760 torrs is atmospheric pressure. 5 torrs is a high vacuum.) (PTR-Precision Technologies, Inc.)

An optical system is used to align the part. The optical system principle is shown in A, Fig. 18-46. Fig. 18-47 shows an operator aligning a piece inside a high vacuum electron beam machine. Fig. 18-48 shows a high production use of a partial vacuum electron beam welder. The parts are moved to the welder because the electron beam welder is too large to move.

## 18-20. LASER BEAM WELDING (LBW)

LASER BEAM WELDING is a high energy density welding process. The term LASER stands for Light Amplification by Stimulated Emission of Radiation. A laser beam is just light. The light beam has one wavelength and is in-phase. In-phase means all the particles or waves move together.

The laser beam is produced by one of three types of lasers:
1. Ruby laser.
2. Nd - YAG laser.
3. $CO_2$ (carbon dioxide) laser.

A ruby laser operates by exciting a ruby crystal. The excited electrons cause additional electrons

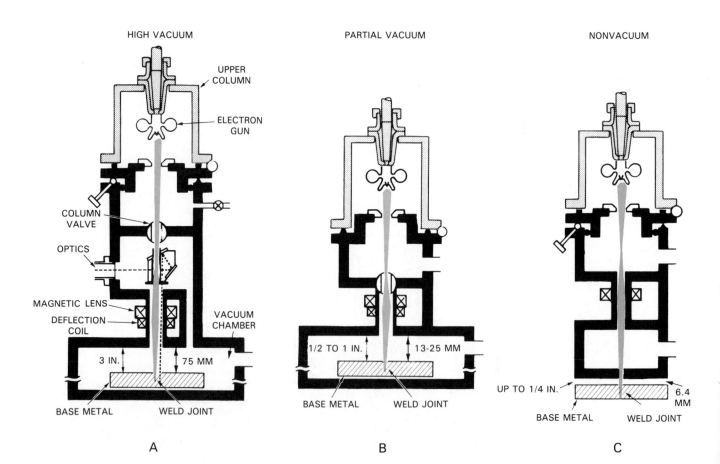

Fig. 18-46. Three types of electron beam welding systems: A—Weld being made under high pressure. B—Weld being made under partial vacuum. C—Weld being made under atmospheric pressure. (PTR-Precision Technologies, Inc.)

Fig. 18-47. Electron beam operator aligning a part for welding. (PTR-Precision Technologies, Inc.)

Fig. 18-48. A partial vacuum electron beam welder used with a four station dial table for high production work. (PTR-Precision Technologies, Inc.)

to become excited. When the excited electrons return to their stable state, they emit light. This light has one wavelength. The light is reflected between mirrors and finally released. A very short laser pulse is emitted. The ruby laser cannot be used continuously, therefore, it is not well suited for some welding applications. A ruby laser is shown schematically in Fig. 1-31. Refer to Heading 1-26 for additional information.

The Nd stands for Neodymium, a bright silvery rare earth metal element. The YAG is Yttrium Aluminum Garnet and acts as a crystal medium and host for the Nd. The combination creates a solid state laser crystal whose electrons are stimulated as are the $CO_2$ and ruby crystals.

A $CO_2$ (carbon dioxide) laser can supply a continuous laser beam. A $CO_2$ laser is the common laser used for welding. A $CO_2$ laser operates on the same principle as the ruby laser. A long tube is filled with $CO_2$ gas. The electrons in the carbon dioxide gas are excited by electrodes which surround the gas tubes. When the excited electrons return to their normal (ground) state, they give off light energy. This energy excites additional electrons which give off additional energy. With the light energy being produced continuously, the laser can be used to weld continuously. The main parts of the $CO_2$ laser are shown in Fig. 18-49. A

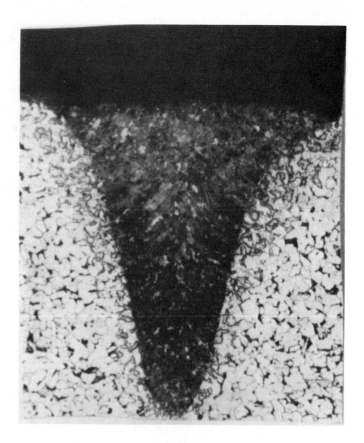

Fig. 18-50. Macrograph of laser weld. The weld is narrower than an arc weld. (Charles Albright, Ohio State University)

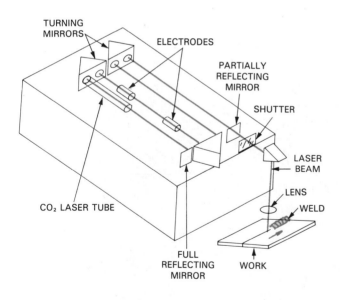

Fig. 18-49. A $CO_2$ laser. Mirrors are used to reflect and aim the laser beam. A lens is used to focus the beam.

$CO_2$ gas laser can produce a full penetration weld in 3/4 inch (approximately 19 mm) steel. Most solid state lasers can only penetrate 1/4 in. (6.4 mm) of steel. A macrograph of a laser weld is shown in Fig. 18-50.

The laser beam is light. It is focused and aimed using lenses and mirrors. Lasers are operated at atmospheric pressure. A shielding gas is used to protect the metal being welded from oxidation. The parts to be welded must have straight edges and be very close together. The travel speeds used for laser welding are much higher than arc welding.

The laser is also used for cutting and for heat treating. In heat treating, the laser beam is not focused to a spot. The beam is slightly divergent so it covers and heat treats a small area.

## 18-21. SELF-GENERATING OXYHYDROGEN GAS WELDING

The production of oxygen gas and hydrogen gas by the electrolysis of water has been done for many years. The burning of hydrogen in air or in pure oxygen has been used as a welding flame or gas welding process for several decades.

However, only recently have these two processes been combined. Distilled water is changed into its two gases by electrolysis. These gases are then fed to a torch and burned.

A self-generating oxyhydrogen gas welding unit is shown in Fig. 18-51. The unit operates on 115

Fig. 18-51. A self-generating oxyhydrogen gas welding outfit. Note the size of the torch tip in the upper corner.
(E. Spirig, supplied by Solder Absorbing Tech, Inc.)

volts AC. An AC to DC converter provides power for the electrolytic reaction. The mixed gases travel along a single hose to the torch. The unit also has a booster which contains methyl alcohol. As the gas mixture bubbles through the booster, alcohol vapor is mixed with the gas and it also burns in the torch flame. The alcohol vapor provides a slightly reducing flame for soldering, brazing, or welding. Fig. 18-52 shows a self-generating gas welding station in use without the methyl alcohol booster.

All the safety precautions as recommended in the previous chapters on oxyfuel gas welding, soldering, and brazing should be followed.

Fig. 18-52. A self-generating oxyhydrogen welding outfit being used to solder a small copper box.
(E. Spirig, supplied by Solder Absorbing Tech, Inc.)

## 18-22. SOLID PELLET OXYFUEL GAS WELDING

This convenient oxyfuel gas welding system uses oxygen rich solid pellets. It has been specially developed to use with a propane cylinder as the fuel source.

The advantage of this system is that it does not require an oxygen cylinder. The propane cylinder and a supply of pellets is all that is required to do a welding, brazing, or cutting job anywhere. See Fig. 18-53.

This oxypropane outfit is designed to use different sizes of pellets, depending on the torch tip size. For larger welding jobs, the torch uses a larger tip and a larger pellet. The oxygen rich solid pellets are burned to release their oxygen. Once the pellet is lighted, it is placed in a special container or cylinder.

The manufacturers of this system supply a very thorough guide for its operation and use. If the guide is carefully followed, satisfactory welds may be made and the operation of the system is quite safe.

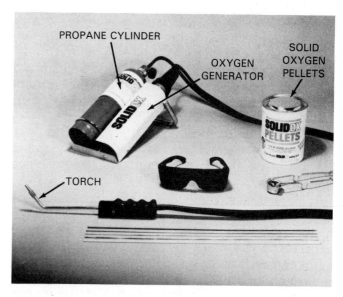

Fig. 18-53. An oxypropane welding system. The oxygen is produced by heating a solid oxygen bearing pellet.
(Cleanweld Products, Inc.)

## 18-23. REVIEW OF SAFETY

1. In submerged arc welding, the flux remaining over the completed weld acts as a heat insulator. The weld, as a result, remains very hot for some time after it is completed.
2. The usual eye protection and protective clothing should be worn when doing electroslag or electrogas welding.

3. There is some flash and sparking from stud arc welding operations. Eyes, face, and hands must be protected from this sparking. Similar protection is necessary when performing arc spot welding.
4. Laser welding uses light energy. The beam of light from the laser welding equipment must never strike any object except the parts to be welded. Always wear eye protection.
5. Plasma arc welding and cutting produce intense sound. In addition to eye, face, and body protection, it is usually necessary to wear ear muffs during such welding or cutting operations.
6. There is usually considerable sparking during the hammering together of parts to be joined in forge welding. Face, eyes, and hands must be adequately protected.
7. Friction welding generates considerable sparking. Protective clothing, goggles, and gloves should be worn.
8. Handling explosives is always dangerous. They must be carefully stored, applied, and detonated. Follow manufacturer's instructions.
9. A considerable quantity of molten metal is formed in the crucible during the thermit welding process. When this molten metal enters the mold, the metal must be carefully managed so that it does not come in contact with moist or wet surfaces. Protect eyes, face, hands, and body.
10. Oxygen and hydrogen gas produced in a self-generating unit may form a very explosive mixture. Hydrogen is a highly explosive gas.
11. The underwater welder should be in touch with the surface operator by means of a communication system. Precautions must be taken if a metal helmet is worn so that no part of it comes in contact with the welding circuit.
12. Solid oxygen pellet welding systems require careful handling. The pellet burns at a very high temperature. A partially burned pellet may be kept for future use; however, handling a burning pellet requires great care. The usual eye protection, clothing, and gloves must be worn. Since this system can be used in so many locations, surrounding areas must be cleared of all combustible material. The flame can produce temperatures of more than 5000°F (2760°).

These precautions, along with common sense will protect you from harm.

## 18-24. TEST YOUR KNOWLEDGE

Write your answers on a separate sheet of paper. Do not write in this book.
1. How are the arc and molten metal protected from oxidation in submerged arc welding?
2. Can more than one electrode wire be used when submerged arc welding? Why?
3. What is the purpose of the copper shoes used during electroslag or electrogas welding?
4. What is the difference between electroslag and electrogas welding?
5. What is the advantage of narrow gap welding?
6. What is the problem with narrow gap welding?
7. In stud welding, what is placed over the stud prior to welding?
8. The tungsten electrode in plasma arc welding is surrounded by an _____ gas. This gas may be an inert gas like _____ or _____, but it cannot contain _____.
9. What is the difference between a transferred and nontransferred arc in plasma arc welding?
10. List three things a welder can do to prevent the rapid heat loss while under water welding.
11. How are the solid state welding processes different from all other welding processes?
12. What is used in cold welding, explosion welding, and forge welding to join two pieces of metal?
13. How is heat generated in inertia welding?
14. What are some advantages (at least 4) of ultrasonic welding?
15. List two advantages of electron beam welding. List the main disadvantages.
16. How high an accelerating voltage may be used in electron beam welding?
17. Why is a $CO_2$ laser better than a ruby laser for welding?
18. How is a laser beam focused on the part being welded?
19. How is the oxygen produced in solid pellet oxyfuel gas welding?
20. What safety precautions should be taken during plasma arc welding or cutting?

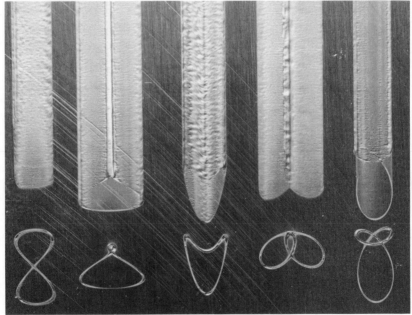

Using oscillatory beam deflection capability with the electron beam. At top, the beam is being moved in a simple ''bow tie'' motion while at bottom, the electron beam is being moved in a variety of other repetitive motion patterns. The view from above shows the effect these beam patterns have on the bead shape produced.
(PTR-Precision Technologies, Inc.)

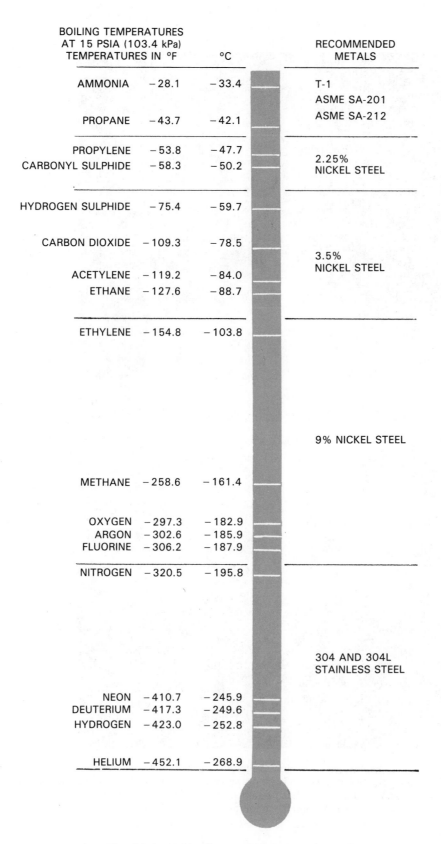

BOILING TEMPERATURES
AT 15 PSIA (103.4 kPa)

| TEMPERATURES IN °F | | °C | RECOMMENDED METALS |
|---|---|---|---|
| AMMONIA | − 28.1 | − 33.4 | T-1 |
| | | | ASME SA-201 |
| PROPANE | − 43.7 | − 42.1 | ASME SA-212 |
| PROPYLENE | − 53.8 | − 47.7 | 2.25% |
| CARBONYL SULPHIDE | − 58.3 | − 50.2 | NICKEL STEEL |
| HYDROGEN SULPHIDE | − 75.4 | − 59.7 | |
| CARBON DIOXIDE | − 109.3 | − 78.5 | 3.5% |
| ACETYLENE | − 119.2 | − 84.0 | NICKEL STEEL |
| ETHANE | − 127.6 | − 88.7 | |
| ETHYLENE | − 154.8 | − 103.8 | |
| | | | 9% NICKEL STEEL |
| METHANE | − 258.6 | − 161.4 | |
| OXYGEN | − 297.3 | − 182.9 | |
| ARGON | − 302.6 | − 185.9 | |
| FLUORINE | − 306.2 | − 187.9 | |
| NITROGEN | − 320.5 | − 195.8 | |
| | | | 304 AND 304L |
| | | | STAINLESS STEEL |
| NEON | − 410.7 | − 245.9 | |
| DEUTERIUM | − 417.3 | − 249.6 | |
| HYDROGEN | − 423.0 | − 252.8 | |
| HELIUM | − 452.1 | − 268.9 | |

Fig. 19-1. Table of recommended steels used to contain low temperature fluids. (Welding Design and Fabrication)

# 19 SPECIAL FERROUS WELDING APPLICATIONS

Practically all metals can be successfully welded. This chapter is devoted to the ferrous metals and alloys for which special welding instruction may be needed. FERROUS METALS and alloys are those which contain large amounts of iron. These include all stainless steels and cast irons. Chapter 20 covers the special welding applications as used on nonferrous metals and alloys.

Most of these special applications necessitate preheat and postweld heat treatment. Chapter 27 describes heat treatment processes. Heading 27-20 discusses the use of temperature indicating crayons, markers, and pellets. These temperature indicators should be used when preheating or postweld heat treating to make sure the correct temperature is reached and maintained during the procedure.

## 19-1. FERROUS METALS AND ALLOYS

The two large classifications of metals are:
1. Ferrous metals.
2. Nonferrous metals.
FERROUS METALS are those bearing a substantial iron content. These are the steels and cast irons. Many alloys are added to steel to obtain desired properties. When welding the alloy steels, special considerations and techniques are often required.

Some examples of uses for alloy steels are to use low alloy steels for high strength, lightweight construction. Use stainless steels for applications where appearance and corrosion resistance are important.

The coating on metals must also be considered when these metals are being welded. Galvanized steel (zinc coated steel) is a very popular metal and it can be successfully welded.

Welding finds an important use in the fabrication and repair of cast iron articles. Cast iron is normally a brittle metal. Under the welding flame

or the electric arc it behaves differently from most of the steels previously described.

## 19-2. ALLOY STEELS

Alloy steels have been produced and used for many decades. New alloys are constantly being developed. Each commercially produced alloy has special properties. Some are more corrosion resistant, some are stronger, some are tougher and able to resist strain without breaking. A variety of alloy steels have been developed to meet existing needs.

In welding these steels, they must be accurately identified so the proper welding procedures may be used. All of the following must be known and carefully followed for the best welding results:
1. The correct preheat temperature.
2. The proper welding process.
3. The proper electrode or filler metal.
4. The correct welding sequence.
5. The postweld heat treatment procedure.
Some of the more popular alloy steels are:
1. Low alloy steels.
2. Stainless steels.
3. Tool and die steels.
4. Galvanized steels.
For low temperature use, the steel alloys shown in Fig. 19-1 are recommended. These steel alloys retain enough strength and ductility to permit their use at these low temperatures. Welds in these metals, for use at these low temperatures, must have similar properties to those of the base metal.

## 19-3. LOW ALLOY STEELS

Many articles are fabricated of rolled low alloy steel. The use of low alloy steel is desirable especially when a reduction in weight is possible. Low alloy steels usually contain less than 3 percent alloying elements. Low alloy steels are 10 to

30 percent stronger than straight carbon steel. These steels are often called high strength low alloy steels. Low alloy steels are desirable where a saving in weight is important. An example of this is a power shovel bucket. Generally speaking, each pound that is saved in the bucket means an extra pound of material may be handled with the same power. One company was able to convert a 5 ton capacity power shovel bucket into a 6 ton carrying capacity by changing from a cast desgin to a fabricated low alloy steel design.

These alloy steels are slightly more expensive than the straight carbon steel. The alloy steels are preferred because of the reduction of the weight and the additional strength. The method of welding the low carbon alloy steels is similar to

| AWS Classification[a] | Type of covering | Capable of producing satisfactory welds in positions shown[b] | Type of current[c] |
|---|---|---|---|
| E70 series — Minimum tensile strength of deposited metal, 70,000 psi (480 MPa) | | | |
| E7010-X | High cellulose sodium | F, V, OH, H | DCEP |
| E7011-X | High cellulose potassium | F, V, OH, H | AC or DCEP |
| E7015-X | Low hydrogen sodium | F, V, OH, H | DCEP |
| E7016-X | Low hydrogen potassium | F, V, OH, H | AC or DCEP |
| E7018-X | Iron powder, low hydrogen | F, V, OH, H | AC or DCEP |
| E7020-X | High iron oxide | H-fillets / F | AC or DCEN / AC or DC, either polarity |
| E7027-X | Iron powder, iron oxide | H-fillets / F | AC or DCEN / AC or DC, either polarity |
| E80 series — Minimum tensile strength of deposited metal, 80,000 psi (550 MPa) | | | |
| E8010-X | High cellulose sodium | F, V, OH, H | DCEP |
| E8011-X | High cellulose potassium | F, V, OH, H | AC or DCEP |
| E8013-X | High titania potassium | F, V, OH, H | AC or DC, either polarity |
| E8015-X | Low hydrogen sodium | F, V, OH, H | DCEP |
| E8016-X | Low hydrogen potassium | F, V, OH, H | AC or DCEP |
| E8018-X | Iron powder, low hydrogen | F, V, OH, H | AC or DCEP |
| E90 series — Minimum tensile strength of deposited metal, 90,000 psi (620 MPa) | | | |
| E9010-X | High cellulose sodium | F, V, OH, H | DCEP |
| E9011-X | High cellulose potassium | F, V, OH, H | AC or DCEP |
| E9013-X | High titania potassium | F, V, OH, H | AC or DC, either polarity |
| E9015-X | Low hydrogen sodium | F, V, OH, H | DCEP |
| E9016-X | Low hydrogen potassium | F, V, OH, H | AC or DCEP |
| E9018-X | Iron powder, low hydrogen | F, V, OH, H | AC or DCEP |
| E100 series — Minimum tensile strength of deposited metal, 100,000 psi (690 MPa) | | | |
| E10010-X | High cellulose sodium | F, V, OH, H | DCEP |
| E10011-X | High cellulose potassium | F, V, OH, H | AC or DCEP |
| E10013-X | High titania potassium | F, V, OH, H | AC or DC, either polarity |
| E10015-X | Low hydrogen sodium | F, V, OH, H | DCEP |
| E10016-X | Low hydrogen potassium | F, V, OH, H | AC or DCEP |
| E10018-X | Iron powder, low hydrogen | F, V, OH, H | AC or DCEP |
| E110 series — Minimum tensile strength of deposited metal, 110,000 psi (760 MPa) | | | |
| E11015-X | Low hydrogen sodium | F, V, OH, H | DCEP |
| E11016-X | Low hydrogen potassium | F, V, OH, H | AC or DCEP |
| E11018-X | Iron powder, low hydrogen | F, V, OH, H | AC or DCEP |
| E120 series — Minimum tensile strength of deposited metal, 120,000 psi (830 MPa) | | | |
| E12015-X | Low hydrogen sodium | F, V, OH, H | DCEP |
| E12016-X | Low hydrogen potassium | F, V, OH, H | AC or DCEP |
| E12018-X | Iron powder, low hydrogen | F, V, OH, H | AC or DCEP |

a. The letter suffix 'X' as used in this table stands for the suffices A1, B1, B2, etc. (see Fig. 9-44) and designates the chemical composition of the deposited weld metal.
b. The abbreviations F, V, OH, H, and H-fillets indicate welding positions as follows: F = Flat; H = Horizontal; H-fillets = Horizontal fillets. V = Vertical } For electodes 3/16 in. (4.8 mm) and under, except 5/32 in. (4.0 mm) and under for classifications OH = Overhead} EXX15-X, EXX16-X, and EXX18-X.
c. DCEP means electrode positive (reverse polarity). DCEN means electrode negative (straight polarity).

Fig. 19-2. AWS Low Alloy Steel Covered Arc Welding Electrodes. This information is from AWS 5.5-81.

that of straight carbon steel.

Oxyfuel gas welding is not recommended for welding low alloy steels. If oxyfuel gas welding is used, a flux should be used to counteract the oxidation of alloying elements. Filler rods of a composition corresponding to the base metal must be used.

All arc welding processes can be used to weld the low alloy steels. When shielded metal arc welding (SMAW), an alloy electrode must be used. The tensile strength and alloy composition of the electrode should match the base metal. Alloy composition is given by a letter and number code after the electrode designation. Examples are:

E7015-A1 (molybdenum steel alloy electrode)

E9018-B1, E8018-B2 (chromium-molybdenum steel electrode)

E8018 - C2 (Nickel steel alloy electrode)

The low alloy SMAW electrodes are discused in Heading 9-16. Fig. 19-2 gives information on the low alloy steel, covered electrodes.

Gas metal arc welding (GMAW) and flux cored arc welding (FCAW) are used to weld low alloy steels. The electrode wire must have a similar alloy composition to the base metal. These electrode alloy designations are similar to those used for SMAW electrodes. Examples are:

ER80S* - B2 (chrome-molybdenum steel)

E90C* - Ni2 (nickel steel)

E80T1* - A1 (carbon-molybdenum steel)

*S - a solid electrode

C - a composite metal cored electrode

T - a tubular electrode

These GMAW and FCAW electrodes are explained in Headings 11-23 and 11-26.

## 19-4. STAINLESS STEELS

A large number of different steels are known as stainless steels. All contain varying amounts of chromium or a combination of chromium and nickel. Other alloying elements are also added.

The minimum chromium content to obtain a stainless steel is about 12 percent. These steels are corrosion resistant, retain a good clean appearance, and have good physical properties.

The American Iron and Steel Institute has classified these alloys by number. A three digit number is used to classify each alloy type. Almost all stainless steels are classified as 3XX or 4XX, such as 304, 316, 410, 430, 446. Some additional stainless steels are classified as 2XX or 5XX. Some classifications have an L following the three numbers like 304L. The L means a low carbon content.

There are three general classifications of Stainless Steels:

1. Austenitic.
2. Martensitic.
3. Ferritic.

These three types of stainless steels are discussed in the following headings.

## 19-5. AUSTENITIC STAINLESS STEEL

When steel is heated to a high temperature, above 1341 °F (727 °C), it is called AUSTENITE. Austenite is a solid solution of iron carbide ($Fe_3C$) in iron.

An austenitic stainless steel remains austenite at room temperature. Chromium is added to make

| AISI Type | % Carbon | % Chromium | % Nickel | % Other |
|-----------|----------|------------|----------|---------|
| 201 | 0.15 | 16 - 18 | 3.5- 5.5 | N, 0.25; Mn, 5.5-7.5; P, 0.06 |
| 202 | 0.15 | 17 - 19 | 4.0- 6.0 | N, 0.25; Mn, 7.5-10.0; P, 0.06 |
| 301 | 0.15 | 16 - 18 | 6.0- 8.0 | — |
| 302 | 0.15 | 17 - 19 | 8.0-10.0 | — |
| 304 | 0.08 | 18 - 20 | 8.0-12.0 | — |
| 304L | 0.03 | 18 - 20 | 8.0-12.0 | — |
| 308 | 0.08 | 19 - 21 | 10.0-12.0 | — |
| 309 | 0.20 | 22 - 24 | 12.0-15.0 | |
| 310 | 0.25 | 24 - 26 | 19.0-22.0 | Si, 1.50 |
| 316 | 0.08 | 16 - 18 | 10.0-14.0 | Mo, 2.0-3.0 |
| 316L | 0.03 | 16 - 18 | 10.0-14.0 | Mo, 2.0-3.0 |
| 317 | 0.08 | 18 - 20 | 11.0-15.0 | Mo, 3.0-4.0 |
| 347 | 0.08 | 17 - 19 | 9.0-13.0 | Cb + Ta, 10 × C min |

Fig. 19-3. Compositions of austenitic stainless steels.

the steel stainless, and nickel is added to allow the steel to remain austenite. Austenitic stainless steels are classified in the 300 series, 3XX. Fig. 19-3 lists the chemical composition of some austenitic stainless steels. The type 201 and 202 are similar to type 301 and 302, except that manganese is added in place of some of the nickel.

Austenitic stainless steels have very good strength and corrosion resistance at elevated temperatures. Molybdenum is added to improve the strength at high temperatures. Austenitic stainless steels are also well suited for low temperature applications.

## 19-6. MARTENSITIC STAINLESS STEEL

A martensitic stainless steel is very hard and not very ductile. MARTENSITE is produced by quenching (rapidly cooling) a steel from the austenite phase. Martensitic stainless steels have enough chromium that when they are cooled in air, they form martensite. These stainless steels have between 12 and 16 percent chromium and very little nickel, usually no nickel. Martensitic stainless steels are part of the 400 series, 4XX. Some of the martensitic stainless steel compositions are shown in Fig. 19-4.

| AISI Type | % Carbon | % Chromium | % Other |
|---|---|---|---|
| 410 | 0.15 max. | 12 | — |
| 414 | 0.15 max. | 12 | 2.0 Nickel |
| 416 | 0.15 max. | 13 | 0.3 Sulphur |
| 420 | 0.15 min. | 13 | — |
| 431 | 0.20 max. | 16 | 2.0 Nickel |

Fig. 19-4. Compositions of martensitic stainless steels.

When welding martensitic stainless steel, the base metal must be preheated and postweld heat treated. Interpass (between weld passes) or continuous heating is often required. The base metal is preheated to 500°F (260°C). After the welding is completed, the weldment is heated to 1300°F (704°C), slow cooled to 1100°F (593°C), and then air cooled. This postweld heat treatment will temper the martensite and prevent cracking. Martensite is tempered to improve its toughness. Martensitic stainless steels are often welded using an austenitic filler metal (300 series). Austenitic filler metal must be used when no postweld heat treatment will be used.

The high carbon martensitic stainless steels are difficult to weld.

## 19-7. FERRITIC STAINLESS STEEL

FERRITIC STAINLESS STEELS are not hardened by heat treating. They contain large amounts of chromium which improves their corrosion resistance. The alloys containing 15-20 percent chromium are easy to weld. The alloys with very high levels of chromium (25 percent) are difficult to weld and have poor mechanical properties. Fig. 19-5 lists the composition of the ferritic stainless steels.

The 15-20 percent chromium alloys do form some martensite in the weld metal. The heat affected zone may lose some of its corrosion resistance. To restore the corrosion resistance of the heat affected zone, the weldment must be postweld annealed at 1300-1550°F (704-843°C)

| AISI Type | % Carbon | % Chromium | % Other |
|---|---|---|---|
| 429 | 0.12 max. | 15 | — |
| 430 | 0.12 max. | 17 | — |
| 434 | 0.12 max. | 17 | 1.0 Molybdenum |
| 442 | 0.20 max. | 20 | — |
| 446 | 0.20 max. | 25 | 0.15 Nitrogen |

Fig. 19-5. Compositions of ferritic stainless steels.

## 19-8. CHROME-MOLY STEELS

Chrome-molybdenum (chrome-moly) steels contain a small amount of molybdenum, .5 to 1 percent, and a medium amount of chromium, 4-10 percent. For example, type 501 or 502 contain 5 percent chromium and .5 percent molybdenum. Type 505 contains 9 percent chromium and 1 percent molybdenum. Although these steels contain less than 12 percent chromium, they still have a good resistance to corrosion. Additional chrome-moly steels have lower chrome contents and less corrosion resistance. Chrome varies between 1 and 3 percent. Molybdenum is still .5-1.0 percent.

Chrome-moly steels are used at high temperatures: for example, with high temperature steam, 1100-1200°F (593-649°C). These steels have good yield strength and creep strength at high temperatures. The YIELD STRENGTH is the point at which the metal assumes a permanent set when a load or stress is released. The CREEP STRENGTH is a metal's ability to resist a slow stretching under a continuous load.

Chrome-moly steels are welded with electrodes

of similar composition. Preheating and interpass heating are required. Postweld heat treating is highly recommended for large, restrained weld joints. The postweld heat treatment is performed before the weld is allowed to cool. The steel can be quenched and tempered for high hardness or annealed for a low hardness, ductile (ability to stretch) steel.

Austenitic electrodes can be used to weld the chrome-moly steels.

## 19-9. STAINLESS STEEL ELECTRODES

Stainless steel electrodes and filler metals are made to join all types of stainless steel. The electrodes and filler metals are designated just like the base metal. A three number classification system is used. The 300 series are austenitic, the 400 series are martensitic or ferritic, and the 500 series are the chrome-moly electrodes.

Shielded metal arc welding (SMAW) electrodes are designated as E308, E316, E410, or E502. Two types of low hydrogen stainless steel electrodes are available. They are designated − 15 and − 16. These designations follow the electrode classification. Examples are: E308 - 15 or E312 - 16. The − 15 type is used with DCEP (DCRP). The − 16 type is used with DCEP (DCRP) or alternating current. The type − 15 gives better penetration.

Electrodes are available with a low carbon or high carbon content. The letter L after the electrode classification means low carbon. A letter H means high carbon. Examples are E308L, E308H and E316L. A complete description and listing of all stainless steel SMAW electrodes is included in Heading 9-19.

Gas metal arc welding (GMAW) electrodes are used with DCEP (DCRP). Argon or an argon-helium mixture is used as a shielding gas. Small additions of carbon dioxide ($CO_2$) or oxygen can be made. The electrodes are designated with an ER before the classification. Examples are ER 308, ER 316L, and ER 321. These same wires are used as filler metal when gas tungsten arc welding.

The electrode used can be selected to match the base metal composition, but it does not have to match. Austenitic electrodes can be used to weld martensitic or ferritic stainless steels.

## 19-10. WELDING STAINLESS STEEL

Stainless steel can be welded by any arc welding process. Other welding processes which can be used include resistance, electron beam, laser, plasma, flash, and electroslag welding. Oxyfuel gas welding is rarely used to join stainless steel.

Before welding stainless steel, the base metal must be cleaned. A stainless steel wire brush is commonly used. Fluorides are used to remove any chromium oxides from the surface. Fluorides are also present in SMAW electrode coverings. The fluorides must be removed when the welding is completed. Fluorides are not used when an inert gas process is used.

Stainless steel is very susceptible to porosity when welding. To prevent porosity, a short arc length must be used. When gas metal or gas tungsten arc welding, shielding gas must be provided for the back side of the weld. This is called a BACKING GAS. The shielding gas or flux covering prevents the chromium from becoming oxidized.

Another problem with stainless steel is grain growth. When too much heat is used on a ferritic or austenitic stainless steel, the grains will grow to a large size. Large grains decrease the mechanical properties of strength and ductility. A stainless steel with large grains may also lose its resistance to corrosion. To prevent this, a lower voltage or current is used to reduce the heat input. Short circuit transfer or a pulsed arc can be used when GMAW instead of a spray transfer. A straight bead, without any weave, should be used. This also keeps the heat input low. Oxyfuel gas welding is not used because it puts too much heat into the base metal.

Other than using a short arc length, reducing the current and voltage, and not using a weave technique, the welding of stainless steel is very similar to welding mild steel. The weld must be thoroughly cleaned after welding and between weld passes.

## 19-11. HEAT TREATING STAINLESS STEEL WELDS

Almost all stainless steel is preheated before welding. Temperatures between 300 and 600 °F (149-316 °C) are common. The martensitic and chrome-moly steels require interpass heating and postweld heat treating to prevent cracking and to obtain the desired properties.

Ferritic and austenitic stainless steels often require postweld heat treating to obtain the desired properties. They are also heat treated to reduce the stresses in the weld. The postweld heat treatment also improves the corrosion resistance of the weld metal and the heat affected zone. The metal should not be held at high temperatures for long periods of time as this will cause grain growth.

The metal is heated to the proper temperature. It is held at the temperature for a period of time, depending on its thickness and then quenched. Ferritic stainless steel is heated to 1600-1650°F (871-899°C). Austenitic stainless steel is annealed by heating to 1850 to 2050°F (1010-1121°C). Stress relieving is done below 800°F (427°C).

## 19-12. NICKEL BASED ALLOYS

The nickel based alloys are better known as Inconel, Monel, or Hastelloy. These alloys contain between 40 and 80 percent nickel. Other major alloying elements include chromium, iron, molybdenum, copper, and cobalt. A number of additional alloying elements include carbon, aluminum, tungsten, and columbium. Fig. 19-6 lists the composition of some of the nickel base alloys.

welding. Preheat or postweld heat treatments are usually not required. A few alloys require postweld heat treating to restore the corrosion resistance of the alloy.

## 19-13. TOOL AND DIE STEEL

Stamping, drawing, and forging presses are widely used in production. The dies and tools used in these machines are capable of producing thousands of duplicate parts. Obviously these stamping, drawing, and forging tools and dies must be strong and must resist wear. They must be made very accurately, and they must hold their accuracy. These tools are usually made of a high carbon alloy steel. They are accurately shaped, heat treated, and ground.

If a part of the tool or die wears or breaks, the

| ALLOY | %Ni | %Cr | %Fe | %Cu | %Mo | %Co | %C | %Al | %W | %Cb |
|---|---|---|---|---|---|---|---|---|---|---|
| Monel alloy 400 | 66.5 | — | 1.25 | 31.5 | — | — | — | — | — | — |
| Monel alloy 502 | 66.5 | — | 1.00 | 28.00 | — | — | 0.05 | 3.00 | — | — |
| Inconel alloy 600 | 76.0 | 15.5 | 8.00 | 0.25 | — | — | 0.08 | — | — | — |
| Inconel alloy 601 | 60.5 | 23.0 | 14.1 | 0.50 | — | — | 0.05 | 1.35 | — | — |
| Inconel alloy 718 | 52.5 | 19.0 | 18.5 | 0.15 | 3.05 | — | 0.04 | 0.5 | — | 5.13 |
| Hastelloy alloy B | 61.0 | 1.0 | 5.0 | — | 28.0 | 2.5 | 0.05 | — | — | — |
| Hastelloy alloy C | 54.0 | 15.5 | 5.0 | — | 16.0 | 2.5 | 0.08 | — | 4.0 | — |
| Hastelloy alloy D | 82.0 | 1.0 | 2.0 | 3.0 | — | 1.5 | 0.12 | — | — | — |
| Hastelloy alloy G | 46.0 | 22.25 | 19.5 | 2.0 | 6.5 | — | 0.03 | — | 0.5 | 2.12 |
| Hastelloy alloy X | 47.0 | — | 18.0 | — | 9.0 | 1.5 | 0.10 | — | 0.6 | — |

Fig. 19-6. Compositions of nickel base alloys.

These high nickel alloys are used for high temperature applications. Inconel alloy 601 and Hastelloy alloy X are used to resist oxidation up to 2200°F (1204°C).

Nickel alloys can be welded by most arc welding processes. Before welding, it is very important that the base metal be thoroughly cleaned.

The electrode or filler metal should be selected to match the base metal composition. Examples of covered SMAW electrodes are ENiCu-2, ENiCrFe-3, and ENiMo-2. The major alloying elements are given in the electrode designation. The compositions must be obtained from the manufacturer. GMAW, GTAW, and SAW electrodes are listed similarily as ERNi-3 and ERNiCrFe-6. Nickel alloys can be oxyacetylene welded, but it is not recommended. Fuel gases other than acetylene cannot be used. A slightly reducing flame is used.

DCEP (DCRP) is used when shielded metal arc

production of a duplicate one is slow and expensive. Tool and die welding makes it possible to reclaim many of these tools by rebuilding the worn

Fig. 19-7. Die repaired by rebuilding the worn areas with overlapping beads. (Eureka Welding Alloys, Inc.)

surface. One example of a rebuilt die is shown in Fig. 19-7. Examples of salvaging valuable tools by welding are shown in Figs. 19-8 and 19-9.

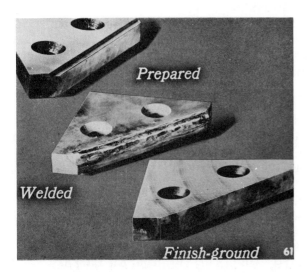

Fig. 19-8. Cutting tool prepared for welding, as welded, and as a finished ground part. (Eureka Welding Alloys, Inc.)

## 19-14. TYPES OF TOOL STEELS

There are many different alloys and trade brands of tool steel. When welding tool steels, it is necessary for the welder to select the electrode or welding rod to match the composition and the heat treatment of the particular tool. There are four general heat treatment classifications of tool steels:
1. Water hardening tool steel.
2. Oil hardening tool steel.
3. Air hardening tool steel.
4. Hot working tool steel (Ni-chrome steel base).

The type of tool steel is not obvious from inspecting the part. The manufacturer should be consulted to determine the type of tool steel.

## 19-15. TOOL STEEL WELDING PROCEDURE

To weld tool steels, the surface must be clean and occasionally shaped (ground) to best receive the electrode deposit. It is also good practice to preheat the tool or die. It is very important to deposit a minimum of metal and then peen the metal to relieve the stresses. Finally, the tool should be heat treated (quenched and tempered) according to the original specifications of the tool or die.

Some of the basic steps to be observed when arc welding tool steels are:
1. Know the type of tool steel. Obtain this data from the drawing or the manufacturer.
2. Use an electrode recommended for the type of

Fig. 19-9. Boring mill cutter. A—Before welding showing cutting edges damaged. B—After welding. When the build-up is completed, the cutter is reground to the proper shape. (Eureka Welding Alloys, Inc.)

steel to be welded. Refer to electrode manufacturers' catalogs.
3. Make the proper joint preparation.
4. Preheat the steel.
5. Weld using DCEP (DCRP).
6. Heat treat the weld.

The electrode deposit should completely cover the worn area. The weld deposit should be built up so it can later be ground to proper dimensions. Tool steel parts that have been welded are shown in Figs. 19-7, 19-8, and 19-9. The weld beads are overlapped and built up.

The weld quality depends upon several factors:
1. Type of steel.
2. Amount that the base metal mixes with the weld deposit.
3. Rate of cooling.
4. Preheat treatment.
5. Technique of welding.
6. Heat treatment after welding.

## 19-16. MARAGING STEELS

MARAGING STEELS contain very low amounts of carbon. The maximum carbon content is .03

percent or 3 point carbon. ONE POINT of carbon content equals .01 percent carbon. (Low carbon steels are usually .10 to .20 percent or 10 to 20 points of carbon. Maraging steels contain 18 percent nickel, 8 percent cobalt, 4 or 5 percent molybdenum, and small amounts of titanium and aluminum. The remainder is iron.

The maraging steels have extremely high strength. Ultimate tensile strength can reach 300 Ksi (2068 MPa). These steels still have good toughness.

The word maraging is a combination of two words, martensite and aging. Martensite forms when these alloys are cooled from the austenite phase. The martensite is then aged for up to 12 hours at 900°F (482°C). This aging greatly increases the strength, hardness, and toughness of the steel.

## 19-17. WELDING MARAGING STEELS

These steels are weldable using SMAW, GTAW, GMAW, or SAW processes. GTAW produces the best results. The filler metal should be of about the same composition as the base metal. The filler wire must have extremely high purity to produce a good weld. Preheat is not required but postweld heat treating is necessary.

When welding, a very low heat input should be used. The filler wire or electrodes must be very clean and pure. The base metal must be thoroughly cleaned prior to welding. Any impurities in the weld will produce a poor weld.

## 19-18. GALVANIZED STEEL

A popular way to protect mild steel against corrosion is to plate it with zinc (galvanizing). The zinc is deposited electrically or the steel is dipped in molten zinc.

When joining pieces of galvanized steel, it is desirable that the galvanizing not be destroyed.

Braze welding is a common method of joining galvanized steel. The metal is heated only to the melting temperature of the brazing rod and the galvanized coating is affected very little. This method may be used on pipe that is galvanized inside as well as outside. Common processes for braze welding include SMAW, GMAW, and GTAW. A brazing electrode or filler metal is used. Very low current settings should be used so that the brazing filler metal is melted but not the base metal. The oxyfuel method can also be used. A small tip should be used with a neutral or slightly oxidizing flame.

Galvanized steel may also be arc welded. The temperatures reached result in some destruction of the galvanized coating in the areas adjacent to the weld. When SMAW galvanized steel, an E6012, E6013, or E7016 electrode is used. When GMAW, an ER70S-3 electrode provides good results.

After the welding is completed, the zinc coating must be replaced. Three methods are used to replace the zinc. The metal must first be cleaned. One method is to apply a zinc rich paint over the damaged area. Another method is to flame spray zinc onto the weld area. The third method is to use a zinc or zinc alloy stick. After the weld solidifies, but is still quite hot, the zinc stick is placed onto the hot metal. The zinc stick melts and covers the weld with a new zinc coating. If the weld has cooled so the zinc will not melt, an oxyfuel gas torch can be used to reheat the metal. Then the zinc stick is used to restore the galvanized coating.

## 19-19. WELDING DISSIMILAR METALS (FERROUS BASE)

There are many iron and steel fabrications which require welding metals together even though their compositions are different. Welding stainless steels to mild steels, low alloy steels to mild steels, and nickel base alloys to stainless steel are typical examples.

It is very important that the composition and the properties of each metal be known. The welding rod or electrode composition should be chosen carefully. The composition should be similar to the metal which has the least amount of alloying metals. The electrode should be a low carbon electrode.

It is advisable to preheat the joint to 200°-300°F (93°-149°C) to produce better welding results. Practice on sample pieces, if possible.

## 19-20. TYPES OF CAST IRON

Cast iron contains between 1.8 and 4 percent carbon. For certain purposes, alloy metals may be added to the cast iron. Cast iron is used extensively for heavy machine parts. Its melting temperature is approximately 2600°F (1427°C).

There are four principal types of cast iron:
1. White cast iron.
2. Gray cast iron.
3. Malleable cast iron or malleable iron.
4. Nodular or ductile cast iron.

Each of these cast iron types is fully described in Chapter 26. A brief description is also given here.

WHITE CAST IRON is a casting that has been cooled rapidly (chilled) after it has been poured.

The name white cast iron is due to the appearance of the fracture. The metal is extremely hard and is very difficult to machine. White cast iron is considered to be unweldable.

GRAY CAST IRON is cast iron that has been cooled very slowly from its critical temperature in sand or in a furnace. The name is derived from the gray appearance of the fracture. The gray is the result of graphite flakes in a matrix of iron (ferrite) and iron carbide. This metal is easy to machine. A welder usually heat treats all cast iron welds, and allows them to cool slowly in order to produce gray cast iron.

MALLEABLE CAST IRON or MALLEABLE IRON is a white cast iron that is heat treated. The casting is heated to 1400°F (760°C) for 24 hours for each inch of thickness and then cooled slowly. This heat treatment permits the carbon to form small rosettes or spheres of carbon in a matrix of low carbon iron.

NODULAR or DUCTILE CAST IRON is produced by adding a small amount of magnesium to the cast iron. The carbon then forms nodules or little spheres in the metal. This metal has much greater ductility than other forms of cast iron.

Gray cast iron, malleable cast iron, and nodular cast iron are all weldable. Preheating is required and these cast irons usually require postweld heat treating.

## 19-21. PREPARING CAST IRON FOR WELDING

Cast iron may be prepared for welding much the same as steel. Thin pieces should be cleaned by grinding. Thick sections, 1/4 in. (approximately 6 mm) or more, should be beveled at a 60 degree angle, leaving a 1/16 in. (1.6 mm) root face. Fig. 19-10 shows a cast iron joint prepared for welding and a cross section of the finished weld.

Cast iron is frequently brazed or braze welded. The process of brazing cast iron is explained in Heading 8-14.

## 19-22. METHODS OF STRENGTHENING A CAST IRON JOINT

Occasionally the welder may prepare the cast iron seam in special ways to obtain a stronger joint. The methods used are:
1. Insert studs in the edges to be joined.
2. Cut round bottom grooves in the cast iron surfaces to be welded. See Fig. 19-11. These methods are especially useful when braze welding cast iron.

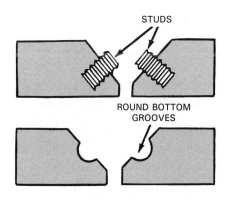

Fig. 19-11. Methods of strengthening a cast iron joint.

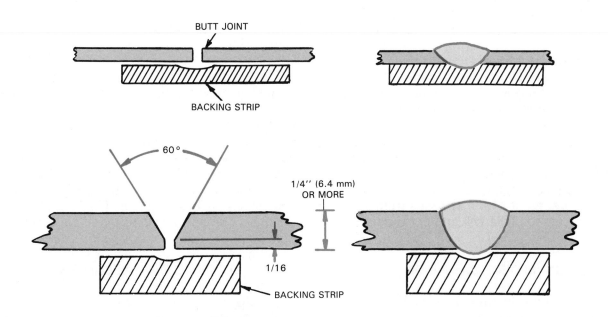

Fig. 19-10. Preparing cast iron joints for welding and cross sections of finished welds.

## 19-23. CAST IRON ELECTRODES AND FILLER METAL

The electrodes used to SMAW cast iron are nickel, nickel alloy, mild steel, or nickel-copper-iron alloy. Nickel alloy electrodes, such as ENi-CI and ENiFe-CI were designed to weld cast iron. The electrodes containing copper, such as ENiCu-A and ENiCu-B, are also used to weld cast iron. The CI stands for cast iron. Cast iron electrodes are described in the American Welding Society specification AWS A5.15.

When oxyfuel welding, two filler metal types are available:

    RCI

    RCI-A.

The RCI-A filler rod has a higher tensile strength of 35 to 40 Ksi (241 to 276 MPa). The oxyfuel gas filler metal is available in round or square cast rods. The round rods are 1/16 to 3/8 inch (1.6 to 10 mm) in diameter and 24 or 36 inches (610 or 914 mm) in length.

## 19-24. PREHEATING CAST IRON

Most cast iron objects must be preheated before welding. This is to prevent cracking of the metal as the weld cools. Cracking is due to the brittleness of the metal and to the fact that most cast iron welds are on complicated frames and structures, for example, cast iron wheels or frames.

PREHEATING is used to reduce the cracking tendency of cast iron. Preheating should be done in a furnace. The structure to be welded should be heated to a temperature of 400 to 600 °F (204 to 316 °C) when shielded metal arc welding will be used. When oxyfuel gas welding, the structure should be heated to 1100-1200 °F (593-649 °C).

These preheat temperatures should be maintained until the welding is completed. Fire bricks are usually built around the structure to be welded to retain the heat and permit quicker and more economical preheating. These high preheat temperatures may tend to make welding unpleasant, but they are needed to obtain a good quality weld.

If a furnace is not available, preheating may be done with an oxyfuel gas torch. By preheating certain members, most of the stresses from welding can be eliminated. This will prevent the part from cracking. See Fig. 19-12.

## 19-25. ARC WELDING CAST IRON

Cast iron is usually arc welded with the SMAW process. GMAW does not produce good welds in cast iron.

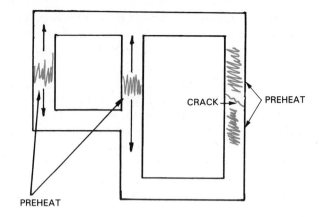

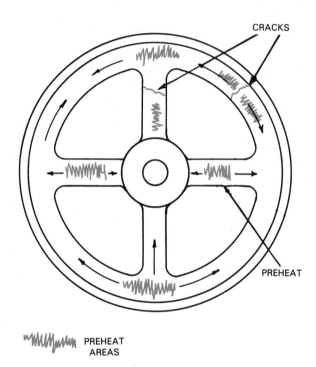

Fig. 19-12. Local preheating of structural cast iron parts.

The welding of cast iron is similar to the welding of mild steel. Electrode size, electrode motion, and arc length are all similar to welding the same thickness of mild steel. AC or DCEP (DCRP) is used.

Deposited weld metal containing copper is, however, weakened by cracks when it is mixed with the cast iron base metal. Care must be taken to prevent unnecessary mixing of the electrode and cast iron base metals. Fig. 19-13 shows a cross section of a shielded metal arc weld on cast iron.

## 19-26. OXYFUEL GAS WELDING CAST IRON

The tip size used for welding cast iron should be similar to the size used for welding steel of the same thickness. A neutral flame should be used

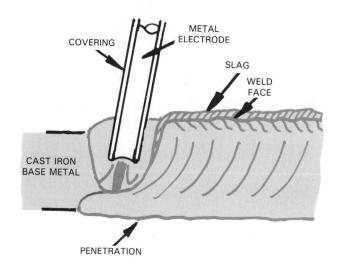

Fig. 19-13. Welding cast iron by the shielded metal arc welding process.

along with a cast iron welding rod of the proper size. A flux is also required.

When welding cast iron, the backhand technique is usually used. The torch should be held at a 60 degree angle with the plate. The inner cone should not touch the metal.

Flux is added by coating the welding rod with flux. The flux enters the weld as the welding rod is consumed. The flux must have the correct constituents. It must be fresh, clean, and be free from moisture.

In cast iron welding, the molten pool is not very fluid. It is important that gas pockets and oxides be worked to the surface of the weld. This may be done by stirring the molten pool with the filler rod. The oxides or gases will then be removed by the flux. The weld must have thorough penetration, and a slight crown in preferred. An oscillating torch and filler rod motion is usually used. Fig. 19-14 shows cast iron being welded by the oxyfuel gas process. Notice how the flux is added, and that the backhand technique is being used.

The oxyfuel gas process is also used to braze and braze weld cast iron. Brazing and braze welding cast iron are described in Heading 8-14.

## 19-27. HEAT TREATING CAST IRON WELDS

A cast iron weld should be cooled slowly. Slow cooling will form gray cast iron and prevent brittle white cast iron from forming.

As soon as the welding is completed, the cast iron is stress relieved. The casting is placed in a furnace before the metal cools. The welded structure is heated to 1100°F (593°C). This temperature is held for one hour for each inch (25.4 mm) of thickness and then slow cooled in the furnace. This will relieve the stresses from welding and prevent white cast iron from forming.

This process of stress relieving is sometimes

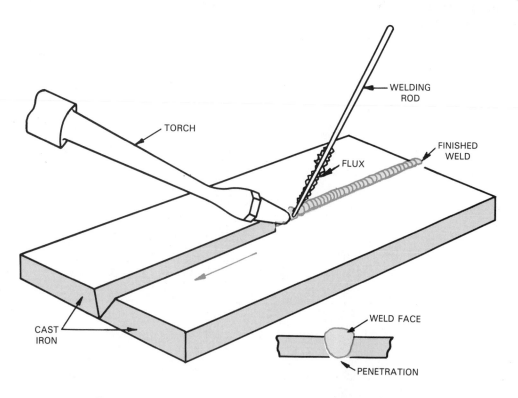

Fig. 19-14. Welding cast iron with the oxyfuel gas process.

more expensive than the welding operation, but it is necessary to insure a good weld.

## 19-28. INSPECTING AND TESTING CAST IRON WELDS

The finished cast iron weld should be slightly crowned without undercutting. It should have proper penetration. No pits or bubbles should appear on the surface of the metal.

To test cast iron welds they should be broken along the length of the joint. The fractured surface is examined for gas pockets, cracks, and hard spots of white cast iron. A good weld will not have any of these defects.

## 19-29. TEST YOUR KNOWLEDGE

Write your answers on a separate sheet of paper. Do not write in this book.

1. What is a ferrous based metal or a ferrous base alloy?
2. List three popular alloy steels.
3. Low alloy steels contain less than _____ percent alloying elements. They are _____ to _____ percent stronger than straight carbon steel.
4. How is the alloy composition of an alloy electrode designated?
5. What letter is used to designate a molybdenum alloy electrode? A chrome-molybdenum electrode?
6. List the three general classifications of stainless steel and give an example of each.
7. List four electrodes used to weld stainless steel. One electrode classification should be a low hydrogen electrode, one should be a low carbon, and one should be a GMAW electrode.
8. Two common problems with the welding of stainless steel are _____ and _____ _____.
9. What is done to prevent the problems listed in question 8?
10. A typical stainless steel used in industry is sometimes called an 18-8 stainless steel. This stainless steel which contains 18 percent chromium and 8 percent nickel is a _____ type stainless steel. See Figs. 19-3, 19-4, and 19-5.
11. What are nickel based alloys used for?
12. What heat treatment is given to a tool steel after welding?
13. Maraging steels contain _____ _____ amounts of carbon. They have _____ _____ strengths.
14. Why is a maraging steel aged?
15. What is the problem with arc welding galvanized steel?
16. How can the strength of a cast iron groove weld be improved?
17. Give two examples of electrodes used to weld cast iron.
18. Why is cast iron preheated prior to welding?
19. What preheat temperature is used when SMAW cast iron?
20. When should a cast iron weld be placed in a furnace for stress relieving?

# 20 SPECIAL NONFERROUS WELDING APPLICATIONS

As stated in Chapter 19, practically all metals can be welded. Chapter 19 covers the welding of the ferrous alloys. This chapter is devoted to welding nonferrous metals and alloys. Nonferrous metals and alloys do not contain large amounts of iron. They may contain small amounts of iron as alloying elements.

Some of the special welding applications used on nonferrous metals and alloys require pre-heat treatment and postweld heat treatment. See Chapter 27 for descriptions of heat treatment processes.

Many types of plastics are being used in industry, some of which are weldable. This chapter also shows how plastics can be welded.

## 20-1. NONFERROUS METALS AND ALLOYS

There are two large classifications of metals. These are:
1. Ferrous metals.
2. Nonferrous metals.

Ferrous metals are those bearing a substantial iron content.

The NONFERROUS METALS consist of all metals which do not contain much iron. Many metals are classified as nonferrous. Some of the more popular of the nonferrous metals are:
1. Copper and its alloys.
2. Aluminum and it alloys.
3. Lead and its alloys.
4. Zinc and its alloys.

Additional nonferrous metals include:
1. Titanium.
2. Beryllium.
3. Zirconium.

## 20-2. ALUMINUM

Aluminum is one of the more common and popular metals. It is available in all standard shapes and forms. It is shaped and formed by all standard methods. Parts made of aluminum are joined by all conventional methods.

Aluminum is lightweight and has good strength. It is well suited for low temperature applications. Aluminum has good resistance to corrosion. (Additional information on aluminum may be obtained in Heading 26-19.)

Aluminum is available in its commercially pure form. It is also alloyed with many other metals. Aluminum can be obtained in the rolled, stamped, drawn, extruded, forged, or cast form. All forms of aluminum, other than cast aluminum, are called WROUGHT ALUMINUM. The most common forms are plates, sheets, rolled forms, extruded forms, and aluminum pipe.

The Aluminum Association has classified aluminum alloys. Some are three digit classifications; others are four digit. The three digit code number (xxx) are castings while the four digit code numbers (xxxx) are wrought aluminum alloys. The three digit series has seven main classifications (0XX through 7XX). The four digit code number (xxxx) has 8 main classifications ranging from 1XXX through 8XXX. Wrought aluminum designations are listed in Fig. 20-1.

| THE ALUMINUM ASSOCIATION ALLOY GROUP DESIGNATION | MAJOR ALLOYING ELEMENT | EXAMPLE |
|---|---|---|
| 1XXX | 99% Aluminum (minimum) | 1100 |
| 2XXX | Copper | 2014, 2017, 2024 |
| 3XXX | Manganese | 3003 |
| 4XXX | Silicon | 4043 |
| 5XXX | Magnesium | 5052, 5056 |
| 6XXX | Magnesium and Silicon | 6061 |
| 7XXX | Zinc | 7075 |
| 8XXX | Other elements | |

Fig. 20-1. Alloys added to aluminum and the designation for each. Group 1XXX contains at least 99 percent pure aluminum.

## 20-3. PREPARING ALUMINUM FOR WELDING

Aluminum is used commercially in two principal forms: wrought aluminum and cast aluminum. The welding procedure is similar for the two forms but some differences do exist. Certain characteristics of aluminum make it rather difficult to weld:
1. The ease with which the aluminum oxidizes at high temperatures.
2. The aluminum melts before it changes color.
3. The oxide melts at a much higher temperature than the metal.
4. The oxide is heavier than the metal (more dense).

Despite these difficulties, aluminum welds can be made which are just as strong and ductile as the original metal. The filler metal, if used, should be of proper aluminum composition. Fig. 20-2 lists the characteristics of various filler metals used to weld different base metal combinations. From this chart, the best filler metal can be selected to join two pieces of aluminum. The welder must know the composition of the alloys being joined and should check the American Welding Society's recommendations for the best welding filler metal to use. See Fig. 20-2.

The metal must be cleaned before welding. It can be mechanically cleaned (clean stainless steel wire brush or clean stainless steel wool), or it can be chemically cleaned (dipped in cleaning solution and then rinsed.) Follow all safety precautions when using cleaning solutions.

Aluminum MUST BE SUPPORTED before, during, and immediately after the weld is made.

When welding aluminum, the welder must obtain the same welding results as when welding steel:
1. Good fusion.
2. Good penetration.
3. Straight weld.
4. Build up over the seam.
5. No defects.

When oxyfuel gas welding aluminum, the base metal is preheated to 300-400 °F (149-204 °C). Preheating is not necessary when arc welding aluminum.

## 20-4. ARC WELDING WROUGHT ALUMINUM

The most common processes for joining aluminum and its alloys are the gas metal arc welding (GMAW) and the gas tungsten arc welding (GTAW) processes. These gas shielded processes are desired because the inert gas protects the aluminum from oxidation.

Gas tungsten arc welding uses AC or DCEP (DCRP). When using AC or DCEP, the oxides on the aluminum are broken up. Aluminum is usually welded using alternating current. The finished weld has a clean, shiny appearance. DCEN (DCSP) can also be used, but there is no cleaning action. The aluminum oxide on the surface must be removed before welding. The finished weld appears dull, but stainless steel brushing will brighten the appearance.

The tungsten electrode should be pure or contain zirconium. Thoriated tungsten electrodes are not recommended for AC welding. The electrode tip should form into a hemisphere (ball shape) when AC welding. Argon or an argon and helium mixture is recommended as the shielding gas. GTAW produces high quality welds on aluminum.

Gas metal arc welding is also used to weld aluminum. DCEP (DCRP) is used. This current provides the necessary cleaning action to remove the aluminum oxide from the base metal. Either pure argon or an argon-helium mixture is used.

Gas metal arc welding (GMAW) is usually used with spray transfer. This can only be used in the flat position. High welding currents (wire feed speed) and high welding speeds are used to produce very high quality welds in aluminum.

Shielded metal arc welding (SMAW) can also be used to weld aluminum. A flux coated aluminum electrode is used. The current is DCEP (DCRP). The quality of a shielded metal arc weld is not as high as produced with GMAW or GTAW. SMAW of aluminum is usually used for repair work when GMAW or GTAW are not available.

## 20-5. OXYFUEL GAS WELDING WROUGHT ALUMINUM

Oxyfuel gas welding aluminum can be done successfully. However, gas tungsten arc and gas metal arc welding are preferred.

Flux is always used when oxyfuel gas welding aluminum. Both the metals being joined and the welding rod, if used, are flux coated. The operator usually mixes the flux with water to form a paste.

Due to the high heat conductivity of aluminum, a larger torch tip is needed for aluminum than for a corresponding thickness of steel. Acetylene is preferred as the fuel gas because it produces the most heat when burned. A neutral or reducing (slightly carburizing) flame should be used.

The appearance and color of the aluminum changes very little before becoming molten. The aluminum may be solid one instant, and then suddenly melt and sag without any apparent change in appearance or color. You may "feel" the surface of the metal with the welding rod by lightly touching the metal with the rod as the metal is

| Base Alloys | Filler Alloys | 1060, 1350 | 1100 | 2014, 2036 | 2219 | 3003, ALCLAD 3003 | 3004 | ALCLAD 3004 | 5005, 5050 | 5052, 5652 | 5083, 5456 | 5086, 5356 | 514.0, A514.0, B514.0, F514.0, 5154, 5254 | 5454 | 6005, 6063, 6101, 6151, 6201, 6351, 6951 | 6061, 6070 | 7005, 7021, 7039, 7046, 7146, A712.0, C712.0 | 413.0, 443.0, 444.0, 356.0, A356.0, A357.0, 359.0 | 319.0, 333.0, 354.0, 355.0, C355.0, 380.0 |
|---|---|---|---|---|---|---|---|---|---|---|---|---|---|---|---|---|---|---|---|
| | Characteristics | WSDCTM | WSDCTM | WSDCTM | WSDCTM | WSDCTM | WSDCTM | WSDCTM | WSDCTM | WSDCTM | WSDCTM | WSDCTM | WSDCTM | WSDCTM | WSDCTM | WSDCTM | WSDCTM | WSDCTM | WSDCTM |
| 319.0, 333.0, 354.0, 355.0, C355.0, 380.0 | 2319 | | | BAAAAA | BAAAAA | | | | | | | | | | | | | | BAAAAA |
| | 4043 | BAAAAA | BAAAAA | CCBCAA | CCBCAA | BBAAAA | BBAAAA | BBAAAA | BBAAAA | AAAAAA | AAAA A | AAAA A | AAAA A | AAAAAA | BBAAAA | BBAAAA | BBAAAA | BBAAAA | |
| | 4145 | AABAAA | AABAAA | ABCBAA | ABCBAA | AABAAA | AABAAA | AABAAA | AABAAA | | | | | | AABAAA | AABAAA | AABAAA | AABAAA | ABBBAA |
| 413.0, 443.0, 444.0, 356.0, A356.0, A357.0, 359.0 | 4043 | AAAAAA | AAAAAA | BBAAAA | BBAAAA | AAAAAA | AAAAAA | AAAAAA | AAAAAA | ABAAAA | ABBA A | ABBA A | ABBA A | ABBAAA | ABAAAA | ABAAAA | ABBAAA | ABAAAA | |
| | 4145 | AABBA | AABBA | AABAA | AABAA | AABBA | | | | | | | | | AABBA | AABBA | AABBA | AABBA | |
| | 5356 | | | | | | | | | BABB A | AAAA A | AAAA A | AAAB A | AAAB A | | | AAAA B | | |
| 7005, 7021, 7039, 7046, 7146, A712.0, C712.0 | 4043 | AACAA | AACAA | BBAAA | BBAAA | ABCAA | ADCBA | ADCBA | ABCBA | BDCBA | | | | | ADCBA | ADCBA | BDCBA | | |
| | 4145 | | | ABAA | AABAA | | | | | | | | | | | | | | |
| | 5183 | BABA A | BABA A | | | BABA A | BABA A | BABA A | BABA A | AABA A | AABA A | AABA A | AABA A | AABA A | AABA A | AABA A | AABA A | | |
| | 5356 | BAAA A | BAAA A | | | BAAA A | BBAA A | BBAA A | BAAA A | ABAA A | ABAA A | ABAA A | ABAA A | ABAA A | ABAA A | ABAA A | ABAAAA | | |
| | 5554 | | | | | | CCAAAA | CCAAAA | CAAAAA | BCAAAA | | | BCAA A | BCAAAA | BCAAAA | BCAAAA | BCAA A | | |
| | 5556 | BABA A | BABA A | | | BABA A | BABA A | BABA A | BABA A | AABA A | AABA A | AABA A | AABA A | AABA A | AABA A | AABA A | AABA A | | |
| | 5654 | | | | | | CCAA B | CCAA B | CAAA A | BCAA A | | | BCAA A | BCAA A | BCAA A | BCAA A | BCAA A | | |
| 6061, 6070 | 4043 | AACAA | AACAA | BBAAA | BBAAA | ABCAA | ADCAA | ADCAA | ABCAA | ADCAA | ADCA | ADCA | ADCA | ADCBA | ACBAA | ACBAA | | | |
| | 4145 | AADBA | AADBA | AABAA | AABAA | AADBA | BCDBA | BCDBA | ABDBA | | | | | | | | | | |
| | 5183 | BAB A | BAB A | | | BAB A | BAB A | BAB A | BAB A | BABC B | AABA A | AABA A | BABC B | BABC A | BAAC A | BAAC B | | | |
| | 5356 | BAA A | BAA A | | | BAA A | BBA A | BBA A | BAA A | BBAC A | ABAA A | ABAA A | BBAC A | BBAC A | BAAC A | BBAC A | | | |
| | 5554 | | | | | | | | | CCABAB | BCAA A | BCAA A | CCAB B | CCAAAA | CBABBA | CBABBB | | | |
| | 5556 | BAB A | BAB A | | | BAB A | BAB A | BAB A | BAB A | BABC B | AABA A | AABA A | BABC B | BABC A | BAAC A | BAAC B | | | |
| | 5654 | | | | | | | | | CCAB A | BCAA B | BCAA B | CCAB A | CCAB B | CBAB B | CBAB B | | | |
| 6005, 6063, 6101, 6151, 6201, 6351, 6951 | 4043 | AACAA | AACAA | BBAAA | BBAAA | ABCAA | ADCAA | ADCAA | ABCAA | ADCAA | ABCA | ABCA | ABCA | ABCBA | ACBAA | | | | |
| | 4145 | AADBA | AADBA | AABAA | AABAA | AADBA | BCDBA | BCDBA | ABDBA | | | | | | | | | | |
| | 5183 | BAB A | BAB A | | | BAB A | BAB A | BAB A | BAB A | BABC B | AABA A | AABA A | BABC A | BABC A | BAAC A | | | | |
| | 5356 | BAA A | BAA A | | | BAA A | BBA A | BBA A | BAA A | BBAC A | AAAA A | AAAA A | BAAC A | BAAC A | BAAC A | | | | |
| | 5554 | | | | | | | | | CCABAB | BAAA A | BAAA A | CAAB A | CAAAAA | CBABBA | | | | |
| | 5556 | BAB A | BAB A | | | BAB A | BAB A | BAB A | BAB A | BABC B | AABA A | AABA A | BABC A | BABC A | BAAC A | | | | |
| | 5654 | | | | | | | | | CCAB A | BAAA B | BAAA B | CAAB B | CAAB B | CBAB B | | | | |
| 5454 | 4043 | ABCCA | ABCCA | | AAAAA | ABCCA | ADCCA | ADCCA | ABCCA | ADCCA | | | | | | | | | |
| | 5183 | BABB A | BABB A | | | BABB A | BABB A | BABB A | BABB A | AAAB A | AABB A | AABB A | AABB A | AABB A | | | | | |
| | 5356 | BAAB A | BAAB A | | | BAAB A | BBAB A | BBAB A | BAAB A | ABAB A | ABAB A | ABAB A | ABAB A | ABAB A | | | | | |
| | 5554 | CAAAAA | CAAAAA | | | CAAAAA | CCAAAA | CCAAAA | CAAAAA | CCAAAA | BCAA A | BCAA A | BCAA A | BCAAAA | | | | | |
| | 5556 | BABB A | BABB A | | | BABB A | BABB A | BABB A | BABB A | AABB A | AABB A | AABB A | AABB A | AABB A | | | | | |
| | 5654 | | | | | | | | | BCAB B | | | BCAA B | BCAB B | | | | | |
| 514.0, A514.0, B514.0, F514.0, 5154, 5254 | 4043 | ABCC | ABCC | | AAAA | ABCC | ADCC | ADCC | ABCC | ADCC | | | | | | | | | |
| | 5183 | BABB A | BABB A | | | BABB A | BABB A | BABB A | BABB A | AABB B | AABA A | AABA A | AABB B | | | | | | |
| | 5356 | BAAB A | BAAB A | | | BAAB A | BBAB A | BBAB A | BAAB A | ABAB A | ABAA A | ABAA A | ABAB A | | | | | | |
| | 5554 | CAAA A | CAAA A | | | CAAA A | CCAA A | CCAA A | CAAA A | CCAA B | BCAA A | BCAA A | BCAA B | | | | | | |
| | 5556 | BABB A | BABB A | | | BABB A | BABB A | BABB A | BABB A | AABB B | AABA A | AABA A | AABB B | | | | | | |
| | 5654 | CAAA B | CAAA B | | | CAAA B | CCAA B | CCAA B | CAAA B | BCAA A | BCAA B | BCAA B | BCAA A | | | | | | |
| 5086, 5356 | 4043 | ABCB | ABCB | | AAAA | ABCB | ACCB | ACCB | ABCB | | | | | | | | | | |
| | 5183 | AABA A | AABA A | | | AABA A | AABA A | AABA A | AABA A | AABA A | AABA A | AABA A | | | | | | | |
| | 5356 | AAAA A | AAAA A | | | AAAA A | ABAA A | ABAA A | AAAA A | ABAA A | ABAA A | ABAA A | | | | | | | |
| | 5554 | | | | | | | | | CCAA A | | | | | | | | | |
| | 5556 | AABA A | AABA A | | | AABA A | AABA A | AABA A | AABA A | AABA A | AABA A | AABA A | | | | | | | |
| | 5654 | | | | | | | | | BCAA B | | | | | | | | | |
| 5083, 5456 | 4043 | ABCB | ABCB | | AAAA | ABCB | ACCB | ACCB | ABCB | | | | | | | | | | |
| | 5183 | AABA A | AABA A | | | AABA A | AABA A | AABA A | AABA A | AABA A | AABA A | | | | | | | | |
| | 5356 | AAAA A | AAAA A | | | AAAA A | ABAA A | ABAA A | AAAA A | ABAA A | A AA A | | | | | | | | |
| | 5554 | | | | | | | | | CCAA A | | | | | | | | | |
| | 5556 | AABA A | AABA A | | | AABA A | AABA A | AABA A | AABA A | AABA A | AABA A | | | | | | | | |
| | 5654 | | | | | | | | | BCAA B | | | | | | | | | |
| 5052, 5652 | 4043 | ABCAA | ABCAA | AAAAA | AAAAA | ABCAA | ABCAA | ACCAA | ABCAA | ADCBA | | | | | | | | | |
| | 5183 | BAB A | BAB A | | | BAB A | BAB A | BAB A | BAB A | AABC B | | | | | | | | | |
| | 5356 | BAA A | BAA A | | | BAA A | BAA A | BBA A | BAA A | ABAC A | | | | | | | | | |
| | 5554 | | | | | | | | | CCAAAB | | | | | | | | | |
| | 5556 | BAB A | BAB A | | | BAB A | BAB A | BAB A | BAB A | AABC B | | | | | | | | | |
| | 5654 | | | | | | | | | BCAB A | | | | | | | | | |
| 5005, 5050 | 1100 | CBAAAA | CBAAAA | | | CCAAAA | | | B AAAA | | | | | | | | | | |
| | 4043 | AACAA | AACAA | BBAAA | BBAAA | ABCAA | ABCAA | ABCAA | ABDAA | | | | | | | | | | |
| | 4145 | BADBA | BADBA | AABAA | AABAA | BBDBA | | | | | | | | | | | | | |
| | 5183 | CAB B | CAB B | | | CABC B | BAB A | BABB A | BAC B | | | | | | | | | | |
| | 5356 | CAB B | CAB B | | | CABC B | BAA A | BAAB A | BAB B | | | | | | | | | | |
| | 5556 | CAB B | CAB B | | | CABC B | BAB A | BABB A | BAC B | | | | | | | | | | |
| ALCLAD 3004 | 1100 | DBAAAA | DBAAAA | | | CCAAAA | | | | | | | | | | | | | |
| | 4043 | AACAA | AACAA | BBAAA | BBAAA | ABCAA | ADDAA | ADDAA | | | | | | | | | | | |
| | 4145 | BADBA | BADBA | AADAA | AABAA | BBDBA | | | | | | | | | | | | | |
| | 5183 | CABC B | CABC B | | | CABC A | BACC A | BACC A | | | | | | | | | | | |
| | 5356 | CABC B | CABC B | | | CABC A | BBBC A | BBBC A | | | | | | | | | | | |
| | 5554 | | | | | | CCABAA | CCABAA | | | | | | | | | | | |
| | 5556 | CABC B | CABC B | | | CABC A | BACC A | BACC A | | | | | | | | | | | |
| 3004 | 1100 | DBAAAA | DBAAAA | | | CCAAAA | | | | | | | | | | | | | |
| | 4043 | AACAA | AACAA | BBAAA | BBAAA | ABCAA | ABDAA | | | | | | | | | | | | |
| | 4145 | BADBA | BADBA | AABAA | AABAA | BBDBA | | | | | | | | | | | | | |
| | 5183 | CAB B | CAB B | | | C BC A | BACC A | | | | | | | | | | | | |
| | 5356 | CAB B | CAB B | | | CABC A | BBBC A | | | | | | | | | | | | |
| | 5554 | | | | | | CCABAA | | | | | | | | | | | | |
| | 5556 | CAB B | CAB B | | | C BC A | BACC A | | | | | | | | | | | | |
| 3003, ALCLAD 3003 | 1100 | BBAAAA | BBAAAA | | | BBAAAA | | | | | | | | | | | | | |
| | 4043 | AABAA | AABAA | BAAAA | BAAAA | AABAA | | | | | | | | | | | | | |
| | 4145 | AACBA | AACBA | AABAA | AABAA | AACBA | | | | | | | | | | | | | |
| 2219 | 2319 | | | BAAAAA | AAAAAA | | | | | | | | | | | | | | |
| | 4043 | BAAAA | BAAAA | BCBCA | BCBCA | | | | | | | | | | | | | | |
| | 4145 | AABAA | AABAA | ABCBA | ABCBA | | | | | | | | | | | | | | |
| 2014, 2036 | 2319 | | | CAAAAA | | | | | | | | | | | | | | | |
| | 4043 | BAAAA | BAAAA | BCBCA | | | | | | | | | | | | | | | |
| | 4145 | AABAA | AABAA | ABCBA | | | | | | | | | | | | | | | |
| 1100 | 1100 | BBAAAA | BBAAA | | | | | | | | | | | | | | | | |
| | 4043 | AABAA | AABAA | | | | | | | | | | | | | | | | |
| | 5356 | | | | | | | | | | | | | | | | | | |
| 1060, 1350 | 1100 | BBAAAB | | | | | | | | | | | | | | | | | |
| | 1060 | CCAAAA | | | | | | | | | | | | | | | | | |
| | 4043 | AABAA | | | | | | | | | | | | | | | | | |

Fig. 20-2. Chart used to select the best filler metal when welding two aluminum alloys.

# Aluminum Filler Alloy Chart

## How To Use The Filler Alloy Chart

First, select the base alloys to be joined, one from the light gray column on the extreme left and the other from the light gray row running along the top of the chart. Then, from the base alloy in the left column, move to the right until you reach the block directly under the base alloy from the top row. Or, from the base alloy in the top row, move down until you reach the block directly across from the base alloy in the left column.

This intersecting block contains horizontal lines of letters (A thru D) which represent the filler alloys directly across from them in the dark gray box at the ends of each row. The letters in each line give the A to D rating of the characteristics tested at the top of each column—W, S, D, C, T, and M.

You will find that by choosing the different filler alloys in each block, you can vary the characteristics of the weld, that is, "trade-off" one characteristic for another until you find the filler that best meets your needs.

*For example*, when joining base alloys 3003 and 1100, find the intersecting block. Now, note that filler alloy 1100 provides excellent ductility (D), corrosion resistance (C), performance at elevated temperatures (T) and color match after anodizing (M), with good ease of welding (W) and strength (S). However, if ease of welding and shear strength are *utmost* in importance, and ductility and color match can be sacrificed slightly, filler alloy 4043 can be used advantageously.

## Legend

Filler alloys are rated on the following characteristics:

| Symbol | Characteristic |
|---|---|
| W | Ease of welding (relative freedom from weld cracking.) |
| S | Strength of welded joint ("as-welded" condition.) (Rating applies particularly to fillet welds. All rods and electrodes rated will develop presently specified minimum strengths for butt welds.) |
| D | Ductility. (Rating is based upon free bend elongation of the weld.) |
| C | Corrosion resistance in continuous or alternate immersion in fresh or salt water. |
| T | Recommended for service at sustained temperatures above 150°F (65.5°C). |
| M | Color match after anodizing. |

- A, B, C and D are relative ratings in decreasing order of merit. The ratings have relative meaning only within a given block.
- Combinations having no rating are not usually recommended.
- Ratings do not cover these alloys when heat-treated after welding.

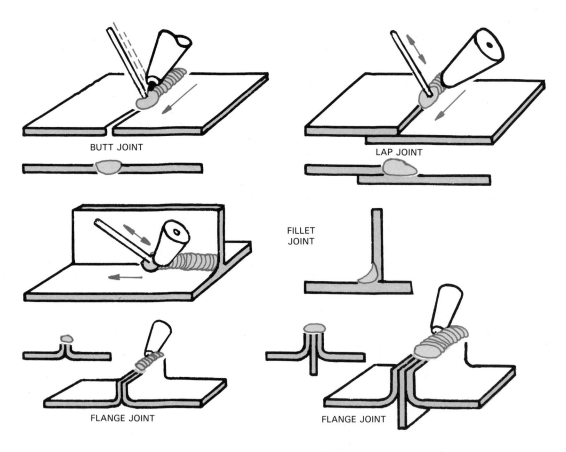

Fig. 20-3. Aluminum joints shown during and after welding. Note the position of the torch and the welding rod. No welding rod is required when welding a flange joint on thin metal.

heated. The welder will feel a change in the metal surface. The aluminum will feel soft or plastic under the welding rod just before the metal melts. When the metal is about to melt, the welding rod should not be withdrawn from the metal. The tip of the welding rod should be allowed to melt with the metal. Fig. 20-3 shows various joints in aluminum sheet being welded. Fig. 20-4 shows aluminum plates being welded.

The flux used for welding aluminum contains chlorides and occasionally fluorides. The fumes from the flux are irritating. Oxyfuel gas welding of aluminum should be done in a well ventilated place. This flux is also irritating to the skin and harmful to clothing. It must be carefully handled. Keep the flux in airtight containers when in storage. It is important to use clean, fresh, aluminum flux at all times.

The finished weld, in addition to all the usual appearances of a good weld, should be of a color similar to the welding rod material with a bright, shiny surface. If the metal is overheated and oxidized, the color becomes darker and the metal has a dull white appearance with a rough surface.

Flux should be washed from a weld with water or with a mild sulphuric acid water solution as soon as possible. Safety goggles and rubber

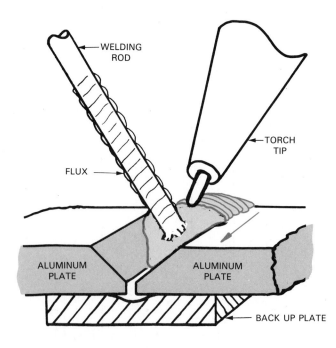

Fig. 20-4. Oxyacetylene welding aluminum plate.

gloves should be worn when using acid solutions. If flux is left on the weld, it will have a corrosive effect.

When learning how to butt weld sheet or plate aluminum, it is convenient to use a backing material.

## 20-6. WELDING CAST ALUMINUM

Welding of cast aluminum and aluminum alloys is not significantly different than welding wrought aluminum. Both sand castings and permanent mold castings are weldable. Aluminum die castings are not usually weldable because of certain ingredients in the aluminum alloys.

Castings that have been heat treated lose the heat treatment properties when welded. If a casting requires heat treatment, it should be welded prior to heat treatment, if possible.

The arc welding processes are preferred to oxyfuel gas welding for cast aluminum, but both can be used. When oxyfuel gas welding thin sections of metal, a stainless steel rod can be used to control and shape the weld pool. Fig. 20-5 shows two rods. These rods are usually made from 1/4 in. stainless steel flattened at one end into a flat, spoonlike shape.

When welding large and/or thick section aluminum castings, the castings should be preheated to 400-500 °F (204-260 °C). Thicker sections should be beveled as is done with steel.

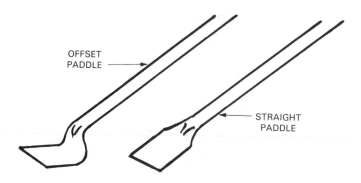

Fig. 20-5. Rods used for weld pool control when oxyfuel gas welding cast aluminum.

## 20-7. WELDING MAGNESIUM

Magnesium and its alloys are often confused with aluminum and welders often try unsuccessfully to use aluminum welding techniques on magnesium. The metal is shaped into various wrought forms and is also cast in sand and in permanent molds.

There are several magnesium alloys on the market. These metals can usually be arc welded. Gas tungsten arc welding (GTAW) and gas metal arc welding (GMAW) are the most widely used processes. Oxyfuel gas welding is recommended only as an emergency repair. Brazing and soldering methods can be used to join magnesium. Fig. 20-6 lists some of the weldable magnesium alloys and their weld strength.

Magnesium oxidizes very rapidly when heated to its melting point. In fact, when small shavings are heated to the melting point, the magnesium will burn spontaneously and leave a white ash. This burning and white ash is one way to identify the metal. BE CAREFUL TO USE ONLY A VERY SMALL AMOUNT OF MAGNESIUM SCRAPINGS OR SHAVINGS TO TRY THIS TEST.

The most popular magnesium alloys are those with aluminum and zinc added. These are the AZ series (A for aluminum and Z for zinc).

The high zinc alloys are weldable by the resistance spot and resistance seam welding processes. These alloys are the ZE, ZH, and ZK series. The usual arc processes cannot be used on all these alloys.

## 20-8. WELDING DIE CASTINGS (WHITE METAL)

DIE CASTINGS are metal alloy castings (sometimes called white metal) cast in iron and steel molds (dies) under pressure or by gravity. The same die (or mold) may be used many times. The finished article requires very little machining or finishing. Die castings may be accurate to .001

| ASTM ALLOY | FILLER ROD ALLOY | BASE METAL STRENGTH | | WELDED STRENGTH | |
|---|---|---|---|---|---|
| | | 1000 psi | MPa | 1000 psi | MPa |
| AZ31B-H24 | AZ61A | 42 | 290 | 37 | 255 |
| AZ10A-F | AZ92A | 35 | 241 | 33 | 228 |
| AZ61A-F | AZ61A | 45 | 310 | 40 | 276 |
| AZ92A-T6 | AZ92A | 40 | 276 | 35 | 241 |
| HK31A-H24 | EZ33A | 38 | 262 | 31 | 214 |
| HM21A-T8 | EZ33A | 35 | 241 | 28 | 193 |
| ZE10A-H24 | AZ61A | 38 | 262 | 33 | 228 |
| ZK21A-F | AZ61A | 42 | 290 | 32 | 221 |

Fig. 20-6 Table of weldable magnesium alloys, recommended filler metal, and weld strengths.

inch (.025 mm). Die casting alloys are usually alloys of high zinc, high aluminum, or high magnesium content. Other alloys include tin and lead alloys. These castings are brittle and break easily, but because of the ease of manufacture, many articles are made using this method. Failure due to brittleness produces considerable demand for repair welding of these castings.

To weld a die casting successfully, you should know the constituent metals in the alloy. The zinc casting is heaviest, the magnesium casting lightest, while the aluminum casting is in between in weight. Zinc die castings are the most common. The zinc die castings melt at about 800°F (427°C) while the magnesium and aluminum castings melt at about 1100 to 1200°F (593 to 649°C). Zinc alloy die castings are very difficult metals to weld because of their low melting temperature and high rate of oxidation. Some typical die casting alloys are shown in Fig. 20-7.

The welding rod should be of a similar composition as the original metal, if possible. Welding rods may be purchased for general repair of die castings.

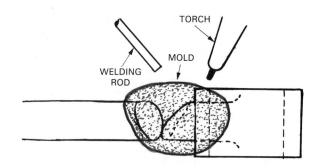

Fig. 20-8. Preparing a die casting for welding by forming a mold of carbon paste around the fracture.

metal melts before it changes color. Use the welding rod to break the surface oxides as the welding metal is being added. The welder may use a stainless steel or brass paddle to smooth the surface of the weld and to remove oxides.

| METAL | Alloy NO.1 | Alloy NO.2* | Alloy NO. 3* |
|---|---|---|---|
| Zinc, percent | 85 | — | — |
| Aluminum, percent | 3 | — | — |
| Copper, percent | 4 | 1 | 4 - 15 |
| Tin, percent | 8 | 4 | 70 - 90 |
| Antimony, percent | — | 15 | Trace |
| Lead, percent | — | 80 | — |
| Melting Temperature, °F | 852 | 550 | 675 |
| Melting Temperature, °C | 456 | 288 | 357 |

*Soldering is recommended for repairing these alloys rather than welding.

Fig. 20-7. Table of three die casting alloys.

A die casting is prepared for welding just as other metals. It must be beveled if there is a thick section. It must be thoroughly cleaned. Any plating must be ground away in the area to be welded. The parts must be firmly supported before, during, and after welding.

These die casting alloys are very fluid. To control the weld metal, carbon paste or blocks are used. Fig. 20-8 shows a die casting with a carbon paste mold formed around the fracture. Fig. 20-9 shows carbon paste and carbon plates used to form molds for controlling weld metal.

The oxyacetylene or oxyfuel gas method is usually used to repair die castings. A heavy carburizing flame should be used and the usual welding procedure followed. A very small tip is recommended. You must be careful because the

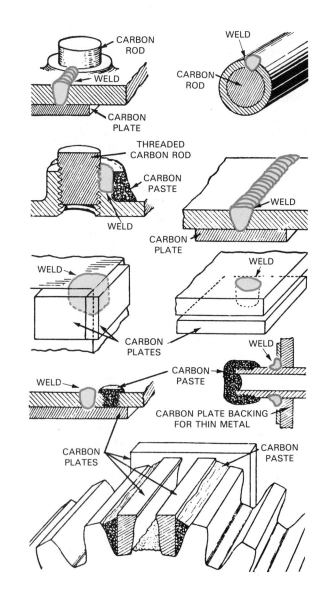

Fig. 20-9. Carbon block and paste backing as used in typical welding applications.

Another successful method of repairing die castings is to use a soldering iron or soldering copper to melt the metal. An oxyfuel gas torch is used to keep the body of the copper at a red heat while the point of the copper heats the die casting and fuses the two pieces together. The torch flame is not put on the die casting at all unless the casting requires preheating. This method is especially successful when repairing small sections.

## 20-9. COPPER AND COPPER ALLOYS

Copper and most of its alloys can be welded. There are two groups of copper:
1. Oxygen-bearing - 99.90 percent copper minimum, Oxygen .04 percent.
2. Deoxidized - 99.50 percent copper minimum, Phosphorous .015 to .040 percent.

The oxygen-bearing copper is difficult to fusion weld. Copper has a high specific heat and consequently it heats at about half the rate of aluminum. Both copper and aluminum have higher heat conductivity that steel. (Additional information on copper may be studied at Heading 26-16.)

Copper is alloyed with many different elements. The two most common are zinc and tin. Copper-zinc alloys are called BRASS. (Refer to Heading 26-17.) Copper-tin alloys are called BRONZE. (Refer to Heading 26-18.) Additional elements can be added to form different brasses and bronzes.

1. COPPER AND ZINC (BRASS) ALLOY

Gilding - 94 to 96 percent Copper-Zinc remainder

Commercial Bronze - 89 to 91 percent Copper-Zinc remainder

Red Brass - 84 to 86 percent Copper-Zinc remainder

Low Brass - 80 percent Copper-Zinc remainder

Cartridge Brass - 70 percent Copper-Zinc remainder

Yellow Brass - 65 percent Copper-Zinc remainder

Admiralty - 71.5 percent Copper-1.1 percent Tin-Zinc remainder

Naval Brass - 61 percent Copper-0.75 percent Tin-Zinc remainder

Manganese Bronze - 58.5 percent Copper-1.00 percent Tin-1.4 percent Iron-0.5 percent Manganese (Maximum)-Zinc remainder

Aluminum Brass - 77.5 percent Copper-2.2 percent Aluminum-Zinc remainder

2. COPPER AND TIN BRONZE

Grade A - .19 percent Phosphorus-94.0 percent Copper-3.6 percent Tin

Grade C - .15 percent Phosphorus-90.5 percent Copper-8.0 percent Tin

Grade D - .15 percent Phosphorus-88.5 percent Copper-10.0 percent Tin

Grade E - .25 percent Phosphorus-95.5 percent Copper-1.25 percent Tin

Alloys of copper and tin (bronze) may crack if not welded properly.

Alloys of copper are used a great deal, mainly because of their ductility, which enables them to be worked easily and shaped into many complicated patterns. These alloys are used in both cast and wrought forms. They are resistant to certain kinds of corrosion and are excellent conductors of heat and electricity. Copper alloys may be recognized by their characteristic red or yellow color.

Pure deoxidized copper is comparatively easy to weld, while some alloys of copper are difficult to weld. In order to test a specimen to determine if it may be easily welded, quickly heat a sample of the metal with a torch to the molten state. If the puddle remains quiet, clear, and shiny, it indicates the metal is comparatively pure copper and that it will be easy to weld. However, if the puddle boils vigorously and gives off a quantity of gaseous fumes, this indicates ingredients in the copper that make it difficult to weld.

Annealed deoxidized copper has a tensile strength of 30-35 ksi (207-241 MPa) and may be welded with arc or oxyfuel gas processes to produce full strength welds.

The tensile strength of annealed oxygen-bearing copper is also 30 to 35 ksi (207-241 MPa). When welding oxygen-bearing copper, the cuprous oxide redistributes itself in the heat affected zone and weakens it. It is difficult to obtain a weld with a strength greater than 70-85 percent of the annealed base metal strength. The heat affected zone also suffers a loss of ductility and corrosion resistance.

If a copper sample is brittle and breaks easily, this is an indication of alloying elements or impurities. One alloy which makes copper susceptible to cracking, and thus difficult to weld, is lead. Phosphorus in small quantities makes welding copper easier.

Most copper, copper alloys, brass, and bronze can be joined by brazing and soldering.

## 20-10. ARC WELDING COPPER

Gas tungsten arc welding (GTAW) and gas metal arc welding (GMAW) processes are preferred when welding copper. Shielded metal arc welding is only used for repair welding or to weld thin sections when GTAW or GMAW is not available.

Gas tungsten arc welding (GTAW) is used with direct current electrode negative (DCSP). A pure tungsten or a 2 percent thoriated tungsten electrode is used. Argon, helium, or nitrogen can be used as a shielding gas. Helium provides a higher arc voltage which gives better penetration. A helium-argon mixture of 2:1 or 3:1 is often used. Stringer beads are recommended.

When gas metal arc welding (GMAW), direct current electrode positive (DCRP) is used. The electrode must be a deoxidized copper electrode. A helium-argon mixture of 3:1 is used as a shielding gas. With this composition, or a higher argon percentage, spray transfer can be used. Wide weave techniques should be avoided. Stringer beads are recommended.

Since copper has a high thermal conductivity, the heat from welding is removed quickly. To prevent the high heat loss, a high preheat temperature is used. Preheat temperatures up to 800 °F (427 °C) are used with a preheat of 400-500 °F (204-260 °C) being the most common.

## 20-11. OXYFUEL GAS WELDING COPPER

Copper and its alloys may be welded with oxyacetylene or other oxyfuel gas processes. However, the high heat of the oxyacetylene flame is preferred. The general procedure is similar to that followed in welding steel.

Because of the rapid heat conduction of copper, a larger tip must be used. A tip one or two sizes larger than would be used on the same thickness of steel is sufficient.

When welding copper and its alloys, a flux is required to prevent oxidation. The backhand method is used. This prevents molten metal from running ahead of the torch, becoming oxidized, and getting trapped in the weld.

## 20-12. ARC WELDING BRASS

Arc welding brass is similar to welding pure copper. One major difference is caused by the presence of zinc. When the zinc is heated by the arc it becomes a gas. This zinc gas forms fumes around the weld and can cause porosity in the weld. Shielding gas must be used to remove any zinc fumes.

The electrode or filler metal used does not contain zinc. A copper-tin or copper-silicon alloy electrode is used. The most common electrodes are classified as ECuSn-C (copper-tin) and ECuSi-A (copper-silicon). The base metal is preheated to 200-600 °F (93-316 °C).

When gas tungsten arc welding, the zinc fumes can be reduced by directing the arc onto the filler

rod. The filler rod is kept in contact with the base metal. The arc heats and melts the filler rod as it is used to fill the weld joint. Using this method, the arc is not directed onto the base metal and the zinc fumes are greatly reduced. This process is illustrated in Fig. 20-10.

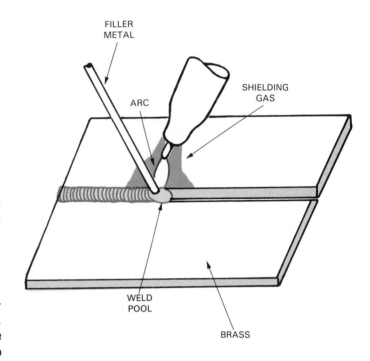

Fig. 20-10. Gas tungsten arc welding brass. The arc is directed at the filler rod to reduce zinc fumes.

## 20-13. OXYFUEL GAS WELDING BRASS

The methods used to oxyfuel gas weld brass are similar to those used when welding deoxidized copper. However, filler metals with matching compositions are not available. A copper silicon filler rod RCuSi-A is used.

Brass can also be joined by braze welding. Brass brazing rods RCuZn-A and RCuZn-C are used. A bronze weld will not be as corrosion resistant.

A good flux is required when welding or braze welding brass.

## 20-14. ARC WELDING BRONZE

The most common arc processes for welding bronze are SMAW and GMAW. Gas tungsten arc welding is used for small repairs. Covered electrodes are available as ECuSn-A and ECuSn-C. They are used with direct current electrode positive (DCEP or DCRP). Gas metal arc welding electrodes are available as ECuSn-C and used with DCEP (DCRP). These filler metals contain some phosphorus to help prevent cracking.

Stringer beads are recommended for both processes. Argon gas or an argon-helium mixture is usually used for shielding with GMAW and GTAW. The bronze base metal should be preheated to 400 °F (204 °C).

## 20-15. OXYFUEL GAS WELDING BRONZE

Oxyfuel gas processes are not recommended for welding bronze. The high heat input can cause cracking and porosity. Arc welding is preferred.

Good quality welds can, however, be produced with practice. Filler rods are available as RCuSn-A and RCuSn-C. If a good color match is not necessary, the RCuZn-C filler metal can be used.

## 20-16. WELDING COPPER—NICKEL ALLOYS

The copper-nickel alloys are very corrosion resistant and have good hot strength. These alloys have many uses. A primary use is to contain sea water or in other salt water applications.

The alloys vary from 5 to 30 percent nickel with the 10 and 30 percent nickel alloys being most common.

These alloys can be soldered, brazed, and welded. A 70 - 30 copper nickel electrode or filler rod is used for most applications.

The following processes are used to weld the copper-nickel alloys: GTAW, GMAW, and SMAW. A short arc length and a slight weave are recommended. Arc welding these alloys are preferred to oxyfuel gas welding.

## 20-17. TITANIUM

Titanium and titanium alloys have very good strength and ductility. Titanium has excellent resistance to corrosion. (Additional information may be obtained in Heading 26-23.)

Titanium is heavier than aluminum but lighter than steel. It is available in many forms such as forgings, bar, sheet, or plate. When titanium is annealed, it is relatively soft and ductile. It obtains its maximum strength when it is solution heat treated and then aged. Strengths can exceed 200 ksi (1380 MPa).

Titanium is available as four different commercially pure grades and alloyed with many different elements. The four commerically pure grades are classified as ASTM Grade 1, Grade 2, Grade 3, and Grade 4. When titanium is alloyed with other elements, the major alloying elements are included as part of the classification. Examples are: Ti-5Al-2.5Sn, Ti-2Al-11Sn-5Zr-1Mo, Ti-6Al-4V,

Ti-8Mn, and Ti-8Mo-8V-3Al-2Fe. The numbers indicate the percentage of each alloying elements. Titanium filler metal and welding electrodes are designated in the same way.

## 20-18. WELDING TITANIUM

The biggest problem with titanium and titanium alloys is that at elevated temperatures, titanium will dissolve oxygen, nitrogen, and hydrogen. Small amounts of oxygen and nitrogen increase the strength and hardness. Large amounts of these elements or any impurities like grease, oil, or moisture cause titanium to become brittle.

To prevent titanium from being contaminated, an inert gas shielding must be used. The inert gas must be very pure with no added moisture. The weld requires shielding gas on both sides. The molten metal and heat affected zone must be protected until they have cooled to at least 800 °F (427 °C).

One method used to shield the weld metal as it cools is to cover the weld and surrounding area with shielding gas. Fig. 20-11 illustrates two side shields around the weld area. Shielding gas completely covers the weld and heat affected zone. In

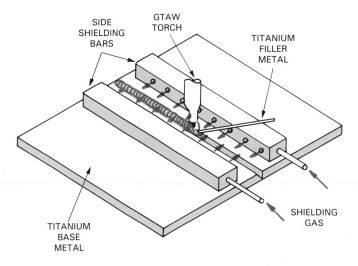

Fig. 20-11. Side shielding is used to cover the weld metal and heat affected zone as they cool.

addition to shielding the top surface, the bottom surface must also be protected with a gas called a BACKING GAS. Fig. 20-12 shows a channel used to protect the root side of the weld. When an automatic machine is used, a trailing shield protects the titanium as it cools. The trailing shield is attached to the welding torch and moves with the torch. A trailing screen is illustrated in Fig. 20-13. The welding torch always provides the primary gas shielding of the molten metal. The shielding

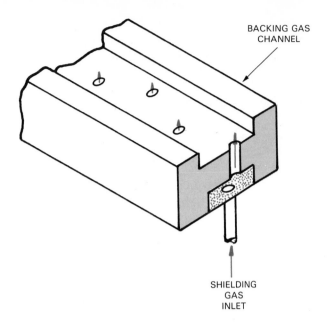

Fig. 20-12. A backing gas channel to protect the root of a weld.

gas used to protect the weld metal and heat affected zone as they cool is called SECONDARY GAS SHIELDING.

Another method used to protect the metal from contamination is to weld in an inert gas atmosphere. The welding is performed in a chamber completely filled with an inert gas. Two types of chambers are used:

1. Small units where the welder is outside the chamber with arm and glove manipulators to handle the electrode holder.
2. Large units with gas locks. The chamber is full of shielding gas and the welder enters the chamber to do the welding. These workers must wear breathing apparatus.

Whenever you enter a gas filled area, you must wear a supplied air or self-contained breathing apparatus. Unconsciousness and deaths have occured in less than ten seconds when workers without breathing equipment entered a chamber filled with a shielding gas, such as argon.

Fig. 20-14 shows a small reach-in cabinet. The titanium is placed inside and then the cabinet is filled with an inert gas. This system has a refrigerating unit to cool and dry the shielding gas atmosphere. It also has a recirculating system to filter the shielding gas.

An inflatable, flexible bag type of shielding gas chamber is shown in Fig. 20-15. The bag is inflated with an inert gas. This bag is designed to allow three operators to weld at the same time.

The filler metal or electrode used when welding titanium should match the base metal composition. Classification of filler metals and electrodes is the same as for the base metal. The major alloying elements and their percentage are listed as part of the designation. Examples are: ERTi-1, ERTi-2, (commercially pure grades 1 and 2), ERTi-3Al-2.5V, ERTi-8Al-1Mo-1V, and ERTi-6Al-4V.

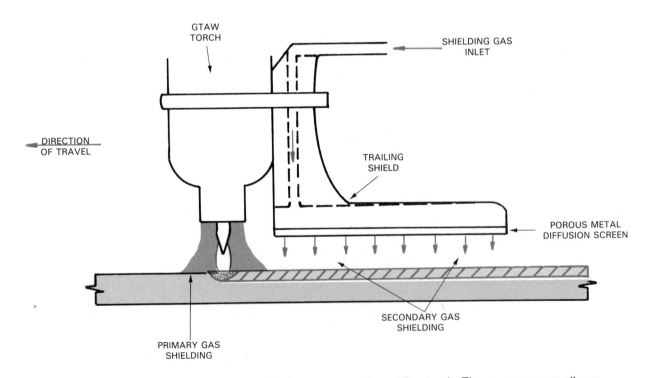

Fig. 20-13. A trailing shield attached to an automatic welding torch. The porous screen allows the shielding gas to cover the entire weld area.

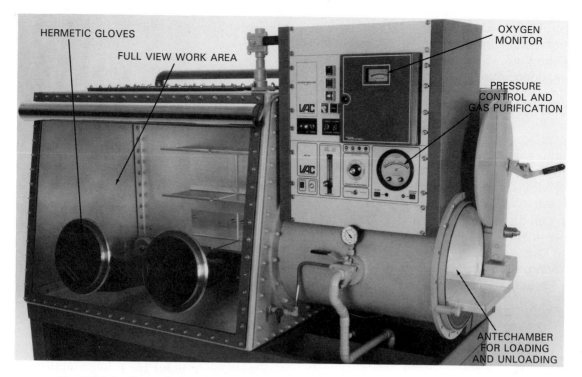

Fig. 20-14. Controlled atmosphere shielding gas cabinet. Difficult to weld metals may be welded within this cabinet. Oxygen and moisture levels are held to less than 1 ppm (part/million).   (Vacuum Atmospheres Co.)

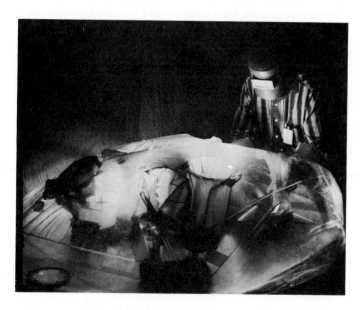

Fig. 20-15. Welding titanium in an atmosphere of inert gas within an inflated plastic container.
(Boeing Co.)

E-electrode; R-solid rod; Ti-titanium; Al-aluminum, Mo-molybdenum; V-vanadium.

The most common methods of welding titanium are GMAW and GTAW. Other processes include plasma arc welding, electron beam, laser, resistance, friction welding, and brazing.

When gas tungsten arc welding (GTAW), a 2

percent thoriated electrode is used (EWTh-2) with direct current electrode negative (DCSP). The electrode should not stick out too far from the gas nozzle. This will help prevent tungsten inclusions. The arc should NOT be touch started on the titanium. A starting tab can be used to strike the arc on. A starting tab is not part of the weld. It is an extra piece of metal placed next to the weld on which the weld is started. The starting tab is removed when the weld is completed.

Whenever welding titanium, cleanliness is very important. The base metal and the weld metal must be free from dirt, grease, and moisture. The weld must be protected by a shielding gas while molten and while it cools. This applies to most welding processes used to weld titanium. Electron beam welding which is done in a high vacuum does not require a shielding gas.

## 20-19. WELDING ZIRCONIUM

Zirconium is a rare metal used for some commercial applications. Its unique properties make it useful for astronautic (travel beyond the earth's atmosphere) applications and for chemical and nuclear energy uses. (Additional information may be obtained in Heading 26-25.)

Zirconium is weldable using the gas tungsten arc welding (GTAW) process. The metal must be both chemically and mechanically cleaned. It must be very carefully fixtured. The weld is best done in a 100 percent shielding gas atmosphere, similar to welding titanium. Zirconium can also be electron beam welded in a high vacuum.

## 20-20. WELDING BERYLLIUM

Beryllium is a commercial metal that is lighter than aluminum and has a fairly high melting temperature. Its tensile strength is approximately 55 ksi (379 MPa). (Additional information may be found in Heading 26-26.)

Beryllium can be gas tungsten arc welded, electron beam welded, solid state welded, resistance spot welded, and brazed. Beryllium is very susceptible to cracking. High heat inputs and high cooling rates must be avoided.

Beryllium is very toxic. Do not allow any beryllium shavings or dust to be taken internally (by breathing or eating).

## 20-21. WELDING DISSIMILAR METALS

It is possible to weld together dissimilar nonferrous metals such as aluminum to copper, aluminum to lead, copper to lead, and many other combinations.

Steel and some of the stainless steels can be welded or brazed to such metals as aluminum, copper, and others.

In most cases, arc welding and oxyfuel gas welding have not proven as successful as resistance welding, friction welding, and/or the solid-state welding processes for welding dissimilar metals.

## 20-22. HOT SHORTNESS

Some nonferrous metals, like aluminum, when heated to the melting point for welding, become extremely weak in the weld pool area. During this time, the metal has a tendency to suddenly fall away from the weld pool, leaving a hole in the metal. This tendency is called HOT SHORTNESS.

Metals which have hot shortness MUST BE SUPPORTED during welding usually with backing strips or backing rings.

## 20-23. USE OF A BACKING STRIP

Metal which extends beyond the back side (side opposite the bead) of a weld is called PENETRATION. When it is necessary to control the shape and size of this penetration, a backing strip is often used.

BACKING STRIPS are used to control the molten metal in the weld pool and to reduce the possibility

of "hot shortness" failure occurring. See Heading 20-22.

A groove is often machined into the metal backing strip. This groove insures that the penetration has the desired shape and size when it cools.

Preformed backing strips can be purchased in roll form. These preformed strips are taped onto the metal.

Backing strips may be removed after welding is completed. However, they are occasionally welded in place and become part of the joint. The backing strips are made of metals that conduct heat readily. The backing strip may be made of a dissimilar metal from the metal being welded. This is to prevent them from being welded into the joint.

When round parts such as pipes are welded, a backing ring may be used. See Heading 21-19 concerning the use of a backing ring.

## 20-24. PLASTICS

Plastics are used for a large variety of component parts in manufactured articles. These plastics have a great variation of chemical compositions and physical characteristics. However, regardless of their properties or form, plastics all fall into one of two groups. The two groups are the thermoplastic and the thermosetting types.

The THERMOPLASTIC group become soft and are capable of being formed when heated. These plastics harden when cooled. Thermoplastics may be reheated and reformed repeatedly.

The following plastics fall into the thermoplastic group classification: acrylic, cellulosic, acetal resin, nylon, vinyl, polyethylene, polystyrene, polyvinylidene, polypropylene, polycarbonate, polyfluorocarbon.

The THERMOSETTING plastics, once formed and cooled, cannot be reheated and reformed.

Thermosetting plastics include: aminoplastic, phenolic, polyester, silicone, alkyd, epoxy, casein, allylic.

## 20-25. PLASTIC WELDING PRINCIPLES

Thermoplastics may be welded in the same manner as metal. All joint designs may be used and successful welds are possible in various positions. Figs. 20-16 and 20-17 show a completed butt weld and a fillet weld on plastics.

The welding rod must be of the same composition as the plastic being welded. Plastic welding rods may be obtained in round, oval, triangular, or flat strip forms.

The welder's choice of welding rod shape is affected by the shape of the joint, the thickness of the plastics to be welded, and the welding equip-

Fig. 20-16. Cross section of plastic butt joint welded on both sides.

Fig. 20-17. Cross section of a plastic T-joint welded on both sides. (Laramy Products Co., Inc.)

ment to be used.

The joints and the welding rod must be clean to produce a high quality weld.

## 20-26. WELDING PLASTICS

Heat for plastic welding is supplied by a heated gas, either compressed air or a shielding gas. The gas is heated as it passes through the welding torch and then is directed onto the surface of the joint. No flame touches the joint, only the heated gas.

Electric heating coils are usually used to heat the gas as it flows through the torch. Fig. 20-18 illustrates a typical torch used for plastic welding. This torch uses an electric coil element to heat the welding gas. Fig. 20-19 shows a complete plastic welding outfit.

When welding, the torch is held in one hand and the plastic welding rod is fed to the weld area with the other hand. This is similar to oxyfuel gas welding. Fig. 20-20 illustrates hand welding of a butt joint on plastic material.

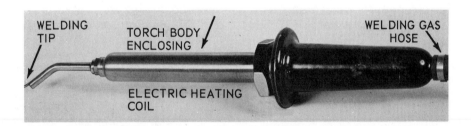

Fig. 20-18. A welding torch designed for welding plastics. It uses an electrical heating coil to heat the welding gas. (Laramy Products Co., Inc.)

Fig. 20-19. A complete plastic welding outfit. (Laramy Products Co., Inc.)

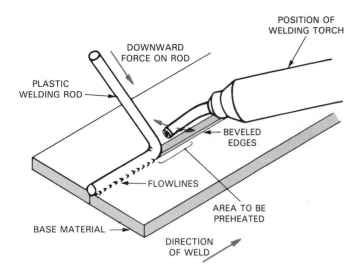

Fig. 20-20. Correct procedure for hand welding a plastics material butt joint. Note the position of the plastic filler rod and the torch tip. (Kamweld Products Co., Inc.)

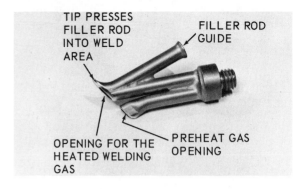

Fig. 20-21. Speed welding tip for welding plastics. (Laramy Products Co., Inc.)

Fig. 20-21 shows a tip which is called a SPEED WELDING TIP. The plastic filler rod is inserted in the guide tube. There are two openings for the heated gas. One opening is used to preheat the plastic, the second opening heats and melts the plastic and the filler rod. An electric welding torch with a speed welding tip installed is shown in Fig. 20-22. A light uniform pressure on the plastic rod is required. A light downward pressure is also exerted on the torch to press the rod into the heated weld area.

If the weld seam warrants it, the welding rod may be in a strip form. The strip is applied in a manner similar to the rod shape using a special speed welding tip. See Fig. 20-23.

The welding gas may be a shielding gas or compressed air. To regulate the welding gas temperature, adjust the gas flow. If the gas flow is slow over the heating element, it picks up more heat. If the gas flows quickly over the heating element, it will pick up less heat. Therefore, the welder decreases the welding gas flow to increase the temperature, and increases the welding gas flow to decrease its temperature.

To change the heating capacity of the torch, the tip size is changed. The torch heats the welding gas to 450-800°F (232-427°C). This heated gas is directed to the weld area where it heats the joint and the welding rod to a soft state. While both the

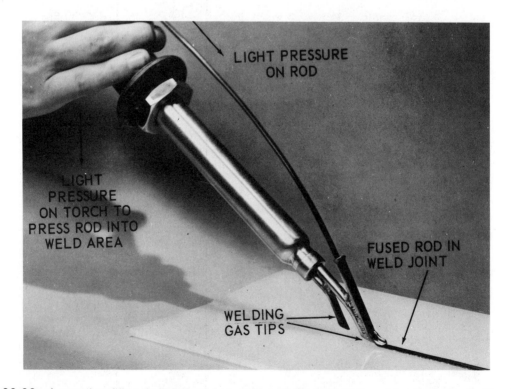

Fig. 20-22. A speed welding tip being used to weld a plastic butt joint. (Laramy Products Co., Inc.)

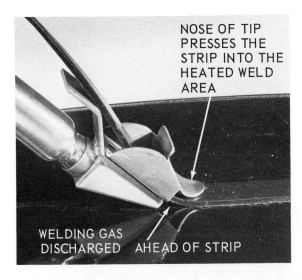

NOSE OF TIP PRESSES THE STRIP INTO THE HEATED WELD AREA

WELDING GAS DISCHARGED AHEAD OF STRIP

Fig. 20-23. Plastic butt joint being welded using a plastic strip as the joining material. Note the special shape of the speed welding tip. (Laramy Products Co, Inc.)

joint and the rod are soft (plastic), they are forced together with light pressure. The surfaces of the joint and welding rod bond. As they cool, they harden, and a good weld results.

Four things are necessary to accomplish a good weld on plastics:
1. Correct welding temperature.
2. Correct pressure on welding rod.
3. Correct welding rod angle.
4. Correct welding speed.

## 20-27. TEST YOUR KNOWLEDGE

Write your answers on a separate sheet of paper. Do not write in this book.

1. Nonferrous metals and alloys can contain iron as an _____ element.
2. All forms of aluminum, other than cast aluminum, are called _____ aluminum.
3. What organization has standardized a system for classifying aluminum alloys?
4. What is the major alloying element in a 4XXX series? A 5XXX series alloy?
5. What filler metal should be used to weld a piece of 5052 aluminum to a piece of 3003 aluminum? See Fig. 20-1.
6. What preheat is used when oxyfuel gas welding aluminum? When arc welding aluminum?
7. Aluminum is usually welded using the _____ or _____ process.
8. What types of welding currents produce cleaning action of the aluminum?
9. What type of flame is recommended when oxyfuel gas welding aluminum?
10. Flux left on an aluminum weld will have a _____ effect.
11. What alloys are added to magnesium to form the most popular magnesium alloy?
12. Why are carbon blocks and carbon paste often used when welding die castings?
13. Which type of copper is comparatively easy to weld?
14. Why does a weld in oxygen-bearing copper have less strength than the base metal?
15. Does the heat in a copper weld leave the weld faster or slower than the heat in a steel weld?
16. What is the biggest problem when welding titanium? Why?
17. List two ways to help eliminate porosity and contamination when welding titanium.
18. When GTAW titanium, what type of electrode and current is used?
19. What type of plastics can be welded?
20. How is the plastic heated when welding?

This diver is using a motorized pipe cutting machine to remove a damaged section of pipe. After the bad section is removed, a new piece will be welded in place.

# 21 PIPE AND TUBE WELDING

Pipes and tubes are mediums used to carry fluids from one point to another. The term FLUID includes substances in either the liquid or gaseous form, or both. The term PIPE usually refers to cylinders of hollow metal, or other material, of substantial wall thickness. Pipes are made by casting, extruding (seamless), or by rolling flat stock into cylinder form and welding the seam. The thickness of the pipe wall is usually thick enough that the pipe may be threaded.

The term TUBE usually refers to hollow cylinders, with a thinner wall than pipe, which are of either seamless or seamed construction. The wall thickness of tubing is usually too thin for threading. Therefore, tube joints must be soldered, brazed, or welded.

## 21-1. TYPES OF PIPES

The size of a pipe is measured according to its inside diameter (ID). The wall thickness usually varies according to the diameter of the pipe. Generally, the larger the pipe, the greater the wall thickness. There are several standard types of metal pipe:

1. Cast iron.
2. Low-carbon steel pipe.
3. Low alloy steel pipe.
4. Stainless steel pipe.
5. Copper and brass pipe.

CAST IRON PIPE is seamless pipe formed in a mold. The properties of cast iron are such that the sections of the pipes are difficult to thread. Cast iron pipe is usually mechanically joined. This limits the use of cast iron pipe to low pressure installations, such as drainage work and the like. Fig. 21-1 shows two types of mechanical pipe butt joints.

LOW CARBON STEEL PIPE or LOW ALLOY STEEL PIPE may either be seamed or seamless pipe. The seamed pipe is fabricated by arc welding (often SAW) or high frequency resistance welding. The

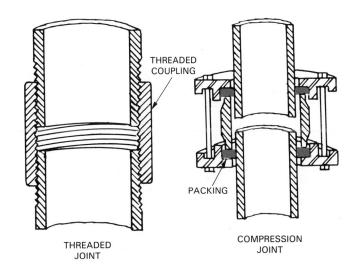

Fig. 21-1. Mechanical pipe butt joints.

seamless pipe is fabricated from solid bar stock, and the pipe is drawn through special dies. Seamless pipe has considerably more strength and can be used for much higher pressure work than seamed pipe.

STAINLESS STEEL PIPE and other noncorrosive steel alloys are used whenever corrosive chemicals are carried in a piping system.

COPPER PIPE and BRASS PIPE are used for water and refrigerant lines. This pipe is usually connected by soldering. Fig. 21-2 shows three types of welded, brazed, and soldered pipe butt joints.

Plastic pipe and tubes are being used where highly corrosive chemicals are being transported at low pressure. Plastic pipes are used in low pressure systems which are laid in water or in the ground since the plastic used is free from corrosion.

Aluminum alloy pipe and tubing is also finding wide use. Aluminum pipe and tubing may be soldered, brazed, or welded relatively easily. Aluminum pipes can be welded using automatic machines.

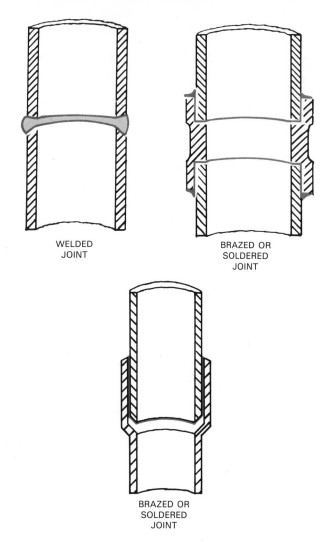

WELDED JOINT

BRAZED OR SOLDERED JOINT

BRAZED OR SOLDERED JOINT

Fig. 21-2. Cross sections of pipe and tube butt joints joined by welding, brazing, and soldering.

## 21-2. TYPES OF TUBES

Tubing is thin wall pipe. One of the most popular metals used in the construction of tubing is copper. Steel and aluminum are also used as tubing materials. In the fabrication of automatic machinery, in refrigerating systems, in automobiles, and in many other applications, tubing is found to be the most satisfactory method of transmitting fluids. Tubing is available in the flexible (soft) form and in the rigid (hard) form. The size of tubing is measured by its outside diameter (OD).

When a change in direction of the tubing occurs, soft tubing may be easily bent to conform with the change in direction. Hard tubing and pipe necessitate the use of appropriate fittings to change direction. Tubing is obtainable in different strengths, and is also manufactured in both the seamed and seamless construction.

Seamless tubing is made by piercing a piece of metal and drawing the metal through a die and over

a mandrel. Another similar method is to force a pierced piece of metal over a mandrel by means of rollers as shown in Fig. 21-3.

Seamed tubing is made by:

1. Rolling flat stock into a cylinder and welding the seam.

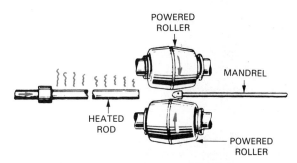

START OF OPERATION

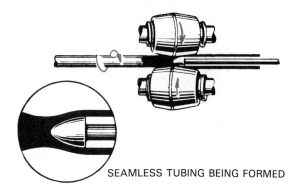

SEAMLESS TUBING BEING FORMED

VARIOUS STEPS IN OPERATION

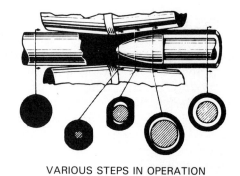

CHANGING SIZE OF TUBING

Fig. 21-3. Forming seamless tubing from a solid rod. (Michigan Seamless Tube Co.)

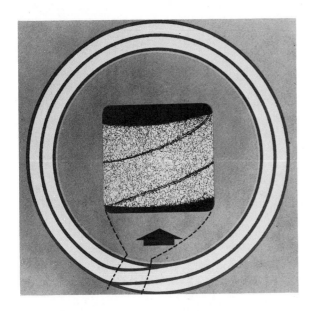

Fig. 21-4. Cross section of seamed tubing which has been furnace brazed. Note macrograph of brazed tubing at center. (Bundy Tubing Div., Bundy Corp.)

2. Rolling thin flat stock into a two metal layer cylinder and then furnace-brazing the lapped surfaces as shown in Fig. 21-4. A brazed joint is shown in Fig. 21-5.

The aircraft industry uses seamless steel tubing of special alloy. Aircraft tubing is seldom bent. It is joined by welding.

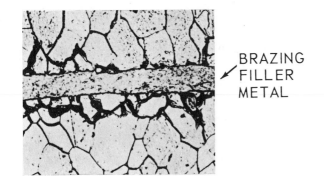

BRAZING FILLER METAL

Fig. 21-5. Microphotograph showing adhesion in brazed tubing joint.

Tubing standards have been established by the American Society of Mechanical Engineers (ASME), the American Society of Testing Materials (ASTM), the Society of Automotive Engineers (SAE), and the U.S. Government (MIL specifications).

Tubing is available in various cross section shapes:
1. Round.
2. Square.
3. Rectangular.
4. Hexagonal.
5. Octagonal.
6. Streamline.
7. Oval.
8. Irregular shapes.

## 21-3. METHODS OF JOINING PIPE

Steel pipes may be joined together or joined to fittings by:
1. Pipe threads.
2. Flange fittings.
3. Welded joints.

The method of joining by threads has been previously explained.

The FLANGE METHOD of joining pipe incorporates the use of threads, but the final connection of the two sections is by means of flange bolts. A gasket is placed between the two flanges and they are bolted together. This construction is popular in threaded pipe installations where it is thought that dismantling might occasionally be necessary.

The WELDING of pipe joints is a popular means of making a pipe installation. The advantages of welding piping are neatness, compactness, rapidity, and low cost. A variety of welding joints are shown in Fig. 21-6. Pipe welding may be performed by using either arc or oxyfuel gas welding. Arc welding is usually used. The process may be manual, semiautomatic, or fully automatic.

When pipe is butt welded in a production plant, flash welding, inertia welding, or friction welding may be used.

## 21-4. METHODS OF JOINING TUBING

Tubing joints may be made by:
1. Compression fitting.
2. Flared fitting.

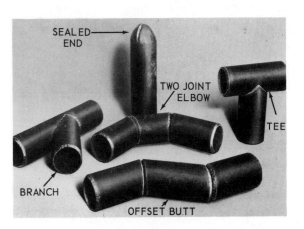

Fig. 21-6. A comparison of pipe joint designs.

3. Quick couplers.
4. Soldering.
5. Brazing.
6. Welding.

COMPRESSION FITTINGS and FLARED FIT-TINGS are used on small copper tubing connections. These fittings are commonly used with the seamed type of tubing (automobiles) or low pressure systems. The flared connection is used where seamless tubing is used in high pressure systems such as some refrigeration systems.

QUICK COUPLERS are fittings that are spring loaded, gasketed assemblies that enable quick assembly and dismantling of tubing systems without loss of the fluid. Fig. 21-7 shows a quick coupler fitting.

SOLDERED CONNECTIONS utilize special fittings, having receptacles (holes) accurately sized for the tubing insertion. The solder is admitted between the tubing and the fitting, where it forms a thin film and strongly binds the two pieces of metal together. The soldered connection may be used for seamed tubing, seamless tubing, or for copper pipe.

## 21-5. PREPARING PIPE JOINTS

Piping must be carefully prepared to insure strong and neat welds. Many companies now carry ready-to-use weld fittings which have bevel-

Fig. 21-7. Quick-disconnect coupling used for hose lines that need easy and fast connecting and disconnecting, such as hydraulic, pneumatic, and water lines. (Aeroquip Corp.)

ed edges ready for welding. The fittings available to the pipe welding trade include tees, 90 and 45 degree elbows, welding neck flanges, concentric

Fig. 21-8. Pipe joint fittings. Such fittings are commercially made in various sizes and thicknesses of steels and steel alloys. The edges are beveled and ready for welding.

reducers, lateral nipples, straight nipples, eccentric reducers, saddle caps, reducing tees, and 180 degree return bends. See Fig. 21-8. These fittings may be obtained in various sizes, and are stocked by wholesale pipe establishments. You must be sure to specify the proper kind and size of fitting when ordering. Fig. 21-9 illustrates a method used to identify a particular fitting.

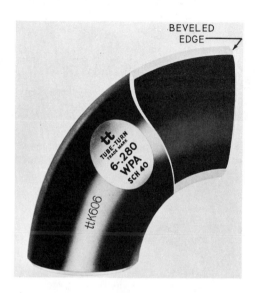

Fig. 21-9. Commercially fabricated pipe elbow fitting designed for welded joints. The 6-.280 means 6-in. pipe with a .280 wall thickness.

When preparing pipe for welding, the contour of the bevel is important. The bevel angle is recommended to be approximately 30 degrees from the vertical. The depth of the bevel extends almost completely through the pipe. A 1/16 inch (approximately 1.6 mm) to a 1/8 inch (approximately 3.2 mm) root face is left on the pipe. The root opening is also 1/16 inch to 1/8 inch (approximately 1.6 to 3.2 mm). These dimensions are shown in Fig. 21-10.

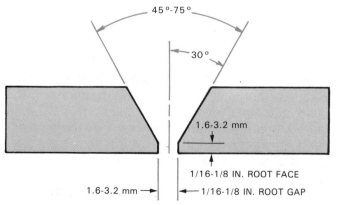

Fig. 21-10. Pipe joint preparation. The bevel angle is usually 30° ± 7.50°. The root opening and root face vary with different applications.

Another method for preparing a butt pipe joint for welding is to use a backing ring which spaces the pipes and helps to control penetration. A backing ring is shown in Fig. 21-11. The backing ring is shown being put in place in Fig. 21-12.

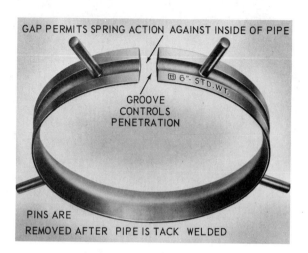

Fig. 21-11. Backing ring for pipe butt joints. This device aligns the pipe and also helps control the penetration. This ring is for 6 in. ID pipe.

Fig. 21-12. A pipe and an elbow being assembled prior to arc welding. Note backing ring used here for alignment and to control penetration.

The beveling of the pipe ends may be done with an oxyfuel gas torch, by machining, or by grinding. A hand operated oxyfuel gas beveling machine is illustrated in Fig. 21-13. A motorized pipe beveling machine with a device for cutting shapes such as saddles is shown in Fig. 21-14. If the flame cut surface is rough, the surface must be ground smooth.

As in all welding, pipes must be accurately aligned, properly spaced, and firmly held in place if the welds are to be successful. Fig. 21-15 shows a fixture for holding two pipes at any desired angle to each other as long as the pipes are in the same plane. The fixture itself can be clamped or bolted to a bench or some other support. If a fixture is not available, V blocks or short

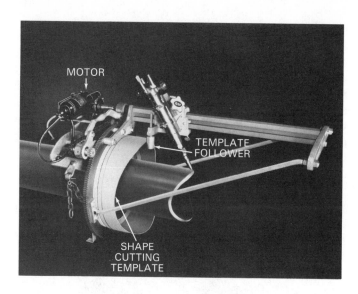

Fig. 21-13. Hand operated pipe beveling machine. Note the hand crack and gearing and angle adjustment for flame cutting head. The pipe being cut is 22 in. (559 mm) in diameter.

Fig. 21-14. Motor driven pipe beveling machine with shape cutting attachment. (H and M Pipe Beveling Machine Co.)

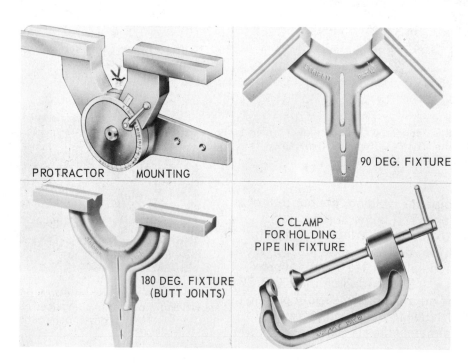

Fig. 21-15. Fixtures for holding pipes while they are being welded. (Strippit, Div. of Houdialle Industries, Inc.)

pieces of angle iron can be assembled to serve as fixtures. Other devices used to help accurately align and hold the pipe during welding are shown in Figs. 21-16 and 21-17. A device for centering a flange on a pipe prior to welding is shown in Fig. 21-18.

One difficulty encountered with all pipe and tube welding is the distortion or alignment of the pipe or tube after the welding is completed. One method to prevent or reduce distortion is to clamp the pipe or tube in a fixture while welding, and allow it to cool before removing the clamps. Another method allows for contraction of the weld metal by providing approximately a 1/4 inch (approximately 6 mm) movement of the pipe for each 24 inches (610 mm) of length. The preset out-of-alignment can be adjusted until the correct results are obtained. Fig. 21-19 shows pipe mounted in a fixture to prevent pipe from being out of line after welding. Fig. 21-20 also shows a pipe mounted out of line with the intent that the weld metal will pull the pipe and/or tube into cor-

Fig. 21-16. Fixture to provide extremely accurate fit-up for pipe welding.
(Dearman System Div./Cogsdill Tool Products, Inc.)

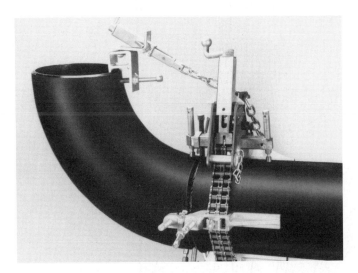

Fig. 21-17. Chain clamp aligns pipe sections for welding.
(Dearman System Div./Cogsdill Tool Products, Inc.)

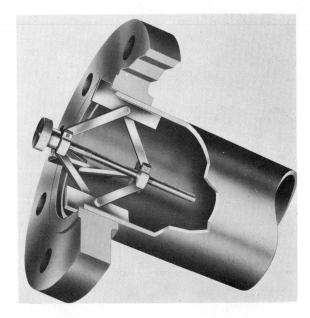

Fig. 21-18. Device for centering flanges and for accurately aligning the flanges 90 degrees to the axis of the pipe.

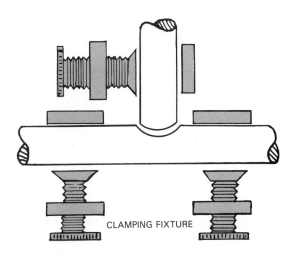

CLAMPING FIXTURE

Fig. 21-19. Clamping fixture used to prevent pipe from being out of line after welding.

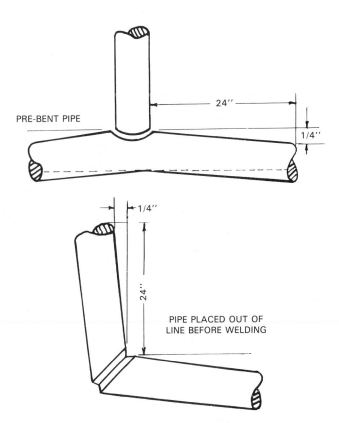

PRE-BENT PIPE

24"

1/4"

1/4"

24"

PIPE PLACED OUT OF LINE BEFORE WELDING

Fig. 21-20. Pipe mounted out of line before welding to correct alignment after welding (angles exaggerated).

rect alignment. One way to reduce distortion is to tack weld the pipes together before welding.

## 21-6. LAYOUT OF PIPE AND TUBE JOINTS

Cutting pieces of pipe to fit one another when the pipe is placed at various angles requires knowing the irregular curves of the contact.

Many times a pipe welder is forced to fabricate or manufacture his own fittings or to prepare

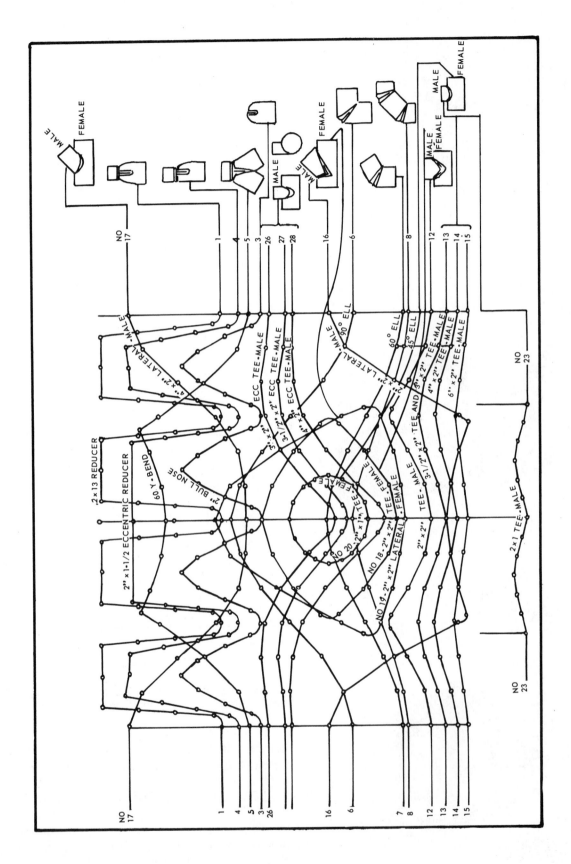

Fig. 21-21. Geometric curves for 2 in. (50.8 mm) diameter pipe joints.

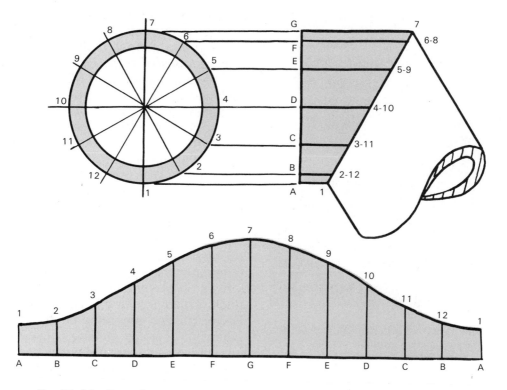

Fig. 21-22. Preparing paper or metal template (layout) for 60° angle pipe joint.

special joints when connecting pipes. Several practical methods have been developed which solve most of the problems encountered.

Paper or sheet metal templets laid out on a drafting board and cut out for shop use are in common use. Examples are shown in Figs. 21-21 and 21-22. Special devices are also used to determine the shape of the curves. One simple method is shown in Fig. 21-23. Another method used is to place two or more pipes in their proper alignment. Lay a soapstone pencil on one pipe and slide it around the pipe while keeping the soapstone in line with the center line of the pipe. Both the opening shape and the shape of the pipe end are marked.

Many special mechanisms have been developed to aid in producing accurate layouts quickly and easily. Figs. 21-24 and 21-25 illustrate a device for marking pipe at an angle. Note the protractor and the freedom of the arm to travel all around the pipe.

## 21-7. ARC WELDING PIPE JOINTS

Pipe may be joined in a number of different configurations. The most common joint is the butt

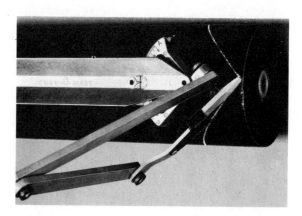

Fig. 21-23. Method used to determine the shape of the end of a pipe for a T-joint. (L-TEC Welding & Cutting Systems)

Fig. 21-24. Special device used to mark pipe prior to cutting.

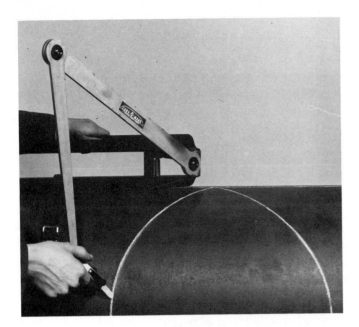

Fig. 21-25. A pipe is being marked prior to cutting. Note the freedom of the arm to mark all around the pipe. (Curv-O-Mark)

joint or butt weld in pipe. The orientation of the pipe can vary widely. The most common position or orientation is when the pipe is horizontal and the weld is vertical. The American Welding Society calls this the 5G position.

Pipe and weld orientations are shown in Fig. 21-26. Before any welding is possible, the pipes must be properly laid out and prepared. The pipe edges must be beveled and ground smooth. The pipes are then placed in proper alignment and held firmly in place.

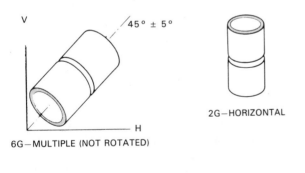

GROOVE WELDS IN PIPE

Fig. 21-26. Pipe joint configurations and their AWS position designations.

After the joint has been prepared, the pipes are tack welded together. This involves striking an arc and forming a small molten pool which joins the two sections of pipe. Pipe diameters less than 6 inches (approximately 150 mm) only need to be tack welded in three locations. Pipes 6 inches (approximately 150 mm) or larger in diameter are tacked in four or more locations.

A vertical butt weld can be welded in one of two ways: vertical-up or vertical-down. VERTICAL-DOWN WELDING is when the weld is started at the top of the pipe and progresses down the joint to the bottom of the pipe. Vertical-down is used for pipes with wall thicknesses less than 1/2 inch (approximately 13 mm). This includes most cross country pipe lines. VERTICAL-UP WELDING is when the weld starts at the bottom of the pipe and moves up toward the top. Vertical-up is used on thick walled pipe and in power plant applications. Fewer passes are required to complete a weld when vertical-up welding. Fig. 21-27 shows a comparison of the number of passes required to complete a joint using these two techniques. The

| TYPICAL NUMBER OF WELD PASSES REQUIRED TO COMPLETE A WELD. | | | |
|---|---|---|---|
| Wall Thickness of Pipe | | Vertical-down | Vertical-up |
| in. | mm | | |
| 1/4 | 6.4 | 3 | 2 |
| 5/16 | 7.9 | 4 | 2 |
| 3/8 | 9.5 | 5 | 3 |
| 1/2 | 12.7 | 7 | 3 |
| 5/8 | 15.9 | — | 4 |
| 3/4 | 19.1 | — | 6 |
| 1 | 25.4 | — | 7 |

Fig. 21-27. Table shows fewer weld passes are required to complete a weld joint when vertical-up welding. Vertical-down is not used on wall thicknesses greater than 1/2 in. (12.7 mm).

methods used to make veritical-up and vertical-down welds are shown in Fig. 21-28.

The different passes in a pipe weld are given names. The first pass in a pipe is called the ROOT PASS. After the root pass is completed, the slag must be removed. The root pass is usually ground out partially to remove any crown in the root pass. This will prevent slag from getting trapped in the weld. The root pass is usually made with an E6010 or E7010 electrode. These electrodes have good penetration. The root pass is sometimes done with GTAW to produce the highest quality root pass possible.

After the root pass is completed and ground, the

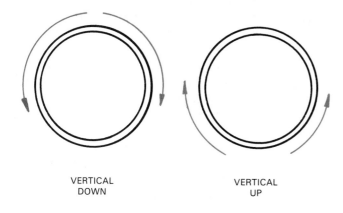

Fig. 21-28. Pipe can be welded vertical-down or vertical-up. Vertical-down welding begins at the top of the pipe and progresses towards the bottom. In vertical-up welding, the weld starts at the bottom and progresses up the pipe to the top.

second pass is welded. This second pass is called the HOT PASS. The hot pass uses more current than the root pass. This hot pass must fuse well with the root pass and the pipe walls. The hot pass must also melt any slag left from the root pass. The hot pass should be welded as soon as possible after the root pass is completed. The hot pass is usually welded within five minutes after the root pass is completed.

FILLER PASSES are used to fill the weld joint. A number of filler passes are usually required. Filler pass welds may be stringer beads (straight beads) or a slight weave may be used. Each filler pass must fuse (melt) into the previous weld passes and into the pipe walls.

The final pass is the COVER PASS. This pass is used to cover the weld joint. A weaving motion is used to produce a wide bead. Some specifications do not allow large cover passes, so smaller passes are used. The final pass or passes should have 1/32 to 1/16 inch (approximately 1 to 1.5 mm) buildup above the pipe. There must be good fusion into all previous weld passes and into the pipe walls and pipe surface. These weld passes are shown and labeled in Fig. 21-29.

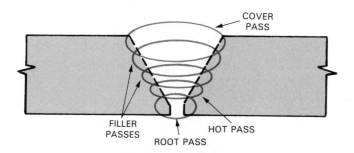

Fig. 21-29. The specific names for the various passes in a pipe weld.

The hot pass, filler passes, and cover pass are made with E6010 or E7010 electrodes or an E7018 electrode. E6010 or E7010 electrodes are used when vertical-down welding; E7018 electrodes are usually used when vertical-up welding. When an alloy pipe is welded, an alloy electrode must also be used. When a backup ring is used, an E7018 electrode can be used for the root pass.

Before attempting to vertical pipe weld, you must be able to weld satisfactorily in flat, vertical, and overhead position. Horizontal pipe welding is similar to horizontal welding on plate.

When large diameter pipe is used, (like cross country pipe), two welders work on the same pipe. Each welder works on one side of the pipe. When two (or more) welders work on the same pipe joint, the weld joint is completed faster.

When welding alloy steel pipe, stainless steel pipe, or aluminum pipe, the technique used is the same as described above. The proper procedures for welding these base metals (electrode selection, preheat, postweld heat treat, etc.) are described earlier in Chapters 19 and 20, (alloy steel in Headings 19-2 and 19-3; stainless steel in Headings 19-4 through 19-11; aluminum in Headings 20-1 through 20-6).

## 21-8. GTAW AND GMAW PIPE JOINTS

Gas tungsten arc welding is used to weld pipe. The GTAW process produces a very high quality weld, but very slowly. This process is used to obtain an excellent quality root pass. Shielding gas

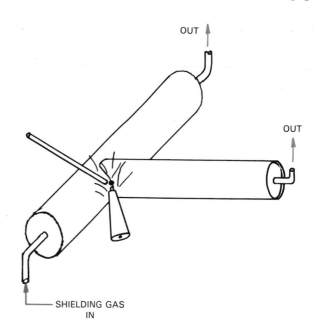

Fig. 21-30. Shielding gas being used to protect the root pass of a weld. Note how the gas enters one end of one pipe and exits each of the other two open ends.

must be used to protect the weld joint. Shielding gas can also be used to protect the weld metal from the inside of the pipe. Gas enters the pipe from one end and leaves from the other end of the pipe, or from each pipe being joined. An example of passing shielding gas through a pipe being welded is shown in Fig. 21-30.

Gas metal arc welding is also used to weld pipe joints. The electrode filler metal should match the base metal composition. GMAW can fill a weld joint faster than GTAW.

The procedure for welding alloy pipe or aluminum pipe is the same as described for welding alloy steel or aluminum base metal. GTAW is commonly used to weld aluminum. Shielding gas is passed through the pipes when welding aluminum or its alloys.

## 21-9. OXYFUEL GAS WELDING PIPE JOINTS

Two techniques are used to weld vertical pipe joints.
1. The easiest and best way, if possible, is to roll the pipe or turn it as the welding progresses.
2. The pipe may be held stationary, and the welding done by starting at the bottom and working up both sides of the pipe. This method is shown in Fig. 21-31.

Of the two techniques, the first usually insures a better weld. The weld is begun at the three o'clock position and welded to the top of the pipe. The pipe is then rotated 90 degrees. Welding continues at the three o'clock position and ends at the top. This continues until the joint is finished.

Fig. 21-32 illustrates this method of rolling the pipe. When using this technique, no overhead and only small amounts of vertical welding are required. Most of the welding is performed near flat position. This technique can only be used when the pipe is not fixed or secured.

The size of torch tip to be used is determined by the thickness of the pipe wall. It is practically the same as for the same thickness of flat sheet or plate. The weld may be done either by forehand welding or, as many companies recommend, backhand welding. The root pass is extremely important. Excessive penetration will restrict and turbulate the fluid flow. Insufficient penetration will leave cracks and crevices on the inside surface of the pipe and may cause failures especially in high pressure pipe. The existence of either of these faults usually necessitates a reject of the joint and it will have to be redone.

On T joints, angle joints, and cluster joints, use a sequence welding technique. This prevents weld metal contraction from pulling the pipe out of line. Fig. 21-33 illustrates one satisfactory sequence for a T-pipe assembly. Whenever possible, do all the welding in a downhand position. As in all pipe welding, the weld metal must be well fused into

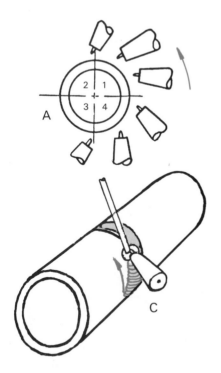
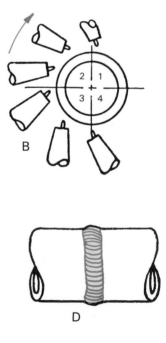

Fig. 21-31. Fixed method of welding pipe butt joints. A-Starting underneath and progressing to the top of one side. B—Starting underneath and progressing to top of other side. C—Position of torch and welding rod. D—Appearance of finished weld. This is vertical-up welding.

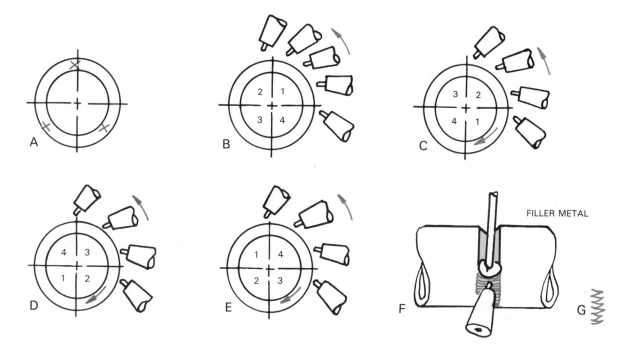

Fig. 21-32. Rolling method of welding pipe butt joints. A—Three tacks 120 degrees apart. B—First part welded. C—Pipe rolled and second quarter welded. D—Pipe rolled and third quarter welded. E—Pipe rolled and fourth quarter welded. F—Position of torch and welding rod. G—Weaving motion used.

the base metal. There must be good penetration and no undercutting.

## 21-10. TUBE WELDING

Tube welding is similar to thin sheet steel welding except the weld joint is a three dimensional curve, as in pipe welding. Also, since the root of the weld is not accessible and because the inner surface is in contact with flowing fluids, the penetration standards are high. Two common tube welding faults are too much penetration and lack of penetration. These two faults are shown in Fig. 21-34. These faults must be repaired before the tubing can be used. X-ray or ultrasonic testing can be used to locate the faults.

Tube materials commonly welded are:
1. Low carbon steel.
2. Chrome-molybdenum steel.
3. Stainless steel.
4. Aluminum.

Fig. 21-33. A method of welding a T-joint pipe assembly with one pipe in the vertical position. The recommended sequence is to prevent the contraction from pulling the pipes out of alignment as they cool.

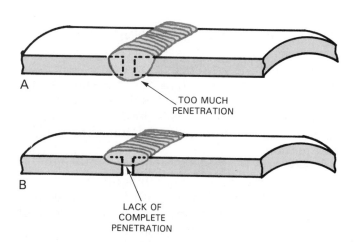

Fig. 21-34. Tube joint weld showing two common penetration defects. A—Too much penetration. B—Not enough penetration.

## 21-11. ARC WELDING TUBE JOINTS

Since the wall thickness on tubing is thin, considerable skill is required to SMAW tube joints. Low amperages must be used and the weld puddle kept small. Expert manipulation of the electrode is necessary to produce high quality welds.

DC welding using covered electrodes of 1/16 and 3/32 inch (1.6 and 2.4 mm) diameter is quite common. Almost all tube metals are weldable. The electrode material and the welding procedure varies with tube metals and tube wall thicknesses.

The same tacking procedures, aligning procedures, and welding sequences are used for tube joints as for pipe joints.

## 21-12. GTAW AND GMAW TUBE JOINTS

Both the GTAW method and the GMAW method may be used for tube welding. Using these methods, it is possible to produce quality welds on all weldable tube joints. The procedures and the electrodes or filler metal used will vary depending on the base metal. Low current settings are used to prevent melting through the thin walls.

Frequently the same kind of shielding gas used for welding is also passed through the inside of the tube or tubes as shown in Fig. 21-30. The gas flow is kept at a low pressure to save gas and to prevent forcing the plastic, molten metal out of position.

A special automatic GTAW tool is used to weld tubing. A tungsten electrode is placed in a rotating head. The automatic welder is clamped onto the tubes to be joined. The tungsten strikes an arc and forms a molten pool. The tungsten electrode in the rotating head rotates completely around the tubing. Shielding gas is used to protect the molten metal. An automatic tube welding machine is shown in Fig. 21-35.

## 21-13. OXYFUEL GAS WELDING TUBE JOINTS

Control of the weld puddle as a tube weld is being made is very important. The thickness of the tube wall determines the weld technique to be used. When the tube wall is thin, best results can be obtained by coordinating the torch and welding rod motion as follows:
1. Heat the base metal edges to a temperature just below the melting temperature.
2. Move the welding rod into the weld area to preheat it. As this action is taken, lift the torch flame enough so the flame is concentrated on the welding rod. Melt some of the welding rod and allow it to settle on the weld (only a very small amount).

3. Remove the welding rod a short distance, lower the flame, and fuse the added metal to the two tube edges.

You should first practice this technique on flat metal and in all positions before attempting tube welding. The complex three dimensional joints require expert handling of the torch and welding rod.

Oxyfuel gas welding is usually used to weld mild steel tubing. It is not used to weld stainless steel, aluminum, or other metals.

## 21-14. AIRCRAFT TUBE WELDING

A special division of tube welding has been created by the aircraft industry. The Federal Aircraft Administration (FAA) requires that one be a licensed airplane mechanic before being allowed to weld aircraft structures. Special alloy tubes are used in the fabrication of the fuselages, the tail, landing gears, and the wing sections of two and four passenger airplanes. This tubing is always of seamless construction, and is made of a very high quality steel with some strengthening alloy metals added. Such steels as chrome-molybdenum (1 percent chromium, 8 1/4 percent molybdenum), and the like are quite commonly used for this type of work. The tube has a relatively thin wall as compared to the ordinary steel pipe. Good fusing through the thickness of the metal is required with an absolute minimum of visible sag penetration.

Because a maximum of strength must be obtained when welding the joints of aircraft tubing, special fabrication designs have been developed. The most common types of special aircraft tubing welded joints are:
1. Fish-tail weld.
2. Telescope tubing weld.
3. Gusset weld.
4. 90 degree T-weld.
5. 45 degree T-weld.
6. Cluster tube weld.

Oxyfuel gas welding can be used to join aircraft tubing. Tubing is usually welded using SMAW, GTAW, or GMAW. Each process and procedure is certified by the FAA. Each aircraft mechanic is also certified to do the welding using these processes. The certification is usually obtained by actual performance tests under the direction of FAA officials.

A FISH-TAIL FITTING is a type of repair joint. A broken or damaged tube is repaired by slipping a new tube over the damaged tube. The ends of the larger tube are cut into a V-shape and then slipped over the small tube which it is to strengthen. The weld is a lap joint. The V-shape allows the weld to be spread over a considerable length of tubing, distributing the load along a greater length of the

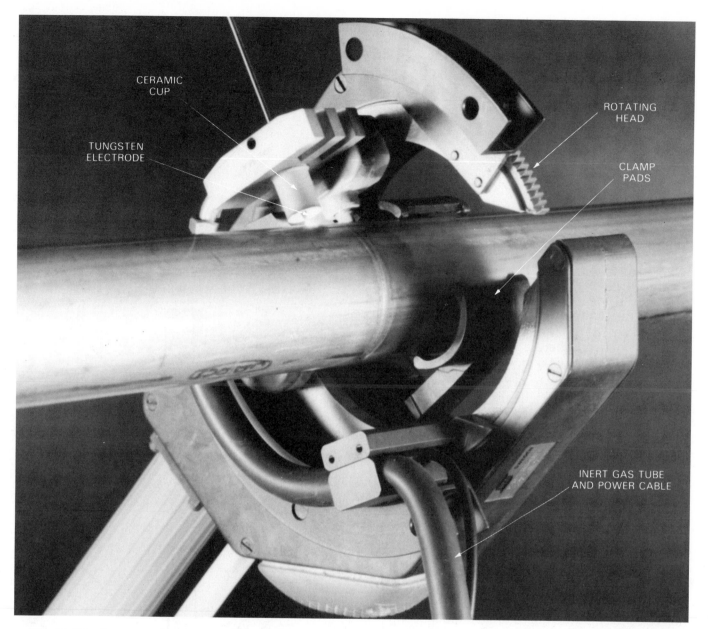

CERAMIC CUP

TUNGSTEN ELECTRODE

ROTATING HEAD

CLAMP PADS

INERT GAS TUBE AND POWER CABLE

Fig. 21-35. An automatic rotating gas tungsten arc welding machine for tubing from 0 - .71 in. (0 - 18 mm) in diameter. The tubing is clamped in place as the electrode is rotated about the weld joint. The weld is completed without the addition of welding wire.   (ESAB Group)

tubing. Fig. 21-36 shows a tube repair with fishtail fittings. The ROSETTE WELDS shown in Fig. 21-36 are plug welds done by drilling the outer tube before assembly and then welding the inner tube to the outer tube.

The CLUSTER TUBE JOINT is used where three or more tubes come together at an angle to one another, Fig. 21-37. The tubing ends are welded together where they come into contact. This cluster joint forms an extremely strong joint to

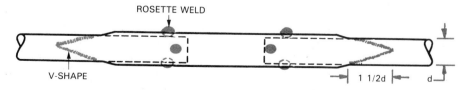

ROSETTE WELD

V-SHAPE

1 1/2d

d

Fig. 21-36. Straight aircraft tubing repair. Note the V-shape at the end of each larger diameter tube.

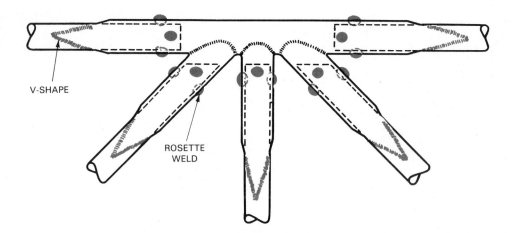

Fig. 21-37. Cluster repair. Note the V-shape at the end of each larger diameter tube.

V-SHAPE

ROSETTE WELD

better withstand tension, compression, vibration, and bending loads. In addition to welding the tubes to one another, a reinforcement is often welded over the cluster joint. Fig. 21-38 shows a type of reinforcement weld used on a cluster joint.

Aircraft tubing welds that are going into service are often inspected by X-rays. Many companies also periodically test samples of the welds made by their welding operators. These samples are tested to destruction to determine both the quality of the weld, and the welder's ability to make a good weld.

## 21-15. HEAT TREATING PIPE AND TUBE WELDED JOINTS

Heat treating a pipe or tube joint presents special problems. The assembly is usually too awkward to put into a furnace. Gas flame heating has been used but temperature accuracy along the weld is difficult. Two methods which have been successfully used are electric heating methods:
1. Resistance heating.
2. Induction heating.
Resistance heating can be used for preheating, for heating during the welding operation, or for postheating. The process involves wrapping resistance wire around the pipes or tubes at the proper places. Current is passed through the wire which heats the wire and the pipe. The temperature can be controlled with a thermostat. Special units have been developed that are easily installed (quick clamp and toggle method). Chapter 27 describes additional methods used to heat metals for heat treatment.

Preheating is often required. Temperatures up to 600 °F (316 °C) may be required. Postweld heat treating can be as high as 1500 °F (816 °C).

The time that the weld is held at the specified

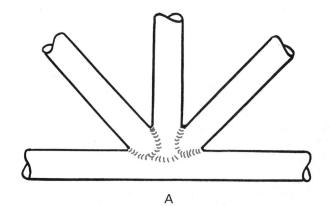

A

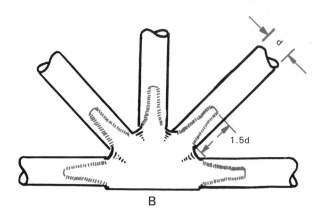

B

Fig. 21-38. Aircraft tubing weld. A—First the tubes are welded to each other. B—Reinforcement is welded to this assembly.

temperature, and the rate of cooling are important factors when postweld heat treating. Some alloys require a two step postweld heat treatment. You should always check with the manufacturer or the supplier of the metal, to determine the exact preheat and postweld heat treatments. Refer to Chapter 27 for additional information.

## 21-16. INSPECTING PIPE AND TUBE WELDS

It is important to know whether a completed pipe or tube weld has sufficient strength, ductility, and toughness. The welding procedure specification (WPS) must be carefully established and checked by inspection and testing facilities. The welding operators must be qualified to make sure they can conform to the accepted procedure. In determining the ability of an individual welder, standard procedures have been developed for testing sample welds produced by a welder. Methods have been devised to determine the quality of the weld joint without destroying it. One example of this type of testing is called visual inspecting. This is just one of many tests classified as nondestructive evaluation. See Chapter 28.

Some of the common faults that must be avoided in pipe welding and all welding are:
1. Too much penetration.
2. Lack of penetration.
3. Slag inclusions.
4. Porosity.
5. Lack of fusion.
6. Undercutting.

Practically all pipe welds are tested either under gas pressure (air, nitrogen, or carbon dioxide), under water pressure, or by using a halogen gas. The water or hydrostatic pressure imposed is usually one and a half times the pressure the weld is expected to be under in actual use. The water method (HYDROSTATIC) checks the integrity of the complete assembly. This test is known as a PROOF TEST.

The GAS PRESSURE METHOD is used mainly to detect leaks. A soap solution is placed around the joint and a leak is indicated by the appearance of bubbles.

The HALOGEN GAS LEAK TEST will locate smaller leaks than any other known method. A halogen gas, such as Refrigerant 12 or Refrigerant 22, is put into the pipe or tube under pressure. Any leak is detected by a very sensitive electronic ionizing sniffer. It is indicated by a meter reading or by a buzzer sound. This method is an adaptation of leak detecting methods used in the Air Conditioning and Refrigeration industry. It will signal a leak of an extremely small size.

## 21-17. CODE REQUIREMENTS

Almost all pipe and tube welding is covered by a welding code or codes. All the requirements of the code or codes must be met. The most commonly used code is the ASME (American Society of Mechanical Engineers) code.

The codes require the welding procedure and the welders be qualified. Refer to Chapter 29 on codes and procedures for more information.

## 21-18. HOT SHORTNESS

Some metals, such as cast iron and aluminum, when heated to the melting point for welding, become extremely weak in the weld puddle area. During this time, the metal has a tendency to suddenly fall away from the weld pool, leaving a hole in the metal. This tendency is called HOT SHORTNESS.

Metals which have hot shortness must be supported during welding usually with backing strips or backing rings.

## 21-19. USE OF BACKING RINGS

Metal which extends beyond the back side (side opposite the bead) of a weld is called ROOT REINFORCEMENT. When it is necessary to control the shape and size of this root reinforcement, a BACKING RING is often used. This is quite common in pipe and tubing welding.

A groove is often machined into the metal backing ring. This groove insures that the reinforcement has the desired shape and size when it cools.

Preformed backing rings are purchased to fit the inside diameter of a pipe. The backing ring usually is used to align the pipes being welded. Pins are attached to the backing ring which aligns the pipes and provides the proper root opening. See Figs. 21-11 and 21-12.

Backing rings may be removed after welding is completed. However, backing rings are occasionally welded in place and become part of the joint. The backing rings are made of metals which conduct heat readily. The backing rings may be made of a dissimilar metal from the metal being welded. This is to prevent them from being welded into the joint.

## 21-20. REVIEW OF SAFETY IN PIPE AND TUBE WELDING

All of the safety rules explained in the previous chapters also apply to pipe and tube welding.

Of particular importance to the pipe and tube welding operator are the following points.

The piping and tubing must be absolutely free of flammable liquids or gases to prevent an explosion. If the pipes or tubes have been used to carry explosives, the lines should be steam cleaned. A shielding gas should flow through the piping system while the weld is being made.

The welding of pipes and tubes requires that the

operator be in an awkward position or climb into almost inaccessible places. A firm footing, excellent ventilation, and a quick escape route are essential.

Because pipes and tubes are frequently installed in occupied buildings, a thorough investigation must be made to locate and remove any and all flammables from near the place of work. Fire fighting equipment must be handy and a fire protection person should be standing by during the welding and/or cutting.

Whenever using a gas under high pressure, always install a pressure relief valve in the test line. It should be adjusted to 5-10 psi (34.5-68.9 kPa) above the testing pressure.

Pipe welds should be vigorously inspected and tested. Sufficient penetration is absolutely essential to a successful weld. It requires considerable skill to weld pipe correctly. A pipe welder's ability should be periodically checked by checking and testing samples to destruction. Pipe welding, when properly done, is stronger than the threaded assembly method.

Most pipe and tube welding procedures are subjected to code regulations. Always inquire of the Local, County, and State Building and Safety Departments before accepting and undertaking a structural, pipe or tube welding project.

## 21-21. TEST YOUR KNOWLEDGE

1. List three ways to produce pipe.
2. For typical use, does seamed pipe or seamless pipe have the greater strength?
3. List a common metal used to make tubes.
4. The size of a tube is measured by its _____ diameter.
5. What are two ways to prevent or reduce distortion in pipe welding?
6. What are the two ways a butt weld in a horizontal pipe can be welded?
7. Vertical-down welding is only used when the pipe wall thickness is less than_____.
8. What are the names given to the first, second, and last passes in a pipe weld?
9. What type of electrodes are commonly used when welding the root pass? Why?
10. How soon after the root pass is welded should the hot pass be welded?
11. What are two common faults in pipe and tube welding that must be avoided?
12. When GTAW or GMAW tube joints, how is the root of the weld protected from oxidation?
13. Is oxyfuel gas welding used to weld stainless steel or aluminum tubes?
14. Tubing used by the aircraft industry is always of _____ construction.
15. What processes can be used to weld aircraft tubing?
16. When is a cluster tube joint used?
17. How are some aircraft welds inspected?
18. List two electric heating methods used to heat treat pipe and tube welds.
19. What is a proof test used for?
20. List two reasons for using a backing ring.

# 22 SPECIAL CUTTING PROCESSES

This chapter will discuss several cutting processes which are considered special or unusual. Such processes include:

1. Cutting through metal sections several feet thick with the oxygen lance cutting (LOC) process.
2. Cutting metal underwater using the oxyfuel gas cutting (OFC) process in construction or ship salvaging operations.
3. Underwater oxygen arc cutting (AOC).
4. Cutting through concrete walls using metal powder cutting (POC).
5. Laser beam cutting (LBC).
6. Cutting difficult-to-cut metal by means of chemical flux cutting (FOC).

## 22-1. OXYGEN LANCE CUTTING (LOC) PRINCIPLES

The principle of oxyfuel gas cutting is that once the metal is heated and the oxygen for cutting is turned on, the metal becomes a fuel and burns (oxidizes). The addition of oxygen supports the oxidation (burning) of the metal. With the ordinary oxyfuel gas cutting torch and tip, the preheating flame and the oxygen blast for cutting is combined in one torch. The maximum thickness which the regular hand held cutting torch will cut is about 24 in. (610 mm).

Steel up to 8 ft. (approximately 2.5 m) thick may be cut using an oxygen lance and a heating torch. The oxygen lance is a long length of ordinary steel tubing or pipe. This tubing is connected to an oxygen hose and an on-off control valve. The metal to be cut is heated to a white hot temperature by means of an oxyfuel gas cutting torch. Normally an oxyacetylene cutting torch is used. However, other fuel gases may be used. Once the metal is heated up to its kindling (burning) temperature, the air valve on the oxygen lance

Fig. 22-1. A large casting being cut using a standard oxyacetylene cutting torch and an oxygen lance.

is turned on and the cutting begins. See Fig. 22-1.

Generally two people are required to do oxygen lance cutting. One person keeps the metal at the kindling temperature with the cutting torch. The second person controls the oxygen lance.

The oxygen lance is used for such operations as cutting risers from large castings. It may be used to cut through thick pieces of steel in scrapping operations or opening the pouring gates in large furnaces.

Refer to Heading 1-29 and to Fig. 1-34 for basic information on the oxygen lance cutting (LOC) process.

## 22-2. OXYGEN LANCE CUTTING (LOC) EQUIPMENT

The equipment required for the oxygen lance operation is listed below.

1. A complete oxyfuel gas welding or cutting outfit to provide the preheat for lancing. For extremely thick sections, a special long preheating torch is required to provide a constant preheat temperature. These special cutting torches do not have a cutting head and a tip set at 90 degrees to the torch body. The torch body, head, and cutting tip are in a straight line. This type torch may be several feet long.
2. One or more oxygen cylinders. For large volume cutting, several cylinders may be manifolded to the regulator.
3. A heavy duty high volume, two stage oxygen regulator.
4. Oxygen hose of an appropriate length complete with fittings for the oxygen lance.
5. Lengths of 1/8 or 1/4 inch (3.2 - 6.4 mm) internal diameter pipe. This pipe must be long enough to penetrate the desired thickness. It must be sufficiently long to allow for consumption of the tubing as the cutting progresses.
6. Safety equipment will be needed. This equipment includes goggles, helmet, leather jacket, leggings, apron, and gloves.

Considerable sparking accompanies the oxygen lance cutting operation. The workers as well as the equipment and surroundings must be protected against hot particles.

## 22-3. OXYGEN LANCE CUTTING (LOC) PROCEDURE

To start a cut with the oxygen lance, it is first necessary to preheat the edge of the piece with a heavy-duty welding or cutting torch. After the metal reaches a cherry red or white hot temperature, the oxygen for the lance is turned on. On extremely thick pieces, the preheat temperature is maintained with a special long torch which can reach well down into the kerf. Generally the cut is started at an angle and gradually brought to a vertical position by manipulating the lance. The cut is much easier to start and maintain using this principle. Fig. 22-2 is a schematic drawing showing the oxygen lance principle.

Lancing may also be used to pierce holes through pieces up to 8 ft. (approximately 2.5 m) thick. This makes it possible to pierce holes in machinery which would be expensive if not impossible to machine.

As the pipe lance extends down into the cut, it is subjected to excessive temperature in the presence of oxygen. The end of the pipe will be burned away continually. As the pipe is consumed, lengths of new pipe are added. To join the new pipe to the old, standard threaded pipe couplings are used. The cutting lance has been known to pierce the center of a freight car axle shaft throughout its length without burning through the side of the shaft. Considerable expense is eliminated, when cutting large steel or cast iron sections, by using this method.

Fig. 22-3 shows operators at work using a standard cutting torch and an oxygen lance to cut a large casting.

## 22-4. PRINCIPLE OF OXYFUEL GAS CUTTING (OFC) UNDERWATER

The oxyfuel gas method of cutting ferrous metals underwater is much the same as the method used on the surface with one exception. In order to keep the flame lighted at extreme depths, the cutting tip and combustion area is surrounded by compressed air. This keeps the water away from the combustion area. Refer to Heading 1-30 and to Fig. 1-35.

The use of acetylene below 15 feet (4.6 m) of water is not recommended since acetylene is not safe to use at pressure over 15 psig (103.4 kPa). Hydrogen should be used as the fuel gas below a depth of 15 feet (4.6 m) of water. A proprietory fuel mixture called MAPP gas may also be used for this purpose. MAPP gas is stable under high pressures. The hydrogen flame has been used to a depth of 200 feet (61 m) under water.

The underwater torch has a special high pressure air system. The purpose of the high pressure air is to keep the water away from the flame by maintaining an air pocket at the point of cutting. The construction of an underwater oxyfuel gas torch is shown in Fig. 22-4. Notice how the pressurized air, shown in red, surrounds the cutting tip #4. Special training in underwater welding and cutting techniques is required and the operator should also be a qualified diver.

Oxyfuel gas torches are usually constructed as compactly as possible. The compressed air sheathing around the tip is adjustable to enable the operator to hold the torch head against the metal being cut. Too high an air pressure will force the air jacket and torch away from the metal being cut. Slots in the air jacket, #1 in Fig. 22-4, permit the escape of the products of combustion and air. The cutting tip must have exceptional preheating capacity to overcome excessive heat loss by con-

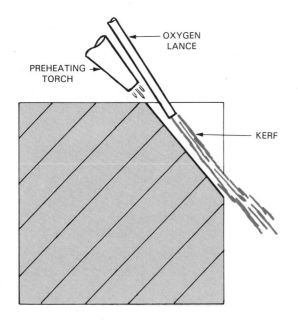

Fig. 22-2. Schematic of oxygen lance cutting (LOC) being used to cut thick metal.

ductivity and convection underwater. The cutting tip orifice must be large in order to overcome corrosion on the surface of the metal. Cutting tips must be large to save the operator time and to reduce the need to remain at great depths for long intervals. The air pressure, and therefore the amount of air, is controlled by a separate air valve designed to be a part of the cutting torch. See Fig. 22-4, part #42. Fig. 22-5 illustrates a typical underwater oxyhydrogen cutting torch.

## 22-5. OXYFUEL GAS UNDERWATER CUTTING EQUIPMENT

The equipment needed for underwater cutting is the same as for surface cutting with some exceptions. A special torch is needed which has a special slotted air jacket surrounding the cutting tip. It also has an additional connection and passages for compressed air, as shown in Fig. 22-4. A compressor and long lengths of air hose are required.

The fuel gas and oxygen cylinders and the air compressor are usually carried on a boat or barge. The welder's air supply compressor, if used, is also carried aboard the surface vessel.

The oxyfuel gas underwater cutting torch has four control valves:
1. The fuel gas valve.
2. The preheat oxygen valve.
3. The cutting oxygen valve (lever operated).
4. The compressed air valve.

Vision is generally very poor underwater. Therefore, the operator must know the position of the valves well enough to operate them skillfully without being able to see them.

## 22-6. OXYFUEL GAS UNDERWATER CUTTING PROCEDURES

A diver who is thoroughly familiar with surface cutting will not find underwater cutting too difficult. The oxygen and fuel gas working pressures are set at the cutting site. Because of the long

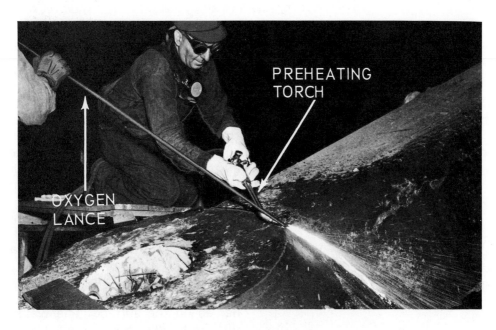

Fig. 22-3. The oxygen lance, with a regular cutting torch, being used to cut a large riser from a steel casting.
(L-TEC Welding & Cutting Systems)

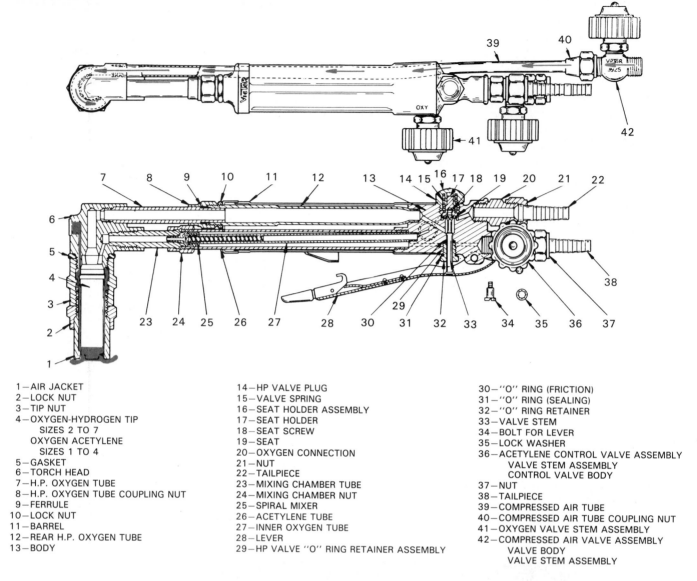

1—AIR JACKET
2—LOCK NUT
3—TIP NUT
4—OXYGEN-HYDROGEN TIP
    SIZES 2 TO 7
    OXYGEN ACETYLENE
    SIZES 1 TO 4
5—GASKET
6—TORCH HEAD
7—H.P. OXYGEN TUBE
8—H.P. OXYGEN TUBE COUPLING NUT
9—FERRULE
10—LOCK NUT
11—BARREL
12—REAR H.P. OXYGEN TUBE
13—BODY

14—HP VALVE PLUG
15—VALVE SPRING
16—SEAT HOLDER ASSEMBLY
17—SEAT HOLDER
18—SEAT SCREW
19—SEAT
20—OXYGEN CONNECTION
21—NUT
22—TAILPIECE
23—MIXING CHAMBER TUBE
24—MIXING CHAMBER NUT
25—SPIRAL MIXER
26—ACETYLENE TUBE
27—INNER OXYGEN TUBE
28—LEVER
29—HP VALVE "O" RING RETAINER ASSEMBLY

30—"O" RING (FRICTION)
31—"O" RING (SEALING)
32—"O" RING RETAINER
33—VALVE STEM
34—BOLT FOR LEVER
35—LOCK WASHER
36—ACETYLENE CONTROL VALVE ASSEMBLY
    VALVE STEM ASSEMBLY
    CONTROL VALVE BODY
37—NUT
38—TAILPIECE
39—COMPRESSED AIR TUBE
40—COMPRESSED AIR TUBE COUPLING NUT
41—OXYGEN VALVE STEM ASSEMBLY
42—COMPRESSED AIR VALVE ASSEMBLY
    VALVE BODY
    VALVE STEM ASSEMBLY

Fig. 22-4. A cross section of an underwater oxyfuel gas cutting torch. The air valve, part #42, controls the air flow to the air jacket, part #1. The air flow and air pocket around the cutting tip are shown in red. (Victor Equipment Co.)

hose lengths, the gauge pressures will be much higher than the working pressures at the torch. The diver welder should be in contact with the surface by radio or rope signal. This is necessary for safety and for gas and oxygen pressure changes if

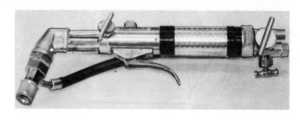

Fig. 22-5. Oxyhydrogen underwater cutting torch. Ellsberg Model. (Craftsweld Equipment Corp.)

necessary. The compressed air pressure is controlled on the surface tender vessel. The volume is controlled by the welder on the torch. A special electric lighter is built into the air jacket. This electric lighter will light the fuel gas and oxygen mixture under water.

To start a cut, the operator holds the torch against the metal and watches the color of the flames around the torch. When a bright yellow flame is seen the preheat temperature has been reached. The cut is then started by turning on the cutting oxygen and proceeding in the same manner as on the surface. The welder is, however, unable to see the cutting area, because it is covered by the air jacket. Because of the pressure created by the compressed air within the air jacket on the torch, an extra effort must be made by the operator to keep the torch against the work. The air tends to push the torch away from the metal being cut. A distinct rumbling sound is heard when the torch is operating. When the sound stops, the diver knows the flame is out.

The torch angle used is 90 degrees. Once the fuel gas and oxygen pressure are set, the welder must move the torch at a uniform speed and keep the torch against the metal. Since the welder cannot see the cut, the forward speed is all that can be controlled.

## 22-7. PRINCIPLES OF OXYGEN ARC CUTTING (AOC) AND SHIELDED METAL ARC CUTTING (SMAC) UNDERWATER

There are two basic principles of cutting metal underwater with an arc. These are the shielded metal arc and the oxygen arc methods. In the oxygen arc process, an arc is struck on the metal to be cut, using a hollow steel electrode which has high pressure oxygen flowing through it, as shown in Fig. 22-6. The arc heats the metal and the oxygen jet quickly oxidizes the metal to perform the cut.

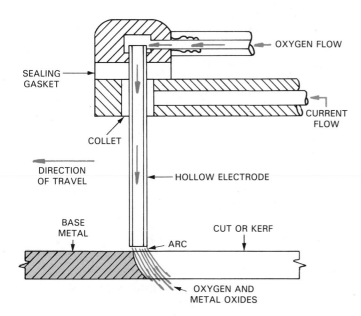

Fig. 22-6. A schematic of an oxygen arc cutting torch. Note the collet which holds the hollow electrode and where the electrical contact is made.

Oxygen arc underwater cutting has proven to be very successful and is widely used in industry. The electrode is usually 5/16 inch (7.9 mm) tubular steel, flux-covered and waterproofed. A fully insulated, special cutting torch (electrode holder) is shown in Fig. 22-7.

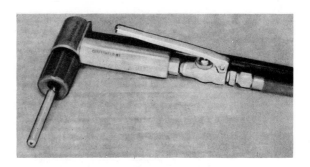

Fig. 22-7. Arc-oxygen underwater cutting torch, Model UA-10. (Craftsweld Equipment Corp.)

This system works well because the moment the arc is struck, a very high temperature spot is produced on the metal. With the oxygen turned on, the cutting starts instantly.

The operator then draws the cutting electrode along the line of cut while holding an arc. This method of cutting is usable at all depths. The electrode is consumed quite rapidly. The welding current is usually shut off when electrodes are changed. All metals may be cut with this system including steels, alloy steels, cast iron, and all the nonferrous metals.

Direct current electrode negative (DCEN) (DCSP) is the polarity used. Fig. 22-8 illustrates the internal construction of an oxygen arc underwater cutting torch.

Tubular (hollow) carbon electrodes and ceramic (silicon carbide) electrodes have been used, but their brittleness has resulted in their replacement by steel electrodes.

In the shielded metal arc cutting (SMAC) process solid electrodes are used. The process can be used to cut both ferrous and nonferrous metal under 1/4 in. (approximately 6 mm) thickness. The base metal is melted by the process and falls from the kerf by gravity. The resulting cut is quite wide and is not as smooth as a cut obtained using the oxygen arc cutting process.

## 22-8. OXYGEN ARC UNDERWATER CUTTING EQUIPMENT

The equipment used in an oxygen arc underwater cutting outfit uses some standard arc welding equipment and some standard oxygen equipment.

The standard arc welding equipment used is a DC arc welding machine.

The standard oxygen equipment used includes oxygen cylinders; a high volume, two stage, oxygen regulator; extra heavy oxygen hose; a welding lens 4 - 6 shades lighter than required on the surface.

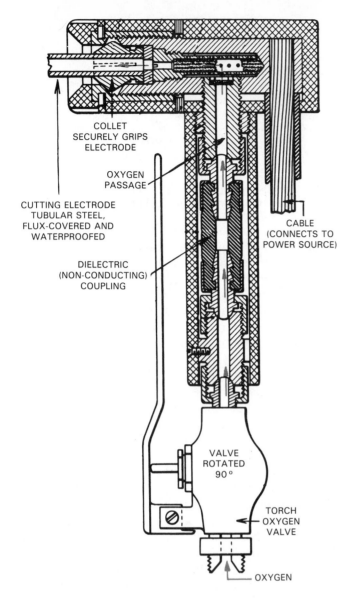

Fig. 22-8. Cross sectional sketch of an arc-oxygen underwater cutting torch. (Craftsweld Equipment Corp.)

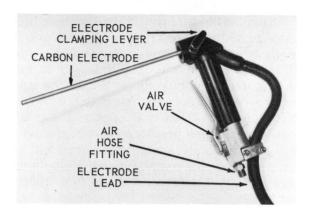

Fig. 22-9. An electrode holder designed for hollow electrodes used when oxygen arc cutting. Note the air hose fitting and the air control lever.

## 22-9. OXYGEN ARC UNDERWATER CUTTING PROCEDURE

With the DC welder and the oxygen turned on, the operator strikes the arc as one would on the surface. The operator then depresses the oxygen valve handle to start the cut. The technique used for underwater arc cutting is to drag the electrode along the line to be cut. A constant arc length is maintained by this method due to the thickness of the flux covering.

Skill is required to move the electrode at a constant speed. The welder must move the electrode slowly enough to insure the cutting of the complete thickness of the metal. If the movement is too slow, however, the cut may be stopped due to the lack of proper preheating of the advance metal. A slow forward movement will also cause too many electrodes and too much oxygen to be consumed during the cutting operation.

## 22-10. METAL POWDER CUTTING (POC) PRINCIPLES

Metal powder cutting is used to cut alloy steels, cast iron, bronzes, nickel, and aluminum. Most materials may be cut with the metal powder cutting technique—even reinforced concrete.

In metal powder cutting, iron powder is introduced into a standard oxyfuel gas cutting torch flame. The heat of combustion of the iron powder increases the total heat of the flame.

When cutting stainless steel with an oxyfuel gas torch, a refractory (difficult to melt) chromium oxide is formed on the surface of the metal in the cutting zone. This oxide insulates the base metal preventing further oxidation by the cutting oxygen. By introducing iron powder into the oxyfuel gas flame, insulating chromium oxides are kept from forming. The cut is able to proceed

The underwater welder is serviced from a tender ship on the surface. Some special equipment needed includes an oxygen arc underwater cutting outfit; an electrical off and on switch for the welding leads on the tender; a waterproofed ground lead; a waterproofed electrode lead; an insulated oxygen arc torch, see Fig. 22-9. Waterproofed, covered, tubular electrodes and rubber gloves are also required.

Electrodes used are tubular iron or steel with a waterproofed flux covering. They are 5/16 inch (7.9 mm) in diameter and 14 inches (356 mm) long.

The hole in the electrode through which the oxygen travels is 1/8 inch (3.2 mm) in diameter.

Oxygen arc cutting electrodes last only a few minutes and may be changed underwater.

unhindered. The high-in-iron particles oxidize or burn in the oxygen stream at the torch tip producing a very high temperature and concentrated heat. The powder also produces a reducing action on the metal, and a high velocity gouging action. Fig. 22-10 shows a large casting being cut using the metal powder cutting process.

The extremely high temperature that results from the ignition of the powder practically eliminates the need of preheating the metal prior to the cutting action. Eliminating the requirement of preheating saves a great deal of time. The metal powder cutting process has a considerable penetrating action or carry over. The cutting action

Fig. 22-10.  A large metal powder cutting torch being used to cut a riser from a large casting.   (L-TEC Welding & Cutting Systems)

Cutting speeds comparable to the conventional oxyfuel gas cutting of low carbon steels may be obtained on stainless steels using the metal powder cutting torch.

When cutting nonferrous metal, it is helpful to add a more active ingredient than iron powder to the oxyfuel gas flame. Ten to thirty percent aluminum powder is used with iron powder when cutting nonferrous metals.

Ferrous metals up to 5 feet (approximately 1.5 m) thick have been cut using metal powder cutting torches mounted on automatic cutting machines.

Reinforced concrete up to 18 inches (approximately 450 mm) thick may be cut with speeds of 1 to 2 1/2 inches (approximately 25-64 mm) per minute, and walls 12 feet (approximately 3.7 m) thick may be cut with a metal powder cutting lance as shown in Fig. 22-11.

Several methods have been used to feed the iron powder to the flame:

1. Introduction of the powder into the oxygen orifice of the tip.
2. Introduction of the powder into a separate orifice in or near the tip. See Fig. 22-12.
3. Introduction of the powder into the preheating orifices of the tip.

If the oxygen orifice method is used, the oxygen pressure must be less than the air or nitrogen pressure feeding the metal powder to the tip. If a separate orifice is used, there is a delay in ignition of the iron powder and the tip must therefore be held 3 or 4 inches (75-100 mm) from the work.

Fig. 22-11.  Cutting concrete using the metal powder cutting process.   (L-TEC Welding & Cutting Systems)

easily jumps slag pockets and carries over from one plate to another.

### 22-11.  METAL POWDER CUTTING (POC) EQUIPMENT

The equipment required for metal powder cutting includes a cutting torch or lance, oxygen and a fuel gas supply, and a method of introducing the metal powder into the flame.

Some special metal powder cutting torches have an additional tube to carry the powder to the torch head. The metal powder is added to the preheating flames at the torch head. Such a torch is shown in Fig. 22-12. The oxygen and metal powder tubes on the torch are both opened when the oxygen control lever is depressed. Normally there is a slight delay in the opening of the oxygen valve. This allows the iron oxide to coat the surface before the oxygen reaches the surface. The delay helps to start the cut faster, particularly on nonferrous metals.

ting oxygen stream. Since the powder is directed into the oxygen stream from all sides, the multiple opening attachment is suitable for form cutting and straight cutting. When powder cutting nonferrous metals, or when cutting extra thick ferrous metals, an attachment with a single powder tube is used. The powder is carried to a point near the conventional tip where it is discharged into the

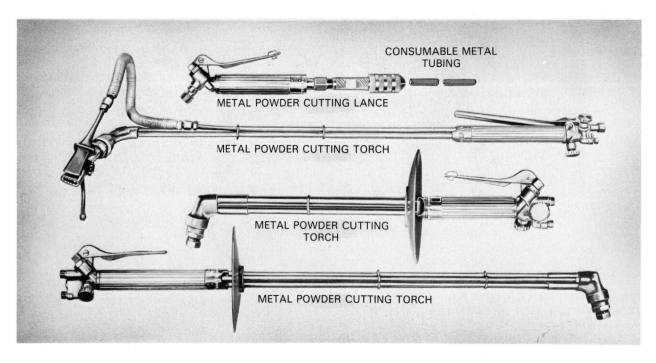

Fig. 22-12. Several types of metal powder cutting torches. Consumable metal tubing is inserted into the end of the metal powder cutting lance.
(L-TEC Welding & Cutting Systems)

Powder cutting adapters are made which may be attached to the conventional machine type cutting torch as shown in Fig. 22-13. Powder is carried to the adapter through tubing, and escapes through multiple openings around the conventional cutting tip. It is melted by the preheat flames and carried to the molten metal by the cut-

flame. More powder may be discharged from the single tube opening than can be released from the small openings in the multiple opening attachment. The single tube discharge is usually considered to be better suited for thick sections than the multiple opening attachment. The single tube attachment works best on straight cuts. It is not as well suited for shape cutting as the multiple opening attachment.

Two types of powder dispensers are presently in use. In one, an enclosed hopper is used as shown in Fig. 22-14. The hopper is filled and then pressurized. Powder is then fed under constant pressure to the metal powder cutting torch or adapter. The metal powder cutting control valve on the torch controls the flow of the metal powder. Air or nitrogen may be used to supply the pressure on the metal powder dispenser. CAUTION—never use oxygen to pressurize the powder dispenser as it may form an explosive mixture with the metal powder.

The second type of dispenser uses vibration ac-

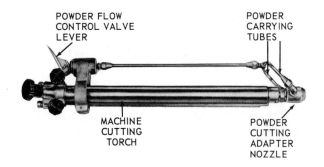

Fig. 22-13. A machine cutting torch with a metal powder cutting nozzle adapter, a powder valve, and manifold tubing attached. (L-TEC Welding & Cutting Systems)

Fig. 22-14. A pressurized powder dispenser for use with metal powder cutting equipment. Notice the pressure regulator used to control the pressure for the powder feed. Note also that only air or nitrogen are to be used to pressurize the powder.

tion to keep the powder agitated and moving along a trough. The powder falls from the vibrating trough into a hopper and is fed to the torch by air pressure. Vibrating type dispensers are used where a more constant and accurately adjusted powder flow is required.

When the thickness of metal to be cut is greater than 18 inches (approximately 450 mm), it is advisable to use the powder cutting lance. Use a specially designed mixing handle to which long lengths of hollow iron pipe are attached, as shown in Fig. 22-15. Oxygen and powder are mixed in the torch and carried to the cutting area through the hollow pipe. The pipe is slowly consumed dur-

ing the operation. Reinforced concrete walls up to 12 feet (about 3.6 m) thick may be cut using the metal powder lance.

## 22-12. METAL POWDER CUTTING (POC) PROCEDURES

The techniques and torch angles used for metal powder cutting are approximately the same as those used for oxyfuel gas cutting. An oscillating motion is sometimes required when cutting cast iron with an oxyfuel gas torch. This motion overcomes the insulating effects of the graphite particles on the metal. No torch motion is required when metal powder cutting cast iron.

To allow the metal powder to reach its ignition temperature before it hits the work, the torch should be held about 1 1/2 inches (about 38 mm) away from the work.

A technique for starting a cut faster on carbon steels has evolved because of powder cutting. This technique is called a FLYING START.

To use the flying start method, the oxyfuel gas torch is equipped with a metal powder cutting attachment. Iron oxide or a mixture of iron oxide and aluminum oxide powder is used. The powder and oxygen are turned on at the start of the cut for a few seconds and the powder is then turned off. Due to the additional heat created by the combustion of the oxide powder, the cut is started almost immediately. Prior to the use of this method, 15-25 percent of the cutting time was spent bringing the base metal up to a temperature where it would oxidize or be cut.

## 22-13. CHEMICAL FLUX CUTTING (FOC)

During the usual oxyfuel gas cutting process, oxides are formed on the base metal surface and

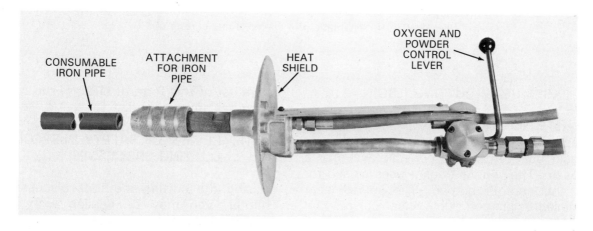

Fig. 22-15. A metal powder cutting lance handle. Consumable black iron pipe is attached to the handle. Clean cuts in reinforced concrete over 12 feet (about 3.6 m) thick can be made with 70 percent iron and 30 percent aluminum powder. (L-TEC Welding & Cutting Systems)

in the kerf. Some of these oxides melt at much higher temperature than the base metal. They make cutting the base metal very difficult.

When chemical flux cutting, a chemical compound is added to the flux rather than iron or aluminum powder as in metal powder cutting. These chemical fluxes have a reducing reaction on oxides which tend to form on the base metal. In other words, the chemical fluxes cut down the amount of these undesirable oxide formations.

Fig. 22-16 is a schematic of a chemical flux cutting outfit. Note that a separate supply line feeds the flux into the work.

In the chemical flux cutting process, the chemical flux is fed into the oxygen stream. This causes a reducing action on the alloy metal oxides, permitting their easy removal from the kerf. The carry-over action of the chemical flux action enables the cutting of multiple layers of metal plates without tight clamping.

in laser beam cutting.

3. A laser beam creates an extremely narrow kerf and a small heat-affected zone.
4. Extremely deep, small diameter holes can be pierced by focusing the laser beam.

The disadvantages are the cost of the equipment and the fact that it is only suited for metals up to 1/2 inch (approximately 13 mm) thick.

Laser beams can be reflected. Through the use of mirrors, laser beam cutting and piercing can be done in relatively hard to reach places. Oxygen, air, helium, argon, or carbon dioxide are sometimes used to blow the molten metal away when using the constant wave laser beam for cutting.

As the metal thickness increases, the required laser power increases and the cutting speed decreases.

Laser cutting and piercing is done with automatic type equipment. Because of the precise

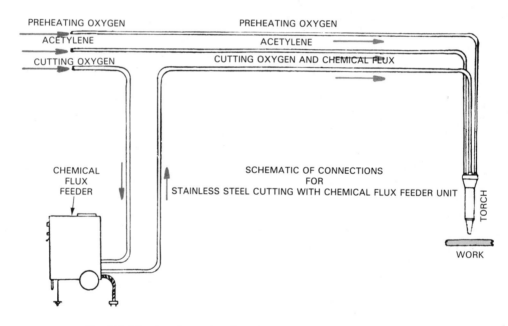

Fig. 22-16. A schematic of a chemical flux cutting outfit.

## 22-14. LASER BEAM CUTTING (LBC)

Chapters 1 and 18 describe the fundamental principles of the laser beam.

The laser beam may be a constant wave or a pulsed beam. The constant beam laser is used for straight and contour cutting. The pulsed laser beam is used to pierce or drill holes.

Some advantages of laser beam cutting are:
1. The laser beam generator does not have to be located near the weld operation.
2. The work piece is not part of the electrical circuit

character of this form of cutting, the cut is often computer programmed.

## 22-15. REVIEW OF SAFETY IN SPECIAL CUTTING PROCESSES

Using the cutting methods discussed in this chapter, you may be dealing with very high temperatures, high gas pressures, high currents, and high voltages.

The operator must be particularly alert for the types of hazards which these conditions may

produce.

Burns, flying sparks, and electrical shock are ever-present hazards.

Reread the review of safety for Chapters 6 and 15. In addition, be sure to observe the following precautions.

1. Clothing, face, eye, and hand protection must be worn.
2. Due to the rather high oxygen pressures carried in cutting hoses, be sure all hoses used are in good condition and that all fittings are tight.
3. Never weld or cut on a container which may have an air or oxygen fuel gas mixture in it, as it may explode!
4. Check with the local fire marshal for recommended procedures before cutting on tanks, cylinders, or other containers which may at some time have held flammable materials.
5. Some of the procedures recommended by the fire marshall for preparing a container which may have held a combustible liquid or gas may include:
   A. Using live steam under pressure to continuously steam out the container for at least a half hour until the entire container is heated to the boiling temperature before beginning the cutting operation.
   B. If it is an edge or dome that is to be cut, fill the container with water up to the point where the cut is to be made.
   C. In some cases after washing out with live steam, argon or nitrogen is flowed through the container. The gas is continually flowed for some time before the cut is made in order to replace all of the air in the container with this inert gas. Argon is preferred to helium for this purpose as it is heavier and will not flow away as readily as helium.
6. Be sure that the area in which all cutting operations are performed is well ventilated. There will be both small metal particles and metal oxides produced in the cutting operations.
7. When performing underwater cutting operations, be sure approved diving equipment is provided. If possible, provide a two-way telephone communicating system between the underwater welder and the surface staff assisting the welder.
8. With most cutting operations there is considerable sparking and throwing of molten slag. Therefore, the area in which the cutting is performed must be of fireproof construction. Combustibles which cannot be moved must be covered with fireproof materials. A

firewatch should also be posted.
9. Always have an approved fire extinguisher at hand when performing cutting operations.
10. If performing heavy oxyfuel gas cutting operations for a considerable length of time, be sure to manifold two or more acetylene cylinders together. In cold weather particularly, the acetylene cylinders cannot provide a heavy flow of acetylene over a long time because of low vapor pressures.
11. Due to the great amount of splatter when flame and arc cutting, be sure the eyes are well protected with goggles or a helmet, even though rays from the flame or arc may not seem to require such protection.
12. Use an inert gas to pressurize and carry the metal powder to the cutting oxygen discharge area when metal powder cutting. Fine powder when confined with a high concentration of oxygen can be explosive.
13. Oxygen should not be used for pressurizing any operating or test procedure. Use carbon dioxide, nitrogen, helium, or argon. Use a pressure relief valve on pressurized powder dispensers. Adjust the relief valve to the maximum pressure to be used.

## 22-16. TEST YOUR KNOWLEDGE

Write your answers on a separate sheet of paper. Do not write in this book.

1. Name the procedures listed below as AWS abbreviations:
   a. SMAC.
   b. OFC.
   c. LOC.
   d. AOC.
   e. POC.
   f. FOC.
   g. LBC.
2. What method can be used to cut steel 8 feet (2.44 m) thick?
3. What is the oxygen lance made of for LOC?
4. For large volume oxygen lance cutting, several oxygen cylinders may be _____ to the pressure regulator.
5. When cutting underwater using oxyfuel gas, what fuel gas is generally used below 15 feet (4.57 m)?
6. What keeps the water away from the cutting flame when OFC underwater?
7. Why are large cutting tips used for OFC underwater?
8. Name the four control valves used on an underwater oxyfuel gas cutting torch.
9. How is the flame ignited underwater?
10. What two arc cutting processes are generally

used to cut underwater?

11. Which polarity is used when oxygen arc cutting?

12. When cutting underwater, what number filter lens is recommended?

13. What type gloves are suggested for an underwater arc welder?

14. How is a constant arc length maintained when doing oxygen arc underwater cutting?

15. Why is metal powder used with the oxyfuel gas flame in the POC process?

16. What flame or arc cutting process can be used to cut concrete walls?

17. Never use oxygen to pressurize the powder dispenser when metal powder cutting or an _____ mixture may result.

18. Describe the "flying start" cutting process.

19. How does chemical flux cutting differ from metal powder cutting?

20. How is it possible to cut and pierce holes in hard to reach places using laser beam cutting?

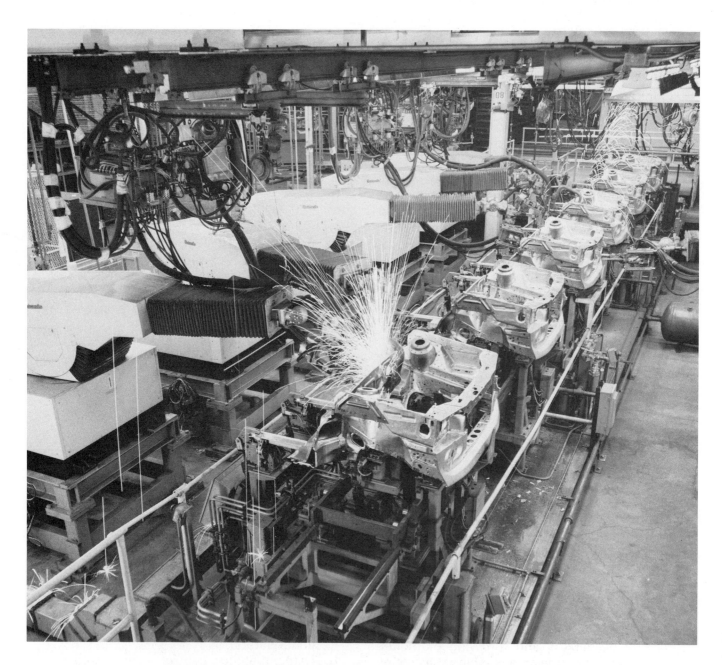

Industrial robots perform welding assignments on an auto assembly line.
(The Ford Motor Co.)

# 23 AUTOMATIC WELDING AND ROBOTS

Virtually every industry and business is moving more towards using automation and robots. The welding industry is converting various manual welding operations to either semi-automatic or fully automatic processes.

The American Welding Society defines AUTOMATIC WELDING as follows: "Welding with equipment which performs the welding operation without adjustment of the controls by a welding operator. The equipment may or may not perform the loading and unloading of the work." This definition indicates that a welding operator is not needed when welding automatically. However, a very knowledgeable welding operator is needed to set up the process.

A welding operator must select the parameters or guidelines to be used for the automatic welding operation. The parameters include voltage, amperage, wire feed speed, travel speed, the proper electrode diameter and alloy, and the correct shielding gas, to name a few.

In addition, various controls, power devices, timers, gauges, and instruments are needed to monitor an automated process. Automatic welding has many forms and can be adapted to perform almost any welding task or operation.

The robot is modernizing the welding industry. The Robot Institute of America defines a ROBOT as "A reprogrammable, multifunctional manipulator, designed to move material, parts, tools, or specialized devices, through variable programmed motions to accomplish a variety of tasks."

Stated another way, a robot is a programmable machine, which moves material or tools like a welding torch, through a set of motions to accomplish a desired task. In welding, the robot simply holds the welding torch and moves it along the weld joint. An automatic welding process is used to make the weld.

## 23-1. ADVANTAGES OF AUTOMATIC WELDING

The two most common automatic welding processes are RESISTANCE WELDING and GAS METAL ARC WELDING. Resistance welding is done automatically because the process happens too fast for a welding operator to control. Resistance welding is covered in Chapters 16 and 17.

Some advantages of automatic gas metal arc welding over manual SMAW arc welding are:
1. No electrode stub loss due to the continuous feed from a reel of welding wire.
2. Relief from the labor of concentrating on the arc length, speed, and other variables which must be controlled for a good quality weld.
3. Much higher currents may be maintained with any given electrode size as compared to manual welding.
4. Weld height, width, fusion, and penetration will be uniform once the automatic controls are adjusted correctly.
5. Weld rates are higher in automatic welding than in manual welding.

Gas metal arc welding is covered in Chapters 11 and 13. The advantages of automatic welding are:
1. Increased productivity.
2. Improved quality.
3. Reduced cost.

## 23-2. FEEDBACK CONTROLS

As a welder prepares to weld, he or she uses certain feedback controls to make sure the equipment is in good working order. A welder turns on the power supply and can hear it make noise. The welder turns on the shielding gas and sees the floating ball in the flowmeter rise. When the trigger on a GMAW torch is pressed, the wire feeds out and the shielding gas flows. These types of feedback are seen or heard by the welder.

A machine or a robot cannot see if the shielding gas is flowing. It cannot hear if the power source is on. A machine or a robot must also know if the cooling water is on. In addition, a machine or robot must know if the air pressure system, hydraulic fluid, or electric motor is working properly to make it move.

An automatic welding machine or robot needs some FEEDBACK CONTROL to determine if everything is working properly. All feedback controls for machines are in the form of an electrical signal. As an example, a robot checks to determine if cooling water is flowing. If an electrical feedback signal is received, this means the cooling water is flowing. If an electrical feedback signal is not received, the cooling water is not flowing, so the robot will turn itself off to prevent any damage.

Another type of control is used to turn something on, like the wire feeder or the shielding gas. This control is a solenoid operated valve or relay. These electrical controls are discussed in the following headings.

### 23-3. FLOW SWITCHES

FLOW SWITCHES are used to check if gas or fluid is flowing through a pipe or tube. A flow switch can be used to light a warning light to tell the welding operator that something is wrong. A switch can also be used to turn off the equipment if fluid stops flowing.

An example of a fluid flow switch is shown in Fig. 23-1. This switch uses a flexible reed which is

Fig. 23-1. A fluid flow switch. This switch may be wired to turn on a warning light and/or turn off the electrical power to the welding machine if the water flow stops or becomes insufficient for cooling. (McDonnell & Miller, Inc.)

placed in the fluid flow. When fluid is flowing, the reed will bend. The bending movement is electrically connected to a safety switch or to an indicator light. The same principle is used for an air or gas flow switch. Larger paddles are used to detect gas flow. An air or gas flow switch is shown in Fig. 23-2.

Another type of switch depends on fluid pressure to activate the switch. The fluid pressure keeps the

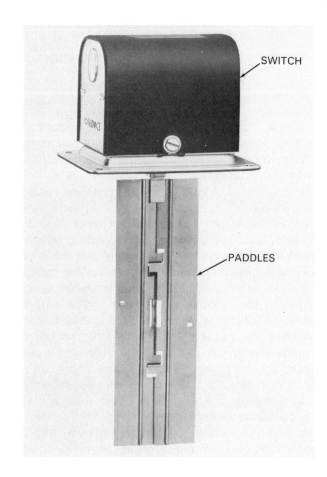

Fig. 23-2. An air flow switch with paddles that fold back when air or gas is flowing. (McDonnell & Miller, Inc.)

electrical circuit closed. If the pressure falls, the switch will open and illuminate a warning light, sound an alarm, or turn off the equipment. An example of a fluid pressure switch is shown in Fig. 23-3.

### 23-4. SOLENOID OPERATED VALVES AND RELAYS

Solenoid operated valves and relays are very similar. A SOLENOID OPERATED VALVE is used to start and stop the flow of a gas or liquid. A SOLENOID OPERATED RELAY is used to start and stop the flow of electrical current. These solenoid operated valves and relays are essential to the operation of semi-automatic and automatic welding equipment.

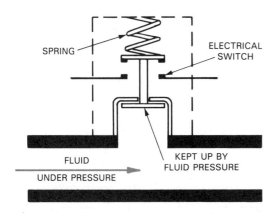

Fig. 23-3. A safety switch which depends on fluid pressure. If the fluid pressure decreases, the electrical circuit will close. The welding machine may be turned off or an alarm activated when the switch closes.

A SOLENOID OPERATED VALVE is shown in Fig. 23-4. Gas or fluid-carrying hoses are connected to the valve at the inlet and outlet. The electrical coil is connected to an electrical circuit. A solenoid may be wired to operate on AC or DC current.

A solenoid operated valve will not allow gas or fluid to pass through the valve unless a certain voltage is supplied to the coil. The plunger, which is also called the valve, is forced against the valve seat by a spring. The plunger and valve seat are the same shape. When the plunger and valve seat are forced together, they form a tight seal so no gas or fluid will flow.

To allow gas or fluid to flow, the plunger must be lifted off the valve seat. The plunger is lifted by supplying the required voltage to the coil. When the coil receives the proper voltage and current, a magnetic field is established. The magnetic field pulls the plunger up, off the valve seat. Gas or fluid will now pass through the valve.

To stop the flow of gas or fluid, the power to the coil is shut off. The magnetic field which holds the plunger up will no longer exist. Then, the spring will force the plunger against the valve seat and no gas or fluid will pass through the valve.

One common use of a solenoid operated valve is to turn the shielding gas on and off during gas tungsten and gas metal arc welding. The shielding gas can also be made to flow by manually pushing the purge switch. Pressing the purge switch will send power to the solenoid gas valve which will allow shielding gas to flow. See Fig. 23-5.

When GMAW or GTAW, the shielding gas is required to flow for a short period after the welding is stopped. This additional gas flow protects the electrode while it cools.

The additional gas flow is provided electronically. The solenoid operated gas valve is connected to a second electrical circuit. This circuit will send electricity to the solenoid, after the electricity in the main circuit is stopped.

This additional gas flow is set by adjusting a knob on a GMAW or GTAW machine. The knob can be used to vary the resistance of the second electrical circuit. The POSTFLOW (additional gas flow) will vary as the resistance of the circuit is changed. Some welding machines control the postflow by counting the cycles in the AC wave. Alternating current has 60 cycles per second. When the postflow is set for 10 seconds, the machine electronically counts 60 cycles per second. After 600 cycles (60 cycles/second x 10 seconds) the current in the second circuit stops and so does the postflow. This electronic circuit timing is very exact.

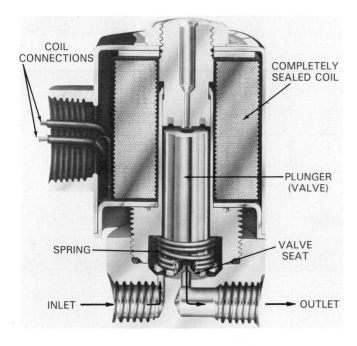

Fig. 23-4. A solenoid valve used to control shielding gas flow. The plunger is forced against the valve seat by the spring. The valve shown is in the closed position. (Airmatic-Allied, Inc.)

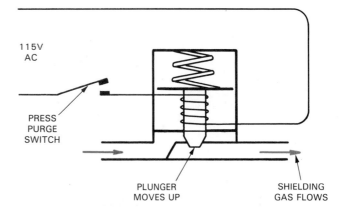

Fig. 23-5. This circuit is used to purge the lines on a GMAW or GTAW torch. When the purge switch is pressed, the solenoid lifts the plunger and allows shielding gas to flow.

Another use for solenoid operated valves is to turn cooling water on and off. Water is used to cool some GTAW and GMAW torches. Water is also used to cool resistance welding equipment.

A SOLENOID OPERATED RELAY is like a solenoid operated valve. In a solenoid operated relay, an electrical coil is wrapped around an iron core. One set of electrical contacts is on the base of the relay. The second set of contacts is on a movable arm. The contacts on the arm are kept away from the contacts on the base by a spring. To operate the relay, proper voltage is sent to the electrical coil. Current passing through the coil will set up a magnetic field. The magnetic field pulls the arm down to close the electrical contacts. When voltage to the coil is turned off, the magnetic field stops. The spring pulls the arm up and the electrical switch opens. A relay is shown in Fig. 23-6.

The big advantage of a relay is that a small voltage (6, 12, 24, or 115 volts) can be used to control a larger voltage (like 115 or 230 volts).

One common example for the use of a relay is in a GMAW machine. A relay is used to operate the wire feeder. When the trigger on the GMAW torch is pressed, the solenoid receives a small voltage and the relay switch closes. After the relay switch closes, the wire feeder receives 115 volts AC and begins to feed wire. See Fig. 23-7. Thus a small voltage of 24 volts, can control a large voltage of 115 volts. Another way to operate the wire feeder

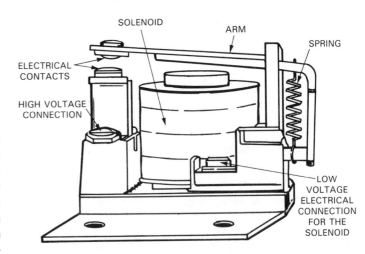

Fig. 23-6. The parts of a solenoid operated relay. When the solenoid receives power, it will pull the arm down and connect the contacts of the switch.

is to push the inch switch. This will also supply 115 volts AC to the wire feeder as seen in Fig. 23-7.

All automatic and semiautomatic welding operations depend on the use of solenoid operated valves and relays. Fig. 23-8 shows a gas metal arc welding circuit. The circuit contains a relay, a solenoid operated gas valve, a transformer, and three switches. The transformer is used to reduce the voltage from 115 volts AC to 24 volts AC. Only 24 volts goes to the gun switch. It is dangerous to have 115 volts in a welder's hand.

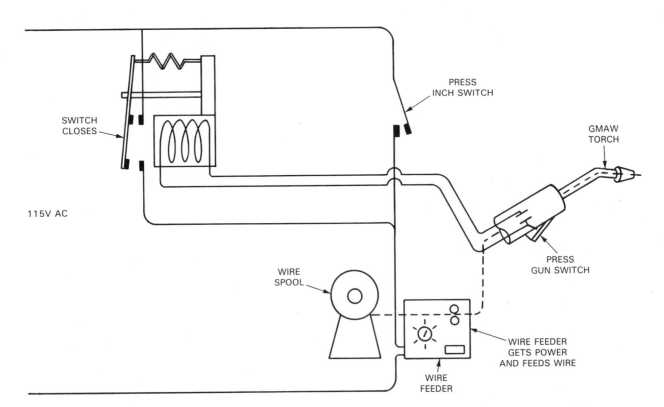

Fig. 23-7. To operate the wire feeder, a welder can press the inch switch or press the gun switch. The gun switch will close the relay which will allow electricity to flow to the wire feeder.

A welder can check to see if everything is operating properly by pressing the purge switch and inch switch. When the welder presses the purge switch, electricity goes to the solenoid operated gas valve. The plunger raises and allows shielding gas to flow. See Fig. 23-5. In the same way, when the inch switch is pressed, the wire feeder receives 115 volts AC and feeds electrode wire.

When the welder is ready to weld, the gun switch is pressed. The 24 volt circuit, shown in red in Fig. 23-8, will cause the relay switch to close. When the relay closes, both the solenoid operated gas valve and wire feeder will get 115 volts AC. Both shielding gas and electrode wire will be delivered to the welding torch. The welder can continue welding until the gun switch is released. When the gun switch is released, the solenoid operated relay will no longer receive 24 volts. The relay spring will open the electrical switch. The gas valve will close and the wire feeder will stop. Note that the postflow gas circuitry is not shown in Fig. 23-8.

## 23-5. THREE-WAY VALVES FOR HYDRAULIC OR PNEUMATIC SYSTEMS

Three-way solenoid operated valves are often used to open and close the operating passages of a hydraulically or pneumatically operated cylinder on an automatic welding machine. Fig. 23-9 shows a three-way valve.

This valve is normally in the closed position. The passage of hydraulic pressure from port A on the inlet to port C on the outlet side of the valve is blocked by the main valve.

Follow the action in Fig. 23-9. When the solenoid is energized, the solenoid pilot valve is moved down. A passage is then completed from port D through port E to the top of the main valve. This pressure on top of the main valve forces it down, thus opening the passage from A to C and to the operating cylinder connected to passage C.

When the solenoid is de-energized, the solenoid pilot valve is moved up by spring pressure blocking the passage from port D to port E. Pressure is relieved from above the main valve by passing through ports E, F, G, and to release port at B. When pressure is relieved from the top of the main valve, it is moved up by pressure from below. This closes the passage from A to C, and opens the passage from C to B, allowing the fluid in the C passages to discharge through opening B.

## 23-6. VARIABLE SPEED DRIVE MOTORS

Variable speed drive motors may be used in automatic equipment to:
1. Move a torch or electrode along the work.

2. Move the work under the torch.
3. Feed electrode wire at a controlled rate.

A variable speed motor is often used to move an oxyfuel cutting torch along the work to be cut. The cutting torch is attached to a carriage. A knob on the carriage is used to control the travel speed of the carriage. The knob is attached to and controls a variable resistor. When the resistance of the field winding circuit is increased, the current in the field winding will decrease. The current through the armature will increase and this will cause the armature of the motor to turn faster. Thus the travel speed of the cutting operation increases.

A variable speed motor can be used to control the speed of a welding, cutting, gouging, flame spraying, and other operations or processes. Sometimes it is necessary to move the work piece instead of the torch. A variable speed motor can be used to control the speed of the work in the same way it controls the speed of the torch. A schematic of a variable speed motor is shown in Fig. 23-10.

## 23-7. AUTOMATIC WELDING PROCESSES

Gas metal arc welding and flux cored arc welding are semiautomatic processes. The welding power source maintains a constant voltage. The wire feeder delivers the electrode wire continuously. A welder must control the manipulation of the torch.

If the GMAW or FCAW torch were attached to a carriage, the welder would not have to hold the torch. By allowing a machine to control the torch, instead of a welder controlling it, the process becomes mechanized. An example of a mechanized flux cored arc welding machine is shown in Fig. 23-11. Only when feedback controls are used does a process become automatic.

Many automatic and mechanized welding processes have already been described in this book. The chart below lists some automatic processes and where they are discussed in this book.

| PROCESS | |
|---|---|
| Cutting | Heading 6-20 |
| Electron beam welding | Heading 18-19 |
| Electroslag, Electrogas welding | Heading 18-3 |
| Friction welding | Heading 18-13 |
| Furnace brazing | Heading 8-15 |
| Gas Tungsten arc welding | Heading 12-17 |
| Gouging | Fig. 14-3 |
| Inertia welding | Heading 18-14 |
| Laser welding | Heading 18-20 |
| Projection welding | Heading 17-9 |
| Resistance welding | Chaps. 16 & 17 |
| Submerged arc welding | Heading 18-2 |

Automatic and mechanized processes which are discussed in other parts of this book.

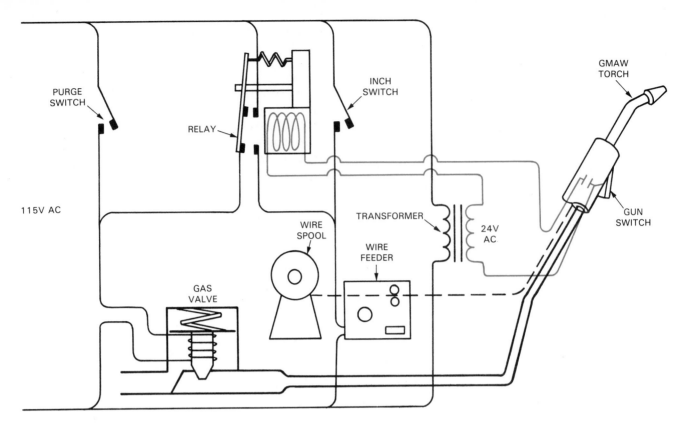

Fig. 23-8. A gas metal arc welding circuit which uses a transformer, a solenoid operated relay, a solenoid operated gas valve, and three switches. When the purge switch is pressed, the gas valve allows shielding gas to flow. When the inch switch is pressed, the wire feeder supplies electrode wire. When the gun switch is pressed, the relay closes. The gas valve supplies shielding gas and the wire feeder supplies electrode wire.

Even though each of the processes listed above is an automatic process, the processes are quite different. In submerged arc welding, the torch moves over the workpiece. In laser and electron beam welding, the equipment is too large to move. See Figs. 18-43 through 18-49. In these two processes, the work is moved beneath the laser or electron beam. A resistance spot weld is made by placing the two pieces of metal between the electrodes. The electrodes come together, a spot weld is made, and the electrodes separate. For a friction or inertia weld, the parts to be joined are clamped in fixtures. After this clamping operation, the process is completely automatic.

From these brief examples, the variety of processes and applications can be seen. Almost every process can be automated. Fig. 23-12 shows a machine which uses two GTAW torches to make two seam welds.

One process which is widely used in industry is automatic gas metal arc welding. Most arc welding robots use the automatic GMAW process. Other robots may use automatic GTAW or resistance welding.

All automatic welding or cutting machines rely on feedback controls, solenoids, relays, and variable speed motors to control and direct the automatic welding process.

## 23-8. INTRODUCTION TO ROBOTS

The major reasons to use an automatic welding process are to:
1. Increase productivity.
2. Improve quality.
3. Reduce cost.

Industrial robots, which are usually just called robots, are used to accomplish all three of these objectives. A robot can make a weld faster than can be done manually. This increases productivity. A robot will perform the same sequence of moves over and over. The robot will maintain the same arc length, same voltage, and same travel speed every time. This will improve the quality of the welding operation and tend to produce fewer defects. Since the robot welds faster, has better quality, and welds continuously, this reduces the cost of the welding operation.

Robots are not suited for every welding application. A robot is a complex, expensive piece of machinery. They require special maintenance and care. But when used properly, a robot can be an

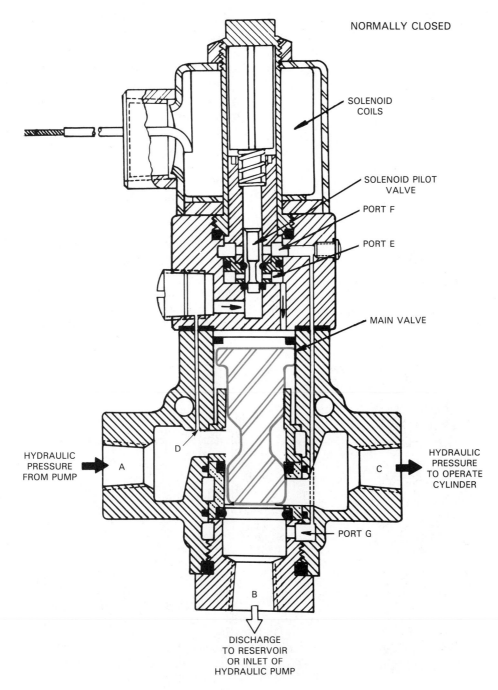

NORMALLY CLOSED

SOLENOID COILS

SOLENOID PILOT VALVE

PORT F

PORT E

MAIN VALVE

HYDRAULIC PRESSURE FROM PUMP

A

D

C

HYDRAULIC PRESSURE TO OPERATE CYLINDER

PORT G

B

DISCHARGE TO RESERVOIR OR INLET OF HYDRAULIC PUMP

Fig. 23-9.  A three-way solenoid valve for controlling hydraulic fluid flow.   (Airmatic/Beckett-Harcum)

asset to large production companies and to the smaller welding job shop.

Refer to Heading 23-18 on Safety Practices in Automatic Welding before proceeding.

## 23-9.  ROBOT SYSTEM

The robot itself is only one part of a robot system. The major parts of a robot system are:
1. Industrial robot.
2. Robot controller.
3. Automatic welding equipment.

4. Positioner.
5. Operator's control.
Each of these components is shown in Fig. 23-13.

## 23-10.  INDUSTRIAL ROBOTS

An industrial robot is required to move the welding torch through a set of motions. The robot needs some system to enable it to move.

Most robots are connected to a 480 volt electrical service. Some others may use 240 volt service. This electricity can be used to operate electric motors

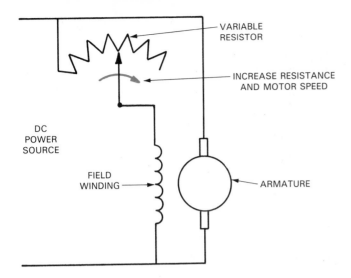

Fig. 23-10. A schematic wiring diagram of a variable speed motor. A manually controlled variable resistor in series with the field winding, controls the speed of the motor. As the resistance is increased, the motor turns faster.

used to move the robot.

Another type of robot is hydraulic powered, but 240 volt or 480 volt electricity is still required to operate the robot. The electricity is used to operate the hydraulic pump. Hydraulic fluid in hydraulic cylinders and hydraulic motors is used to move the robot.

A robot can only move or rotate about a mechanical joint. Each mechanical joint allows the robot to move in one plane, horizontal or vertical. Each joint moves about an AXIS. Many robots are five axis robots. This means the robot has five mechanical joints about which it can move. Some robots have six axes.

Each axis is given a name. Four of the axes are named after similar moving parts of the human body. A six axis robot with the name and movement of each axis is shown in Fig. 23-14. The axis at the base is called the base rotation. This axis allows the robot to move in a horizontal plane, left

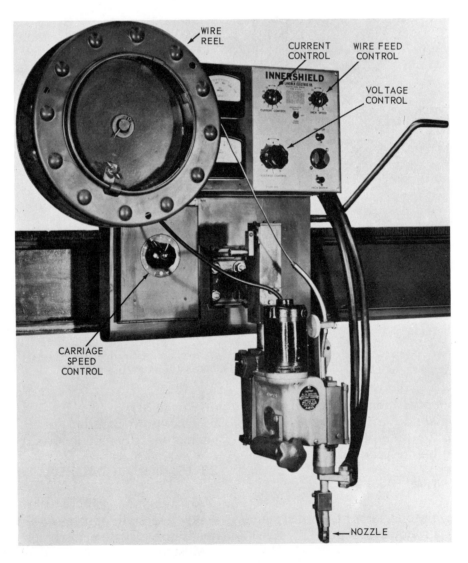

Fig. 23-11. An automatic welding head for flux cored arc welding. (The Lincoln Electric Co.)

Fig. 23-12. Two welds are made at the same time with automatic GMAW equipment   (Cecil C. Peck Co.)

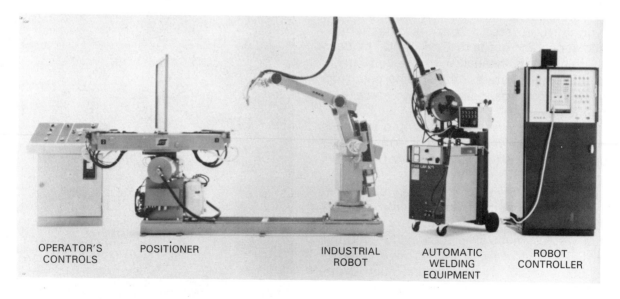

Fig. 23-13. A complete robot welding system showing the major parts that make up the system.   (ESAB Automation, Inc.)

and right. The waist, shoulder, arm or elbow, wrist pitch, and roll axes allow the robot to move in all planes. These axes work to position and manipulate the welding torch.

Each axis or joint must have its own mechanism to make it move. An electric robot must have an electric motor for each axis. The motors for the waist, shoulder, and elbow are usually larger than the motors used to move the wrist and roll axes. In the same way, a hydraulic robot must have an actuator for each axis. A HYDRAULIC ACTUATOR is used to change hydraulic energy into mechanical

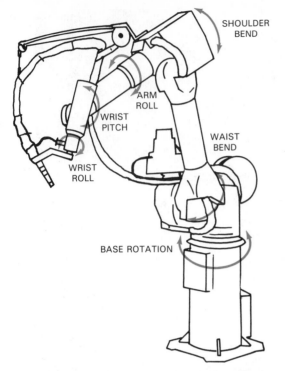

Fig. 23-14. A six axis robot. Note the name and movement of each axis. (GMFanuc Robotics Corp.)

energy. Two common hydraulic actuators are a hydraulic cylinder and a hydraulic motor.

A HYDRAULIC MOTOR operates like a water wheel. Hydraulic fluid causes a turbine wheel with blades on it to rotate. A gear is attached or machined at one end of the turbine shaft. The gear on the turbine shaft meshes with a second gear. When the first gear rotates, the second gear also rotates. This second gear causes some part of the robot to move.

The hydraulic motors in the waist and shoulder are larger than the other hydraulic motors. A hydraulic cylinder is often used to move the elbow axis. See Fig. 23-15.

The robot has a limited area that it can reach. This area in which the robot can move and work is called the ROBOT'S WORKING VOLUME. Any parts that are to be welded must be placed in the working volume. Each type of robot has its own working volume. Some working volumes are much larger than others. A working volume for one type of robot is shown in Fig. 23-16.

Never enter the working volume of a robot when it is in operation.

## 23-11. ROBOT CONTROLLER

A robot controller controls the entire robot system. It is the "brains" of the operations. The ROBOT CONTROLLER controls the movement of the robot, the movement of the positioner, and con-

trols the welding equipment. The controller requires 480 volts or 240 volts to operate.

The main part of the robot controller is a computer. Every part of the robot system is connected to this computer. In addition to controlling the operation of the system, the computer monitors all the parts to make sure everything is running properly. The computer also stores the program or sequence of actions that the robot system will follow to perform the welding operation.

All the responsibilities of the robot controller can be classified as follows:
1. Monitor the robot system.
2. Store the program.
3. Execute the program.

The robot controller uses feedback control to monitor the robot system. As discussed in Headings 23-2 and 23-3, feedback controls and flow switches are used to monitor the operation of the system. Checks are constantly being made to make sure the motors or hydraulic pumps do not overheat. Checks are made to make sure cooling water is flowing. If something is not operating properly, the controller will illuminate a warning light. If the problem is serious, the controller may shut down the robot system to prevent any damage.

A PROGRAM is a series of operations. Through this series of operations the robot system performs

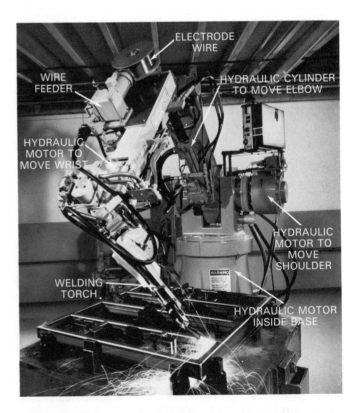

Fig. 23-15. A six axis hydraulic robot. Note the hydraulic motor used to move the shoulder and wrist. Also note the hydraulic cylinder used to move the elbow.
(Cincinnati Milicron)

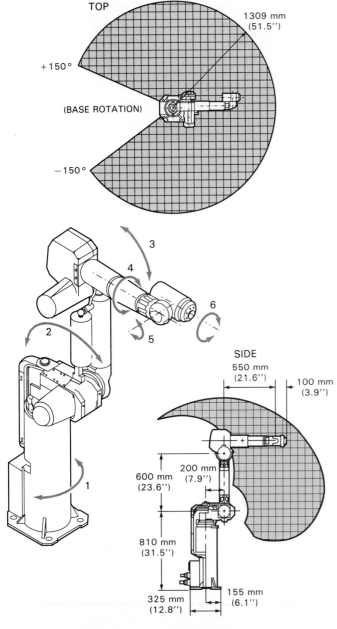

TOP

1309 mm
(51.5")

+150°

(BASE ROTATION)

−150°

3

4

6

2

5

1

SIDE

550 mm
(21.6")

100 mm
(3.9")

200 mm
(7.9")

600 mm
(23.6")

810 mm
(31.5")

155 mm
(6.1")

325 mm
(12.8")

Fig. 23-16. An example of a robot's working volume and axis descriptions. The robot cannot reach beyond its working volume. (GMFanuc Robotics Corp.)

An example of the use of solenoid operated valves is a hydraulic robot. To move the robot, hydraulic fluid must be allowed to enter a hydraulic cylinder or flow through a hydraulic motor. Solenoid operated valves control the flow of hydraulic fluid; see Fig. 23-9. In order to move one axis of the robot, the controller sends a signal to the solenoid operated hydraulic valve. The solenoid lifts the plunger and allows hydraulic fluid to flow. When hydraulic fluid flows, the robot moves. When the controller determines that the robot has moved to the desired location, the electrical signal to the solenoid is stopped. No more hydraulic fluid will flow and the robot's movement stops.

The operation of an electric robot is similar. To move one axis of the robot, the controller sends an electric signal to a solenoid operated relay. The solenoid will move the plunger and close the electrical switch. Now 240 volt or 480 volt electricity will operate an electric motor to move a part of the robot. When the controller turns off the signal to the solenoid operated relay, the relay switch will open. No current will flow to the electric motor and the robot will not move.

It must be understood that each axis of a robot has its own electric motor, hydraulic motor, or hydraulic cylinder. Each electric motor has a relay. Each relay is controlled by the robot controller. Likewise, each hydraulic motor and hydraulic cylinder has its own solenoid operated hydraulic valve. The robot controller controls each valve. Through the use of solenoid operated valves and relays, the robot controller controls the movements of the robot.

In addition to controlling the movements of the robot, the controller also controls the movement of the positioner. The weld positioner is moved by means of electric or hydraulic motors or hydraulic cylinders. These are turned on and off by solenoid operated valves or relays also. The controller also turns on the automatic welding equipment. A robot controller operates the entire robot welding system.

## 23-12. AUTOMATIC WELDING EQUIPMENT

Common welding processes used with robots are gas metal arc welding, gas tungsten arc welding, and resistance spot welding. Gas metal arc and resistance spot welding are the most widely used. Automatic welding equipment used in robot welding is connected to the robot controller. As stated in Heading 23-11, a robot controller operates the entire robot system. This includes the welding equipment.

The controller can set the desired welding voltage and wire feed speed. This is done through electrical signals and electric circuits. When resistance spot

the welding operation. A controller stores the program so that the exact same series of operations can be performed over and over. Programming the robot is explained in Heading 23-15.

The most important function of the robot controller is to execute the program. Through the use of solenoid operated valves and relays, the controller can execute the program.

A robot controller uses electrical signals to perform the sequence of operations called a program. A controller usually sends small electrical signals to solenoid operated valves and relays. Through these valves and relays, the controller controls every part of the robot system.

welding, the controller can set the current level and electrode force.

When it is time for the robot system to begin welding, an electrical signal is sent to the welding equipment by the controller. This signal closes a relay switch and the welding equipment begins to operate. Electrode wire, shielding gas, and the proper welding voltage and current are delivered to the welding torch. When resistance spot welding, the electrodes close with the proper electrode force, and the welding current flows. The resistance welding sequence of squeeze time, weld time, and hold time, is also controlled by the robot controller.

Automatic welding equipment used for robot welding is exactly the same as the manual or semi-automatic equipment described in other headings in this book. The only difference is that the equipment is controlled by the robot controller. A robot controller sets the voltage and wire feed speed, or current and electrode force, depending on the process. A controller also turns the equipment on and off when required in the program.

## 23-13. POSITIONER

A POSITIONER is used to hold and move the parts to be welded. Parts are often clamped into fixtures on a positioner. When clamped into fixtures, the parts will not move during welding.

Another use of a positioner is to allow the robot to reach all areas of a part. A large piece to be welded can be placed on a positioner. Part of the piece will be in the robot's working volume, part may be outside. Remember that a robot cannot reach outside of its working volume.

The robot can weld all the joints that are within its working volume. The area outside of the working volume is moved inside by rotating the positioner. The portion that was inside and was welded is moved outside. The robot can now complete the welding operation. By using a positioner, the robot's working volume can be increased.

If welding must be done on the top and bottom of a part, a positioner can help. Welding can be done on the top, then the positioner can rotate so the bottom becomes the top. See Fig. 23-17.

Positioners are available that can hold two or more parts to be welded. When one part is being welded, a second part can be placed in fixtures on the positioner. After the robot completes welding one part or assembly, the positioner moves the second part into the robot's working volume. A welding operator removes the part that was welded and installs another part to be welded. By using a positioner, the robot can do more welding which increases productivity. See Fig. 23-18.

Some positioners have two or more rotating tables. A metal screen can be used to separate the

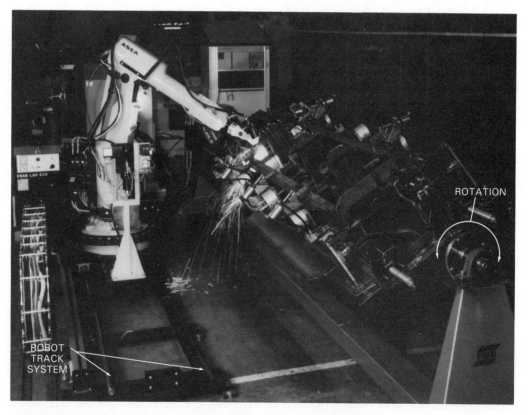

Fig. 23-17. The positioner rotates to allow this six axis robot better access to all welds.
(ESAB Automation, Inc.)

Fig. 23-18. A rotating positioner used with an industrial robot. Stations "A" and "B" rotate 360° to provide the robot with access to all weld areas. When the part on station "A" is welded, the entire positioner rotates so that the part on station "B" is positioned for welding. (ESAB Automation, Inc.)

tables. An operator can set up parts on one side of the screen while welding is being done on the other side. By using a screen, the operator is protected from the arc and spatter. See Fig. 23-19.

Positioners usually have two axes about which they rotate. Some positioners have as many as five axes. A positioner is not a required part of a robot system. Parts or assemblies can be clamped to a stationary table. The advantage of positioners is that they are able to effectively increase the working volume of a robot. Positioners also increase the productivity of a robot system.

## 23-14. OPERATOR'S CONTROLS

The person directing a robot controls the operation of the robot system by using an OPERATOR'S CONTROL PANEL. An operator uses the operator's control panel to begin the robot program and to turn off the robot in an emergency.

An operator's control panel is usually used when an operator is required to load and unload parts. When a robot completes the welding operation on one part, the robot moves back out of the way. It will not move from this position until the welding operator pushes a button.

While the robot is out of the way, the part that was just welded is removed. A new part to be welded is placed in the welding fixture. The operator then presses the "Cycle Start" button on the operator's control panel. When the cycle start button is pressed, the robot will perform the complete welding operation. When the robot finishes welding, it will again move out of the way and wait for the cycle start button to be pressed again.

If something goes wrong during the operation of the robot system, the operator can stop the process. A red "Emergency Stop" button is on the operator's control panel. To stop the program, the "Emergency Stop" button is pressed.

Not all robot systems use an operator's control panel. Some robot systems have the emergency stop and cycle start buttons on the robot controller. An operator's control panel is shown in Fig. 23-19. Also see the operator's control panel in Fig. 23-18.

Refer to Heading 23-18 on Safety Practices in Automatic Welding.

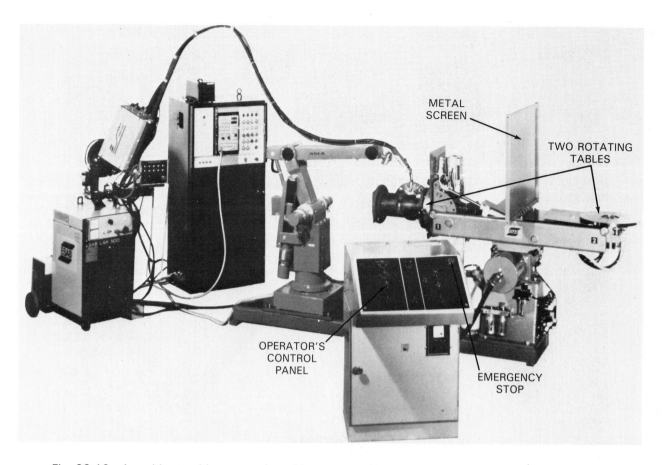

Fig. 23-19. A positioner with two rotating tables separated by a metal screen. The screen protects the operator from viewing the arc and from the metal spatter. The operator's control panel is used to begin the welding operation. An operator can stop the operation by pressing the emergency stop button.    (ESAB)

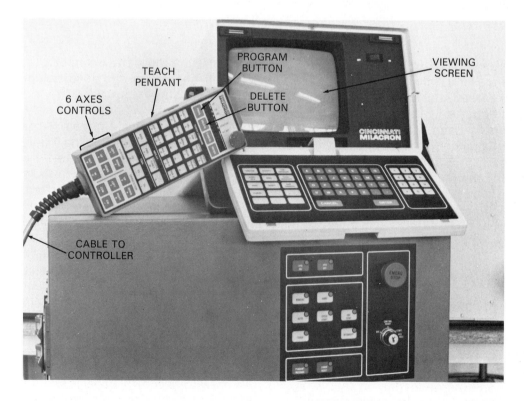

Fig. 23-20. A teach pendant for a six axes robot. Note the "Program" and "Delete" buttons. A viewing screen is used to display some variables used by the robot, like travel speed and welding parameters.    (Cincinnati Milicron)

## 23-15. PROGRAMMING THE ROBOT SYSTEM

A robot system is controlled by a robot controller. The robot controller stores the sequence of steps and functions that the robot will perform. This sequence of steps and functions is called a program. Before the controller can store a program, it must be taught each step and function in the program. This process is called PROGRAMMING or teaching the robot system.

When teaching or programming the robot system, the welding operator uses a teach pendant. A teach pendant is shown in Fig. 23-20. The TEACH PENDANT has control over each axis of the robot and each axis of the positioner (if used). The welding operator can move the robot using the teach pendant. One axis is moved at a time by pressing the proper axis button.

There are many steps and functions within one program. Some of the more important steps and functions are listed below.
1. The movement of the robot.
2. The speed of movement.
3. When to begin welding.
4. When to stop welding.

One of the most important parts of a program is the movement of the robot. There are two ways to program the motion of the robot: POINT TO POINT or CONTINUOUS PATH. Point to point motion involves programming a series of points. The robot will move in a straight line from one point to the next point.

In POINT TO POINT MOVEMENT, the welding operator uses the teach pendant and moves the robot to a desired location. After the new location is reached, the "Program" button is pressed. This point becomes one point in the program. The welding operator then moves the robot to the next point to be in the program. Again the "Program" button is pressed. This process is continued until the robot's movements are completed.

It may be necessary to delete a point in a program. This is done the same way as when programming a point. The robot is moved to the point to be deleted. The "Delete" button is pressed. That point is no longer a part of the program.

An example of point to point movement is shown in Fig. 23-21. A robot operator wants to move the welding torch on the end of the robot arm from point 1 to point 4. Four points are programmed so the robot's movements will miss the obstacle between points 1 and 4. The robot will move from point 1 to point 2, from point 2 to 3, and then from point 3 to point 4. Each movement will be a straight line.

If the robot operator decides to delete point 2, the robot will move from point 1, to point 3, and then to point 4. This movement is faster than mov-

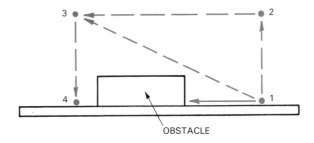

Fig. 23-21. Point to point movement. To move a robot from point 1 to point 4, two paths can be followed. The path 1-2-3-4, or the path 1-3-4 can be followed. A robot cannot move directly from point 1 to point 4 or it will hit the obstacle.

ing from 1 to 2 to 3 to 4. The obstacle is still missed. If point 3 were deleted, the robot would attempt to move directly from point 1 to point 4. During this movement the welding torch on the robot arm would hit the obstacle and the welding torch may be damaged. The robot may also require maintenance after such a collision.

The other type of robot movement is continuous path. In CONTINUOUS PATH MOVEMENT the robot will follow the exact path that it was taught. If the robot moved along a curve while being programmed, it will repeat the curve. There are no programmed points. Every movement of the robot will be part of the program.

If the robot was taught to follow the path 1-2-3-4 in Fig. 23-21, it would always perform the same move. Point 2 cannot be deleted. To change the program, the robot must be reprogrammed. This involves deleting or erasing the old program, and reprogramming the path.

Point to point movement is usually preferred.

Another important function that must be programmed is the speed at which the robot moves. Speeds can be relatively slow, like welding speeds of 10 to 30 inches per minute (approx. 4 to 13 mm/sec). Maximum speeds depend on the robot being used. Speeds can exceed 2000 inches per minute (approx. 850 mm/sec) with speeds of 800 inches/min (340 mm/sec) common.

Each point to point movement can be made at a different speed. When a robot is moving through open areas away from any obstacles, high travel speeds can be used. When the robot moves toward the workpiece, or near an obstacle, the travel speed is usually reduced. The travel speed can be reduced in steps. See Fig. 23-22. If the movement from point 3 to point 5 was all done at 60 inches/minute (25.4 mm/sec), the movement would be quite slow. Moving from 3 to 4 at 400 inches/minute (169 mm/sec) requires less time. Reducing the travel speed in steps allows a faster speed to be used, except for the last few inches (millimeters).

In order for the robot system to know when to

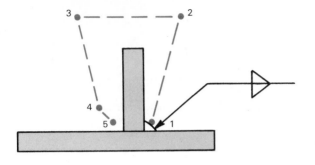

| FROM | TO | SPEED | |
|------|-----|-------|--------|
|      |     | in/mm | mm/sec |
| 1    | 2   | 400   | 169    |
| 2    | 3   | 800   | 339    |
| 3    | 4   | 400   | 169    |
| 4    | 5   | 60    | 25.4   |

Fig. 23-22. Robot speeds used to go from point 1 to point 5. Note the speed is faster from point 3 to 4, than from 4 to 5.

begin welding, it must be programmed. In addition to knowing when to weld, the controller must be programmed to turn on the shielding gas and wire feeder when used. The controller must set the current and electrode force in resistance welding.

Each of these operations is a step in the program. A typical set of functions required to begin GMAW is described below. The first function is to set the voltage and wire feed speed. Welding parameters are set by entering the values while programming the robot. The second function is to turn on the shielding gas. Next, add a delay function for .5 seconds. This allows the shielding gas to cover the welding area. The next function is to begin feeding electrode wire. Then, add another delay function of .5 seconds. This is to allow the arc to stabilize and a proper sized molten pool to develop. The final function is for the robot to begin moving along the joint to be welded. A similar set of functions is required to stop welding.

Setting the welding parameters is usually done using the teach pendant.

Fig. 23-23 shows two pendants. One pendant is used to program the robot system. The second pendant is used to set the welding parameters.

The robot system has two modes of operation, manual and automatic. All programming is done in the manual mode. Any change to the robot's movements or change in the welding parameters is done in the manual mode.

To execute the program, the robot system is placed in the automatic mode. The robot system is then completely automatic. It will perform the program exactly. The program can be repeated over and over.

## 23-16. ROBOT SYSTEM OPERATION

When the programming of a robot system is completed, it can then be used to perform the desired welding operations.

Before making any welds, the welding operator must inspect the robot system. This inspection includes the following:
1. Check if the shielding gas cylinder has gas.
2. Check if the electrode wire spool has wire.

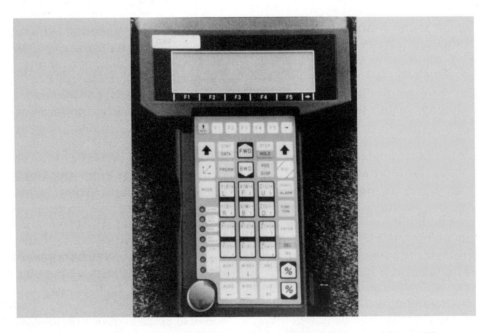

Fig. 23-23. A teach pendant used to program a six axis robot and its positioner. (GMFanuc Robotics Corp.)

3. Check all electrical connections.
4. Check hydraulic lines for leaks.

In addition to checking the equipment, the welding operator should check the welding parameters. For GMAW these include:
1. Voltage setting.
2. Wire feed speed.
3. Proper electrode wire composition.
4. Proper electrode diameter.
5. Proper contact tube (or nozzle) to work distance.
6. Proper travel speed for the robot when it is doing the welding.

Finally, the robot operator places the parts to be welded in their proper place. This usually involves placing the parts in a fixture. See Fig. 23-24.

The robot system must be in the automatic mode. To begin the program, the ''Cycle Start'' button is pressed. The robot controller will direct the robot system through the program steps. Remember that each part of the robot system, the robot, the positioner, and the automatic welding equipment, is controlled by the controller.

When a robot system has completed a program,

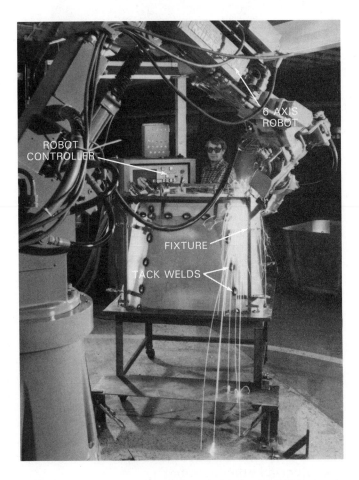

Fig. 23-24. A fixture is used to hold the aluminum parts together while tack welds are made. After tacking is completed, the fixturing is removed and the seams are welded completely. (Cincinnati Milicron)

the robot will usually move back out of the way. The system will not move again until the ''Cycle Start'' button is pressed again.

Never let any part of your body be near a moving welding torch. Never enter the working volume of a robot when it is in operation.

Industry uses many robots on assembly lines. The robots shown in Fig. 23-25 are resistance welding robots.

## 23-17. COMPARING ROBOT SYSTEMS

There are many manufacturers producing different robots made for different purposes. Not all robots are made for welding.

Before any robot system is purchased, the manufacturer should be consulted. Each robot system has advantages and disadvantages. The two types of robot systems available for welding are the hydraulic robot and the electric robot.

A hydraulic robot is much larger and often weighs over 5000 pounds (approx. 2300 kg). A hydraulic robot is capable of lifting more weight than an electric robot. Hydraulic robots usually cost more than electric robots.

Electric robots can be large or small. Small electric robots often weigh less than 1000 pounds (approx. 450 kg); some weigh under 300 pounds (approx. 140 kg). Large electric robots can weigh as much as or more than a hydraulic robot.

One advantage of an electric robot is that it usually has better accuracy. Robot accuracy is also called REPEATABILITY. Robot accuracy is the ability of a robot to go from one point to another point exactly. There is always some chance of error involved. The robot may not go exactly to the programmed point, but will be very close. An electric robot usually has an accuracy or repeatability of ±.010 inches (.254 mm). This means that the robot will move to within .010 inch (.254 mm) of the programmed point. Hydraulic robots have repeatability of .030 to .040 inches (.762 to 1.016 mm).

## 23-18. REVIEW OF SAFETY PRACTICES IN AUTOMATIC WELDING

All of the safety practices discussed in previous chapters relative to GMAW, GTAW, resistance welding, cutting processes, and the like also apply to automatic welding and cutting processes.

Be very alert when around power driven devices and machines. Never let any part of your body be near a moving welding torch. Never enter the working volume of a robot when it is in operation. Some type of fencing should be placed around the working volume of a robot so no one enters the area.

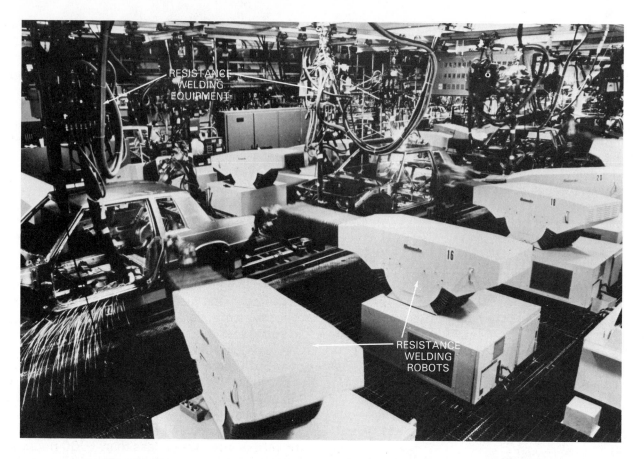

Fig. 23-25. Robots used to resistance spot weld automobile bodies. (Unimation Inc.)

In addition, check the electrical leads and gas hoses so they do not get caught during the operation of automatic equipment or robots. Before operating a system at full speed, the program, or sequence of moves, should be executed at slow speeds. During this test run, any interference of cables or hoses can be seen and corrected.

If the automatic welding equipment or robot system is not functioning properly, the welding operator should press the emergency stop button. Do not attempt to fix the equipment while it is operating. An automatic machine or robot has a lot more power and strength than a person.

## 23-20. TEST YOUR KNOWLEDGE

Write your answers on a separate sheet of paper. Do not write in this book.

1. How does automatic welding differ from manual welding or semiautomatic welding? (Refer to the definition of automatic welding.)
2. What is a welding operator required to do in automatic GMAW?
3. List the advantages of automatic welding.
4. What is the use of flow switches?
5. What type of electrical current is required to operate a solenoid?
6. How is fluid or gas prevented from flowing through a solenoid operated gas valve?
7. How is a relay switch made to close?
8. What is the advantage of using a relay?
9. What happens when the gun switch on a GMAW torch is pressed?
10. How does a knob on a welding or cutting carriage vary the speed of the operation?
11. What do automatic welding and cutting machines rely on to perform and to control the automatic process? List at least three items.
12. List the five major parts of a robot system.
13. What does it mean when a robot has five axes?
14. What is a hydraulic actuator used for?
15. What is the working volume of a robot?
16. The main part of a robot controller is a _____.
17. List two advantages of using a positioner.
18. What does a welding operator use to teach or program a robot system?
19. When comparing hydraulic and electric robots, what is the big advantage of an electric robot over a hydraulic robot?
20. What should be done to prevent people from entering the working volume of a robot while it is in operation?

# 24 METAL SURFACING

Bearing surfaces, metal rock crusher surfaces, and the surfaces of earth moving equipment are examples of machine areas which wear rapidly. Surfaces wear because of abrasion, fatigue, chemical or atmospheric corrosion, and metal transfer or adhesion. These large and expensive parts would have to be replaced if they could not be repaired.

Metal surfaces may be rebuilt by adding such items as metals, ceramics, or metal alloys. These materials may be added as powders or solids. The methods used to apply these surfacing materials are hardfacing and thermal spraying.

Hardfacing is done using:
1. Oxyfuel gas.
2. The electric arc.
   Methods of thermal spraying are:
1. Flame spraying.
2. Electric arc spraying.
3. Detonation spraying.
4. Plasma arc spraying.

Today, as equipment costs are rising, more and more industries are having worn parts repaired using hard surfacing or thermal spraying. Manufacturers are making new parts which have hard surfaces applied by means of thermal spraying to increase their service life.

## 24-1. HARDFACING AND METAL SPRAYING

When the surfacing material applied to a part is hard like a rod or an electrode, the process is called HARDFACING.

If the surfacing material is sprayed on at extremely high temperatures, the process is called THERMAL SPRAYING. When thermal spraying the surfacing material may be in a powdered, metal droplet, or plasma form.

Hardfacing and thermal spraying may be defined as processes by which a layer of material is bonded or fused to the surface of a base material.

The base material is usually metal.

Concrete, stone, and many synthetic materials have had metal applied by thermal spraying. The surfacing metal may be applied to improve the appearance, to improve the physical strength, or to protect the base material from chemical corrosion.

By applying a layer of the correct alloy to the surface of a piece of equipment, the part may be made more resistant to corrosion by chemicals. They may also be made more resistant to wear and abrasion from contact with abrasive materials, or to cracks or breakage due to shock.

Some of the advantages of surfacing a part with a desired metal alloy or other material are:
1. Certain dimensions may be maintained under adverse conditions, abrasion, corrosion, or impact shocks.
2. The service life of a part may be greatly increased.
3. Production costs may be lowered. Less expensive, low alloy materials may be used to make the part. Only the areas of high wear on such a part are coated with a more expensive alloy required to stand up under the service conditions. Low alloy steel surfaces which contact chemically corrosive gases or liquids may be surfaced with stainless steel to resist the corrosive action. The resulting cost per square foot would be appreciably lower than if stainless steel was used to manufacture the entire item.
4. Costs would be lowered also, since fewer replacement parts would need to be carried in stock because of the increased service life of each part.

Surfacing treatments may improve the surface's resistance to corrosion, impact breaks, or abrasive wear. However, no single surface treatment will give the maximum resistance to all of these types of deterioration at the same time.

When consideration is given to the surfacing of

a particular part, four factors should be considered:

1. The nature and cause of the wear problem.
2. The surfacing material needed to reduce this wearing condition.
3. The surfacing process which will most economically apply the selected surfacing material.
4. The proper technique to be used for depositing the surfacing material.

## 24-2. THE NATURE OF WEAR PROBLEMS

As mentioned previously, wear, or the deterioration of metals, may be caused by chemical action. The gradual consumption of the metal may be by external rusting or corrosion. It may also be by chemical etching of the parts (pipes or tanks) carrying certain chemicals.

It is possible to combat chemical wear by putting a layer of material on the surface which will resist oxidation and chemical etching, as shown in Fig. 24-1. Such materials as nickel base, copper base, cobalt base alloys, and a variety of tungsten carbide mixtures are available. Lead and zinc metals may also be applied. The surfacing material used will depend on the type of corrosion expected.

Parts may wear due to constant impacts with hard materials. Road grader blades, tool and die surfaces, power shovel buckets, and engine valve faces are examples of parts which suffer impact wear. Impact wear tends to change the original dimensions of a part by pounding it out of shape. Impact forces may also cause a part to break.

To reduce impact wear or breakage, a base material which has greater toughness should be used. Harder surfacing material will permit a softer inner base metal to withstand surface wear. The softer inner metal will permit the hard surface to withstand impact loads.

Another cause of wear is contact with abrasive materials. Bulldozer blades, rock crushing rollers and housings, and rock quarry conveyors are examples of parts which suffer abrasive wear.

Hard materials such as tungsten carbide or ceramics such as chromium and aluminum oxide compounds may be used to reduce abrasive wear.

Soft metals such as copper, brass, bronze, and aluminum have been sprayed on steel parts to prevent abrasive wear where two surfaces come into contact.

The material used to prevent a wear problem must be chosen for its ability to prevent wear from one or more of the following causes:

1. Abrasion—a rubbing or scraping action.
2. Surface fatigue—the loss of areas of the metal surface from pitting or flaking.
3. Corrosion—usually chemical, heat, or atmospheric attacks on the surface.
4. Adhesion—loss of surface material from one surface to another.

## 24-3. HARDNESS DETERMINATION

Various characteristics of metals and their importance are explained in Chapter 26.

In the metal surfacing field, the property usually given the most attention is hardness. HARDNESS is the property of the metal to resist indentation or the action of cutting tools (or materials). It is important that the hardness of the surfacing material be known in order to insure the requirements of a particular job. The three most common methods of measuring hardness are the Brinell scale, the Rockwell scale and the Shore Scleroscope scale. Test procedures for these methods are given in Chapter 28.

Fig. 24-1. A flame spraying torch being used to spray a hard material on a wearing surface.　(Metallizing Co. of America)

A reasonably accurate test for hardness may be conducted by using a mill file if testing machines are not available. Follow this procedure:

1. If metal is removed easily the hardness is about a 100 Brinell or 60 Rockwell B. An example of this hardness would be low carbon steel.
2. If metal is readily cut with moderate pressure exerted the hardness is about 200 Brinell or 15 Rockwell C. Medium carbon steel is an example of this hardness.
3. If metal is difficult to cut, though possible, the hardness is approximately 300 Brinell or 30 Rockwell C. An example is high alloy steel.
4. If metal is cut with great pressure only, the hardness is probably about 400 Brinell or 40 Rockwell C. Tool steel has this hardness.

5. If metal is nearly impossible to cut, the hardness is about 500 Brinell or 50 Rockwell C. An example of this hardness is tool steel.
6. If metal cannot be cut, the hardness is about 600 Brinell or 60 Rockwell C. Hardened tool steel has this hardness.

## 24-4. SELECTION OF A SURFACING PROCESS

The OXYFUEL GAS PROCESS for depositing hardfacing material has many desirable features:
1. The oxyfuel gas process is used where a surface is to be applied which will require a minimum of final surface finishing.
2. The carburizing or reducing flame, normally used to apply hardfacing material, adds carbon to the surface and improves the abrasive resistance of the metal.
3. Preheating and postheating may have to be carefully controlled with the oxyfuel gas flame to prevent cracking on the surfacing material.
4. Nonferrous metals are more easily applied by this method.
5. Oxyfuel gas is well adapted to fusing hardfacing materials.

When a flawless surface deposit is required, THERMAL or ARC SPRAYING PROCESSES are preferred. However, the metal being surfaced must be very clean. This requires additional equipment and labor costs. Because of the high quality of the deposit, the arc and thermal spraying processes are being used more and more in the aircraft and aerospace fields for surfacing new parts subjected to some form of extreme wear. One application for surfacing is in building up larger dies which become broken or require changes.

The MANUAL GAS METAL ARC PROCESS is often used. Larger volumes of surfacing material may be deposited. This process has these advantages:
1. The equipment is highly portable and is common to most welding shops.
2. Surfacing materials may be laid on more easily in various positions and locations due to the great variety of electrodes available.
3. Great varieties of surfacing materials may be laid because of the large variety of electrodes available.
4. Austenitic stainless steels, austenitic manganese steels, and nickel-chromium-ferritic steels may be surfaced with the gas metal arc process.

AUTOMATIC and SEMIAUTOMATIC SURFACING PROCESSES normally will give a higher quality of surfacing deposit. These processes are not as portable as the manual process, nor are they as adaptable to position welding. Automatic and semiautomatic processes are best used in the flat position. The advantages of speed and accuracy of the deposit are best utilized in the hardfacing of long straight or curved surfaces in new production. The automatic and semiautomatic processes work best for repairs when parts can be dismantled and brought to the automatic machines.

Thermal spraying may be used to deposit almost all virgin metals and many alloys to other metal surfaces. The deposit is usually thin, and the thickness may be held to close tolerances. The surfaces being treated must be cleaned and slightly roughened. Sandblasting may be used for this purpose. In some instances multiple passes may be made to build up a surface with hardfacing material without danger of cracking the previous layers.

Virtually any material may be bonded to any other with the PLASMA ARC PROCESS. Ceramics may be sprayed on the surfaces of metals to increase their resistance to corrosion from heat and/or chemicals. Stainless steel may be sprayed on low alloy steel to increase resistance to corrosion and abrasive wear.

The ELECTRIC ARC SPRAYING PROCESS is the most economical thermal spraying process. It is capable of depositing a large variety of coatings with high efficiency.

## 24-5. OXYFUEL GAS SURFACING PROCESSES

To obtain the best results, the surface should be cleaned and the metal preheated to eliminate warpage. Hardfacing, using a solid rod, and thermal spraying may be done with the OXYFUEL GAS PROCESS. When surfacing, use a tip one to two sizes larger than used with the same diameter rod when welding. A slightly reducing (carburizing) flame is desired. Refer to Fig. 1-2. Any carbon added to the surface will aid in the hardfacing process. The reducing flame will reduce the oxides on the surface of the work. The angles for the torch and rod are the same as for welding, as shown in Fig. 24-2. When surfacing steel, the metal should be brought up to approximately 2200 °F (1204 °C) before the rod is touched to the work. This step is shown in Fig. 24-3. The preheat temperature is important. It is advisable to use a temperature indicating crayon, or some other means for determining the temperature. The process of surfacing is very similar to brazing as it applies to the welder's techniques and manipulation. Fig. 24-4 shows steel being surfaced using the forehand process. Fig. 24-5 shows the backhand technique.

The techniques used for holding the torch and the surfacing rod are shown in Figs. 24-6 and 24-7.

When surfacing cast iron the surfacing material

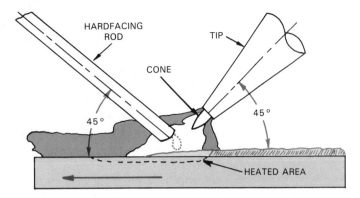

Fig. 24-2. Hardfacing material being applied with the use of an oxyfuel gas torch. The outer flame area preheats the base metal in advance of the deposited surfacing metal. The base metal melts the surfacing rod as in brazing. A carburizing flame is used when hardfacing.

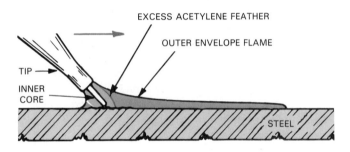

Fig. 24-3. Flame and tip position to use when preheating a metal prior to hardfacing.   (Haynes Stellite Co.)

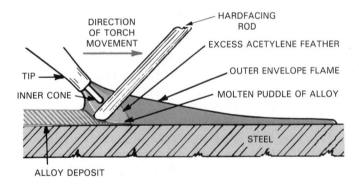

Fig. 24-4. Forehand method of hardfacing.

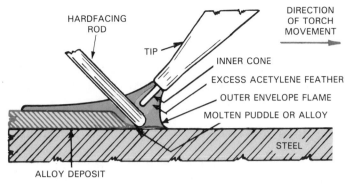

Fig. 24-5. Backhand method of hardfacing. Note that the torch points in the opposite direction to its forward movement.

Fig. 24-6. Hardfacing the wear areas of an automobile engine camshaft.   (Wall Colmonoy Corp.)

Fig. 24-7. Helix type conveyer being rebuilt using a hardfacing rod and oxyacetylene flame.   (Wall Colmonoy Corp.)

will not flow as easily. To produce a good job, the rod must be used to break through the surface crust of the iron by rubbing the heated surface with the rod.

An oxyfuel gas torch may be fitted with a hopper assembly as shown in Fig. 24-8 to do flame spraying. A variety of surface hardening and cladding materials are available for use with this flame spraying torch combination. Surfacing or cladding powder is placed into the hopper on the torch. When the base metal is preheated to the proper temperature, the lever on the torch is pressed to release the powder in the hopper. The surfacing powder is released into the oxygen and fuel gas tube and melted in the flame at the torch tip. The molten surfacing or cladding powder is transferred to the base metal where it fuses with the surface.

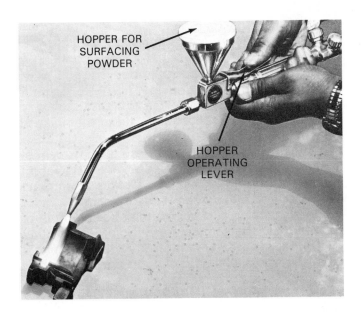

HOPPER FOR SURFACING POWDER

HOPPER OPERATING LEVER

Fig. 24-8. An oxyfuel gas torch equipped with a hopper to hold surfacing material. A silicon-boron-nickel powder is being used to repair a glass container mold.
(Wall Colmonoy Corp.)

## 24-6. MANUAL SHIELDED METAL ARC SURFACING PROCESS

Direct current is normally used with electrode negative (DCEN) polarity. A higher current setting is used than is normal for the electrode diameter. The high current setting will permit a long arc which will preheat the metal ahead of the arc. This will keep the surfacing material molten long enough for the impurities to float to the surface. An angle of about 45° is correct for the electrode. A long arc is necessary to spread the heat over a large area and thus prevent deep penetration, localized heating and warpage.

Preheating is not necessary before applying surfacing materials to all metals, but with alloy steels it is advisable.

When in doubt about preheating, the following guide rules may be helpful:
1. Except for heavy sections, low and mild carbon steels do not require preheating.
2. Preheat high carbon and medium carbon high alloy steels.
3. 12-14 percent manganese steel parts should be preheated to 200°F (93°C) to relieve stresses. However, the part temperature should never exceed 500°F (260°C).
4. Cast iron should be preheated to 500-700°F (260-371°C) at a slow and uniform rate.
5. When in doubt, preheat to 500-700°F (260-371°C), but be certain the part being preheated is not a manganese steel.

If narrow beads are desired, an oscillating motion should be used in the direction of travel with no sideways motion. A good technique is to lay a bead about 1 in. (about 25 mm) long and then start the next stroke of the same length 3/4 in. (about 20 mm) back on the first stroke and extend it 1/4 in. (about 6 mm) past the first stroke. The third stroke should start 3/4 in. (about 20 mm) back on the second stroke and extend 1/4 in. (6 mm) beyond the second stroke. This procedure is continously repeated until the desired length bead is laid.

Wider beads require a circular motion. A circular motion is started at the center and continued in spirals until the desired width of bead is obtained. The operator then continues to make circular motions and extend each circle about 1/4 in. (about 6 mm) beyond the last circle in the direction of the desired path. Each time a circular motion is made the weld moves forward 1/4 in. (6 mm), as shown in Fig. 24-9.

Fig. 24-9. Circular motion being used to obtain bead widths from 3/4 to 1 1/4 in. (about 20 to 30 mm).

The recommended height of a deposit is 1/16 to 1/4 in. (1.5-6 mm), and the recommended bead width is 3/4 to 1 1/4 in. (20-30 mm).

When laying a second bead, the first and second bead should overlap 1/4 to 1/3 the width of the first bead. This process will insure a fairly uniform surface to the deposited metal as shown in Fig. 24-10.

If more than one layer of surfacing material is required, the beads of the first layer should be cleaned before depositing the second layer. This will minimize slag inclusion. Each bead should be made in the same manner as described above for laying beads from 3/4 to 1 1/4 in. (20-30 mm) wide.

An interesting technique for depositing hardfacing beads is to lay the beads in a basket weave pattern. This leaves low spots between the beads. If the part being hardfaced is used in abrasive

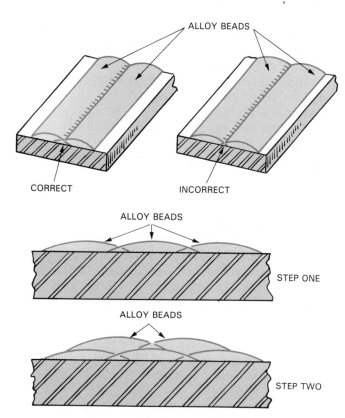

Fig. 24-10. The proper way to overlap adjacent surfacing beads.

material such as sand, the abrasive material will build up in the low spots between the beads and act as an added protection for the base metal as shown in Fig. 24-11.

It should be noted here that small beads made with small diameter rods cool fastest and mix the least with the base metal. Wide beads made with high current and large electrodes cool slowly and mix with the base metal to a greater extent.

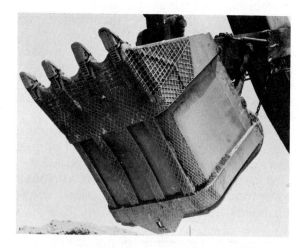

Fig. 24-11. Basket weave pattern for laying beads. The abrasive material will build up between the beads and help protect the base metal.   (Stoody Deloro STellite, Inc.)

Before the rod diameter, current setting, and surfacing procedure are decided, the welder must determine:
1. How wide the bead should be.
2. How fast the bead should cool.
3. How harmful the mixing of the surfacing material and base metal will be to the finished job.

In many situations better results may be obtained by using superimposed layers of different hard surfacing alloys. Fig. 24-12 shows a hammer mill head which has been hardsurfaced.

The automatic and semiautomatic methods of surfacing are the same as for welding except that a different type of wire is used. See Chapter 23 for automatic welding procedures and equipment.

## 24-7. FLAME SPRAYING PRINCIPLES

The flame spraying process may be used to spray pure or alloyed metals onto a surface which requires a coating buildup. Flame spraying is accomplished by melting surfacing or coating materials in an oxyfuel gas flame. These molten materials are then transferred to the surface of the material being coated. The surfacing or coating material may be in the form of wire, rod, or powder. The coating materials may be ceramics, hardening metals and alloys, or soft metals like brass or bronze. The fuel gas used in the oxyfuel gas process may be acetylene, propane, natural gas, hydrogen, or methylacetylene-propadiene. The sprayed coating may be used "as sprayed." It may also be fused to the base metal by reheating with an oxyfuel gas torch, or by using electric induction heating coils.

Worn or inaccurately machined parts may be repaired by having the surfaces built up to the acceptable dimensional limits by one of the flame

Fig. 24-12.  A hammer mill head after it has been rebuilt with hardfacing electrodes.   (Stoody Deloro Stellite, Inc.)

spraying processes. Usually, this process is workable with most inorganic (not containing carbon) materials which melt without decomposition.

Fig. 24-13. Metal surfacing a shaft using a flame spraying torch which has a built-in wire feed motor. The shaft is being built up to the required diameter.

Flame sprayed coatings, for example, can be used to build up scored surfaces on engine crankshaft journals, so they can be machined to accept standard size main bearing inserts.

Wire flame spraying may use any material which can be formed into a wire shape. The wire is usually a metal or metal alloy. Ceramic material may also be formed into wire and flame sprayed. Powders may also be placed inside hollow metal wire or thin plastic tubes.

Fig. 24-13 shows a shaft being metal surfaced, using a wire feed torch held in the operator's hand. A similar operation, with the torch attached to the lathe cross feed mechanism, which insures even metal spraying, is shown in Fig. 24-14. Occasionally, in original production, surfaces are flame sprayed with a harder alloy material to reduce wear on these surfaces.

Surfacing materials used may also be in a powder form. Fig. 24-15 shows a part being flame sprayed with a torch which is supplied with a high pressure carrier gas to feed the powder to the torch. Oxygen is not used as the carrier gas. Oxygen, in combination with a fine powder, can create an explosive mixture. Another method used to supply the powdered metal for flame spraying is shown in Fig. 24-16. The powder is fed to the flame by gravity from the hopper mounted on top

Fig. 24-14. A flame spraying torch with a built-in wire feed mechanism being used to surface a crankshaft journal. The torch is mounted on the lathe cross-feed for better spray control.
(Metallizing Co. of America, Inc.)

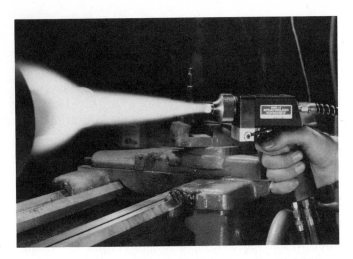

Fig. 24-15. Metal surfacing a part as it revolves in a lathe. Metal for spraying is fed to the torch in a powder form. A carrier gas feeds the powder metal to the torch. This flame spraying process is followed up with an oxyacetylene flame fusing operation.    (Wall Colmonoy Corp.)

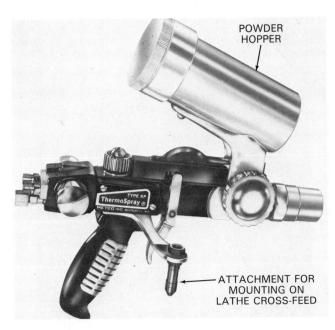

Fig. 24-16. The powder is gravity fed to this flame spraying torch. This torch is equipped with an electric vibrator to improve the powder flow to the torch.    (Metco, Inc.)

of the torch. Powder flame spraying may be done with any material which melts below the flame temperature of about 4000-5500 °F (2205-3038 °C). Some close fitting parts of jet engines, for instance, require airtight seals and operate at extremely high temperatures. The contacting surfaces of these parts are metal sprayed with a soft metal. As the rotating parts expand and contract, the soft metals contract and conform to the rotating parts to produce the required airtight seal.

In production brazing of certain alloys, the flow of brazing material between the parts to be joined (capillary flow) is difficult to obtain. To insure that the brazing material will cover the entire surface to be joined, the surfaces may be first flame sprayed with brazing metal. They are assembled and then the assembly is heated in an oven to complete the brazing operation.

Flame spraying may also be used on a surface to improve its wear or corrosive resistance. The surface produced by flame spraying normally requires very little finishing prior to use. Flame spraying is often used to refinish worn shaft bearing journals.

An interesting application of flame spraying is to use layers of blended metals. This blending allows a material which would normally not adhere to another material to be applied successfully. One material could be applied to the other in several

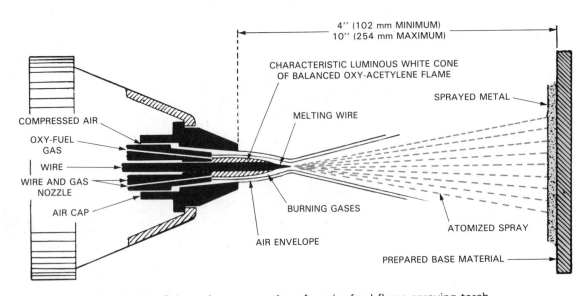

Fig. 24-17. Schematic cross section of a wire feed flame spraying torch.

graded layers. The beginning layer would have little of the final coating material in the powder mixture. As following layers are applied, the content of the final coating material is increased until the pure material is applied in the final coatings.

## 24-8. FLAME SPRAYING TORCH

The flame spraying torch must feed oxygen, fuel gas, a high pressure carrier gas (not oxygen), and the coating wire or powder to the torch nozzle. Each must be in the correct amount in order that the flame spraying process may be successful.

Fig. 24-17 shows a schematic of the wire feed type flame spraying torch. The oxyfuel gas flame at the nozzle melts the coating material. The carrier gas pressure propels the molten powder or metal to the surface being coated. The wire feed mechanism on the torch in Fig. 24-18 is driven by an air turbine motor. A wire feed, flame spraying torch which uses an electric motor to drive the coating wire is shown in Fig. 24-19. Fig. 24-20 shows the location and names of the essential parts of a typical wire feed, flame spraying torch.

The high temperature oxyfuel gas flames completely surround the wire as it is fed through the torch nozzle. The pressurized carrier gas atomizes the molten wire material and sprays it on the surface to be coated.

Fig. 24-21 schematically illustrates the operation of a powder feed, flame spraying torch. The coating material may be fed to the torch nozzle by gravity as shown in Fig. 24-22. Fig. 24-23 illustrates a flame spraying torch which uses a

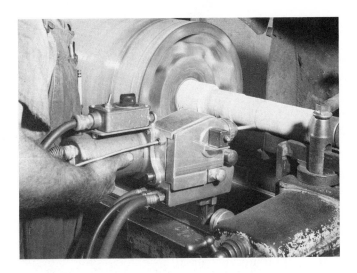

Fig. 24-19. An automated metal spraying operation. The electrically powered wire feed is electronically controlled. (Metallizing Co. of America, Inc.)

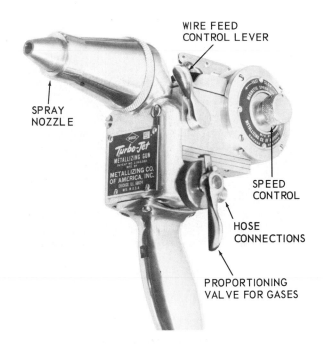

Fig. 24-20. Flame spraying torch with wire feed. (Metallizing Co. of America, Inc.)

pressurized carrier gas to carry the coating powder to the torch. The carrier gas also propels the molten powder to the surface being coated.

## 24-9. FLAME SPRAYING PROCESS

Flame spraying is a process in which metals, alloys, or ceramic materials in a wire or powdered form are fed through a special torch (spray gun). They are melted in an oxyfuel gas flame and atomized by high pressure gases. These gases carry the atomized particles to the cleaned and

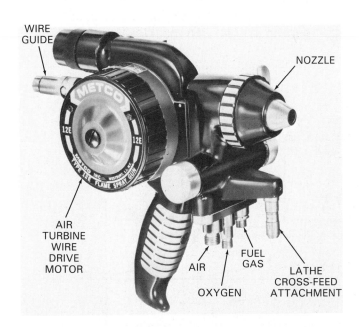

Fig. 24-18. A flame spraying torch. An air turbine wire drive motor is built into this torch. (Metco, Inc.)

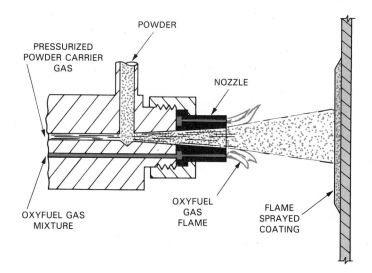

Fig. 24-21. A schematic of a flame spraying torch which is using powdered coating material. A carrier gas (not oxygen) is used to move the powder to the torch. The same gas propels the molten powder to the surface being coated.

Fig. 24-22. A gravity feed flame spraying torch in operation.

Fig. 24-23. A flame spraying torch designed for use on a lathe.

prepared surface. Fig. 24-24 illustrates a flame spraying outfit used to spray metal in a wire form.

The equipment necessary for wire feed flame spraying is:

1. Air Compressor—Capacity of at least 30 cu. ft. per minute (850 L/min.) of free air required.
2. Compressed Air Drying Unit—If there is an excess amount of moisture in the air supply, it is generally advisable to use an air drying unit. Any amount of moisture will interfere with the metallizing bond.
3. Air Receiver Tank—This unit smooths out the flow of compressed air to the compressed air regulator by compensating for compressor pumping pulsations.
4. Fuel Gas and Oxygen Regulators—Two-stage regulators recommended.
5. Fuel Gas and Oxygen Cylinders.
6. Gas Flow Meters—Gas flow meters keep a constant gas volume flow to the gun nozzle.
7. Compressed Air Control—A single stage regulator may be used to control the air pressure. Air filters are normally used to further purify the air. If the air supply is used to supply air to operator's mask, an air filter is required.

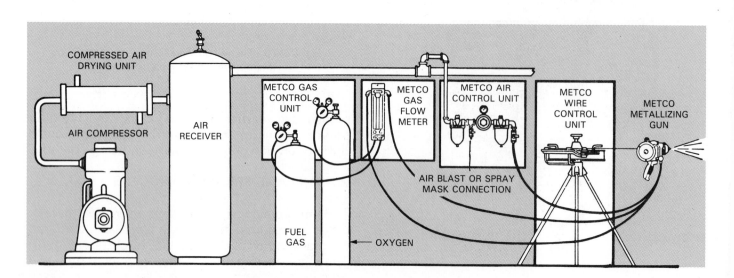

Fig. 24-24. A schematic of the equipment used in a typical wire feed flame spraying outfit.    (Metco, Inc.)

8. Wire Feed Control—This unit will straighten the wire as it comes from the coil and feed it at a uniform rate to the gun nozzle.
9. Flame Spraying Gun—The flame spraying gun mixes the fuel gas, oxygen, and the desired powder or alloy wire. Some guns contain mechanisms to draw the wire from the wire coils.

Fig. 24-25 illustrates a combination flow control for oxygen, fuel gas, and powder. The hopper holds the powder.

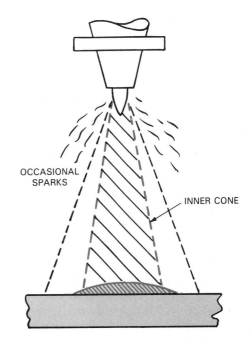

Fig. 24-26. Schematic showing that approximately 95 percent of the sprayed metal is in the inner cone, and is deposited on the surface being sprayed.

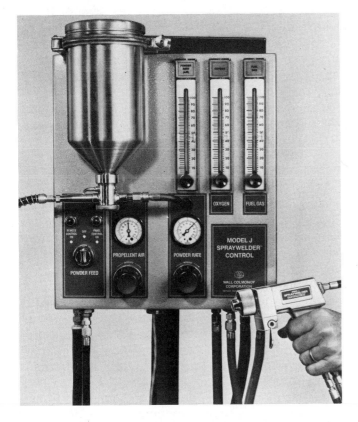

Fig. 24-25. A combination flow control for oxygen, fuel gas, and powder. The powder is stored in the hopper.
(Wall Colmonoy Corp.)

The width of the spray cone varies with the wire diameter, the type of metal being sprayed, and the spraying speed (wire feed rate). The spray cone, which is visible, is considerably larger than the actual cone of spray metal due to the sparks which travel around the actual cone of metal. About 95 percent of the metal being sprayed is in a smaller inner cone not visible when spraying, as indicated in Fig. 24-26.

Different types of nozzles are used for various fuel gases and these should not be interchanged. If a propane nozzle, with a cupped end, is used with acetylene gas, backfiring may occur.

A complete automated flame spraying installation in operation is shown in Fig. 24-27. A solid

ceramic rod is used in this torch to spray a ceramic coating onto a part.

Another process used to flame spray surfaces is by means of a powdered surfacing material carried to the work by an inert gas, compressed air, or gravity. The powder in the pressure fed units is carried to the torch by a third hose from a hopper where the powdered material is stored under pressure. When this torch is used with powdered

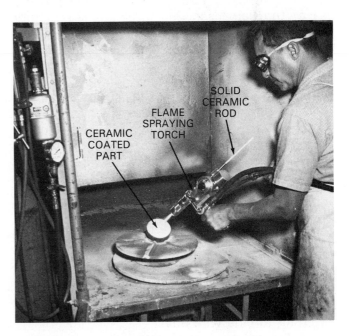

Fig. 24-27. A complete flame spraying installation. A solid ceramic rod is being used in this torch to spray a ceramic coating on a small part. (Metallizing Co. of America)

ceramics, an electric vibrator is used to help the ceramic powder flow. One advantage of using powdered material for flame spraying is that flux may be mixed with the coating material. Flux will help the coating to adhere better.

With powders, the spraying rate is critical. If the spray rate is too slow, the small particles stay in the flame chamber too long and may vaporize. If the spray rate is too fast, the material may not melt completely. If this happens, it will not bond with the surface or other particles properly. The surface of the material being coated is sometimes preheated by using an oxyacetylene flame to improve the bonding of the material with the surface.

The spray must be applied in smooth even layers. The gun is held perpendicular to the surface being sprayed. Material layers about .001 to .002 in. (.025 - .050 mm) are sprayed. Layer after layer may be sprayed until the total desired coating thickness is obtained.

The spray gun to surface distance should be about 4 to 10 in. (about 100 - 250 mm). This distance will depend on the process, coating material, and rate of deposit used. In the case of thin parts or small diameter shafts, the parts may have to be cooled by an air stream to prevent overheating.

## 24-10. STARTING AND ADJUSTING THE FLAME SPRAYING STATION

The safety precautions recommended for use with all gas welding equipment apply also to the use of the flame spraying torch. Review safety precautions in Headings 4-34, 4-35, and 4-36. Check all hoses, connections, and gauges for leaks before using the equipment. The procedures for adjusting and lighting the wire feed type torch are as follows:
1. Open the air valve and adjust the air pressure regulator to the correct pressure. Refer to the torch manufacturer's data sheets for recommendations for wire size, air cap size, air pressure, oxygen and fuel gas pressures, and deposit or spraying speeds.
2. Close the air valve.
3. With the drive roll knob loose, insert the wire into the rear wire guide and through the gun to the nozzle.
4. Tighten the drive roll knob until the wire begins to feed.
5. Adjust the wire feed speed. If equipped with an air turbine wire drive:
   a. Adjust the speed control ring which controls the speed of the turbine in the air motor. When the speed control ring is

moved clockwise, the wire will speed up. When the control ring is turned counterclockwise, the wire will slow down.
   If equipped with an electric motor wire drive:
   a. Adjust the motor speed controls to feed the wire at the recommended rate.
6. Open the valve handle to the run position and adjust the fuel gas and oxygen pressures to the values specified by the torch manufacturer. Close the gas proportioning valve handle.

Before actually spraying, the wire feed rate should be set accurately on a test piece. If the wire feeds too fast, the tip of wire will extend beyond the hottest part of the flame. If this occurs, the surfacing material will leave the torch in large droplets. This will form a coarse, bumpy surface on the part being sprayed.

If the wire is fed too slowly, the coating material will oxidize more easily. The sprayed surface will be highly oxidized and weak.

## 24-11. ELECTRIC ARC SPRAY SURFACING

Another process used to spray materials is the ELECTRIC ARC SPRAY method. This method consists of maintaining an arc between two current carrying wires. The wires are automatically fed to the arc position in the spray gun by electric drive mechanisms similar to those used in gas metal arc welding. The wire material melted by the arc is then sprayed onto the surface to be coated. An air jet or an inert gas jet directed across the arc propels the molten wire to the surface to be coated, as shown in Fig. 24-28. Fig. 24-29 illustrates a commercial arc spraying torch.

It is claimed that production rates can be tripled using the arc spraying process. Arc spray guns can spray between 3 and 200 lbs. (1.4 - 90 kg) of

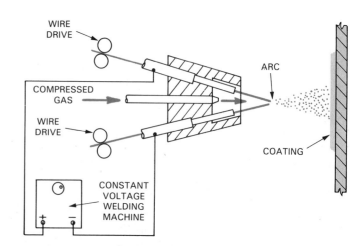

Fig. 24-28. A schematic of the electric arc spray method of coating the surface of a part.

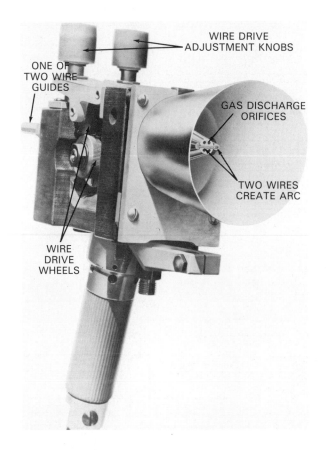

WIRE DRIVE
ADJUSTMENT KNOBS

ONE OF
TWO WIRE
GUIDES

GAS DISCHARGE
ORIFICES

TWO WIRES
CREATE ARC

WIRE
DRIVE
WHEELS

Fig. 24-29. An arc spraying torch. Notice the wire drive unit. The upper drive wheel is adjustable. Two electrode wires create an arc. The molten material is propelled to the coated surface by air or inert gas.
(Metallizing Co. of America)

wire per hour. Different materials may be used in each of the wires fed to the arc. Using two different wires allows an alloying of materials at the torch. This alloy or material mixture is then sprayed onto the surface to be coated.

Because of the low cost of electricity as an energy source, arc spraying is more economical than any other surface coating process. Fig. 24-30 is a table which compares the cost of energy for various thermal spraying processes.

Industrial ear muffs, goggles, and protective clothing should be worn when using this process.

## 24-12. DETONATION SPRAYING

Detonation spraying is done with a gun which resembles a small cannon. See Fig. 24-31.

In the DETONATION SPRAYING gun, air or oxygen, a fuel gas, and a charge of coating material mix in the combustion chamber. An ignition system ignites the mixture, setting off the charge several times per second. The coating material in a molten condition is propelled from the end of the gun at 2500 ft/sec (762 m/sec). Nitrogen is

sometimes used to purge (clean) the chamber after each explosion. Temperatures within the detonation spray gun reach 6000°F (3315°C).

Very high bond strength and coating densities are obtained with this process.

Industrial type ear muffs must be worn when using this process. Goggles and other protective clothing should be worn to protect the operator from sparks and ultraviolet burns.

## 24-13. PLASMA ARC SPRAYING PROCESS

The PLASMA ARC or jet is the result of forcing gas, such as nitrogen, hydrogen, argon, and the like into an enclosed electric arc. The gas is heated to such a high temperature by the electric arc that its molecules become ionized atoms containing a great deal of energy. Plasma is considered by physicists as a fourth state of matter (neither gas, liquid, or solid). Chapter 18 explains the process in detail.

Temperatures up to 30,000°F (16 650°C) may be obtained with plasma arc equipment. This process can be very successfully used for surfacing by spraying. In general, any inorganic material

| Powder flame spray | | | |
|---|---|---|---|
| Oxygen | 95ft³/h($0.0204/ft³) | = | $ 1.94 |
| Acetylene | 59ft³/h($0.0810/ft³) | = + | 4.78 |
| Total | | | $ 6.72 |
| **Wire flame spray, acetylene fuel** | | | |
| Oxygen | 90ft³/h($0.0204/ft³) | = | $ 1.84 |
| Acetylene | 32ft³/h($0.0810/ft³) | = | 2.59 |
| Air | 2100ft³/h($0.00015/ft³) | = + | 0.32 |
| Total | | | $ 4.75 |
| **Wire flame spray, propane fuel** | | | |
| Oxygen | 185ft³/h($0.0204/ft³) | = | $ 3.77 |
| Propane | 30ft³/h($0.0187/ft³) | = | 0.56 |
| Air | 2100ft³/h($0.00015/ft³) | = + | 0.32 |
| Total | | | $ 4.65 |
| **Rod flame spray** | | | |
| Oxygen | 194ft³/h($0.0204/ft³) | = | $ 3.96 |
| Acetylene | 57ft³/h($0.0810/ft³) | = | 4.62 |
| Air | 2100ft³/h($0.00015/ft³) | = + | 0.32 |
| Total | | | $ 8.90 |
| **Electric arc spray** | | | |
| Electricity | 3 kW($0.06/kWh) | = | $ 0.18 |
| Air | 2100ft³/h($0.00015ft³) | = + | 0.32 |
| Total | | | $ 0.50 |
| **Plasma spray[a]** | | | |
| Argon | 75ft³/h($0.08/ft³) | = | $ 6.00 |
| Electricity | 40 kW($0.06/kWh) | = + | 2.40 |
| Total | | | $ 8.40 |

[a]Plasma spray with nitrogen results in costs similar to that of argon; nitrogen costs less per cubic foot but spray nozzles wear out faster.

Data supplied by Tafa Metallisation

Fig. 24-30. Comparison of typical energy cost for one hour of operation at typical spray rate for various processes. Note the cost of the arc spray method compared to the other methods.

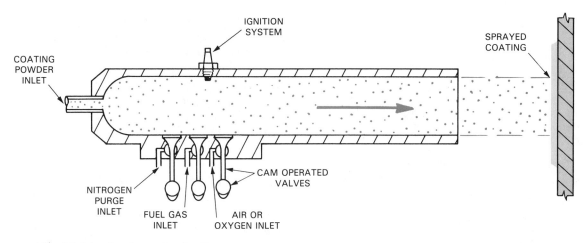

Fig. 24-31. A schematic of a detonation spray gun in use. The cam operated valves allow nitrogen for purging and oxygen and fuel gas for combustion to enter the chamber several times per second. The powdered coating material enters at the back of the gun.

which will not decompose may be sprayed. Examples of materials being applied to a base metal are: ferrous metals, ceramics, tungsten, tungsten carbides, tantalum, zirconium diboride, platinum, columbium, hafnium, vanadium carbides and others.

Coating densities applied with the plasma arc process are up to 98 percent of the theoretical density. Pure tungsten and tungsten carbides may also be applied to almost any base material with up to 95 percent densities.

A schematic of a typical plasma arc torch is shown in Fig. 24-32. The principle of the process is explained below.

Using direct current, an arc is formed between the internal electrode and the nozzle. The gas is ionized within the nozzle and becomes plasma. This is known as a non-transferred plasma arc. The material to be sprayed is carried to the arc area in the nozzle. The coating material is melted and atomized by the heat of the plasma. The atomized surfacing material is then carried to the surface to be sprayed by the high velocity of the plasma jet. Speeds of 20,000 ft. per second (approximately 6100 m/sec) may be reached with the plasma jet. Normally much lower speeds are required. Fig. 24-33 shows a cross section of a plasma arc material spraying torch.

Nitrogen with from 5 - 10 percent hydrogen gas is generally used for the plasma gas. Pure nitrogen is used as the carrier gas to propel the powdered surfacing material to the melting zone within the nozzle.

The entire plasma jet nozzle is usually water cooled. Distilled water is often used in the cooling circuit to prevent mineral deposits and thereby lengthen the life of the equipment. Fig. 24-34 shows the external view of a typical torch.

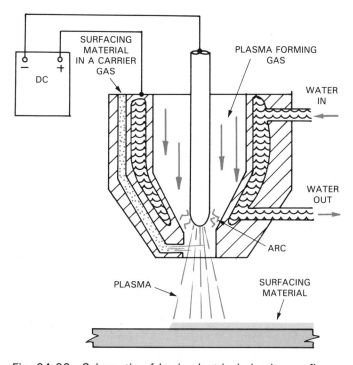

Fig. 24-32. Schematic of basic electrical circuit, gas flow, surfacing material flow, and cooling of plasma arc spraying torch.

A complete plasma arc station would include:
1. A plasma jet spray gun.
2. A control unit with an on-off switch, starter arc controls, gas flow controls, ammeter, and voltmeter.
3. A heat exchanger in which city water is used to cool the distilled water circulating in the gun. Water flow controls are also included in this unit.
4. A powder control unit to control the pressure and quantity of the powder fed to the gun.
5. A power supply unit. The current supplied to

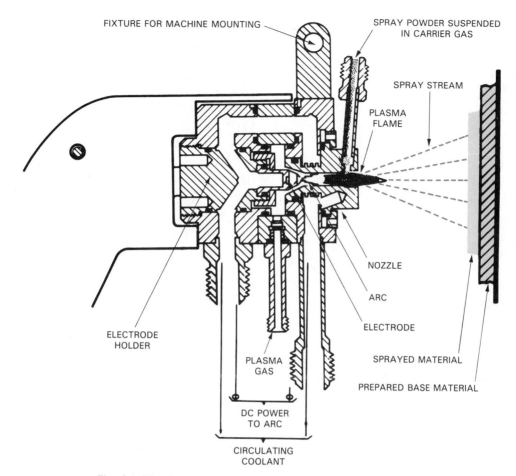

FIXTURE FOR MACHINE MOUNTING

SPRAY POWDER SUSPENDED IN CARRIER GAS

SPRAY STREAM

PLASMA FLAME

NOZZLE

ARC

ELECTRODE

SPRAYED MATERIAL

PREPARED BASE MATERIAL

ELECTRODE HOLDER

PLASMA GAS

DC POWER TO ARC

CIRCULATING COOLANT

Fig. 24-33. Schematic of plasma arc material spraying torch.

the electrodes is controlled and stabilized by this unit.

6. Gas and water hoses and electrical cables.
7. Safety equipment. Face shield, heat resistant and reflecting clothing, gloves, and ear plugs or muffs to protect the operator's ears from the intense noise caused by the plasma jet.
8. A water wash spray booth is used to protect other workers from overspray of both flame and material.

An interesting application of the plasma arc material spraying process is used when making aerospace parts such as rocket nozzles. The parts are exposed to extremes of temperature and erosion wear due to the high velocities and temperatures to which they are exposed. Rocket nozzles are made from zirconium, platinum, or other rare and expensive metals. These materials are too rare in fact to be machined to shape where the machining chips would have to be salvaged and remelted.

In the plasma arc spraying process, the desired metal is sprayed onto an aluminum or brass form (mandrel) to the desired thickness. The mandrel or form is then removed from the shell of rare metal by dissolving it with acid.

## 24-14. SURFACE PREPARATION

The bonding of coating materials relies on the mechanical interlocking of the cooled sprayed particles. Fig. 24-35 shows how an arc or flame sprayed surface might appear under a powerful microscope.

Surface preparation is essential to any thermal spray coating application. Before the surface is coated, it must first be cleaned thoroughly. Nonporous surfaces may be cleaned by steam, vapor degreasing, hot detergent washing, or cleaning with industrial solvents. Porous materials such as cast iron may need to be baked at 400 - 600 °F (about 200 - 320 °C) to vaporize any grease or oil that is in the porous areas.

Abrasive cleaning methods, such as grit blasting, may be used to remove mill scale. White metals such as aluminum and magnesium are blasted using aluminum oxide or quartz, as in Fig. 24-36. Fig. 24-37 shows a comparison of the bond strengths of materials and the cleaning methods used.

Machine cutting is used to remove contaminated (dirty) material. Material is also removed to provide for the thickness of the new coating.

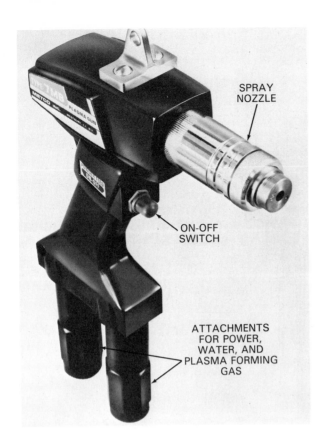

Fig. 24-34. An external view of plasma arc material spraying torch. (Metco, Inc.)

Fig. 24-36. Cleaning a gear by grit blasting. A special enclosed chamber is used.

| Material | Deposition Method | Bond Strength lb/in.² | Surface* Preparation |
|---|---|---|---|
| Molybdenum | Flame | 5,500 | Grit blast |
| 9 Al-bronze | Electric arc | 6,000 | Clean only |
| 9 Al-bronze | Electric arc | 6,700 | Grit blast |
| 95 Ni-5 Al | Electric arc | 9,100 | Clean only |
| 95 Ni-5 Al | Electric arc | 9,700 | Grit blast |
| 95 Ni-5 Al | Plasma | 9,500 | Clean only |
| Aluminum | Flame | 1,400 | Grit blast |
| Aluminum | Electric arc | 4,900 | Grit blast |

*Special bond coats are sometimes needed.

Fig. 24-37. How bond strengths compare. Note the beneficial effect or grit blasting. (Hobart/Tafa)

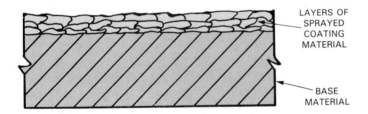

Fig. 24-35. A cross section of an arc or flame sprayed coating on a surface. The sprayed particles interlock and bond to the surface and to each other as they cool.

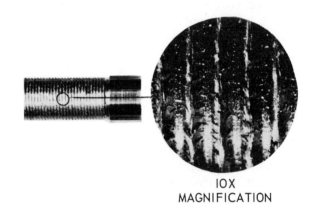

IOX MAGNIFICATION

Fig. 24-38. Appearance of steel stock after machine cutting the surface prior to flame or arc spraying. The 10-X magnification shows the roughness needed for good spraying results.

On cylindrical parts, like shafts and pistons, the parts are turned on a lathe. Material is removed to level out a worn surface and to provide for the thickness of the spray coating. Cylindrical parts are often rough threaded to a depth of .003 to .006 in. (approximately .08 - .15 mm) after they are straight turned. Surfacing materials generally adhere or bond better to cylindrical parts that have been rough threaded to the shallow depths stated above. See Fig. 24-38.

Surfaces to be coated are often preheated to drive off moisture. Any moisture on the surface will interfere with the sprayed materials bonding process. Preheating at 200 - 250°F (93 - 121°C) is helpful when flame spraying materials.

Some sprayed materials will not readily bond with some surfaces. To insure good bonds for such coatings, a bond coat is applied before the final coating is sprayed. The bond coat is an intermediate coating between the base material and the finished coating. Materials such as molybdenum, 9% aluminum-91% bronze, and 95% nickel-5%aluminum, are used as bond coats for other sprayed materials.

## 24-15. SELECTING THERMAL SPRAY COATINGS

When selecting the best material to use for a surface coating, the following variables must be considered:
1. The material being coated. Is the material metal, stone, ceramic, or plastic?
2. Is the metal hard or soft?
3. Is the metal porous or non-porous?
4. Is the coating material to be soft or hard?
5. Is the coating material to be in the wire or powder form?
6. Should the coating material be corrosion resistant?

Fig. 24-39 is a useful guide for the selection of thermal spray coatings.

## 24-16. TESTING AND INSPECTING THERMAL COATINGS

Finish thermal sprayed coatings are generally inspected using visual inspection. The operator should inspect each layer of material before continuing. The operator should look for uniformly applied, fine to medium sized granules. There should be no blisters, cracks, chips, or loose particles. No

## A GUIDE TO THE SELECTION OF THERMAL SPRAY COATINGS

| Surface material | Use and characteristics | Recommended deposit thickness after finishing, in. (mm) | Finishing method |
|---|---|---|---|
| Aluminum | Reclamation of light-alloy components. Corrosion resistance on ships, structures, marine, and industrial applications. | 0.010-0.080 (.254-2.03 mm) | Grind, turn, or as-sprayed |
| Aluminum bronze | Excellent general purpose bronze for all reclamation operations; high wear resistance, good machining properties. | 0.015-0.150 (.381-3.81 mm) | Grind or turn |
| Brass | Base coating for rubber bond. | 0.010-0.020 (.254-.508 mm) | As-sprayed |
| Copper | For resurfacing and reclaiming printing rollers, electrical contacts, slip rings, other. | 0.010-0.150 (.254-3.81 mm) | Turn or hand finish |
| Low-carbon steel | Economical for worn or mis-machined bearing surfaces and shafts where high hardness is not required. | 0.015-0.080 (.381-2.03 mm) | Grind or turn |
| Molybdenum | Correct machining error. Rectify worn bearing housings. Rectify press and interference fits. Hardface new components. Bond coat. | 0.005-0.015 (.127-.381 mm) | Grind |
| Monel | For high corrosion resistance. Low contraction rate permits heavy buildup. Takes a superfine turned finish. | 0.015-0.150 (.381-3.81 mm) | Grind or turn |
| 95 Ni-5 Al | A superior bonding and one step coating for all materials except copper based alloys. | 0.003-0.050 (.076-1.27 mm) | Grind or turn |
| 80 Ni-20 Cr | High temperature (to 1,800 °F or 982 °C) corrosive atmosphere. Base coat for ceramics and metals. Good self-bonding characteristics. | 0.010-0.015 (.254-.381 mm) | Grind or turn |
| Nickel-chromium-iron | Economical base coat. Less corrosion resistant than 80 Ni-20 Cr or 90 Ni-5 Al. Good bond strength. | 0.010-0.070 (.254-1.78 mm) | Grind or turn |
| Phosphor bronze | Use on components of similar material. | 0.015-0.080 (.381-2.03 mm) | Turn |
| 13 Cr stainless steel | Reclaim all types of bearing surface i.d. and o.d. faces. Where hard wearing tough deposit required. | 0.015-0.150 (.381-3.81 mm) | Grind |
| 18 Cr-5 Ni stainless steel | Where stainless properties are required and a turned finish may be necessary. Less shrinkage than 18 Cr-8 Ni, permitting heavy buildup. | 0.015-0.150 (.381-3.81 mm) | Grind or turn |
| 18 Cr-8 Ni stainless steel | Reclaim worn 18 Cr-8 Ni stainless components. Stainless steel surfacing of new components of cast iron or mild steel to replace manufacture from solid stainless steel. | 0.015-0.080 (.381-2.03 mm) | Grind |
| Zinc | Corrosion resistance. High deposition rates allow quick buildup. | 0.006-0.015 (.152-.381 mm) | As-sprayed |

Fig. 24-39. A guide to the selection of thermal spray coatings.

oil or other contaminates should be visible on the surface. If defects are visible they must be removed and recoated. Defects are generally removed by grit blasting.

The coating bond strength may be tested by scratching two lines on the coating surface. The two lines are ten times the coating thickness apart. The amount of coating material which comes off between these two lines is compared with other samples. The more coating that comes off the weaker the bond strength.

The 90 degree bend test may also be used to test the bond strength. Coatings that flake off indicate a weak or poor bond. In this test it is acceptable if the coatings crack or craze since they generally are hard and have little flexibility.

Dye penetrants and magnetic particle inspections may be used to reveal cracks and porosity.

Eddy current instruments can be used in production to measure the coating thickness. This test is used to insure uniformity of coating thicknesses.

## 24-17. REVIEW OF SURFACING SAFETY

All of the safety precautions specified for gas welding, arc welding, welding on containers and the like, also apply to metal surfacing operations. Review all safety precautions before attempting any metal surfacing operations.

Excellent ventilation is essential at all times because of the fluxes used, and because of the toxic effects of some of the alloys in the surfacing materials. The best protection is to use a supplied air respirator. Breathing masks with filters may be used as a second choice. Thermal spraying stations should have ventilation systems capable of exhausting 200-300 cu. ft. of air per minute (approximately 5600-8500 L/min) from the spray area. This exhaust should be passed through a wet collector to capture all dust which contains metal and coating material.

Protective clothing should be kept clean and in good condition. Your eyes should be protected at all times. Ear plugs or muffs are necessary when using the arc spraying, detonation spraying, and plasma arc spraying torch as they create an ear damaging sound pitch and intensity.

You must be constantly alert to avoid spraying metal on flammables or on another person.

## 24-18. TEST YOUR KNOWLEDGE

Write your answers on a separate sheet of paper. Do not write in this book.
1. List at least four causes of surface wear.
2. List the methods used to perform hard surfacing and thermal spraying.
3. List two advantages of surfacing a part.
4. What is meant by hardfacing?
5. Name two unusual materials which are not metals that have been flame sprayed.
6. The loss of areas of a metal surface by flaking or pitting is called _____ _____.
7. If a metal is difficult to cut, though possible, the hardness is probably _____ Brinell or _____ Rockwell C.

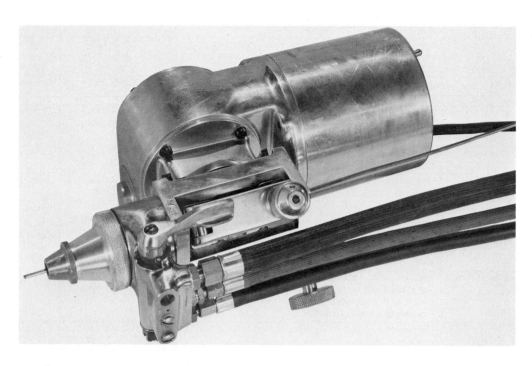

A large motor driven wire feed flame spraying torch.    (Metallizing Co. of America)

8. When using powder in flame spraying, name two ways of delivering the powder to the flame.

9. When flame spraying with powdered coating materials, _____ should not be used as the carrier gas. It may cause an explosion if combined with fine powder.

10. When thermal spraying, how thick is each layer of the coating material?

11. When flame spraying, if the surface of the applied coating is bumpy and coarse, what is the probable cause?

12. Which thermal spraying process is the most economical to use? What does it cost per hour?

13. Name six materials which may be applied to a base metal by plasma arc spraying.

14. When thermal spraying using the plasma arc, which type electrical circuit is recommended: the transferred or non-transferred arc?

15. Which thermal spray process propels the coating material to the surface with the highest velocity? What is the velocity in ft./sec. (m/sec)?

16. What is the temperature of the plasma arc spraying process?

17. When a cylindrical part is rough threaded, how deep are these threads cut?

18. What materials are used as bonding coats as a base for other thermal spray coatings? Name three.

19. When visually inspecting thermal sprayed coating, what should not appear? Name four things.

20. Industrial ear muffs should be worn when using which three thermal spraying methods?

Industry photo. A large chamber, high vacuum, electron beam welding machine. (PTR-Precision Technologies, Inc.)

# METAL TECHNOLOGY

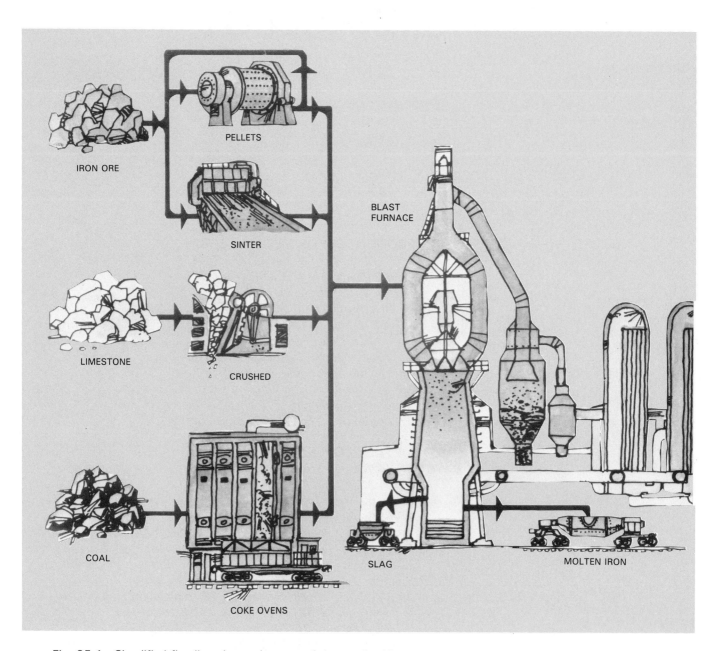

IRON ORE

PELLETS

SINTER

LIMESTONE

CRUSHED

COAL

COKE OVENS

BLAST
FURNACE

SLAG

MOLTEN IRON

Fig. 25-1. Simplified flowline shows the start of the steelmaking process. (American Iron and Steel Institute)

# 25 PRODUCTION OF METALS

It is important that a welder understand the origin, production, and refinement of metals in use today. Having a thorough understanding of metals is an asset to the welding technician. At this time review Figs. 1-33 through 1-40 on pages 48 to 61. This will provide an overview of the production of steel and iron. The processes used to change ore to commercial metals, and then to produce alloys, are constantly being improved.

Iron and steel are the most common metals in industrial use. Great quantities are produced each year. Aluminum is also being produced in large quantities. Magnesium and titanium have many applications, particularly in the aerospace industries. Recently, many exotic metals have been successfully produced, mainly due to the needs of our nuclear and space age.

This chapter will discuss several metals that have industrial applications.

## 25-1. METHODS OF MANUFACTURING STEEL

Steel is produced by adding controlled amounts of carbon to iron which is quite pure. Most, if not all, of the carbon and other impurities have been removed from the iron. This alloy (mixture) of iron and carbon is called STRAIGHT or PLAIN CARBON STEEL.

ALLOY STEELS are produced by adding additional elements, such as nickel, chromium, and manganese to plain carbon steel. The alloying elements are added in controlled amounts.

Follow the steps in steelmaking in Fig. 25-1. The first step in the production of steel is to refine the IRON ORE. The iron ore contains many unwanted impurities. When the iron ore is refined it is called PIG IRON. The refinement of iron ore is usually done in the reducing atmosphere of a BLAST FURNACE. Pig iron contains some impurities and a relatively large amount of carbon. The high carbon content makes pig iron brittle. In order to produce steel, the carbon in the iron must be removed in the oxidizing atmosphere of a STEELMAKING FURNACE.

The oxidizing atmosphere in these furnaces "burns off" and eliminates the carbon and impurities in the iron. Once the impurities and carbon have been eliminated or reduced to a minimum, controlled amounts of carbon and other alloying elements may be added to the iron to produce the type of steel desired.

The furnaces and processes used to produce steel and alloy steel are as follows:
1. Basic oxygen furnace (Heading 25-10).
2. Open hearth furnace (Heading 25-11).
3. Electric furnace (Heading 25-12).
4. Crucible furnace (Heading 25-13).
5. Induction furnace (Heading 25-14).
6. Vacuum furnace (Heading 25-15).

When it has been determined that the steel in the steelmaking furnace contains the desired amounts of carbon and other elements, the molten steel is formed into ingots, slabs, blooms, or billets. All these shapes differ in cross-sectional shape or area. See Fig. 25-2. From these rough

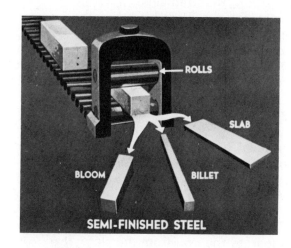

Fig. 25-2. An ingot being rolled into the semi-finished shapes generally produced.

stock shapes, the steel may later be formed into more exact shapes and products. Refer to Fig. 1-33.

## 25-2. MATERIAL USED BY THE BLAST FURNACE

Iron seldom exists free in nature. It is mined from the earth in the form of IRON ORE or IRON OXIDES mixed with impurities in the form of clay, sand, and rock. The most important types of iron ore are:

| | | | |
|---|---|---|---|
| Hematite (red iron) | $Fe_2O_3$ | 70 | % iron |
| Magnetite (black) | $Fe_3O_4$ | 72.4 | % iron |
| Limonite (brown) | $Fe_2O_3H_2O$ | 63 | % iron |
| Siderite (iron carbonite) | $FeCO_3$ | 48.3 | % iron |

Taconite ($Fe_3O_3$) has an iron content of 25-35 percent. It is made commercially useful by refining it to 70 percent iron prior to shipping.

A good FLUX that will melt and combine with the impurities in the molten iron ore must be used in the blast furnace. One flux that is commonly used is limestone. The LIMESTONE combines with the impurities and floats them in the combined state (SLAG) above the molten iron. The molten slag is drained from above the pig iron just before the pig iron is tapped or removed from the furnace.

One of the best fuels for the blast furnace is COKE. The coke furnishes enough heat to reduce the iron ore. Coke is low in such impurities as sulphur and phosphorus. Some modern blast furnaces use gas injection and solid fuel injection.

COKE is produced by heating soft coal (Bituminous) in a closed container until the gases and impurities are driven off. Coke, which is practically pure carbon, then remains.

To operate a modern blast furnace for one day requires approximately 2,000 tons (1,800 Mg) of iron ore, 1,000 tons (900 Mg) of coke, 500 tons (450 Mg) of limestone, and 4,000 tons (3,600 Mg) of air.

## 25-3. BLAST FURNACE

The blast furnace has five major operations to perform:
1. Deoxidize iron ore.
2. Melt the slag.
3. Melt the iron.
4. Carbonize the iron.
5. Separate the iron from the slag.

The modern blast furnace is a huge tubular furnace made of steel and lined with firebrick. The average size is approximately 100 feet (30 m) high and 25 feet (8 m) in diameter. Some are much larger. Fig. 25-3 shows a typical blast furnace in cross section. Around the bottom of the furnace are openings (TUYERES) through which hot air may be blown. An opening near the top allows gases to escape. The iron ore, limestone, and coke are carried up to the top of the furnace and dumped down into the furnace through a bell-shaped opening (hopper).

The coke burns and produces enough heat to melt the iron. The excess carbon from the coke unites with the iron and lowers its melting temperature. The melted iron forms at the bottom of the furnace. It is drawn off when a sufficient quantity has collected. The flux melts and collects the impurities. It floats on top of the molten iron. The flux can be drawn off the furnace through an opening higher than the one through which the iron is taken out. The operation of the furnace is continuous. The right proportions of iron ore, limestone, and coke are regularly dumped in at the top of the furnace. Every few hours a batch of blast furnace iron (pig iron) is drawn off from the bottom of the furnace. See Fig. 1-34.

The iron coming out of the blast furnace is called PIG IRON because it was formerly cast into bars called pigs. In modern practice, some of the iron from the blast furnace is not allowed to cool. The molten iron is taken directly to another furnace such as the open hearth furnace or the cupola furnace.

(For further information on blast furnace chemistry see Chapter 31.)

## 25-4. CAST IRON

Approximately 30 percent of all the pig iron produced is used in the manufacture of gray cast iron. Gray cast iron is the most common form of cast iron. GRAY CAST IRON is simply a casting that has been cooled slowly thus allowing some of the carbon to separate, forming free graphite (carbon) flakes. This graphite causes the gray appearance. Gray cast iron can be machined. WHITE CAST IRON is made by cooling the casting quickly. White cast iron is very hard and brittle. It is very difficult to machine white cast iron. CAST IRON is usually made by melting and oxidizing pig iron in a cupola furnace.

## 25-5. CUPOLA FURNACE

The CUPOLA FURNACE is used to produce cast iron. The cupola furnace resembles a small blast furnace. Coke is used as fuel to heat the furnace. Limestone is used as a flux. Pig iron is added along with scrap cast iron and steel.

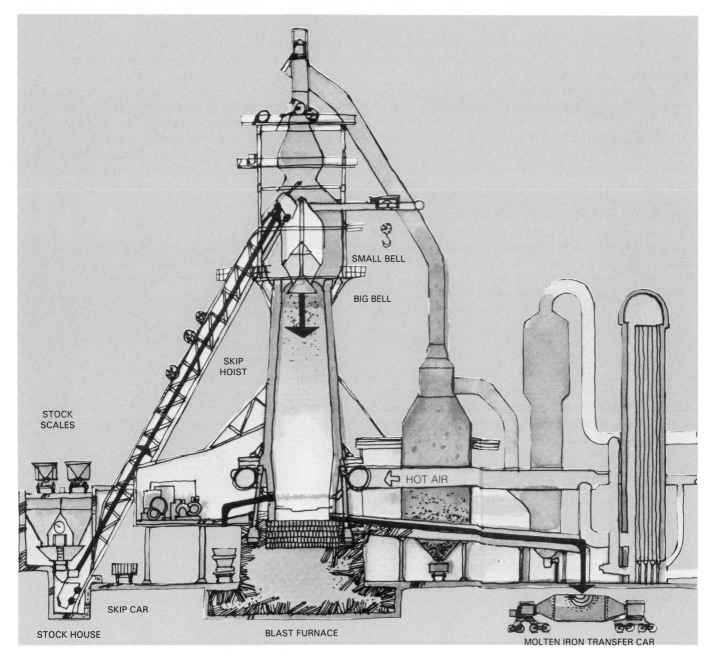

Fig. 25-3. Cross section of a blast furnace. Iron ore, limestone, and coke form iron at the bottom of the furnace. (American Iron and Steel Institute)

Labels in figure: SMALL BELL, BIG BELL, SKIP HOIST, STOCK SCALES, HOT AIR, SKIP CAR, STOCK HOUSE, BLAST FURNACE, MOLTEN IRON TRANSFER CAR

The cupola furnace eliminates the excess carbon and impurities as the metal and flux melt. When a large quantity of molten metal is formed, the furnace is ready to be tapped. TAPPING is the term used to draw off the molten metal from the furnace.

First the molten slag is drawn off. Then the furnace is tapped. The molten metal from the cupola furnace is ready to be cast. It is then called cast iron.

## 25-6. MALLEABLE IRON

Cast iron is a desirable metal from which intricate metal parts may be cast. Cast iron is very fluid when molten and flows freely to all parts of a mold. It may be machined relatively easily. However, it has several undesirable characteristics such as brittleness and lack of MALLEABILITY (the ability to be pounded into shape). In parts demanding malleability or resistance to shock, malleable iron may be used. MALLEABLE IRON is made by prolonged heating or ANNEALING of white cast iron at a temperature of approximately 1650 °F (900 °C). The castings are held at this temperature for about 50 hours. Then they are allowed to cool slowly.

The casting is heated to allow the carbon to diffuse (move) within the structure. The carbon gathers and forms small rosette shapes. This leaves the remaining iron low in carbon which has good ductility (ability to change shape without cracking or breaking) and toughness.

By this heat treating process, brittle white cast iron is transformed into a soft, malleable cast iron.

## 25-7. DUCTILE CAST IRON

Gray cast iron has free graphite (carbon) present in the metal. The graphite is in a flake form and the flakes cause the iron to be brittle. To prevent the graphite from forming flakes, magnesium is added to the liquid iron. The iron is then poured and cast. The magnesium attracts the carbon and forms graphite spheres. The remaining iron is low in carbon and therefore very ductile. DUCTILE CAST IRON has good strength, hardness, and ductility, as shown in Fig. 25-4. Ductile iron is also known as NODULAR IRON.

## 25-8. WROUGHT IRON

WROUGHT IRON contains the least amount of carbon of any of the ferrous metals used commercially. It is normally manufactured in a puddling furnace. (A puddling furnace or a reverberatory furnace is used to oxidize or burn off all the impurities and carbon in pig iron. These furnaces are used to produce wrought iron which is almost pure iron.) Wrought iron is no longer produced in large quantities. Steel has taken its place in most applications.

To make wrought iron, pig iron is melted on the hearth of the reverberatory or puddling furnace which is lined with iron oxide. This process almost completely removes all of the carbon, silicon, and manganese from the pig iron. As the carbon is removed, the fusion temperature of the iron rises. The iron becomes pasty and can be rolled up in balls and removed from the furnace. It is then squeezed through rollers to remove most of the excess slag. The wrought iron is rolled into muck bars and finally into commercial forms.

Wrought iron is soft, tough, and malleable. It is ideal for ornamental work, as it is rust resisting, easily shaped, and may be easily welded.

## 25-9. STEEL

STEEL may be defined as iron combined with 0.1 to 1.7 percent of carbon. See Chapter 26 for more technical specifications of steel.

Steel is produced by reducing cast iron and scrap steel in one of several types of furnaces. Several types of steel producing furnaces are:
1. Basic oxygen.
2. Open hearth.
3. Electric.
4. Crucible.
5. Induction.
6. Vacuum.

## 25-10. BASIC OXYGEN PROCESS

The basic oxygen furnace, shown in Fig. 25-5, is tipped on its side and molten iron and scrap steel are poured into the mouth of the furnace. The furnace is then rotated to a vertical position under an exhaust hood. A water-cooled oxygen lance is lowered to a position above the molten metal and oxygen is blown into the furnace. Refer to Fig. 1-35. Oxygen from above burns off the im-

| DUCTILE IRON TYPE | TENSILE STRENGTH | | YIELD STRENGTH | | ELONGA-TION % | HARDNESS BRINELL |
|---|---|---|---|---|---|---|
| | PSI | MPa | PSI | MPa | | |
| *60-45-10 | 60,000 to 80,000 | 414 to 552 | 45,000 to 60,000 | 310 to 414 | 10 to 25 | 140 to 200 |
| 80-60-03 | 80,000 to 100,000 | 552 to 689 | 60,000 to 75,000 | 414 to 517 | 3 to 10 | 200 to 275 |
| 100-70-03 | 100,000 to 120,000 | 689 to 827 | 70,000 to 90,000 | 483 to 621 | 3 to 10 | 240 to 300 |
| 120-90-02 | 120,000 to 150,000 | 827 to 1,034 | 90,000 to 125,000 | 621 to 862 | 2 to 7 | 270 to 350 |
| xHeat Resistant | 60,000 to 100,000 | 414 to 689 | 45,000 to 75,000 | 310 to 517 | 0 to 20 | 140 to 300 |

*This number is decoded as follows:
  60 means 60,000 psi tensile strength, 45,000 psi yield strength,
  and 10% elongation.

Fig. 25-4. Properties for five types of ductile cast iron.

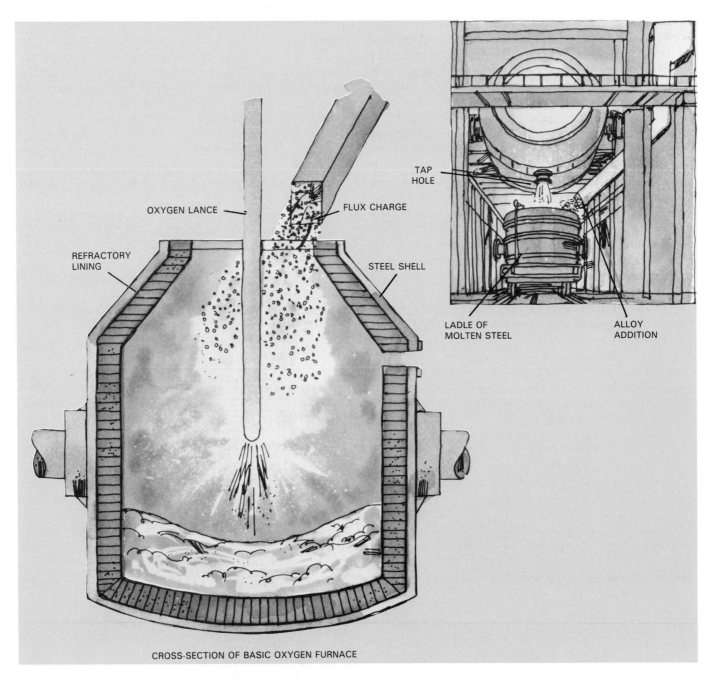

OXYGEN LANCE

FLUX CHARGE

TAP HOLE

REFRACTORY LINING

STEEL SHELL

LADLE OF MOLTEN STEEL

ALLOY ADDITION

CROSS-SECTION OF BASIC OXYGEN FURNACE

Fig. 25-5. Schematic of a basic oxygen furnace. Oxygen is blown into the furnace and the products of combustion are exhausted into a pollution control system. (American Iron and Steel Institute)

purities in the molten metal and produces steel. In 40 to 60 minutes, about 80 tons (72,575 kg) of quality steel can be produced. At the end of the prescribed time, the oxygen is turned off. The oxygen furnace is then tipped to pour off the steel produced. A new charge of molten iron and scrap steel is poured in to begin a new cycle.

## 25-11. OPEN HEARTH PROCESS

The open hearth furnace method of making steel, Fig. 1-37, is sometimes called the Siemens-Martin Open Hearth Process. In such a furnace, Fig. 25-6, the metal is contained in a large shallow basin, holding from 150 to 300 tons (136 to 272 Mg) of metal. The furnace is charged with molten and solid pig iron, scrap, and fluxes. Up to 50% scrap can be used.

At each end of the basin or metal container there is a preheating stove. The preheating stove is made of firebricks arranged in a checkerboard pattern. Air and a fuel gas enter the furnace at one end after passing through one of the preheating stoves. As the fuel gas and air burn above the metal in the furnace, they heat the metal. When the metal is heated, the unwanted elements and impurities oxidize or burn off. As exhaust gases exit from the furnace, they travel through the unused preheating furnace on the exhaust side of the furnace. The hot gases heat the bricks in this stove as they travel out of the furnace. Fig. 25-6 shows the furnace to illustrate the method of preheating the incoming gas and air. Periodically the direction of the incoming fuel gas and air is reversed. The cold input gases travel over the hot bricks in the preheat stove which were previously heated by the exhaust gases. This is called a REGENERATIVE PROCESS OF HEATING. The preheating of the incoming fuel gas and air results in higher temperatures inside the open hearth furnace. The total time for processing each charge is 20 to 30 hours.

Pure oxygen is now being used in some open hearth furnaces to speed the production of steel. For example, an 82.5 ton (74.8 Mg) heat in a fur-nace using oxygen may be completed in about 60 minutes.

Since more steel can be made in a given time, the use of oxygen is economically feasible and is growing in popularity. The high temperatures obtained by using oxygen in the steel making process air in burning out the carbon and impurities in the iron.

Today, less than 1 percent of the steel produced in the U.S.A. is made in the open hearth furnace.

Some advantages of the open hearth process are:

1. A minimum quantity of steel is lost in the process.
2. Better control over the alloying elements is obtained.
3. Steel is cleaner because it contains fewer oxides.
4. Larger batches of steel may be made at one time.
5. Pig iron and scrap unsuitable for the basic

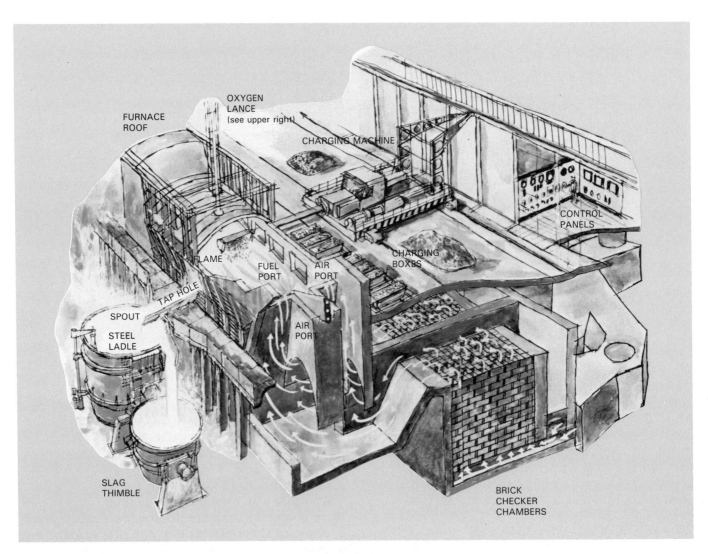

Fig. 25-6. Schematic drawing of an open hearth furnace. Note how the materials are charged into the furnace and also how they are removed from the furnace.

oxygen process may be made into steel by the open hearth method.

6. Steel manufactured by the basic oxygen process may be further refined in the open hearth furnace.

By taking periodical chemical analysis of the metal in the heat, the exact composition may be determined. By adding alloying ingredients, the heat may be held to very close chemical tolerances and a better quality steel may be produced than is possible in the basic oxygen process.

## 25-12. ELECTRIC FURNACE

A popular method used to produce special, high quality, steel alloys is the electric furnace, as shown in Fig. 25-7. In this type furnace, the chemical constituents of the metal may be closely controlled. Various alloying elements may be added to create a steel with predetermined characteristics. Refer to Fig. 1-38.

Generally, pig iron and selected scrap steel are used to charge the electric furnace. The furnace is then closed to the atmosphere.

The heat required to melt the metal in this fur-

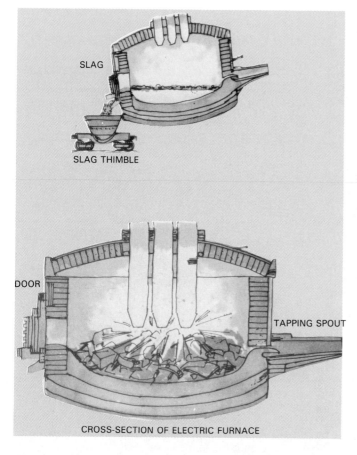

CROSS-SECTION OF ELECTRIC FURNACE

Fig. 25-7. An electric furnace. Note the carbon electrodes at the top. This furnace is tilted to pour molten metal into the ladle.

nace is produced by an electric arc.

Large diameter movable electrodes are installed in the top of the furnace above the metal. An arc is struck between the carbon electrodes and the metal to furnish the heat required to melt the metal. As the electrodes are consumed they are moved down to a given distance above the molten metal. The carbon electrodes may be changed as required from the top of the furnace.

Test samples of molten metal may be removed through small inspection ports. These samples are analyzed and alloying elements are added to the furnace as required.

Electric furnaces vary in capacity from 5 to 50 tons (4.54 to 45.36 metric tons). The time required to complete a heat varies from three to six hours. The entire furnace is generally built to tilt to make loading and unloading easy.

## 25-13. CRUCIBLE FURNACE

One of the oldest methods of refining steel is the crucible furnace. The heat for this furnace is produced by burning fuel gas. A crucible or covered pot made of ceramics and graphite is usually used to hold the charge. In this process, wrought iron or wrought iron and scrap steel are loaded into a crucible. The correct amounts of carbon and other alloying elements required to produce the desired finished steel are then charged into the crucible. The use of scrap steel generally lowers the quality of the finished product.

After the crucible is charged, it is covered and sealed. The crucible is then placed into the furnace where hot gases heat the crucible and its contents.

The quality of the steel produced in a crucible furnace is generally considered to be higher than that produced in an electric furnace, but the process is much slower and more expensive. The charge in a crucible may vary from a few pounds (kilograms) to several tons (megagrams or metric tons).

## 25-14. INDUCTION HEATING PROCESS

Induction heating is defined as "raising the temperature of a material by means of electrical generation of heat within the material and not by any other heating method such as convection, conduction, or radiation." In an induction furnace, the metal to be heated is contained in a vessel and electrical conductors are wound around the vessel to form a coil. Alternating current is passed through the coil and induction heating of the vessel and metal in the vessel takes place.

Induction heating is caused when an alternating magnetic field is set up in a metal. The alternating magnetic field is created by passing alternating current through the electrical conductors.

When magnetic materials, such as iron and steel, are placed within the area of an alternating magnetic field, they are heated. The heating is by both hysteresis and eddy current losses. See the schematic in Fig. 25-8.

HYSTERESIS LOSS is caused by friction among the molecules of the molten steel. The action of the magnetic field causes the molecules in the molten steel to move around. The magnitude of this hysteresis loss and the heat created are proportional to the frequency of the electric current.

EDDY CURRENT LOSSES are losses caused by electric resistance. The resistance is a result of small circulating currents within a material placed in an alternating magnetic field. These resistance losses cause heat which is absorbed by the metal being heated. The amount of heat created by eddy currents is proportional to the square of the alternating frequency of the current. It is also proportional to the square of the amperage flowing in the conductor, which produces the magnetic field.

By accurately controlling the frequency and amperage of the alternating current passing through the induction coil, it is possible to accurately control the temperature of the metal be-

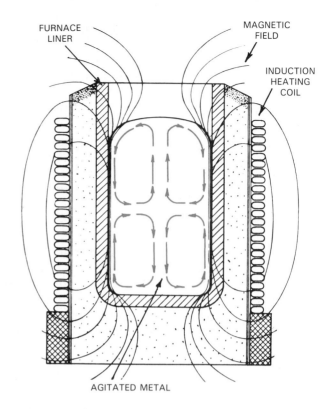

Fig. 25-8. Schematic cross section of a small induction furnace. Metal in the furnace is agitated and mixed by magnetic eddy currents, to produce a more homogeneous metal.

Fig. 25-9. A complete induction melting furnace installation. (Ajax Magnethermic Corp.)

ing heated. A small induction furnace unit is shown in Fig. 25-9.

A low-frequency induction furnace with a 4 ton (3.63 metric ton) capacity is being used in conjunction with a cupola furnace to further refine cupola iron. Tests show that iron from the cupola-induction furnace combination shows superior tensile strength, fluidity, and other favorable casting metal characteristics. It also has desirable machining characteristics. Fig. 25-10 shows a schematic cross section of a large induction furnace.

The design of the furnace and its low frequency induction coil sets up a controlled stirring action of the molten metal in the furnace. In addition, the stirring action keeps the slag on the surface agitated providing openings in the slag layer through which entrapped gases may escape.

## 25-15. VACUUM FURNACES

The melting of steel, steel alloys, titanium, and other pure metals in a vacuum greatly reduces the amount of gas in the metal. There is only a small amount of gas present in the furnace. Since there is very little gas present, the absorption of gases by the molten metal is very small. Steels, melted by more conventional methods, absorb gases which cause porosity and inclusions in the metal when it solidifies. Gases formed in a vacuum furnace are pulled away from the molten metal by the vacuum pumps. The absence of these gases improves many of the qualities of the steel including ductility, magnetic properties, impact strength, and fatigue strength. See Fig. 1-39.

There are two main types of vacuum furnaces used today. The two types are the vacuum arc and vacuum induction-type furnaces.

In the VACUUM ARC FURNACE, the metal is first produced by another method. The metal is then formed into long round or square cylinders. These long metal cylinders are then melted as huge consumable electrodes in the furnace. The metal electrodes are fed into the furnace at a controlled rate of speed to control the arc length.

As the metal drops from the end of the electrode, it falls into a water-cooled steel crucible. The steel crucible is the grounded part of the electrical circuit. Air and contaminating gases are constantly pumped out of the furnace by vacuum pumps. Fig. 25-11 shows such a furnace

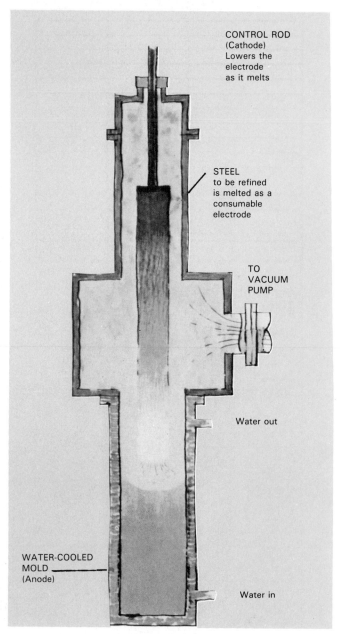

Fig. 25-11. Consumable metal electrode vacuum arc furnace shown schematically.

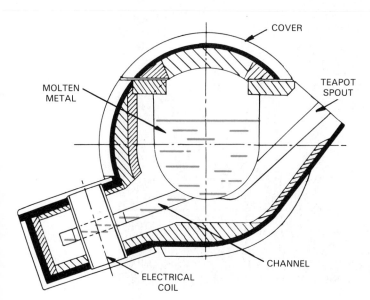

Fig. 25-10. Schematic cross section of a large induction furnace.

schematically. The copper crucible is shown being installed in the furnace in Fig. 25-12.

This process is frequently used when the uniformity and purity of the metal is very important. No provision is made in this furnace for adding alloying elements.

The VACUUM INDUCTION FURNACE is used when close control of the chemistry of the metal is of prime importance. The metal is melted in an electric vacuum induction furnace. The induction furnace is airtight and attached to vacuum pumps so that contaminating gases are constantly

Fig. 25-12. A copper crucible for a consumable electrode vacuum melting furnace being lowered into position. (Allegheny Ludlum Steel Corp.)

removed. Provisions are made to allow alloying materials to be added to the furnace without destroying the vacuum.

The heating of the metal, the pouring of the metal into ingots, and the cooling of ingots are done under vacuum conditions to prevent contamination. Fig. 25-13 shows a large induction vacuum melting furnace installation.

## 25-16. CONTINUOUS CASTING PROCESS

The continuous casting process for the manufacture of steel is shown in Fig. 25-14. Liquid steel is poured into a reservoir or TUNDISH. From the tundish the metal flows vertically into a water-cooled mold. The molten metal in contact with the sides of the mold cools quickly and

HOT INGOTS

Fig. 25-13. Removing hot ingots from vacuum chamber of induction vacuum melting furnace. The melting furnace is located in the tank on the left. Vacuum chambers are arranged to permit continuous operation of the unit.

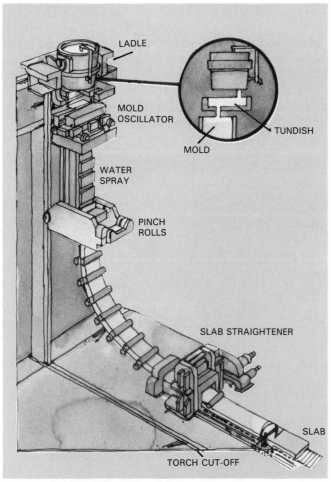

Fig. 25-14. Schematic drawing of the continuous casting process for making steel.

ING rolls as the column of steel is pulled from the mold. As the column of steel leaves the mold, jets of water are sprayed on the metal to cool and solidify the entire column. The metal, as it comes from the mold, may be in the form of a slab or a square bar. As the metal leaves the withdrawing rolls it is cut to desired lengths for further processing. Refer to Fig. 1-40.

This process has the advantages of eliminating ingot pouring, removing of the ingots from their molds, the use of soaking pits and reheating furnaces, and the rough rolling of ingots into semi-finished forms.

## 25-17. MANUFACTURING STAINLESS STEELS

STAINLESS STEEL, the most common of the alloy steels, may be made in the open hearth furnaces in the same manner as straight carbon steels. Alloy metals are added during the time the metal is in the furnace. Stainless steels may also be made or refined in the electric furnaces. Various types of stainless steels are described in Chapter 19.

## 25-18. MANUFACTURING COPPER

A great deal of copper ore contains a high percentage of pure copper. This ore is crushed and then washed in water to remove the lighter weight earth particles from the heavier copper ore. The ore is then mixed with coke and limestone and placed into a small blast furnace. The liquid copper settles to the bottom of the furnace. Impurities in the copper ore are floated above the molten copper in the form of slag. This is the same process as when refining iron ore in a blast furnace.

shrinks away from the side of the mold. This forms a shell around the molten metal in the center of the mold. This shell is supported by the WITHDRAW-

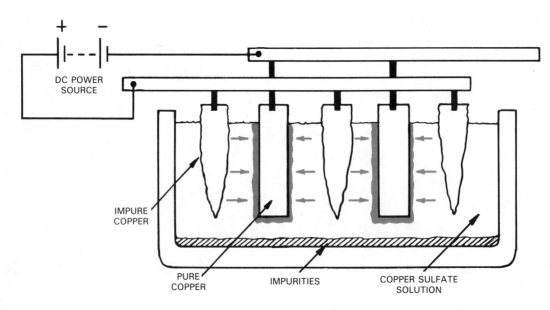

Fig. 25-15. Schematic drawing of an electrolytic cell used to refine copper.

The copper removed from the blast furnace is further refined by the electrolysis process.

ELECTROLYSIS may be defined as a chemical change, or decomposition. The decomposition is created in a material by passing electricity through a solution of the material or through the substance while it is in a molten state.

The electrolytic cell used to refine copper is shown in Fig. 25-15. Pure copper bars are used as the cathodes (negative). Impure copper to be refined forms the anodes (positive) in the cell. Copper sulfate with some sulphuric acid is used as the electrolyte, or fluid, in the cell.

When the electric current is turned on, pure copper leaves the anode and deposits on or ''plates'' the cathode. The impurities fall to the bottom of the electrolytic cell. When the impure copper anode is consumed, it is replaced. The heavily plated pure copper cathodes are removed and replaced with thinner bars. The sediment in the bottom of the cell often contains small quantities of silver and gold.

## 25-19. MANUFACTURING BRASS AND BRONZE

BRASS is an alloy of copper and zinc. BRONZE is an alloy of copper and tin. When a third or fourth element is added to brass or bronze to improve the physical properties, an alloy brass or an alloy bronze is created. Some of the alloying elements added to brass are: tin, manganese, iron, silicon, nickel, lead, and aluminum. Alloying elements added to bronze are: nickel, lead, phosphorous, silicon, and aluminum.

To make brass or bronze, the alloying elements are heated in a crucible or an electric cupola.

The term bronze is often used with copper alloys which contain no tin or small quantities of tin in the alloy.

Hardware bronze, as an example, contains approximately 90 percent copper, 8 percent zinc, and 2 percent lead. Manganese bronze contains approximately 58.5 percent copper, 1.0 percent tin, about 39 percent zinc, .28 percent manganese, and 1.4 percent iron.

## 25-20. MANUFACTURING ALUMINUM

Aluminum is normally produced by separating it from the oxide ($Al_2O_3$) found in bauxite ore. It may also be found in many other forms. After the aluminum oxide is removed from the ore, it is dissolved in a molten bath of sodium-aluminum-fluoride (cryolite). An electric current is passed through the molten bath and pure aluminum is obtained by an electrolysis process, called the Hall Process.

The electrolytic cell used in this process is a carbon lined, open-top furnace. The electrolytic cell also has carbon electrodes suspended in a solution of aluminum oxide and cryolite. As the current passes through the solution, the aluminum oxide is reduced. Pure aluminum is deposited at the cathode, or negative terminal, which is the lining of the furnace. The pure aluminum collects at the bottom of the cell as shown in Fig. 25-16. Such a furnace is in continuous operation. Periodically the molten aluminum is poured from the cell into ingot molds which are stored for further processing.

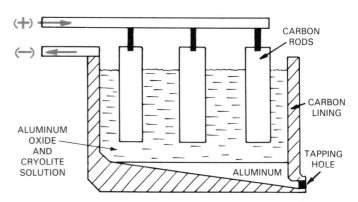

Fig. 25-16. Schematic drawing of the Hall process for producing aluminum.

## 25-21. MANUFACTURING ZINC

Zinc is principally produced by a distilling process. The zinc ore is heated with coke in a clay crucible and the zinc vapor is then condensed in a clay condenser. It may also be refined by an electrolytic process. It is used mainly as an alloying metal and for galvanizing.

## 25-22. PROCESSING METALS

Most metals as they come from the furnace are originally cast into ingots or molds for further processing. As needed, the castings are reheated to a definite temperature depending on the metal. The metal is then formed into a finished or semifinished shape by one of the following methods:
1. Casting.
2. Rolling (hot and cold).
3. Forging.
4. Extruding.
5. Drawing.
CASTING a metal in a sand or permanent mold is a popular method of producing objects with intricate shapes. A stationary or spinning (centrifugal) mold may also be used.

To improve the physical properties of a metal, and to form them into more usable rough stock

Fig. 25-17. A macrograph (4X) of etched cold rolled steel. The upper figure is across the grain of rolling, and the lower figure is along the grain of rolling.

shapes, they are often ROLLED. The dense structure of steel after rolling is shown in Fig. 25-17.

In a steelmaking plant, the large ingots are rolled between large powerful rollers in a rolling mill. The ingots are reduced to blooms, billets, slabs, and even sheet stock as required. These blooms, billets, slabs, and sheet stock may be further processed in a rolling mill to produce rails, T beams, I beams, angles, bar stock, etc. Fig. 25-18 illustrates some typical shapes which are formed by rolling. Numerous operations are required to form some of the shapes.

FORGING, either drop or press, is used to obtain shapes stronger than castings and which are not easily rolled into shape. Forge hammers and/or forming dies are used to pound the metal into the shape desired.

EXTRUDING metal is a process by which metal normally in its plastic state is pushed with great force through dies which are cut in the shape of the desired cross section. This process produces long lengths of metal with uniform cross section.

DRAWING is a process of pulling metal through dies to form wires, tubing, and moldings. Drawing is a popular method of shaping metal to meet a certain requirement.

Many intricate shapes which would be difficult or impossible to make by any other method are now being produced by the POWDERED METALS PROCESS. The metal to be used is reduced to a powder texture and is forced under pressure into a heated steel mold. The properties of the finished part approximate those of the original solid metal. Fig. 25-19 shows a die set being used to form gears from powdered metals.

These parts are made to very close tolerances. Most of them can be used without any additional machining or grinding.

Metal may also be reduced to actual fibers. The fibers are then laid down or woven to form mats. Mats of fibrous metal are finding increasing use in resistance welding where the mats are placed between parts to be welded.

## 25-23. REVIEW OF SAFETY

All furnaces which contain molten metal should be handled with great caution. Careless handling can cause great bodily harm and damage to equipment. Spilled molten metal will spread with great speed and will burn almost anything combustible. Special clothing, spats, and special shoes must be worn. Special face guards should be worn. The eyes and face must be protected from flying particles and glare.

Gloves are needed at all times.

OSHA regulations for workers in metal refining plants and foundries should be obeyed.

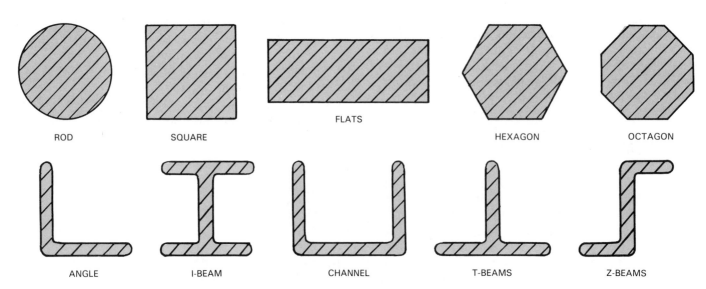

ROD    SQUARE    FLATS    HEXAGON    OCTAGON

ANGLE    I-BEAM    CHANNEL    T-BEAMS    Z-BEAMS

Fig. 25-18. Standard cross sectional shapes of metals. The welding process may be used to combine the principal forms into complex structures.

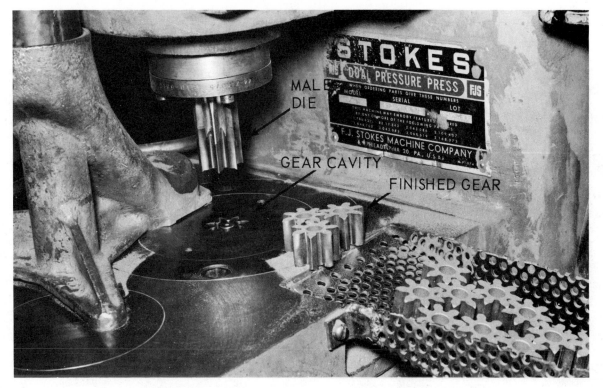

Fig. 25-19. Pump rotors made from powdered metals in a male-female die set.

## 25-24. TEST YOUR KNOWLEDGE

Write your answers on a separate sheet of paper. Do not write in this book.

1. What is coke?
2. What is the purpose of limestone?
3. Is pig iron ductile or brittle?
4. What materials are used in a blast furnace to produce pig iron?
5. Why is a flux needed in a blast furnace?
6. The most common form of cast iron is _____ cast iron.
7. What is done with the molten metal produced in a cupola furnace?
8. a. How is malleable iron produced?
   b. How is ductile iron produced?
9. Steel contains between _____ and _____ percent carbon.
10. Why is oxygen used in place of air in some open hearth furnaces?
11. Describe how the electric arc furnace is used to produce steel.
12. Which type of furnace produces the highest quality steel? (Open hearth, electric, or crucible)
13. List four advantages of the continuous casting process.
14. How is a magnetic material in an induction furnace heated?
15. What are the advantages of a vacuum furnace?
16. The production of copper requires two processes. First, copper ore is reduced in a _____ _____. It is further refined by the _____ _____.
17. When producing brass or bronze, what type of furnace is most often used?
18. How is aluminum produced?
19. What forming processes can be used to produce long pieces with a constant cross section?
20. What is done to a metal during a forging process?

Industry photo. A gear elevated welding positioner used in heavy applications. Unit is able to position up to 500,000 pound workpieces. (K.N. Aronson, Inc.)

# 26 METAL PROPERTIES AND IDENTIFICATION

It is necessary for a welder to have some accurate means of identifying metals. The welder must also have a good understanding of the constituents of metals in order to intelligently solve welding problems.

Metals are divided into two major fields. These are:

1. Ferrous metals.
2. Nonferrous metals.

Iron and its alloys are classified as ferrous metals.

The NONFERROUS METALS include metals and alloys which contain either no iron or insignificant amounts of iron. Some of the more popular nonferrous metals a welder encounters are copper, brass, zinc, bronze, lead, and aluminum.

A great amount of welding is done on FERROUS METALS ranging from wrought iron to low, medium, and high carbon steel, tool steel, stainless steel, and cast iron.

The various welding processes now make it possible to satisfactorily weld practically all nonferrous metals.

## 26-1. IRON AND STEEL

Iron is produced by reducing iron oxide, commonly called iron ore, to pig iron by means of the blast furnace. Review Chapter 25.

Many types of furnaces are used to change pig iron into the various steels. Some of these furnaces are: the open hearth furnace, the basic oxygen furnace, the electric furnace, and the cupola furnace. See Chapter 25 for more information on these furnaces.

CARBON STEEL is an alloy of iron and controlled amounts of carbon. ALLOY STEEL is a combination of carbon steel and controlled amounts of other desirable metal elements. The percentage of carbon content determines the type of carbon steel. For example, WROUGHT IRON has .003 percent carbon, meaning three thousandths of one percent. LOW CARBON STEEL contains less than .30 percent carbon. MEDIUM CARBON STEEL varies between .30 and .55 percent carbon content. HIGH CARBON STEEL contains approximately .55 to .80 percent carbon, and VERY HIGH CARBON STEEL contains between .80 and 1.70 percent carbon. CAST IRON contains 1.8 to 4 percent carbon.

The carbon generally combines with the iron to form CEMENTITE, a very hard, brittle substance. Cementite is also known as IRON CARBIDE. This action means that as the carbon content of the steel increases, the hardness, the strength, and the brittleness of the steel also tend to increase.

Various heat treatments are used to enable steel to retain its strength at the higher carbon contents, and yet not have the extreme brittleness usually associated with high carbon steels. Also, certain other substances such as nickel, chromium, manganese, vanadium, and other alloying metals may be added to steel to improve certain physical properties.

A welder must also have an understanding of the impurities occasionally found in metals and their effect upon the weldability of the metal.

Two of the detrimental IMPURITIES sometimes found in steels are phosphorus and sulphur. Their presence in the steel is due to their presence in the ore, or to the method of manufacture. Both of these impurities are detrimental to the welding qualities of steel. Therefore, during the manufacturing process, extreme care is always taken to keep the impurities at a minimum (.05 percent or less). Sulphur improves the machining qualities of steel, but it is detrimental to its hot forming properties.

During a welding operation, sulphur or phosphorus tends to form gas in the molten metal. The resulting gas pockets in the welds cause brittleness. Another impurity is dirt or slag (iron oxide). The dirt or slag is imbedded in the metal during rolling. Some of the dirt may come from the by-products of the process of refining the metal. These impurities may also produce blow holes in the weld and reduce the physical properties of the metal in general. See Heading 26-8 for a procedure used to test a steel for impurities.

## 26-2. PHYSICAL PROPERTIES OF IRON AND STEEL

A PHYSICAL PROPERTY is a characteristic of a metal which may be observed or measured. As mentioned previously, the physical properties of steel are affected by the following:
1. Carbon content.
2. Impurities.
3. Addition of various alloying metals.
4. Heat treatment.

Chapter 28 describes various machines used for determining a metal's physical properties. Modified testing machines are being introduced into the welding shop to enable operators to identify metals and to check on the physical properties of weldments.

Some of the more important physical properties of steels are:
1. Tensile strength.
2. Compressive strength.
3. Hardness.
4. Elongation.
5. Ductility.
6. Brittleness.
7. Toughness.
8. Grain size.

TENSILE STRENGTH is the ability of a metal to resist being pulled apart. This property may be measured on a tensile testing machine which puts a stretching load on the metal. Fig. 26-1 illustrates the types of loads imposed on structures. Fig. 26-2 shows how the tensile strength, elongation (explained below) and yield point are affected by the carbon content of steel. As the carbon content increases, the tensile strength and yield point first increase then decrease. (The YIELD POINT is the point on the stress strain curve when the metal begins to plastically deform.)

The COMPRESSIVE STRENGTH of a metal is a measure of how much squeezing force it can withstand before it fails. The metal to be tested is mounted in a tensile tester, but instead of pulling on the metal, a squeezing (compression) force is applied.

HARDNESS is the quality which allows a metal to resist penetration. The Rockwell, Brinell, or Scleroscope may be used to test hardness in a metal. See Heading 28-14 concerning hardness testing.

ELONGATION is a measure of how much a

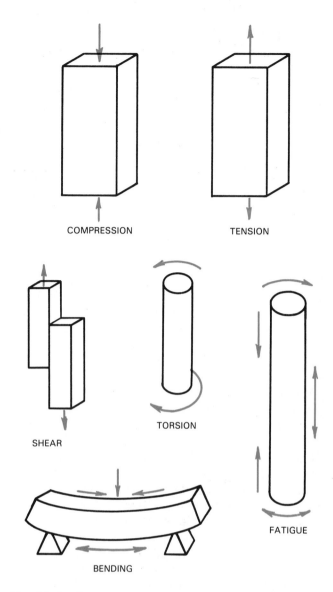

Fig. 26-1. Types of stresses or loads imposed on structures: Compression, Tension, Shear, Torsion, Bending, and Fatigue. In bending, the lower surface is in tension and the upper surface is in compression. In fatigue, a vibration or repeated reversal of the load is applied. The fatigue load may be either compression, tension, torsion, shear, or a combination of these loads.

metal will stretch before it breaks. Elongation, or the percent of elongation, is measured during a tensile test. Two marks, exactly two inches (50.8 mm) apart are placed on the tensile sample and the distance between the marks is again measured after the sample breaks. The original distance between the marks is compared to the increase in length to determine the percent of elongation. A metal that has over 5 percent elongation is said to be DUCTILE. One with less than 5 percent elongation is considered BRITTLE. The formula for determining percent elongation is:

$$\% \text{ Elongation} = \frac{L_f - L_i}{L_i} \times 100.$$

$L_f$ is the final length between marks. $L_i$ is the initial length (2 inches or 50.8 mm). Fig. 26-2 shows how the elongation of a steel decreases as the carbon content increases.

DUCTILITY is the ability of a metal to be stretched. Other terms which refer to this same property are formability, malleability, and workability. A very ductile metal such as copper or aluminum may be pulled through dies to form wire.

BRITTLENESS is the opposite of ductility. A brittle metal will fracture if it is bent or struck a sharp blow. Most cast iron is very brittle.

TOUGHNESS is the ability to prevent a crack from propagating (growing). A metal is said to be tough if it can withstand an impact or shock loading.

The GRAIN SIZE and microstructure of a metal can be viewed under a microscope. The MICRO-STRUCTURE of a metal is the structure observed through a microscope. Microscopic views give a good indication of a metal's heat treatment, tensile strength, and ductility.

Before these properties are studied in detail, the welder should have an understanding of the effect of carbon on the properties of steel and a knowledge of alloys in general.

## 26-3. ALLOY METALS

As mentioned previously, an alloy metal may be defined as an intimate mixture of two or more elements. Any ferrous or nonferrous metal may be alloyed to form an alloy metal with new and desirable characteristics.

Steel is a combination of iron and controlled amounts of carbon. Alloy steels are created by adding other elements to plain carbon steel. Some elements which are alloyed with carbon steel and the qualities imparted to steel by each are:

CHROMIUM—increases resistance to corrosion; improves hardness and toughness; improves the responsiveness to heat treatment.

MANGANESE—increases strength and responsiveness to heat treatment.

MOLYBDENUM—increases toughness and improves the strength of steel at higher temperatures.

NICKEL—increases strength, ductility, and toughness.

TUNGSTEN—produces dense, fine grains; helps steel to retain its hardness; helps steel to retain strength at high temperatures.

VANADIUM—retards grain growth and improves toughness.

The melting temperature of a metal is changed somewhat whenever an alloying metal is added. This may be illustrated by examining the cooling curve for a simple alloy. See Fig. 26-3.

A simple alloy consists of two metals in any proportion. An example of a simple alloy is the combination of tin and lead which is called solder. The melting temperature of the lead is 621 °F (327 °C). Tin has a melting temperature of 450 °F (232 °C). However, as the two metals are mixed, any combination of the two results in a lower melting temperature than 621 °F (327 °C). At a certain proportion of the metals, the lowest melting temperature is reached. This point is called the EUTECTIC POINT, as shown in Fig. 26-3.

| SAE AISI No. | Carbon Content in Percentages | Tensile Strength | | Yield Point | | Elongation in Percentages |
|---|---|---|---|---|---|---|
| | | Lbs/sq.in. | MPa | Lbs/sq.in. | MPa | |
| 1006 | .06 | 43,000 | 300 | 24,000 | 170 | 30 |
| 1010 | .10 | 47,000 | 320 | 26,000 | 180 | 28 |
| 1020 | .20 | 55,000 | 380 | 30,000 | 210 | 25 |
| 1030 | .30 | 68,000 | 470 | 37,500 | 260 | 20 |
| 1040 | .40 | 76,000 | 520 | 42,000 | 290 | 18 |
| 1050 | .50 | 90,000 | 620 | 49,500 | 340 | 15 |
| 1060 | .60 | 98,000 | 680 | 54,000 | 370 | 12 |
| 1070 | .70 | 102,000 | 700 | 56,000 | 390 | 12 |
| 1080 | .80 | 112,000 | 770 | 56,000 | 390 | 12 |
| 1090 | .90 | 122,000 | 840 | 67,000 | 460 | 10 |
| 1095 | .95 | 120,000 | 830 | 66,000 | 460 | 10 |

Fig. 26-2. Approximate physical property changes of carbon steel as the carbon content changes.

In this diagram, at temperatures below line 1-2 generally the lead is solid and the tin is liquid. As the alloy cools, all of the tin becomes solid at line 4-2. Likewise, under line 2-3 the tin solidifies and the lead remains a liquid. At line 2-5 all the lead becomes solid. Between points 4 and 5 the alloys have the same SOLIDUS (starting-to-melt) temperature. In the area to the left of 1-4 the alloy remains solid up to line 1-4 and is called ALPHA. In the area to the right of 3-5 the alloy remains solid up to the line 3-5 and is called BETA. The other regions on the diagram are mixtures of alpha, beta, and liquid. All the regions are labeled on Fig. 26-3.

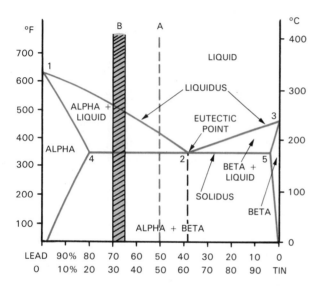

Fig. 26-3. Lead-tin diagram. Line A represents the popular 50-50 solder; Line B represents the range of body solders. An alloy solidifies over a range of temperatures. The alloy begins to solidify at the liquidus. It is completely solid at the solidus.

## 26-4. COOLING CURVES

The temperature of a virgin (pure) metal (not an alloy) at atmospheric pressure remains constant during the change from the solid to the liquid phase. As virgin metals are heated to their melting temperature, a thermometer will show a constant rise in temperature per unit of time with a constant heat input. The temperature increases until the metal reaches its melting temperature. As the metal melts, the temperature will remain constant for a length of time. During this time, the metal absorbs heat energy. This heat energy is required to give the metal atoms the energy required to break away from the solid and become liquid. After the metal is melted, the temperature will again rise as the metal is heated.

As a metal cools from the liquid phase, the temperature may be observed to drop until the

point where the metal solidifies. At this point, the temperature will remain constant while the atoms in the metal molecules change back to the solid state molecular structure. During this period of time while the atoms are changing position, the metal releases heat. After the metal has solidified, the temperature will again drop until it reaches room temperature.

The melting or solidifying of an alloy differs slightly from that of a pure metal. The alloy does not solidify at one specific temperature. It solidifes over a range of temperatures. This is shown in Fig. 26-3. The alloy represented by line A begins to solidify when it crosses the liquidus line 1-2-3. It is not completely solid until it crosses the solidus line 4-5.

A change other than solid to liquid, or liquid to solid is possible. One form of a solid may change to another form of a solid. This is true for iron and steel. There are three typical structures.

One structure is a BODY CENTERED CUBIC STRUCTURE. There is one atom at each corner of a cube and one in the center. See Fig. 26-4. This is the structure of steel at room temperature. A second structure is the FACE CENTERED CUBIC STRUCTURE. There is again one atom at each corner of a cube and one in the center of each face. This is the structure of austenite, a form of steel above 1341°F (727°C). Aluminum also has a face centered cubic structure. The final structure is a HEXAGONAL CLOSE PACKED STRUCTURE. All three of these structures are illustrated in Fig. 26-4.

When an alloy changes structure, there is a delay in the heating or cooling similar to a phase change.

The temperatures at which the phase changes or structural changes occur, greatly depend on the alloy content. These temperatures are important when heat treating a metal and are often called CRITICAL TEMPERATURES. The critical temperatures for steel are shown on the iron-carbon diagram, Fig. 26-5.

## 26-5. IRON-CARBON DIAGRAM

The iron-carbon diagram, Fig. 26-5, shows the important temperatures for steel and cast iron. The iron-carbon diagram is very important in producing, forming, welding, and heat treating steel and cast iron. Before the diagram can be used, each region on the diagram must be understood.

Each of the regions on the diagram is labeled. There are also red capital letters on the diagram to help locate the various regions as they are discussed in the text. These red letters are not found on other iron-carbon diagrams.

FERRITE, the region marked A, contains very small amounts of carbon. Ferrite is also known as alpha ferrite or alpha iron. Ferrite can contain a maximum of .02 percent of carbon at 1341 °F (727 °C). As the temperature is increased to 1674 °F (912 °C), the amount of carbon in ferrite decreases to zero. Also, as the temperature of ferrite decreases to room temperature the amount of carbon in ferrite decreases.

Ferrite is present in all steel and cast iron. It always contains the same amount of carbon even though the carbon content of the steel or cast iron varies. Ferrite forms alone in regions A and E. Ferrite also forms along with cementite and is called pearlite in regions E, F, and G. Ferrite is formed when steel or cast iron is cooled below 1341 °F (727 °C). Ferrite has a body centered cubic structure and is both ductile and tough.

Pure CEMENTITE has a molecular formula of $Fe_3C$. Cementite contains 6.69 percent carbon. Because it has such a high carbon content, pure cementite is not shown on Fig. 26-5. As mentioned previously, as the carbon content in steel increases, the hardness and brittleness also increase. Cementite, with its high carbon content is very hard and brittle. Cementite is also known as iron carbide.

Cementite is found on the iron-carbon diagram, but in combination with austenite, region K, or pearlite, regions E, F, and G. Wherever the cementite is on the iron-carbon diagram, it always contains 6.69 percent carbon. Between 2106 °F (1152 °C) and 1341 °F (727 °C), region K, the cementite is mixed with austenite. Below 1341 °F (727 °C), the cementite can be found alone, regions F and G, or in combination with ferrite in the form of pearlite, regions E, F, and G. Cementite is thus formed in all steels and cast irons.

PEARLITE is a combination of ferrite and cementite as mentioned previously. The ferrite and cementite occur in alternating layers in the microstructure. The microstructure is shown in Fig. 26-6. Pure pearlite, line B, is formed at 1341 °F (727 °C) and .77 percent carbon. Pearlite always contains .77 percent carbon. If a steel contains less than .77 percent carbon, the pearlite is found along with ferrite, region E. If a steel or cast iron contains more than .77 percent carbon, cementite and pearlite form, in regions F and G.

AUSTENITE, region C, is the final important region on the iron-carbon diagram. Austenite is stable above 1341 °F (727 °C) and below 2800 °F (1538 °C). Austenite can contain up to 2.11 percent carbon. Even though austenite does not occur at room temperature, it is an important region when heat treating steels. Austenite has a

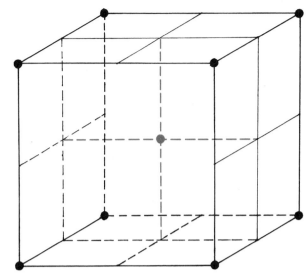

A. EXAMPLES: IRON, CHROMIUM, TUNGSTEN, AND TITANIUM.

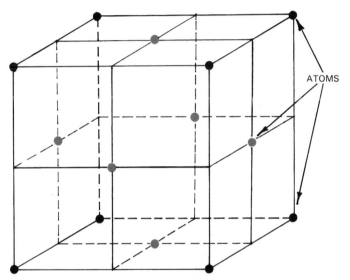

ATOMS

B. EXAMPLES: AUSTENITE (IRON AT HIGH TEMPERATURES), ALUMINUM, COPPER, NICKEL, SILVER, GOLD, AND LEAD.

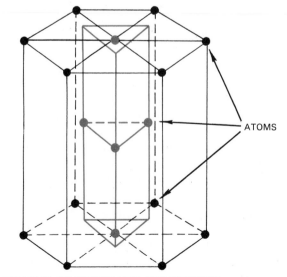

ATOMS

C. EXAMPLES: MAGNESIUM, ZINC, AND ZIRCONIUM.

Fig. 26-4. Three types of crystal structure. A—Body centered cubic. B—Face centered cubic. C—Hexagonal close-packed.

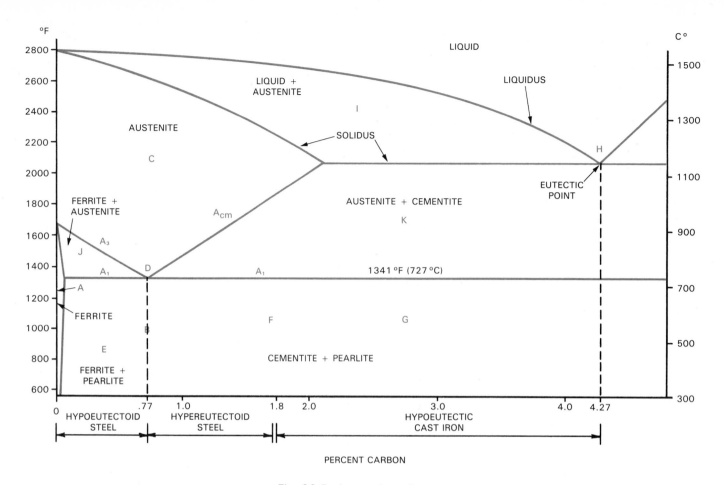

Fig. 26-5. Iron-carbon diagram.

face centered cubic structure which is different from the ferrite, body centered cubic structure. See Fig. 26-4 for the different structures. Austenite is also called GAMMA IRON.

The EUTECTOID POINT, point D, is an important point on the iron-carbon diagram. It occurs at .77 percent carbon and at 1341 °F (727 °C). When a steel is cooled through this point, pearlite is formed.

Very few steels have exactly the eutectoid composition. Most steels contain less than the eutectoid composition and are called HYPOEUTECTOID STEELS, region E. The microstructure of these steels has a combination of ferrite and pearlite. Those steels which have a higher carbon content than the eutectoid point are called HYPEREUTECTOID STEELS, region F. The microstructure of a hypereutectoid steel has a combination of cementite and pearlite.

The EUTECTIC POINT, point H, is another important point on the diagram. It occurs at 4.27 percent carbon and 2095 °F (1146 °C). The reaction at the eutectic point is similar to the reaction at the eutectoid point. At the EUTECTIC POINT, a liquid transforms into two solids, austenite and cementite. At the eutectoid point, a solid

(austenite) transforms into two new solids, ferrite and cementite, called pearlite.

Cast iron is formed when the amount of carbon is between 1.8 percent and 4.0 percent. These cast irons have less than the eutectic composition and are called HYPOEUTECTIC CAST IRONS, region G.

There are three remaining regions on the iron-carbon diagram. These are the liquid plus austenite region, the ferrite plus austenite region and the austenite plus cementite region. As a steel or cast iron is cooled through the LIQUID PLUS AUSTENITE REGION, region I, the liquid transforms into austenite. As a steel is cooled through the FERRITE PLUS AUSTENITE REGION, region J, the austenite transforms into ferrite. Also, as a steel or cast iron is cooled through the AUSTENITE PLUS CEMENTITE REGION, region K, the austenite transforms into cementite.

The important lines on the iron-carbon diagram are also labeled. The LIQUIDUS begins at 2800 °F (1538 °C) and goes down to 2095 °F (1146 °C) at 4.27 percent carbon, then increases again. The SOLIDUS also begins at 2800 °F (1538 °C) and slopes down to 2095 °F (1146 °C) at 2.11 percent carbon. The solidus remains at 2095 °F (1146 °C) as the carbon content increases above 2.11 percent. The horizontal line at 1341 °F (727 °C), the EUTECTOID TEMPERATURE, is the LOWEST CRITICAL TEMPERATURE for steels. This temperature is known as the $A_1$. The UPPER CRITICAL TEMPERATURE for hypoeutectoid steel is called the $A_3$. It begins at 1674 °F (912 °C) and slopes down to the eutectoid point. The upper critical temperature for a hypereutectoid steel is called the $A_{cm}$. This line extends from the eutectoid point up to the solidus line at 2.11 percent carbon. The lines $A_1$, $A_3$, and $A_{cm}$ are found on most iron-carbon diagrams.

Most steels contain less than 1 percent carbon. A small change in carbon can greatly change the characteristics of a steel. The carbon content is specified in 1/100 of 1 percent. Each 1/100 is called one point of carbon. Thus a steel containing .10 percent carbon is called a 10 point steel. One that has .85 percent is called an 85 point steel.

As an example of how a steel solidifies, follow the changes when a 20 point carbon steel is cooled through the complete temperature range. Begin with the steel as a liquid above 2800 °F (1538 °C). The molten metal cools until it reaches the liquidus line. At this point, solid austenite begins to form. The steel is now in the liquid plus austenite region, region I. It continues to cool until the solidus, where the steel is completely solid. The solid steel cools through the austenite region, region C. The steel crosses the $A_3$ line and begins to transform to ferrite. Ferrite continues to form until the eutectoid temperature is reached, the $A_1$. Now the remaining austenite forms pearlite. No further changes occur upon cooling to room temperature. The final structure has grains of ferrite and colonies of pearlite. The microstructure is seen in Fig. 26-7.

The iron-carbon diagram is also very important when heat treating steel. The strength and ductility of a steel can vary over a wide range depending on the heat treatment. See Chapter 27.

## 26-6. IDENTIFICATION OF IRON AND STEEL

Because of the effect on the properties of steel caused by the three variables of carbon content, temperature, and time, a welder must be able to determine quite accurately the composition of the steel being handled. The manufacturers' specifications for the particular steel are most desirable. Whenever possible, the welder should obtain these specifications and keep them on file. At the same time, the welder should mark the metal to correspond with filed information. Where manufacturers' specifications are not available, other methods may be used to determine the nature of the metal.

Many tests have been developed to identify iron and steel. The following tests are the most common for shop use:
1. Spark test (with the power grinder).
2. Oxyacetylene torch test.
3. Fracture test.
4. Color test.
5. Density or specific gravity test.
6. Ring or sound of the metal upon impacting with some other metal.
7. Magnetic test.
8. Chip test.

Of these eight tests, numbers three through eight should take place almost subconsciously, over a period of time, in the welder's mind, while working on various metals. The spark test and the gas torch test must be done under carefully prepared conditions. These tests indicate to a remarkably accurate degree the properties and constituents of the metal.

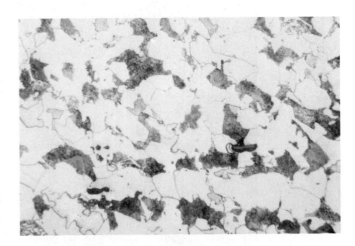

Fig. 26-6. Microstructure of pearlite. The dark regions are cementite, the light regions are ferrite.

Fig. 26-7. Microstructure of a 1020 steel. The white areas are ferrite grains. The other areas are pearlite.

## 26-7. SPARK TEST

A SPARK TEST METHOD of identifying metals is extensively used by welders to identify irons and steels. A power grinder is used as the test equipment. WHEN GRINDING YOU MUST ALWAYS WEAR GOGGLES. The grinder must be inspected to see that it is in good condition before proceeding with the test. A sample is tested by lightly touching the rim of the revolving wheel with the metal. The friction of the wheel surface as it contacts the metal, heats the particles removed to the incandescent (glowing) and burning temperature. The sparks resulting from the contact are found to differ in character for different steels. The lighter the contact, the better. One should use a black background to better identify the spark.

The theory of the spark test is when a metal is heated the different parts of the metal oxidize at different rates and the oxidation colors are different. Relatively pure iron when heated by the grinding wheel does not oxidize quickly. The sparks are long and fade out on cooling. The varying compounds of carbon and iron have different ignition temperatures when touched to the grinder. Therefore, the characteristics of the sparks differ as the carbon content of steel or cast iron increases. Four characteristics of the spark generally denote the nature and condition of the steel. The characteristics are:

1. Color of spark.
2. Length of spark.
3. Number of explosions (spurts) along the length of the individual sparks.
4. Shape of the explosions (forking or repeating).

For example, a mild steel containing 20 point carbon (.20 percent) will show a long white spark which will jump approximately 70 inches (1.78 m) from the power grinder (using a 12 in. [.30 m] wheel). Some of these sparks will suddenly explode, shooting off smaller sparks at approximately 45 degree angles to the direction of travel of the original spark. A 30 point carbon steel will have sparks almost identical to the 20 point carbon steel, with the exception that more spark lines will explode. The total length of these spark lines will decrease slightly. This serves to show that as the carbon content of steel increases, the explosions of the sparks become more frequent. Also, as the carbon content increases, the length of the spark is decreased. In Fig. 26-8, for example, a high carbon tool steel, containing 80 points of carbon, has a very short spark with explosions occurring very

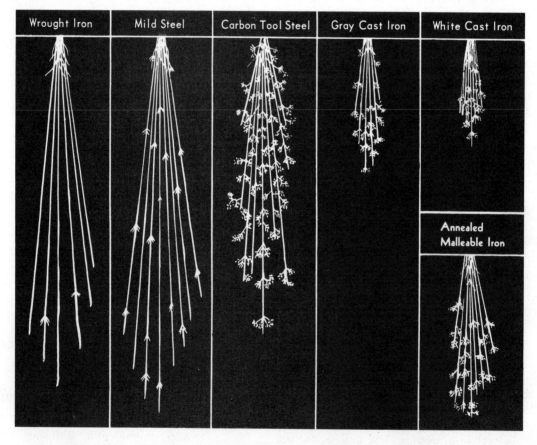

Fig. 26-8. Spark test for common cast irons and steels. (Norton Co.)

| Metal | Volume of Stream | Relative Length of Stream* | Color of Stream Close To Wheel | Color of Streaks Near End of Stream | Quantity of Spurts | Nature of Spurts |
|---|---|---|---|---|---|---|
| Wrought iron | Large | 65 in. (1.65 m) | Straw | White | Very few | Forked |
| Machine steel (AISI 1020) | Large | 70 in. (1.78 m) | White | White | Few | Forked |
| Carbon tool steel | Moderately large | 55 in. (1.40 m) | White | White | Very many | Fine, repeating |
| Gray cast iron | Small | 25 in. (0.64 m) | Red | Straw | Many | Fine, repeating |
| White cast iron | Very small | 20 in. (0.51 m) | Red | Straw | Few | Fine, repeating |
| Annealed mall. iron | Moderate | 30 in. (0.76 m) | Red | Straw | Many | Fine, repeating |

*Figures obtained with 12 in. (.31 m) wheel on bench stand are relative only. Actual length in each instance will vary with grinding wheel, pressure, etc.

Fig. 26-9. Table of spark test characteristics of common cast irons and steels.    (Norton Co.)

rapidly. The sparks dissipate themselves very quickly. See Fig. 26-9 for a table of spark test characteristics.

When higher carbon content metals are tested, you may become confused between extremely high carbon steel and cast iron because the spark is somewhat the same. Heat treatments have some effect on the nature of the spark. The cast iron spark, when leaving the point of contact, is a dull red and the spark jumps only 20 to 25 in. (508-635 mm) from the wheel. The spark from a high carbon steel of approximately 130 point carbon is white when it leaves the grinding wheel. The length of the spark is usually a little longer than the cast iron spark. Both of these metals show a small spark explosion at the end of the spark flight. The number of spark explosions in these two cases is considerably less than one would expect from such a high carbon content.

## 26-8. OXYACETYLENE TORCH TEST

Even if you know the physical composition and the chemical composition of a metal, you must also know whether the metal has good welding properties. For example, some cold rolled sheet steels may show very good physical and chemical properties. However, during some part of the manufacturing process, impurities have been added to it or certain work was done to the metal affecting its properties. The metal will be found not to melt and fuse readily. The final weld may be unsatisfactory. The usual cause of this condition is that there are impurities imbedded in the metal. The impurities are usually slag and roller dirt or excessive sulphur and phosphorus. For these reasons a welder should subject steel to the torch test.

The actual test consists of melting a puddle in the steel. If the metal is thin, the puddle penetrates through the thickness of the steel until a hole is formed. This puddling should be done with a neutral flame, held at the proper distance from the metal. The puddle should not spark excessively or boil. The puddle should be fluid and should possess good surface tension. The appearance on the edge of the puddle or hole indicates the weldability of the steel. If the metal that was melted has an even, shiny appearance upon solidification, the metal is generally considered as having good welding properties. However, if the molten metal surface is dull or has a colored surface, the steel is unsatisfactory for welding. The steel is also considered unsatisfactory for welding if the surface is rough, perhaps even broken up into small pits or porous spots.

This test is accurate enough for most welding. The test is very easily applied with the equipment on the job. The test determines the one thing that is fundamentally necessary in any welding job, that is, the weldability of the metal. Fig. 26-10 shows how this test is conducted. While performing the weldability test of the metal, it is important to note the amount of sparking emitted from the molten metal. A metal which emits few sparks has good welding qualities.

## 26-9. MISCELLANEOUS IDENTIFICATION TESTS

As an added precaution to the oxyacetylene torch tests, six other tests are occasionally used in the welding shop. The FRACTURE TEST is used extensively and consists of breaking a portion of the metal in two. If it is a repair job, the fractured surface may be inspected. The appearance of the surface where the metal is cracked shows the grain structure of the metal. If the grains are large, the metal is ductile and weak. If the grains are small, the metal is usually strong and has better ductility. Small grains are usually preferred. The fracture shows the color of the metal which is a good means of identifying one metal from another. The test also indicates the type of metal by the ease with which it may be fractured. Fig.

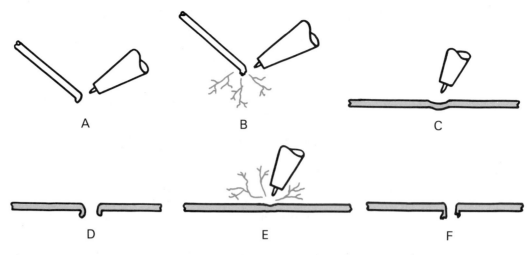

Fig. 26-10. Torch flame test: A—Good quality filler rod. B—Poor quality filler rod. C and D—Good quality base metal. E and F—Poor quality base metal.

26-11 shows the appearance of the fractured surface of several different metals.

The COLOR TEST separates two main divisions of metals. The irons and steels are indicated by their typical gray white color. Nonferrous metals come in two general color classifications of yellow and white. Copper may be rather easily identified by a welder and the same applies to brass and bronze. Aluminum is a white metal. Aluminum alloys, zinc, and the like, are all of somewhat the same silver-gray color although they may vary in shade.

The metals may also be differentiated by means of the weight or DENSITY TEST of the specimen. A good example of identification by density or specific gravity is identifying aluminum and lead. Roughly speaking, their colors are somewhat similar, but anyone may readily distinguish between the two metals because of their respective weights. Lead weighs about three times as much as aluminum.

The RING TEST, or the sound of the metal test, is an easy means of identifying certain metals after some experience with this method. It is used extensively for identifying heat-treated steels from annealed steels. It is also used to detect alloys from the virgin metal. An example is the difference between aluminum and duralumin, an alloy of aluminum and copper (2017). The pure aluminum sheet has a duller sound, or ring, than the duralumin which is somewhat harder and has a more distinct ringing sound.

The MAGNETIC TEST is an elementary test used to separate iron and steel metals from the nonferrous metals. Generally speaking, all steels are affected by magnetism while the nonferrous metals are not. However, some stainless steels are not magnetic.

Fig. 26-11. Fractured surfaces of several different metals.
(L-TEC Welding & Cutting Systems)

Another test which must be accompanied by considerable experience is the CHIPPING TEST of a metal. In this test, the cutting action of the chisel indicates the structure and heat treatment of the metal. Cast iron, for example, when being chip-tested, breaks off in small particles, whereas a mild steel chip tends to curl and cling to the original piece. Higher-carbon, heat-treated steels cannot be tested this way because of hardness.

A rough test between mild carbon steel and chrome-moly steel may be indicated by the relative hardness of the metals while being hacksawed.

## 26-10. IDENTIFICATION OF IRON AND STEEL ALLOYS

Extensive research has been carried on with iron and carbon steels in which other elements have been added to improve or bring out certain properties. The two main fields in which this development has taken place are in the stainless and low carbon alloy steels. These metals are sometimes difficult to identify from clean ordinary steels. The type of metal being welded must be known since its composition greatly affects its welding properties.

Alloys are added to plain iron-carbon steel to develop or improve the various physical properties of the metal. A common example includes the stainless steels. Alloying elements are used in these steels to make them corrosion proof, and to develop other desired physical properties such as toughness and strength.

These metals consist of various combinations of chromium and nickel, along with molybdenum, titanium, zirconium, aluminum, or tungsten.

These other elements are added to the steel to increase its strength, hardness, and/or toughness. A good example of this type of alloy is the use of tungsten. The addition of a small amount of tungsten to steel produces an extremely hard metal without sacrificing its other properties to any appreciable extent. Low carbon steels are improved by adding various alloying elements. These alloys serve the purpose of producing stronger steels at a minimum cost. They are called high strength, low alloy (HSLA) steels. Heat treatment is often of great importance when obtaining the best physical properties of alloy steels.

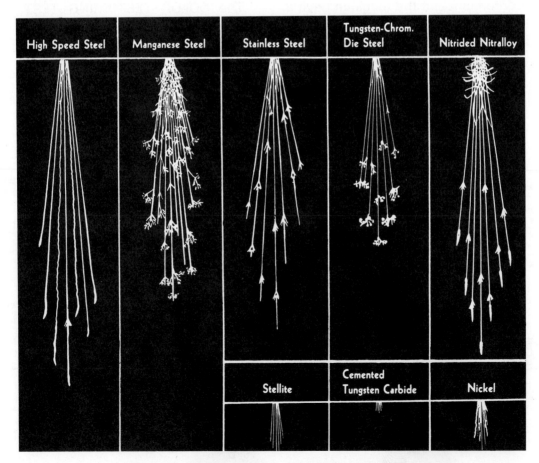

Fig. 26-12. Spark test for some alloy steels and alloying metals. (Norton Co.)

## 26-11. SPARK TEST FOR ALLOY STEELS

The appearance of the power grinder spark for alloy steels varies considerably, depending on the alloys included in the metal. The basic iron-carbon spark test is identical with the straight, iron-carbon steels. The number of the spark explosions increases as the length of the spark flight decreases. The color of the spark becomes brighter and brighter as the carbon content increases. The principle variations the alloy elements give to a particular steel are shown in the multitude of angled, branching sparks. These branching sparks are the result of the different alloys in the metal. These alloys also change the color of the spark.

As an example, a high-speed steel has only a slight indication of the spark explosion. Manganese in steel causes the sparks to shoot out at 45 degree angles to the flight of the original spark. These sparks also tend to explode (repeating sparks) producing the appearance of a leafless tree branch, as shown in Fig. 26-12.

Tungsten-chromium steels, which are used for high speed work, show the typical, high-speed steel spark with the exception that the spark turns to a chromium yellow (straw color) at the end of its flight. The chromium and the tungsten have a deteriorating effect on the carbon spark, causing it to be very fine or thin with repeating spurts. Fig. 26-13 lists the spark characteristics for alloy steels. The chromium and tungsten spark is also an interrupted one. That is, it disappears for a portion of its flight and then appears again.

## 26-12. TORCH TEST FOR ALLOY STEELS

It is difficult to give rigid specifications of the torch test for all alloy steels. However, it is generally known that the higher the alloy contents, the more difficult it is to weld the metal. Under the influence of the torch flame, the metals have a tendency to boil because of the action of the alloys in them. Special fluxes are often used when welding them.

Stainless steels are difficult to oxyfuel gas weld, and a flux is needed to produce good welds. Without the flux, the metal boils and a porous puddle is produced.

Manganese steel is often used for surfaces which are exposed to abrasive wear, such as power shovel buckets. The usual application of welding in this case is to build up the worn surfaces. The steel melts readily under the torch flame and presents few difficulties in welding.

Nickel steels can be identified under the torch flame by the metal boiling action. Nickel-chrome steels behave in somewhat the same manner.

## 26-13. MISCELLANEOUS TESTS FOR ALLOY STEELS

The COLOR TEST, RING TEST, MAGNETIC TEST, FRACTURE TEST, and the CHIP TEST are all applicable to alloy steels, but perhaps the easiest one is the color test. The different alloy constituents tend to change the color of the metal, which in certain compositions is very noticeable and characteristic.

| Metal | Volume of Stream | Relative Length of Stream* | Color of Stream Close To Wheel | Color of Streaks Near End of Stream | Quantity of Spurts | Nature of Spurts |
|---|---|---|---|---|---|---|
| High-speed steel (18-4-1) | Small | 60 in. (1.52 m) | Red | Straw | Extremely few | Forked |
| Austenitic manganese steel | Moderately large | 45 in. (1.14 m) | White | White | Many | Fine, repeating |
| Stainless steel (Type 410) | Moderate | 50 in. (1.27 m) | Straw | White | Moderate | Forked |
| Tungsten-chromium die steel | Small | 35 in. (0.89 m) | Red | Straw** | Many | Fine, repeating |
| Nitrided Nitralloy | Large (curved) | 55 in. (1.40 m) | White | White | Moderate | Forked |
| Stellite | Very small | 10 in. (0.25 m) | Orange | Orange | None | |
| Cemented tungsten carbide | Extremely small | 2 in. (0.05 m) | Light orange | Light orange | None | |
| Nickel | Very small*** | 10 in. (0.25 m) | Orange | Orange | None | |
| Copper, brass, aluminum | None | | | | | |

*Figures obtained with 12 in. (.31 m) wheel on bench stand and are relative only. Actual length in each instance will vary with grinding wheel, pressure, etc.
**Blue-white spurts.    ***Some wavy streaks.

Fig. 26-13. Table of spark test characteristics of alloy steels.
(Norton Co.)

Stainless steels, for example, have a distinct silvery color, which sets them apart from the other steel alloys. Some of the metals are magnetic, whereas others are not. Identification can be accomplished to a certain extent by using this test. The chip test is not used extensively for identifications because of the similarity in the appearance of practically all the stainless steel alloys. However, some of these alloys are considerably harder than others, and the chip test will bring this out quite clearly.

The ring or metallic sound of the metal is another test which can be applied to steel alloys with considerable accuracy. However, it is recommended that this test be used in conjunction with other tests.

## 26-14. NUMBERING SYSTEMS FOR STEELS

The Society of Automotive Engineers (SAE) has long been a leader in standardizing in industry. The Society has developed a numbering system for identifying practically all steels. The code is based on a number of four digits, for example, 2315. The first digit 2 classifies the steel, the number 2 representing nickel steels. The second digit represents the percent of the alloy metal in the steel, that is, 2315 represents 3.25 to 3.75 percent nickel.

The last two digits represent the carbon content of the metal in points (hundredths of one percent), as for example, .10 to .20 percent carbon, Fig. 26-14 is a table of sample SAE steels. SAE first number designations for steel and steel alloys are as follows:

1XXX Carbon steels.
11XX Special, sulphur-carbon steels that have free cutting properties.
12XX Phosphorus-carbon steels
13XX Manganese steels.
2XXX Nickel steels.
3XXX Nickel-chromium steels.
4XXX Molybdenum steels.
5XXX Chromium steels.
6XXX Chromium-vanadium steels.
7XXX Tunsten steels.
9XXX Silicon-manganese steels.

Note: The X represents the space for the second, third, and fourth number of the SAE designation.

SAE-AISI CARBON STEEL COMPOSITION NUMBERS

| SAE | AISI | CARBON RANGE | MANGANESE RANGE | PHOSPHORUS MAX. | SULFUR MAX. |
|---|---|---|---|---|---|
| 1010 | C1010 | 0.08-0.18 | 0.30-0.60 | 0.040 | 0.050 |
| 1015 | C1015 | 0.13-0.18 | 0.30-0.60 | 0.040 | 0.050 |
| 1020 | C1020 | 0.18-0.23 | 0.30-0.60 | 0.040 | 0.050 |
| 1025 | C1025 | 0.22-0.28 | 0.30-0.60 | 0.040 | 0.050 |
| 1030 | C1030 | 0.28-0.34 | 0.60-0.90 | 0.040 | 0.050 |
| 1035 | C1035 | 0.32-0.38 | 0.60-0.90 | 0.040 | 0.050 |
| 1040 | C1040 | 0.37-0.44 | 0.60-0.90 | 0.040 | 0.050 |
| 1045 | C1045 | 0.43-0.50 | 0.60-0.90 | 0.040 | 0.050 |
| 1050 | C1050 | 0.48-0.55 | 0.60-0.90 | 0.040 | 0.050 |
| 1055 | C1055 | 0.50-0.60 | 0.60-0.90 | 0.040 | 0.050 |
| 1060 | C1060 | 0.55-0.65 | 0.60-0.90 | 0.040 | 0.050 |
| 1065 | C1065 | 0.60-0.70 | 0.60-0.90 | 0.040 | 0.050 |
| 1070 | C1070 | 0.65-0.75 | 0.60-0.90 | 0.040 | 0.050 |
| 1075 | C1075 | 0.70-0.80 | 0.40-0.70 | 0.040 | 0.050 |
| 1080 | C1080 | 0.75-0.88 | 0.60-0.90 | 0.040 | 0.050 |
| 1085 | C1085 | 0.80-0.93 | 0.70-1.00 | 0.040 | 0.050 |
| 1090 | C1090 | 0.85-0.98 | 0.60-0.90 | 0.040 | 0.050 |
| 1095 | C1095 | 0.90-1.03 | 0.30-0.50 | 0.040 | 0.050 |

RESULPHURIZED CARBON STEELS

| SAE | AISI | CARBON RANGE | MANGANESE RANGE | PHOSPHORUS MAX. | SULFUR MAX. |
|---|---|---|---|---|---|
| 1115 | C1115 | 0.13-0.18 | 0.60-0.90 | 0.040 max. | 0.08-0.13 |
| 1120 | C1120 | 0.18-0.23 | 0.70-1.00 | 0.040 | 0.08-0.13 |
| 1125 | C1125 | 0.22-0.28 | 0.60-0.90 | 0.040 | 0.08-0.13 |
| 1140 | C1140 | 0.37-0.44 | 0.70-1.00 | 0.040 | 0.08-0.13 |

Fig. 26-14. Typical SAE-AISI carbon steel composition numbers.

## REPHOSPHURIZED AND RESULPHURIZED CARBON STEELS

| SAE | AISI | CARBON RANGE | MANGANESE RANGE | PHOSPHORUS MAX. | SULFUR MAX. |
|------|-------|------------|----------------|-----------------|-------------|
| 1211 | C1211 | 0.13 max. | 0.60-0.90 | 0.07-0.12 | 0.08-0.15 |
| 1213 | C1213 | 0.13 max. | 0.70-1.00 | 0.07-0.12 | 0.24-0.33 |

## MANGANESE STEELS

| SAE | AISI | CARBON RANGE | MANGANESE RANGE | PHOSPHORUS MAX. | SULFUR MAX. | SILICON MAX. |
|------|------|------------|----------------|-----------------|-------------|--------------|
| 1330 | 1330 | 0.28-0.33 | 1.60-1.90 | 0.040 | 0.040 | 0.20-0.35 |
| 1335 | 1335 | 0.33-0.38 | 1.60-1.90 | 0.040 | 0.040 | 0.20-0.35 |
| 1340 | 1340 | 0.38-0.43 | 1.60-1.90 | 0.040 | 0.040 | 0.20-0.35 |
| 1345 | 1345 | 0.43-0.48 | 1.60-1.90 | 0.040 | 0.040 | 0.20-0.35 |

## NICKEL STEELS

| SAE | AISI | CARBON RANGE | MANGANESE RANGE | PHOSPHORUS MAX. | SULFUR MAX. | NICKEL RANGE |
|------|------|------------|----------------|-----------------|-------------|--------------|
| 2315 | ---- | 0.10-0.20 | 0.30-0.60 | 0.040 | 0.050 | 3.25-3.75 |
| 2330 | ---- | 0.25-0.35 | 0.50-0.80 | 0.040 | 0.050 | 3.25-3.75 |
| 2340 | ---- | 0.35-0.45 | 0.60-0.90 | 0.040 | 0.050 | 3.25-3.75 |
| 2345 | ---- | 0.40-0.50 | 0.60-0.90 | 0.040 | 0.050 | 3.25-3.75 |
| ---- | 2515 | 0.10-0.20 | 0.30-0.60 | 0.040 | 0.050 | 4.75-5.25 |

## NICKEL CHROMIUM STEELS

| SAE | AISI | CARBON RANGE | MANGANESE RANGE | PHOSPHORUS MAX. | SULFUR MAX. | NICKEL RANGE | CHROMIUM RANGE | SILICON |
|------|-------|------------|----------------|-----------------|-------------|--------------|----------------|---------|
| 3140 | 3140 | 0.38-0.43 | 0.70-0.90 | 0.040 | 0.040 | 1.10-1.40 | 0.55-0.75 | 0.20-0.35 |
| 3310 | E3310 | 0.08-0.13 | 0.45-0.60 | 0.025 | 0.025 | 3.25-3.75 | 1.40-1.75 | 0.20-0.35 |

## MOLYBDENUM STEELS

| SAE | AISI | CARBON RANGE | MANGANESE RANGE | PHOSPHORUS MAX. | SULFUR MAX. | CHROMIUM RANGE | NICKEL RANGE | MOLYBDENUM RANGE | SILICON |
|------|------|------------|----------------|-----------------|-------------|----------------|--------------|------------------|---------|
| 4130 | 4130 | 0.28-0.33 | 0.40-0.60 | 0.040 | 0.040 | 0.80-1.10 | ---- | 0.15-0.25 | 0.20-0.35 |
| 4140 | 4140 | 0.38-0.43 | 0.75-1.00 | 0.040 | 0.040 | 0.80-1.10 | ---- | 0.15-0.25 | 0.20-0.35 |
| 4150 | 4150 | 0.48-0.53 | 0.75-1.00 | 0.040 | 0.040 | 0.80-1.10 | ---- | 0.15-0.25 | 0.20-0.35 |
| 4320 | 4320 | 0.17-0.22 | 0.45-0.65 | 0.040 | 0.040 | 0.40-0.60 | 1.65-2.00 | 0.20-0.30 | 0.20-0.35 |
| 4340 | 4340 | 0.38-0.43 | 0.60-0.80 | 0.040 | 0.040 | 0.70-0.90 | 1.65-2.00 | 0.20-0.30 | 0.20-0.35 |
| 4615 | 4615 | 0.13-0.18 | 0.45-0.65 | 0.040 | 0.040 | ---- | 1.65-2.00 | 0.20-0.30 | 0.20-0.35 |
| 4620 | 4620 | 0.17-0.22 | 0.45-0.65 | 0.040 | 0.040 | ---- | 1.65-2.00 | 0.20-0.30 | 0.20-0.35 |
| 4815 | 4815 | 0.13-0.18 | 0.40-0.60 | 0.040 | 0.040 | ---- | 3.25-3.75 | 0.20-0.30 | 0.20-0.35 |
| 4820 | 4820 | 0.18-0.23 | 0.50-0.70 | 0.040 | 0.040 | ---- | 3.25-3.75 | 0.20-0.30 | 0.20-0.35 |

## CHROMIUM STEELS

| SAE | AISI | CARBON RANGE | MANGANESE RANGE | PHOSPHORUS MAX. | SULFUR MAX. | CHROMIUM RANGE | SILICON |
|-------|--------|------------|----------------|-----------------|-------------|----------------|---------|
| 5120 | 5120 | 0.17-0.22 | 0.70-0.90 | 0.040 | 0.040 | 0.70-0.90 | 0.20-0.35 |
| 5140 | 5140 | 0.38-0.43 | 0.60-0.90 | 0.040 | 0.040 | 0.70-0.90 | 0.20-0.35 |
| 5150 | 5150 | 0.48-0.53 | 0.60-0.90 | 0.040 | 0.040 | 0.70-0.90 | 0.20-0.35 |
| 52100 | E52100 | 0.95-1.10 | 0.25-0.45 | 0.025 | 0.025 | 1.30-1.60 | 0.20-0.35 |

Fig. 26-14 (Continued)

| SAE | AISI | CARBON RANGE | MANGANESE RANGE | PHOSPHORUS MAX. | SULFUR MAX. | CHROMIUM RANGE | VANADIUM | SILICON |
|---|---|---|---|---|---|---|---|---|
| 6118 | 6118 | 0.16-0.21 | 0.50-0.70 | 0.040 | 0.040 | 0.50-0.70 | 0.10-0.15 | 0.20-0.35 |
| 6120 | 6120 | 0.17-0.22 | 0.70-0.90 | 0.040 | 0.040 | 0.70-0.90 | 0.10 min. | 0.20-0.35 |
| 6150 | 6150 | 0.48-0.53 | 0.70-0.90 | 0.040 | 0.040 | 0.80-1.10 | 0.15 min. | 0.20-0.35 |

NICKEL CHROMIUM MOLYBDENUM STEELS

| SAE | AISI | CARBON RANGE | MANGANESE RANGE | PHOSPHORUS MAX. | SULFUR MAX. | CHROMIUM RANGE | MOLYBDENUM RANGE | NICKEL RANGE | SILICON |
|---|---|---|---|---|---|---|---|---|---|
| 8115 | 8115 | 0.13-0.18 | 0.70-0.90 | 0.040 | 0.040 | 0.30-0.50 | 0.08-0.15 | 0.20-0.40 | 0.20-0.35 |
| 8615 | 8615 | 0.13-0.18 | 0.70-0.90 | 0.040 | 0.040 | 0.40-0.60 | 0.15-0.25 | 0.40-0.70 | 0.20-0.35 |
| 8720 | 8720 | 0.18-0.23 | 0.70-0.90 | 0.040 | 0.040 | 0.40-0.60 | 0.20-0.30 | 0.40-0.70 | 0.20-0.35 |
| 8822 | 8822 | 0.20-0.25 | 0.75-1.00 | 0.040 | 0.040 | 0.40-0.60 | 0.30-0.40 | 0.40-0.70 | 0.20-0.35 |

SILICON MANGANESE STEELS

| SAE | AISI | CARBON RANGE | MANGANESE RANGE | PHOSPHORUS MAX. | SULFUR MAX. | SILICON RANGE | NICKEL RANGE | CHROMIUM RANGE | MOLYBDENUM RANGE |
|---|---|---|---|---|---|---|---|---|---|
| 9260 | 9260 | 0.55-0.65 | 0.70-1.00 | 0.040 | 0.040 | 1.80-2.20 | - - - - | - - - - | - - - - |
| 9840 | 9840 | 0.38-0.43 | 0.70-0.90 | 0.040 | 0.040 | 0.20-0.35 | 0.85-1.15 | 0.70-0.90 | 0.20-0.30 |

CHROMIUM NICKEL STEELS

| SAE | AISI | CARBON | MANGANESE MAX. | SILICON MAX. | PHOSPHORUS | SULFUR | CHROMIUM RANGE | NICKEL RANGE | MOLYBDENUM |
|---|---|---|---|---|---|---|---|---|---|
| 30304 | 304 | 0.08 max. | 2.00 | 1.00 | 0.040 | 0.040 | 18.00 min. | 8.00 min. | - - - - |
| 30317 | 317 | 0.10 max. | 2.00 | 1.00 | 0.040 max. | 0.040 max. | 16.00-18.00 | 10.00 min. | 2.00 max. |
| 51410 | 410 | 0.15 max. | 1.00 | 1.00 | 0.040 max. | 0.040 max. | 11.50-13.50 | 0.60 max. | 0.60 max. |

Fig. 26-14. (Continued)

Some special divisions of corrosion and heat resistant steels are as follows:

30XXX   Nickel-Chromium steels.
51XXX   Chromium steels.

The American Iron and Steel Institute (AISI) has also prepared standards for steels and steel alloys. In general their code system is identical with that of the SAE code system.

The American Society for Testing and Materials (ASTM) has standardized a code system for identifying and labeling steels.

## 26-15. NONFERROUS METALS

NONFERROUS METALS and their alloys are metals which do not contain iron (ferrous). They may contain small quantities of iron, as an alloying element, but they are not considered ferrous metals. This group includes copper, brass, bronze, aluminum, solder, stellite, lead, zinc, nickel, etc.

These metals can be identified by various means. They have distinctive colors. They are nonmagnetic and are usually relatively soft metals. Nonferrous metals will not spark when touched to the grinding wheel.

CAUTION - Grinding of nonferrous metals is not recommended except for the extremely short intervals of time required for spark testing. The oxides of some nonferrous metals are toxic and therefore the operator must wear an air filtering breathing apparatus and protective clothing. The grinder must be equipped with an adequate exhaust system. Nonferrous metal will generally clog the grinding wheel and prevent it from grinding properly.

## 26-16. COPPER

Copper is an element of the metal family which has many uses because of its electrical and thermal conductivity and its ability to resist corrosion. Most copper has a reddish brown color. Copper melts at a temperature of approximately 1980°F (1083°C). This is higher than the melting temperature of silver and considerably lower than the melting temperature of iron, as shown in the chart in Fig. 26-15.

Manufacturing processes are such that commercial copper usually contains sulphur, phosphorus, and silicon as its impurities. Each of these impurities has a tendency to make the copper more brittle and to reduce its weldability. However, a very small amount of phosphorus in copper is an aid to the welding of the metal. It is helpful because the dissolving property that phosphorus has for copper oxides permits it to act as a flux.

The only copper recommended for fusion welding purposes is deoxidized copper. This copper has had a very small amount of silicon added to it. Silicon has the property of dissolving whatever cupric oxides are present in the metal. Enough silicon is added to the copper during its manufacture so that an excess is left in the copper after deoxidizing action takes place. If this amount is too much, as mentioned previously, the copper tends to become brittle.

Another feature of copper, which is typical of practically all nonferrous metals, is its behavior called HOT SHORTNESS. As copper is heated to its melting temperature, the copper becomes very weak at a certain temperature even though it is still a solid. The slightest shock or weight will tend to distort the metal unless it is very firmly supported and firmly clamped. This support prevents distortion while the metal is passing through this "hot shortness" temperature. The approximate point at which hot shortness will occur may be determined by the welder by the color of the metal as it is heated. When the color of the metal becomes a medium cherry red, the copper is at the hot shortness temperature.

## 26-17. BRASS

Brass is an alloy of copper and zinc, although small amounts of other metals are frequently added. The amount of the zinc in the alloy may vary from 10 to 40 percent. A very common alloy of copper and zinc to form brass is 70 percent copper and 30 percent zinc. This metal is used principally because of its acid resistance qualities, its appearance, and because it is a good brazing alloy. There are two common types of brass: one type is called MACHINE BRASS, which contains 32 to 40 percent zinc, and RED BRASS which contains from 15 to 25 percent zinc. Some additional metals which are added to improve the physical properties of brass are tin, manganese, iron, and lead. These make brass a triple alloy.

Brass may be identified by its color which is an opaque yellow.

## 26-18. BRONZE

Bronze is an alloy of copper and tin. A common ratio is 90 percent copper and 10 percent tin. Bronze is more coppery in color than brass. It behaves much like brass when being welded. Generally speaking, the welder uses the same filler rod and the same flux for both bronze and brass.

Like brass, bronze is highly resistant to corrosion. Because of its attractive appearance, it is often used for decorative parts and objects.

## 26-19. ALUMINUM

Aluminum is an element of the metal family known for its electrical conductivity, heat conductivity, resistance to corrosion, and light weight. It is obtainable either in rolled (wrought) or cast form. Aluminum may be combined with many other metals to form alloys. The different alloys that are added to aluminum are listed in Fig. 26-16 along with the Aluminum Association designation for the alloys.

In its pure form, the metal has a white color. It is very ductile in the rolled aluminum sheet form. Cast aluminum is very brittle. The strength of the pure metal is considerably less than that of steel. Its melting temperature is approximately 1220°F (660°C) as shown in Fig. 26-15. This metal also has a critical point called HOT SHORTNESS. For that reason, it must be carefully supported when being welded. An element which may be added to aluminum to decrease its "hot shortness" is silicon.

| Metal | Melting Temperatures | |
|---|---|---|
| | °F | °C |
| Aluminum | 1217 | 659 |
| Armco iron | 2795 | 1535 |
| Bronze 90 Cu Bronze 10 Sn | 1562-1832 | 850-1000 |
| Brass 90 Cu 10 Zn | 1868-1886 | 1020-1030 |
| Brass 70 Cu 30 Zn | 1652-1724 | 900-940 |
| Copper | 1981 | 1083 |
| Iron | 2786 | 1530 |
| Lead | 621 | 327 |
| Mild Steel | 2462-2786 | 1350-1530 |
| Nickel | 2646 | 1452 |
| Silver | 1761 | 960 |
| Tin | 450 | 232 |
| Zinc | 786 | 419 |

Fig. 26-15. Melting temperatures of some common metals.

Aluminum sheet metal may be obtained in several qualities and grades. The purest commercial aluminum contains 99 1/2 percent aluminum, while the more popular commercial grades contain 99 percent aluminum. Two metals which are added to aluminum to increase its desirable physical qualities are manganese and magnesium. Amounts of these are relatively small, varying from one to five percent.

Special alloys of aluminum have been developed for aircraft work. These alloys are noted for their high tensile strength, but they are usually very difficult to weld satisfactorily. One of these alloys is called Duralumin (2017). Any mechanical working of the metals tends to increase their brittleness.

Due to the activity of the metal, aluminum oxidizes very readily upon being heated. Aluminum oxide melts at about 5000°F (2760°C). Therefore, wire brushing or chemicals must be used to remove the oxide. The oxide is more dense that the molten metal and settles into the weld pool, causing a porous weld. Therefore, special fluxes must be used at all times during the welding process unless an inert gas process is used.

Another feature of aluminum which adds to the difficulty of welding is that it does not change in color before it reaches the melting temperature. In other words, the metal upon being heated maintains the same color, but when reaching the melting point, it suddenly becomes liquid. When welding aluminum, the welder can determine the melting temperature of the metal by using the filler rod to scratch the surface to reveal any softening. If the welder uses the ANSI lens shade recommendation of Shade No. 4 for gas welding aluminum, the weld is more visible and a slight change in color can be noted as the melting point is reached.

Aluminum may be identified by its silvery white color when fractured, and by a comparison of weights with other metals. Aluminum is about one third the weight of iron for a given volume.

Another test for aluminum is to burn some chips of the metal. Aluminum will burn to a black ash.

Magnesium may be mistaken for aluminum. However, magnesium chips when heated will actually ignite and will form a white ash. CAUTION MUST BE TAKEN WHEN IGNITING MAGNESIUM CHIPS SO THAT ONLY A VERY SMALL QUANTITY OF THE CHIPS ARE IGNITED, SINCE MAGNESIUM CHIPS AND POWDER BURN VIOLENTLY.

## 26-20. HARD SURFACING METALS

Certain metal alloys have the property of extreme hardness. There are several different types:
1. Ferrous metal with up to 20 percent alloying elements such as chromium, tungsten, and manganese.
2. Ferrous metal with over 20 percent alloying elements such as chromium, tungsten, and manganese. (Cobalt and nickel are sometimes added.)
3. Nonferrous metal with alloying elements such as cobalt, chromium, and tungsten.
4. Tungsten carbide (fused), with some other alloying elements.
5. Granules of tungsten carbide.

See Chapter 28 for information on metal surfacing.

These metals are difficult to identify. It is best to keep them in their identifying packages. Fig. 26-12 shows the characteristic spark of stellite and for cemented tungsten carbide, two hard surfacing materials.

These metals range from extremely hard to very tough. Their best application, therefore, is as a thin coating on metal of a more ductile nature. This combination produces a long wearing surface and also one of great strength.

## 26-21. SOFT SURFACING METALS

The soft surfacing metals are usually the nonferrous metals. These may be identified by the methods described earlier in this chapter.

The properties of the deposited surface metal depend upon the surfacing metal and the method used to apply the metal to the surface of the parent metal. Surfacing metals can be applied by:
1. Dipping.
2. Electroplating.
3. Spraying.
4. Brazing or soldering.

## 26-22. NEW METALS

Many of the formerly exotic metals are used as alloying elements to improve the properties of the more common metals. A few of these metals are: COLUMBIUM, TITANIUM, LITHIUM, BARIUM, ZIRCONIUM, TANTALUM, BERYLLIUM, NOBELIUM, and PLUTONIUM.

The manufacture and use of these metals is increasing. Titanium, zirconium, and columbium are widely used as alloys in steel. Welding engineers are constantly developing and improving methods to weld and braze these metals. See Chapter 20 for information on the welding of these metals.

## 26-23. TITANIUM

Titanium is an important structural metal. It is the fourth most abundant metal exceeded only

| The Aluminum Association Alloy Group Designation | Major Alloying Element | Examples |
|---|---|---|
| 1XXX | 99% Aluminum | 1100 |
| 2XXX | Copper | 2014, 2017, 2024 |
| 3XXX | Manganese | 3003 |
| 4XXX | Silicon | 4043 |
| 5XXX | Magnesium | 5052, 5056 |
| 6XXX | Magnesium and Silicon | 6061 |
| 7XXX | Zinc | 7075 |
| 8XXX | Other elements | |

Fig. 26-16. Alloys added to aluminum and the designation for each. The 1XXX group contains at least 99 percent pure aluminum.

by aluminum, iron, and magnesium. The commercially pure titanium melts at about 3035 °F (1668 °C). Many of the alloys are weldable under certain conditions. This metal has good weight to strength ratio, high temperature properties, and it is very corrosion resistant. The aircraft, chemical, and transportation industries use titanium in various applications. It is about 67 percent heavier than aluminum and about 40 percent lighter than stainless steels. It retains its strength very well up to 1000 °F (538 °C).

Titanium-carbon alloys find wide acceptance. The carbon content varies from .015 to 1.1 percent. The titanium-carbon alloys become more brittle as the carbon content increases, but they also become more corrosion resistant. The maximum tensile strength is reached at about .4 percent carbon and is approximately 112,000 psi (pounds per square inch) (772.2 MPa). However, .04 percent carbon alloy has a strength of 92,000 psi (634.3 MPa).

Titanium alloys using tungsten and carbon have a tensile strength of about 130,000 psi (896.3 MPa). In combination with aluminum, it has a tensile strength of about 114,000 psi (786 MPa). Some of the other alloys added to titanium and the resulting tensile strength are given below:

Chromium-nickel
195,000 psi (1344.5 MPa)

Chromium-molybdenum
175,000 psi (1206.6 MPa)

Chromium-tungsten
150,000 psi (1034.2 MPa)

Manganese-aluminum
160,000 psi (1103.2 MPa)

Titanium-manganese alloys, such as 8 percent manganese, are popular in aircraft structure. Some other alloys consist of 3 percent aluminum and 1/2 percent of manganese, 6 percent alu-minum and 4 percent vanadium, and 5 percent aluminum and 2 1/2 percent tin.

## 26-24. LITHIUM

The welding industry is interested in lithium. When this metal is alloyed with some other brazing metal, the brazing can usually be performed without flux. For example, titanium can be successfully brazed in an inert atmosphere using an alloy of about 98 percent silver and 2 percent lithium. Lithium has a melting temperature of 367 °F (186 °C).

## 26-25. ZIRCONIUM

Zirconium is a rare metal which is found in nature combined with silicon as zirconium silicate ($ZrSiO_4$) commonly known as the semiprecious jewel zircon. Zirconium oxide melts at 4892 °F (2700 °C) and is used as linings for high temperature furnaces.

This metal oxidizes easily. Its strength is affected by oxygen, nitrogen, and hydrogen with which it may combine. Zirconium is used as an alloying element in alloy steels.

Zircalbys are a family of zirconium alloys containing about 1 1/2 percent tin, nickel, chromium and up to 1/2 percent iron. These alloys have improved corrosion resistance and higher strength than unalloyed zirconium.

Because zirconium readily unites with oxygen and nitrogen when heated, it is welded best in an inert gas filled chamber. The shielding gases normally used are argon or helium.

## 26-26. BERYLLIUM

Beryllium is a light metal with a density (1.845 g/cc) slightly higher than the density of magnesium. It is often used as an alloying element with other metals.

It has been used as an alloying element in copper and nickel to increase elasticity and strength characteristics.

Beryllium has a high thermal and electrical conductivity and a high heat absorption rate. Because of these qualities, it is difficult to weld.

Some beryllium copper alloys are very strong and are used to replace forged steel tools in places when an explosive atmosphere may be present since these alloys are non-sparking.

Beryllium has been welded by the inert-gas, resistance, ultrasonic, electron-beam, and diffusion welding processes. Brazing and soldering methods of joining have also been used.

Beryllium is used with magnesium to reduce its tendency to burn during melting and casting.

This metal and its compounds have dangerously poisonous (toxic) properties, and special precautions must be taken when working with the metal and alloys which use this metal.

## 26-27.  REVIEW OF SAFETY

Spark testing metals for identification requires the tester to wear goggles and/or face shield. The grinder must be in good condition, including balanced grinding wheels and wheel rest clearance not to be greater than 1/16 to 1/8 in. (1.59 - 3.18 mm).

When a torch flame or an arc is used to identify the metal by its melting and sparking behavior, all the safety precautions for gas welding and/or arc welding must be followed. See Chapters 3, 4, 9, and 10.

## 26-28.  TEST YOUR KNOWLEDGE

Write your answers on a separate sheet of paper. Do not write in this book.

1. What is meant by an alloy steel?
2. How much carbon does a medium carbon steel contain?
3. Carbon usually combines with iron to form _____, which is also called _____.
4. What is elongation?
5. What characteristics does nickel give to steel?
6. Does an alloy solidify at one temperature, or over a range of temperatures?
7. What is meant by a critical temperature of a metal?
8. How do the properties of steel change as carbon content increases?
9. Pearlite is a combination of _____ and _____.
10. The carbon content of ferrite is _____; the carbon content of cementite is _____; the carbon content of pearlite is _____.
11. Using the iron-carbon diagram, describe the different regions a 1050 steel passes through as it cools from a liquid to a solid at room temperature.
12. What is present in the microstructure of a 1050 steel at room temperature?
13. Name five methods of identifying metals.
14. What can be learned from the spark test?
15. What can be learned from the oxyacetylene torch test?
16. What is the percent of carbon or points of carbon in an SAE 1045 steel?
17. What do the initials ASTM, SAE, and AISI represent?
18. Describe the term "hot shortness."
19. What element may be added to aluminum to reduce hot shortness?
20. What alloying element is added to the aluminum series 2XXX? 5XXX?

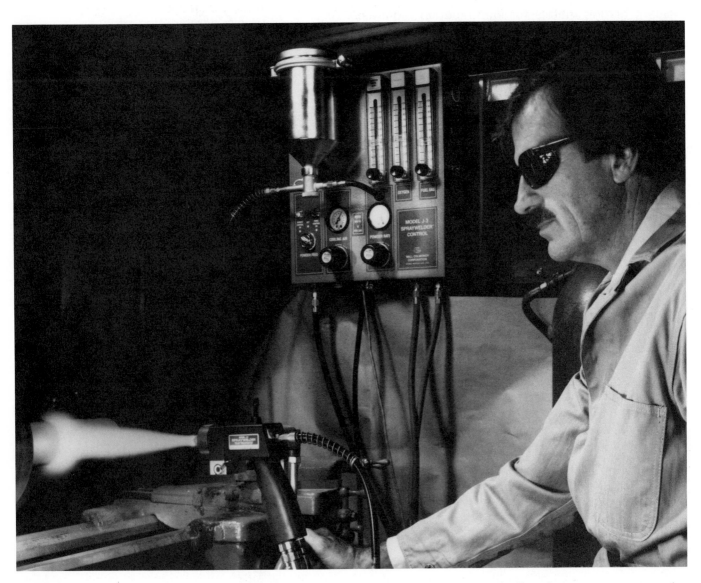

Industry photo. Spraywelder is used to apply and fuse powdered alloys. Fingertip controls allow the operator to start and stop the flow of combustible gases and powder flow.    (Wall Colmonoy Corp.)

# 27 HEAT TREATMENT OF METALS

The welder should be particularly interested in the heat treatment of metals. It is important to know what welding does to the heat treatment of the metal. It is good to understand the effect of welding on the physical properties of the metal. It is necessary to know if a metal must be preheated for welding, and if the metal should be heated during the welding operation. It is useful to learn what heat treatment procedure to use to bring back, as nearly as possible, the original properties of the metal (postweld heat treatment).

The most common application of postweld heat treating in a welding shop is STRESS RELIEVING. It relieves the metal of internal stresses and strains, caused by the expansion and contraction during welding. It also improves the properties of the metal in the weld and in the heat affected zone. Most structural welding involves only the knowledge of how a metal may be annealed or stress relieved. There are actually three types of heating: preheating, concurrent or interpass heating, and postweld heat treating of a metal. In large shops and in manufacturing plants these are a part of the procedure determined by the engineering staff.

In repair welding an extensive and accurate knowledge of all phases of heat treating is required by the welder.

## 27-1. THE PURPOSES OF HEAT TREATMENT

All metals can be heat treated. Some metals are affected very little by heat treating, but some, particularly most steels, are greatly affected. Heat treating may serve the following purposes:
1. Develop ductility.
2. Improve machining qualities.
3. Relieve stresses.
4. Change grain size.
5. Increase hardness or tensile strength.
6. Change chemical composition of metal surface as in casehardening.
7. Alter magnetic properties.
8. Modify electrical conduction properties.
9. Induce toughness.
10. Recrystallize metal which has been cold worked.

When heat treating there are four factors of great importance:
1. The temperature to which the metal is heated.
2. The length of time that the metal is held at that temperature.
3. The rate (speed) at which the metal is cooled.
4. The material surrounding the metal when it is heated, as in casehardening.

When a weld is made, the metal in and around the weld joint is heated to a variety of temperatures depending on its distance from the weld joint. This drop in the metal temperature from the weld joint outward is called a TEMPERATURE GRADIENT. In steels, some of the metal in the area of the weld is heated through only one critical point, some two, and some may be heated to only 500°F (260°C) as shown in Fig. 27-1.

Because of the uneven heating, the strength, ductility, grain size, and other metal properties may vary greatly in the weld and in the heat affected zone (HAZ).

The welding operator may use PREHEATING and/or CONCURRENT HEATING (added heating while welding) to reduce temperature gradients in the weld area. Heating the metal before welding

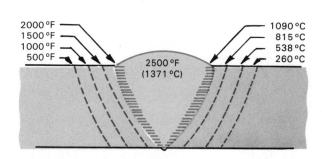

Fig. 27-1. Temperature zones in the weld area.

will help to prevent internal stresses and strains which may cause weld failure. The welder may use CONTINUOUS HEATING for the same purpose. POSTWELD HEATING is used to relieve stresses and bring all the metal in the area of the weld to the same heat treat condition.

## 27-2. METHODS OF HEATING

Heating metals is a complex operation. All metals expand when heated and contract on cooling. A rather drastic change in volume takes place as metals are heated above their critical temperatures, or are cooled below their critical temperatures. Recall that a CRITICAL TEMPERATURE is where a phase change occurs or a crystalline structure change takes place. Refer to Heading 26-4.

Unequal heating or cooling causes unequal expansion and contraction. Expansion and contraction often cause a warping of the structure. The ability to warp metals by heating them is not always a disadvantage. Warped articles can be straightened by the proper local heating of the metal.

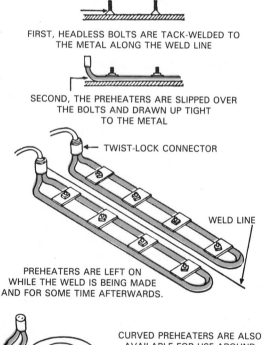

HOW CHROMALOX TUBULAR PREHEAT UNITS ARE INSTALLED AND OPERATED

FIRST, HEADLESS BOLTS ARE TACK-WELDED TO THE METAL ALONG THE WELD LINE

SECOND, THE PREHEATERS ARE SLIPPED OVER THE BOLTS AND DRAWN UP TIGHT TO THE METAL

TWIST-LOCK CONNECTOR

WELD LINE

PREHEATERS ARE LEFT ON WHILE THE WELD IS BEING MADE AND FOR SOME TIME AFTERWARDS.

CURVED PREHEATERS ARE ALSO AVAILABLE FOR USE AROUND CIRCULAR WELDS SUCH AS FLANGE PLATES, PATCHES, ETC.

Fig. 27-2. Electrical resistance units for preheating. (Edwin L. Wiegand Div., Emerson Electric Co.)

Heating can be done by using a flame such as:
1. Air-fuel gas.
2. Oxyfuel gas.
Or, it can be done by:
1. Electrical resistance heating.
2. Induction heating.
3. Furnace heating.

Electrical resistance heating may be used to heat a part by putting electrical resistance units against the parts to be heated as shown in Fig. 27-2. Electric resistance units may also be used to heat a furnace.

Induction heating is a means of heating using high frequency alternating current to create eddy currents in the metal. This method heats a metal throughout its thickness. See Fig. 27-3.

General heating is usually done in a furnace. The furnace can be heated with direct heat or indirect heat. The furnace may be heated by air-fuel gas, by electrical induction, or by electrical resistance units.

## 27-3. METHODS OF COOLING

The speed and evenness of cooling a metal object determines to a great extent the physical properties of the metal.

Cooling may be accomplished in various ways. The heat loss in cooling is usually a combination of:
1. Convection.
2. Conduction.
3. Radiation.

Cooling in a gas occurs mainly by convection and radiation. Cooling in a liquid occurs mainly by convection and conduction. Cooling can only be

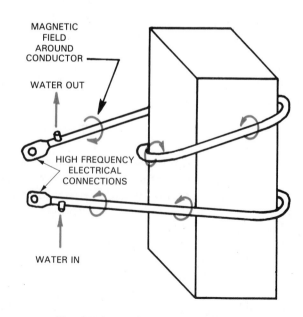

MAGNETIC FIELD AROUND CONDUCTOR

WATER OUT

HIGH FREQUENCY ELECTRICAL CONNECTIONS

WATER IN

Fig. 27-3. An induction heating coil.

done to approximately the temperature of the cooling medium. The colder the cooling medium the faster the cooling will take place. Cooling in a gas (by convection) is the slowest. Cooling in a cold liquid is fastest. The liquid can be any desired temperature. For example: water at room temperature (70 °F, 21 °C), ice water (32 °F, 0 °C), or liquid nitrogen (−230 °F, −146 °C) (cryogenics).

Air is a common gas cooling medium. Two methods of cooling in air are:
1. Thermal convection cooling in still air.
2. Forced convection cooling with fans.

Water is a common liquid cooling medium. Two methods of cooling in water are:
1. Submerging in a water tank.
2. Spraying water on the metal to be cooled.

Oil, molten low temperature metals, liquid air, or liquid nitrogen are other cooling liquids.

Another means of cooling and controlling the cooling of metals is to clamp the part in either a water cooled or refrigerant cooled fixture.

One problem in cooling a metal is that the surface of the metal cools faster than the inside of the metal.

## 27-4. CARBON CONTENT OF STEEL

Steel may be obtained with various carbon contents and alloying metals. Fig. 27-4 shows the carbon content of some common items made of steel. The family of steels starts with low carbon steel (almost wrought iron). As the carbon content increases, the steel becomes harder, stronger, and more brittle until approximately 1.70 percent carbon is reached. This is the maximum percent of carbon that will combine with iron to form steel. Iron with carbon contents over 1.70 percent forms cast iron.

Heading 26-2 describes the physical properties of steel.

## 27-5. CRYSTALLINE STRUCTURE OF STEEL

Practically all heat treatments deal with the crystalline structure of steel or the grain size. The heat treatments are also concerned with the distribution of cementite and ferrite in the metal. Cementite is commonly called iron carbide. In 10 point carbon steel, which contains 1/10 of one percent carbon, most of the carbon is in chemical combination with iron, forming cementite, $Fe_3C$. Since there is not much carbon in a 10 point steel, there is very little cementite. There is, however, a considerable amount of free iron which is called ferrite. Ferrite is very ductile. Steel containing a large amount of ferrite cannot be hardened. The only heat treatment possible is to alter the grain structure and size. Heating this steel to the upper transformation temperature, the $A_3$ critical temperature, does not greatly affect the steel's hardness. This is because it is the amount of cementite which determines the hardness. The $A_3$ critical temperature may be seen on the iron-carbon diagram, Fig. 26-5.

Steels with a carbon content above 35 points can be hardened. The final product and hardness of the steel greatly depends on the cooling rate from the austenite region. The products which result from different cooling rates are shown in a Time-Temperature-Transformation (T-T-T) diagram. See Fig. 27-5.

To understand the Time-Temperature-Transformation diagram three different cooling rates for steel will be examined. In Fig. 27-5, the steel is cooled from the austenite region. It is cooled at the desired rate to give the desired product. The first line crossed as the metal cools is where the product begins to form. The second line crossed indicates where the transformation is complete.

| ARTICLE | CARBON CONTENT |
|---|---|
| Axles | .40 |
| Boiler Plate | .12 |
| Boiler Tubes | .10 |
| Castings, Low Carbon Steel | Less Than .20 |
| Casehardening Steel | .12 |
| Cold Chisels | .75 |
| Files | 1.25 |
| Forgings | .30 |
| Gears | .35 |
| Hammers | .65 |
| Lathe Tools | 1.10 |
| Machinery Steel | .35 |
| Metal Tools | .95 |
| Nails | .10 |
| Pipe, Steel | .10 |
| Piano Wire | .90 |
| Rails | .60 |
| Rivets | .05 |
| Set Screws | .65 |
| Saws for Wood | .80 |
| Saws for Steel | 1.55 |
| Shaft | .50 |
| Springs | 1.00 |
| Steel for Stamping | .90 |
| Tubing | .08 |
| Wire, Soft | .10 |
| Wood Cutting Tools | 1.10 |
| Wood Screws | .10 |

Fig. 27-4. Steels and how the carbon content varies depending on the use of the steel.

If a steel is cooled slowly (line A on Fig. 27-5) from the austenite region, pearlite forms. The microstructure is shown in Fig. 27-6. If the steel is quenched in ice water (line C) martensite is formed. See Fig. 27-7. If the steel is initially quenched to a moderate temperature of 675°F (357°C) and then slow cooled, bainite is formed. Each of these products has different characteristics. They are listed in Fig. 27-8. Martensite, which is cooled the fastest, is the hardest structure. Pearlite, which is cooled slowly, has the best ductility.

The steels with higher carbon contents and higher alloy contents form martensite easier than low carbon, low alloy steels. This is true also for weldments as the weld metal cools.

The weld metal may contain pearlite or martensite or both. The grains may be large or small, and the carbides may be large or they may be small and dispersed.

The heat affected zone has large and small

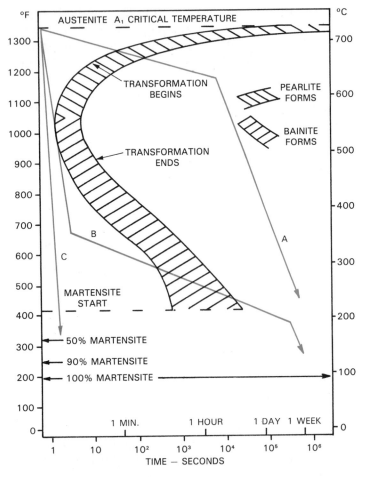

Fig. 27-5. Time-Temperature-Transformation (T-T-T) diagram for a eutectoid steel. If the steel is cooled slowly (line A), pearlite forms. If the steel is first quenched and then cooled slowly, bainite forms (line B). If the steel is fully quenched (line C), martensite forms. Note the time is in powers of ten. $10^4$ means $10 \times 10 \times 10 \times 10 = 10,000$.

Fig. 27-7. Microstructure of martensite magnified 500 diameters.

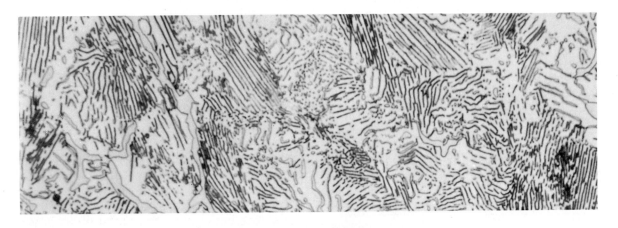

Fig. 27-6. Microstructure of a high carbon steel with pearlite grain structure. This is magnified 1000 diameters.

| STRUCTURE | CHARACTERISTICS |
|-----------|-----------------|
| Martensite | Extremely hard, strong, and brittle. |
| Bainite | Good strength, not as hard as martensite, ductile and tough. |
| Pearlite | Good ductility, not as hard as bainite, fairly strong and tough. |

Fig. 27-8. A table of the characteristics for the various types of steel structures.

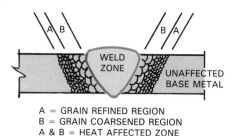

A = GRAIN REFINED REGION
B = GRAIN COARSENED REGION
A & B = HEAT AFFECTED ZONE

Fig. 27-9. The heat affected zone (HAZ) has two regions: the grain refined region and the grain coarsened region. The grain coarsened region was heated to a higher temperature than the grain refined region.

grains. The small grain region is next to the unaffected base metal. This area was heated to just above the $A_3$ critical temperature. New refined grains were formed. The region cools and is called the GRAIN REFINED REGION. The next region is the GRAIN COARSENED REGION. See Fig. 27-9.

The temperature here was well above the $A_3$ critical temperature. The grains had time to grow to a large size. Grain growth is a function of time and temperature. The higher the temperature and the longer the time, the larger the grains will be. The grains are the smallest as they cross the $A_3$ critical temperature upon heating. This is illustrated in Fig. 27-10. Examples 2 and 3 show a steel which is heated to a temperature well above the $A_3$ critical temperature. The grain size is large. In example 4, the steel is heated to a temperature just above the $A_3$ critical temperature. This produces a very small grain size. Steel upon cooling below the $A_3$ critical temperature will retain the largest grain size achieved.

The various regions in the heat affected zone and the weld metal have different properties. Postweld heat treating is used to make the different regions more uniform.

There are a number of different heat treatments used in industry today. These include annealing, normalizing, quench and tempering, thermal stress relieving, and spheroidizing.

## 27-6. ANNEALING STEEL

The term annealing is a common term in heat treating and includes several types of heat treating operation. Generally speaking, ANNEALING is considered to be that type of heat treatment

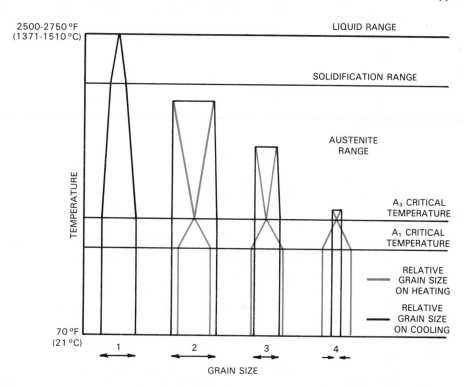

Fig. 27-10. Influence of heating on the grain size of steel. A steel that is heated to a high temperature has large grains. A steel that is heated to just above the $A_3$ critical temperature and then cooled has small grains.

which leaves the metal in a soft condition.

An annealing operation may be performed for several reasons:

1. To soften the metal so it may be cold worked.
2. To make the steel machinable.
3. To relieve the internal stresses and strains produced while the metal was being shaped or welded.

The general procedure for fully annealing most steels is as follows:

1. Heat the steel 50-100 °F (28-56 °C) above the $A_3$ critical temperature. See Fig. 27-11.
2. Hold for a period of time. A general rule is to allow one hour at the desired temperature for every inch of thickness.
3. Furnace cool the steel at a controlled rate. Cool to 50 °F (28 °C) below the $A_1$ critical temperature.
4. Air cool to room temperature. This process is shown graphically in Fig. 27-12, view A.

As mentioned previously, the final grain size depends on time and temperature. The grain size also affects the physical properties of the steel. As seen in Fig. 27-13, the elastic limit (yield strength) and ductility increase as the grain size decreases. Note that as the number of grains per inch increases the grain size decreases. (The ELASTIC LIMIT is the maximum load that will still produce elastic [linear] deformation in the steel. The YIELD STRENGTH is the load [ksi or MPa] where plastic deformation begins.)

Another type of annealing is the PROCESS ANNEAL. It is usually performed on low carbon steels. The steel is heated but remains below the $A_1$ critical temperature. It is held at this temperature long enough for softening to occur. Because of the lower temperatures involved, it is faster and more economical than a full anneal.

## 27-7. NORMALIZING STEEL

If the extremely soft, full annealed structure is not required, the normalizing heat treatment can be used. It saves time and money compared to annealing.

NORMALIZING is used to make the internal structure of a steel more uniform. A normalized structure will usually have more uniform mechanical properties and better ductility.

The procedure for normalizing is as follows:

1. Heat the steel to 100 °F (56 °C) above the upper critical temperature. See Fig. 27-11.
2. Hold the steel at the desired temperature until the temperature is uniform throughout.
3. Cool in still air to room temperature.

The normalizing process is shown in Fig. 27-12, view B.

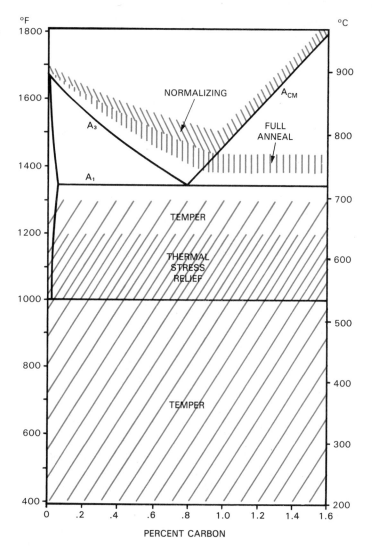

Fig. 27-11. The heat treatment temperatures for various carbon steels. To use the chart, find the percent carbon of the steel to be heat treated on the lower horizontal axis. Go directly up to the desired heat treatment. The proper heat treatment temperature is found on the vertical axis.

## 27-8. QUENCH AND TEMPERING STEEL

A steel that is quenched very fast from the austenite region forms martensite. Martensite is very hard and strong, but it is brittle.

TEMPERING is used to improve the ductility and toughness of martensite. Some strength and hardness is lost. The final structure has very good strength and hardness, along with good toughness and ductility.

A quench and tempered structure is produced as follows:

1. Heat the steel into the austenite region. (The steel may already be in this region due to welding.)
2. Quench very rapidly to form martensite.
3. Reheat to 400 ° - 1300 °F (204 ° - 704 °C) to temper the martensite.

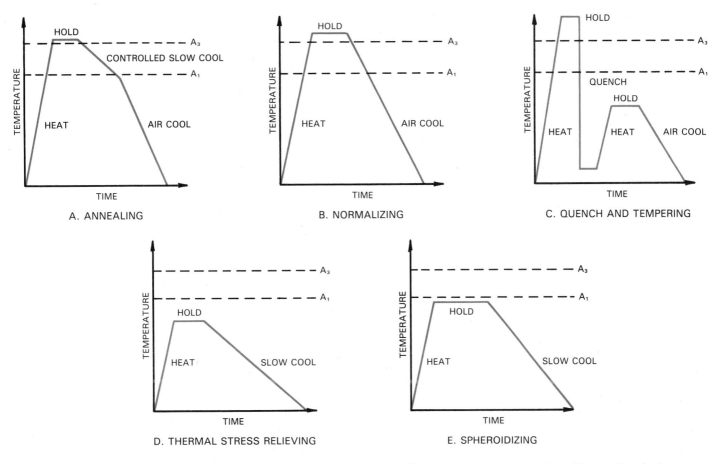

Fig. 27-12. Diagrams showing the five heat treatment processes. The hold time is long enough for the steel to obtain a uniform temperature. The hold time for spheroidizing is much longer to allow the cementite to form spheres.

4. Cool in air to room temperature.

This process is shown in Fig. 27-12, view C. Notice that the tempering temperatures are below the $A_1$ critical temperature. This is shown also on Fig. 27-11.

## 27-9. THERMAL STRESS RELIEVING

THERMAL STRESS RELIEVING is used to reduce the residual stresses in a steel. These stresses may be the result of cold working or welding.

Thermal stress relieving is similar to tempering. The steel is heated to 1000° - 1200°F (537° - 649°C). This is below the $A_1$ critical temperature. The results of this heat treatment are that the residual stresses are reduced and ductility is improved.

The procedure for thermal stress relieving is as follows:
1. Heat the steel to the desired temperature.
2. Hold at this temperature.
3. Slowly cool back to room temperature.

This process is shown in Fig. 27-12, view D. Fig. 27-14 shows resistance coils in place on a large tank. These will be used to preheat the metal

before welding and to stress relieve it after welding.

## 27-10. SPHEROIDIZING

SPHEROIDIZING is a process to improve the ductility of a high carbon steel. The cementite forms into small, separate spheres. The majority of the structure is ferrite. The microstructure is shown in Fig. 27-15. The result of spheroidizing a steel is that a high carbon steel has good ductility and can be formed or machined.

The procedure for spheroidizing is as follows:
1. Heat the steel to a temperature just below the $A_1$ critical temperature.
2. Hold at this temperature to allow the cementite to form spheres.
3. Slowly cool back to room temperature.

This process is shown in Fig. 27-12, view E.

## 27-11. HARDENING STEEL

Steels of less than 35 point carbon content cannot be successfully hardened by heat treatment. The maximum amount of carbon a steel contains is 170 points of carbon. Within these

| NO. OF GRAINS PER LINEAR INCH (25.4 mm) | ELASTIC LIMIT | | REDUCTION OF AREA, PERCENT |
|---|---|---|---|
| | psi | MPa | |
| 210 | 44,000 | 303 | 20 |
| 222 | 44,500 | 307 | 22 |
| 322 | 47,000 | 324 | 35 |

Fig. 27-13. Effects of grain size on elastic limit and ductility. The reduction in area is an indicator of a metal's ductility or its ability to stretch before breaking. The higher the percent of area reduction the higher the ductility of the metal.

extremes it is possible to obtain almost any degree of hardness by using the proper carbon content and the proper heat treatment. It is also common practice to obtain a refinement of the grain structure along with the hardness.

To obtain the maximum hardness, the steel is heated just above its $A_3$ critical temperature and then quenched rapidly. The critical temperatures for steel with various carbon contents are shown in Fig. 27-16. As mentioned in Heading 27-5, the rate of cooling determines to a great extent the hardness and brittleness of the metal. Cold water, air, and molten metal baths are used as mediums for cooling metal. The water cooling results in extreme brittleness along with maximum hardness. Air cooling and molten metal bath cooling tends to relieve much of the brittleness but at a sacrifice of some hardness.

Fig. 27-15. Microphotograph of the grain structure of spheroidized high carbon steel. The spheres are cementite; the background is ferrite.

Hardness of steel depends on the distribution and structure of the cementite throughout the steel. When cooled rapidly there is no time for the cementite to clump. As the rate of cooling is slowed, the cementite has time to clump leaving ductile ferrite which is softer than cementite.

Fig. 27-14. Resistance heating coils used on a large storage tank. The heating coils are used to preheat the joint and to thermally stress relieve the welded structure after welding. (Duraline Div. of J.B. Nottingham & Co., Inc.)

| PERCENT CARBON | CRITICAL TEMPERATURE FOR HARDENING AND FULL ANNEALING | | COMMERCIAL ANNEALING | |
|---|---|---|---|---|
| | °F | °C | °F | °C |
| .10 | 1675-1760 | 913-960 | | |
| .20 | 1625-1700 | 885-927 | | |
| .30 | 1560-1650 | 849-899 | | |
| .40 | 1500-1600 | 816-871 | | |
| .50 | 1450-1560 | 788-849 | 1020 | (549) |
| .60 | 1440-1520 | 782-827 | to | to |
| .70 | 1400-1490 | 760-810 | 1200 | (649) |
| .80 | 1370-1450 | 743-788 | | |
| .90 | 1350-1440 | 732-782 | | |
| 1.00 | 1350-1440 | 732-782 | | |
| 1.10 | 1350-1440 | 732-782 | | |
| 1.30 | 1350-1440 | 732-782 | | |
| 1.50 | 1350-1440 | 732-782 | | |
| 1.70 | 1350-1440 | 732-782 | | |
| 1.90 | 1350-1440 | 732-782 | | |
| 2.00 | 1350-1440 | 732-782 | | |
| 3.00 | 1350-1440 | 732-782 | | |
| 4.00 | 1350-1440 | 732-782 | | |

Fig. 27-16. Critical temperatures for various carbon steels.

## 27-12. TEMPERING A COLD CHISEL

A common tempering operation in a laboratory is to heat treat a cold chisel. Most cold chisels are made of approximately 75 point carbon steel and must have the following properties:
1. The cutting edge must be extremely hard.
2. The metal just back of the cutting edge must be hard and tough, not brittle.
3. The body of the chisel and the hammering end must also be tough.

To obtain these various properties from the same carbon content steel, heat treat as follows:
1. Heat the cold chisel slowly to just above its critical temperature (1350°F, 732°C, cherry red).
2. Allow enough time for the body of the chisel to assume this temperature throughout its thickness; then quench about one inch of the cutting edge end of the chisel in cold water.
3. Wait until the end in the water turns dark; then withdraw the chisel quickly from the water while the other end is still cherry red.
4. Heat from the cherry red end will travel to the chilled cutting edge, reheating it slowly. With a small pad faced with emery cloth, polish the flat surfaces near the cutting end of the chisel. Be careful of burns.

Now watch the polished surface carefully. As the heat from the shank of the chisel travels into this part, the polished surface will gradually change color. It will first become straw, then

bronze. As the temperature rises the surface will turn purple because of the oxidation of the metal. The purple color indicates the steel has been reheated to approximately 600°F (316°C). Now place the cutting edge in the water slowly. If the shank of the chisel has cooled below a cherry red color as it should have done, the complete chisel may be immersed in a bucket of cold water.

If the timing is correct on the heat treatment, and if the original temperature is not too high, the chisel will have a hard cutting edge, with a fairly hard and tough body. If the original temperature was too high, the chisel will be brittle and will crack easily because of the large crystalline structure. The cutting edge will be too hard and will chip off when used. If the cutting edge was requenched before it should have been, the cutting edge will be hard and the edge will break. If the shank of the chisel is quenched too soon (before its color has become a dark red), the shank will be too hard and brittle and will crack when being used. A chisel with brittle qualities is an extremely dangerous tool because if it cracks or chips the flying particles of metal may inflict injuries.

## 27-13. SURFACE HARDENING

Surface hardening is widely used today in many manufacturing operations. In many applications, such as gears, shafts, connection rods, and the like, it is advisable to produce as hard a wearing surface as possible. The internal portion of the structure must remain ductile and tough. The hard surface provides long wear and maintains an exact contour while the tough interior insures that the part will withstand shocks without breaking.

One method which is commonly used to surface harden a metal is FLAME HARDENING. A part that is to be flame hardened is first heat treated to produce a tough, ductile structure throughout. The surfaces to be hardened are then put in a flame hardening machine. A multiple-tipped oxyfuel flame is passed over the surface to heat it quickly to a high temperature. Following the flame, a heavy stream of water quickly cools the entire surface, Fig. 27-17. The depth of the hardness can be controlled by controlling the temperature of the surface and the cooling rate. Flame hardening is a rapid and economical means of hardening a metal surface.

Two additional methods of surface hardening a metal are to use a laser or an electron beam gun. The process is similar to flame hardening except a laser or an electron beam is used instead of a flame. The laser or electron beam is directed at the surface to be surface hardened. Both systems can be accurately focused so that small or difficult

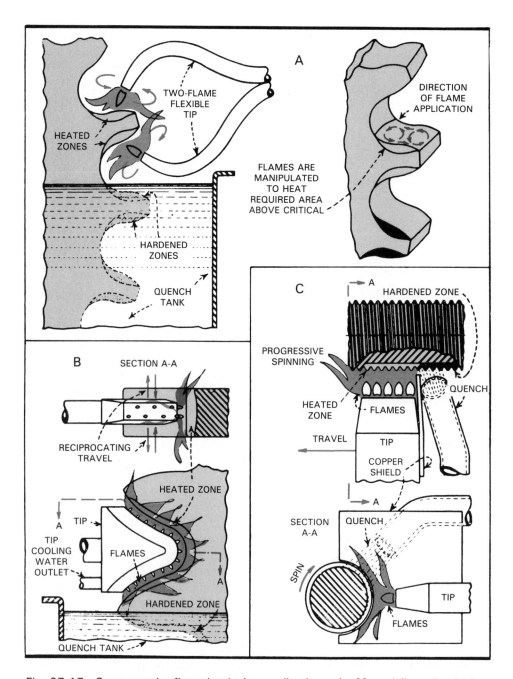

Fig. 27-17. Some popular flame hardening applications. A—Manual flame hardening allows concentration of heat at areas requiring maximum depth of hardness. B—Root and opposing faces of adjacent teeth are heated with contour tip. C—Progressive spinning is employed for fine threads, using water cooled nonquenching tip.

areas can be surface hardened. The depth of hardening can be controlled better with a laser or electron beam than with oxyfuel heating. These processes are much faster than oxyfuel heating, but very expensive.

## 27-14. CASEHARDENING

Another solution to the problem of producing a metal article with a tough interior and a hard, wear resistant surface is known as CASEHARDENING. The article is made out of low carbon steel, so any machining required may be done easily. The surfaces to be casehardened are exposed to carbon while they are at a high temperature. This is achieved by placing the part that is to be casehardened in a furnace. The furnace contains a carbon atmosphere, usually carbon monoxide (CO). In the high temperature of the furnace, the low carbon steel absorbs some of the carbon into its surface. The carbon penetrates into the part at about 1/64 in. (.4 mm) per hour. The surface of the steel part now contains more carbon than the body of the part.

In some cases it is not desirable to caseharden a particular portion of the part that is placed in the furnace. To prevent an area from absorbing carbon, the area is plated with copper.

The casehardened part is heat treated using the quench and temper process. The steel is heated to its critical temperature, then quenched. This produces a very hard surface. The interior of the part is still low carbon steel. It has refined (small) grains. As stated in Heading 27-6 and Fig. 27-13, as the grain size decreases, the strength and ductility of a steel increases. The interior of a casehardened part is thus strong and tough.

The part is finally tempered to improve the toughness of the hard outer surface. A photograph of a casehardened gear tooth is shown in Fig. 27-18.

## 27-15. HEAT TREATING TOOL STEELS

Tool steels are all high carbon steels. Some examples of the use of tool steels are chisels, hammers, saws, springs, and lathe tools. The carbon content of these steels is shown in Fig. 27-4.

When these tool steels are formed, the metal must be ductile. To obtain good ductility the tool steel is either annealed or the spheroidizing process is used. Spheroidized steel is more ductile than an annealed tool steel.

These tool steels must all be very hard when used. To produce the hardness the steel is heat treated using the quench and tempering process. Different tempering temperatures are used to obtain the desired combination of hardness and toughness.

To properly heat treat a tool steel, the welder must know the carbon content of the steel and the heat treatment process to be used. Fig. 27-11 can be used to determine the correct heat treating temperature for the steel.

To determine the correct heat treatment temperature in Fig. 27-11, find the carbon content of the steel on the lower (horizontal) axis. Follow a vertical line up to the desired process. Then follow across horizontally to the temperature scale on the left or right. For example, determine the proper temperature to fully anneal a .90 percent carbon metal tool. Find .90 on the horizontal scale. Go directly up to the full anneal region. Go directly across to the temperature scale. The correct temperature to fully anneal a .90 percent carbon tool is 1400 °F (760 °C).

## 27-16. HEAT TREATING ALLOY STEELS

As alloying elements are added to steel, the critical temperature of the steel is changed. Alloy-

Fig. 27-18. A cross section of a casehardened steel gear tooth, showing casehardened surface. (General Motors Corp.)

ing elements usually lower the critical temperatures. When heat treating an alloy steel, the welder cannot just consider the amount of carbon in the steel. The alloy content of the steel must also be known.

It is almost impossible for a welder to know what alloys are in a steel. The only sure way to know is to obtain the percentages from the steel manufacturer. Even with the alloying elements known, it is very difficult to select the correct temperature for a heat treatment. The best way to heat treat an alloy steel is to request and follow the manufacturer's recommendation.

## 27-17. HEAT TREATING CAST IRON

There are four main types of cast irons. These are white cast iron, gray cast iron, ductile cast iron, and malleable cast iron. These four types are similar in reference to the carbon content of each, but their properties are quite different. Ductile cast iron has magnesium added as an alloying element. The reason for the differences in the other three cast irons are the different cooling rates and heat treatments.

WHITE CAST IRON is formed when a casting is cooled quickly. It can also be formed by heat treating a cast iron. The cast iron is heated to a temperature of 1650 to 2050 °F (899 to 1121 °C) and quenched (rapidly cooled).

GRAY CAST IRON is formed when a casting is cooled slowly. It can also be formed by heating the casting to a temperature of 1850 to 2050° (1010 to 1121 °C) and cooling the casting slowly. The heat treatment is used after a gray cast iron is welded. The weld metal and the heat affected zone are usually much harder than the rest of the casting. This heat treatment makes the structure and the properties of the casting more uniform.

The heat treatment that is most often used is heat treating a white cast iron to form MALLEABLE IRON. Malleable iron is ductile and also machinable whereas white cast iron is not. To form malleable iron, a white cast iron is heated to 1450 to 1650 °F (788 to 899 °C). It is held at this temperature for 24 hours for each inch of casting thickness. While the casting is at this temperature, the carbon separates into irregular spheres of carbon. The casting is slow cooled to room temperature. The structure of malleable iron contains small spheres of carbon and the remainder of the casting is low carbon ferrite. This structure is ductile or malleable.

If MALLEABLE CAST IRON is welded, the weld metal and the heat affected zone will lose their ductile properties and become white cast iron. The welded casting would have to be heat treated again which is a long process. One way to prevent the casting from forming white cast iron is to braze or braze weld the casting. This process keeps the casting at a temperature low enough to prevent the formation of white cast iron.

## 27-18. HEAT TREATING COPPER

Copper and copper alloys become hard and brittle when they are mechanically worked. However, the metal can be made ductile again by annealing. To anneal copper, it should be heated to a temperature between 1400 and 1800 °F (760 and 982 °C). It is then water quenched. This heat treatment brings back the copper's original ductile qualities.

It is important to be careful when heating copper to its annealing temperature because copper undergoes a physical phenomenon or change, called HOT SHORTNESS. At this temperature the copper suddenly loses its tensile strength. If the copper is not adequately supported or if it is subjected to strain, it will fracture easily.

Some copper alloys are known for their high tensile strength and high hardness. To obtain the high strengths the copper alloy is first annealed. It is then cold worked, usually rolled. The alloy is then age hardened. This age hardening involves heating the alloy to 600 to 950 °F (316 to 510 °C) for a number of hours. One copper-beryllium alloy has a tensile strength of 173,000 psi (1193 MPa).

## 27-19. HEAT TREATING ALUMINUM

Aluminum is similar to copper. Aluminum becomes harder and more brittle when it is cold worked. Aluminum also has the characteristic of hot shortness so one must support the part as it is being heated.

There are a number of heat treatments used for aluminum and its alloys. A letter-number code is used to signify the various heat treatments. The

| H | | strain hardening |
| O | | annealed |
| W | | solution heat treated only |
| T | | thermally heat treated |
| | T1 | naturally aged |
| | T2 | annealed (for cast products only) |
| | T3 | solution heat treated and cold worked |
| | T4 | solution heat treated and naturally aged |
| | T5 | artificially aged (from a cast condition) |
| | T6 | solution heat treated and artificially aged |
| | T7 | solution heat treated and stabilized |
| | T8 | solution heat treated, cold worked and artificially aged |
| | T9 | solution heat treated, artificially aged and then cold worked |
| | T10 | artifically aged and then cold worked |

Fig. 27-19. Letter-number code for the heat treatments used on aluminum alloys. Each 'T' number is a subcategory of the broad classification 'T'.

common heat treatments and their letter-number codes are shown in Fig. 27-19. The letter-number code follows the alloy designation. See Fig. 26-16 for the aluminum alloy designations. Some examples are as follows: 1100-O, 2017-T4; 6061-T6; 7075-W.

STRAIN HARDENING is a cold working process. Some examples of strain hardening are rolling, stretching, and forming. Strain hardening increases the hardness of an aluminum alloy.

Annealing an aluminum alloy is done by heating the alloy to 650 °F (343 °C) or 775 °F (413 °C) depending on the alloy. The alloy is heated for two

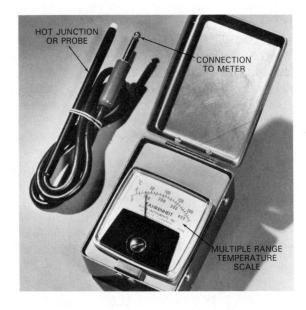

Fig. 27-20. Thermoelectric temperature measuring instrument. The electric probe draws small quantities of heat energy from the surface to be measured. (Royco Instruments, Inc.)

Fig. 27-21. Temperature indicating pellets. The 275°F (135°C) pellet is melting; the 325°F (163°C) pellet is not. This indicates that the temperature is between 275 and 325°F (135 and 163°C). (Tempil®, Big Three Industries, Inc.)

Fig. 27-22. This liquid, used to determine temperature, is applied to the surface of the metal. After the liquid dries, it will melt at the temperature marked on the bottle. (Tempil®, Big Three Industries, Inc.)

to three hours and then cooled. The cooling rate is not important as it will not affect the final structure.

A SOLUTION HEAT TREATMENT is similar to annealing. The procedure to solution heat treat an aluminum alloy is to heat the alloy to 940-970°F (504-521°C). Hold at this temperature and then quench in water. This produces a uniform distribution of the alloys in the aluminum. The aluminum alloy is then cold worked or aged to achieve the desired strength and hardness.

Aging can be done naturally or artificially. Natural aging occurs at room temperature. Artificial aging is done by heating the aluminum alloy to an elevated temperature. An aluminum alloy that is aged has better strength and hardness than when annealed or solution heat treated. The temperature for T6 artificial aging is between 320-360°F (160-182°C) for 6 to 12 hours.

## 27-20. TEMPERATURE MEASUREMENTS

Correct heat treatments can be obtained only if the temperature of the metal can be accurately measured.

One of the more difficult measurements is to determine if the surface temperature is also the inner temperature of the structure.

Several ways used to measure high temperatures are:
1. Optical pyrometers.
2. Gas thermometers.
3. Thermoelectric pyrometers, Fig. 27-20.
4. Resistance thermometers.
5. Radiation pyrometers.
6. Color change of the metal.
7. Temperature indicating crayons, cones, pellets, and liquids. The use of a temperature indicating pellet is shown in Fig. 27-21.

One of the most important factors in welding is the proper preheating and postheating of the metal. The temperature to which the metal is heated is of vital importance as a few degrees too little or too much can give undersirable results. Automatic furnaces with heat source controls and thermocouple temperature indicators are

desirable. However, their expense makes them impractical for many small shops. An excellent solution to the problem for small shops is the use of temperature indicators such as crayons, pellets, or liquids. The operation of liquid temperature indicator is shown in Fig. 27-22.

Fig. 27-23 shows a 250°F (121°C) crayon. Fig. 25-24 shows a crayon in use. These materials

Fig. 27-23. Temperature indicating crayon which will melt at 300°F (149°C). (Tempil®, Big Three Industries, Inc.)

Fig. 27-24. Temperature indicating crayon in use. The 250°F (121°C) mark will melt when surface temperatures are above 250°F (121°C). (Tempil®, Big Three Industries, Inc.)

indicate a great variety of temperatures. They indicate temperatures in 12 1/2 °F (7 °C) steps in the 113 to 400 °F (45 to 204 °C) range. They are in 50 °F (28 °C) steps in the 400 to 2000 °F (204 to 1093 °C) range. Pellets exist with 100 °F (56 °C) steps from 2000 to 2500 °F (1093 to 1371 °C). Fig. 27-25 shows a test kit which contains a number of temperature indicating crayons. These crayons are used to mark the metal surface. When the correct temperature crayon is applied, the crayon will melt and change color. Fig. 27-26 shows one way a welder can use such a crayon to measure the preheat temperature of a pipe prior to welding.

## 27-21. REVIEW OF SAFETY

Heat treatment of metals requires one to use safe working habits with the hot metal. Gloves are required and goggles are very strongly recommended, in most cases. Hot metals should be safeguarded and marked to prevent other persons from injuring themselves.

Quenching metals requires a safety shield to protect the face from high temperature flying fluids. Since the heat source used for heat treating varies, one should investigate the required safety steps for each heat source.

## 27-22. TEST YOUR KNOWLEDGE

Write your answers on a separate sheet of paper. Do not write in this book.
1. The most common postweld heat treatment is _____.
2. Heating can occur at three different times while welding: before, during, and after.

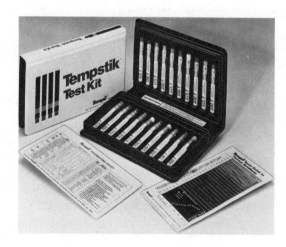

Fig. 27-25. A temperature test kit. This test kit is portable and covers up to 800 °F (427 °C) which is the range for preheating for welding and some heat treatments.
(Tempil°, Big Three Industries, Inc.)

Fig. 27-26. A welder preheating a pipe joint to 300 °F (149 °C) prior to welding. Note at A the crayon mark and how it has melted in the area where the torch is pointing.
(Tempil°, Big Three Industries, Inc.)

What name is given to each period of heating?
3. What four factors are important when heat treating a metal?
4. What structure in steel determines the hardness of the steel? What structure in steel determines the ductility of a steel?
5. How is martensite formed? How is pearlite formed?
6. Describe the regions in the heat affected zone and how they were formed.
7. What is the purpose of annealing?
8. How does normalizing differ from annealing?
9. Describe the quench and tempering process and the properties of a steel that is quench and tempered.
10. Thermal stress relieving reduces the _____ _____ and improves the _____ of a steel.
11. What happens when a high carbon steel receives a spheroidizing heat treatment?
12. What does the hardness of a steel depend on?
13. Describe one way to produce a gear with a hard surface and a tough interior.
14. Describe the heat treatment that is most often used on cast irons.
15. What heat treatment is used to soften copper?
16. What process is used to give copper high strengths and hardness?
17. How is an aluminum alloy hardened?
18. What heat treatment did an aluminum alloy receive that has a designation T6?
19. What does a solution heat treatment produce in an aluminum alloy?
20. How can temperature indicating crayons, pellets, and liquids be used to properly identify the temperature of a metal?

The largest friction welder ever built. This welding machine has a maximum thrust capacity of 4,500,000 lbs. (2000 metric tons) and is capable of joining tubular steel sections of up to 225 in.² (145,161 mm²). (Manufacturing Technology, Inc.)

# 28 INSPECTING AND TESTING WELDS

All welds have discontinuities (flaws). DISCONTINUITIES are interruptions in the normal crystalline lattice structure of the metal. However, a discontinuity is not always a defect. They are only DEFECTS when the weld becomes unsuitable for the service for which it was intended.

The role of WELD INSPECTION is to locate and determine the size of discontinuities (flaws). This nondestructive evaluation (NDE) must be done without damaging the weld. Nondestructive testing (NDT) is also used to check a weld's quality without destroying the weld.

The role of destructive testing is to determine the effects of various discontinuities. Destructive testing is used to determine the physical properties of a weld and to predict the service life of a weld. To perform this type of test, it is necessary to destroy the weld.

## 28-1. NONDESTRUCTIVE EVALUATION (NDE)

NONDESTRUCTIVE EVALUATION (NDE) is used on a weld without reducing its continued serviceability.

Manufacturers are interested in the soundness of welds or their fitness for service. This has made NDE an important phase of welding.

FITNESS FOR SERVICE is basically the concept that all welds contain discontinuities (flaws), but not all discontinuities are defects. Ideally, when a weld joint is designed, it is welded and then evaluated nondestructively. After all discontinuities are located, the weld is tested to destruction. This is done on several samples. The purpose of these tests is to determine how large a discontinuity must be to become a defect. When this defect size or critical size is determined, all discontinuities of this size and above will be considered defects.

On all similar welds, nondestructive evaluations may be made to locate all discontinuities above the critical defect size.

The methods used for nondestructive evaluation (NDE) include:
1. Visual inspection (VT).
2. Magnetic particle inspection (MT).
3. Liquid penetrant inspection (PT).
4. Ultrasonic testing (UT).
5. X-ray inspection (RT).
6. Eddy current inspections (ET).
7. Mass spectrometer detection (LT).
8. Air pressure leak inspection (LT).
9. Halogen gas leak inspection (LT).

The most popular NDE methods are: visual, magnetic particle, liquid penetrant, ultrasonic, and X ray.

## 28-2. DESTRUCTIVE TESTS

DESTRUCTIVE TESTS are used to determine the physical properties of a weld. For every type of weld and material used, a weld procedure qualification is normally written. Refer to Heading 29-8. This weld qualification procedure will specify the type of test or tests to be used to check the properties of the weld. The procedure for making each test is also specified in the weld qualification procedure for each weld.

Destructive tests render the weld unfit for further service. Destructive type tests include:
1. Tensile test.
2. Chemical analysis.
3. Bend test.
4. Microscopic test.
5. Macroscopic test.
6. Hardness texst.
7. Impact test.
8. Fatigue.
9. Hydrostatic test to destruction.
10. Peel test (used on spot and projection welds).
11. Tensile shear test (used on spot and projection welds).

## 28-3. VISUAL INSPECTION (VT)

A weld which is not required to have high physical strength may be inspected visually for cracks, inclusions, contour, and certain other qualities. This type of inspection is subjective in nature, and usually has no definite and rigidly held limits for acceptability. A template may be used to check the contour of the weld bead. Figs. 28-1 and 28-2 shows such templates. Using the visual inspection method, an inspector may compare a finished weld with an accepted standard, and pass or reject a weld by the comparison method only.

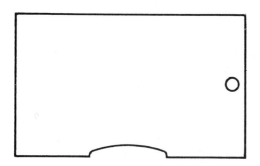

Fig. 28-1. Template for testing the bead contour of welds.

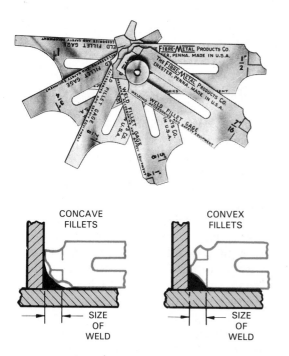

CONCAVE FILLETS CONVEX FILLETS

SIZE OF WELD SIZE OF WELD

Fig. 28-2. Above. Gauges for measuring and inspecting both convex and concave fillet welds from 1/4 to 1 in. (6.4-25.4 mm). Below. Blades of gauge used to check concave and convex weld contours. (Fibre Metal Products Co.)

This test is effective only when appearance is the most important quality of the weld.

To inspect the inside of hard to reach places,

miniature television cameras and lenses have been developed. These miniature TV inspection units can be placed inside pipes, cylinders, and tanks to make visual inspections. What the camera lens sees is viewed by the inspector on a TV screen. See Figs. 28-3 and 28-4.

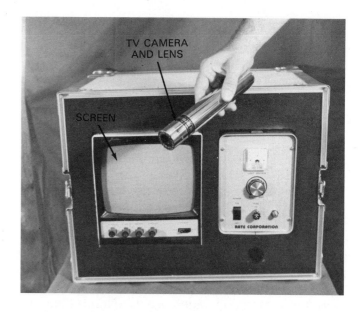

Fig. 28-3. A miniature TV camera and lens used to visual inspect hard to reach areas. (Flo-Max Corp.)

Fig. 28-4. A number of different lenses used with the TV camera shown in Fig. 22-3. (Flo-Max Corp.)

## 28-4. MAGNETIC PARTICLE INSPECTION (MT)

MAGNETIC PARTICLE INSPECTION is most effective in checking a weld for surface or near surface flaws. It is used only on materials which can be magnetized. Prior to applying the magnetic particle material, the surface must be cleaned with a special cleaner furnished with the magnetic particle kit. A material containing fine magnetic particles is then applied to the weld being inspected.

Fluorescent magnetic particles may also be used. These magnetic particles are applied as a liquid or in a dry form.

The metal is then subjected to a strong magnetic field. Because of the magnetic field, the sides of any small cracks and pits become the north and south poles of miniature magnets. The magnetic particles in the applied powder or liquid are attracted to the cracks and pits by their magnetism. These particles are colored red, black, or gray and may be suspended in a liquid. The choice of black, gray, or red is dependent on which color gives the best color contrast against the test weld. When the magnetic field is removed, the inspector will find a concentration of magnetic particles in the area of every flaw. If fluorescent magnetic particles were used, a black (ultra violet) light is used to show the location of flaws. If defects are found, they are ground or chipped away. The part is then rewelded and retested. Fig. 28-5 shows the magnetic field which is created around the weld area being tested. Fig. 28-11 shows how defects look when using a dye penetrant. Flaws will look much the same during a magnetic particle inspection.

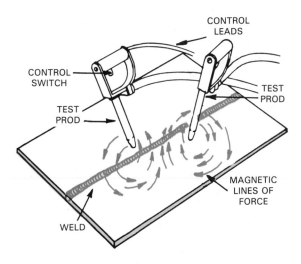

Fig. 28-5. A magnetic field is created around a weld as current is passed through the weld between the two test prods. (Magnaflux Corp.)

Two methods are used to create the magnetic field:
1. Passing current through the object being inspected by means of test prods.
2. Placing a powerful electromagnetic yoke or permanent magnet against the object to allow the magnetic field to pass through the material being inspected.

The flux lines in the magnetic field should be as perpendicular to the crack as possible. Because the location of the crack is not known, the article should be magnetized twice. One inspection should be made with the test prods or yoke in one direction, and one inspection with the yoke at 90 degrees to the first inspection. Fig. 28-6 shows magnetic powder being used to detect flaws in a weld which is being magnetized by electrical test prods.

Fig. 28-6. Magnetic powder and electrical test prods are used to locate surface defects in a weld. A special dual prod assembly frees one of the inspector's hands and allows the application of powder while current is flowing. (Magnaflux Corp.)

An electromagnetic yoke, the cleaner, and magnetic powders are shown in Fig. 28-7. The use of this method is shown in Fig. 28-8.

The magnetic field can also be produced by using a permanent magnet as shown in Fig. 28-9. The electro and permanent magnet systems are light and convenient. Permanent magnets are also used where an electrical spark may be dangerous. Another type of electromagnet, shown in Fig. 28-10, is used when inspecting pipes and shafts.

The material to be tested should be as clean and bright as possible prior to the test.

The magnetic particle test is very popular for checking for cracks in many service trades (aircraft, automotive, truck, marine, and the like).

Locating cracks in axles, shafts, gears,

Fig. 28-7. A kit for magnetic particle inspection. (Magnaflux Corp.)

Fig. 28-9. A magnetic particle test kit with a permanent magnet. The squeeze bottles contain red, black, and gray powder containing magnet particles. (Magnaflux Corp.)

Fig. 28-8. Magnetic particle inspection being used to check an engine block for cracks. Note the dark lines indicating cracks. (Magnaflux Corp.)

crankshafts, and landing gear parts before performance failure occurs is of considerable importance.

## 28-5. LIQUID PENETRANT INSPECTION (PT)

The LIQUID PENETRATION INSPECTION method uses colored liquid dyes and fluorescent liquid penetrants to check for surface flaws. This system can be used to detect surface flaws in metals, plastics, ceramics, and glass. This method will not detect subsurface discontinuities or defects.

The steps required to perform a satisfactory colored liquid dye or fluorescent liquid penetrant inspection are:
1. Clean the part with the special cleaner provided in the kit.
2. Apply the liquid penetrant and allow time for it to penetrate into any cracks or pits.
3. Depending on the type of penetrant used, the excess liquid penetrant is removed with:
   a. water.
   b. solvent.
   c. an emulsifier.
4. Apply the developer liquid.
5. Inspect for discontinuities (flaws) and defects.
6. Clean the surface after inspection is complete.

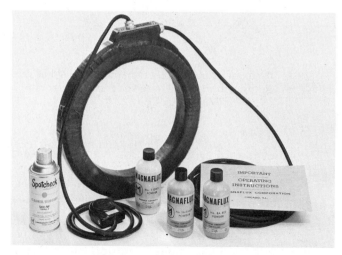

Fig. 28-10. Ring type or torus shaped electromagnet used to magnetize pipes or shafts to be inspected with magnetic particles. (Magnaflux Corp.)

Prior to applying the developer, the penetrant must be removed. Some penetrants are removed by washing with water; others are removed by using a solvent. A third type of penetrant uses an emulsifier. The emulsifier makes the penetrant water soluble. Water is used to wash away the penetrant and emulsifier.

The developer is then applied. Some of the penetrant which is in a flaw will be drawn out. The dye can be seen and the discontinuity located. Surface discontinuities (flaws) and defects will show up as shown in Fig. 28-11.

LARGE CRACK OR OPENING—
a continuous streak which bleeds up rapidly.

CRACK OR COLD SHUT—
a broken line of dots which takes several minutes to come up.

FATIGUE CRACK, PARTIAL WELD OR LAP—
a series of red dots forming an irregular line.

PITS AND POROSITY—
a concentration of red spots.

Fig. 28-11. Indications of surface flaws found by the liquid dye or fluorescent penetrant method of inspection.

Fluorescent liquids are used in a manner similar to dye penetrants. A fluorescent liquid is applied to the surface being inspected. After a short time, the excess fluorescent liquid is removed with a cleaner. The developer is applied and an ultraviolet (blacklight) source is then brought to the surface. All areas where the fluorescent liquid has penetrated will show up clearly under the ultraviolet light, as shown in Fig. 28-11.

The dye, the cleaner, and the developer are available in aerosol spray cans for convenience. Some solvents used in the cleaners and developers contain high percentages of chlorine, a known health hazard, to make the liquids nonflammable. Solvents and developers containing chlorine should be used with great care.

The penetration ability of the dye varies according to the materials being tested. The penetrant activity varies with the temperature. It is important to allow sufficient time to permit accurate inspection. This time may vary from 3 to 60 minutes. At room temperatures, the time usually recommended is from 3 to 10 minutes.

## 28-6. ULTRASONIC TESTING (UT)

ULTRASONIC TESTING is a nondestructive testing (NDT) method of locating and determining the size of discontinuities (flaws) by means of sound waves. It can be used on virtually any type of material. These sound waves are passed through the material. Ultrasonic testing (UT) uses high frequency sound waves at more than one million hertz (MHz). One hertz is one cycle per second.

An electronic device called a TRANSDUCER is placed on the surface of the material being tested. The ultrasonic waves are sent into the material through the transducer. The ultrasonic wave will not travel through the air. Therefore, an excellent contact must be made between the transducer and the surface of the material. To insure a good contact, a material called a COUPLANT is used. This couplant can be water, oil, or glycerol. The only specification for a couplant is that it is not harmful to personnel or the part being tested.

The ultrasonic wave is sent into the material for very short periods of time. This time period is only one to three millionth of a second (1-3 microseconds). Once the UT wave is sent through the material, it is reflected from the surface boundaries or flaws where density changes occur. The reflected wave is then received by the same transducer which started the wave. A second transducer with the same frequency rating may be used. After a wave is sent and received back, another wave is sent for 1-3 microseconds. This process is repeated approximately 500,000 times per second until the test is completed.

Each reflected wave signal is amplified and displayed on a cathode ray tube (CRT) on an oscilloscope. See Fig. 28-12. The oscilloscope is calibrated to show the distance between the transducer and any flaw found. The equipment must be calibrated for each type of material and thickness tested. Operators must be skilled to obtain consistent results. See Fig. 28-13.

The equipment used for ultrasonic testing is light and portable. Ultrasonic testing is done easily on the job site. The advantages of UT are:
1. Fast.
2. Immediate results.
3. Usable on most materials.
4. Access from two sides is not required.
The disadvantages of UT are:
1. A couplant is required.
2. Defects parallel to the sound beam are difficult to detect.

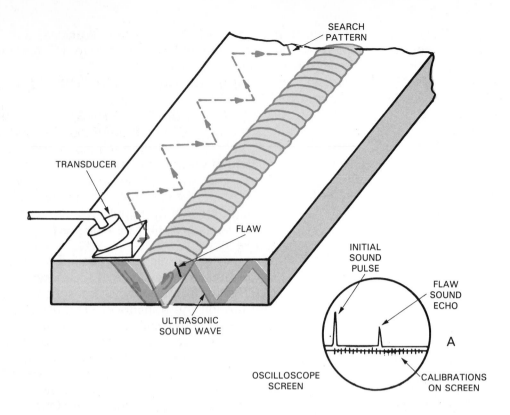

Fig. 28-12. A schematic drawing showing the path followed by the ultrasonic search unit and the path of the sound waves as they move through the metal being tested. The initial, and echo indicators, are shown on the oscilloscope (CRT) in a manner similar to that shown at A. The distance between peaks is an indication of how far the flaw is from the searching unit head.

3. CRT displays require training to interpret accurately.
4. Equipment must be calibrated regularly.

## 28.7. EDDY CURRENT INSPECTION (ET)

EDDY CURRENT INSPECTION uses an AC coil to induce eddy currents into the part being inspected. An AC coil produces a magnetic field. The magnetic field induces an AC current into the part. This induced current flows in a closed circular path. These induced currents are known as EDDY CURRENTS.

To perform an eddy current test, the coil is placed on the part to be tested. The coil is calibrated or adjusted to obtain a set value of impedance. Impedance is the resistance to the flow of alternating current. The set value of impedance is viewed on an oscilloscope.

The coil is moved over the surface of the part to be inspected. If a discontinuity is present in the material it will interrupt the flow of the eddy currents. The change in the eddy currents affects the impedance of the coil. This change can be observed on the oscilloscope.

Eddy currents only detect discontinuities near the surface of the part. The depth of inspection depends on the alternating current frequency. With common frequencies, the depth of inspection for common metals will not exceed 1/8 in. (3.2 mm). Eddy currents can be used to inspect flat and circular parts like pipes. Eddy currents are used to check for porosity and cracks. They can also be used to inspect for proper heat treatment.

## 28-8. X-RAY INSPECTION (RT)

Welds may be checked for internal discontinuities (flaws) by means of X-RAY INSPECTION. The X ray is a wave of energy which will pass through most materials and reproduce their image on film (radiography). X rays may also be displayed on a fluorescent screen (fluoroscopy), or on a television screen for viewing at a remote spot (TVX).

Fig. 28-14 diagrammatically illustrates the principle of the fluoroscopic system of examining a pipe weld. The radioactive X-ray energy may be produced electronically in an X-ray machine, or by means of radioactive isotopes. Equipment using

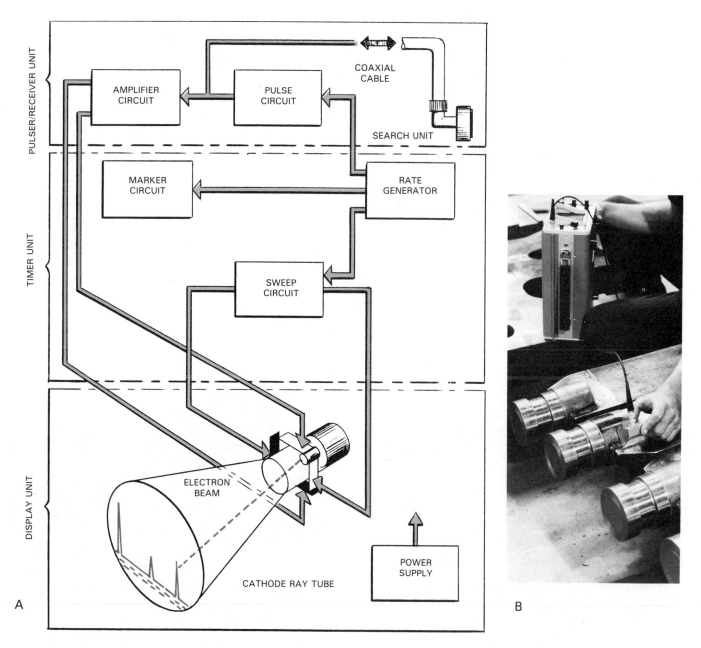

A

B

Fig. 28-13. A—Block diagram shows the component functions of an ultrasonic testing (UT) unit. Ultrasonic sound is sent and received by the transducer search unit. The reflected signal is displayed on the cathode ray tube (CRT). The pattern indicates any defects and their location. B—Ultrasonic unit in use testing a weld on an equalizer bar for a railroad car.   (Sperry Div., Automation Industries, Inc.)

radioactive isotopes is portable and may be used to check welds made in the field. Portable X-ray equipment is shown in Fig. 28-15.

Certain safety precautions must be observed while using radioactive materials, to prevent illness and even death from overexposure. Radiographers inspect completed welds before any further work is done on the weldments. A radiograph (X-ray picture), if made, is a permanent record of welds made on critical construction such as pipelines, ships, aircraft, and the like. Radiographs are often required on critical welds. Fig. 28-16

shows equipment being readied for taking a radiograph.

Several popular radioactive isotopes are:
Cesium 137
Cobalt 60
Iridium 192
Samarium 153
Thulium 170

These isotopes give off gamma rays. The radioactive isotope X-ray machines are small and need no electrical source. They are more flexible in their use than standard X-ray equipment.

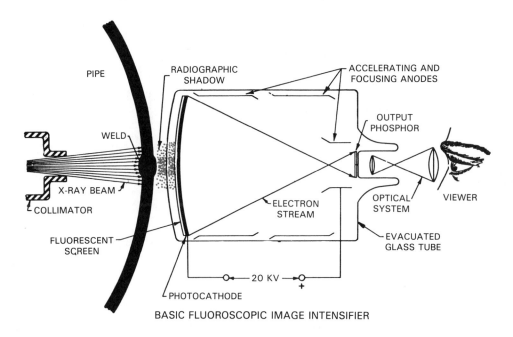

BASIC FLUOROSCOPIC IMAGE INTENSIFIER

Fig. 28-14. Diagrammatic explanation of a fluoroscopic examination of a pipe weld. (National Tube Div., U.S. Steel)

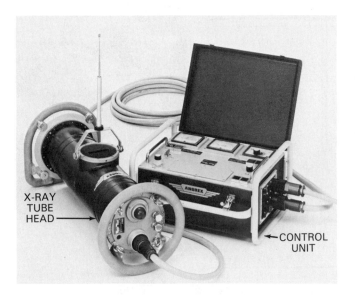

Fig. 28-15. A portable radioactive isotope type X-ray machine. (TFI Corp.)

Isotopes must be handled with great caution. Only specially trained personnel should handle them.

The energies needed to x-ray vary considerably. Equipment is available from 50 KV (K = 1000, V = volts) to 24,000 KV (24,000,000 Volts). 140 KV will x-ray 2 inch thick steel, while 24,000 KV will x-ray 20 inch thick steel. Fig. 28-17 illustrates the principle of radiography.

Flaws in a weld usually are easily seen in an X-ray radiograph, as shown in Fig. 28-18. However, the depth at which the flaw occurs cannot be determined with an X ray made from only one

Fig. 28-16. A portable X-ray unit used to inspect the welds on a large cylinder. Note the stand used to hold the X-ray tube head in position. (TFI Corp.)

direction.

The equipment required when inspecting by means of X rays depends on the:
1. Kind of material.
2. Thickness of material.
3. Accessibility of part to be tested.
4. The geometry of part to be tested. For example, a flat plate is easier to inspect than a pipe cluster.

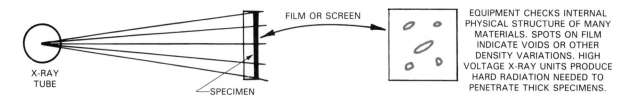

EQUIPMENT CHECKS INTERNAL PHYSICAL STRUCTURE OF MANY MATERIALS. SPOTS ON FILM INDICATE VOIDS OR OTHER DENSITY VARIATIONS. HIGH VOLTAGE X-RAY UNITS PRODUCE HARD RADIATION NEEDED TO PENETRATE THICK SPECIMENS.

Fig. 28-17. A schematic of an X-ray radiograph (photograph) of a weld.

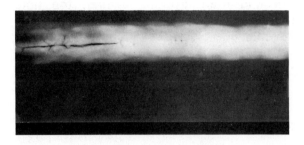

Fig. 28-18. An X-ray radiograph of a weld in 1 1/2 in. (38.1 mm) steel plate. Note how the undersurface crack at the left shows up in the X-ray radiograph.

## 28-9. INSPECTING WELDS USING PNEUMATIC OR HYDROSTATIC PRESSURE

A common method of testing pressure-vessel welds (tanks and pipe lines) for leaks is to use gas or air pressure. Carbon dioxide gas pressure is well suited for the purpose. It is nonexplosive when in contact with oils or greases. A small pressure of 25 psi to 100 psi (172.4 kPa - 689.5 kPa) is applied in the vessel or pipe, and a soap and water solution put on the outside of each weld. Leaks will be indicated by the formation of bubbles. The ability of a vessel to hold pressure is also an indication of the soundness of its welds. A vessel to be inspected may be pressurized and the pressure noted on a gauge. The pressure is then turned off and after 24 hours the gauge is again checked. Any drop in pressure indicates a leak. This test is easily applied and is safe since the pressures used are usually under 100 psig (689.5 kPa).

Another test for pressure vessels is to coat the surface with a lime solution. After the lime has dried, pressure is built up in the vessel. Where the lime flakes from the metal, a flaw is indicated as being present. Hydrostatic pressure, using water as the fluid, is the usual medium used in this test. This test may also be used to reveal the weakest portion of a welded vessel if enough pressure is created without destroying the vessel. Strain gages which measure the slightest movement may be used to test for any flex points in a weld.

To inspect pressure vessels for leaks, water is still a popular method. However, water is not a reliable check for extremely small leaks.

Pressure vessels may be filled with chlorine, fluorine, helium, or other non-oxygenated gases. These gases will flow through extremely small holes. A pickup mass spectrometer tube is placed on a suspected leakage area. A flow of one part of these gases in 1,000,000 parts of air may be detected to indicate a very small leak in a pressure vessel. Follow all proper safety precautions when using chlorine gas.

## 28-10. BEND TESTS

A popular method of testing a weld which does not require elaborate equipment is the destructive BEND TEST. The method is fast and shows most weld faults quite accurately. A sample specimen may be tested to destruction to determine:
1. The physical condition of the weld, and thus check on the weld procedure.
2. The welder's qualifications.

This method is particularly appropriate where large numbers of identical pieces are to be fabricated. The usual method of testing is to take one piece out of a predetermined quantity and test

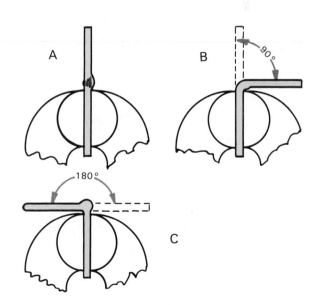

Fig. 28-19. Testing a weld sample in a vise. A—Sample before bending. B—Sample after bending 90 deg. (note the weld is closed in on itself, to test the penetration). C—Sample after bending it back 180 deg.

it to destruction. The destruction test may show up such qualities as tensile strength, ductility, fusion, penetration, and crystalline structure.

The equipment used for a test of this kind depends on the shape and type of articles to be tested. A common method is to clamp the piece to be tested in a vise and by means of a bending bar, bend the metal at the welded joint. Fig. 28-19 illustrates a bend test. This method of testing quickly gives the approximate strength of the weld, while the stretching of the metal determines to some extent its ductility. Any cracking of the metal will show a lack of fusion or defective penetration.

Many fixtures have been devised to help test weld specimens to destruction. After the weld has been broken, the appearance of the fracture will show the crystalline structure. Large crystals will usually indicate an incorrect welding procedure, or poor heat treatment after welding. Small crystals will usually indicate a good weld.

The guided root or face bend test is described in Chapter 29. It is a bend test made under specified conditions to determine certain physical properties of a weld. The radius of the bend made on the test sample must meet AWS test standards. See Fig. 29-13. As the bend is made, the metal is guided or controlled at all times by the bending fixture so that all samples are bent in exactly the same manner. Fig. 28-20 illustrates a hydraulically operated guided bend tester. Free bend tests are also made on specially prepared weld samples as described in Chapter 29.

## 28-11. IMPACT TESTS

It has been found that it is possible for a weld to show good results by a variety of tests, but fail

under a rapidly applied, impact load. The impact test may be made by either the Izod or Charpy method. These methods are similar, but the shape and position of the notch varies. Test samples are taken from the weld, the heat affected zone (HAZ), and the base metal.

A test piece is notched in a specified manner and clamped in the jaws of an impact testing machine. A heavy pendulum is lifted to a given height and then dropped against the notched specimen to determine what impact force it can withstand. See Figs. 28-21 and 28-22.

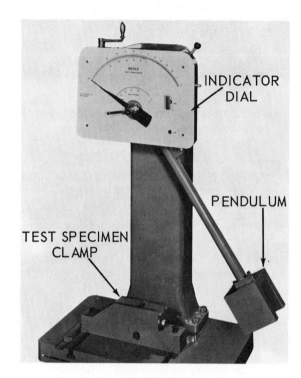

Fig. 28-21. A Charpy impact testing machine. In this tester the pendulum is lifted and dropped against the test specimen which is held in the clamp. The impact force is registered by the dial indicator.

The indicator needle on the testing machine shows what force was exerted on the test specimen. This test determines the impact strength of the weld specimen.

## 28-12. LABORATORY METHODS OF TESTING WELDS

Most large companies that do welding have laboratories for determining the weld strength of welds. These laboratories are equipped with modern weld testing equipment. This equipment can be used to determine the physical and chemical properties of weld or base metal samples. Occasionally, the testing is performed by

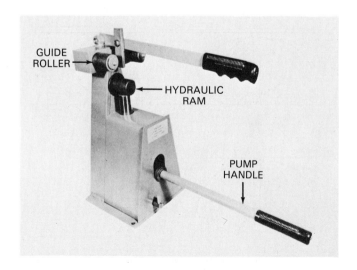

Fig. 28-20. A hydraulically operated guided bend test machine. (Vega Enterprises)

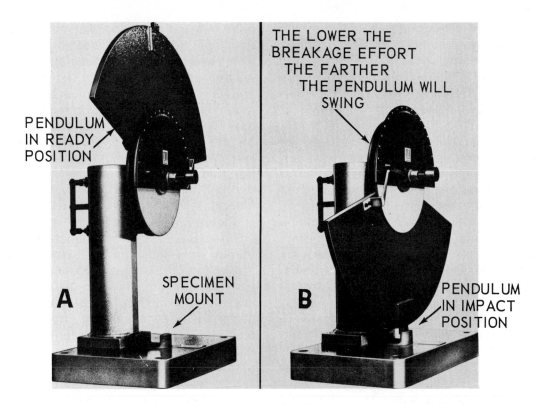

Fig. 28-22. A Charpy impact testing machine. A—Shows the pendulum in the ready position. B—Shows the pendulum in the impact position. (Physmet Corp., Div. of Manlabs, Inc.)

a metallurgical department. In some cases, a part of the shop is set aside for testing purposes.

Some metal properties to be determined in a laboratory or shop test are:
1. Tensile strength.
2. Ductility.
3. Hardness.
4. Microstructure.*
5. Macrostructure (deep etch).*
6. Chemical constituents.*
   *Laboratory tests only.

The conditions under which the specimens are tested are kept identical. The specimens are all of a standard size specified by the welding code used. The length of the specimens need not be the same, but the cross section area of the specimens must be the same. Samples must be taken from specified positions on the weld sample being tested.

The Society of Automotive Engineers (SAE), the American Society of Mechanical Engineers (ASME), and American Society of Testing Materials (ASTM), have adopted standards for laboratory tests of metals. These standards specify the physical qualities to be maintained for different types of welds and base metals. They also specify the size and location of various test specimens or samples. Some sample test pieces are shown in Fig. 28-23. Specifying the location

Fig. 28-23. Samples of weld test specimens tested to destruction. These were machined to standardized measurements before testing. All pieces shown are tensile test specimens except the pair on the right. This pair is a Charpy impact test specimen after it has been fractured.

and size of the metal sample is necessary in order to compare test results. Test results of completed welds and base metal samples may be used to compare welds, welding operators, and welding procedures. Standardizing tests make it possible to build standard testing machines.

## 28-13. TENSILE DUCTILITY METHOD OF TESTING WELDS

To TENSILE TEST a weld, a specimen of the weld is mounted in a machine and stretched. The machine determines three values for the weld:

1. Tensile strength of the metal.
2. Yield point of the metal.
3. Ductility of the metal.

The TENSILE STRENGTH is recorded as the number of pounds per square inch or kilopascals required to break or fail a metal.

The ELASTIC LIMIT of the metal means that it can be stretched just so far, and will return to its original length after the load is released. However, when more load is applied to the specimen after the elastic limit has been reached, the specimen loses its elasticity and the metal will not return to its original condition or size. When the specimen stretches instantaneously, or gives at a certain loading, but does not break, this is called the YIELD POINT or YIELD STRENGTH. Fig. 28-24 is a graph of the tensile load imposed on a metal sample.

The yield point is important. It is not desirable to load metal to the point where it will stretch and not return to its former shape. When a metal stretches without returning to its original shape it is said to take a PERMANENT SET. Metal structures are designed strong enough so the yield

point is never reached. Many machines have been developed to test tensile strength and yield point. Some machines record directly, even automatically, the tensile strength of the metal. Fig. 28-25 shows a hydraulic universal tester. Some machines are large, while others are portable and may be taken to field locations for testing metals on the premises. Fig. 28-26 shows a portable screw-operated machine. Fig. 28-27 illustrates a portable universal tensile tester which is hydraulically powered. A hydraulically powered laboratory type universal tensile testing machine is shown in Fig. 28-28.

Fig. 28-29 shows a laboratory type tensile tester. This machine uses motor driven screw power to pull test the specimen. These machines can test the tensile strength of metals, fabrics, assembly strengths, welds, etc. Some machines designed for tensile testing may also be used to test the compressive strength of brittle substances by applying a crushing force on the material being tested.

It is desirable for these testers to be universal, so they can test metals of different cross sections

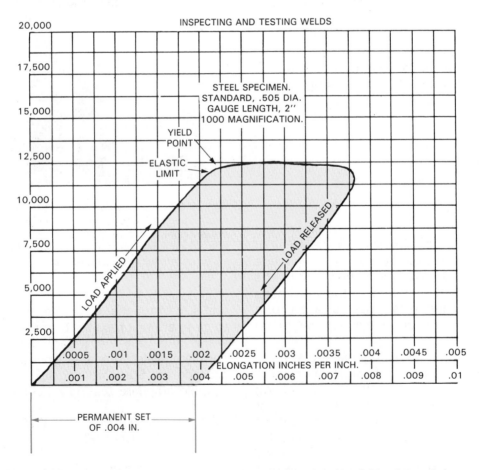

Fig. 28-24. Strength graph of a metal sample showing the yield point and the permanent set (lengthening) left in the steel when returned to a no-load condition. (Tinius Olsen Testing Machine Co.)

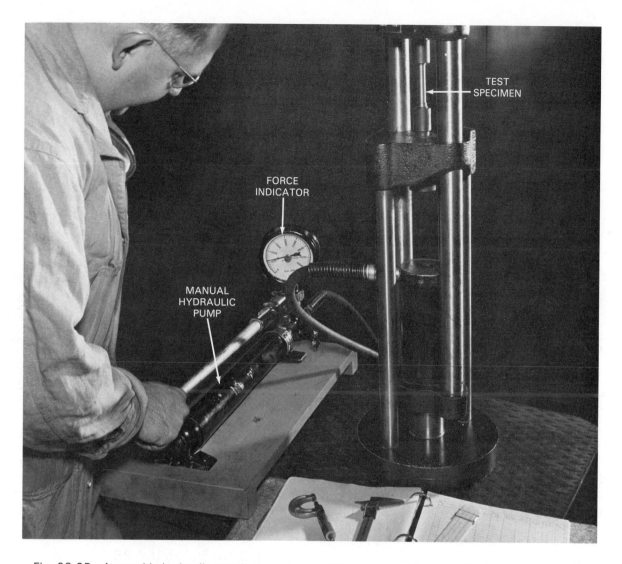

Fig. 28-25. A portable hydraulic tensile tester in use. This same machine may also be used to perform a guided bend test on a weld.

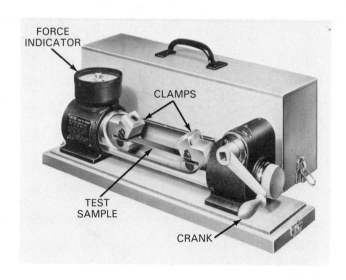

Fig. 28-26. A portable tensile tester. The force applied to the test sample is obtained through the mechanical advantage of a threaded shaft which is moved by turning the operating crank.   (Detroit Testing Machine Co.)

and shapes; that is, the machine will hold and test round, oval, square, and rectangular specimens. The machine tests the tensile strength of the

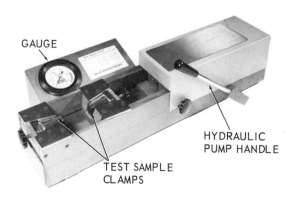

Fig. 28-27. A portable universal tensile testing machine. Force applied to pull the test sample apart is obtained by operating a hydraulic jack built into the tester housing. The force applied is read on the gauge.   (Vega Enterprises)

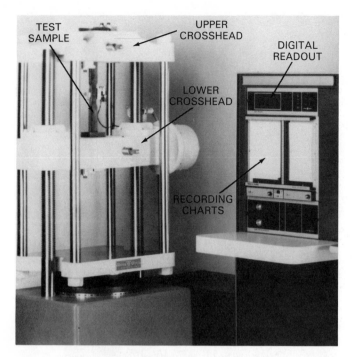

Fig. 28-28. A 120,000 lb. capacity universal tensile testing machine. The text specimen is clamped into the upper and lower crossheads. A hydraulic tensile (pulling) force is applied to test the specimen. (Tinius Olsen Testing Machine Co., Inc.)

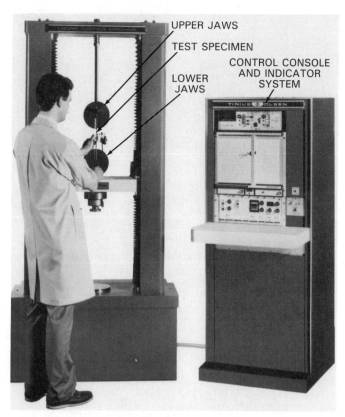

Fig. 28-29. A 30,000 lb. capacity universal tensile testing machine. The test specimen is clamped into the upper and lower jaws. Motor-driven screws move the upper crosshead up to pull the specimen. Test results are read on a digital readout or on a charted graph.
(Tinius Olsen Testing Machine, Co., Inc.)

metal. It also tests the DUCTILITY, meaning the ELONGATION (stretchability) of the metal before it fails.

To measure the elongation of a weld, prick-punch or scribe two points on the weld specimen, 2 inches (50 mm) apart. Then measure the change in this distance after the specimen breaks.

The elongation is determined in percent by dividing the difference between the two readings by the original distance. For example: Let 2 inches (50 mm) be the original distance between the punch marks. The distance between the punch marks after the metal is stretched to its elastic limit is 2 1/4 inches (56.25 mm). Then 2 1/4 in. − 2 in. = 1/4 in. and 1/4 in. ÷ 2 in. = 12.5 percent. (Likewise in metric measurements: 56.25 mm − 50 mm = 6.25 mm and 6.25 mm ÷ 50 mm = 12.5 percent.) Fig. 28-30 is a table showing the tensile strength and percent of elongation for some common metals.

| METAL | TENSILE STRENGTH ANNEALED | | TENSILE STRENGTH HEAT TREATED OR HARDENED | | PERCENT ELONGATION |
|---|---|---|---|---|---|
| | psi | MPa | psi | MPA | |
| Low Carbon Steel | 55,000 | 379 | — | — | 25 |
| Medium Carbon Steel | 76,000 | 524 | — | — | 18 |
| Stainless Steel | 75,000 | 517 | 90,000 | 621 | — |
| Crome-Moly Steel | — | — | 128,000 | 886 | 17 |
| High Tensile Steel | — | — | 262,000 | 1806 | 4.5 |
| Duralumin-2017 | 26,000 | 179 | 62,000 | 427 | 22 |
| Aluminum-6061 | 18,000 | 124 | 45,000 | 310 | 17 |
| Commercial Bronze | 38,000 | 262 | 45,000 | 310 | 25-45 |

Fig. 28-30. Tensile strength and percent of elongation values for various metals.

## 28-14. TESTING THE HARDNESS OF WELDS

Another important factor which should be determined when testing a weld is the hardness of the weld metal. HARDNESS may be defined as a resistance to permanent indentation. The hardness of welds is particularly important if the welds must be machined. There are many special metal alloy electrodes used to produce strong, hard surfaced welds on machine tools.

Methods used to determine metal hardness have been standardized. One of the most popular methods is to use what is called a ROCKWELL hardness testing machine, Fig. 28-31. This machine works somewhat like a press and is provided with a platform for holding the specimen. A point is pressed into the metal which is being tested by means of fixed weights operating through leverage, as shown in Fig. 28-32. The point used may be a 1/16 in. (1.6 mm) steel ball. It may also be a diamond ground to a 120 degree

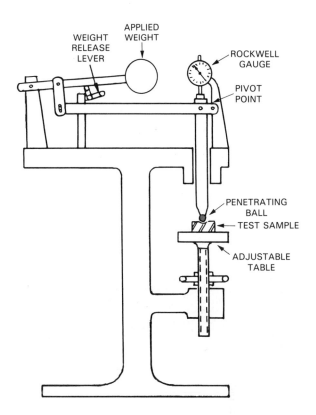

Fig. 28-32. Diagrammatic drawing showing how weight is applied to penetrator through leverage in a Rockwell hardness tester.

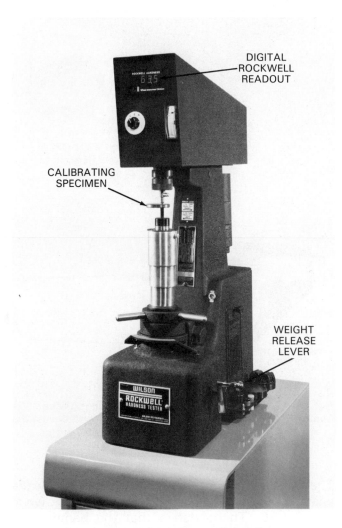

Fig. 28-31. Laboratory type digital ROCKWELL® hardness tester. (Page-Wilson Corporation)

point. When the ball-point is used, two loads are applied to the metal sample being tested. The machine is calibrated and the gauge set to zero during the first loading. The first load is 10 kg (22 lb.). The distance the ball penetrates the metal between the first load and the final or major load of 100 kilograms (220.5 lb.), indicates the hardness on a gauge registering from 0-100.

The hardness is indicated on the Rockwell B scale, when the 1/16 in. (1.6 mm) steel ball is used. When using the diamond cone penetrator, as shown in Fig. 28-33, the weights of 10 kilograms (22 lb.) and 150 kilograms (330.7 lb.) are used. The hardness is read on the Rockwell C scale.

Portable Rockwell hardness testers are available and may be used to check metals on the job. These units are easily calibrated and give accurate readings. Fig. 28-34 shows a portable tester as it is used in production work.

An equivalent hardness testing device is shown in Fig. 28-35. The tungsten-carbide ball indentor used is spring loaded. A piezoelectric crystal is attached in the indentor rod. When the indentor is released it strikes the metal being tested. The deceleration of the indentor is sensed by the piezoelectric crystal which generates a proportional voltage. This voltage is displayed on a

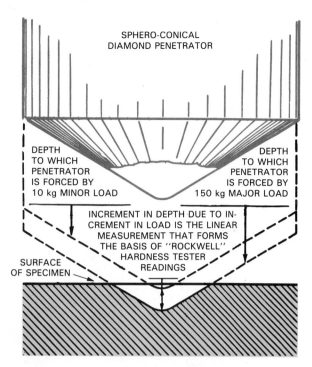

Fig. 28-33. Diamond penetrator used in the Rockwell hardness tester. Note the depressions left in the metal by the diamond under the 10 kg (22 lb.) and the 150 kg (331 lb.) load.

meter. The equivalent tester is calibrated by using a test sample of known hardness.

A second method of testing hardness is to use a direct reading SCLEROSCOPE. This machine is

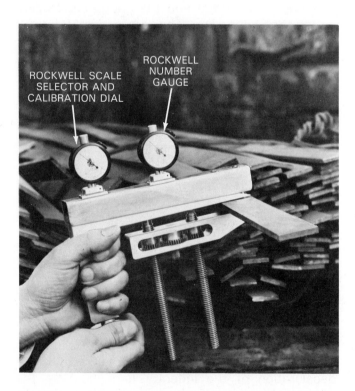

Fig. 28-34. A portable Rockwell hardness tester being used to check the hardness of steel stock.

based on the impact or rebound of a ball or hammer from a test specimen. The machine consists of a vertical glass tube channel of a certain height. At the top of this is mounted a steel hammer, having a diamond tip of a certain diameter and size.

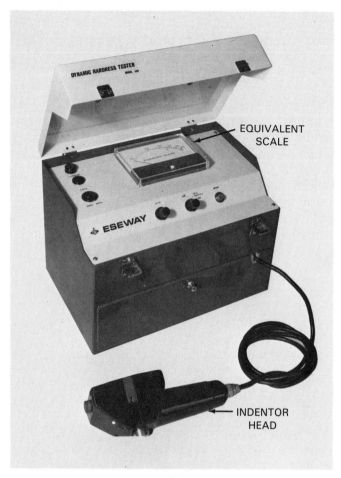

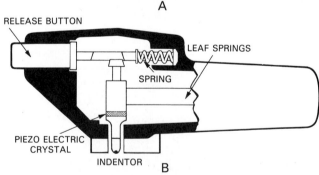

Fig. 28-35. An equivalent hardness testing device. A—Complete unit is shown. B—Schematic of the indentor head.   (Engineering & Scientific Equip., Ltd.)

The specimen to be tested is placed below the channel, and the hammer is released, as shown in Fig. 28-36. The distance that the hammer rebounds after it contacts the metal may be read on the scale beside the channel. The hardness of the

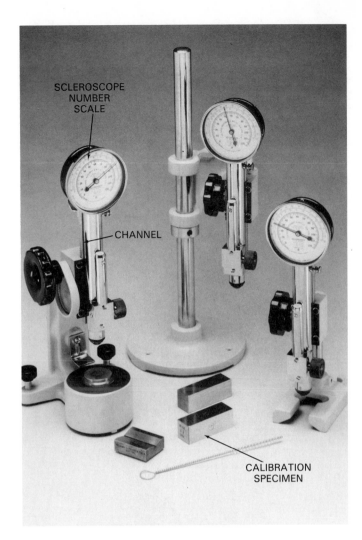

Fig. 28-36. Scleroscope hardness testing machine. The distance the small hammer rebounds up the channel after it drops on the test specimen is read on the dial gauge. Three different mounting arrangements are shown here.
(Shore Instrument & Mfg. Co., Inc.)

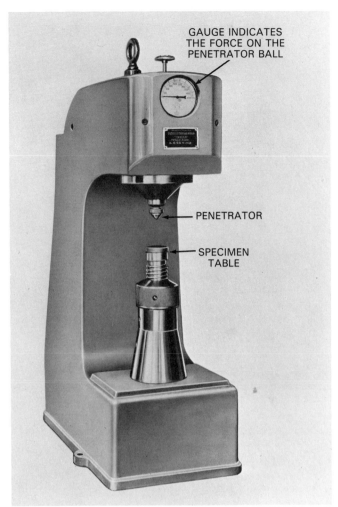

Fig. 28-37. A hydraulically operated Brinell hardness tester.
(Detroit Testing Machine Co.)

metal, as indicated by the scale number with this tester, will range from 0 to 140. The higher the number, the harder the metal. A high carbon steel will indicate approximately 95 points on the scale. A rubber tube-and-bulb arrangement manipulates the hammer for testing purposes.

A third method of testing for hardness is to use a BRINELL machine having a penetrator ball built into a press, as shown in Fig. 28-37. The penetrator ball is moved by hydraulic pressure indicated on a dial scale. The dial indicates the force exerted on the specimen. The specimen is mounted below the ball, and the 10 mm (.394 in.) diameter ball is pressed (not dropped) into the specimen under a load of 29,420 newtons (6,614 lb.) for 10 seconds. A microscope is then used to measure the diameter of the indentation in millimeters. The area of the depression divided by the load gives the Brinell-hardness number. A table is supplied with the machine to determine the hardness number, once the diameter of the indentation is known.

This method is often used for testing softer metals like copper and aluminum. The construction of the machine is shown in Fig. 28-38. The table in Fig. 28-39 shows a comparison of hardness numbers in the Rockwell, Brinell, and Scleroscope scales.

MICROHARDNESS TESTERS have been developed which make it possible to test the hardness of a part without appreciably injuring the part. An extremely small diamond penetrator is used to penetrate the surface to be tested. The loads on the diamond penetrator may be varied from .25 newtons to 490 newtons (.06 lb. to 110 lb.). After the surface has been penetrated, the size of the indentation is measured using a powerful microscope. This measurement indicates the hardness of the surface tested. Because the metal surface is marked microscopically, the tests can be performed more frequently over the surface of

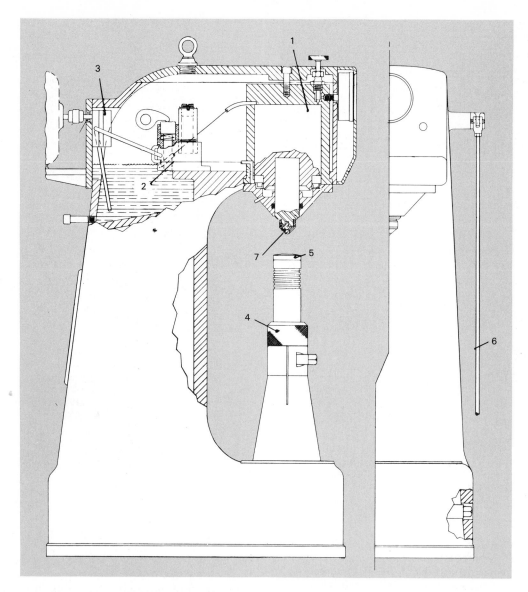

Fig. 28-38. Interior construction of a Brinell hardness tester. The main parts are: 1—Hydraulic cylinder, 2—Hydraulic valve and pressure regulator, 3—Hydraulic pump, 4—Manual adjustment, 5—Anvil, 6—Hand or foot operated control lever, 7—Ball (penetrator).

| ROCKWELL C | ROCKWELL B | BRINELL | SCLEROSCOPE |
|---|---|---|---|
| 69 | — | 755 | 98 |
| 60 | — | 631 | 84 |
| 50 | — | 497 | 68 |
| 40 | — | 380 | 53 |
| 30 | — | 288 | 41 |
| 24 | 100 | 245 | 34 |
| 20 | 97 | 224 | 31 |
| 10 | 89 | 179 | 25 |
| 0 | 79 | 143 | 21 |

Fig. 28-39. Table showing comparison of hardness numbers in Rockwell, Brinell, and Scleroscope scales.

a metal object. Fig. 28-40 shows a microhardness testing machine.

The length of an indentation made on hardened steel with the diamond penetrator under a load of .98 newtons (.22 lb.) is about .0015 in. (.038 mm) and the depth of the penetration is about .00005 in. (.001 mm). Fig. 28-41 illustrates the indentation made with the Knoop diamond penetrator.

## 28-15. MICROSCOPIC METHOD OF TESTING WELDS

A test commonly used in the metallurgical laboratory for testing a weld is to procure a sample of the weld and polish it to a very smooth, mirror-

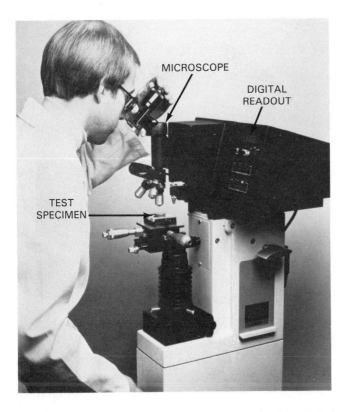

Fig. 28-40. Microhardness testing machine with digital reading. (Page-Wilson Corporation)

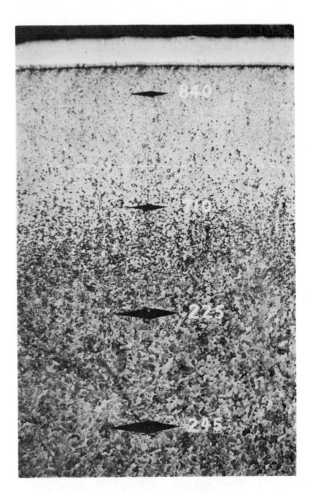

Fig. 28-41. The microhardness indentations made by a Knoop diamond penetration. The white layer at the top of the photograph is chromium (75X).

like finish. After polishing, the sample should show absolutely no scratches on the surface. The sample is then placed under a microscope which magnifies the surface of the metal from 50-5,000 diameters. The usual magnification is 100 or 500 diameters.

The appearance of the metal under the microscope reveals such things as the amount of impurities in the metal, heat treatment, and grain size. In most steels, a microscopic study of a sample can accurately determine the carbon content. Fig. 28-42 shows a microscope used to inspect metal specimens.

Usually, the metal being worked on is treated with acid or etched. To etch a sample after it has been polished, it is wiped with a weak acid solution. The acid solution used for steel is usually 4 percent nitric acid, and 96 percent alcohol, called nitrol. The acid is allowed to remain on the metal for a time. It is then washed off. The metal is then studied under a microscope. Certain acids bring out features, such as the grain boundaries in the metal, the kind of impurities, and slag spots. Poor fusion and shrinkage cracks are easily seen under a microscope. A type of weld frequently given this test is the sample that is removed from boiler seams.

The most important feature of the micrographic

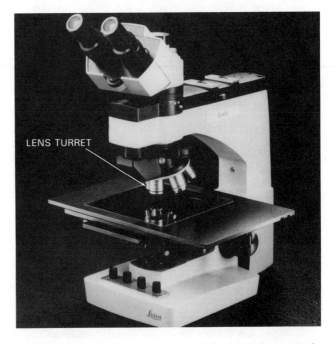

Fig. 28-42. A laboratory type microscope used to examine polished metal specimens. This microscope has four different lenses in a rotating turret. (Leica, Inc., Buffalo, NY)

study of the metal is that photographs may be taken. Photographs are taken of the metal by means of a specially adapted camera attached to the microscope. By taking photographs of each specimen, a very accurate comparison may be made between the samples studied. See Chapter 26 for microphotographs of common metals.

Sections for testing and examination may be cut from the base metal or the weld by a number of methods, one of which is shown in Fig. 28-43. The machine shown uses a circular blade to cut boat-shaped sections of any depth, as shown in Fig. 28-44.

The sections removed may be polished for micro and macroscopic examination. They may also be used for hardness tests, or chemical analysis. Defective areas of a weld may be removed by this method and the cavity filled with a sound weld.

Methods are being developed to perform MICROMETAL ANALYSIS. By this method, extremely small specimens may be removed from the metal to be analyzed. This method enables an analysis to be made of metal parts that are to be placed in service.

## 28-16. MACROSCOPIC METHOD OF TESTING WELDS

A microscopic view of a weld does not cover enough area to obtain a picture of an entire weld

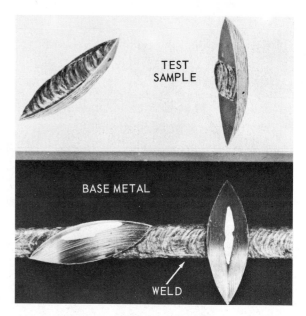

Fig. 28-44. Typical samples removed from a weld for inspection and testing using the machine shown in Fig. 28-43. The samples may be cut at any angle to the weld bead.

for inspection purposes. MACROSCOPIC pictures are only 10 to 40 magnifications (10X to 40X), and are better suited to this purpose. When the sample is deeply etched with hot nitric acid, the structure of the weld stands out more clearly. Fig. 28-45 shows a macroscope with 40X magnifications. The crystalline structure of the metal is not so clearly revealed, but cracks, pits, and pin holes are clearly seen, as shown in Fig. 28-46. Scale inclusions are easily detected by this method. This test also shows up the crystal grain size. A large grain size indicates improper heat treatment after or during welding.

## 28-17. CHEMICAL ANALYSIS METHOD OF TESTING WELDS

The complete investigation of welded material consists of a thorough chemical analysis. This type test is made in a metallurgical laboratory. Most companies are not equipped to perform this test. The chemical analysis may be both qualitative and quantitative. QUALITATIVE ANALYSIS determines the different kinds of chemicals in the metals. The QUANTITATIVE ANALYSIS determines the kind and amount of each chemical in the metal. This type of investigation is necessarily tedious and expensive. The tests are not of direct value to a welder or to a welding company. A chemical analysis is usually used when large quantities of metal are believed to be of the wrong composition. The weldability of a metal is dependent to a great extent on the impurities in the metal.

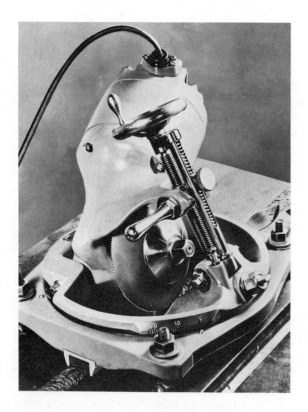

Fig. 28-43. Machine for removing test samples of metal from welds. (Fibre Metal Products Co.)

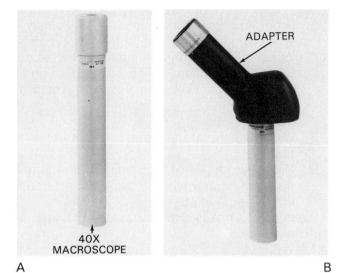

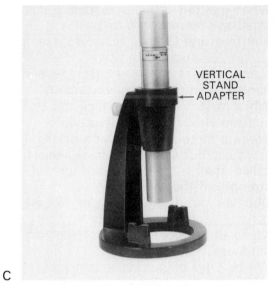

A

B

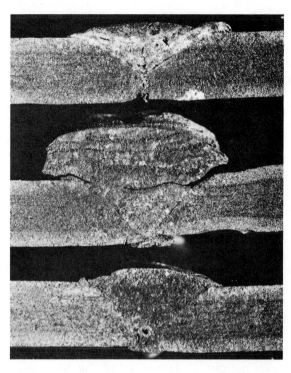

C

D

Fig. 28-45. A small pocket size 40X macroscope is shown at A. At B, C, and D, various adapters for the macroscope are shown. (Leica, Inc., Buffalo, NY)

Fig. 28-46. A macroscope photograph of etched welds in cross section. The quality of the weld is more easily judged than with the naked eye.

Usually the manufacturer of the metal can supply complete data on the physical and chemical properties of the metal.

## 28-18. THE PEEL TEST

Lap joints may be tested to destruction by means of the peel test. The PEEL TEST is most often used to check the strength of a resistance spot weld. All recommended machine settings are made on the spot welding machine before several

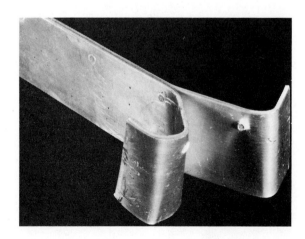

Fig. 28-47. Destructive peel test. Test welds are made to determine weld strength, size, and proper machine settings required for a given job.

spot welds are completed on test pieces.

The test pieces are then peeled apart. See Fig. 28-47. If the spot weld nugget is of the correct diameter and is torn out of one piece, the spot weld is considered to be properly made. All machine settings are considered to be correct for the parts to be welded when the peel test is made satisfactorily.

Spot welds may also be inspected in a nondestructive manner using ultrasonic test methods.

## 28-19. REVIEW OF SAFETY

Proper inspecting and testing of welds requires attention to certain safety factors. There is a danger of parts flying when the weld is broken. A face shield should be worn to protect the operator. The operator should never stand in the likely trajectory of a flying test piece.

One should avoid exposure to the types of radiation used to inspect metals. Only trained experts should handle isotopes used as the radiation source for X-ray inspection. Thick concrete wall, lead, and thick water sections are some of the shielding techniques used to protect operators from X-ray radiation.

## 28-20. TEST YOUR KNOWLEDGE

Write your answers on a separate sheet of paper. Do not write in this book.
1. What is a discontinuity?
2. When does a discontinuity become a defect?
3. What does NDE mean? What does NDT mean?
4. List nine types of NDE.
5. What are the most popular methods of NDE?
6. What equipment may be used to visually inspect welds inside small pipes or cylinders?
7. When making a magnetic particle inspection, the test should be done twice. During the second inspection, the yoke or test prods should be at _____ ° to the direction of the first inspection.
8. What are the steps used in a dye penetrant test?
9. Name two types of NDE tests that require little or no equipment and are most often used in the small shop?
10. In UT, what is the couplant and what is it used for?
11. Which type of X-ray inspection equipment is most portable?
12. What method is used to test for leaks as small as one part in a million?
13. The Izod and Charpy tests are examples of _____ tests.
14. Name three laboratory tests generally not done in a small shop or company.
15. Which test is made to check the elongation and the yield point of a material?
16. When a metal stretches at a certain loading or force, but does not break, this point is called the _____ point or _____ strength of the metal.
17. Ductility is the ability of a metal to _____ before it breaks.
18. Name three types of tests used to test for hardness.
19. Which type of hardness test may be used with minimum damage to the metal surface?
20. _____ pictures are 10 to 40 magnifications (10X – 40X).

# 29 PROCEDURE AND WELDER QUALIFICATIONS

Whenever a building, bridge, ship, or pressure vessel is welded, it is necessary that the manufacturer and buyer reach an agreement on how each weld will be made.

The agreement includes the welding method, base metal specifications and thickness, filler metal composition, preheat and postheat treatment, and other variables which will affect a WELDING PROCEDURE. This agreement also includes the PERFORMANCE QUALIFICATION TESTS for the welders who are to make the welds agreed upon in the welding procedure. All these agreements are written in the form of a CODE.

To eliminate the necessity of writing a new code for each new job, several government agencies, societies and associations have developed codes which may be used. Federal, state, and local governmental agencies, for example, have written building and safety codes. Insurance companies which must insure welded structures have also prepared procedures and welder performance qualification codes. Codes are also written for brazed joints.

## 29-1. GOVERNMENT AGENCIES

Governmental codes and standards are established by the following partial list of agencies:

Bureau of Mines
Federal Aviation Administration (FAA)—aircraft welding
Interstate Commerce Commission (ICC)—cylinders; design, construction, content, etc.
Military Specifications (MIL)
NASA—National Aeronautics and Space Administration
National Bureau of Standards
Occupational Safety and Health Administration (OSHA)
U.S. Air Force
U.S. Army

U.S. Coast Guard
U.S. Navy

## 29-2. ASSOCIATIONS AND SOCIETIES

A sample of the associations and societies that have established codes, standards and/or tentative standards are:

ASTM—formerly the American Society for Testing Materials
American Institute of Steel Construction (AISC)
American National Standards Institute—Voluntary safety, engineering, and industrial standards
American Petroleum Institute (API)—Standard for Field Welding for Pipe Lines (1104)
American Society for Metals (ASM)
American Society of Mechanical Engineers (ASME)—Boiler and Pressure Vessel Code: Section I-Power Boilers, Section VIII-Pressure Vessels, Section IX-Welding and Brazing Qualifications
American Welding Society (AWS)—Structural Welding Code (D1.1-82)
Association of American Railroads (AAR), Operation and Maintenance Division
Institute of Electrical and Electronics Engineers (IEEE)
Mechanical Contractors Association of America
Resistance Welder Manufacturers' Association (RWMA)
SAE—formerly the Society of Automotive Engineers
Tubular Exchanger Manufacturers' Association, Inc. (TEMA)

## 29-3. INSURANCE COMPANIES AND ASSOCIATIONS

Insurance companies or associations having established standards for eligibility for insurance where welding is used include:

Factory Assurance Corporation
The Hartford Steam Boiler Inspection and
Insurance Co.
Lloyd's of London
Lloyd's Register of Shipping Rules and
Regulations
National Board of Fire Underwriters
National Fire Protection Association
Underwriters Laboratories

## 29-4. IMPORTANCE OF CODES AND SPECIFICATIONS

The codes and specifications written by the various organizations have been written to aid the manufacturer in producing a safe product. The codes and specifications that are available are only guide lines to a manufacturer. The codes and specifications do not have to be followed unless they are part of a contract agreement. The manufacturer or builder of a product and the buyer must reach an agreement as to how the product will be built. The agreement is called a CONTRACT.

The contract will refer to the codes and specifications to be used to construct the product. The manufacturer or builder is required to follow those codes and specifications. If the manufacturer or builder does not follow the specifications in the contract, the firm has broken the contract. Companies usually use the established codes rather than write new codes of their own.

There are other examples of when a code must be followed. The Interstate Commerce Commission (ICC) regulates the transportation of all gas cylinders in the United States. Any manufacturer which produces cylinders for use in the United States must follow ICC specifications.

Some states have adopted certain codes as state laws. Any product produced or sold in that state must be manufactured according to the adopted code.

Some jobs may require the builder to follow a number of codes. If a bridge were constructed which also carried oil and pressure pipes over a span of water, several codes may be involved. The American Welding Society (AWS) "Structural Welding Code" may be used for the bridge structure. The oil pipe lines would be welded using the American Petroleum Institute (API) "Standard for Field Welding of Pipe Lines." Pressure pipes may be welded using the American Society of Mechanical Engineers (ASME) "Boiler and Pressure Vessel Code."

All codes applying to a welding job must be available to the welders and assigned inspector on the job. Copies of any welding code are available, for a fee, from the agency, society, or association which publishes them.

## 29-5. PROCEDURE QUALIFICATIONS

PROCEDURE QUALIFICATIONS are limiting instructions written by a manufacturer or contractor which explain how welding will be done. The welding must be done in accordance with a code. These limiting instructions are listed in a document known as a WELDING PROCEDURE SPECIFICATION (WPS).

A welding procedure specification must list in detail:
1. The various base metals to be joined by welding.
2. The filler metal to be used.
3. The range of preheat and postweld heat treatment.
4. Thickness, and other variables described for each welding process.

The variables are listed as essential or nonessential. These variables will be covered in Heading 29-6.

Each manufacturer or contractor must qualify the Welding Procedure Specification (WPS) by welding test coupons and by testing the coupons in accordance with the code. A test weld is made and test coupons (samples) are cut from it. The test coupons are used to make tensile tests, root bends, and face bends as required by the code. The results of these tests are recorded on a document known as a PROCEDURE QUALIFICATION RECORD (PQR).

A different welding procedure specification is required for each change in an essential variable. Each Welding Procedure Specification (WPS) must have a Procedure Qualification Record (PQR) to document the quality of the weld produced. Each welding procedure must be written and the PQR performed before the welding procedure can be used. The WPS and the PQR must be kept on file by the manufacturer.

## 29-6. WELDING PROCEDURE VARIABLES

Most welding codes list the WELDING PROCEDURE VARIABLES which are essential and nonessential.

ESSENTIAL VARIABLES are those which, when changed, will affect the mechanical properties of the weldment. Changes of essential variables require that the welding procedure specifications (WPS) be requalified.

Some of the essential variables for welds made with the shielded metal-arc welding (SMAW) process are changes in:

1. Base metal thickness (beyond accepted limits).
2. Base metal strength or composition.
3. Filler metal used (strength or composition).
4. Preheat temperature.
5. Postheat temperature.
6. Thickness of postheat treatment sample.

NONESSENTIAL VARIABLES are those which, when changed from the approved welding procedure specifications, do not require requalification of the procedure. The changes made must be recorded on the WPS.

For SMAW, some of the nonessential variables are changes in:

1. Type of groove.
2. Deletion of backing in single welded butt joints.
3. Size of the electrode.
4. Addition of other welding positions to those welding positions which are already qualified.
5. Maintenance or reduction of preheat prior to postheat treatment.
6. Current or polarity or the range of amperage or voltage.

## 29-7. P NUMBERS

To reduce the number of welding procedure qualifications required, the American Society of Mechanical Engineers (ASME) has developed a system of grouping metals with similar characteristics. The grouping is done by P NUMBERS. The P number grouping is based on comparable base metal characteristics. The P numbers are further divided into subgroup numbers. These subgroup numbers further refine and divide the P number metals, based on comparable base metal characteristics, such as composition, weldability, and mechanical properties. There are 61 P groups in the ASME Code and many subgroup numbers within each P group. Fig. 29-1 lists metal classified as P1, with subgroups 1, 2, 3, and 4.

Each welding procedure specification (WPS) must have a procedure qualification record (PQR) to document the quality of the weld. When a WPS has a PQR to document the weld quality, it is a qualified procedure. The WPS is only qualified to weld one subgroup number under a specific P number. If a WPS is qualified to weld a P1, group 1 base metal, then any metal listed as a P1, group 1 metal in Fig. 29-1 can be welded using the same WPS.

If the P number or subgroup number is changed, a new WPS must be written and requalified.

## 29-8. PROCEDURE QUALIFICATION SPECIMENS

To qualify a welding procedure specification (WPS), a contractor or manufacturer must use the WPS and weld a TEST PLATE. Several test specimens (samples) are taken from the test plate for testing and approval.

A tension test and a bend test are required. The number and thickness of the required tension test and transverse bend test specimens are shown in Fig. 29-2. Also shown are the requirements for tension and longitudinal bend test specimens.

When a test plate is qualified, it will serve to qualify a range of thicknesses. If a test plate 1/8 in. thick is qualified, it will serve to qualify all thicknesses from 1/16 to 1/4 in., as shown in Fig. 29-2. The maximum thickness in the qualified range is "2t" or twice the test plate thickness.

## 29-9. WELDING PERFORMANCE QUALIFICATIONS

Every welding procedure must be qualified. In addition, every welder and welding operator must be qualified to perform each welding procedure. The welder or welding operator is qualified by passing a welding performance qualification test.

The WELDING PERFORMANCE QUALIFICATION TEST requires the welder or welding operator to complete a weld made in accordance with the welding procedure specification (WPS). Some departures from the exact WPS may be allowed by the code used. Test specimens are cut from the test plate.

The type and number of test specimens required for mechanical testing is specified by the code used. Also specified is the manner by which the specimens are removed from the weld sample. Specimens may also be tested by radiographic examination (X-ray).

The welder or welding operator is qualified by welding position. Welders may be qualified to perform a welding procedure in only one or possibly all positions. A welder qualified only in the flat position is not qualified to make welds in any other position.

Each welder or welding operator must be qualified for each process used. A welder qualified to weld in accordance with one qualified WPS may also be qualified to weld in accordance with another WPS. This is only true when the second WPS and the essential variables are within the limits of the code.

Welders may be required to REQUALIFY if they have not used the specific process for three

| | | | | P-NUMBERS | | |
|---|---|---|---|---|---|---|
| | | | Grouping of Base Metals for Qualification | | | |
| QW | P-No. | Group No. | Base Metal Specification | Minimum Specified Tensile, ksi | Type of Base Metal (Nominal Composition) | |
| | | | Steel and Steel Alloys | | | |
| 422.1 | 1 | 1 | SA-31 Grade A | 45 | Carbon Steel Rivets (C) | |
| | | | Grade B | 58 | Carbon Steel Rivets (C) | |
| | | | SA-36 — | 58 | Carbon Steel Plate (C-Mn-Si) | |
| | | | SA-53 Acid Bessemer | 50 | Carbon Steel Furnace Welded Pipe | |
| | | | Open Hearth | 45 | Carbon Steel Furnace Welded Pipe | |
| | | | Grade A | 48 | Carbon Steel Smls. or Welded Pipe (C) | |
| | | | SA-106 Grade A | 48 | Carbon Steel Pipe (C-Si) | |
| | | | Grade B | 60 | Carbon Steel Pipe (C-Si) | |
| | | | SA-135 Grade A | 48 | Carbon Steel Electric-Resistance-Welded Pipe (C) | |
| | | | Grade B | 60 | Carbon Steel Electric-Resistance-Welded Pipe (C-Mn) | |
| | | | SA-155 Grade C45 | 45 | Carbon Steel Pipe (C) | |
| | | | SA-155 Grade C50 | 50 | Carbon Steel Pipe (C) | |
| | | | Grade C55 | 55 | Carbon Steel Pipe (C) | |
| | | | SA-155 Grade KC60 | 60 | C-Si Steel Pipe (C-Si) | |
| | | | Grade KCF60 | 60 | C-Si Steel Pipe (C-Si) | |
| | | | SA-178 Grade A | 47[1] | Carbon Steel Electric Welded Boiler Tube (C) | |
| | | | Grade C | 60 | Carbon Steel Electric Welded Boiler Tube (C) | |
| | | | SA-179 — | — | Carbon Steel Smls. Low-Carbon Steel Tubes (C) | |
| | | | SA-181 Class 60 | 60 | Carbon Steel Pipe Flanges (C-Si) | |
| | | | SA-192 | 47 | Carbon Steel Boiler Tubes, Smls. (C-Si) | |
| | | | SA-210 Grade A-1 | 60 | Carbon Steel Tubes (C-Si) | |
| | | | SA-216 Grade WCA | 60 | Carbon Steel Castings (C-Si) | |
| | | | SA-226 | 47 | Carbon Steel Electric-Welded Tubes (C-Si) | |
| 422.1 | 1 | 2 | SA-105 | 70 | Carbon Steel Pipe Flanges (C-Si) | |
| | | | SA-106 Grade C | 70 | Carbon Steel Pipe (C-Si) | |
| | | | SA-155 Grade KC70 | 70 | C-Si Steel Pipe (C-Si) | |
| | | | SA-181 Class 70 | 70 | Carbon Steel Pipe Flanges (C-Si) | |
| | | | SA-210 Grade C | 70 | Carbon Steel Electric-Resistance-Welded Steel Tubes (C-Mn-Si) | |
| | | | SA-234 Marking WPB | 70 | Carbon Steel Piping Fittings (C-Mn-Si) | |
| | | | Marking WPC | 70 | Carbon Steel Piping Fittings (C-Mn) | |
| | | | SA-299 | 75 | C-Mn-Si Steel Plates (C-Mn-Si) | |
| | | | SA-350 Grade LF2 | 70 | Carbon Steel Forgings (C-Mn-Si) | |
| | | | SA-372 Type II | 75 | Carbon Steel Forgings (C-Mn-Si) | |
| | | | SA-414 Grade F | 70 | Carbon Steel Sheet (C-Mn) | |
| | | | SA-420 Grade WPL6 | 70 | Carbon Steel Piping Fittings (C-Mn-Si) | |
| | | | SA-455 Type I | 75 | Carbon Manganese Steel Plates (C-Mn) | |
| | | | SA-487 Class AQ | 70 | Carbon Steel Castings (C) | |
| | | | SA-508 Class 1 | 70 | Forgings (C-Si Steel 0.35 max. C) (C-Si) | |
| | | | SA-515 Grade 70 | 70 | C-Si Steel Plates (C-Si) | |
| | | | SA-516 Grade 70 | 70 | C-Mn-Si Steel Plates (C-Mn-Si) | |
| 422.1 | 1 | 3 | SA-487 Class BQ | 80 | Carbon Steel Castings (C) | |
| | | | SA-537 Class 2 | 80 | C-Mn-Si Steel Plates (C-Mn-Si) | |
| | | | SA-671 Grade CD80 | 80 | C-Mn-Si Steel Pipe (C-Mn-Si) | |
| | | | SA-672 Grade D80 | 80 | C-Mn-Si Steel Pipe (C-Mn-Si) | |
| | | | SA-691 Grade CMSH-80 | 80 | C-Mn-Si Steel Pipe, Fusion Welded (C-Mn-Si) | |
| | | | SA-737 Grade C | 80 | C-Mn-V-N Plates | |
| | | 4 | SA-724 Grade A (5/8 in. max. T) | 90 | Carbon Steel Plate (C-Mn-Si) | |
| | | | Grade B (5/8 in. max. T) | 95 | Carbon Steel Plate (C-Mn-Si) | |

Note: (1) Tensile value is expected minimum.

Fig. 29-1. A partial listing of P1 groups 1, 2, 3, 4 base metals. To find a complete listing of all P numbers and groups, refer to ASME section IX Welding & Brazing Qualifications.

## TENSION TEST AND TRANSVERSE BEND TESTS

| Thickness T of Test Coupon Welded, in. [Note (1)] | Range of Thickness T of Base Metal Qualified, in. [Note (2)] | | Range of Thickness t of Deposited Weld Metal Qualified, in. [Note (2)] | | Type and Number of Tests Required (Tension and Guided Bend Tests) — Note (6) | | | |
|---|---|---|---|---|---|---|---|---|
| | Min. | Max. | Min. | Max. | Tension QW-462.1[4] | Side Bend QW-462.2 | Face Bend QW-462.3(a) | Root Bend QW-462.3(a) |
| Less than 1/16 | T | 2T | t | 2t | 2 | — | 2 | 2 |
| 1/16 to 3/8, incl. | 1/16 | 2T | 1/16 | 2t | 2 | — | 2 (5) | 2 (5) |
| Over 3/8, but less than 3/4 | 3/16 (8) | 2T | 3/16 (8) | 2t | 2 | Note(3) | 2 (5) | 2 (5) |
| 3/4 to less than 1 1/2 | 3/16 (8) | 2T | 3/16 (8) | 2t when t < 3/4 | 2 | 4 | — | — |
| 3/4 to less than 1 1/2 | 3/16 (8) | 2T | 3/16 (8) | 2t when t ≥ 3/4 | 2 | 4 | — | — |
| 1 1/2 and over | 3/16 (8) | 8 (7) | 3/16 (8) | 2t when t < 3/4 | 2 | 4 | — | — |
| 1 1/2 and over | 3/16 (8) | 8 (7) | 3/16 (8) | 8 (7) when t ≥ 3/4 | 2 | 4 | — | — |

NOTES:

(1) When the groove weld is filled using two or more welding processes, the thickness $t$ of the deposited weld metal for each welding process shall be determined and used in the "Range of thickness $t$" column. The test coupon thickness $T$ is applicable for each welding process.

(2) See QW-403 (.2, .3, .6, .7, .9, .10) and QW-407.4 for further limits on range of thicknesses qualified.

(3) Four side bend tests may be substituted for the required face and root bend tests.

(4) The deposited weld metal of each welding process shall be included in the tension test of QW-462.1(a), (b), (c), or (e); and in the event turned specimens of QW-462.1(d) are used, the deposited weld metal of each welding process shall be included in the reduced section insofar as possible.

(5) Applicable for a combination of welding processes only when the deposited weld metal of each welding process is on the tension side of either the face or root bend.

(6) When toughness testing is a requirement of other Sections, it shall be applied with respect to each welding process.

(7) For the welding processes of QW-403.7 only; otherwise per Note (2) or 2T, whichever is applicable.

(8) When the weld metal thickness deposited by a process is 3/8 in. or less, this minimum shall be 1/16 in. for that process.

## TENSION TESTS AND LONGITUDINAL BEND TESTS

| Thickness T of Test Coupon Welded, in. [Note (1)] | Range of Thickness T of Base Metal Qualified, in. [Note (2)] | | Range of Thickness t of Deposited Weld Metal Qualified, in. [Note (2)] | | Type and Number of Tests Required (Tension and Guided Bend Tests) - Note (5) | | |
|---|---|---|---|---|---|---|---|
| | Min. | Max. | Min. | Max. | Tension QW-462.1[3] | Face Bend QW-462.3(b) | Root Bend QW-462.3(b) |
| Less than 1/16 | T | 2T | t | 2t | 2 | 2 | 2 |
| 1/16 to 3/8, incl. | 1/16 | 2T | 1/16 | 2t | 2 | 2 (4) | 2 (4) |
| Over 3/8 | 3/16 (6) | 2T | 3/16 (6) | 2t | 2 | 2 (4) | 2 (4) |

NOTES:

(1) When the groove weld is filled using two or more welding processes, the thickness $t$ of the deposited weld metal for each welding process shall be determined and used in the "Range of thickness $t$" column. The test coupon thickness $T$ is applicable for each welding process.

(2) See QW-403 (.2, .3, .6, .7, .9, .10) and QW-407.4 for further limits on range of thicknesses qualified.

(3) The deposited weld metal of each welding process shall be included in the tension test of QW-462.1(a), (b), (c), or (e); and in the event turned specimens of QW-462.1(d) are used, the deposited weld metal of each welding process shall be included in the reduced section insofar as possible.

(4) Applicable for a combination of welding processes only when the deposited weld metal of each welding process is on the tension side of either the face or root bend.

(5) When toughness testing is a requirement of other Sections, it shall be applied with respect to each welding process.

(6) When the weld metal thickness deposited by a process is 3/8 in. or less, this minimum shall be 1/16 in. for that process.

Fig. 29-2. ASME procedure qualification specimens. The QW numbers are article numbers in the ASME Code.

months. They may also be required to requalify if there is a reason to question their ability to make welds which meet the welding procedure specification.

A welder shall be requalified for a process whenever a change is made in one or more of the essential variables. The essential variables for shielded metal-arc welding (SMAW) listed in the ASME Boiler and Pressure Vessel Code are:

1. The deletion of backing in single welded butt joints.
2. A change in the ASME electrode specification number or AWS electrode classification number.
3. The addition of other welding positions than those in which the welder has already qualified.
4. A change from upward to downward, or from downward to upward, in the progress of the weld.

TABULATION OF POSITIONS
OF WELDS

| Position | Diagram Reference | Inclination of Axis | Rotation of Face |
|---|---|---|---|
| Flat | A | 0° to 15° | 150° to 210° |
| Horizontal | B | 0° to 15° | 80° to 150° 210° to 280° |
| Overhead | C | 0° to 80° | 0° to 80° 280° to 360° |
| Vertical | D E | 15° to 80° 80° to 90° | 80° to 280° 0° to 360° |

Fig. 29-3. Welding test positions. The horizontal reference plane is taken to lie always below the weld under consideration. Inclination of axis is measured from the horizontal reference plane toward the vertical. Angle of rotation of face is measured from a line perpendicular to the axis of the weld and lying in a vertical plane containing this axis. The reference position (0 deg.) of rotation of the face invariably points in the direction opposite to that in which the axis angle increases. The angle of rotation of the face of the weld is measured in a clockwise direction from this reference position (0 deg.) when looking at point P.   (AWS A3.0)

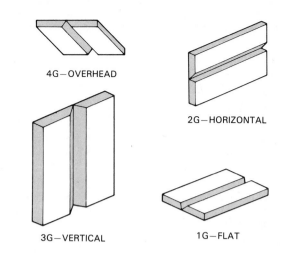

**GROOVE WELDS IN PLATE**

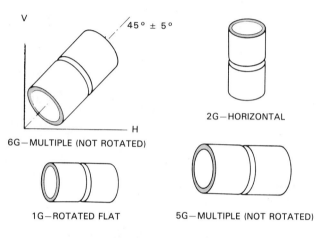

**GROOVE WELDS IN PIPE**

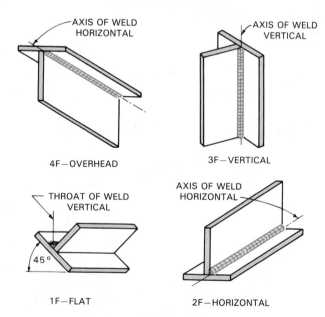

**FILLET WELDS IN PLATE**

Fig. 29-4. Test positions for various joints. G stands for groove weld, F stands for fillet weld. Note that test position 5G and 6G are not rotated and the weld position changes as the weld is made.   (AWS A3.0)

## 29-10.  WELDING QUALIFICATION SPECIMENS

Welding positions and test specimen specifications illustrated in the following headings are taken from the American Society of Mechanical Engineers (ASME) Boiler and Pressure Vessel Code, Section IX.

This information is provided for those who may wish to practice making specimens similar to those which may be required for welder qualification tests. This information is only a small part of the complete code and should not be used as a code. The complete code should be obtained from the agency, society, or association which writes the code.

## 29-11.  WELDING TEST POSITIONS

The weld positions shown in Figs. 29-3, 29-4, 29-5, and 29-6 are typical of those used for many welding codes. A welder or welding operator must be tested to become qualified for each of the positions required in a welding procedure specification

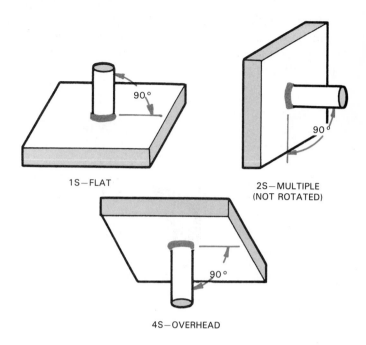

Fig. 29-6.  Test positions for stud welds. (ASME—Section IX)

(WPS). A qualification in one position does not automatically qualify the welder to weld in any other position. When the welding position changes, the welder must be requalified for that position.

## 29-12.  METHODS OF TESTING SPECIMENS

There are six different weld specimen tests:
1. Tension-reduced section (Heading 29-13).
2. Side bend (Heading 29-14).
3. Face and root bend - transverse (Heading 29-15).
4. Face and root bend - longitudinal (Heading 29-15).
5. Fillet weld - procedure (Heading 29-16).

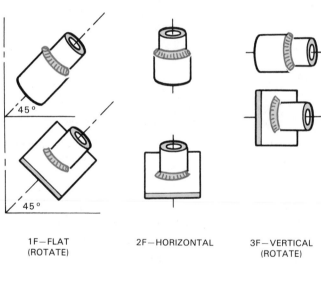

1F—FLAT (ROTATE)  2F—HORIZONTAL  3F—VERTICAL (ROTATE)

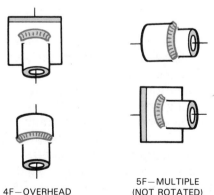

4F—OVERHEAD  5F—MULTIPLE (NOT ROTATED)

Fig. 29-5.  Test positions for fillet welds in pipe joints. (ASME—Section IX)

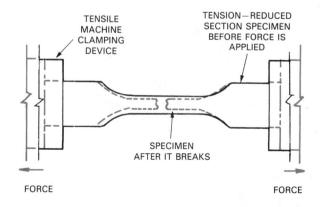

Fig. 29-7.  Tension-reduced section test. When force is applied, the specimen is stretched until it breaks. The maximum force is noted and the tensile strength is calculated.

6. Fillet weld - performance (Heading 29-16).

Fig. 29-7 shows how the tension-reduced section specimen is tested. Fig. 29-8 demonstrates the side bend test.

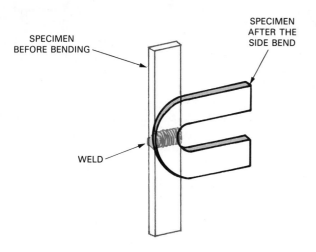

Fig. 29-8. The side bend. Note the direction of the bend.

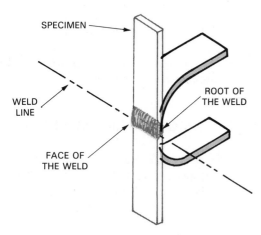

A — TRANSVERSE FACE BEND
The weld face is on the outside of the bend

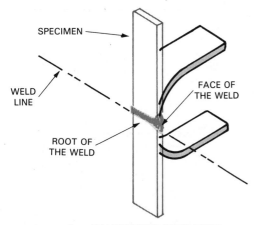

B — TRANSVERSE ROOT BEND
The root of the weld is on the outside of the bend

Fig. 29-9. Transverse face and root bend. The transverse sample is cut across the weld line.

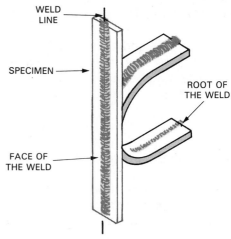

A — LONGITUDINAL FACE BEND
The weld face is on the outside of the bend

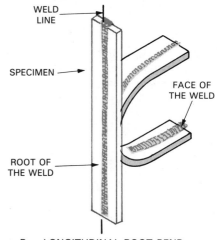

B — LONGITUDINAL ROOT BEND
The root of the weld is on the outside of the curve

Fig. 29-10. Longitudinal face and root bend. The longitudinal sample is cut in the same direction as the weld line.

Fig. 29-9 shows the transverse face and root bend samples before and after bending. The longitudinal face and root bend samples are shown before and after testing in Fig. 29-10.

## 29-13. TENSION-REDUCED SECTION TESTING

The tension-reduced section test sample should be cut from the test weld from the area shown in Fig. 29-11, A, B, and C.

The sample must then be prepared according to the dimensions shown in Fig. 29-12, A or B.

Before subjecting the weld sample to a tension load, the thickness and width of the weld at the narrowest point must be measured.

The cross-sectional area of the weld is obtained by multiplying the weld width times its thickness. The tensile strength is found by dividing the maximum tensile load by the cross-sectional area of the weld.

| Discard | | This Piece |
|---|---|---|
| Reduced Section | | Tensile Specimen |
| Root Bend | | Specimen |
| Face Bend | | Specimen |
| Root Bend | | Specimen |
| Face Bend | | Specimen |
| Reduced Section | | Tensile Specimen |
| Discard | | This Piece |

| Discard | | This Piece |
|---|---|---|
| Side Bend | | Specimen |
| Reduced Section | | Tensile Specimen |
| Side Bend | | Specimen |
| Side Bend | | Specimen |
| Reduced Section | | Tensile Specimen |
| Side Bend | | Specimen |
| Discard | | This Piece |

**PLATES—¹/₁₆ TO ³/₄ IN.
PROCEDURE QUALIFICATION**

A

**PLATES—OVER ³/₄ AND
ALTERNATE ³/₈ TO ³/₄ IN.
PROCEDURE QUALIFICATION**

B

| Discard |
|---|
| Reduced-Section Tensile Specimen |
| |
| |
| Reduced-Section Tensile Specimen |
| |
| Discard |

—Longitudinal Face-Bend Specimen

—Longitudinal Root-Bend Specimen

—Longitudinal Face-Bend Specimen

—Longitudinal Root-Bend Specimen

C

**PLATES—LONGITUDINAL    PROCEDURE QUALIFICATION**

Fig. 29-11. Order of removal of test specimens from the test weld.
(ASME—Section IX)

Example:

    1/4 in. (6.4 mm) thickness
    × 3/4 in. (19.1 mm) width
    = 3/16 in.² (122.2 mm²) area.

If the tensile load is 11,250 lb. (50 042 N) then:

$$\text{Tensile strength} = \frac{\text{tensile load}}{\text{area}}$$

$$\text{Tensile strength} = \frac{11250 \ (50042)}{3/16 \ (122.2)}$$

$$= 60,000 \text{ lbs./in.}^2$$
$$(409.5 \text{ N/mm}^2)$$

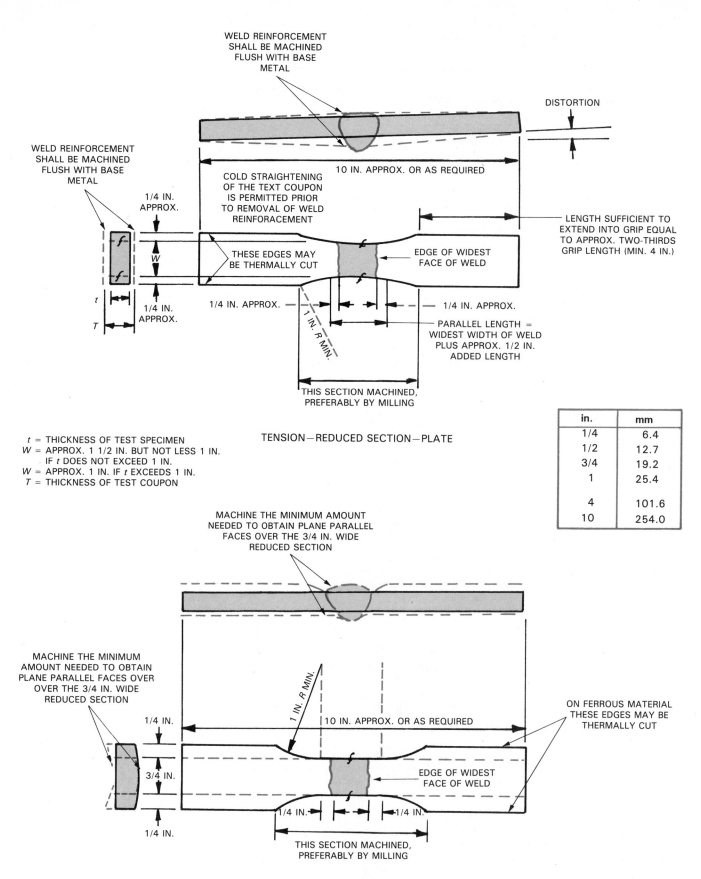

WELD REINFORCEMENT
SHALL BE MACHINED
FLUSH WITH BASE
METAL

DISTORTION

WELD REINFORCEMENT
SHALL BE MACHINED
FLUSH WITH BASE
METAL

1/4 IN.
APPROX.

10 IN. APPROX. OR AS REQUIRED

COLD STRAIGHTENING
OF THE TEXT COUPON
IS PERMITTED PRIOR
TO REMOVAL OF WELD
REINFORACEMENT

LENGTH SUFFICIENT TO
EXTEND INTO GRIP EQUAL
TO APPROX. TWO-THIRDS
GRIP LENGTH (MIN. 4 IN.)

THESE EDGES MAY
BE THERMALLY CUT

EDGE OF WIDEST
FACE OF WELD

1/4 IN.
APPROX.

1/4 IN. APPROX.

1 IN. R MIN.

1/4 IN. APPROX.

PARALLEL LENGTH =
WIDEST WIDTH OF WELD
PLUS APPROX. 1/2 IN.
ADDED LENGTH

THIS SECTION MACHINED,
PREFERABLY BY MILLING

$t$ = THICKNESS OF TEST SPECIMEN
$W$ = APPROX. 1 1/2 IN. BUT NOT LESS 1 IN.
     IF $t$ DOES NOT EXCEED 1 IN.
$W$ = APPROX. 1 IN. IF $t$ EXCEEDS 1 IN.
$T$ = THICKNESS OF TEST COUPON

TENSION—REDUCED SECTION—PLATE

| in. | mm |
|-----|-------|
| 1/4 | 6.4 |
| 1/2 | 12.7 |
| 3/4 | 19.2 |
| 1 | 25.4 |
| 4 | 101.6 |
| 10 | 254.0 |

MACHINE THE MINIMUM AMOUNT
NEEDED TO OBTAIN PLANE PARALLEL
FACES OVER THE 3/4 IN. WIDE
REDUCED SECTION

MACHINE THE MINIMUM
AMOUNT NEEDED TO OBTAIN
PLANE PARALLEL FACES OVER
OVER THE 3/4 IN. WIDE
REDUCED SECTION

ON FERROUS MATERIAL
THESE EDGES MAY BE
THERMALLY CUT

1/4 IN.

3/4 IN.

1 IN. R MIN.

10 IN. APPROX. OR AS REQUIRED

EDGE OF WIDEST
FACE OF WELD

1/4 IN.

1/4 IN.

1/4 IN.

THIS SECTION MACHINED,
PREFERABLY BY MILLING

TENSION—REDUCED SECTION—PIPE

Fig. 29-12. Specifications for preparing tension-reduced section specimens from plate and pipe samples.
(ASME—Section IX)

## 29-14. SIDE BEND TEST

The side bend test sample is removed from the test weld as indicated in Fig. 29-11, part B. Each sample must be bent in a jig with the dimensions as shown in Fig. 29-13.

To be acceptable, the side bend test and all other bend tests must not break. According to the ASME, the bend test may not show defects larger than those indicated in the code article QW-163:

*The weld and heat-affected zone of a transverse-weld bend specimen shall be completely within the bent portion of the specimen after testing.*

*The guided-bend specimens shall have no open defects exceeding 1/8 in. (3.2 mm), measured in any direction on the convex surface of the specimen after bending, except*

| Thickness of Specimens | | A | | B | | C | | D | | Material | Refer To |
|---|---|---|---|---|---|---|---|---|---|---|---|
| in. | mm | in. | mm | in. | mm | in. | mm | in. | mm | | |
| 3/8 | 9.5 | 1 1/2 | 38.1 | 3/4 | 19.1 | 2 3/8 | 60.3 | 1 3/16 | 30.2 | All | |
| $t$ | $t$ | $4t$ | $4t$ | $2t$ | $2t$ | $6t + 1/8$ | $6t + 3.2$ | $3t + 1/16$ | $3t + 1.6$ | Others | |
| 1/8 | 3.2 | 2 1/16 | 52.4 | 1 1/32 | 26.2 | 2 3/8 | 60.3 | 1 3/16 | 30.2 | P-No. 23 and P-No. 35 except as shown below | QW-422.23 QW-422.35 |
| 3/8 | 9.5 | 2 1/2 | 63.5 | 1 1/4 | 31.8 | 3 3/8 | 85.7 | 1 11/16 | 42.9 | P-No. 11, P-No. 25, SB-148 Alloys CDA-952 & 954 and SB-271 Alloy CDA-952 & 954 | QW-422.11 QW-422.25 QW-422.35 |
| $t$ | $t$ | $6 2/3t$ | $169.3t$ | $3 1/3t$ | $84.7t$ | $8 2/3t + 1/8$ | $220t + 3.2$ | $4 1/2t + 1/16$ | $114.3t + 1.6$ | | |
| 1/16-3/8 in. incl | 1.6-9.5 | $8t$ | $8t$ | $4t$ | $4t$ | $10t + 1/8$ | $10t + 3.2$ | $5t + 1/16$ | $5t + 1.6$ | P-51 | QW-422.51 |
| 1/16-3/8 in. incl | 1.6-9.5 | $10t$ | $10t$ | $5t$ | $5t$ | $12t + 1/8$ | $12t + 3.2$ | $6t + 1/16$ | $6t + 1.6$ | P-No. 52 | QW-422.52 |
| 1/16-3/8 in. incl | 1.6-9.5 | $10t$ | $10t$ | $5t$ | $5t$ | $12t + 1/8$ | $12t + 3.2$ | $6t + 1/16$ | $6t + 1.6$ | P-No. 61 | QW-422.61 |

| in. | mm |
|---|---|
| 1/8 | 3.2 |
| 1/4 | 6.4 |
| 1/2 | 12.7 |
| 3/4 | 19.1 |
| 1 1/8 | 28.6 |
| 1 1/2 | 38.1 |
| 2 | 50.8 |
| 3 | 76.2 |
| 3 7/8 | 98.4 |
| 6 3/4 | 171.5 |
| 7 1/2 | 190.5 |
| 9 | 228.6 |

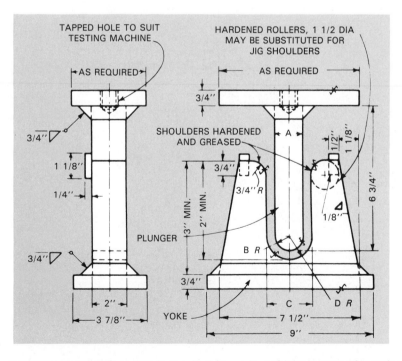

Fig. 29-13. Guided-bend jig specifications. Notice that a different jig is required for various thicknesses and metals. The QW-422 numbers referred to are ASME P number metal specifications. (ASME—Section IX). The millimeter equivalents are not shown as a part of the ASME-Section IX code. They are shown here for reference.

*that cracks occurring on the corners of the specimen during testing shall not be considered, unless there is definite evidence that they result from slag inclusions or other internal defects. For corrosion resistant weld overlay cladding, no open defect exceeding 1/16 in. (1.6 mm) measured in any direction shall be permitted in the cladding, and no open defects exceeding 1/8 in. (3.2 mm) shall be permitted in the bond line.*

## 29-15. FACE AND ROOT BENDS

Weld samples are generally welded from one side only. The FACE of the weld is the surface of the joint where the weld bead is applied. The ROOT or bottom of the weld is the surface opposite the weld bead.

After a test weld is completed, test specimens (samples) are cut from the test weld as shown in Fig. 29-11.

When the test specimen is cut across the weld, the specimen is called a TRANSVERSE SPECIMEN. The test specimen may be cut parallel (in the same direction) to the weld. In this case, it is a LONGITUDINAL SPECIMEN.

There are four types of face and root bends. They are:
1. Transverse face bend.
2. Transverse root bend.
3. Longitudinal face bend.

4. Longitudinal root bend.

Each face and root bend specimen must be bent in a jig constructed as shown in Fig. 29-13.

All bends, to be acceptable in the ASME Code, must not have defects larger than 1/8 in. (3.2 mm) as stated in article QW-163. Refer to Heading 29-14.

## 29-16. FILLET WELD PROCEDURE TEST

A weld sample must be prepared according to the dimensions shown in Fig. 29-14. This illustration also shows how specimens for the procedure test are cut from the test weld.

In order to pass the ASME test, a specimen must satisfy article QW-183:
*QW-183 Macro-Examination - Procedure Specimens*

*One face of each cross section shall be smoothed and etched with a suitable etchant (See QW-470) to give a clear definition of the weld metal and heat-affected zone. In order to pass the test:*

*Visual examination of the cross section of the weld metal and heat-affected zone shall show complete fusion and freedom from cracks; and*

*There shall not be more than 1/8 in. (3.2 mm) difference in the length of the legs of the fillet.*

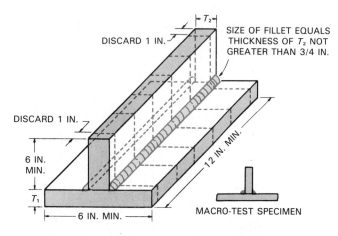

MACRO TEST: THE FILLET SHALL SHOW FUSION AT THE ROOT OF THE WELD BUT NOT NECESSARILY BEYOND THE ROOT. THE WELD METAL AND HEAT AFFECTED ZONE SHALL BE FREE OF CRACKS.

QW-462.4(a) FILLET WELDS — PROCEDURE

| $T_1$ | $T_2$ |
|---|---|
| 1/8 IN. AND LESS OVER 1/8 IN. | $T_1$ EQUAL TO OR LESS THAN $T_1$ BUT NOT LESS THAN 1/8 IN. |

| in. | mm |
|---|---|
| 1/8 | 3.2 |
| 3/4 | 19.1 |
| 1 | 25.4 |
| 6 | 152.4 |
| 12 | 304.8 |

Fig. 29-14. Specifications for preparing a fillet weld procedure qualification weld. This figure also shows how the test specimen is removed from the weld sample. (ASME—Section IX). The millimeter equivalents are not shown as part of the ASME—Section IX code. They are shown here for reference.

## 29-17. FILLET WELD PERFORMANCE TEST

The performance test sample for the fillet weld must be prepared and made as shown in Fig. 29-15, A or B. Test specimens are removed from the weld sample as shown in the illustration.

The test specimens are placed under a load to bend the two parts (stems) flat on each other. To pass this performance test, the specimen must agree with the ASME article QW-182:

*QW-182 Fracture Tests*

*The stem of the 4 in. (102 mm) performance specimen center section in QW-462.4 (b) or the stem of the quarter section in QW-462.4 (c), as applicable, shall be loaded laterally in such a way that the root of the weld is in tension. The load shall be steadily increased until the specimen fractures or bends flat upon itself.*

*If the specimen fractures, the fractured surface shall show no evidence of cracks or incomplete root fusion, and the sum of the lengths of inclusions and gas pockets visible on the fractured surface shall not exceed 3/4 in. (19.1 mm) to pass the test.*

A macro-examination is then made as shown in article QW-184:

*QW-184 Macro-Examination - Performance Specimens*

*The cut end of one of the end sections from the plate or the cut end of the quarter section from the pipe, as applicable, shall be smoothed and etched with a suitable etchant (See QW-470) to give a clear definition of the weld metal and heat-affected zone. In order to pass the test:*

*Visual examination of the cross section of the*

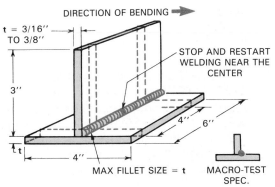

### A  PERFORMANCE TEST SPECIMEN FOR A FILLET WELD IN PLATE

| in. | mm |
|------|-------|
| 3/16 | 4.8 |
| 3/8 | 9.5 |
| 2 | 50.8 |
| 3 | 76.2 |
| 4 | 101.6 |
| 6 | 152.4 |

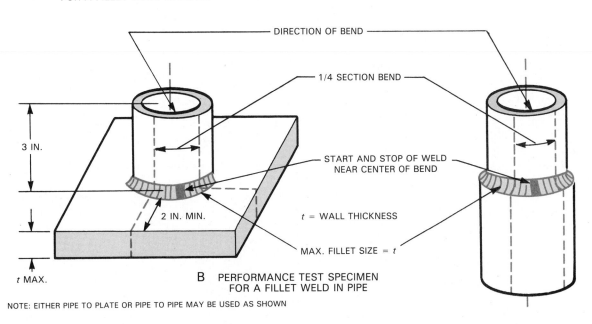

### B  PERFORMANCE TEST SPECIMEN FOR A FILLET WELD IN PIPE

NOTE: EITHER PIPE TO PLATE OR PIPE TO PIPE MAY BE USED AS SHOWN

Fig. 29-15.  Specifications for preparing a fillet weld performance qualification weld. This figure also shows how the test specimen is removed from the weld sample.   (ASME—Section IX). The millimeter equivalents are not shown as a part of the ASME—Section IX code. They are shown here for reference.

*weld metal and heat-affected zone shall show complete fusion and freedom from cracks, except that linear indications at the root, not exceeding 1/32 in. (0.8 mm) shall be acceptable; and*

*The weld shall not have a concavity or convexity greater than 1/16 in. (1.6 mm); and*

*There shall be not more than 1/8 in. (3.2 mm) difference in the lengths of the legs of the fillet.*

## 29-18. TEST YOUR KNOWLEDGE

Write your answers on a separate sheet of paper. Do not write in this book.

1. When a structure is to be built, the _____ and _____ must reach an agreement on how each weld will be made.
2. Name four organizations that have prepared welding codes.
3. What organization has prepared a code for structural welding, such as for bridges and similar structures?
4. What organization has prepared a welding code for pressure vessels?
5. Why have different codes and specifications been written?
6. When must a code or specification be followed?
7. Must an inspector on a job have all the codes for that job available to him?
8. What do the letters WPS mean?
9. What do the letters PQR stand for?
10. How is a WPS qualified?
11. What must be done if an essential variable on a WPS is changed?
12. What must be done if a nonessential variable on a WPS is changed?
13. What is the meaning of P numbers as used in connection with the ASME code?
14. If a test plate of 1/8 in. (3.2 mm) thickness is qualified, what is the range of thicknesses that may be brought in under this qualification?
15. Under what conditions is it necessary to requalify a welder qualified in a given procedure?
16. Is it necessary that a welder be qualified for different welding positions?
17. What are three of the essential welding procedure variables for shielded metal arc welding?
18. What position of welding does a 3F signify? a 5G?
19. Name four specimen tests for qualifying welders.
20. An E7018 electrode is used to butt weld two 1/2 in. (12.7 mm) plates together. A tension reduced section specimen is prepared with a width of 3/4 in. (19.1 mm). The specimen was tested. The maximum load is 27,000 lbs. (120,100 N). What is the tensile strength of this weld sample?

# 30 THE WELDING SHOP

Welding shops may be divided into three principal groups:
1. The independent shop which specializes in repairing and fabricating metal structures by the various methods of welding.
2. The welding shop, or department, attached to a factory or manufacturing establishment.
3. The welding equipment shop which overhauls and repairs welding equipment and sells equipment and supplies.

The equipment found in each shop will vary depending on the type and amount of repair or production being done.

Various types of supplies and equipment found in a welding shop will be explained and illustrated in this chapter. Welding machines and stations have been explained in other chapters, and will not be included in this chapter.

## 30-1. WELDING SHOP DESIGN

The architectural design of the welding shop is important. A typical shop would incorporate these features:
1. Heavy-duty load bearing floors, preferably of concrete.
2. A fire resistant building structure.
3. A building which is well ventilated and has provision for localized exhaust ventilation.
4. Large doors and some means of moving heavy equipment and material into and out of the shop.
5. Heavy-duty electrical service readily available.

The concrete floors should be at the ground level. The utilities should be arranged around the outside of the room and overhead. A monorail or a double rail crane system should be installed to provide easy movement of equipment and material to any spot in the room. If the business is of sufficient size to justify accessory rooms such as paint booths, storerooms, toilets, showers,

locker rooms, and offices, these should be in an annex to the shop for safety, cleanliness, and to reduce the noise level.

## 30-2. WELDING SHOP EQUIPMENT

Equipment used in the welding shop depends to a considerable extent upon the kind of work handled. Some of the common equipment used is:
1. Oxyfuel gas welding stations.
2. AC and DC arc welding stations.
3. GMAW and GTAW arc welding stations.
4. Resistance welding machines.
5. Oxyfuel gas cutting equipment.
6. Preheating and postheating furnaces.
7. Overhead crane and/or heavy-duty hoists.
8. Benches, vises, anvils.
9. Jigs and fixtures.
10. Heavy-duty power tools such as lathes, drill presses, grinders, nibblers, shears, brakes.
11. Grit blasting equipment.
12. Weld inspecting and testing equipment.
13. A paint booth.
14. Weld positioners.
15. Stationary and portable ventilation units.
16. Portable screens to set up around shop floor welding jobs, see Fig. 30-1.

Local and state Building and Safety Codes must be followed when designing, building, and using all welding shops. The shop must be adequately equipped with personnel safety equipment, first aid supplies, and fire safety equipment such as fire extinguishers and blankets.

## 30-3. AIR VENTILATION AND CONDITIONING EQUIPMENT

Air contamination in the shop must be kept to a minimum for health and good housekeeping. The operators must have sufficient clean air and oxygen to eliminate respiratory problems, and to be

Fig. 30-1. A portable welding booth with filter quality plastic screening material. This type screen can be set up anywhere to protect other workers from ultraviolet rays. (Singer Safety Company)

Fig. 30-2. Shop-wide fume extractor system. The pickup hoods can be positioned as desired. (Nederman, Inc.)

Fig. 30-3. This portable fume extractor has an electrostatic precipitator to clean and filter the fumes that pass through it. It can be set up at the site of large shop floor welding jobs. (Nederman, Inc.)

comfortable. The temperatures and humidity conditions should be comfortable. Toxic gases and toxic dust particles must be held to an absolute minimum by the proper selection of the materials used. If dust and gases occur, they must be withdrawn and filtered. Welders who work in highly toxic areas should wear supplied air breathing equipment or at least filtered air gear.

Welding stations which produce dust particles and gases should have adequate exhaust ventilation. The amount of dust produced varies with the size of the equipment, kind of metal being heated, fluxes used and the like. It is desirable to have an industrial hygiene technician take air samples under operating conditions and to adjust or modify the ventilating system until safe working conditions prevail.

Venting exhaust fumes from a welding station is shown in Fig. 30-2.

Fig. 30-3 illustrates a portable fume extractor. This unit can be set up on the site of a large shop floor welding job.

Air which is exhausted must be replaced. When the outdoor temperatures are comfortable, open doors and windows provide the air replacement. However, during cold weather conditions, the replacement air must be both heated and humidified. Provisions must be included in the heating plant for seasonal shop requirements.

## 30-4. MECHANICAL METAL CUTTING EQUIPMENT

Metal may be cut to size and beveled to specifications prior to welding by several different processes:
1. Abrasive cutting wheel.
2. Power hacksaw.
3. Metal bandsaw.
4. Shears.
5. Nibblers.

Power hacksaws, shears, and power cutoff saws are needed to cut standard size stock to needed lengths. Fig. 30-4 illustrates an abrasive wheel cutoff machine. The machine illustrated is enclosed to prevent accidents.

Where 90 deg. or other standard angle cuts are needed, a reciprocating type power hacksaw or a band type power hacksaw are commonly used. Metal cutting bandsaws are used for both straight and contour cuts.

Metal shears of all types, both manual and power operated, find extensive use in welding shops. Floor mounted, manually operated floor shears are sometimes used. The powered floor shear is more often found in shops. Shears are often equipped with different blades which make it possible to cut flat stock, angle iron, and bar stock to length. Squaring shears are used for thin metal while alligator jaw shears (some with

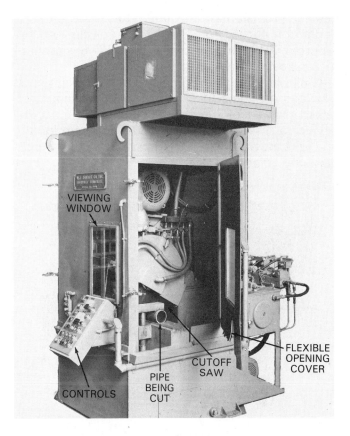

Fig. 30-4. An abrasive wheel cutoff saw. The unit is enclosed to provide safety. The part being cut can extend out of the machine through the flexible covers in the sides of the cabinet. (W.J. Savage Co., Inc.)

Fig. 30-5. This hydraulic shear will cut 3/4 in. (approx. 20 mm) thick mild steel plate up to 20 ft. (6.1 m) long. (Niagara Machine & Tool Works)

special provisions for round stock cutting and angle iron cutting) are used for stock up to 8 in. (approx. 200 mm) in width. Fig. 30-5 shows a hydraulically powered squaring shear.

A nibbler machine complete with attachments provides a means of cutting irregular shapes for fabricating sheet metal and plate. This machine uses two small, sharp steel blades, one stationary and one powered moving blade. The cutting is done by shearing as shown in Fig. 30-6. It is

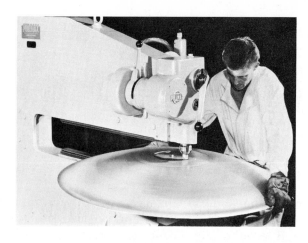

Fig. 30-6. Enlarged view of a cut made by a nibbler machine.

Fig. 30-7. Electrically powered floor model nibbler machine making a circular shearing cut. (American Pullmax Co.)

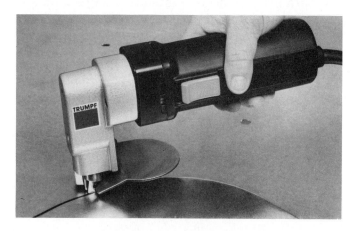

Fig. 30-8. A light duty portable electric nibbler. The tool can cut on straight or curved lines. (Trumpf, Inc.)

possible to cut straight or curved lines with this type machine. Fig. 30-7 shows a floor model and Fig. 30-8 a portable model. Portable nibblers will cut metal up to 1/4 in. (6.4 mm) thick. The nibbler is an excellent tool for cutting curved shapes out of sheet metal. The internal construction of a nibbler showing the mechanism which produces the

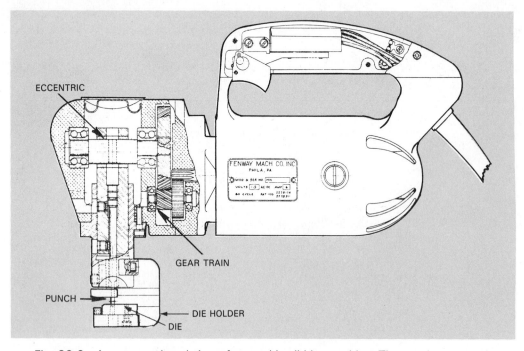

Fig. 30-9. A cross sectioned view of a portable nibbler machine. The rotating eccentric makes the upper punch or cutter go up and down. (Fenway Machine Co., Inc.)

shearing action is illustrated in Fig. 30-9.

Tools have been developed for cutting tubing and pipe of all diameters and thicknesses to produce the special notched ends needed to weld or braze the joints. Fig. 30-10 shows a manually operated notching machine. The different shaped ends of tubing which may be obtained are shown in Fig. 30-11.

When metal exceeds 1/8 in. (3.2 mm) thickness, metal to be gas or arc welded is usually beveled. The bevels may be made by grinding, by flame or arc cutting, or may be made by a beveling machine as shown in Fig. 30-12.

### 30-5. METAL FORMING EQUIPMENT

After metal has been cut to size and shape it must often be bent at various angles. This bending may be done in a shop vise if the parts are small and thin. However, most metal bending is done in a manual or power brake. Fig. 30-13 illustrates

Fig. 30-10. A tube and angle notching machine. The shape of the upper and lower dies can be changed to cut tubes and angles of various sizes.   (Vogel Tool & Die Corp.)

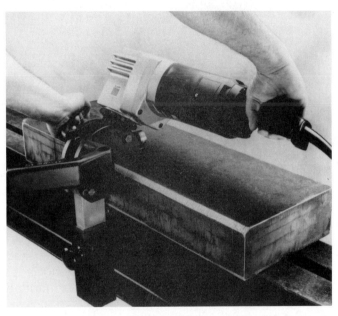

Fig. 30-12. A heavy duty portable electric edge beveling machine. This is a nibbler with a special cutter and guide. (Trumpf, Inc.)

Fig. 30-11. Metal angle and both square and round tubing may be notched as shown.   (Vogel Tool & Die Corp.)

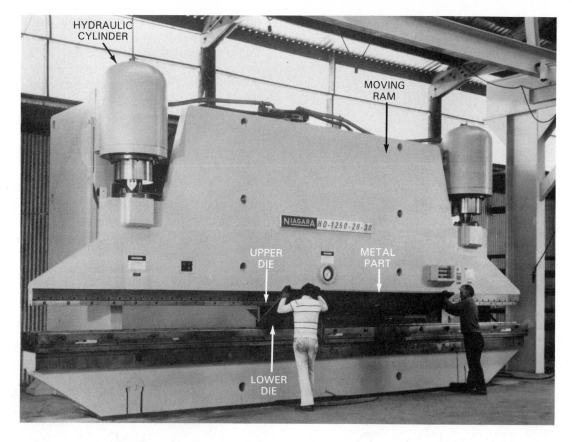

Fig. 30-13. This hydraulic brake press will bend up to 1 in. (25.4 mm) thick mild steel with 1250 tons (11 121 kN) of force. (Niagara Machine & Tool Works)

a large, 1250 ton (11 121 kN) powered brake press.

The upper and lower brake dies may be changed as a set to change the radius of the bend made. See Fig. 30-14. The angle of bend can be changed by adjusting how far the upper die moves down. As the upper die continues to move down, the angle of the bend is increased. See Fig. 30-15 for examples of how the angle is changed.

## 30-6. FURNACES

A preheating furnace is a necessity for practically all welding shops. Some metals and practically all complicated metal structures are subject to excessive straining and perhaps breakage if heated or cooled unevenly. To minimize these conditions, the metal must be heated gradually to the correct temperature before the necessary welding is performed. The structure is then allowed to cool slowly and evenly after the welding is completed. This practice makes possible a minimum of warpage. It also prevents excessive stresses and cracking of the metal as it cools to room temperature. Three styles of preheaters are:
1. Portable torch type.

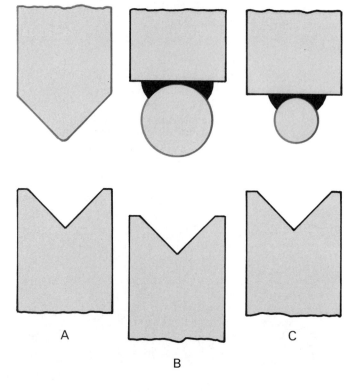

Fig. 30-14. Three upper and lower dies for a hydraulic brake press. A—90° angles with minimum bend radius. B and C—Round bars on the upper die vary the bend radius on the part.

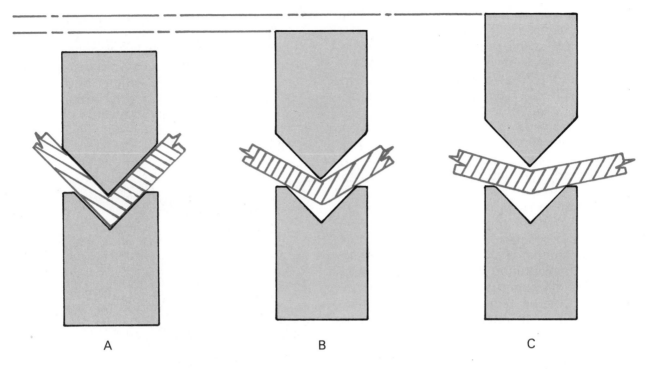

Fig. 30-15. Hydraulic brake press dies at various full down positions. Note how the angle of the bend on the metal is varied by changing the travel distance of the upper die. A—The upper die is adjusted for a 90° bend. B and C—The upper die is stopped at a higher setting for less than a 90° bend.

2. Flat top open type.
3. Enclosed furnace.

The torch type may be a large portable blowtorch such as shown in Fig. 30-16. A preheat furnace may use city gas, oil, gasoline, kerosene, or propane as the fuel. It may use air or oxygen as the combustion supporting gas. These furnaces, or heaters, may be used for the smaller metal structures or for localized preheating.

The open flat top preheater consists of a grate, or series of bars, with a gas burner underneath. The article to be preheated is placed on top of the bars or grates. Occasionally a part too large for any existing shop furnace may be brought in for repair. In such cases a preheat furnace may be built around the part using firebrick and heat resistant materials. Asbestos materials should not be used. Heat is then produced under the part. In some cases, large gas fired industrial space heaters are used to provide heat for improvised preheat furnaces such as shown in Fig. 30-17.

Clay and firebricks may be built up around the article to be preheated in order to enclose it. This shield may also protect the welder from the furnace's heat which is dissipated into the room.

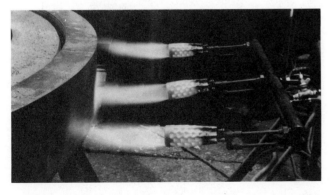

Fig. 30-16. A bank of three high output preheating torches used to heat a large steel ring prior to welding. Such torches may be used for preheating, post-heating, or heat treating. (Belchfire Corp.)

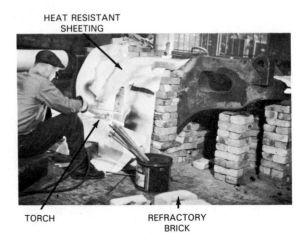

HEAT RESISTANT SHEETING

TORCH

REFRACTORY BRICK

Fig. 30-17. A shop built furnace made of firebrick and heat resistant sheeting.

Such heat may make the working conditions uncomfortable, especially in warm weather.

An advantage of this type of furnace is its flexibility. The article being preheated, or welded, is readily accessible if certain bricks or asbestos sheets are removed. The article may be welded while it is still located in the furnace.

The enclosed type of preheating furnace is a typical furnace which may be used for heat treating, carburizing, or preheating. It consists of a firebrick lined, steel structure and usually uses gas for fuel. A forced air blower is generally used for supporting the combustion and for raising the temperature. An advantage of this type furnace is its economy. It is able to cool the article slowly, for annealing purposes, and vent the heat outdoors rather than into the shop. The enclosed furnace has the disadvantage of requiring that the preheated article be removed from the furnace before it is welded. If continual heating is required for welding, this furnace cannot be used.

## 30-7. OVERHEAD CRANE

The size of the work to be handled by the welding shop may necessitate the use of a crane

for moving the article to be welded from place to place.

In the small shop, a chain fall, either hand driven or motor driven, may be suspended from rails or beams, which run the length of the shop. A movable hoist can be used for lifting jobs which weigh only a few hundred pounds. Fig. 30-18 illustrates a powered cable hoist. Fig. 30-19 shows a similar hoist in use.

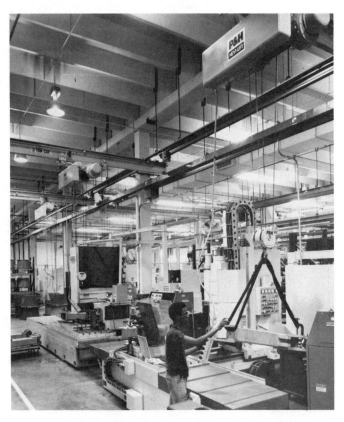

Fig. 30-19. A hoist in use in a large shop. (Harnischfeger Corp.)

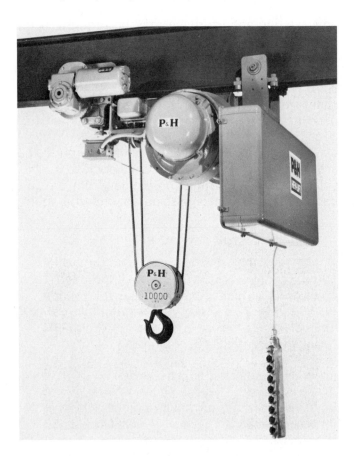

Fig. 30-18. A 10,000 lb. (4536 kg) capacity hoist. The buttons control the up and down motion, and travel along the track. (Harnischfeger Corp.)

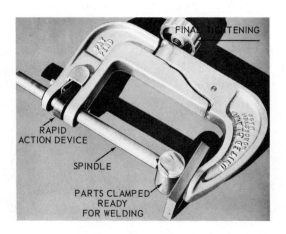

Fig. 30-20. Fast acting clamp. The shaft or spindle is made of beryllium copper alloy. This permits arc welding within 1/4 in. (6.4 mm) of the spindle without injury from arc spatter. (United Clamp Mfg. Co.)

## 30-8. JIGS AND FIXTURES

One of the greatest problems encountered in a welding shop is to prevent the metal from warping or buckling during or after the welding operation. Many devices have been used to hold the metal during welding so that the warpage and bending of the metal will be reduced to a minimum. Clamps, V-blocks, vises, and special holding jigs are used extensively for this purpose. Fig. 30-20 shows a fast action clamp. This clamp has a beryllium copper alloy spindle which resists injury from arc welding spatter. Fig. 30-21 shows a

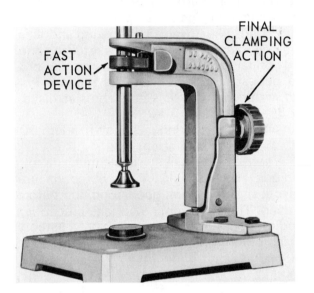

Fig. 30-21. A table mounted fast action clamp used for production work. (United Clamp Mfg. Co.)

table mounted type fast action clamp. Fig. 30-22 shows how close one may arc weld to the clamp without injury to the clamp. When a large quantity

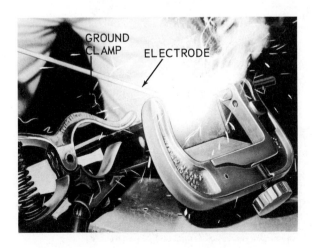

Fig. 30-22. An assembly being arc welded while the parts are held together with a fast action clamp. Note the ground clamp. (United Clamp Mfg. Co.)

of articles of the same type are to be made, it is economical to design and make jigs to hold the articles so the finished products will be identical in shape and size as shown in Fig. 30-23.

Fig. 30-23. An assembly of toggle clamps mounted on a steel plate to form a clamping fixturel. (De-Sta-Co)

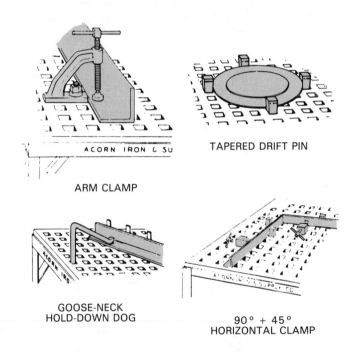

ARM CLAMP

TAPERED DRIFT PIN

GOOSE-NECK HOLD-DOWN DOG

90° + 45° HORIZONTAL CLAMP

Fig. 30-24. A faceplate (platen) used with various clamping devices. (Acorn Iron and Supply Co.)

Angle iron, cast iron, or steel grooved face plates may be used to advantage for fixtures of this kind. Fig. 30-24 shows a cast iron platen with various accessories used to clamp the metal.

Welding positioners are popular for production work in shops. They help to turn the work so that flat position welding is possible on all joints. Fig. 30-25 shows a typical powered welding positioner with a rotary turntable. Also see the illustration on page 8.

Fig. 30-26 illustrates a computerized welding positioner. This positioner moves the weldment into position as required to permit automatic robot welding.

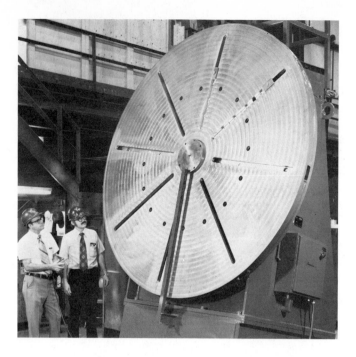

Fig. 30-25. A large rotary welding positioner.
(K.N. Aronson, Inc.)

## 30-9. POWER TOOLS

Power tools of various types are a necessity in high production welding. They help to cut overall production costs.

The drill press or portable drill is used to make holes in parts of weldments. They are also used for drilling holes for reinforcing rods in large cast iron welds. Electrically or air powered wrenches can be used when disassembling large jobs. Power chisels save a great deal of labor when opening up a crack for a welding repair, or in removing scale from ferrous or nonferrous castings prior to welding. The power chisel may also be used when removing excessive welding material from a surface to be ground.

The pedestal grinder may be used for snag (rough) grinding, for finishing parts, and for preparing small parts for welding. Fig. 30-27 illustrates a large pedestal grinder. The portable grinder finds wide usage in preparing and finishing a welded joint. It is especially useful for grinding welds on large weldments that cannot be brought to a pedestal grinder. Portable grinders may be used with rigid or flexible type abrasive discs. See Fig. 30-28 and Fig. 30-29.

An abrasive belt grinder is shown in Fig. 30-30.

In all grinding operations you should wear eye protection. The abrasive wheels must be in good condition. They should be inspected for cracks, looseness, and general good condition before being used. The operator should take precautions to stand in a safe and secure position. Fasten small articles before grinding or sanding them.

A flexible shaft grinder is shown in Fig. 30-31.

Fig. 30-26. A computerized welding positioner which moves as it is programmed to permit robot welding in the flat position. (K.N. Aronson, Inc.)

Fig. 30-27. A large electric pedestal grinder. (Cincinnati Electric Tool, Inc.)

Fig. 30-29. A portable grinder being used with a flexible grinding disc. (Norton Co.)

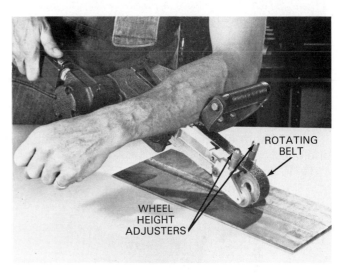

ROTATING BELT

WHEEL HEIGHT ADJUSTERS

Fig. 30-30. A rotating belt grinder. The wheels on each side of the grinding belt are adjustable. This grinder can be adjusted to remove a weld bead without cutting into the base metal. The arm resting on the protective belt guard is the recommended position for most efficient grinding. (Dynabrade, Inc.)

Fig. 30-28. A portable electrical grinder. The machine is being used with a rigid abrasive grinding disc. Note the face shield. (Norton Co.)

## 30-10. BLAST CLEANING EQUIPMENT

A blast cleaning unit is handy when dirty or rusty metal stock must be cleaned before and/or after the welding has been done.

Air pressure is used to drive steel grit, shot, sand, artificial abrasives, or even walnut shells against the surface to be cleaned. The material used depends on the metal and the condition of the surface being cleaned. The smoothness of the finished surface will be affected by the material used for blasting. The blasting operation is performed in a special cabinet with a suction unit and a recovery chamber for the abrasive. The operator stands outside the cabinet and uses specially built-in gloves to handle the blast nozzle and the parts, as shown in Fig. 30-32.

The air pressure varies from 20 to 100 psig (138 to 689 kPa). An exhaust system is used to remove the abrasive from the booth. Filters are mounted in the exhaust system to trap and remove the blasting material before the material can reach the exhaust fan.

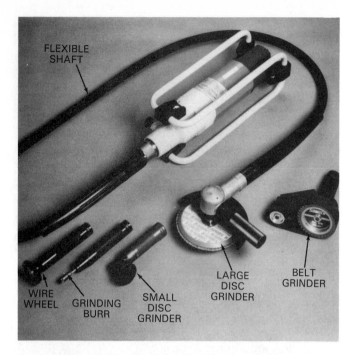

Fig. 30-31. Universal electric flexible shaft drive grinder with accessories. (Suhner Industrial Products Corp.)

## 30-11. WELDING SHOP TOOLS

The hand tools needed in a welding shop should include:
1. Wrenches.
   a. Welding equipment wrenches.
   b. Socket wrench sets up to a 1 in. drive size

Fig. 30-32. A blast cleaning cabinet. A nozzle which uses compressed air and grit is used to clean the metal before and after welding. (Ruemelin Mfg. Co.)

for large jobs. Metric sizes are useful in many applications.
   c. Box and open end wrenches for dismantling and assembling various articles.
   d. Cylinder wrenches for acetylene cylinder.
   e. Pipe wrenches for preparing pipe material for welding.
2. Hammers of several sizes and types. Sledge hammers for straightening and bending heavy stock, and chipping hammers.
3. Chisels (manual and power).
4. Files of all types and sizes.
5. Screwdrivers.
6. Wire brushes (manual and power).
7. Power grinder dresser.
8. Grinder safety goggles.
9. Squares.
10. Levels.
11. Clamps of all types.
12. Mallets.
13. Soldering irons.
14. Hacksaws and blades (hand and power).
15. Pedestal drill press.
16. Portable hand drills (electric and air).

## 30-12. WELDING SHOP SUPPLIES

It is recommended that the reader refer to Chapters 3, 5, 9, 11, 14, and 16 for information on welding supplies. A list of the more common supplies is as follows:
1. GASES.
   a. Oxygen.
   b. Acetylene.
   c. Argon, Helium, or $CO_2$.
   d. Occasionally other gases such as natural, propane, methylacetylene-propadiene (MAPP), etc.
2. FILLER METALS.
   a. Steel.
      1/16, 3/32, 1/8, 3/16, and 1/4 in. dia. (1.59, 2.38, 3.18, 4.76, 6.35 mm)
   b. Cast iron.
      1/8 and 1/4 in. (3.18 and 6.35 mm) round.
   c. Aluminum.
      1/16, 3/32, and 1/8 in. (1.59, 2.38, and 3.18 mm) dia. rods.
      1/4 in. (6.38 mm) square cast rod.
   d. Miscellaneous welding rods for special welding tasks.
   e. Brazing and soldering rods for various metals.
   f. Various GMAW electrode wire for the types of jobs normally done.
   g. Various types tungsten electrode for GTAW.

3. ELECTRODES

The shop operator must keep a stock of electrodes as required by the types of metals being welded. Today, numerous electrodes are available for all types of metal and thicknesses to be welded.

4. METAL STOCK

a. Sheet and plate metal. A welding shop should have on hand at all times a quantity of the various standard sizes of sheet metal. See Fig. 31-8 for a table of Brown and Sharpe & Steel Sheet Manufacturers Standard gauge sizes for wire and metal thicknesses. Since a great variety of plain carbon and alloy steel may be stored, it is essential that stored metal be identified. Most suppliers ship bar and formed stock with one end painted. A specific color is used to identify the alloy and carbon content of the metal. If the supplier does not paint the metal, the shop should paint the ends. A shop developed color code should be kept on file. All employees should be instructed never to cut the end of the stock which is painted. If the type of metal becomes unknown, it is virtually worthless. Most jobs require a specific metal to be used.

b. Pipe. A quantity of pipe stock should be kept on hand for standard fabricating use.

c. Angle Iron. Angle iron of various sizes is also an important item.

d. Round and other forms of solid rod. Solid rod is often used in a welding shop, and a limited quantity should be kept on hand for special jobs.

5. FLUXES.

Fluxes are used principally when oxyfuel gas welding and brazing are done. The common fluxes which should be kept on hand are:

a. Brazing flux.
b. Cast iron welding flux.
c. Cast iron brazing flux.
d. Aluminum flux.
e. Silver soldering flux.

It is important that these fluxes be kept in sealed containers and in a cool, dry place when not in use. It is further recommended that when using the flux, enough should be transferred to a smaller container to handle the job at hand. This will keep the larger supply clean and fresh.

6. VITRIFIED FIREBRICK.

Firebrick is used for bench tops and for enclosing articles that are to be preheated. They are also used to slow the cooling rate of heated articles.

7. MISCELLANEOUS.

a. Glycerine for lubricating oxyacetylene moving parts, and wiping cloths for cleaning purposes.
b. Welding backing rings, and adhesive backing forms.
c. First aid materials.
d. Safety goggles.
e. Gloves.
f. Parts for normal repairs and replacements.
g. Miscellaneous fastening devices like bolts, nuts, and washers for assembly jobs.

## 30-13. WELD TEST EQUIPMENT

Welding shops may employ one or more of the following methods to check the quality of the welds and weldments produced:

1. Visual inspection.
2. Dye penetrants for surface cracks.
3. Magnaflux for surface cracks.
4. X ray to detect internal flaws.
5. Ultrasonic testers to detail internal flaws.
6. Brinell, Rockwell, Scleroscope testers to test metal hardness.
7. Tensile testers to check yield strengths in production welds. Fig. 30-33 illustrates a small, shop type tensile tester.
8. Bend testers to determine qualifications of the welders.

These tests are described in Chapter 28 on Inspecting and Testing Welds.

## 30-14. REPAIR OF WELDING EQUIPMENT

There are many specialty firms in business to repair oxyfuel gas regulators, gauges, and

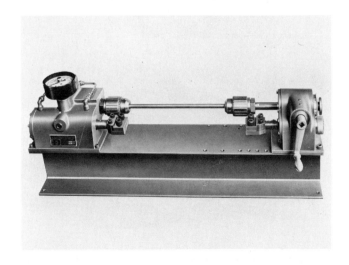

Fig. 30-33. This tensile test machine applies the pulling force by means of a threaded shaft. The gauge reads the yield strength of the test sample. (Detroit Testing Machine Co.)

torches. Similar firms specialize in the repair of arc and resistance welding machines, flow gauges, and electronic timers, and wire drives.

The specialty tools, gauges, and meters required to repair the wide variety of makes and styles of equipment are very expensive.

It is recommended that repairs, other than very minor maintenance types of repair, be done only by trained specialists. Large companies may employ such experts. Smaller jobs may have to send small equipment items out for repair. For large equipment repair, the expert will have to be called in.

Oxyfuel gas tips can be cleaned out with the correct size drill, see Fig. 4-5. The end of the tip may be reshaped with a tip dressing tool. The packing or seal around the torch valve stem can usually be replaced with manufacturer's replacement parts.

Hose nipples, hose nuts, and hoses may be replaced easily and safely with some training by shop personnel.

Arc welding electrode and work leads can be replaced, and end connections or lugs can also be easily installed. Fig. 30-34 and Fig. 30-35 show cable connections and one method used to install them.

Fig. 30-35. A metal forming method for making cable connections. (TWECO Products, Inc.)

GMAW wire drives may be repaired in the shop, but major repairs may require a service call.

Calling in a repair specialist may be the most economical way of repairing an item when safety and down time are considered.

## 30-15. REVIEW OF SAFETY

All the safety procedures described throughout the text also apply to the various types of welding shops explained in this chapter.

Adequate ventilation, eye protection, protective clothing are all necessary. Safety shoes should be worn at all times when handling weldments and when repairing equipment.

Some of the precautions one must take when working in a welding shop are as follows:
1. Cranes and hoists must be periodically inspected for condition of cables, hooks, clamps, rails, and the like.
2. Ventilation ducts and air moving equipment must be periodically inspected, cleaned, and moving parts lubricated.
3. Aisles in the shop should be clearly marked and kept clear of material and equipment.
4. When heavy equipment is being moved, all personnel in the vicinity should be alerted and should be moved out of potentially dangerous positions.
5. Periodic inspections should be made of all safety equipment and supplies in the shop.

## 30-16. WELDING SHOP POLICY

It is difficult to list all the policies that a welding shop should pursue in order to maintain correct business relationships. Work should all be done under a legal contract basis with clearly and definitely defined provisions for all emergencies. The welding shop should be extremely careful to make certain that all metal supplied meets the specifications required for the job. The welding shop should list the specifications of the metal when purchasing it. They should hold the supplier

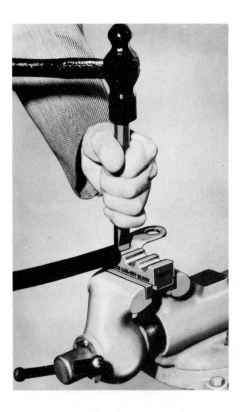

Fig. 30-34. Forming the barrel of a cable lug to obtain a connection between the cable and the lug. The completely assembled lug and cable is shown in Fig. 30-35. (TWECO Products, Inc.)

of the material to these specifications.

The welding shop must estimate the final cost of fabricating an article prior to starting work. This takes considerable training and experience. To estimate it properly, these items must be considered:

1. Length, depth, and shape of weld.
2. Type of weld—straight or curved.
3. Position of the weld—flat, vertical, or overhead weld.
4. Type of metal to be used.
5. Type of electrode or welding rod required.
6. Labor costs involved.
7. Energy costs—oxygen, fuel gases, electricity.
8. Shielding gas costs.

Depreciation of equipment and other overhead costs also affect the final cost of the jobs. By calculation, the estimator can determine the operating costs. Accurate information may be obtained from welding supply houses. They may provide information on how much welding rod, or electrode to use for a specific thickness and length of weld. They may also furnish information on how much gas or electricity may be involved in making welds of certain lengths and on various thicknesses of metal.

With this basic information, the estimator can determine the labor cost, the cost of material, and the power cost. Another item that should not be neglected in estimating the cost is the matter of handling the material to be welded. When dealing with small articles, this may be neglected. With large cumbersome forms which have to be moved from one part of the shop to another and turned in various positions, this item is important.

There should be a clear understanding about transportation costs. Either party may take this responsibility. Transportation costs involve moving the articles to the shop to be welded, and returning them to the person contracting for the work.

As with many other kinds of work, estimators, after gaining considerable experience, can calculate the cost of a weld job. This cost estimation of a job is vital to a shop's financial survival. An error in estimating the real cost of a job can cost a shop a great deal of money. Inspection and quality testing costs should be included in any agreement. A welding shop should use carefully worded contract forms. A lawyer who specializes in drawing up industrial contracts should be employed to draw up a standard form that may be applied to various types of welding jobs handled by the shop.

Many skilled welders have failed in their own welding businesses because they neglected the business aspects of the business. Accurate records must be kept at all times. The total cost of operating a shop includes not only the actual operating cost but the overhead cost as well. Consultation with an experienced accountant is highly recommended.

It is important to keep an accurate record of each job. These records can then be used as a base for estimating future jobs similar in nature.

## 30-17. TEST YOUR KNOWLEDGE

Write your answers on a separate sheet of paper. Do not write in this book.

1. List four important features which must be included in a well designed welding shop.
2. For what reasons should the offices, paint booths, toilets, shower and locker rooms be kept separated from the welding shop?
3. List three reasons for having adequate ventilation in a shop.
4. During the winter, replacement air for the ventilation system must be _____ and humidified.
5. List four different mechanical processes used to cut metal in a welding shop.
6. Name three types of shears available.
7. What is a notching machine?
8. What type of tool or machine is used to make straight bends in metal? Name two.
9. What adjustment is made to change the angle of the bend in a hydraulic brake press?
10. Name two reasons why weldments may be preheated.
11. Name one disadvantage of an enclosed furnace.
12. Why is a jig or fixture used in welding?
13. Welding _____ are used to rotate the part so that all welds may be made in the flat position.
14. In all grinding operations, you should wear _____ _____.
15. List five materials used as cleaning abrasives when blast cleaning.
16. Why do smaller welding shops send most welding equipment out to specialty shops for repair?
17. When heavy equipment is being moved all _____ in the vicinity should be alerted and moved out of potentially _____ positions.
18. What is the cause of failure for most new welding shops?
19. Why is it a good practice to identify all metal stock which is kept in shop inventory?
20. Which type of mechanical metal cutting machine mentioned in this chapter is best for cutting curved lines on metal sheet?

A portable fume extractor. This lightweight, portable exhaust unit is ideal for use when welding in close areas, such as inside a large pipe.   (Nederman, Inc.)

Material covered in this chapter will deal with technical understanding and useful reference tables.

Information in the chapter will include:

1. The effect of hose diameter and length on gas flow and pressure.
2. Various oxyfuel gas flame temperatures.
3. The chemistry of welding.
4. The chemical composition of pickling and metal cleaning solutions.
5. Properties of metals.
6. Sheet metal gauges.
7. Judging the temperature of mild steel by its color.
8. Drill, metal gauge, and tap sizes.
9. Metric practices and conversions for use in welding.
10. Understanding the meaning of temperature and heat.

## 31-1. THE EFFECT OF TEMPERATURE ON CYLINDER PRESSURE

The pressure within a cylinder or tank will vary as the outside temperature changes. Gas in a confined space, like a cylinder, cannot expand. Therefore, if it is heated, the pressure inside the cylinder will rise. If the temperature of the cylinder cools, the pressure of the gas in the cylinder will go down.

Fig. 31-1 uses a full 244 cu. ft. (6.92 m³) cylinder at 2200 psig (15 168 kPa) and 70°F (21.1°C) to illustrate how the pressure changes as the temperature changes.

## 31-2. THE EFFECT OF HOSE DIAMETER AND LENGTH ON THE GAS FLOW AND PRESSURE AT THE TORCH

The volume of gas which reaches the torch valves is determined by the inside diameter of the

| TEMPERATURE OF OXYGEN | | GAUGE READING | |
|---|---|---|---|
| °F | °C | psig | kPa |
| 100 | 37.8 | 2325 | 16030 |
| 90 | 32.2 | 2283 | 15741 |
| 80 | 26.7 | 2242 | 15458 |
| 70 | 21.1 | 2200 | 15168 |
| 60 | 15.6 | 2158 | 14879 |
| 50 | 10.0 | 2117 | 14596 |
| 40 | 4.4 | 2075 | 14307 |
| 30 | −1.1 | 2034 | 14024 |
| 20 | −6.7 | 1992 | 13734 |
| 10 | −12.2 | 1951 | 13452 |
| 0 | −17.8 | 1909 | 13162 |

Fig. 31-1. The effect of temperature on the oxygen pressure within a fully charged 244 ft.³ (6.92 m³) oxygen cylinder at 2200 psig and 70°F (15 168 kPa @ 21.1°C).

hose, the length of the hose, and the gas pressure.

In oxyfuel gas welding and cutting, an insufficient volume of oxygen or fuel gas will affect the energy output of the flame. It may also cause a flashback accident to occur. An insufficient supply of oxygen for cutting will cause a poor cut to occur.

Insufficient volumes of shielding gases when GTAW, GMAW, GTAC, GMAC, or other gas-using processes will cause poor welds and cuts.

Therefore, the volume of gas delivered to the torch valves is very important.

Pressure on the gas is what propels the gas. A low pressure will deliver lower volumes of gas. Three factors will affect the gas pressure that reaches the torch valves. One factor is the pressure set on the regulator. The second is the length of the hose used in the system. The third is the inside diameter of the hose. The inside surface of the hose offers resistance to the flow of the gases.

As the length of hose increases, a higher regulator setting is required to obtain the desired pressure to the torch valve. Two pressures will register on the regulator. The regulator gauge will read a pressure higher when the gas is not flowing (static) than when the gas is flowing (dynamic). Careful thought must be given to the hose diameter, hose length, and regulator pressure settings when an outfit is prepared for welding. Fig. 31-2 illustrates the effects of hose diameter and length on the flow and pressure at the torch valves.

## 31-3. FLAME CHARACTERISTICS

Not all gases burn at the same temperature. Some gases are well suited for preheating and others, because of their high flame temperatures, are best for welding. Fig. 31-3 lists the flame temperatures for a number of fuel gases.

## 31-4. THE CHEMISTRY OF OXYACETYLENE GAS WELDING

The chemical formula for the products involved in the burning of the oxyacetylene flame are:

| Hose Diameter | Hose Length | Cutting Tip Size | Reg. PSI Static | Reg. PSI Flowing | Inlet PSI Torch | PSI Drop In Hose | Flow CFH |
|---|---|---|---|---|---|---|---|
| 3/16 | 50 | 3 | 50 | 47 | 37½ | 9½ | 169 |
| 3/16 | 100* | 3 | 51 | 47 | 26 | 21 | 129 |
| 3/16 | 50 | 5 | 84½ | 78 | 44 | 34 | 370 |
| 3/16 | 100* | 5 | 83½ | 78 | 22 | 56 | 215 |
| 3/16 | 50 | 7 | 108 | 100 | 24 | 76 | 510 |
| 3/16 | 100* | 7 | 106½ | 100 | 9 | 91 | 270 |
| 3/16 | 50 | 9 | 138½ | 130 | 19½ | 110½ | 735 |
| 3/16 | 100* | 9 | 136½ | 130 | 7 | 123 | 405 |
| 1/4 | 50 | 3 | 50½ | 47 | 44½ | 2½ | 194 |
| 1/4 | 100* | 3 | 50 | 47 | 42½ | 4½ | 188 |
| 1/4 | 50 | 5 | 86 | 78 | 68½ | 9½ | 540 |
| 1/4 | 100* | 5 | 85 | 78 | 58½ | 19½ | 470 |
| 1/4 | 50 | 7 | 114 | 100 | 68 | 32 | 1140 |
| 1/4 | 100* | 7 | 110 | 100 | 49 | 51 | 870 |
| 1/4 | 50 | 9 | 149½ | 130 | 65 | 65 | 2010 |
| 1/4 | 100* | 9 | 144 | 130 | 36½ | 93½ | 1290 |
| 1/4 | 100** | 3 | 50 | 47 | 36 | 11 | 164 |
| 1/4 | 100** | 5 | 84½ | 78 | 42 | 36 | 360 |
| 1/4 | 100** | 7 | 108 | 100 | 25 | 75 | 560 |
| 1/4 | 100** | 9 | 140 | 130 | 18 | 112 | 795 |
| 3/8 | 50 | 3 | 51 | 47 | 46 | 1 | 190 |
| 3/8 | 50 | 5 | 86 | 78 | 74½ | 3½ | 580 |
| 3/8 | 50 | 7 | 117 | 100 | 86 | 14 | 1400 |
| 3/8 | 50 | 9 | 163½ | 130 | 89½ | 40½ | 2700 |
| 3/8 | 100* | 3 | 51 | 47 | 46 | 1 | 198 |
| 3/8 | 100* | 5 | 86 | 78 | 72 | 6 | 570 |
| 3/8 | 100* | 7 | 115 | 100 | 77 | 23 | 1280 |
| 3/8 | 100* | 9 | 155 | 130 | 75 | 55 | 2280 |

\*—Two 50 ft. lengths of hose connected together with standard hose unions
\*\*—Four 25 ft. lengths of hose connected together with standard hose unions

Caution: Do not exceed the maximum working pressures (regulator setting) listed for the hoses during welding and cutting operations.
Oxygen . . . . . . . . . . . . . . . . . . . . . . . . . . . . . . . . . . . . . . . . . . . . . . . . . . . . . . . . . . . . . . . . . . . . . . . . . . 150 psig Maximum
Acetylene . . . . . . . . . . . . . . . . . . . . . . . . . . . . . . . . . . . . . . . . . . . . . . . . . . . . . . . . . . . . . . . . . . . . . . . 15 psig Maximum
Propane, Propylene, MAPP and all other fuel gases . . . . . . . . . . . . . . . . . . . . . . . . . . . . 40 psig Maximum

Caution: HOSES. Do not use excessively long hoses or hoses with many hose unions. Either will restrict gas flow and pressure causing lower cutting efficiency and possibly leading to dangerous operating conditions.

MANIFOLDING CYLINDERS. When required flows in cubic feet per hour exceed the recommended withdrawal rate from one cylinder then additional cylinders must be manifolded to provide safe and efficient operation. Acetylene must not be withdrawn at more than 1/7 of the cylinder capacity (47 cubic feet per hour for a 330 cu. ft. cylinder). Consult your gas supplier for manifolding instructions for the gases and cylinders supplied to you.

Fig. 31-2. The effect of hose diameter and length on the flow and pressure at the torch valves.
(Smith Equipment, Div. of Tescom Corp.)

Acetylene = $C_2H_2$
Oxygen = $O_2$
Carbon Monoxide = CO
Carbon Dioxide = $CO_2$
Water (Vapor) = $H_2O$

The chemical formula for the conbustion of oxygen and acetylene is: $2\ C_2H_2 + 3\ O_2 \rightarrow 4\ CO + 2\ H_2O + Heat$. In this equation, a chemical change takes place as the gas welding gases are ignited at the torch tip. This is what happens: two molecules of acetylene ($C_2H_2$) combine with three molecules of oxygen ($O_2$) and are ignited. The result of the burning of this combination of molecules is four molecules of carbon monoxide (CO), plus two molecules of water vapor ($H_2O$) plus heat.

| Gas | Chemical Formula | BTU/ Ft³ | MJ/m³ | Neutral Flame Temperature with Oxygen | |
|---|---|---|---|---|---|
| | | | | °F | °C |
| Acetylene | $C_2H_2$ | 1470 | 55 | 5612 | 3100 |
| MPS* | $C_3H_4$ | 2406 | 90 | 4712 | 2600 |
| Propylene | $C_3H_6$ | 2371 | 88 | 4532 | 2500 |
| Propane | $C_3H_8$ | 2498 | 93 | 4442 | 2450 |
| Hydrogen | $H_2$ | 275 | 10 | 4334 | 2390 |
| Natural gas | $CH_4 + H_2$ | 1000 | 37 | 4260 | 2350 |

*methylacetylene — propadiene (stabilized)

Fig. 31-3. The heat units and flame temperatures of various fuel gases.

Carbon monoxide (CO) is a very unstable gas. It will unite readily with oxygen to form carbon dioxide ($CO_2$). The carbon dioxide accounts for the fact that surrounding the cone of the welding flame is an area where the flame is of lesser intensity.

It is in this area that the carbon monoxide is mixing with the atmospheric oxygen to form carbon dioxide. The layer of carbon monoxide tends to keep the molten weld metal from oxidizing. The carbon monoxide absorbs any free oxygen present.

The chemical action in the outer flame becomes:

$$2\ CO + O_2 \rightarrow 2\ CO_2 + heat$$

The oxygen, in this case, comes from the atmosphere surrounding the welding flame.

This principle must be remembered when welding in a confined space in which a free movement of air does not exist around the torch tip. Under these conditions more oxygen will need to be fed to the torch tip in order that a neutral flame may be maintained.

## 31-5. CHEMICAL CLEANING AND PICKLING SOLUTIONS

Chemical cleaning, degreasing, and pickling solutions are often required prior to brazing some nonferrous metals.

Magnesium, magnesium alloys, and copper and copper alloys require chemical cleaning prior to brazing. The following solutions are used on magnesium. Solutions for preparing copper are also given.

DEGREASING is generally done with an alkaline solution of sodium bicarbonate, sodium hydroxide, and water. The solution is mixed at a temperature of 190-212 °F (88-100 °C) as follows:

3 oz. (.085 kg) sodium bicarbonate.
2 oz. (.057 kg) sodium hydroxide.
1 gal. (3.79 L) water.

Bright chrome pickling or modified chrome pickling is done to remove oxides and scale from the metal surfaces.

BRIGHT PICKLE SOLUTION may be made by mixing these chemicals at 60-100 °F (15.6-37.8 °C):

24 oz. (.680 kg) chromic acid.
5.3 oz. (.150 kg) ferric nitrate.
.47 oz. (.013 kg) potassium fluoride.
1 gal. (3.79 L) water.

The CHROME PICKLING SOLUTION is made by mixing the following chemicals at 70-90 °F (21.1-32.2 °C):

24 oz. (.680 kg) sodium dichromate.
24 fl. oz. (.710 L) concentrated nitric acid.
1 gal. (3.79 L) water.

A MODIFIED CHROME PICKLING MIXTURE is made by combining the following at 70-100 °F (21.1-37.8 °C):

2 oz. (.057 kg) sodium acid fluoride.
24 oz. (.680 kg) sodium dichromate.
1.3 oz. (.037 kg) aluminum sulfate.
16 fl. oz. (.473 L) nitric acid, 70%.
1 gal. (3.79 L) water.

To remove oil, grease, and dirt from copper and its alloys, the following solution can be prepared. The solution should be used at 125-180 °F (51.7-82.2 °C).

6.8 oz. (.193 kg) sodium orthosilicate.
.8 oz. (.023 kg) sodium carbonate.
.4 oz. (.011 kg) sodium resinate.
1 gal. (3.79 L) water.

A pickling solution for copper is a 10 percent solution of sulfuric acid used at 125-150 °F (51.6-65.6 °C). Scale can be removed by using the following solution at room temperature:

40 percent concentrated nitric acid.
30 percent concentrated sulfuric acid.

.5 percent concentrated hydrochloric acid.
29.5 percent water
(Percent by volume)

## 31-6. THERMIT REACTION CHEMISTRY

The welding of parts or making of castings using the exothermic Thermit process has been used for many years. The chemical reaction of this steel welding or casting process is:

$$8 Al + 3 Fe_3O_4 \rightarrow 9 Fe + 4 Al_2O_3 + heat$$

The aluminum and iron oxide mixture must be heated to approximately 2200°F (1204°C) to start the reaction described above.

Copper, nickel, and manganese have also been welded or cast using this process. The word "Thermit" is a registered trade name commonly used to identify this process.

## 31-7. BLAST FURNACE OPERATIONS

The blast furnace is used to convert iron ores to pig iron. See Heading 25-1. The blast furnace, in production of pig iron, must: (1) deoxidize the iron ore; (2) melt the iron; (3) melt the slag; (4) carburize the iron; (5) separate the iron from slag.

The following raw materials are fed to the furnace:

| ORES | % IRON |
|---|---|
| Hematite (red iron) $Fe_2O_3$ | 70 |
| Magnetite (black) $Fe_3O_4$ | 72.4 |
| Limonite (brown) $Fe_2O_3H_2O$ | 63 |
| Siderite Iron Carbonate $FeCO_3$ | 48.3 |

FLUX

A good flux must melt and unite with impurities, called gangue, and carry them away in the form of slag. Sand and Alumina are chief impurities. The basic flux used in the blast furnace is limestone.

FUEL

The fuel used must melt the charge and furnish heat for the reactions in the furnace. It must be low in phosphorus and sulphur. Coke is ideal for this purpose.

AIR

Preheated compressed air is forced into the lower part of the furnace.

The main chemical reactions in the blast furnace are:

1. $C + O_2 \rightarrow CO_2 +$ heat
2. $CO_2 + C \rightarrow 2CO +$ heat
3. $Fe_2O_3 + 3CO \rightarrow 3CO_2 + 2Fe +$ heat
4. $MnO + CO \rightarrow CO_2 + Mn +$ heat

(1-3) In the lower zone of the furnace the CO acts as a reducing agent.

(4-5) In the upper zone of the furnace.

5. $SO_2 + 2CO \rightarrow 2CO_2 + S +$ heat
6. $Fe_2O_3 + 3C \rightarrow 3CO_2 + 2Fe +$ heat
7. $CaCO_3 + heat \rightarrow CaO + CO_2$

(6-7) In the lower zone as much heat is required as in the upper part and the process goes practically to completion at about 1500°F (816°C).

8. $MgCO_3 + heat \rightarrow MgO + CO_2$
9. $CaSO_4 + 2C \rightarrow CaS + 2CO_2$

(8-11) In the lower zone and only partly complete.

10. $CaO + Al_2O_3 \rightarrow CaO + Al_2O_3$
11. $CaO + SiO_2 \rightarrow CaO + SiO_2$

The $Al_2O_3$, CaO, and MnO go through both zones unchanged as not enough heat is furnished to cause the reactions in (4) and (5) to go to completion.

Pure iron melts at 2700°F (1482°C). With impurities, it melts at lower temperatures.

Iron and slag separate in the bottom of the blast furnace because the slag is lighter and floats on top of the molten iron. Impurities such as silicon, manganese, and carbon are soluble in iron and remain in the iron. Iron sulphide and iron phosphide are also soluble in iron so their quantities must be kept small.

## 31-8. PROPERTIES OF METALS

Fig. 31-4 lists most metals used in the welding industry. Their chemical symbol is given along with their melting temperatures and some other properties.

## 31-9. WEIGHT AND EXPANSION PROPERTIES OF VARIOUS METALS

When designing a structure, the type of material to be used and its weight must be considered.

To prevent distortions in the completed weldment, the expansion rate of metals must be considered. This is especially true when jigs and fixtures are used. Fig. 31-5 is a table which lists the weight and expansion rate of several metals.

## 31-10. STRESSES CAUSED BY WELDING

STRESS is a force which causes or attempts to cause a movement or change in the shape of parts being welded (strain).

During welding, the heat created by the welding process causes the metal to expand. When the welded metal cools, it contracts (shrinks). Usually, it does not return to the original shape or position. If it does not return to its original shape, DISTORTION has occurred. Distortion can be

| ELEMENTS | SYMBOL | MELTING TEMPERATURE | | SPECIFIC GRAVITY | WEIGHT PER CU. FOOT | GRAMS PER CU. Cm | SPECIFIC HEAT | |
|---|---|---|---|---|---|---|---|---|
| | | °F | °C | | | | BTU/lb/°F | Cal/g/°C |
| Aluminum | Al | 1,218 | 659 | 2.7 | 166.7 | 2.67 | 0.212 | 0.226 |
| Antimony | Sb | 1,166 | 630 | 6.69 | 418.3 | 6.6 | 0.049 | 0.049 |
| Armco Iron | .. | 2,795 | 1,535 | 7.9 | 490.0 | 7.85 | 0.115 | 0.108 |
| Barium | Ba | 1,600 | 850 | 3.6 | 219.0 | .... | ... | 0.068 |
| Beryllium | Be | 2,348 | 1,285 | 1.84 | .... | 1.845 | ... | 0.46 |
| Bismuth | Bi | 520 | 271 | 9.75 | 612.0 | .... | ... | 0.029 |
| Boron | B | 3,990 | 2,200 | 2.29 | 143.0 | .... | ... | 0.309 |
| Brass (70Cu 30Zn) | .. | 1652-1724 | 900-940 | 8.44 | 527.0 | .... | 0.092 | ... |
| Brass (90Cu 10Zn) | .. | 1868-1886 | 1020-1030 | 8.60 | 540.0 | .... | 0.092 | ... |
| Bronze (90Cu 10sn) | .. | 1562-1832 | 850-1000 | 8.78 | 548.0 | .... | 0.092 | ... |
| Cadmium | Cd | 610 | 321 | 8.64 | 550.0 | .... | ... | 0.055 |
| Carbon | C | 6,510 | 3,600 | 2.34 | 219.1 | 3.51 | 0.113 | 0.165 |
| Cast Pig Iron | .. | 2012-2282 | 1100-1250 | 7.1 | 443.2 | .... | 0.13 | ... |
| Cerium | Ce | 1,184 | 640 | 6.8 | 432.0 | .... | ... | 0.05 |
| Chromium | Cr | 2,770 | 1,520 | 6.92 | 431.9 | 6.92 | 0.104 | 0.12 |
| Cobalt | Co | 2,700 | 1,480 | 8.71 | 555.0 | .... | ... | 0.099 |
| Columbium | Cb | 3,124 | 1,700 | 7.06 | 452.54 | 7.25 | ... | ... |
| Copper | Cu | 1,980 | 1,100 | 8.89 | 555.6 | 8.9 | 0.092 | ... |
| Gold | Au | 1,900 | 1,060 | 19.33 | 1205.0 | 19.2 | 0.032 | 0.031 |
| Hydrogen | H | −434.2 | −259 | 0.070 | 0.00533 | .... | ... | 3.415 |
| Iridium | Ir | 4,260 | 2,350 | 22.42 | 1400.0 | 22.4 | 0.032 | 0.032 |
| Iron | Fe | 2,790 | 1,530 | 7.865 | 490.0 | 7.85 | 0.115 | 0.108 |
| Lead | Pb | 621 | 327 | 11.37 | 708.5 | 11.32 | 0.030 | 0.030 |
| Lithium | Li | 367 | 186 | .534 | .... | 32.8 | ... | 0.79 |
| Magnesium | Mg | 1,204 | 651 | 1.74 | 108.5 | .... | ... | 0.249 |
| Manganese | Mn | 2,300 | 1,260 | 7.4 | 463.2 | 7.40 | 0.111 | 0.107 |
| Mercury | Hg | −38 | −39 | 13.55 | 848.84 | 13.6 | 0.033 | 0.033 |
| Molybdenum | Mo | 4,530 | 2,500 | 10.3 | 638.0 | .... | ... | 0.065 |
| Nickel | Ni | 2,650 | 1,450 | 8.80 | 555.6 | 8.9 | 0.109 | 0.112 |
| Open Hearth Steel | .. | 2462-2786 | 350-1530 | 7.8 | 486.9 | .... | 0.115 | ... |
| Osmium | Os | 4,890 | 2,700 | 22.48 | 1405.0 | .... | ... | 0.031 |
| Palladium | Pd | 2,820 | 1,550 | 12.16 | 750.0 | .... | ... | 0.059 |
| Platinum | Pt | 3,190 | 1,750 | 21.45 | 1336.0 | 21.4 | 0.032 | 0.032 |
| Rhodium | Rh | 3,540 | 1,950 | 12.4 | 776.0 | .... | ... | 0.060 |
| Ruthenium | Ru | 4,440 | 2,450 | 12.2 | 762.0 | .... | ... | 0.061 |
| Silenium | Se | 424 | 218 | 4.8 | 300.0 | .... | ... | 0.084 |
| Silicon | Si | 2,590 | 1,420 | 2.49 | 131.1 | 2.10 | 0.175 | 0.176 |
| Silver | Ag | 1,800 | 960 | 10.5 | 655.5 | 10.5 | 0.055 | 0.056 |
| Tantalum | Ta | 5,160 | 2,800 | 16.6 | 1037.0 | .... | ... | 0.036 |
| Tellurium | Te | 846 | 452 | 6.23 | 389.0 | .... | ... | 0.047 |
| Thallium | Ti | 576 | 302 | 11.85 | 740.0 | .... | ... | 0.031 |
| Thorium | Th | 3,090 | 1,700 | 11.5 | 717.0 | .... | ... | 0.028 |
| Tin | Sn | 450 | 232 | 7.30 | 455.7 | 7.30 | 0.054 | 0.054 |
| Titanium | Ti | 3,270 | 1,800 | 5.3 | 218.5 | 3.50 | 0.110 | 0.142 |
| Tungsten | W | 5,430 | 3,000 | 17.5 | 1186.0 | 19.0 | 0.034 | 0.034 |
| Uranium | U | .... | .... | 18.7 | 1167.0 | 18.7 | 0.028 | 0.028 |
| Vanadium | V | 3,130 | 1,720 | 6.0 | 343.3 | .... | 0.115 | ... |
| Wrought Iron Bars | .. | 2,786 | 1,530 | 7.8 | 486.9 | .... | 0.11 | ... |
| Zinc | Zn | 787 | 419 | 7.19 | 443.2 | .... | 0.093 | ... |
| Zirconium | Zr | 3,090 | 1,700 | 6.38 | 398.0 | .... | ... | 0.066 |

Fig. 31-4. Metal Properties. This table lists the metals used in welding, their chemical symbol, and other useful information.

| Metal | Weight per ft³ (lbs.) | Weight per m³ (kg) | Expansion per °F rise in temperature (.0001 in.) | Expansion per °C rise in temperature (.0001 mm) |
|---|---|---|---|---|
| Aluminum | 165 | 2643 | 1.360 | 62.18 |
| Brass | 520 | 8330 | 1.052 | 48.10 |
| Bronze | 555 | 8890 | .986 | 45.08 |
| Copper | 555 | 8890 | .887 | 40.55 |
| Gold | 1200 | 19222 | .786 | 35.94 |
| Iron (Cast) | 460 | 7369 | .556 | 25.42 |
| Lead | 710 | 11373 | 1.571 | 71.83 |
| Nickel | 550 | 8810 | .695 | 31.78 |
| Platinum | 1350 | 21625 | .479 | 21.90 |
| Silver | 655 | 10492 | 1.079 | 49.33 |
| Steel | 490 | 7849 | .689 | 31.50 |

Fig. 31-5. Weight and expansion properties of various metals.

minimized by clamping the parts into a fixture while welding.

When unclamped parts cool down, they are often distorted (changed in shape). This is due to the stresses remaining in them. Stresses which remain in parts after welding is completed are known as RESIDUAL STRESSES. See Fig. 31-6.

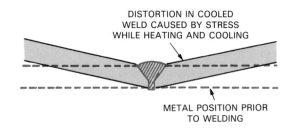

Fig. 31-6. Distortion caused by welding stresses in an unclamped part. Residual stress after distortion may be zero.

When parts are clamped into a jig or fixture for welding, expansion, contraction, and resulting distortion is held to a minimum. Even though the distortion has been reduced, residual stresses remain in the metal after cooling. See Fig. 31-7. These residual stresses may cause the metal to distort at any time if not removed. To reduce the residual stresses caused by welding, parts should be given a stress relieving heat treatment.

## 31-11. WIRE AND SHEET METAL GAUGES

Two systems are in use in the United States to measure the thickness of metal. The thickness of metal is measured in gauge sizes. The two systems are the BROWN AND SHARPE GAUGE

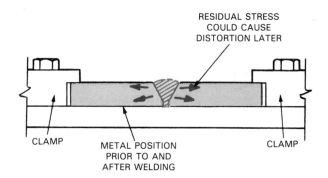

Fig. 31-7. Residual stress remaining in a clamped part after welding. After the weld has cooled, the distortion may be near zero, but the internal residual stress may be considerable. Residual stress may be removed by stress relieving.

(B & S) and the STEEL SHEETS MANUFACTURE STANDARD (SSMS). The B & S gauge system is usually used to measure nonferrous sheet and wire. Sheet steels are measured using the Steel Sheets Manufacture Standard gauge system.

Fig. 31-8 lists the B & S and SSMS gauge sizes.

## 31-12. JUDGING TEMPERATURE BY COLOR

Chapter 27 explains the need and procedure for the proper heat treatment of a metal after welding or brazing. Proper heat treatment and/or mechanical working will maximize the desired physical properties of a metal.

For proper heat treatment, the metal must be heated to specific temperatures and then cooled. Steel as it is heated to higher and higher temperatures, turns to different colors. These colors give a fairly accurate indication of the temperature of the steel and may be used for tempering. Refer to Fig. 1-37.

| Gauge | Brown and Sharpe Gauge (inch) | Steel Sheet Mfrs. Std. (inch) | | Gauge | Brown and Sharpe Gauge (inch) | Steel Sheet Mfrs. Std. (inch) |
|---|---|---|---|---|---|---|
| 6-0 | 0.5800 | — | | 17 | 0.0453 | 0.0538 |
| 5-0 | 0.5165 | — | | 18 | 0.0403 | 0.0478 |
| 4-0 | 0.4600 | — | | 19 | 0.0359 | 0.0418 |
| 3-0 | 0.4096 | — | | 20 | 0.0320 | 0.0359 |
| 2-0 | 0.3648 | — | | 21 | 0.0285 | 0.0329 |
| 0 | 0.3249 | — | | 22 | 0.0253 | 0.0299 |
| 1 | 0.2893 | — | | 23 | 0.0226 | 0.0269 |
| 2 | 0.2576 | — | | 24 | 0.0201 | 0.0239 |
| 3 | 0.2294 | 0.2391 | | 25 | 0.0179 | 0.0209 |
| 4 | 0.2043 | 0.2242 | | 26 | 0.0159 | 0.0179 |
| 5 | 0.1819 | 0.2092 | | 27 | 0.0142 | 0.0164 |
| 6 | 0.1620 | 0.1943 | | 28 | 0.0126 | 0.0149 |
| 7 | 0.1443 | 0.1793 | | 29 | 0.0113 | 0.0135 |
| 8 | 0.1285 | 0.1644 | | 30 | 0.0100 | 0.0120 |
| 9 | 0.1144 | 0.1495 | | 31 | 0.0089 | 0.0105 |
| 10 | 0.1019 | 0.1345 | | 32 | 0.0080 | 0.0097 |
| 11 | 0.0907 | 0.1196 | | 33 | 0.0071 | 0.0090 |
| 12 | 0.0808 | 0.1046 | | 34 | 0.0063 | 0.0082 |
| 13 | 0.0720 | 0.0897 | | 35 | 0.0056 | 0.0075 |
| 14 | 0.0641 | 0.0747 | | 36 | 0.0050 | 0.0067 |
| 15 | 0.0571 | 0.0673 | | 37 | 0.0045 | 0.0064 |
| 16 | 0.0508 | 0.0598 | | 38 | 0.0040 | 0.0060 |

Fig. 31-8. A table of the Brown and Sharpe and Steel Sheet Manufacturers Standard gauge sizes for metal thickness. The decimal equivalents of the customary U.S. systems are also shown.

Fig. 31-9 shows the approximate temperature at which each color occurs.

| Color | °F | °C |
|---|---|---|
| Faint straw | 400 | 205 |
| Straw | 440 | 225 |
| Deep straw | 475 | 245 |
| Bronze | 520 | 270 |
| Purple | 540 | 280 |
| Light blue | 590 | 310 |
| Blue | 640 | 340 |
| Black | 700 | 370 |
| Dark red | 1000 | 600 |
| Dull cherry red | 1200 | 650 |
| Cherry red | 1300 | 700 |
| Bright cherry red | 1400 | 750 |
| Orange red | 1500 | 800 |
| Orange yellow | 2200 | 1200 |
| Yellow white | 2370 | 1300 |
| White heat | 2550 | 1400 |

Fig. 31-9. The approximate temperature at which various color changes occur when clean, carbon steel is heated to the melting temperature.

## 31-13. DRILL SETS AND SIZES

Drill bits are produced in a variety of sizes and marked using four different size marking systems. These four sizing systems are:
1. Fractional drills.
2. Number drills.
3. Letter drills.
4. Metric drills.
The shank of each drill, when large enough, carries a stamped identification of the drill size. The size is given as a number, fraction, or letter. Metric drills are marked with a number. It is important to know if the drill set being used is a number drill set or a metric drill set. The diameters are different even though the number used may be the same.

Depending on the quality and use, drill bits are available in either high carbon steel (least expensive) or alloy steel, marked HSS (high speed steel), for high speed use.

Fractional drill sets may begin with size 1/64 in. and go to 1/2 in. in steps of 1/64 in. Larger fractional sizes are available; see Fig. 31-10.

Number drill sets begin with No. 1 (0.228 in.) and go to No. 80 (0.0135 in.). Most commonly used number drill sets include No. 1 through No.

| 1/64 | .015625 | 17/64 | .265625 |
|---|---|---|---|
| 1/32 | .03125 | 9/32 | .28125 |
| 3/64 | .046875 | 19/64 | .296875 |
| 1/16 | .0625 | 5/16 | .3125 |
| 5/64 | .078125 | 21/64 | .328125 |
| 3/32 | .09375 | 11/32 | .34375 |
| 7/64 | .109375 | 23/64 | .359375 |
| 1/8 | .125 | 3/8 | .375 |
| 9/64 | .140625 | 25/64 | .390625 |
| 5/32 | .15625 | 13/32 | .40625 |
| 11/64 | .171875 | 27/64 | .421875 |
| 3/16 | .1875 | 7/16 | .4375 |
| 13/64 | .203125 | 29/64 | .453125 |
| 7/32 | .21875 | 15/32 | .46875 |
| 15/64 | .234375 | 31/64 | .484375 |
| 1/4 | .250 | 1/2 | .500 |

Fig. 31.10. Fractional drills with their decimal equivalents from 1/64 to 1/2 in.

| DRILL NUMBER | DECIMAL SIZE | DRILL NUMBER | DECIMAL SIZE | DRILL NUMBER | DECIMAL SIZE |
|---|---|---|---|---|---|
| 1 | .2280 | 28 | .1405 | 55 | .0520 |
| 2 | .2210 | 29 | .1360 | 56 | .0465 |
| 3 | .2130 | 30 | .1285 | 57 | .0430 |
| 4 | .2090 | 31 | .1200 | 58 | .0420 |
| 5 | .2055 | 32 | .1160 | 59 | .0410 |
| 6 | .2040 | 33 | .1130 | 60 | .0400 |
| 7 | .2010 | 34 | .1110 | 61 | .0390 |
| 8 | .1990 | 35 | .1100 | 62 | .0380 |
| 9 | .1960 | 36 | .1065 | 63 | .0370 |
| 10 | .1935 | 37 | .1040 | 64 | .0360 |
| 11 | .1910 | 38 | .1015 | 65 | .0350 |
| 12 | .1890 | 39 | .0995 | 66 | .0330 |
| 13 | .1850 | 40 | .0980 | 67 | .0320 |
| 14 | .1820 | 41 | .0960 | 68 | .0310 |
| 15 | .1800 | 42 | .0935 | 69 | .02925 |
| 16 | .1770 | 43 | .0890 | 70 | .0280 |
| 17 | .1730 | 44 | .0860 | 71 | .0260 |
| 18 | .1695 | 45 | .0820 | 72 | .0250 |
| 19 | .1660 | 46 | .0810 | 73 | .0240 |
| 20 | .1610 | 47 | .0785 | 74 | .0225 |
| 21 | .1590 | 48 | .0760 | 75 | .0210 |
| 22 | .1570 | 49 | .0730 | 76 | .0200 |
| 23 | .1540 | 50 | .0700 | 77 | .0180 |
| 24 | .1520 | 51 | .0670 | 78 | .0160 |
| 25 | .1495 | 52 | .0635 | 79 | .0145 |
| 26 | .1470 | 53 | .0595 | 80 | .0135 |
| 27 | .1440 | 54 | .0550 | | |

Fig. 31-11. A table of number drills with their decimal sizes. If the size cannot be seen on the drill, a micrometer can be used to measure the size.

60. In number drills the higher the number, the smaller the drill. Fig. 29-11 lists the number drill sizes.

Letter drill sets come in sizes "A" (0.234 in.) to "Z" (0.413 in.); see Fig. 31-12. Metric drill sets include sizes from 0.100 mm (0.0039 in.) to 25.50 mm (1.004 in.); see Fig. 31-13.

| DRILL LETTER | DECIMAL SIZE | DRILL LETTER | DECIMAL SIZE |
|---|---|---|---|
| A | .234 | N | .302 |
| B | .238 | O | .316 |
| C | .242 | P | .323 |
| D | .246 | Q | .332 |
| E | .250 | R | .339 |
| F | .257 | S | .348 |
| G | .261 | T | .358 |
| H | .266 | U | .368 |
| I | .272 | V | .377 |
| J | .277 | W | .386 |
| K | .281 | X | .397 |
| L | .290 | Y | .404 |
| M | .295 | Z | .413 |

Fig. 31-12. A table of letter drills with their decimal equivalents.

### 31-14. TAPPING A THREAD

When a thread is cut in a hole, a tool called a TAP is used. Before the tap can be used, a hole of the correct size must be drilled. Regular letter, fraction, number, or metric drills are used. When these drills are used prior to tapping a hole, they are called TAP DRILLS.

Fig. 31-14 lists a number of different National Fine (N.F.) and National Coarse (N.C.) thread sizes and their proper tap drill size.

If the tap drill used is too small, the worker may break the tap. If the tap is too large, the thread depth will be too shallow.

### 31-15. METRIC PRACTICES FOR WELDING

Most of the world with the exception of the United States uses the metric system for measurement. All countries using metrics however were not using the same metric units for measurements. In weight, for example, the unit might be grams or kilograms; and lengths could be in millimeters or meters.

An international group has developed a metric system agreed to world-wide. This measurement

| Metric Drills | | Metric Drills | | Metric Drills | | Metric Drills | |
|---|---|---|---|---|---|---|---|
| mm | Equiv. Inch | mm | Equiv. Inch | mm | Equiv. Inch | mm | Equiv. Inch |
| 0.40 | 0.0157 | 2.0 | 0.0787 | 4.1 | 0.1614 | 7.3 | 0.2874 |
| 0.45 | 0.0177 | 2.05 | 0.0807 | 4.2 | 0.1654 | 7.4 | 0.2913 |
| 0.50 | 0.0197 | 2.1 | 0.0827 | 4.3 | 0.1693 | 7.5 | 0.2953 |
| 0.58 | 0.0228 | 2.15 | 0.0846 | 4.4 | 0.1732 | 7.6 | 0.2992 |
| 0.60 | 0.0236 | 2.2 | 0.0866 | 4.5 | 0.1772 | 7.7 | 0.3031 |
| 0.65 | 0.0256 | 2.25 | 0.0886 | 4.6 | 0.1811 | 7.8 | 0.3071 |
| 0.70 | 0.0276 | 2.3 | 0.0906 | 4.7 | 0.1850 | 7.9 | 0.3110 |
| 0.75 | 0.0295 | 2.35 | 0.0925 | 4.8 | 0.1890 | 8.0 | 0.3150 |
| 0.80 | 0.0315 | 2.4 | 0.0945 | 4.9 | 0.1929 | 8.1 | 0.3189 |
| 0.85 | 0.0335 | 2.45 | 0.0965 | 5.0 | 0.1968 | 8.2 | 0.3228 |
| 0.90 | 0.0354 | 2.5 | 0.0984 | 5.1 | 0.2008 | 8.3 | 0.3268 |
| 0.95 | 0.0374 | 2.55 | 0.1004 | 5.2 | 0.2047 | 8.4 | 0.3307 |
| 1.00 | 0.0394 | 2.6 | 0.1024 | 5.3 | 0.2087 | 8.5 | 0.3346 |
| 1.05 | 0.0413 | 2.65 | 0.1043 | 5.4 | 0.2126 | 8.6 | 0.3386 |
| 1.10 | 0.0433 | 2.7 | 0.1063 | 5.5 | 0.2165 | 8.7 | 0.3425 |
| 1.15 | 0.0453 | 2.75 | 0.1083 | 5.6 | 0.2205 | 8.8 | 0.3465 |
| 1.20 | 0.0472 | 2.8 | 0.1102 | 5.7 | 0.2244 | 8.9 | 0.3504 |
| 1.25 | 0.0492 | 2.85 | 0.1122 | 5.8 | 0.2283 | 9.0 | 0.3543 |
| 1.30 | 0.0512 | 2.9 | 0.1142 | 5.9 | 0.2323 | 9.1 | 0.3583 |
| 1.35 | 0.0531 | 2.95 | 0.1161 | 6.0 | 0.2362 | 9.2 | 0.3622 |
| 1.4 | 0.0551 | 3.0 | 0.1181 | 6.1 | 0.2402 | 9.3 | 0.3661 |
| 1.45 | 0.0571 | 3.1 | 0.1220 | 6.2 | 0.2441 | 9.4 | 0.3701 |
| 1.5 | 0.0591 | 3.2 | 0.1260 | 6.3 | 0.2480 | 9.5 | 0.3740 |
| 1.55 | 0.0610 | 3.25 | 0.1280 | 6.4 | 0.2520 | 9.6 | 0.3780 |
| 1.6 | 0.0630 | 3.3 | 0.1299 | 6.5 | 0.2559 | 9.7 | 0.3819 |
| 1.6 | 0.0630 | 3.4 | 0.1339 | 6.6 | 0.2598 | 9.8 | 0.3868 |
| 1.65 | 0.0650 | 3.5 | 0.1378 | 6.7 | 0.2638 | 9.9 | 0.3898 |
| 1.7 | 0.0669 | 3.6 | 0.1417 | 6.8 | 0.2677 | 10.0 | 0.3937 |
| 1.75 | 0.0689 | 3.7 | 0.1457 | 6.9 | 0.2717 | 10.1 | 0.3976 |
| 1.8 | 0.0709 | 3.75 | 0.1476 | 7.0 | 0.2756 | 10.2 | 0.4016 |
| 1.85 | 0.0728 | 3.8 | 0.1496 | 7.1 | 0.2795 | 10.3 | 0.4055 |
| 1.9 | 0.0748 | 3.9 | 0.1535 | 7.2 | 0.2835 | 10.4 | 0.4094 |
| 1.95 | 0.0768 | 4.0 | 0.1575 | 7.25 | 0.2854 | 10.5 | 0.4134 |

Fig. 31-13. A table of metric drill sizes with their decimal equivalents in U.S. customary units.

system is called the International System of Units or SI. Industry in the United States also uses SI.

The American Welding Society has published a booklet entitled, AWS A1.1-80, ''Metric Practice Guide for the Welding Industry.'' This publication describes the manner in which SI measurements will be used in the welding industry. Metrics is an easily used system since it never deals with fractions. Parts of a unit are expressed as decimals. The SI system uses only one unit for each physical quantity.

The base units used in the International System of Units (SI) are as follows:

| Measure | Unit | Symbol |
|---|---|---|
| Length | meter | m |
| Mass (weight) | kilogram | kg |
| Time | second | s |
| Electric Current | ampere | A |
| Thermodynamic temperature | kelvin | K |
| Luminous Intensity | candela | cd |
| Amount of Substance | mole | mol |
| Temperature | degree Celsius | °C |

In the SI system, the following terms are used to express parts or multiples of a base unit:

| TAP | | TAP DRILL |
| --- | --- | --- |
| Dia. of Screw | Threads per Inch | |
| 6 | 32 | No. 36 |
| 6 | 40 | No. 33 |
| 8 | 32 | No. 29 |
| 8 | 36 | No. 29 |
| 10 | 24 | No. 25 |
| 10 | 32 | No. 21 |
| 12 | 24 | No. 16 |
| 12 | 28 | No. 14 |
| 1/4 | 20 | No. 7 |
| 1/4 | 28 | No. 3 |
| 5/16 | 18 | F |
| 5/16 | 24 | I |
| 3/8 | 16 | 5/16 |
| 3/8 | 24 | Q |
| 7/16 | 14 | U |
| 7/16 | 20 | 25/64 |
| 1/2 | 13 | 27/64 |
| 1/2 | 20 | 29/64 |

Fig. 31-14. A table of NF and NC tap drill sizes.

milli = 1/1000 or .001 of the base unit.
micro = 1/1,000,000 or .000001 of the base unit.
Examples:

millisecond = 1/1000 of a second or .001 seconds.
microampere = 1/1,000,000 of an ampere or .000001 amperes.
These terms are used to express large numbers of the base unit:
kilo = 1000 × 1 unit.
mega = 1,000,000 × 1 unit.
Examples:
kilogram = 1000 grams.
megasecond = 1,000,000 seconds.
Fig. 31-15 is a list of conversion factors for use with the common measurement units used in welding. Fig. 31-16 is a conversion table for changing electrode diameters and fillet weld sizes to SI metric units.

## 31-16. TEMPERATURE SCALES

Many means have been used to measure the temperature level. In the International System of Units (SI) or metrics, the temperature scale is in degrees CELSIUS. In the Celsius scale, ice melts at 0 °C and water boils at 100 °C. The FAHRENHEIT scale is based on a scale of 180 equal points with ice melting at 32 °F and water boils at 212 °F.

One other temperature scale is used in scientific work. This scale measures temperatures in degrees KELVIN. At 0 °K on the kelvin scale, the

| | CONVERTING FROM | TO | MULTIPLY BY (4 DIGIT ACCURACY) |
| --- | --- | --- | --- |
| Length | in. | m | .0254 |
| | in. | mm (millimeters) | 25.40 |
| | ft. | m | .3048 |
| Mass (Weight) | pound | kg | .4536 |
| Pressure | lbs/in² or psi | kPa (kilopascal) | 6.895 |
| Tensile Strength | ksi (kilopounds/in²) | MPa (megapascals) | 6.895 |
| Area | in² | m² | .0006451 |
| | in² | mm² | 645.2 |
| | ft² | m² | .0929 |
| | ft² | mm² | 92903 |
| Volume | in³ | m³ | .00001639 |
| | in³ | mm³ | 16387 |
| | ft³ | m³ | .02832 |
| | ft³ | mm³ | 28316846 |
| Energy | BTU | J (joule) | 1054.35 |
| Force | pound | N (newton) | 4.448 |

Fig. 31-15. A table showing how to convert common U.S. Conventional units into SI units.

| ELECTRODE SIZES | | FILLET SIZES | |
|---|---|---|---|
| in. | mm | in. | mm |
| 0.030 | 0.76 | 1/8 | 3 |
| 0.035 | 0.89 | 5/32 | 4 |
| 0.040 | 1.02 | 3/16 | 5 |
| 0.045 | 1.14 | 1/4 | 6 |
| 1/16 | 1.59 | 5/16 | 8 |
| 5/64 | 1.98 | 3/8 | 10 |
| 3/32 | 2.38 | 7/16 | 11 |
| 1/8 | 3.18 | 1/2 | 13 |
| 5/32 | 3.97 | 5/8 | 16 |
| 3/16 | 4.76 | 3/4 | 19 |
| 1/4 | 6.35 | 7/8 | 22 |
| | | 1 | 25 |

Fig. 31-16. Conversion tables from U.S. conventional to SI metric. The table gives equivalents for electrodes and fillet sizes in metric sizes.

temperature is −273°C. This is the point at which all molecular motion stops. The point at which ice melts is 273°K. Water boils at 373°K.

To convert °F to °C, use the formula below:

$$(°F − 32) × .555 = °C$$

Example:

80°F = ?°C
(80°F − 32) × .555 = ?°C
48 × .555 = 26.6°C or 27°C

To convert °C to °F, use this formula:

$$(°C × 1.8) + 32 = °F$$

Example:

100°C = ?°F
(100°C × 1.8) + 32 = ?°F
180°C + 32 = 212°F

or 40°C = ?°F
(40 × 1.8) + 32 = ?°F
72 + 32 = 104°F

## 31-17. PHYSICS OF ENERGY, TEMPERATURE, AND HEAT

The welding industry is interested in the energy in heating gases. It is interested in turning solids to liquids. Therefore, welders should know some of the theory of molecular energy.

The Molecular Theory of heat is generally accepted by chemists and engineers as the best explanation of heat energy. The three most important forms of energy are HEAT ENERGY, MECHANICAL ENERGY, and ELECTRICAL ENERGY. One form of energy can be converted into another form of energy. For example: an electric motor may be used to turn electrical energy into mechanical energy. The bearings of a motor warm up a little while it is running. This shows that some of the mechanical energy is being turned into heat.

One horse power (mechanical energy) = 2545.6 Btu (2684 kJ) per hour (heat energy).
One horse power (electrical energy) = 746 watts.
746 watts = 2545.6 Btu (2684 kJ) per hour.
1 watt = 3.412 Btu (3.497 kJ) per hour
1 kW = 3412 Btu (3497 kJ) per hour.

The Molecular Theory of heat, briefly explained, is as follows: All matter consists of molecules and atoms. The atoms contain protons, electrons, neutrons, and short life particles. It is believed that molecules and atoms are always in motion.

Molecules in a sheet of paper or a piece of steel are continually in motion. The rapidity with which these molecules move determines the amount of heat energy in the material. The rapidity with which the molecule moves determines the heat level and is known as TEMPERATURE. Whether one atom is moving at a certain speed, or whether a thousand atoms are moving at the same speed, the temperature will be the same.

The number of molecules or atoms in a substance at a given temperature is what determines the heat or HEAT ENERGY of a substance. This, in brief, is the difference between temperature and the amount of heat. The speed of the motion of the molecules determines the temperature. But, the amount of the substance determines the energy.

All substances may exist in four forms, solid, liquid, gas, and plasma. It is usually conceded that there is no change in the chemical composition of the same substance in any of these four forms. Therefore, it is explained that in a particular substance in the solid form, the molecules have a vibrating motion. They stay in the same position relative to each other, but they are vibrating. When energy is applied to the substance, the molecules vibrate faster than before. This is indicated by an increase in temperature. However, only a certain definite amount of heat energy may be put into a solid substance, and only a certain temperature rise can be obtained in a solid.

After this amount has been absorbed by the substance, if any additional energy is added the molecules will travel at such a rate that the molecules cannot stay within their vibrating bonds. At this heat level an internal change of structure occurs within the molecule. This is accompanied by the absorption of a large amount of energy, and the substance changes slowly from the solid to the liquid state. The substance will absorb heat during the change but there is no temperature rise. All the heat being applied, regardless of how fast or slow, produces the internal structural change, rather than an increase of the motion of the molecule.

The theory of the energy in a liquid substance is that the molecule now travels in a straight line until it comes into contact with another molecule, instead of vibrating. This necessarily means that the substance will now have no definite shape and will have no rigidity. It must, therefore, be kept in a container. However, the structure of the molecule is such that the individual molecules still have considerable attraction for one another. One molecule will attract another enough to divert its path and also to prevent it from traveling too far away.

As energy is applied to the liquid molecule, the rate of travel increases and this is indicated by a temperature rise. After a certain amount of energy has been absorbed by the liquid, it will reach a point where energy applied results not in a temperature change, but in an internal structure change of the molecule. The result of this is that the liquid now turns into a gas. While a liquid is being turned into a gas, the temperature cannot rise as the heat being applied results only in a structural change. Upon becoming a gas, the molecules lose their attraction one for the other and travel in straight paths until contacting another molecule or some other substance. This means that gas must be confined to sealed containers. Further heating will change the gas into the plasma state.

Ice, water, and steam are the best examples of the above theory. All three conditions are easily obtained. Changing from one to the other does not reveal any change in chemical composition.

The heat which turns solids to liquids, liquids to solids, liquids to gases, or gases to liquids, is called LATENT HEAT (hidden heat). This is because the thermometer gives no indication of the amount of this heat. For example, it requires 970 Btu to change one pound of water at 212 °F, to steam at 212 °F (or 2255 joules to change one gram of water at 100 °C, to steam at 100 °C).

The oxyacetylene flame temperature is the same for all tip sizes. The amount of energy delivered to the metal, however, increases as the tip size and gas volume increases.

## 31-18. HEALTH HAZARDS

As a part of instruction in welding, it is important to be familiar with, and to be able to recognize conditions that may be hazardous to health. It is important too, to remember that the best way to attack the hazards is to eliminate or control the conditions that are responsible.

The oxyfuel gas flame is generally safe, but there is a carbon monoxide (CO) problem in a poorly ventilated place.

Coatings on metals are a problem when the metal is heated. Nitrogen dioxide, coated electrode fumes, and iron oxide and cadmium and zinc coating fumes are examples.

Red lead paint was often used outdoors for metal finishing and protection. Lead oxide fumes caused by burning lead paint coatings can produce acute lead poisoning.

Cadmium plate is frequently used on small parts. Cadmium oxide fumes at low levels produce a chronic condition. At high levels, the fumes are harmful to lungs and liver.

Silver brazing alloy fumes can be quite dangerous.

Termeplate is a metallic lead coating and is dangerous when heated.

Fluorides in fluxes are common. Fluoride fumes are harmful.

Manganese dioxide is not too toxic, but it may cause trouble if ventilation is poor.

When gas tungsten arc welding aluminum, there is a possibility of real trouble as the ultraviolet frequency is right to form ozone. This gas is most toxic and may cause severe lung and body damage. It is irritating and causes coughing. It is advisable to keep the arc area well ventilated.

Ultraviolet rays are harmful to the welder's vision. Safety goggles with proper lenses will provide protection.

Ionizing radiations markedly increase the possibility of eye cataracts. Even small amounts of ultraviolet may generate ionization and accelerate cataracts. Those operating welding equipment, as well as helpers and other people in the area must take precautions at all times to prevent eye injury by wearing safety glasses.

The use of carbon dioxide creates a carbon monoxide (CO) problem as it breaks down in the welding arc.

Beryllium is toxic in even very small amounts. Therefore, any operation involving beryllium must be contained.

Cobalt should be handled about the same as beryllium.

Thorium is toxic. Therefore, when thoriated electrodes are used, an alpha emission is produced and causes an ionization effect. Ventilate the area well.

Reducing furnaces may have carbon monoxide (CO) emissions. Such furnaces should be well vented at both the charging end and the discharging end.

When vacuum furnaces or welding chambers are used, the pump exhaust must be vented away from people.

Inert gases are dangerous to use in confined spaces. Inert gases displace air and oxygen. Workers entering spaces or tanks that are filled with an inert gas must wear supplied air or self contained breathing equipment.

Oil smoke is a problem. The aromatics produced can be dangerous. Fig. 31-17 shows a table of safe limits for some welding fumes.

NEVER FAIL TO PROTECT THE EYES, SKIN, AND RESPIRATORY SYSTEM. PROVIDE ADEQUATE VENTILATION DURING ALL WELDING CUTTING, BRAZING, AND SOLDERING OPERATIONS.

| MATERIAL | GASES | | MILLION PARTS |
|----------|-------|-------|---------------|
| | ppm | mg/m³ | per Cu. Ft. |
| Acetylene | 1000 | | |
| Beryllium | | .002 | |
| Cadmium Oxide fumes | | .1 | |
| Carbon Dioxide | 5000 | | |
| Copper fumes | | .1 | |
| Iron Oxide fumes | | 10.0 | |
| Lead | | .2 | |
| Manganese | | 5.0 | |
| Nitrogen Dioxide | 5.0 | | |
| Oil Mist | | 5.0 | |
| Ozone | .1 | | |
| Titanium Oxide | | 15.0 | |
| Zinc Oxide fumes | | 5.0 | |
| Silica, crystalline | | | 2.5 |
| Silica, amorphous | | | 20.0 |
| Silicates: Asbestos | | | 5.0 |
| Portland Cement | | | 50.0 |
| Graphite | | | 15.0 |
| Nuisance Dust | | | 50.0 |

Fig. 31-17. A table of safe limits for some welding fumes. All gases tend to reduce oxygen by replacement. Such gases as argon, helium, carbon dioxide, etc., present this danger.

## 31-19. RESOURCES FOR HEALTH AND SAFETY INFORMATION

The Occupational Safety and Health Administration (OSHA) provides many regulations and controls for the safe operation, handling and storage or welding supplies and equipment.

Instructional personnel should be familiar with OSHA requirements. Information concerning safety for any particular trade or industry can be received from the Department of Labor in the state where the program is located.

Publications are available also from the U.S. Department of Health and Human Services, Public Health Service, Center for Disease Control, National Institute of Environmental Health Services, Robt. A. Taft Laboratories, and others.

## 31-20. TEST YOUR KNOWLEDGE

Write your answers on a separate sheet of paper. Do not write in this book.

1. What is the pressure in a 244 ft.³ oxygen cylinder at 30°F.?
2. If an oxyfuel gas torch with a 1/4 in. inside diameter hose and a number 7 tip is used 100 ft. from the cylinders, what pressure drop occurs in the hose?
3. Which fuel gas has the highest flame temperature? What is the temperature?
4. When mixing an alkaline solution for cleaning, _____ ounces of sodium bicarbonate are mixed with one gallon of water.
5. What chemicals are mixed to make a bright pickle solution?
6. A pressure of 250 lbs. per square inch or psi is equal to what SI metric pressure?
7. What material is used in the blast furnace as a flux material?
8. At what temperature does pure iron melt in °F and °C?
9. Pure nickel melts at what temperature?
10. How much heavier than aluminum is steel?
11. Which two metal thickness gauges are used in the United States?
12. At what temperature is clean, carbon steel if it is heated to a color of cherry red?
13. A number 44 drill is how large in diameter in decimals of an inch?
14. What is the decimal size of a 7/32 inch drill?
15. Which letter drill is larger in diameter: a G or an H?
16. Before tapping a 12-28NF thread, what size tap drill should be used?
17. When tapping a hole, what may happen if the tap drill is too small?
18. What do the letters SI mean?
19. In SI, how large is a microampere?
20. What precautions must be taken before entering a space or tank which is filled with an inert gas like argon, helium, or $CO_2$?

# GENERAL CONVERSIONS

| Property | To convert from | To | Multiply by |
|---|---|---|---|
| acceleration (angular) | revolution per minute squared | rad/s² | $1.745\ 329 \times 10^{-3}$ |
| acceleration (linear) | in./min² | m/s² | $7.055\ 556 \times 10^{-6}$ |
| | ft/min² | m/s² | $8.466\ 667 \times 10^{-5}$ |
| | in./min² | mm/s² | $7.055\ 556 \times 10^{-3}$ |
| | ft/min² | mm/s² | $8.466\ 667 \times 10^{-2}$ |
| | ft/s² | m/s² | $3.048\ 000 \times 10^{-1}$ |
| angle, plane | deg | rad | $1.745\ 329 \times 10^{-2}$ |
| | minute | rad | $2.908\ 882 \times 10^{-4}$ |
| | second | rad | $4.848\ 137 \times 10^{-6}$ |
| area | in.² | m² | $6.451\ 600 \times 10^{-4}$ |
| | ft² | m² | $9.290\ 304 \times 10^{-2}$ |
| | yd² | m² | $8.361\ 274 \times 10^{-1}$ |
| | in.² | mm² | $6.451\ 600 \times 10^{2}$ |
| | ft² | mm² | $9.290\ 304 \times 10^{4}$ |
| | acre (U.S. Survey) | m² | $4.046\ 873 \times 10^{3}$ |
| density | pound mass per cubic inch | kg/m³ | $2.767\ 990 \times 10^{4}$ |
| | pound mass per cubic foot | kg/m³ | $1.601\ 846 \times 10$ |
| energy, work, heat, and impact energy | foot pound force | J | $1.355\ 818$ |
| | foot poundal | J | $4.214\ 011 \times 10^{-2}$ |
| | Btu* | J | $1.054\ 350 \times 10^{3}$ |
| | calorie* | J | $4.184\ 000$ |
| | watt hour | J | $3.600\ 000 \times 10^{3}$ |
| force | kilogram-force | N | $9.806\ 650$ |
| | pound-force | N | $4.448\ 222$ |
| impact strength | (see energy) | | |
| length | in. | m | $2.540\ 000 \times 10^{-2}$ |
| | ft | m | $3.048\ 000 \times 10^{-1}$ |
| | yd | m | $9.144\ 000 \times 10^{-1}$ |
| | rod (U.S. Survey) | m | $5.029\ 210$ |
| | mile (U.S. Survey) | km | $1.609\ 347$ |
| mass | pound mass (avdp) | kg | $4.535\ 924 \times 10^{-1}$ |
| | metric ton | kg | $1.000\ 000 \times 10^{3}$ |
| | ton (short, 2000 lbm) | kg | $9.071\ 847 \times 10^{2}$ |
| | slug | kg | $1.459\ 390 \times 10$ |
| power | horsepower (550 ft lbf/s) | W | $7.456\ 999 \times 10^{2}$ |
| | horsepower (electric) | W | $7.460\ 000 \times 10^{2}$ |
| | Btu/min* | W | $1.757\ 250 \times 10$ |
| | calorie per minute* | W | $6.973\ 333 \times 10^{-2}$ |
| | foot pound-force per minute | W | $2.259\ 697 \times 10^{-2}$ |
| pressure | pound force per square inch | kPa | $6.894\ 757$ |
| | bar | kPa | $1.000\ 000 \times 10^{2}$ |
| | atmosphere | kPa | $1.013\ 250 \times 10^{2}$ |
| | kip/in.² | kPa | $6.894\ 757 \times 10^{3}$ |
| temperature | degree Celsius, $t\,^{\circ}C$ | K | $t_K = t\,^{\circ}C + 273.15$ |
| | degree Fahrenheit, $t\,^{\circ}F$ | K | $t_K = (t\,^{\circ}F + 459.67)/1.8$ |
| | degree Rankine, $t\,^{\circ}R$ | | $t_K = t\,^{\circ}R/1.8$ |
| | degree Fahrenheit, $t_F$ | | $t\,^{\circ}C = (t_F - 32)/1.8$ |
| | kelvin, $t_K$ | | $t\,^{\circ}C = t_K - 273.15$ |
| tensile strength (stress) | ksi | MPa | $6.894\ 757$ |
| torque | inch pound force | N·m | $1.129\ 848 \times 10^{-1}$ |
| | foot pound force | N·m | $1.355\ 818$ |

## General Conversions continued

| Property | To convert from | To | Multiply by |
|---|---|---|---|
| velocity (angular) | revolution per minute | rad/s | $1.047\ 198 \times 10^{-1}$ |
| | degree per minute | rad/s | $2.908\ 882 \times 10^{-4}$ |
| | revolution per minute | deg/min | $3.600\ 000 \times 10^{2}$ |
| velocity (linear) | in./min | m/s | $4.233\ 333 \times 10^{-4}$ |
| | ft/min | m/s | $5.080\ 000 \times 10^{-3}$ |
| | in./min | mm/s | $4.233\ 333 \times 10^{-1}$ |
| | ft/min | mm/s | $5.080\ 000$ |
| | mile/hour | km/h | $1.609\ 344$ |
| volume | in.$^3$ | m$^3$ | $1.638\ 706 \times 10^{-5}$ |
| | ft$^3$ | m$^3$ | $2.831\ 685 \times 10^{-2}$ |
| | yd$^3$ | m$^3$ | $7.645\ 549 \times 10^{-1}$ |
| | in.$^3$ | mm$^3$ | $1.638\ 706 \times 10^{4}$ |
| | ft$^3$ | mm$^3$ | $2.831\ 685 \times 10^{7}$ |
| | in.$^3$ | L | $1.638\ 706 \times 10^{-2}$ |
| | ft$^3$ | L | $2.831\ 685 \times 10$ |
| | gallon | L | $3.785\ 412$ |

*thermochemical

## UNITS PERTAINING TO WELDING

| Property | Unit | Symbol |
|---|---|---|
| area dimensions | square millimeter | mm$^2$ |
| current density | ampere per square millimeter | A/mm$^2$ |
| deposition rate | kilogram per hour | kg/h |
| electrical resistivity | ohm meter | $\Omega \cdot m$ |
| electrode force (upset, squeeze, hold) | newton | N |
| flow rate (gas and liquid) | liter per minute | L/min |
| fracture toughness | meganewton meter$^{-3/2}$ | MN·m$^{-3/2}$ |
| impact strength | joule | J = N·m |
| linear dimensions | millimeter | mm |
| power density | watt per square meter | W/m$^2$ |
| pressure (gas and liquid) | kilopascal | kPa = 1000 N/m$^2$ |
| tensile strength | megapascal | MPa = 1000 000 N/m$^2$ |
| thermal conductivity | watt per meter kelvin | W/(m·K) |
| travel speed | millimeter per second | mm/s |
| volume dimensions | cubic millimeter | mm$^3$ |
| wire feed rate | millimeter per second | mm/s |

# CONVERSIONS FOR COMMON WELDING TERMS*

| Property | To convert from | To | Multiply by |
|---|---|---|---|
| area dimensions (mm²) | in² | mm² | $6.451\ 600 \times 10^2$ |
| | mm² | in² | $1.550\ 003 \times 10^{-3}$ |
| current density (A/mm²) | A/in² | A/mm² | $1.550\ 003 \times 10^{-3}$ |
| | A/mm² | A/in² | $6.451\ 600 \times 10^2$ |
| deposition rate** (kg/h) | lb/h | kg/h | .045** |
| | kg/h | lb/h | 2.2** |
| electrical resistivity ($\Omega \cdot m$) | $\Omega \cdot cm$ | $\Omega \cdot m$ | $1.000\ 000 \times 10^{-2}$ |
| | $\Omega \cdot m$ | $\Omega \cdot cm$ | $1.000\ 000 \times 10^2$ |
| electrode force (N) | pound-force | N | 4.448 222 |
| | kilogram-force | N | 9.806 650 |
| | N | lbf | $2.248\ 089 \times 10^{-1}$ |
| flow rate (L/min) | ft³/h | L/min | $4.719\ 475 \times 10^{-1}$ |
| | gallon per hour | L/min | $6.309\ 020 \times 10^{-2}$ |
| | gallon per minute | L/min | 3.785 412 |
| | cm³/min | L/min | $1.000\ 000 \times 10^{-3}$ |
| | L/min | ft³/h | 2.118 880 |
| | cm³/min | ft³/h | $2.118\ 880 \times 10^{-3}$ |
| fracture toughness ($MN \cdot m^{-3/2}$) | $ksi \cdot in.^{1/2}$ | $MN \cdot m^{-3/2}$ | 1.098 855 |
| | $MN \cdot m^{-3/2}$ | $ksi \cdot in.^{1/2}$ | 0.910 038 |
| heat input (J/m) | J/in. | J/m | $3.937\ 008 \times 10$ |
| | J/m | J/in. | $2.540\ 000 \times 10^{-2}$ |
| impact energy | foot pound force | J | 1.355 818 |
| linear measurements (mm) | in. | mm | $2.540\ 000 \times 10$ |
| | ft | mm | $3.048\ 000 \times 10^2$ |
| | mm | in. | $3.937\ 008 \times 10^{-2}$ |
| | mm | ft | $3.280\ 840 \times 10^{-3}$ |
| power density (W/m²) | W/in.² | W/m² | $1.550\ 003 \times 10^3$ |
| | W/m² | W/in.² | $6.451\ 600 \times 10^{-4}$ |
| pressure (gas and liquid) (kPa) | psi | Pa | $6.894\ 757 \times 10^3$ |
| | lb/ft² | Pa | $4.788\ 026 \times 10$ |
| | N/mm² | Pa | $1.000\ 000 \times 10^6$ |
| | kPa | psi | $1.450\ 377 \times 10^{-1}$ |
| | kPa | lb/ft² | $2.088\ 543 \times 10$ |
| | kPa | N/mm² | $1.000\ 000 \times 10^{-3}$ |
| | torr (mm Hg at 0 °C) | kPa | $1.333\ 22 \times 10^{-1}$ |
| | micron (μm Hg at 0 °C) | kPa | $1.333\ 22 \times 10^{-4}$ |
| | kPa | torr | $7.500\ 64 \times 10$ |
| | kPa | micron | $7.500\ 64 \times 10^3$ |
| tensile strength (MPa) | psi | kPa | 6.894 757 |
| | lb/ft² | kPa | $4.788\ 026 \times 10^{-2}$ |
| | N/mm² | MPa | 1.000 000 |
| | MPa | psi | $1.450\ 377 \times 10^2$ |
| | MPa | lb/ft² | $2.088\ 543 \times 10^4$ |
| | MPa | N/mm² | 1.000 000 |
| thermal conductivity (W/[m·K]) | cal/(cm·s· °c) | W/(m·K) | $4.184\ 000 \times 10^2$ |
| travel speed, wire feed speed (mm/s) | in./min | mm/s | $4.233\ 333 \times 10^{-1}$ |
| | mm/s | in./min | 2.362 205 |

*Preferred units are given in parentheses.
**Approximate conversion.

# 32 GLOSSARY OF TERMS

This chapter explains the meaning of terms most used by welders. Technical engineering terms have been simplified. For additional definitions, refer to the AWS WELDING TERMS AND DEFINITIONS (publication No. AWS A3.0-80). It is published by the American Welding Society.

## A

ABRASION: Worn condition produced by rubbing.

AC or ALTERNATING CURRENT: Kind of electricity which reverses its direction of electron flow in regular intervals.

ACETYLENE ($C_2H_2$): Gas composed of two parts of carbon and two parts of hydrogen. When acetylene is burned in an atmosphere of oxygen, it produces one of the highest flame temperatures obtainable.

ACETYLENE CYLINDER: Specially built container manufactured according to ICC standards. Used to store and ship acetylene (Occasionally called "tank" or "bottle").

ACETYLENE HOSE: See HOSE.

ACETYLENE REGULATOR: Automatic valve used to reduce acetylene cylinder pressures to torch pressures and to keep the pressures constant.

ACOUSTIC EMISSION: Sound produced by a material as it undergoes some change.

ACTUAL THROAT: Shortest distance from the root of the weld to the face of the weld.

ADHESION: Act of sticking or clinging.

AIR CARBON ARC CUTTING (AAC): Arc cutting process which uses a carbon arc to heat the metal and an air blast to remove the molten metal and form the cut or gouge.

ALLOY: An alloy is a pure metal which has additional metal or nonmetal elements added while molten. The alloy has different (usually improved) mechanical properties than the pure metal.

AMPERE: Unit of electrical current. One ampere is required to flow through a conductor having a resistance of one ohm at a potential (pressure) of one volt.

AMPERE-TURNS: (A term used with electromagnets.) It is equal to the number of turns of wire in a coil times the number of amperes flowing in the coil.

ANNEALING: Softening metals by heat treatment.

ANODE: Positive terminal of an electrical circuit.

ARC: Flow of electricity through a gaseous space or air gap.

ARC BLOW: Wandering of an electric arc from its normal path because of magnetic forces.

ARC CUTTING: Making a cut in a metal using energy of an electric arc.

ARC GOUGING: Arc cutting process used to cut a groove or bevel.

ARC VOLTAGE: Electrical potential (pressure or voltage) across the arc.

ARC WELDING: Group of welding processes used to melt and weld metal using the heat of an electric arc with or without filler metal.

ATOM: Smallest whole part of an element. Its nucleus is made up of protons and neutrons.

AUSTENITE: High temperature form of steel. It has a face centered cubic structure.

AUTOMATIC WELDING: Welding is mechanically moved while controls govern the speed and/or the direction of travel.

AXIS OF A WELD: Imaginary line along the center of the weld metal from the beginning to the end of the weld.

## B

BACKFIRE: Short "pop" of the torch flame followed by extinguishing of the flame or continued burning of the gases.

BACKHAND WELDING: Moving the weld in opposite direction to which gas flame is pointing.

BACKING: Some material placed on the root side of a weld to aid in the control of penetration.

BACKING RING: Metal ring placed inside of a pipe before butt welding; it insures complete weld penetration and a smooth inside surface.

BACK-STEP WELDING: Welding small sections of a joint in a direction opposite the progression of the weld as a whole.

BACKWARD WELDING: See BACKHAND WELDING.

BASE METAL: Metal to be welded, cut, or brazed.

BAUXITE: Ore from which aluminum is obtained. Consists mostly of hydrated alumina ($Al_2O_3 3H_2O$).

BEAD: Appearance of the finished weld; the metal added in welding.

BEVEL: Angling the metal edge where welding is to take place.

BLACK LIGHT: Light waves below the visible range of violet light. The wavelength reacts with certain dyes causing the dyes to fluoresce in a color range visible to the eye.

BLASTING: Method of cleaning or surface roughening by projecting a stream of sharp angular abrasives against the material.

BLOWHOLE: See preferred term POROSITY.

BLOWPIPE: Another name for an oxyfuel gas torch.

BODY: Main structural part of a regulator.

BOND LINE: Junction between the thermal spray deposit and the base metal.

BRAZE WELDING: Making an adhesion groove, fillet, or plug connection above 840 °F (450 °C). The metal is not distributed by capillary action.

BRAZEMENT: Assembly joined by brazing.

BRAZING: Making an adhesion connection with a minimum of alloy which melts above 840 °F (450 °C) and which flows by capillary action between close fitting parts.

BRINELL HARDNESS: Accurate measure of hardness of metal made with an instrument. Measurement is made as a hard steel ball is pressed into the smooth surface at standard conditions.

BRITTLENESS: Quality of a material which causes it to develop cracks with little bending (deformation) of the material.

BRONZE WELDING: See BRAZE WELDING.

BUILDUP: Amount of a weld face extended above surface of joined metals.

BURNING: See FLAME CUTTING.

BUTT JOINT: Assembly in which the two pieces joined are in the same plane with the edge of one piece touching the edge of the other.

BUTTERING: Surfacing layer on the face of a groove weld. Usually as a transition layer when welding dissimilar metals.

BUTTON: Part of a weld torn out in destructive testing of a spot, seam, or projection welding.

C

CABLE: See LEAD.

CAPILLARY ACTION: Property of a liquid to move into small spaces if it has the ability to ''wet'' these surfaces.

CARBON: Element which, when combined with iron, forms various kinds of steel. In solid form, it is used as an electrode for arc welding. As a mold, it will hold weld metal. Motor brushes are made from carbon.

CARBURIZING: See CASEHARDENING.

CASEHARDENING: Adding carbon to the surface of a mild steel object and heat treating it to produce a hard surface.

CASTINGS: Metallic forms which are produced by pouring molten metal into a shaped container or cavity called a mold.

CAST-WELD ASSEMBLY: Cast parts fixed in an assembly by welding.

CATHODE: Electrical term for negative terminal.

CELSIUS: Temperature scale in SI metric.

CEMENTITE: Compound also known as Iron Carbide, $Fe_3C$. It contains 6.67 percent carbon.

CHAMFER: See preferred term BEVEL.

CHARPY: Impact testing machine which strikes the specimen with a swinging hammer. The specimen is placed against an anvil with supports 40 mm apart.

CHEMICAL FLUX CUTTING (FOC): Oxyfuel gas cutting process which uses a chemical flux to help in cutting certain materials.

CHILL: Cool rapidly.

CHLORINATION: Passing of dry chlorine gas through molten aluminum alloys to remove trapped oxides and dissolved gases.

CIRCUIT: Path of electron flow from the source through components and connections back to its source.

CLADDING: Somewhat thick layer of material applied on a surface to improve resistance to corrosion or other agents which tend to wear away the metal.

CLEARANCE: Gap or space between adjoining or mating surfaces.

COATED ELECTRODE: See COVERED ELECTRODE.

COATING: A relatively thin layer of material applied to a surface to prevent corrosion, wear, or temperature scaling.

COHESION: Sticking together through attraction of molecules.

COLD WELDING (CW): Use of high pressure and no outside heat to force metal parts to fuse.

COLD WORK: Metal part on which a permanent strain has been placed by an outside force while the metal is below its recrystallization temperature.

COLD WORKING: Bending (deforming) metal at a temperature lower than its recrystallization temperature.

COMBINED STRESSES: Stress type more complex than simple tension, compression, or shear.

COMBUSTIBLE: Flammable, easily ignited.

COMPLETE FUSION: Fusion which has occurred over the entire surface of a base metal being welded.

COMPLETE JOINT PENETRATION: When weld metal completely fills the groove and fuses with the base metal through its entire thickness.

COMPRESSIVE STRENGTH: Greatest stress developed in material under compression.

CONCAVE WELD FACE: Weld having the center of its face below the weld edges.

CONDUCTIVITY: Ability of a conductor to carry current.

CONDUCTOR: Substance capable of readily transmitting electricity or heat.

CONE: Inner visible flame shape of a neutral or nearly neutral flame.

CONSTRICTING: Reducing in size or diameter as in a constricted arc or constricting orifice.

CONTINUOUS CASTING: Method of casting metal in an open ended mold so that metal is fed into and cools in the mold in a continuous form.

CONTINUOUS WELD: Making the complete weld in one operation.

CONVEX WELD: Weld with the face above the weld edges.

COOLING STRESSES: Stresses resulting from uneven distribution of heat during cooling.

CORNER JOINT: Junction formed by edges of two pieces of metal touching each other at angle of about 90 degrees (right angles).

CORROSION: Interaction of a metal—chemically and electro-chemically—with its surroundings causing it to deteriorate.

CORROSION EMBRITTLEMENT: Loss of ductility or workability of a metal due to corrosion.

CORROSION FATIGUE: Effect of repeated stress in a corrosive atmosphere characterized by shortened life of the part.

COUPLANT: A material placed between the transducer and metal surface when doing ultrasonic testing.

COUPON: Piece of metal used as a test specimen. Often an extra piece as in a casting or forging.

COVERED ELECTRODE: Metal rod used in arc welding which has a covering of materials to aid arc welding process.

COVER LENS OR PLATE: A removable pane of clear glass or plastic used to protect the expensive filtering welding lens.

CRACK: Break or separation in rigid material running more or less in one direction.

CRACKING: The action of opening a valve slightly and then closing the valve immediately.

CRATER: Depression in the face of a weld, usually at the termination of a weld or under the arc while welding.

CREEP: Permanent deformation caused by stress or heat or both.

CREVICE CORROSION: Deterioration of a metal caused by concentration of dissolved salts, metal ions, oxygen, or other gases in pockets not disturbed by the fluid stream. Buildup eventually causes deep pitting.

CROWN: Curve or convex surface of finished weld.

CRYOGENICS: Study of physical phenomena at temperatures below −50°F (−46°C).

CUP: See preferred term NOZZLE.

CUPOLA: Blast furnace in the shape of a vertical cylinder used in making gray iron.

CUTTING HEAD: The part of a cutting machine or cutting equipment to which a cutting torch or tip is attached.

CUTTING NOZZLE: See CUTTING TIP.

CUTTING PROCESS: Action which causes separating or removal of metal.

CUTTING TIP: Part of an oxygen cutting torch from which the gases are released.

CUTTING TORCH: Nozzle or device which controls and directs the gases and oxygen needed for cutting and removing the metal in oxyfuel gas cutting.

CYCLE: Set of repeating events which occur in order or one complete reversal of alternating current.

CYLINDER: Container holding the supply of high pressure gas used in welding. (See OXYGEN, ACETYLENE).

CYLINDER MANIFOLD: See MANIFOLD.

D

DC or DIRECT CURRENT: Electric current which flows only in one direction.

DCEN: See DIRECT CURRENT ELECTRODE NEGATIVE.

DCEP: See DIRECT CURRENT ELECTRODE POSITIVE.

DCRP: See DIRECT CURRENT REVERSE POLARITY.

DCSP: See DIRECT CURRENT STRAIGHT POLARITY.

DECALESCENCE: Transformation which takes place during superheating of iron or steel. The metal surface darkens due to the sudden decrease of temperature during the rapid absorption of the latent heat of transformation.

DECARBURIZATION: Carbon loss from the surface of a ferrous alloy when the alloy is heated in a medium that reacts with the carbon on the surface.

DEEP ETCHING: Eating away (as with corrosive action of acid) of a metal surface for purpose of examining the surface (under a magnifier) to detect features such as grain flow, cracks, or porosity.

DEFECT: Imperfection which, by its size, shape,

location, or makeup, reduces the useful service of a part.

DEGASIFIER: Substance which can be added to molten metal to draw off soluble gases before they are trapped in the metal.

DEGASSING: Process of removing gases from liquids or solids.

DEGREE: One unit of a temperature scale.

DEMAGNETIZATION: Removal of existing magnetism from a part.

DEMURRAGE: Charge made by a gas supplier to the gas user as rent on the gas cylinder. The user is allowed free use for a number of days. Then a daily charge is made for additional usage.

DENDRITE: Crystal development often found in cast metals as they are slowly cooled through the solidification range. The crystal shows a tree-like branching pattern.

DEOXIDIZER: Substance which, when added to molten metal, removes either free or combined oxygen.

DEOXIDIZING: Process of removing oxygen from molten metals with a deoxidizer. Also: the removal of other desirable elements using elements or compounds that readily react with them. Or: in metal finishing, removing oxide films with chemicals or electrochemical processes.

DEPOSITION RATE: Weight of material applied in a unit of time. Usually expressed in lbs/hr or kg/hr.

DEPTH OF FUSION: Depth to which base metal is melted during welding.

DESTRUCTIVE TESTING (DT): Testing a sample until it fails. Used to determine how large a discontinuity can be before it is considered a flaw.

DETONATION FLAME SPRAYING: Thermal spraying process which uses controlled explosions of a mixture of fuel gas and oxygen to propel a powdered coating material onto the surface of a workpiece.

DIE CASTING: Production of parts by forcing molten metal into a metal die or mold. Also, the part made by this process.

DIE FORGING: Part shaped by use of dies in the forging operation.

DIFFUSION: Spreading of an element throughout a gas, liquid, or solid so that all of it has the same composition.

DIODE: Device used in an electrical circuit which will permit the electricity to flow in only one direction.

DIRECT CURRENT ELECTRODE NEGATIVE (DCEN): Direct current flowing from the electrode (cathode) to the work (anode).

DIRECT CURRENT ELECTRODE POSITIVE (DCEP): Direct current flow from the work (cathode) to the electrode (anode).

DIRECT CURRENT REVERSE POLARITY: See preferred term DIRECT CURRENT ELECTRODE POSITIVE.

DIRECT CURRENT STRAIGHT POLARITY: See preferred term DIRECT CURRENT ELECTRODE NEGATIVE.

DISCONTINUITY: Any abrupt change or break in the shape or structure of a part (cracks, seams, laps, bumps, or changes in density). Usefulness of part may or may not be affected.

DISTORTION: Warping of a part of a structure.

DOWNHAND WELDING: See preferred term FLAT POSITION WELD.

DRAG: Offset distance between the actual and theoretical exit points of the cutting oxygen stream measured on the exit side of the material.

DRAWING: A shop term used mistakenly for the term "TEMPERING" in the heat treatment of metal.

DROSS: Oxidized metal or impurities which may form on molten metals. It often forms on the other side of metal when it is flame or arc cut.

DUCTILE CRACK PROPAGATION: Slow development of cracks and noticeable warping or deformation caused by an outside pressure.

DUCTILITY: Ability of a material to be changed in shape without cracking or breaking.

DUTY CYCLE: Percentage of time in a 10 minute period that a welding machine can be used at its rated output without overloading.

E

EDGE JOINT: Joint formed when two pieces of metal are lapped with at least one edge of each at an edge of the other.

EDGE WELD: Weld produced on an edge joint.

EFFECTIVE PENETRATION: (A term used in ultrasonic testing.) The greatest depth at which ultrasonic transmissions can effectively detect discontinuities.

EFFECTIVE THROAT: The least distance from the root of a weld to the weld face not counting any reinforcements. See also JOINT PENETRATION.

ELASTIC DEFORMATION: Temporary change of dimensions caused by stress; part returns to original dimension when the stress is removed.

ELASTIC LIMIT: Greatest stress to which a structure may be subjected without causing permanent deformation.

ELASTICITY: Ability of a material to regain its original size and shape after deformation.

ELECTRIC ARC SPRAYING (EASP): Process using the heat of the arc and a compressed gas to atomize and propel a coating material onto the base material.

ELECTROCHEMICAL CORROSION: Corrosion caused by current flow between the contact areas of two dissimilar metals.

ELECTRODE: Terminal point to which electricity is brought in the welding operation and from which the arc is produced to do the welding. In most electric arc welding, the electrode is usually melted and becomes a part of the weld.

ELECTRODE EXTENSION: Length of unmelted electrode extending beyond the end of the contact tube in GMAW, FCAW, SAW, GTAW.

ELECTRODE LEAD: Electrical conductor between the welding machine and the electrode holder.

ELECTRODE SKID: Sliding of an electrode along the work surface during spot, seam, or projection welding.

ELECTROMAGNET: Magnet produced by a coil carrying an electric current. Coil surrounds a mass of ferrous material which also becomes magnetized.

ELECTROMOTIVE FORCE: Energy, measured in volts, which causes the flow of electric current.

ELECTRON: Fundamental part of an atom which has a small negative charge.

ELECTRON BEAM WELDING (EBW): Focused stream of electrons which heats and fuses metals.

ELECTRONIC CONTROLLER: A device which controls the resistance welding process.

ELECTROSLAG WELDING (ESW): Process using one or more arcs between continuously fed metal electrodes and base metal. The molten metal, slag, and flux is held in place by moving water cooled shoes as the weld progresses vertically.

ELEMENT: Chemical substance which cannot be divided into simpler substances by chemical action.

ELONGATION: Percentage increase in the length of a specimen when stressed to its yield strength.

EMBRITTLEMENT: Reducing the normal ductility of a metal by a physical or chemical change.

EROSION: Reduction in size of an object because of a liquid or gas impacting on the object.

EUTECTIC ALLOY: Mixture of metals which has a melting point lower than that of any of the metals in the mixture, or of any other mixture of these metals.

EXPLOSIVE WELDING (EXW): Joins metal as powerful shock waves create pressure to cause metal flow and resultant fusion.

EXPULSION WELD: Resistance spot weld which squirts (expels) molten metal.

F

FACE OF A WELD: Exposed surface of the weld.

FAHRENHEIT: Temperature scale once used in most English speaking countries. Symbol is F.

FATIGUE: Condition of metal leading to cracks under repeated stresses below the tensile strength of the material.

FATIGUE LIMIT: Stress limit below which a material can be expected to withstand any number of stress cycles.

FATIGUE STRENGTH: Most stress that a metal will withstand cracking for a certain number of stress cycles.

FEED RATE: Speed at which a material passes through the welding gun in a unit of time.

FERRITE: Iron which contains very small amounts of carbon. It has a body centered cubic structure.

FERRITE BANDING: Bands of free ferrite which line up in the direction the metal was worked. Bands are parallel.

FILLER METAL: Metal added in making welded, brazed, or soldered joints.

FILLER ROD: See preferred term WELDING ROD.

FILLET WELD: Metal fused into a corner formed by two pieces of metal whose welded surfaces are approximately 90 degrees to each other.

FILTER PLATE: An optical material which protects the eyes from ultraviolet, infrared, and visible radiation.

FLAME CUTTING: Cutting performed by an oxyfuel gas torch flame which has a second oxygen jet.

FLAME SPRAYING (FLSP): A thermal spraying process in which an oxyfuel gas flame is the source of heat for melting the coating material. Compressed gas may be used to propel the coating material to the base material.

FLASH: Impact of electric arc rays against the human eye. Also the surplus metal formed at the seam of a resistance weld.

FLASH WELDING (FW): Process using electric arc in combination with resistance and pressure welding.

FLASHBACK ARRESTORS: Check valves usually installed between torch and welding hose to prevent flow of burning fuel gas and oxygen mixture back into hoses and regulators.

FLAT POSITION WELD: Horizontal weld on the upper side of a horizontal surface.

FLAW: Discontinuity in a weld larger than the accepted limit.

FLUORESCENT DYE: Dye which gives off or reflects light when exposed to short wave radiation.

FLUORESCENT PENETRANT: Penetrating fluid with a fluorescent dye added to improve visibility of discontinuities.

FLUX: Material used to prevent, dissolve, or help to remove oxides and other undesirable surface substances.

FLUX CORED ARC WELDING (FCAW): Welding method in which heat is supplied by an arc between a hollow, flux filled electrode and the base metal.

FLUX CORED ARC WELDING ELECTROGAS (FCAW-EG): A type of flux cored arc welding process. Similar to ELECTROSLAG WELDING with shielding gas and flux cored electrode added.

FLUX OXYGEN CUTTING: See CHEMICAL FLUX CUTTING.

FOCAL SPOT (EBW and LBW): Spot where an energy beam's energy level is most concentrated and where it has the smallest cross sectional area.

FORGING: Metallic shapes being made by either hammering or squeezing the original piece of metal.

FOREHAND WELDING: Welding in the same direction that the flame is pointing.

FORMING: Changing the shape of a metal part without changing its thickness.

FREE BEND TEST: Bending the specimen without using a fixture or guide.

FREE CARBON: In steel or cast iron, that part of the total carbon content which is present in the form of graphite or temper carbon.

FRICTION WELDING (FRW): Weld in which welding heat is generated by revolving one part against another part, under very heavy pressure to create friction heat.

FULL-WAVE RECTIFIED SINGLE-PHASE AC: Alternating current in which the reverse half of the cycle is made to travel the same direction as the other half of the wave. It produces pulsating direct current with no interval between pulses.

FULL-WAVE RECTIFIED THREE-PHASE AC: Conversion of alternating current by rectifier in which there is little pulsation in the resulting direct current.

FUSION: Intimate mixing or combining of molten metals.

FUSION FACE: The surface of the base metal which is melted during welding.

FUSION WELDING: Any type of welding using fusion as part of the process.

## G

GAMMA RAYS: Electromagnetic radiation given off by a nucleus. Gamma rays always accompany fission.

GAS HOLES: Holes created by gas escaping from molten metal. These holes are round or elongated, smooth edged dark spots. They appear in clusters or individually.

GAS METAL ARC WELDING (GMAW): Arc welding using a continuously fed consumable electrode and a shielding gas. Sometimes called MIG.

GAS POCKETS: Cavities in weld metal caused by entrapped gas.

GAS TUNGSTEN ARC WELDING (GTAW): Arc welding using a tungsten electrode and a shielding gas. The filler metal is added using a welding rod.

GAUSS: Unit of magnetic flux density or induction.

One gauss is one line of flux per square centimeter.

GENERATOR: Mechanism which generates electricity or produces some substance, for example an electric generator or an acetylene generator.

GOUGING: Cutting a groove in the surface of a metal using an oxyfuel gas air arc cutting outfit.

GRAPHITIZATION: Forming of graphite in iron or steel either during solidification or later during heat treatment.

GROOVE WELD: Welding rod fused into a joint which has the base metal removed to form a V, U, or J trough at the edge of the metals to be joined.

GROSS POROSITY: Pores, gas holes, or globular voids in a weld or casting. Called gross because they are bigger and more numerous than would be found in good practice.

GUIDED BEND TEST: Bending a specimen in a definite way by using a fixture.

## H

HALF-WAVE RECTIFIED AC: Simplest manner of rectification in which the reverse half of the cycle is blocked out completely. Result is a pulsating direct current with intervals when no current is flowing.

HAMMER FORGING: Deforming of workpiece by repeated blows.

HAND SHIELD: See SHIELD and HELMET.

HARDFACING: Filler material placed on a surface to toughen the surface to resist abrasion, erosion, wear, corrosion, galling, or impact wear.

HARDENING: Making metal harder by a process of heating and cooling.

HARDNESS: Ability of metal to resist plastic deformation; same term may refer to stiffness or temper, resistance to scratching or abrading.

HAZ: See HEAT AFFECTED ZONE.

HEAT: Molecular energy of motion.

HEAT AFFECTED ZONE: That part of the base metal altered by heat from welding, brazing, or cutting operations.

HEAT-CHECKING: Crazing of a die surface, especially one subjected to alternate heating and cooling.

HEAT CONDUCTIVITY: Speed and efficiency of heat energy movement through a substance.

HEAT TINTING: Using heat to color a metal surface through oxidation for purpose of showing details of the microstructure.

HEAT TREATABLE ALLOYS: Aluminum alloys that reach maximum strength by solution heating and quenching.

HELIUM (He): Inert, colorless, gaseous element used as a shielding gas in welding.

HELMET: Protecting hood which fits over the arc

welder's head, provided with an approved filtering lens through which the operator may safely observe the electric arc.

HERTZ: Unit of frequency equal to one cycle per second.

HORIZONTAL POSITION: Weld performed on a horizontal seam at least partially on a vertical surface.

HOLD TIME: Time force continues on electrodes in resistance welding after the current is turned off.

HORN: Arm of a resistance welder that carries the current and applies the electrode pressure.

HOSE: Flexible, usually reinforced, rubber tube used to carry pressured gases or water to a torch.

HOT FORMING: Operations performed on metal while it is above the recrystallization temperature of the metal. Operations may include bending, drawing, forging, heading, piercing, and pressing.

HOT SHORTNESS: Weakness of metal which occurs in the hot forming range.

HOT WORKING: Shaping of metal at a temperature and rate which does not cause strain hardening.

HYDROGEN: Gas formed of the single element, hydrogen. When combined with oxygen, it forms a very clean flame which, however, does not produce a very high temperature.

HYDROGEN EMBRITTLEMENT: Low ductility condition in metals due to absorption of hydrogen.

I

IMPACT ENERGY: Amount of energy that must be exerted to fracture a part; measurement is usually made in an Izod or Charpy test.

IMPACT STRENGTH: Material's ability to resist shock.

IMPACT TEST: Carefully measured test of how materials behave under heavy loading such as bending, tension, or torsion. Charpy or Izod tests, for example, measure energy absorbed in breaking a specimen.

IMPEDANCE: Total resistance to flow of alternating current as a result of resistance and reactance.

IMPURITIES: Undesirable elements or compounds in a material.

INCLUSION: Foreign matter introduced into and remaining in welds or castings.

INCOMPLETE FUSION: Less than complete fusion of weld material with base metal or with preceding bead.

INCOMPLETE JOINT PENETRATION: Lack of fusion between metals appearing as elongated darkened lines. May occur in any part of a weld groove.

INCOMPLETE PENETRATION: Incomplete root penetration or failure of two weld beads to fuse.

INDENTATION: Depression left on the surface of the base metal after spot, seam, or projection welding.

INDENTATION HARDNESS: Degree of resistance of a material to indentation. This is the standard test of a material's hardness.

INDICATION, MAGNETIC: Magnetic particle pattern held magnetically on surface of material being tested.

INDICATION, PENETRANT: Visual evidence of a discontinuity, that is, the penetrant can be seen in the crack.

INDICATION, ULTRASONICS: Signal on the ultrasonic equipment indicating a crack or discontinuity in a material being tested.

INDUCED CURRENT: Secondary current which is set in motion as a second conductor in the shape of a closed loop is placed in the magnetic field around another current carrying conductor.

INDUCTANCE: In the presence of a varied current in a circuit, the magnetic field surrounding the conductor generates an electromotive force in the circuit itself. If another circuit is next to the first, the changing magnetic field of the first circuit will cause voltage in the second.

INDUCTION: Magnetism induced in a ferromagnetic material by some outside magnetic force.

INDUCTION HARDENING: Process of quench hardening using electrical induction to produce the heat.

INDUCTIVE REACTANCE: Force which opposes flow of alternating current through a coil. This force is independent of the resistance of a conductor to a flow of current.

INERT GAS: Gas which does not normally combine chemically with the base metal or filler metal.

INFRARED RAYS: Heat rays coming from both the arc and the welding flame.

INSIDE CORNER WELD: Two metals fused together. One metal is held 90 degrees to the other. The fusion is performed inside the vertex of the angle.

INTERFACE: Surface which forms a common boundary between two bodies.

INTERGRANULAR CORROSION: Corrosion occurring, for the most part, between grains or on the edges of the grain in a ferrous material.

INTERMITTENT WELD: Joining two pieces and leaving sections unwelded.

ION: Atom or a group of atoms positively or negatively charged as a result of having gained or lost one or more electrons. Also: sometimes a free electron.

IONIZATION: Adding or removing electrons from atoms or molecules to create ions.

IRON-CARBON DIAGRAM: Diagram which shows the critical temperatures for the varying amounts of carbon in iron which form steels or cast irons.

IZOD TEST: Type of test for impact strength made by striking the test piece with a measured downstroke of a pendulum. The specimen, usually notched, is held by one end in a vise. Energy absorbed, measured by the upward swing of the pendulum, indicates the impact strength of the specimen.

## J

JIG: Fixture or template to accurately position and hold a part during welding or machining operations.
JOINT: Where two pieces are joined in an assembly.
JOINT EFFICIENCY: Strength of a welded joint given as a percentage of the strength of the base metal.
JOINT PENETRATION: Depth of weld metal and base metal fusion in a welded joint.

## K

KERF: Width of cut produced by a cutting operation.
KEYHOLE: Welding technique in which concentrated heat penetrates the workpiece leaving a hole at the leading edge of the weld. As the heat source moves on, molten metal fills the hole and forms the weld bead.
KILLED STEEL: Steel which has been deoxidized with a strong agent to reduce the oxygen content. Deoxidizing is carried to the point where no reaction takes place between carbon and oxygen as the metal solidifies.
KILOPASCAL (kPa): Unit of pressure in SI metrics; one thousand pascals. See PASCALS.

## L

LACK OF FUSION: Defect in a weld caused by lack of union between weld metal and base metal.
LAMINATE: Sheets or bars made up of two or more metal layers built up to form a structural member. Also: forming a metallic product with two or more bonded layers.
LAP: Folded-over section of metal which is then rolled or forged into the surface. Considered a surface defect.
LAP JOINT: Joint in which the edges of the two metals to be joined overlap.
LASER BEAM WELDING (LBW): Process in which single frequency light beam concentrates a small spot of heat to fuse small, light metal materials.
LAYER: Certain weld metal thickness made of one or more passes.
LEAD: Electricity-carrying wire from the power source to the electrode holder or to the ground clamps.
LEG OF FILLET WELD: Distance from the point where the base metals touch to the toe of the fillet.

LENS: Specially treated glass or plastic through which a welder may look at an intense flame without being injured by the harmful rays or glare.
LIGHT METAL: Low density metal such as aluminum, magnesium, titanium, beryllium, or their alloys.
LIQUIDUS: Lowest temperature at which a metal or alloy is completely liquid.

## M

MACRO-ETCH: Eating away of the metal surface to make gross structural details stand out so that they can be observed with the naked eye or with magnification up to 10 times.
MACROGRAPH: Photographic or graphic reproduction of the surface of a prepared specimen which has been magnified up to 10 times normal size.
MACROSTRUCTURE: Physical makeup or structure of metals revealed under magnification of not more than 10 diameters.
MALLEABILITY: Ability of a metal to be deformed without breaking.
MALLEABLE CAST IRON: Cast iron made by annealing white cast iron while the metal undergoes decarburization, graphitization, or both thus eliminating all or most of the cementite.
MALLEABLE CASTINGS: Cast forms of metal which have been heat treated to reduce their brittleness.
MANIFOLD: Pipe or cylinder with several inlet and outlet fittings. It is designed so that several cylinders can be connected together while the gas or oxygen in the cylinders can be piped to several locations or stations.
MAPP: Stabilized methylacetylene-propadiene fuel gas.
MAPS: Fuel gas methylacetylene-propadiene (stabilized).
MARTENSITE: Very hard and brittle form of steel. It is formed by rapid cooling from the austenite phase.
MECHANICAL PROPERTIES: Description of a material's behavior when force is applied for purpose of determining the material's suitability for mechanical usage. Properties described, for example are, modulus of elasticity, elongation, fatigue limit, hardness and tensile strength.
MEGAPASCAL (Mpa): One million pascals; a unit of measure for pressure in SI metrics. See PASCAL.
METAL ARC CUTTING (MAC): Method of cutting metal with an electric arc. Molten metal flows away from the base metal usually by force of gravity.
METAL POWDER CUTTING (POC): Oxygen cutting process which cuts metal using a powder, such as iron or aluminum, to improve the cutting.

METALLIC DISCONTINUITY: Break in the surface of a metal part or a void such as a gas pocket.

METALLOGRAPHY: Scientific study of the constitution and structure of metals and alloys as observed by the naked eye or with the aid of magnification and X ray.

METALLURGY: Science and technology of metal.

MIG: See preferred term GAS METAL ARC WELDING.

MIXING CHAMBER: Part of the welding torch where the welding gases are mixed prior to combustion.

MODULUS OF RUPTURE: In a bend or torsion test, the stress at which fracture occurs expressed as a constant.

**N**

NEUTRAL FLAME: Flame resulting from combustion of perfect proportions of oxygen and the welding gas.

NEWTON: SI metric unit of force. A force of 9.8 newtons is required to lift a mass of 1 kilogram.

NITRIDING: Casehardening process; adding nitrogen to a solid ferrous alloy by keeping the alloy at a suitable temperature while in touch with a material rich in nitrogen.

NODULAR CAST IRON: Cast iron containing primary graphite which is in a ball-like or globular form rather than in flakes as in gray cast iron. Also known as spheroidal graphite iron, it is more ductile and has greater strength than ordinary iron.

NONDESTRUCTIVE TESTING (NDT): Testing for defects using techniques which do not destroy or damage the part.

NORMALIZING: Heating steel above the temperature used for annealing and then cooling it in still air at room temperature; used as a preparation for further heat treatment. Normalized steel has a uniform unstressed condition with a grain size and refinement that makes the metal more suitable for heat treating.

NOTCH BRITTLENESS: Tendency of a material to break at points where stress is concentrated.

NOTCH SENSITIVITY: Measure of the extent of reduction of strength in a metal after introducing stress concentration (by notching).

NOZZLE: Device which directs the shielding media or gas.

**O**

OFFTIME (resistance welding): Time that the electrodes are off the work. The time between repeating cycles.

OPTICAL PYROMETER: Temperature measuring device which compares the incandescence (white, glowing hot) of a heated object with an electrically heated filament whose incandescence can be regulated.

ORIFICE: Opening through which gases flow.

OUTSIDE CORNER WELD: Fusing two pieces of metal together with the fusion taking place on the outer part of the seam.

OVERHEAD POSITION: Weld on the underside of joint with the face of the weld in a horizontal position.

OVERHEATING: Damaging the properties of a metal by too much heat. When original properties cannot be restored, the overheating is known as "burning."

OVERLAP: Extension of the weld face metal beyond the toe of the weld.

OXIDATION: Combining of a substance with oxygen. Rapid oxidation is called "burning."

OXIDIZING: Combining oxygen with any other substance.

OXIDIZING FLAME: Flame produced by an excess of oxygen in the torch mixture, leaving some free oxygen which tends to burn the molten metal.

OXYACETYLENE CUTTING: Oxyfuel gas cutting process which uses an oxygen and acetylene flame for heat and a jet of oxygen to oxidize the molten metal to form a cut.

OXYACETYLENE WELDING (OAW): Method of oxyfuel gas welding in which oxygen and acetylene are combined and burned to provide the heat.

OXYFUEL GAS CUTTING: Cutting metal using an oxygen jet and a preheating flame which combines oxygen and a fuel gas.

OXYFUEL GAS WELDING: Method of welding which combines and burns oxygen and a fuel gas to create the required heat.

OXYGEN ($O_2$): Gas which makes up 21 percent of the composition of air. A gas used to support combustion in oxyfuel gas welding and cutting.

OXYGEN CYLINDER: Specially built container manufactured according to ICC standards and used to store and ship certain quantities of oxygen.

OXYGEN HOSE: Reinforced, multilayered, flexible tube usually of rubber. Used to carry high pressure gases.

OXYGEN LANCE CUTTING (LOC): Oxyfuel gas cutting process which heats base metal and then blows away molten metal with jet of oxygen from an iron pipe.

OXYGEN REGULATOR: Automatic valve used to reduce cylinder pressures to torch pressures and to keep the pressures constant.

OXYGEN-ACETYLENE CUTTING: See preferred term OXYACETYLENE CUTTING.

OXYGEN-ACETYLENE WELDING: See preferred term OXYACETYLENE WELDING.

OXYGEN-FUEL GAS CUTTING: See preferred term OXYFUEL GAS CUTTING.

OXYHYDROGEN FLAME: Chemical combining of oxygen with the fuel gas hydrogen.

OXYPROPANE GAS FLAME: Chemical combining of oxygen with the fuel gas LP (liquified petroleum).

## P

PASCAL: Unit for measuring pressure in the SI metrics.

PASS: See preferred term WELD PASS.

PEARLITE: Form of steel which has alternating layers of ferrite and cementite.

PEEL TEST: Destructive test which mechanically separates a resistance welded lap joint by peeling one piece away from the other.

PENETRANT: Either a liquid or a gas which, when applied to the surface of a metal, enters cracks (discontinuities) to make them visible.

PENETRATION: Extent to which the weld metal combines with the base metal as measured from the surface of the base metal.

PERCUSSION WELDING (PEW): Type of resistance welding in which the heat comes from an arc produced by an electrical discharge and instantaneous pressure applied during or immediately following the heating.

PHYSICAL PROPERTIES: Properties or qualities other than mechanical properties, that have to do with the physics of a material. Examples are density, ability to conduct electricity, ability to conduct heat, and thermal expansion.

PHYSICAL TESTING: Examination of a material to find out its physical properties.

PICKLING: Removing surface oxides from metals by chemical or electro-chemical reaction.

PLASMA: Temporary physical condition of a gas after it has been exposed to and has reacted to an extremely high temperature.

PLASMA ARC CUTTING (PAW): Process of metal cutting using an electric arc and fast flowing ionized gases.

PLASMA SPRAYING (PSP): Thermal spraying process which uses a nontransferred plasma arc to melt and propel the coating material to the base material.

PLASTIC WELDING: Process in which heated air softens and fuses synthetic plastic materials.

PLASTICITY: Ability of a metal to bend without breaking (rupturing).

PLUG WELD: Weld made in a hole of one piece of metal as it is lapped over another piece of metal.

POLARITY: The direction of flow of electrons in a closed direct current welding circuit. When the electrons flow from the electrode to the work the polarity is DCEN (direct current electrode negative) or DCSP. When the electrons flow from the work

to the electrode the polarity is DCEP (direct current electrode positive) or DCRP.

POROSITY: Gas pockets or voids in a metal.

POSTHEATING: Temperature to which a metal is heated after an operation has been performed on the metal.

POWDER METALLURGY: Art and technology of producing powdered metal and of utilizing it in the production of parts.

PREHEATING: Temperature to which a metal is heated before an operation is performed on the metal.

PROCEDURE QUALIFICATION RECORD (PQR): Document containing the actual welding variables used to produce an acceptable test weld for the purposes of qualifying a welding procedure specification.

PRODS: Two hand held electrodes designed to be pressed against the surface of a part for purpose of passing a magnetizing electric current through the part; used to find defects with magnetic particles.

PROGRAM: Series of times, events, and pressures set on the automatic controls of a welding machine.

PROJECTION WELDING (RPW): Type of resistance welding in which current flow is concentrated at predetermined points by projections, embossments or intersections.

PUDDLE: See preferred term WELD POOL.

PULSE ARC WELDING (GMAW-P): Welding arc in which the current is interrupted or pulsed as the welding arc progresses. Proper term is gas metal arc welding-pulsed arc.

PYROMETER: Device for determining temperatures over a wide range.

## Q

QUENCH AGING: Change in metal produced by rapid cooling after heat treating.

QUENCH ANNEALING: Process used to soften austenitic ferrous alloys by solution heat treatment.

QUENCH HARDENING: Hardening an iron alloy by austenitizing followed by rapid cooling so that some or all of the austenite becomes martensite.

QUENCHING: Rapid cooling of metal in a heat treating process.

## R

RADIOGRAPH: Photograph made by passing X rays or gamma rays through the object to be photographed and recording the variations in density on a photographic film.

RADIOGRAPHER: One who performs radiographic

operations.

RADIOGRAPHER'S EXPOSURE DEVICE: Instrument containing the X-ray source for making radiographic records on sensitized film.

RADIOGRAPHIC INTERPRETATION: "Reading" of the films to determine cause and significance of discontinuities below the surface of the material radiographed; one of the determinations of the reading is the suitability of the material.

RADIOGRAPHIC SCREENS: Sheets, either metallic or fluorescent, used to intensify the radiation effect on film.

RADIOGRAPHY: Use of radiant energy found in X rays or gamma rays to examine opaque objects and make a record of the examination.

RAYS: See INFRARED RAYS and ULTRAVIOLET RAYS.

REACTANCE: Opposition to the flow of alternating current as a result of inductance or capacitance.

RECARBURIZE: Adding carbon to molten cast iron or steel. Also: the process of adding more carbon to a surface which has lost some carbon in processing.

RECTIFIED ALTERNATING CURRENT: Alternating current made to flow in one direction only by use of a device like a diode to stop normal reversing of the current.

RECTIFIER: Device such as a diode or a circuit which acts like a one way valve. It converts one half of a waveform of alternating current to useful current flowing in the same direction as the other half of the waveform.

REDUCING FLAME: Oxyfuel gas flame with a slight excess of fuel gas.

REDUCTION OF AREA: Difference in cross sectional area of a specimen after fracture as compared to original cross sectional area.

REFRACTORY: Material resistant to heat or difficult to melt.

REGULATORS: See ACETYLENE REGULATOR, OXYGEN REGULATOR.

REINFORCEMENT OF WELD: Excess metal on the face of a weld.

RESIDUAL STRESS: Stress still present in a body freed of external forces or thermal gradients.

RESISTANCE SEAM WELDING (RSEW): Resistance welding process which usually uses round, rotating electrodes to make a continuous seam or overlapping spot welded seam.

RESISTANCE SPOT WELDING (RSW): Resistance welding process which uses the resistance to the flow of electricity to create the heat for fusion. A small spot is welded on two overlapping pieces between two electrodes.

RESISTANCE WELDING (RW): Process using the resistance of the metals to the flow of electricity as the source of heat.

REVERSED POLARITY: See DIRECT CURRENT ELECTRODE POSITIVE.

ROCKWELL HARDNESS TESTER: Tester which measures the hardness of materials based on depth of penetration of a standardized force.

ROOT CRACK: Crack in either the weld or the heat affected zone at the root of a weld.

ROOT OF JOINT: Point at which metals to be joined by a weld are closest together.

ROOT OF WELD: That part of a weld farthest from source of weld heat and/or from filler metal side.

ROOT PENETRATION: Depth to which weld metal extends into the root of a welded joint.

ROSETTE WELD: See PLUG WELD.

## S

SCLEROSCOPE TEST: Hardness test which uses the height of rebound of a falling piece of metal to determine how much energy the material being tested absorbed.

SCRATCH HARDNESS: Hardness of a metal determined by measuring the width of a scratch made by a cutting point under a known pressure.

SECONDARY HARDENING: Tempering of some alloy steels at a higher temperature than normally used for hardening. Result is a hardness greater than is achieved by tempering at the lower temperature for the same period of time.

SEMI-KILLED STEEL: Incompletely deoxidized steel which contains sufficient dissolved oxygen to react with the carbon so that carbon monoxide is formed. This offsets solidification shrinkage.

SEQUENCE: Order in which operations take place.

SHEAR: Force which causes two parts of the same body touching each other to slide parallel to their contacting surfaces.

SHEAR FRACTURE: Break in which crystalline material separates by sliding under action of shear stress.

SHEAR STRENGTH: Stress required to fracture a part in a cross sectional plane when two forces being applied are parallel and opposite, but are offset a little.

SHIELD: Eye and face protector. It enables a person to look directly at the electric arc through a special lens without being harmed.

SHIELDED METAL ARC CUTTING (SMAC): Arc cutting process in which metal is cut by melting it with the heat of an arc between a covered electrode and the base metal.

SHIELDED METAL ARC WELDING (SMAW): Arc welding process which melts and fuses the metals using the heat of an arc between a covered electrode and the base metal. The electrode wire also acts as the filler metal.

SHORT ARC: Gas metal arc process which uses a

low arc voltage. The arc is continuously interrupted as the molten electrode metal bridges the arc gap.

SHOT PEENING: Working the surface of a metal by bombarding it with metal shot.

SIZE OF WELD: Joint penetration in a groove weld. Also the nominal lengths of the legs of a fillet weld.

SLAG: Nonmetallic byproduct of smelting and refining made up of flux and nonmetallic impurities. Also material which forms on the underside of an oxyfuel gas or arc cut.

SLAG INCLUSIONS: Nonfused, nonmetallic substances in the weld metal.

SOLDERING: Means of fastening metals together by adhering to another metal at a temperature below 840 °F (450 °C). Filler metal only is melted.

SOLID PELLET OXYFUEL GAS WELDING: Portable welding system using fuel gas with pellets. Pellets produce oxygen and eliminate need for bulky oxygen cylinders.

SOLID STATE CONTROLLER: Electronic controller which uses transistors, diodes, and other semiconductor devices.

SOLIDUS: Highest temperature at which a metal or alloy is completely solid.

SPHEROIDIZING: Heating and cooling to produce a spheroidal or globular form of carbide in steel.

SPRAY ARC: Gas metal arc process which has an arc voltage high enough to continuously transfer the electrode metal across the arc in small globules.

STICKOUT: See preferred term ELECTRODE EXTENSION.

STRAIGHT POLARITY: See DIRECT CURRENT ELECTRODE NEGATIVE.

STRAIN: Reaction of an object to a stress.

STRESS: Load imposed on an object.

STRESS RELIEVING: Even heating of a structure to a temperature below the critical temperature followed by a slow, even cooling.

SUBMERGED ARC WELDING (SAW): Process in which electric arc is submerged in granular flux.

SUPERFICIAL ROCKWELL HARDNESS TEST: Test for determining surface hardness of thin sections or small parts, or where a large hardness impression might be harmful.

SURFACING: Depositing material on a surface to obtain desired properties or dimensions.

### T

TACK WELD: Small weld used to temporarily hold components together.

TANK: Thin walled container of fluids or gases. Usually weaker than a cylinder for pressurized gases.

TEMPER: Part of heat treating in which hardened steel or hardened cast iron is heated to a temperature below its melting point for purpose of decreasing the hardness and increasing the toughness.

TEMPERING: Reheating hardened or normalized ferrous alloys and then cooling at any rate desired.

TENSILE STRENGTH: Maximum pull stress in pounds per square inch or mega pascals (newtons per square millimeter) that a specimen will withstand.

THROAT OF A FILLET WELD (actual throat): Distance from the weld root to the weld face.

TIG: See preferred term GAS TUNGSTEN ARC WELDING (GTAW).

TIME, TEMPERATURE, TRANSFORMATION DIA— GRAM: Diagram which shows how time and temperature affect the steel transformation when cooling from the austenite phase.

TINNING: In soldering, a coating of the soldering metal placed on the metals to be soldered.

TIP: End of the torch where the gas burns, producing the high temperature flame. In resistance welding, the electrode ends.

T-JOINT: Joint formed by placing one metal against another at an angle of 90 degrees to form a T-shape.

TOE CRACK: Base metal crack at the toe of a weld.

TOE OF WELD: Junction between the face of a weld and the base metal.

TORCH: Mechanism which the operator holds during gas welding and cutting and from which issue the gases that are burned to produce heat. The mechanism held during some arc welding processes is also known as a torch.

TORSION: Twisting motion resulting in shear stresses and strains.

TOUGHNESS: Metal's ability to absorb energy and deform before breaking.

TRANSDUCER: Device used to send and receive sonic (sound) waves when doing ultrasonic testing.

TTT CURVE: See TIME-TEMPERATURE-TRANSFORMATION CURVE.

TUYERE: Opening in a blast furnace through which air is forced to support combustion.

### U

ULTIMATE COMPRESSIVE STRENGTH: Most compressive stress that a material can stand under a gradual and evenly applied load.

ULTIMATE STRENGTH: Greatest conventional stress, tensile, compressive, or shear that a material can stand.

ULTRASONIC: Mechanical vibrations in a frequency above the range of humanly audible sound.

ULTRASONIC TESTING (UT): Nondestructive testing method which transmits and receives high frequency sound waves through the material being inspected.

ULTRASONIC WELDING (USW): Process using high sound frequencies to produce metal fusion.

ULTRAVIOLET RAYS: Energy waves that come from electric arcs and welding flames at such a frequency that these rays are in the short light wavelength known as the ultraviolet ray light spectrum.

UNDERBEAD CRACK: Crack in the base metal near the weld and beneath the surface.

UNDERCUT: Depression at the toe of the weld which is below the surface of base metal.

## V

VERTICAL POSITION: Type of weld where the welding is done in a vertical seam and on a vertical surface.

## W

WELD BEAD: Deposit or row of filler metal from a single welding pass.

WELD CRACK: Crack in weld metal.

WELD METAL: Fused portion of base metal or fused portion of the base metal and the filler metal.

WELD NUGGET: Weld metal in a spot, seam, or projection weld.

WELD PASS: Single progression of a weld or surfacing operation. The result of a pass is a bead, layer, or spray deposit.

WELD POOL: Portion of weld that is molten due to the heat of welding.

WELDER: One who performs a weld. (Sometimes incorrectly used to describe a welding machine.)

WELDING: Art of fastening metals together by means of flowing the materials together with or without filler material.

WELDING PROCEDURE SPECIFICATION (WPS): Lists in detail the base metal P numbers to be welded, the filler metals used, the preheat or postwelding heat treatment, metal thickness, and other variables for each welding process as essential or nonessential.

WELDING ROD: Wire which is melted into the weld metal.

WELDING SEQUENCE: Order in which the component parts of a structure are welded.

WELDMENT: Assembly of component parts joined together by welding.

WORKLEAD: Electrical conductor which carries the current between the welding machine and the workpiece.

WROUGHT IRON: Commercial iron made up of slag (also known as iron silicate) fibers entrained in a ferrite matrix.

## Y

YIELD POINT: The lowest stress to which a material or body can be subjected and at which strain increases without an appreciable or proportionate increase in stress.

YIELD STRENGTH: Stress value in psi or kPa at which a specimen assumes a specified limiting permanent set.

# ACKNOWLEDGMENTS

The publishing of a book of this nature would not be possible without the assistance of the many segments of the Welding Industry.

The authors gratefully acknowledge the cooperation of the following:

Acorn Iron & Supply Co.; Acro Automation Systems, Inc.; Aeroquip Corp.; Aidlin Automation Corp.; Airmatic/Beckett-Harcum; Air Products & Chemicals, Inc.; Ajax Magnethermic Corp.; Allegheny Ludlum Steel Corp.; Aluminum Co. of America; American Iron and Steel Institute; American Pullmax Co.; The American Society of Mechanical Engineers; American Welding Society; Anchor Swan Corp.; Andrews/Mautner, Inc.; Applied Power, Inc.; Arcair Co.; Arcos Corp.; Argopen; K.N. Aronson Machine, Inc.; Arvin Automation; Atlas; Bausch and Lomb, Inc.; Belchfire Corp.; Bernard Welding Equipment Co.; Boeing Co.; Bundy Tubing Div., Bundy Corp.; Cam-Lok Div. Empire Products, Inc.; Cincinnati Electric Tool, Inc.; Cincinnati Milicron; Cleanweld Products, Inc.; Clements Mfg. Co.; CMW, Inc.; Collis Div., Chamberlain Mfg. Corp.; CONCOA; Craftsweld Equipment Corp.; Curv-O-Mark; Cut-Mark, Inc.; Dearman Systems; De-Sta-Co.; Despatch Industries, Inc.; Detroit Testing Machine Co.; Diamonite Products Mfg., Inc.; Dockson Corp.; Dow Chemical Co.; Duffers Associates, Inc.; Duraline Div. of J.B. Nottingham & Co., Inc.; Dynabrade, Inc.; Engineering & Scientific Equipment. Ltd.; Erico Products, Inc.; ESAB Automation, Inc; ESAB Group; ESAB Welding Products Div.; Eureka Welding Alloys, Inc.; Falstrom Co.; Fenway Machine Co., Inc.; Ferranti Sciaky, Inc.; Fibre-Metal Products Co.; Flo-Max Corp.; Ford Motor Co.; Fusion, Inc.; Gasflux Co.; General Electric Co.; General Motors Corp.; GMFanuc Robotics Corp.; Gor-Vue Corp.; Goss,Inc.; Gullco International; H & M Pipe Beveling Machine Co.; Hacker Instruments, Inc.; Handy & Harman; Harnischfeger Corp.; Harris Calorific Div.; Hercules Welding Products, Div. of Obara Corp.; Hobart Brothers Co.; Ind. Schweisstechnik E. Jankus; Invincible Airflow Systems; Jackson Products; Jewel Mfg. Co.; Kamweld Products Co., Inc.; Kedman Co.; Kelsey-Hayes Co.; Kolene Corp.; L-TEC Welding & Cutting Systems; Laramy Products Co., Inc.; Leica, Inc., Buffalo, NY; The Lincoln Electric Co.; Magnaflux Corp.; Maitlen & Benson; Manufacturing Technology, Inc.; McDonnell & Miller, Inc.; Metallizing Co. of America, Inc.; Metco, Inc.; Michigan Seamless Tube Co.; Miller Electric Mfg. Co.; Modern Engineering Co., Inc.; National Cylinder Gas; National Welding Equipment Co.; Nederman, Inc.; Nelson Stud Welding; Niagara Machine & Tool Works; Norton Co.; NLC, Inc.; The Ohio Nut & Bolt Co.; Omark Industries; Page-Wilson Corp.; Cecil C. Peck Co.; Pertron Controls Corp.; Peterson Mfg. Co., Inc.; Physical Acoustics Corp.; Physmet Corp., Div. of Manlabs,Inc.; Pressed Steel Tank Co.; Prest-O-Lite; Progressive Machine Corp.; PTR-Precision Technologies, Inc.; Racal Health and Safety, Inc.; Royco Instruments, Inc.; Resistance Welding Corp.; Rexarc, Inc.; Robotron Div. of Midland-Ross Corp.; W.J. Savage, Inc.; Shore Instrument & Mfg. Co.; Singer Safety Co.; Smith Welding Equip. Div.; Sonobond Corp.; Sperry Div., Automation Industries, Inc.; E. Spirig (Solder Absorbing Tech., Inc.); Stoody Deloro Stellite, Inc.; Strippit, Div. of Houdaille Industries; Inc.; Suhner Industries Products, Inc.; Swanstrom Tools, USA; Taylor-Winfield Corp.; Tec Torch Co., Inc.; Tempil, Big Three Industries, Inc.; TFI Corp.; Thrall Car Mfg. Co.; Thermal Dynamics Corp.; Thermocote-Welco Co.; Tinius Olsen Testing Machine Co.; Trumpf; Tuffaloy Products, Inc.; Tweco Products, Inc.; Unimation, Westinghouse Automation Div.; United Clamp Mfg. Co.; U.S. Steel Corp.; Vega Enterprises; Vacuum Atmospheres, Inc.; Vickers, Inc.; Victor Equipment Co.; Vogel Tool & Die Corp.; Wall Colmonoy Corp.; Wear Technology Div., Cabot Corp.; Weldex, Inc.; Welding & Fabrication Data Book; Welding Design & Fabrication; Weldma Co.; Edwin L. Wiegand Div., Emerson Electric Co.

Industry photo shows welders completing the bottom of a covered hopper car. The entire railcar has been turned on its side in a positioner to allow the welders to work in the horizontal or vertical position in place of the overhead position.    (Thrall Car Manufacturing Co.)

# INDEX

# SAFETY INDEX

This special index is provided as a quick reference allowing you to look up items of safety in the welding field. By reading the material listed on the pages following each entry, you may learn the general safety guidelines recommended for each procedure. Be sure to follow all safety precautions as you proceed through your study of welding.